ISBN 978-0-265-94909-2
PIBN 10914262

Vol. XXXI No. 1                                    Toronto, January 1, 1922

# Electrical News

**Engineering Contracting** — **Merchandising Transportation**

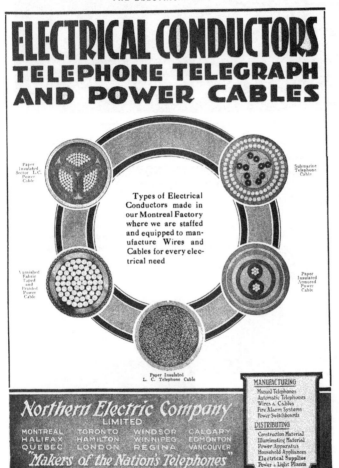

## "The Henderson Business Service helped me--it can help you!"

# e Story of a Contractor-Dealer's Rise to Success

ve months ago I was in exactly the same position as you are.
I all the business I could handle—I knew I was getting more than my share of the business in my district. I worked early and late, employed a large number of men, people even atulated me on my success. Yet I knew—and my books showed that I wasn't making a

lly I went to an old friend who knew the selling game from end to end. I told him every- and asked his advice. It was he who advised the Henderson Business Service. The Hen-n Business Service showed me the exact price to charge for every article I sold—wire, it, condulets, sockets, appliances, everything.

rouble had been that I didn't know what to charge for the merchandise I was selling over unter or sending out on the jobs.

lerson Business Service showed me all this and more. For instance it suggested how much rge for small repairs and house wiring rates per outlet. It brought me almost weekly the d price list that was being used all over Canada. The rest was easy. From then on I bsolutely certain of making a profit on every article I sold.

all this for about ten cents a day!

hat's my advice to you. Get this Henderson System and stick to it. You're bound to come top at the end of the year, if you do."

The Boston Daily Globe

**NEW ENGLAND GRIPPED IN SNARL OF WIRES**
**TRANSPORTATION, POWER, LIGHTS, CRIPPLED**

TELEPHONE POLES DOWN IN WOBURN

Destruction From Storm Increases as Weight of
Ice and Gale Bring Down Thousands of Poles
...Loss Reaches Into the Millions

HOW VERY PUNY MAN IS WHEN
PITTED AGAINST THE ELEMENTS

No Lights in
Towns

# One of many reasons
## for Fibre Conduit

ONE more page-wide head-line that tells of overhead wiring destruction with losses running into the millions. Reading between the lines we find again the story of the ultimate economy of Fibre Conduit.

Had those wires run underground, snugly encased in Johns-Manville Fibre Conduit, it would have been a different tale—the story of continuous service in spite of wind, rain, snow, sleet, or fire. There would have been no frantic repair gangs clearing up wreckage and replacing it at tremendous cost.

Fibre Conduit is free of the service interruptions of overhead wiring. It is permanent, conve-

nient and safe. Its first cost may be a little more but it is cheaper in the end just as inside metal conduit is ultimately cheaper than the out-of-date knob and tube wiring. Its value lies in its ability properly to serve and protect its cables.

Thousands of feet of Johns-Manville Fibre Conduit all over the country bear witness of its greater dependability. Its ever increasing sale proves it a satisfactory and profitable investment.

Send now for a booklet describing its use.

Canadian Johns-Manville Co.,
Limited

Toronto          Montreal          Winnipeg
Vancouver        Windsor           Hamilton
                 Ottawa

*Johns-Manville Fibre Conduit is storm-proof*

*Through*
**ASBESTOS**
*and its allied products*
Electrical Materials
Brake Linings
Insulations
Roofings
Packings
Cements
*Fire Prevention Products*

# JOHNS-MANVILLE
# Fibre Conduit

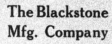

## Saleability

There can be no doubt about the selling possibilities contained in these "Simplex" Ironers.

Their splendid construction and high-grade materials insure a long service. And this, together with the fact that they fill a real need, quickly wins over the busy housewife. In addition we render every possible help to the dealer in the way of window cards, folders and newspaper cuts.

It is an opportunity no electrical merchant should overlook. Write and let us send you full particulars. Prices will appeal to dealer and consumer.

### Canadian Ironing Machine Co., Limited
Woodstock      Ontario

SIMPLEX IRONER
"THE BEST IRONER"

## ALPHABETICAL LIST OF ADVERTISERS

# Contracting Conduit Engineers

### Head Office
## Power Building
## Montreal

*Vancouver*             *Winnipeg*

---

# Electrical Books --- Special Prices
### The following Books are Offered Subject to Previous Sale:

A Laboratory Manual of Alternating Currents, by John H. Morecroft, E.C. Published in 1912 by Renouf Publishing Company, 248 pages, illustrated. Price $1.50.

Baudot Printing Telegraph System, by H. W. Pendry. Published in 1913 by Whittaker & Company. 144 pages, illustrated. Price $2.50.

Direct-Acting Steam Pumps, by Frank F. Nickel. Published in 1915 by McGraw-Hill Book Company. 258 pages, illustrated. Price $2.50.

Direct Current Machinery, Theory and Operation of, by Cyril M. Jansky, B.S., B.A., 1st edition, published in 1917 by McGraw-Hill Co., Inc. 265 pages, illustrated. Price $2.50.

Dynamo-Electric Machinery, by Francis B. Crocker, E. M., Ph.D. Published in 1908 by American School of Correspondence. 236 pages, illustrated. Price 50c.

Electric Railway, by A. Morris Buck, M.E. Published in 915 by McGraw-Hill Book Company, Inc. 390 pages, illustrated.

"Electrical Rates" - by G. P. Watkins, Ph. D., - recently published by D. Van Nostrand Company, 228 pages, price $3.00.

"Engineering Electricity," by Ralph G. Hudson, S.B. 190 pages illustrated. Published in 1920 by the John Wiley & Sons, Incorporated. Price $2.00.

Examples in Alternating-Currents, (Vol. 1), by F. E. Austin. B.S. Published in 1915. 228 pages, illustrated. Price $1.50.

Handbook of Machine Shop Electricity, by C. E. Clewell. Published in 1916 by McGraw-Hill Book Company. 461 pages, illustrated. Price $3.00.

"How to Sell Electrical Labor-Saving Appliances," by Electrical Merchandising first edition. 118 pages, illustrated.

Principles of Alternating Current Machinery, by Ralph R. Lawrence. Published in 1916 by McGraw-Hill Book Company. 614 pages, illustrated. Price $4.50.

Principles of Electrical Design—D.C. and A.C. Generators, by Alfred Still. Published in 1916 by McGraw-Hill Book Co. 365 pages, illustrated. Price $3.00.

Radiodynamics—The Wireless Control of Torpedoes and Other Mechanisms, by B. F. Miessner. Published in 1916 by D. Van Nostrand Company. 206 pages, 112 illustrations. Price $2.00.

Radiation, Light and Illumination, by Steinmetz. Published in 1909 by McGraw-Hill Book Company. 304 pages, illustrated. Price $3.00.

"Storage Batteries," by C. J. Hawkes, 157 pages, illustrated, published in 1920 by The Wm. Hood Dunwoody Industrial Institute. Price $2.00.

Telegraph Practice, a Study of Comparative Method, by John Lee, M.A. Published in 1917 by Longmans-Greene & Co. 102 pages. Price $1.00.

Telegraphy—A Detailed Exposition of the Telegraph System of the British Post Office, by T. E. Herbert, A.M, Inst. E.E. Third edition. Published in 916 by Whittaker & Company. 985 pages, 630 illustrations. Price $2.00.

"Theory and Calculation of Transient Electric Phenomena and Oscillations," by Charles Proteus Steinmetz. Third edition, revised and enlarged. Published in 1920 by McGraw-Hill Book Co., Incorporated. 606 pages, illustrated. Price $6.00.

The Dynamo (2 volume), by Hawkins & Wallis, Published in 1909 by Whittaker & Company. Price $4.00.

The School of Practical Electricity (Electrician's Course in 2

QUIET operation is a very desirable feature in a fan for office service. In homes, hospitals, theatres, and hotels too, silent service is an important characteristic and dealers find this feature of the R & M Fan a big sales help.

Since noise in any machine usually indicates vibration, wear and tear on the machine, the smooth, quiet operation of the R & M Fan is a promise of the long life and untroubled service the customer can expect when he buys an R &M Fan. And likewise it is the dealer's insurance of a satisfied, profitable customer.

### The Robbins & Myers Company of Canada Limited,

Brantford     -     Ontario

# Robbins & Myers Fans

## and Motors

# Appliances
## That Are Judged By
# What They Can Do!

Women judge stores most often by the lines that are carried in stock.

And they judge the lines by the satisfaction they give. Stores that carry the New Canadian Beauty Electrical Appliances (irons, toasters, grills, stoves, percolators, etc.) are better patronized—because Canadian Beauty Appliances have been bought by women, tested thoroughly and have given service absolutely up to their highest expectations.

The New Canadian Beauty Electrical Appliances please purchasers in every way —appearance, performance, length of life and usefulness—in fact, every article in the entire line meets the most exacting requirements of all women.

The public are keenly interested in electrical appliances. The public realize the immense advantages of the electric iron. the

toaster, the grill, the heater, the percolator, etc.

Men and women are beginning to d i s c r i m i n ate among the different brands of appliances, and in this shuffle the Canadian Beauty line has received a very large share of the public preference.

Have you benefitted any? Or are you carrying a less popular line?

As you know, people are no longer buying in a haphazard fashion.

In the new Canadian Beauty Appliances the public get service, durability and smart appearance.

It is a line rich in selling talk—real, truly, honest - to - goodness selling talk.

Order from your nearest jobber or write . us direct—tonight.

*Majestic Heater*

### Renfrew Electric Products, Limited
**Toronto**
29 Richmond St. W.,
**RENFREW, ONT.**
**Winnipeg**
803 Lindsay Bldg.

*The New*
# Canadian Beauty
## Electrical Appliances

Generation, Transmission and Application of Electricity

For nearly thirty years the recognized journal for the Electrical Interests of Canada.

Published Semi-Monthly By

## HUGH C. MACLEAN PUBLICATIONS
LIMITED

THOMAS S. YOUNG, Toronto, Managing Director
W. R. CARR, Ph.D., Toronto, Managing Editor
HEAD OFFICE - 347 Adelaide Street West, TORONTO
Telephone A. 2700

| | | |
|---|---|---|
| MONTREAL | - - | 119 Board of Trade Bldg. |
| WINNIPEG | - - - | Electric Ry. Chambers |
| VANCOUVER | - - - - | Winch Building |
| NEW YORK | - - - - - | 296 Broadway |
| CHICAGO | - - | Room 803, 63 E. Adams St. |
| LONDON, ENG. | - - | 16 Regent Street S. W. |

ADVERTISEMENTS
Orders for advertising should reach the office of publication not later than the 5th and 20th of the month. Changes in advertisements will be made whenever desired, without cost to the advertiser.

SUBSCRIBERS
The "Electrical News" will be mailed to subscribers in Canada and Great Britain, post free, for $2.00 per annum. United States and foreign, $2.50. Remit by currency, registered letter, or postal order payable to Hugh C. MacLean Publications Limited.
Subscribers are requested to promptly notify the publishers of failure or delay in delivery of paper.

Authorized by the Postmaster General for Canada, for transmission as second class matter.

Vol. 32          Toronto, January 1, 1922          No. 1

## Are we "Sold" Ourselves? Let us Start the Year Right

An investigation was recently carried out by the Merchandising Division of the Commercial Section of the N.E.L.A., New England Division, to ascertain the extent to which employees and officials of the member companies are using appliances of various kinds. Replies were received from 45 companies reporting upon 218 officials and 6,492 other employees—a total of 6,710.

From the results it appears that the officers and employees of electrical companies are not themselves sold on the value of electrical appliances and, from the figures submitted, it is no wonder if they are not able to make sales to the general public. If a salesman has not sufficient confidence in the merits of a toaster or iron to install one in his own home it is extremely unlikely that he can arouse the necessary enthusiasm in his customer. It is a case that seems to fit the Scriptural admonition that we must first remove the mote from our own eye before we can see clearly enough to remove it from the eye of others.

In a study of the figures representing 6,710 electrical homes it is shown that only 2,941 own electric irons—only 44 per cent; toasters are used by 1,661 homes—a little less than 25 per cent; vacuum cleaners show the same results as toasters—25 per cent; 18 per cent use fans; 12 per cent use electric washers; 4 per cent use electric ranges, and so on down the list.

If this represented the percentage for the total inhabitants of those towns and cities it would be nothing for the electrical industry to be specially proud of.

We can't say whether these percentages would apply to Canada. Possibly Canadian figures would be somewhat better; we have many reports, for example, of more than 44 per cent of the total household population using electric irons. It will be remembered, however, that the statistics we gave some time ago regarding the general use of appliances indicated that an unbelievably large percentage of our people don't use them as they should.

There is a very real lesson to be learned from these figures, however. We would suggest that every electrical company, whether manufacturing, jobbing, contracting or merchandising, would do well to take an inventory of the electrical equipment in the homes of their employees to see how far the "Do it Electrically" idea has sold itself to the men themselves whose life work it is to sell it. There is no other selling argument that carries so much weight as "I know from my own experience." It looks as if the N.E.L.A. had exposed one of the real weaknesses of the electrical industry.

## Conciliation Board Recommends Wage Reductions

In the application of the B. C. Electric Railway Company for a reduction of wages to employees, alteration of working conditions, abolition of extra pay to motormen and conductors for Sundays and holidays, and for definition of wages and conditions governing the operation of one-man cars, the Board of Conciliation appointed to consider the application, has handed down its findings. The report, which has been released for publication by the Hon. the Minister of Labour for Canada, is not a unanimous one. The majority report, signed by Mr. W. C. Ditmars, chairman, and Mr. A. G. McCandless, company's appointee, finds in favor of the company on the question of the wage reduction and the cutting out of overtime, as well as in altering working conditions. The majority does not, however, recommend the full demands of the company, holding that the present is not an opportune time to make such drastic changes. The minority report handed in by Mr. R. P. Pettiplece, does not agree with the amount of the decrease to be granted.

After reciting the numerous meetings held, and evidence submitted, the majority report recommends a general reduction of ten per cent in all wages, with exceptions later mentioned. After expressing strong opposition to the principle of overtime for work on Sundays and holidays in operating cars, because such work is necessary to the operation of an industry giving continuous service, the majority report recommends that time and a quarter for Sundays and time and a half for holidays be made the new basis. Another recommendation was that extra men be guaranteed a minimum monthly earning of $87.50 instead of the present guarantee of six hours work per day. Spread-over time, arising out of runs not being completed within ten hours, is recommended to be paid at 10 cents per hour, instead of 25 cents, additional to the regular pay, as at present. Reporting time to be paid according to schedule running time between nearest office and relief point, is another recommendation, which would also eliminate extra box time now allowed. For shop men, overtime is recommended to be time and a half except that when men are required to work more than five hours overtime, after having already worked eight hours at straight time, they shall be paid double time for all time worked in excess of such five hours. For the one-man cars that the company proposes to instal on certain lines, the majority recommends that men operating them shall be paid 10 per cent in excess of the proposed schedule, owing to the greater responsiblity attached.

### Minority Report Recommends Less

Mr. Pettipiece, representing the employees, in his minority report does not specifically fall in with the findings of the majority on revised working conditions but does so tacitly. His recommendation of a 5% reduction in wages would, he figures, with the reductions in overtime, and changes in working conditions, make a total actual reduction of 12 per cent all round. That, he comments, would, in his opinion, have been an ample reduction for the Board to make at the present time. As the chairman did not agree with his view, Mr. Pettipiece submitted his minority recommendation.

It is stated by officials of the company that the majority report of the board will be accepted by the company, though it did not concede the full demands. It was remarked that the award definitely stated the majority opinion that the company was justified in asking for a fifteen per cent reduction, although only 10 per cent has been recommended. The decision of the men awaits the outcome of meetings of the various locals of the union at which the two reports will be considered. It is not anticipated that any strike action is likely to follow.

---

## Election Returns by Wireless A Welcome Improvement on Billboards

Election returns by wireless—in your own easy chair? Of course! It is the modern method and the best one yet found. Through the kindness of a local company who own and operate a powerful wireless telephone a wonderfully complete election service was rendered on the evening of December 6th to all Toronto amateurs who cared to tune in their wireless receivers and listen for the returns.

Those who stood for hours in the crowd before bulletin boards or sheets on which figures, names and cartoons were being flashed will doubtless be interested in the wireless method. Its speed exceeds that of the bulletin board. and you can get all the information you want right in your own home and entertain your friends with the results.

It is probable that 500 homes in Toronto alone received returns by wireless. but if it is an advantage to city folks think what it means to farmers or villages and towns at a distance from headquarters.

As one instance of the way it worked out on last Dec. 6, the Anglican Young People's Association of St. John's Church, Weston, Ont., decided to have a wireless entertainment and announced that election returns would be received by wireless.

A temporary aerial was swung between two trees outside the place of meeting, a valve detector and three-step amplifier was installed and the current from the highest step operated a Magnovox which spoke right up and gave the information to all of the one hundred and fifty persons who formed the audience. It is hard to tell whether election interest or wireless curiosity brought out the crowd, which consisted of voters from all over town as well as the non-voting members of the Association. However, they were there and heard, for the first time, a human voice which by wireless had crossed several miles of country.

The service was so fast that there was not any time for music between the announcements, but election returns were the most important matters that evening, and with the different views represented and comments passing back and forth in the audience all those present enjoyed this novel means of entertainment.

How much better was that for the audience than standing before the bulletin boards and then wending their way home at midnight over miles of bad roads?

---

## Government Telephone Commissioner Addresses Members of Manitoba Electrical Ass'n.

"Good service generates the desire to use the service more."

This is the policy which underlies the administration of the Manitoba Government Telephones, according to Commissioner J. E. Lowry. Mr. Lowry addressed the Manitoba Electrical Association on Thursday, the 10th Nov. and in addition to describing the work of the Exchange, spoke of the desire which imbued all telephone employees to give courteous and efficient service. The occasion recorded the largest attendance that has been present at an Association meeting.

Mr. Lowry had demonstration equipment and apparatus to illustrate his talk. Before going into the technical operation of the phone system, however, he made a few references to the "service" given customers.

"The service in Winnipeg cannot be considered as bad," he said, "when one considers that approximately half a million calls each day are handled. The average duration of these calls is 1½ minutes. If one man or woman were to talk into a telephone continuously, 24 hours a day, for 365 days a year, for 37 years, he or she would have equaled the amount of time taken up in phone calls in Winnipeg in one single month. In other words, the phone time consumed by Winnipeg subscribers in talking during one month of 26 working days equals the continuous use of one phone over a period of 37 years.

"We have in Winnipeg practically the last word in automatic telephone apparatus."

Mr. Lowery very lucidly explained the complexities of the phone system, and demonstrated the workings of the automatic system, showing how efficient and time-saving it was over the manual system. He showed how easy it was, too, for errors to be made, and how great and insistent was the effort of the department employees to eliminate errors and cause for complaints. The audience dispersed with a feeling of keener appreciation of the efforts being made to give Winnipeg the "best phone service possible."

---

## New Power Unit at White Rock

The Gaspereau Light, Heat & Power Co., Ltd., have recently placed contracts for the installation of a new power unit to be installed in their White Rock, N. S., plant. The prime mover will consist of an S. Morgan Smith horizontal water turbine with a nominal maximum output of 150 brake h.p., coupled to a Westinghouse, 125 kv.a., 3-phase, 60-cycle, 2,300-volt, alternator, with direct-connected exciter. This unit will add about 50 per cent to the present station capacity when operated in parallel with the main unit already installed, but its chief purpose is for operation during the off-peak period under light load to insure higher station efficiency and resulting economy in the use of water power during dry periods when the river flow requires to be augmented from storage.

This work is in charge of Mr. Kenneth L. Warren, consulting engineer, who states that this plant, when completed, will be one of the most compact. dependable and efficient hydro-electric power stations, for its size, in Nova Scotia.

---

The Ontario Gazette announces the incorporation of the Matachewan Power Company, Ltd., with a capitalization of $1,000,000. Head office, Toronto. It is understood this company will develop water powers on the Montreal River, in the Matachewan area, Northern Ontario.

# Winnipeg Hydro Starts Survey Work at Slave Falls, Man.

The Winnipeg Hydro Electric System have a gang of men making a survey of the power site at Slave Falls, which is situated four and one half miles from their present plant at Point du Bois. Following the policy of the Department to keep their supply of power well ahead of the demand, Mr. J. G. Glassco requested the city council to go ahead with the survey, It is expected the work of survey will extend over two or three years.

By undertaking this work early, and taking plenty of time, considerable saving will result in the ultimate expenditure. Mr. J. A. McGillivray, who was engineer-in-charge of the Point du Bois extension, is taking charge of the present operatons.

## Construction Advantages

There are a number of features which make Slave Falls the logical point for the next power development on the Winnipeg river. As general examples the following are given.

In all hydro-electric developments the cost of transportation facilities and equipment forms a considerable portion of the total cost. With a comparatively negligible expenditure, the Point du Bois railway track will be extended to Slave Falls, thus providing an excellent railway connection to the site of the new power house from the Canadian Pacific Railway.

Excellent gravel pits, already opened and provided with spur trackage are situated within four miles of Slave Falls. With the material for manufacturing the concrete so close at hand the cost of the power house building and river dams will be a minimum.

Point du Bois and Slave Falls are situated so closely together that both plants will be operated together as one large power house. Hence it will only be necessary to increase the Point du Bois staff just sufficiently to operate a combined plant of approximately 170,000 h.p.

Electrical energy for operating the machinery used in the construction of the Slave Falls plant will cost practically nothing, as it will be transmitted the short distance of four miles from the Point du Bois power house on a steel tower line which will be eventually used for the necessary electrical tie between the two power houses.

## Operating Advantages

It has frequently been stated that the new power development at Slave Falls can only be operated satisfactorily in conjunction with the existing power house of the City Hydro. There are several reasons why this is so, and it is not difficult to predict accurately the unfortunate condition that would exist if the two power sites were not combined to form one large development.

In order that the Point du Bois development may be economically operated it will be necessary at certain times of the year that all the water flowing down the river be made to pass through the power house turbines. Hence within certain limits, the amount of water flowing from Point du Bois to Slave Falls will depend on the amount of electrical load which the Point du Bois power house is carrying.

A sudden reduction of the load at Point du Bois can therefore seriously diminish the amount of load that the Slave Falls power house might carry. The natural flow of the river at these times would be held back at Point du Bois for filling up the large pondage area above the power house. In the meanwhile Slave Falls would have to patiently wait

for Point du Bois to carry more load or for the regulating dams to overflow.

In a similar manner it would be possible for the Slave Falls power house to interfere considerably with the efficient operation of the Point du Bois plant. This condition would occur if the amount of load on the Slave Falls caused the elevation of the water between the two power houses to be too high for the most efficient operation of Point du Bois.

It can thus be demonstrated that the highest efficiency in the use of these two falls is obtained by controlling the ratio between the head at Point du Bois and the head at Slave Falls. As the head at each plant bears a certain relation to its load, it follows that the load at the plants must be kept in certain relation to each other to obtain the desired result. At the time when both plants are working at their maximum output, for every horsepower generated at Slave Falls, there should be a corresponding load of one and three quarters horse power at Point du Bois.

The manner in which one development can interfere with the operation of another development on the same river depends to a great extent on the distance between the two plants under discussion and the pondage area between them. With a limited pondage area such as that between Point du Bois and Slave Falls the interference can be considerable, while with plants located forty or fifty miles apart there is not much chance for interference provided the elevation of the dams at the up-stream power house is definitely fixed by some authority such as the Dominion Government.

## Automatic Telephones on Pacific Coast

The Northern Electric Company recently installed a private automatic telephone exchange in the large pulp and paper plant of the Powell River Company, at Powell River, B. C. The P. A. X. (Private Automatic Exchange), installed, is known as the "step by step" type of equipment. It provides for 125 individual lines and common equipment to which 75 lines can be added as required. The power plant consists of motor generator set, storage batteries and power board, equipped with automatic charging switch, ringing machine, automatic circuit breaker and the usual equipment found on an up-to-date power board. Both mill and townsite are now provided with excellent service 24 hours a day, without attendant operator. This "peps" up the organization for night and day operation by giving uniform service at speed and accuracy not attainable by manual operation.

Wiring is provided in the equipment for special features or services which can be added to further increase the efficiency of the plant. Some of these automatic services are "The Night-Watchman Service," wherby telephones in daily use and suitably located are detailed for watchman stations; the watchman on his rounds simply raises the telephone receiver which operates time recording equipment at the central office, thus making a permanent record of all calls. The "Conference" service provides equipment which permits the head of the organization to hold a round table conference by telephone. Each party summoned to the conference calls a predetermined number which connects him with the conference wire. The "Code Call" service locates persons who are about the plant but not at their desk. By dialing the code call number of the person wanted, buzzers, bells, horns, or whistles in various parts of the plant sound the code and the person wanted steps to the nearest telephone and dials a number which will connect him with the party originating the call. Telephones are equipped with a dial by which all the various services are controlled and operated, and a pair of wires connect each telephone to the central office equipment. This installation, while the first of its kind in British Columbia, indicates that telephone development is keeping abreast with the advance of things electrical on the Pacific Coast.

# A Central Heating Plant in a Western Town

### By M. D. CADWELL
Superintendent of Utilities, North Battleford, Sask.

Central steam heating was first introduced into commercial circles in North Battleford in the summer of 1916 when a steam main was laid from the power plant to the new public library which was built that year, located on Main street 750 feet distant. The service was so successful that it was decided to extend the system to serve the business section of the city after the close of the war, provided ways and means could be arranged to finance the initial cost of the installation.

A very comprehensive research was conducted by the writer, for a period covering nearly four years, relative to the merits of central heating and its adaptation to conditions prevailing in North Battleford.

Finally, in the summer of 1920, it was decided that the proposed installation would prove advantageous and profitable to the community and the city council signified its willingness to proceed at once with the installation, provided the patrons would finance the cost of same. An advance deposit, representing the fixed figure on a unit basis, and proportional to the requirements of each consumer, was agreed upon and tenders were called for supply of necessary material.

A portion of the installation was made in the Fall of 1920, but, due to the lateness of the season, less than ten consumers received service during the season of 1920-21.

Meanwhile, materials were received from time to time and in the spring of last year, practically all supplies requisite for the completion of the system were on the ground prior to the date on which excavation could be commenced.

By the end of September last, consumers were receiving steam service and installation of the distribution system was completed.

The popularity of the service and the ever increasing demand for steam, necessitated the installation of a new and larger steam main to supplement the original to a point opposite the public library, from which point the main distribution system was commenced in the Fall of 1920. This installation has recently been completed and has been operating since November 26th last.

The complete installation now comprises over 5000 lineal feet of piping arranged as below:

### Under Ground Mains

| | | |
|---|---|---|
| 743 lineal feet of | 12 | inch pipe |
| 270 lineal feet of | 10 | inch pipe |
| 326 lineal feet of | 8 | inch pipe |
| 920 lineal feet of | 5 | inch pipe |
| 1318 lineal feet of | 4 | inch pipe |
| 90 lineal feet of | 3 | inch pipe |
| 56 lineal feet of | 2½ | inch pipe |

### Under Ground Services

| | | |
|---|---|---|
| 131 lineal feet of | 4 | inch pipe |
| 407 lineal feet of | 3 | inch pipe |
| 522 lineal feet of | 2½ | inch pipe |
| 147 lineal feet of | 2 | inch pipe |
| 98 lineal feet of | 1½ | inch pipe |

### Under Ground Drains

| | | |
|---|---|---|
| 94 lineal feet of | 1½ | inch pipe |
| 55 lineal feet of | 1¼ | inch pipe |

5077 total lineal feet.

Practically all under-ground piping is of genuine wrought iron, a considerable portion of which was imported as it is not made in Canada above certain sizes.

All piping is thoroughly insulated and enclosed in circular casing manufactured in the City of North Battleford and every lineal foot has been graded with an engineer's level, and all lines were set with the transit,

Complete and perfect drainage has been provided for all piping placed below the ground level. The estimated life of the plant is 50 years.

The total cost of the system, as installed, approximates $35,000, and has been financed by the patrons of the plant in addition to certain lines of credit which were arranged with two of the firms who submitted the lowest tender for a considerable quantity of the materials required.

The complete system was designed by the writer and was installed by the employees of his department under his personal supervision.

There are now upwards of 40 consumers, and this figure will exceed 50 in the near future.

During the recent inclement weather, upwards of seventy five tons, or 150,000 lbs. of steam were delivered to the patrons every twenty four hours.

Exhaust steam from the generating units at the power plant is used as the source of heat.

While the City of North Battleford now owns and operates a comprehensive and ideal central heating plant, the first municipally owned plant in Canada to date, it has not cost the rate-payers a single farthing and is greatly augmenting the revenue from the utilities.

Meanwhile, the operation of this new utility is being watched by engineers and other interested municipal officials from coast to coast, and it is sincerely hoped that the results obtained by this municipality, will in the near future, justify similar installations elsewhere, and that the comfort and convenience now being enjoyed by the patrons of the North Battleford heating plant may soon become the privilege of many in other urban centres.

# Carrier Telephony Developments

### By MR. H. J. VENNES
Before the Northern Electric Engineering Society

As a consequence of the rapid development in the art of telephonic communication, it is not surprising that the public, or even those engaged in other branches of the electric industry, are hardly acquainted with the new development, "Carrier Telephony." In general, this development makes it possible to carry on two or more conversations and a telegraphic message simultaneously over one circuit.

Mr. Vennes said that Carrier Telephony is a very broad subject and that it was impossible to cover the entire field adequately in a single lecture, but he wished to point out a few of the factors that have been responsible for the developments and progress which have been done in this field of communication. The possibilities of multiplex transmission were conceived a long time ago, some of the ideas dating back as far as 1890 and since that time various theories have been proposed giving various methods for obtaining several simultaneous conversations over a single telephone line. Until very recently, however, no apparatus has been available which would make carrier telephony a practical realization. The first actual demonstration of this kind was made by Major-General Squier in 1911 when he conducted a multiplex transmission test between Baltimore and Washington. He had at his disposal one of the few high-frequency generators which were available at that time, this being an Alexanderson high-frequency alternator. This apparatus was very elaborate and expensive and could not be considered as being adaptable to the commercial field where carrier telephony should actually be applied.

During the next few years a considerable amount of progress was made in the telephone art in the establishing of the transcontinental telephone line for the 1915 Fair which was held at San Francisco. This development resulted in the perfection of the vacuum tube, a device which has come to be of great importance in carrier telephony. About the same time a practical demonstration in radio telephony was also made upon radio telephone messages which were sent out from Arlington and which were heard as far away as 6,000 miles. This demonstration showed the possibilities of using the vacuum tubes for the generating of high frequency currents and also for combining or modulating the voice current with high frequency current so that the former could be transmitted as a radio current.

### Over Ordinary Telephone Lines

As a result of these developments and demonstrations the attention of telephone engineers was directed towards the possibilities of transmitting these high frequency currents over ordinary telephone wires thereby reducing the amount of power which would be needed and consequently reducing the cost of the equipment . However; before any apparatus could be built, it was necessary to investigate the characteristics of the telephone lines on transmitting these frequency currents. It was necessary to determine what frequency ranges should be used since the vacuum tube had solved the problem of generating alternating currents of any desired frequency; consequently it had to be decided which range of frequencies could most economically be used. The experimental data which was obtained showed that the attenuation or losses in the telephone line when high frequency currents were transmitted increased very rapidly as the frequency was increased and that approximately 30,000 cycles appeared to be the upper limit of frequencies which could economically be used for carrier telephone transmission. Experimental apparatus was then built and practical tests carried out which were very successful. The first real commercial installation was made in 1918 when a four channel carrier telephone system was put into operation between Baltimore and Pittsburg. Since that time many more installations have been made and all are giving very satisfactory results.

### One Installation In Canada

Mr. Vennes stated that a carrier installation has also been made in Canada this summer. The system in now installed and operating very satisfactorily between Calgary and Edmonton. This is the first commercial installation outside the United States.

Carrier Telephony depends on three major factors,—

1. The generating of a high frequency current.

2. The combining or modulation of this high frequency current with the voice.

3. The selecton at the distant end of the high frequency current which pertains to any particular channel.

The first two operations are carried out by the vacuum tube and the third is accomplished by the use of what are known as wave filters. The principal thing to be borne in mind when dealing with modulation of a high frequency current by the voice is that two new frequencies are produced. These are the carrier frequency plus the voice frequency and the carrier frequency minus the voice frequency. In other words, the voice frequency and are really converted to a high frequency current which have the voice characteristics impressed on them. These currents are then transmitted to the distant end where they are selected into their proper channels and again passed through into a circuit where the original voice currents are reproduced. Thus, by combining various voice currents with as many different high frequency currents we would then actually have the conversation of several parties transmitted as a different high frequency current and each one can be separated out at the distant end.

### Wave Filters

The wave filters in carrier telephony are of three different types. One type is known as a low pass filter, and will permit of all currents below a given frequency to pass. The second type is known as a high pass filter and will permit currents of all frequencies above a certain point to pass. The third type is known as a band filter and will permit frequencies within two given limits to be transmitted. The band filters are used for separating the current for each individual channel and the high and low pass filters are used for separating all the carrier currents from the ordinary voice and telegraph currents which may be transmitted over the telephone line simultaneously with the carrier currents.

Mr. Vennes, at the conclusion of his address, referred to several charts. One of these gave a very clear idea of what a complete carrier current telephone circuit involves. He traced the actual transmission through this system so as to show where these various operations actually take place. An interesting chart was used to explain the theory and the varied functions of the electric valve. Still another chart showed the ranges of frequency used for carrier purposes, the voice frequency and the telegraph frequency.

## Canadian Electrical Association, Montreal Section

On December 19 a Montreal Section of the Canadian Electrical Association was formed at a meeting held at the building of the Engineering Institute of Canada. Mr. Julian C. Smith, vice-president of the Shawinigan Water & Power Company, and president of the Canadian Electrical Association, was in the chair. He outlined the objects in forming the Montreal Section. The headquarters of the Association, he pointed out, were in Montreal and it was, therefore, fitting that a Section should be located there. The Section would be a benefit in the direction of mutual assistance to the members; it would also result in extending the activities of the parent association, while the reading of papers on subjects of interest to the members would prove beneficial.

Mr. P. T. Davies, Southern Canada Power Co., was elected chairman; Mr. J. W. Pilcher, Canadian General Electric Co., vice-president; and Mr. E. Vinet, Shawinigan Water & Power Co., secretary-treasurer.

The program committee is composed of Messrs. L. A. Kenyon, Montreal Light, Heat & Power Consolidated, chairman; A. Lee Jones, Canadian General Electric; A. J. Soper, Northern Electric; George K. McDougall, consulting engineer; Charles F. Medbury, Canadian Westinghouse; George R. Atchison, Southern Canada Power, and W. A. Bunyon, Shawinigan Water & Power.

The following are the members of the Membership Committee: Messrs. M. K. Pike, Northern Electric, chairman; E. Playford, Canadian General Electric; A. Anderson, contractor-dealer; P. S. Gregory, Shawinigan Water & Power; H. C. Haskell, Southern Canada Power; G. A. Wendt, Canadian Westinghouse; and W. O'Brien, Montreal Light, Heat & Power Consolidated.

A large number of films were shown at the conclusion of the formal business. These were: "Back of the Button," loaned by the National Electric Light Association; "A Square Deal for his Wife," loaned by the Western Electric Company, and a number of pictures showing the water powers of Manitoba.

# World-wide   Power   Development

## All Parts of the World Involved in an Unprecedented Investigation of the Value of Water Power Resources—Uses of Hydro-Electric Energy Rapidly Increasing ( Continued )

By Prof. A. H. GIBSON, D. Sc., before the British Association.

The figures already quoted indicate that the scope for inland water power development throughout the world, and more particularly throughout the British Empire, is likely to be large for many years to come, and it is gratifying to know 'that British engineers are prepared to play a large part in such development work.

The utilization of this water power is likely to give rise to some economic problems of interest and importance. When industrial conditions have again become stabilized, the competitive ability of the various nations will depend largely on economy in the application of energy to production and transportation, and the possession of cheap water power is likely to counterbalance the posession of such resources as coal and iron as a measure of the industrial capacity of a nation.

While it is probably true in industrial communities that the most attractive water power schemes have already received attention, many of those available in countries which have hitherto been non-industrial are capable of extremely cheap development and will certainly be utilized as soon as a market for their output can be assured.

It is in such countries that the result of these developments is likely to be most marked, and will require most careful consideration. Thus the hydro-electric survey of India now being carried out by the Indian Government indicates that very large water power resources are available in the country, and that, although a few large schemes have been or are being developed, the resources of the country are practically untouched. There can be little doubt that in the course of time a large amount of cheap energy will be available in India for use in industrial processes, and as the country possesses a large and prolific population readily trained to mechanical and industrial processes, along with ample supplies of raw material for many such processes, all the conditions would appear to be favorable for its entry into the rank of manufacturing and industrial nations.

### Modern Tendencies in Water-power Development

The large amount of attention which has been concentrated on the various aspects of water power development during the past ten years has been responsible for great modifications and improvements in the design, arrangement, and construction of the plant.

Broadly speaking, these have been in the direction of increasing the size, capacity, reliability, and efficiency of individual units; of improving the design of the turbine setting and of the head and tail works; of increasing the rotative speed of low head turbines; of detailed modifications in the reaction type of turbine to enable it to operate under higher heads than have hitherto been considered feasible; and of increasing the voltage utilized in transmission.

The capacity of individual units has been increased rapidly during recent years, and at the present time units having a maximum capacity of 55,000 horsepower under a head of 305 ft. are being installed in the Queenston-Chippewa project at Niagara, while units of 100,000 horsepower are projected for an extension of the same plant.

These modern high-power turbines are usually of the vertical shaft, single runner type, with the weight of the shaft, runner, and generator carried from a single-thrust bearing of the Michell type. This type lends itself to a simple and efficient form of setting, while the friction losses in the turbine are extremely low. As a result of careful overall design it has been found possible to build units of this type having an efficiency of approximately 93 per cent.

One of the great drawbacks of the low head turbine in the past has been its relatively slow speed of rotation, which necessitated either a slow speed, and consequently costly generator, or expensive gearing. As a result of experiment it has, however, been possible so to modify the form of the runner as greatly to increase the speed of rotation under a given head without seriously reducing the efficiency.

Investigations in this direction are still in progress and promise to give rise to important results. At the present time, however, turbines are in existence which are capable of working efficiently at speeds at least five times as great as would have been thought feasible ten years ago.

### Welded Pipe for High Heads

The non-provision of a suitable pipe line has, until recent years, tended to retard the development of plants for very high heads. Under such heads the necessary wall thickness, even with a moderate pipe diameter, becomes too great to permit of the use of riveted joints. Recent developments in electric welding and oxy-acetylene welding have, however, rendered it possible to construct suitable welded pipes, and by their aid, and by the use of solid-drawn steel pipes in extreme cases, it has been found possible to harness some very high falls. The highest as yet utilized is at the Fully installation in Switzerland. Here the working head is 5412 ft., corresponding to a working pressure of 2360 lbs per square inch. The pipe line is 19.7 in. in diameter and 1¾ in. thick at its lower end, and each of the three Pelton wheels in the power house develops 3000 horsepower, with an efficiency of 82 per cent.

Until comparatively recently the Pelton wheel was looked upon as the only practicable turbine of high heads, and the use of the Francis turbine was restricted to heads below about 400 ft. This was due partly to the fact that a reaction turbine of comparatively small dimensions gives a large output under a high head, and except in turbines of comparatively large power the water passages become very small and the friction losses in consequence large.

A further and more important reason for the general choice of the Pelton wheel for high heads was the fact that in the earliest Francis turbines, when operating under heads involving high speeds of water flow, corrosion of the runner was very serious. This corrosion is now generally attributed to the liberation of air containing nascent oxygen at points where eddy formation causes regions of low pressure. Careful design of the vanes has enabled this to be largely prevented in modern runners, and in consequence the field of useful application of the Francis turbine has been extended until at present turbines of this type are operating successfully under a head of 850 ft., and this limit will probably be exceeded in the near future.

The great increase in all constructional costs since 1914 has increased the cost of the average hydro-electric plant by something of the order of 150 per cent., and since the cost of energy produced by such a plant is mainly due to fixed charges on the capital expenditure this cost has gone up in an even greater proportion owing to the higher interest charges now demanded.

It is true that the same increased cost applies within narrow limits to the output from every steam plant erected since the war, and the relative position of the two types of power plant with coal at about 25s per ton is much the same as before the war.

The fact remains, however, that a newly constructed hydro-electric plant has often to compete in the market with a steam plant built in pre-war days whose standing charges are comparatively low, and in order to enable it to do so with success the constructional cost must often be reduced to a minimum compatible with safe and efficient operation. With this in view many modifications in design and construction have been introduced in recent plants, but there would still appear to be ample scope for investigation into the possibility of reducing the first cost by modifying many of the details of design and methods of construction now in common use.

### Modification in Plant Design

Among recent modifications in this direction may be mentioned:—

1. The elimination of the dam in storage schemes in which natural lochs or reservoirs are utilized, this water level being drawn down in times of drought instead of being raised in times of flood. This reduces the cost of construction appreciably in favorable circumstances and eliminates the necessity for paying compensation for flooding of the land surrounding the reservoir.

2. The substitution, where feasible, of rockfill dams for those of masonry or monolithic concrete.

3. The introduction of outdoor installations with the minimum of power house construction.

4. The simplification of the power plant.

Some progress has already been made in these directions, and it is probable that experience based on recent installations and experimental investigations will enable considerable further progress to be made.

### Research in Hydro-electric Problems

There are few branches of engineering in which research is more urgently required and in which it might be more directly useful.

Among the many questions still requiring investigation on the civil and mechanical side may be mentioned:

1. Turbines.—Investigation of turbine corrosion as affected by the material and shape of the vanes.

Effects of erosion due to sand and silt.

Resistance to erosion offered by different materials and coatings.

Bucket design in low head high-speed turbines.

Draught tube design.

Investigations of the directions and velocities of flow in modern types of high-speed turbines.

Investigation of the degree of guidance as affected by the number of guide and runner vanes.

2. Conduits and Pressure Tunnels.—The design of large pipe lines under low heads with the view of reducing the weight of metal. The investigation of anti-corrosive coatings, so as to reduce the necessity for additional wall thickness to allow for corrosion.

Methods of strengthening large thin-walled pipes against bending and against external pressures.

Methods of lining open canals and of boring and lining pressure tunnels.

Effects of curvature in a canal or tunnel.

3. Dams.—Most efficient methods of construction and best form of section especially for rockfill and earthen dams. Best methods of producing water tightness.

4. Run-off Data.—Since the possibility of designing an installation to develop the available power efficiently and economically depends in many cases essentially on the accuracy of the run-off data available, the possession of accurate data extending over a long series of years is of great value.

While such data may be obtained either from steam gaugings or from rainfall and evaporation records, the former method is by far the more reliable. For a seasonable degree of accuracy, however, records must be available extending over a long period of years, and at the present moment such data are available only in very few cases.

Where accurate rainfall and evaporation records are available it is possible to obtain what is often a sufficiently close approximation to the run-off, but even rainfall records are not generally at hand where they are most required, and, even in a district where such records are available they are usually confined to easily accessible points, and are seldom extended to the higher levels of a catchment area where the rainfall is greatest. Even throughout the United Kingdom our knowledge of the rainfall at elevations exceeding 500 ft. is not satisfactory, and little definite is known concerning that at elevations exceeding 1000 ft.

In this country evaporation may account for between 20 and 50 per cent. of the annual rainfall, depending on the physical characteristics of the site, its exposure, mean temperature, and the type of surface covering. In some countries evaporation may account for anything up to 100 per cent. of the rainfall. As yet, however, few records are available as to the effect of the many variables involved. An investigation devoted to the question of evaporation from water surfaces and from surfaces covered with bare soil and with various crops, under different conditions of wind, exposure, and mean temperature, would appear to be urgently needed. If this could be combined with an extension of Vermeulle's investigation into the relationship between rainfall, evaporation, and run-off on watersheds of a few characteristic types, it would do much towards enabling an accurate estimate of the water power possibilities of any given site to be predetermined.

Even more useful results would follow the initiation of a systematic scheme of gauging applied to all streams affording potential power sites.

Among other questions which are ripe for investigation may be mentioned:—

1. The combined operation of steam and water power plants to give maximum all-round efficiency.

2. The relative advantages of high-voltage direct-current and alternating-current generation and transmission for short distances.

3. The operation of automatic and semi-automatic generating stations.

### Tidal Power

The question of tidal power has received much attention during the last few years. In this country it has been considered by the Water Power Resources Committee of the Board of Trade, which has issued a special tidal power report dealing more particularly with a suggested scheme on the Severn. The outline of a specific scheme on the same estuary was published by the Ministry of Transport towards the end of 1920.

In France a special commission has been appointed by the Ministry of Public Works to consider the development of tidal power, and it has been decided to erect a 3000-kilowatt experimental plant on the coast of Brittany. With the view of encouraging research, the Government proposes to grant concessions, where required, for the laying down of additional installations.

The tidal rise and fall around our coasts represent an enormous amount of energy, as may be exemplified by the fact that the power obtainable from the suggested Severn installation alone, for a period of eight hours daily throughout the year, would be of the order of 450,000 horsepower.

Many suggestions for utilizing the tides by the use of current motors, float-operated air compressors, and the like, have been made, but the only practicable means of utilizing tidal energy on any large scale would appear to involve the provision of one or more dams, impounding the water in tidal basins, and the use of the impounded water to drive turbines.

The energy thus rendered available is, however, intermittent; the average working head is low and varies daily within very wide limits, while the maximum daily output varies widely as between spring and neap tides.

If some electro-chemical or electro-physical process were available, capable of utilizing an intermittent energy supply subject to variations of this kind, the value of tidal power would be greatly increased. At the moment, however, no such process is commercially available, and in order to utilise any isolated tidal scheme for normal industrial application it is necessary to provide means for converting the variable output into a continuous supply constant throughout the normal working period.

Various schemes have been suggested for obtaining a continuous output by the co-ordinated operation of two or more tidal basins separated from each other and from the sea by dams with appropriate sluice gates. This method, however, can only get over the difficulty of equalizing the outputs of spring and neap tides if it be arranged that the maximum rate of output is that governed by the working head at the lowest neap tide, in which case only a small fraction of the available energy is utilized.

When a single tidal basin is used it is necessary to provide some storage system to absorb a portion of the energy during the daily and fortnightly periods of maximum output, and for this purpose the most promising method at the moment appears to involve the use of an auxiliary high-level reservoir into which water is pumped when excess energy is available, to be used to drive secondary turbines as required. It is, however, possible that better methods may be devised. Storage by the use of electrically heated boilers has been suggested, and the whole field of storage is one which would probably well repay investigation.

If a sufficiently extensive electrical network were available, linking up a number of large steam and inland water power stations, a tidal power scheme might readily be connected into such a network without any storage being necessary, and this would appear to be a possibility which should not be overlooked in the case of our own country.

A tidal power project on any large scale involves a number of special problems for the satisfactory solution of which our present data are inadequate.

Thus the effect of a barrage on the silting of a large estuary, and the exact effect on the level in the estuary and in the tidal basin at any given time, can only be determined by experiment, either on a small installation or preferably on a model of the large scheme.

Many of the hydraulic, mechanical, and electrical problems involved are comparatively new, and there is little practical experience to serve as a basis of their solution.

Among these may be mentioned:

1. The most advantageous cycle of operations as regards working periods, mean head, and variations of head.

2. The methods of control and of sluice gate operation.

3. Effect of changes of level due to wind or waves.

4. The best form of turbine and setting and the most economical turbine capacity.

5. The possibilities of undue corrosion of turbine parts in salt water.

6. The best method of operation; constant or variable speed.

7. Whether the generators shall be geared or direct driven.

8. Whether generation shall be by direct or alternating current.

The questions of interference with navigation and with fisheries, of utilizing the dam for rail or road transport across the estuary, and, above all, economic questions connected with the cost of production, and the disposal of the output of such an installation, also require the most careful consideration before a scheme of any magnitude can be embarked upon with assurance of success.

In view of the magnitude of the interests involved and of the fact that rough preliminary estimates indicate that to-day current even for an ordinary industrial load could be supplied from such an installation at a price lower than from a steam generating station giving the same output with coal at its present price, it would appear desirable that these problems should receive adequate investigation at an early date.

In view of the considerations already outlined, and especially in view of the large part which British engineering will probably play in future water power developments, the provision on an adequate scale at some institution in this country of facilities for research on hydraulic and cognate problems connected with the development of water power is worthy of serious attention.

At present the subject is treated in the curriculum of the engineering schools of one or two of our universities, but in no case is the laboratory equipment really adequate for the purpose in question.

What is required is a research laboratory with facilities for experiments on the flow of water on a fairly large scale; for carrying out turbine tests on models of sufficient capacity to serve as a basis for design; and, if possible, working in conjunction with one or more of the hydro-electric stations already in existence, or to be installed in the country, at which certain large scale work might be carried out.

The provisions of such a laboratory is at the moment under consideration in the United States, and in view of the rapidity with which the designs of hydraulic prime movers and their accessories are being improved at the moment, it would appear most desirable that the British designer, in order that the deservedly high status of his products should be maintained and enhanced, should at least have access to equal facilities, and should, if necessary, be able to submit any outstanding problems to investigation by a specially trained staff.

The extent to which our various heat engine laboratories have been able of recent years to assist in the development of the internal combustion engine, and to which our experimental tanks have assisted in the development of the ship-building industry, is well known to most of us, and the provision of similar facilities to assist in the development of our hydro-electric industry would probably have equally good results in this connection.

# The Development of Wireless—I

## A Series of Short Interesting Articles Covering Wireless Progress to Date

### By F. K. D'ALTON

In the present day, when wireless communication has reached such a high state of efficiency that the human voice can be carried over distances of hundreds of miles through the ether, it is interesting to review the steps in the development of this art and to predict its future as we see it.

There are many methods used in the communication of intelligence and these may be classified under two headings—

(a) Where the transmitted energy is guided to its destination;

(b) Where the transmitted energy is radiated in all directions.

In the former class come such methods as messenger service, postal service, telegraph, telephone and, perhaps the simpler speaking tube; but the latter class includes sound and light and also the subject of this paper—radio communication, whereas any of the systems under the second heading may rightly be called "radio" systems, the term is restricted and now understood to mean communication by radiated waves which are produced electrically.

All systems of communication require three parts,—the transmitter, the conducting medium and the receiver, and the fact that the transmitter and receiver are at some distance from each other gives the system a purpose.

The vocal cords, the air and the ear form a complete communication system but the velocity of propagation of sound waves is low—about 1100 ft. per second and the transmitted energy is rapidly diminished as it travels.

With a system using light where the illuminant, the ether and the eye form the three important parts, a very much greater speed is attained—approximately 188,000 miles per second. The radiated energy is not so rapidly reduced.

Sound waves are impeded by certain materials and conducted more freely by others and in much the same way particular materials are opaque and will not permit the passage of light through them while the transparent and translucent substances will allow it to travel through them with more or less diversion in direction. Both light and sound waves will travel in all directions from the source if so permitted but sound waves will surround an object whereas the light waves travel in straight lines.

If now we take a string of fixed length and cross section and submit it to a given stress in tension, we will find on plucking it that a definite note is emitted and each time we perform this experiment the pitch is the same. A change in any of the above constants will alter the pitch unless two of these quantities be so changed that their effects counteract each other.

The pitch of the sound depends then upon certain constants in the sound producing device, and is a measure of the length of the emitted sound waves. A change of wave length entails a change of pitch; the longer waves giving the lower pitch and vice versa.

In the study of light we find that certain constants in the source determine the color, or mixture of colors of the light radiated and here the wave length is also a definite amount; variations in wave length showing themselves as changes in color.

In wireless communication, waves considerably longer than light waves are used. These are radiated in all directions with fixed wave lengths and follow most of the laws of light with the exception that they will penetrate most of the substances which are opaque to light. The methods used for

generating these waves use electricity as a means to an end and in fact, this is the only known agent. The wave lengths are determined by certain electrical constants in the circuits of the transmitter. Changes in wave length do not appear as differences in pitch or color but necessitate changes in the constants of the receiving device and in this way provide an excellent means of selection of particular stations. Unlike sound or light waves, these wireless waves are not detectable by the human senses until certain changes have been made which result in the emission of sound corresponding to variations in the wireless waves themselves.

While communication by sound and light has been used for centuries, it is only within the last three score years that anything has been known about wireless.

The first electrical methods of communication were, of course, the land telegraph over wire lines, in which a conventional code was used; the operators necessarily had to be educated in this code. The received messages were written down by a marking device which made short and long marks on a moving tape. The operator then more or less at his leisure, could interpret these markings and write them as the letters of the alphabet which they represented.

It was not very long until operators found that by ear they could read and interpret the ticking of the automatic machine with the result that the recording devices were discarded and the messages written down in the actual words for which they stood. Greater speed was in this way obtained, and now in the ordinary railway and telegraph office, the station masters and operators use their sense of hearing in receiving signals.

We cannot leave the land telegraph system without suggesting to the minds of our readers a few late developments. High speed combined with absolute accuracy characterizes the present day automatic receiver which without the aid of any interpreter writes the incoming message in readable form, very much exceeds the speed of translation of the best operators and can run all day at top speed without tiring. Economy is obtained by means of multiplex methods whereby many messages are sent simultaneously over a pair of conductors without the slightest interference or deviation from their intended courses.

Prof. Alexander Graham Bell made the first marked improvement on the land telegraph when he introduced a transmitting device which caused the current in the line to vary in accordance with sound waves, and at the receiving end introduced a device which would regenerate the sound waves from the variations in the current, thus making possible the telephone.

We need hardly hesitate to mention the rapidity with which the telephone has developed, and the manner in which it has changed from a mere experimental accomplishment to a common necessity of life, particularly in the business world.

The elements in the system have been improved to give almost perfect speech over distances of thousands of miles; lines have been extended, freed by many novel schemes of interference from power lines and other electrical circuits until to-day it is a mere matter of a few minutes for one person to get into touch with any other person wherever he be on the continent. It is doubtful if anything will supersede this system, which to-day is augmented by the radio, enabling one to carry on personal conversation with others, at sea or in isolated places as well as in accessible locations on land.

In the next issue we shall commence the story of Radio. Many of the difficulties and obstacles to be overcome in finally producing the now popular wireless telephone may be surmised in considering the limitations of other methods of communication as to both publicity of messages and power used.

# The Electrical Contractor

## Is The Contractor-Dealer an Individual or a Team ?

### By CHARLES L. EIDLITZ*

For several years past we have heard that the contractor should and must turn his attention to merchandising. This has been preached, written about, and exploited loudly by men who, it seems to me, ought to have known better.

I have had twenty-five years of contracting experience, I have met I believe nearly all kinds of contractors in my thirty-three years in the electrical line, and I can say without fear of contradiction that practically none of the men whom I know as contractors have the requisite knowledge, instinct, or experience that goes to make a successful merchant.

The contracting business, except perhaps in the smallest towns, is distinctly an engineering proposition. The training of the men in this line is entirely opposed to the training necessary to make a man a good merchant. I feel I can go even a bit further, i. e., that the very fact that a man selects contracting as his business, proves conclusively that he has no leaning toward the merchant type of business.

When I say that an electrical contractor should not take on merchandising, I do not mean that the small contractor or electrician should not carry a stock of electrical materials, if there is no supply house in his locality. There can be no possible harm that I can see in his having such a stock, nor in his attempting to turn over a few dollars here and there by disposing of it, either for profit, or for the accommodation of his clients, but I do say, and most emphatically, that the electrical contractor in the larger cities should not and cannot successfully make a combination of contracting and merchandising, and that if he tries it he will make a failure of both.

A man who has successfully run a merchandising business may take on a repair work contracting line as a side issue but the reverse of this arrangement will not work out.

The reason for this seems to me to be perfectly plain and simple. The successful contractor is more or less of an engineer, and he comes in contact with his clients in an advisory capacity. It is his function to construct for his purchaser an efficient, economical system, and I defy any man to do this satisfactorily and honestly, if every time he thinks of an outlet on a plan, he is at the same time figuring on selling his client some device which he has in stock, and the sale of which will net him a profit.

If a contractor represents or carries some special line of material in stock he is bound to try and sell this material to his client.

How much faith would you have in a doctor who was known to be interested in some special patent medicine, and whose prescriptions seemed to consist mainly of this particular medicine? Would you not begin to be a little suspicious? When suspicion enters at all in professional matters, confidence immediately begins to sneak for the door.

I am sure that every jobber, manufacturer and contractor will agree with me that a consulting engineer, i. e., one who

*Ex-president N.E.C.A., in Electrical Record

regularly prepares plans and specifications, should not be in the employ of or be retained in any way by any manufacturer, whose materials are required in the lines in which the engineer specifies. Would not an owner justly look with suspicion on one so connected? And so the contractor, when acting as engineer or adviser, is in the same class if he carries a stock of goods.

It is perfectly proper for a contractor doing the construction work on a job to sell the owner the motors, etc., but what kind of a motor do you think the contractor would sell if he had a large stock of Specific Electric Company motors in his place, all paid for and unsold? Wouldn't he try to get rid of his stock? Wouldn't that be absolutely natural? Yet, this particular type of motor might not be the best type for the job in question.

Would the contractor sell him the exact size necessary, if he had a lot of near sizes in stock? In other words, and in blunt words, would the contractor pass up the opportunity to unload his stock?

For this reason if for no other, the contractor must remain a free lance, he must not have a stock of motors, appliances, etc. He must be free to buy what he believes best for the particular job in hand, and must be uninfluenced by anything except the giving to the owner or client, the best judgment that he has.              ....

So much then for the moral side.

As a money-making proposition there can be only one answer. The type of men who have made successes in these two lines, are absolutely dissimilar.

The contracting business being practically a profession, while merchandising is distinctly a business, the entire training of the men is different. In the case of contracting the men are accustomed to buying for a sale already made, while in merchandising the buying is of a speculative nature; the judgment of the buyer as to salability, values, and similar considerations being essential to success.

If the advocates of contractor-dealer schemes think that a contractor should take the profits made in contracting, and sink them in an attempt to popularize the use and sale of devices, made by some special manufacturer, they certainly have not been clever in their attempts to put it over, and I think that I can repeat my statement made at the Buffalo convention, "that I would rather be an old fogy type of electrical contractor than a half-baked merchant."

And to sum up the whole situation it seems to me that there has been a whole lot of wasted effort and energy, that has cost nearly everyone some money, and that has resulted in messing up the entire business without having accomplished anything of a constructive nature.

In the meantime this "would-be-dealer" has tied up money in devices that he does not know how to market, money that he might better have spent for developing his contracting business on which he could have made a profit by selling proper installations. But as it is, he has neither helped the appliance business nor increased his contracting business.

Why not leave things alone? The small contractor was always something of a dealer. The large contractor has always handled certain material sales in conjunction with his work or for the accommodation of his clients. If the industry is looking for a name for the man who makes a business of selling heaters, fans, toasters, washing machines vacuum cleaners, and the like, call him an electrical merchant and let the contractor attend to his own work without this ridiculous and unnecessary excitement, caused by trying to tie a can to his tail filled with electrical merchandise, which may make a lot of noise as he travels around banging it on the pavement, but heavens and earth! this does not make him a dealer, and it never will.

## Changing a Fireproof Warehouse into a Modern Office Building

The Imperial Office Building, Hamilton, is located at the corner of Hughson and Main Sts. This building was formerly a modern fireproof warehouse for storage purposes. The structure was orginally designed for 12 storeys.

The Mills Bros. Ltd., of Hamilton recently purchased this building and made extensive alterations to the interior to provide modern offices for rental. When the alterations were completed the two top floors were equipped for the Canada Business College. The District Income Tax collector's offices are located on the entire fourth floor. The ground floor was equipped for the Chamber of Commerce headquarters.

The alterations to the building included the following: partitions were provided for the subdivision of the several offices, corridors etc; also passenger elevators, toilets, vaults, stairway, heating system and lighting system. One passenger elevator has been installed of the most modern direct-connected type made by the Otis Fensom Elevator Co. The final installation will include an additional elevator. A separate source of power is provided for each to assure continuity of service and minimum delay and inconvenience to the tenants. Alternating current is used for the elevator service and is proving very satisfactory, eliminating any possible breakdown of a motor-generator set which might cause trouble had direct current been used. The initial cost was much lower.

The lighting, which was installed by Culley & Breay of Hamilton, was a complete conduit system throughout. The conduit was exposed throughout as shown in the photographs. The lighting service is provided from two sources to provide for continuous service and prevent inconvenience to the tenants; all offices and space has been rented with light. With an installation of this type it is unnecessary to provide individual meters, the installation is not so complicated and any changing in the offices does not require alterations to wiring. Sufficient capacity has been provided in all branch wiring, risers and mains for extensions and increased loads. The mains have been designed to provide for the additional six storeys to be added in future.

The service equipment is Square D, with large double-throw switch for emergency purposes. The power and lighting service enters on Main St. through a standpipe. This standpipe is only temporary until the underground system of the company is completed. The transformer vault for the underground system is located at the same point outside.

The lighting system required special study as the owners required it to be such that it would not be necessary to use desk lamps and further, they wanted to use three different types of fixtures. The fixtures had to be readily removable without having to make joints in the wiring. The outlet must be such that a lamp could be installed if a fixture were not required or a drop cord could be installed. Pull switches were provided near each outlet for individual

control, using Bryant No. 2387 pull switches on an outlet box for this purpose.

To meet full requirements of the owners it was decided to use the Benjamin fixture connector No. 1412 and Benjamin outlet box receptacle No.1405. The brass cover of the receptacle made a neat appearance and finish. It was possible with this receptacle to use a ball and bands in the halls, toilets and stairways as a fixture.

The tenants installed their own fixtures, two different types of which are shown. Any person could install the fixture. It was even unnecessary to have the services of a wireman as the contractor sold the fixtures complete over the counter ready for installation. This proved a wonderful saving and convenience. The tenants could change fixtures about without altering the wiring in any way. No matter what type of fixture they had it could be connected with the proper fittings sold by the dealer over the counter.

The mobility and flexibility of this system was appreciated by the owners and tenants. The dealer did not have any worry or lose any money with the fixture installation. The Superintendent of the work highly recommends this method to all persons contemplating equipping a building such as this one; in fact it could be used in all office buildings and industrial buildings to advantage. The small extra cost per outlet was very trifling when the convenience is considered. The pull switch at every outlet gave the maximum convenience and control and for office buildings is much more desirable where rooms may be altered to meet different requirements. It eliminates the necessity of moving wall switches when partitions are changed.

The entire electrical installation was designed and supervised by V. K. Stalford, Hamilton, Ontario.

## Was a Prominent Contractor

Mr. Harry A. Rooks, of the Rooks Electric Co., Toronto, died on December 16. Deceased was born at London, Ont., 42 years ago and came to Toronto 18 years ago. He was always active in association work and was a member of the original committee of the Toronto Electrical Contractors' Association, and for the last three years had been a member of the Ontario Association of Electrical Contractors & Dealers. Whatever movement was under way for the good of the industry, Harry could be depended upon to put his hand in his pocket. Among the men in the trade, where he was well known and universally liked, he will be greatly missed. The business will be carried on under the same name.

---

The Canada Electric Co., 175 King St. E., Toronto, has secured the contract for electric wiring on the stables and sheds to be erected at the north end of the city by the Board of Control at an estimated cost of $60,000. Also the contract for electrical work on a new school to be erected on Coxwell & Sammon Avenues by the trustees of S. S. No. 7.

# Off to a Good Start

### Toronto Electrical Interests Leading
### Off with Electrical Home Campaign

The Toronto electrical industry is starting this year 1922 off with an Educative Campaign which doubtless, will be the means of stimulating and creating business for every section of the industry.

The first move takes the form of an "Electrical Home." The plan, indeed, calls for a series of four homes to be shown during the first three or four months of the year. The first home will be opened on Saturday, January 21.

This decision is the result of much preliminary work on the part of a representative committee of electrical men who have been meeting almost daily for several weeks discussing ways and means and, in a general way, "prospecting" the field with a view to locating any unexpected obstacles. The decision has now been definitely reached that the time is opportune to "start something" and it has been deemed advisable to open the campaign, as noted above, with the first Electrical Home exhibit on January 21. This home will remain open for a period of two weeks.

The general committee that has been working on this campaign includes the various interests involved—central stations, manufacturers, jobbers, contractor-dealers. Members of the committee include such well-known names as E. M. Ashworth, acting general manager Toronto Hydro-electric system; J. Herbert Hall, managing director Conduits, Limited; C. A. McLean, general manager Masco Company, Ltd.; W. R. Ostrom, manager Northern Electric Company; A. S. Edgar, manager Supply Department, Canadian General Electric Company; E. A. Drury, chairman Toronto Contractor-Dealers' Association; A. W. J. Stewart, manager Appliance Department, Toronto Hydro, and F. John Bell.

Through the kindness of the Society for Electrical Development, Mr. Kenneth A. McIntyre, Canadian representative, has been delegated to give his entire time, free of charge, to the organizing of the campaign.

Complete details of the campaign are being distributed through the mails and will probably have reached those interested before this issue of the Electrical News is off the press.

# Quebec Contractors and Dealers, English Section, Report Very Successful Year's Work

At a well attended meeting of the English Section of the Electrical Contractor-Dealers Association, Province of Quebec, held in room 305 Drummond Building, the following officers were elected: President, J. A. Anderson, of J. A. Anderson & Co; vice-president, E. J. Gunn, Gunn Electric Company; secretary-treasurer, Louis Kon, secretary manager Electrical Co-operative Association, Province of Quebec.

The executive consists of: Messrs. J. M. Walkley, the retiring president, R. S. Mu_ir, M. R. Henry, H. Vincent, and C. E. Barrett.

The same committees as last year were re-appointed to continue the work of better business relations between the jobbers and the contractor-dealers, the French and English Sections, and the Electrical Co-operative Association, Province of Quebec.

Great appreciation of the work of Messrs. Gunn, Walkley and Anderson, as members of the committee handling the adjustment of problems between the jobbers and contractor-dealers was expressed by all the members present, and a spirit of satisfaction as to results achieved lately and determination to bring about a higher standing of the electrical contractor-dealers in the business communities of the province was very marked in the speeches of all the members.

It was also felt that the work of Messrs. St. Amour, Rochon and Pelletier, who were appointed by the French Section to act jointly with the English Section of the Contractor-Dealers Association in achieving the above is already showing considerable progress.

The Electrical Contractor-Dealers' Association has justified its formation. Previous endeavors to get the contractor-dealers in Montreal to combine failed, partly because of the language difficulty, and partly because of lack of sustained interest. This is now a thing of the past. Mr. J. A. Anderson, the new president, in an interview with a representative of the Electrical News, outlined the work of the association during the past year, and told of the harmony that prevails between the English and French sections. The division into sections has worked out in an admirable way, each doing its work in its own way, but on a co-operative basis.

"It was only during the past few months that the association accomplished anything tangible," said Mr. Anderson. "A committee, composed of Messrs. J. M. Walkley, E. J. Gunn, and myself, interviewed the jobbers with a view to securing concerted action in connection with surplus stocks and also to obtain more stability with regard to the prices quoted by jobbers and by contractor-dealers. While we recognize that competition is desirable and inevitable, yet it was felt that conditions were such as to leave little or no profit to those immediately concerned. Some progress has been made as the result of the negotiations. The jobbers agreed that the position taken by the contractors is reasonable. On our part we believe that jobbers should not deal with consumers at wholesale rates, but that they are entitled to sell to consumers at the full retail prices; at the same time our view is that jobbers should eliminate as far as possible this class of business, which really belongs to the contractor-dealer.

"Another question discussed with the jobbers related to dealing with industrial concerns by the jobbers. An industrial concern coming within the scope of the jobbers was defined as one having a permanent electrical staff. The Federal and Provincial Governments, municipalities, railways, and certain very large companies are within that category, but the contractor-dealers are of opinion that there are many concerns that do not buy heavily and that should be the exclusive customers of the contractor-dealers.

"The French Section gave us the most cordial co-operation in approaching the jobbers. A committee consisting of Messrs. St. Amour, Rochon, and Pelletier, accompanied the English committee. Our relations with the French section are of the most cordial character, and I desire to express our appreciation of the assistance of Mr. Simoneau, the president, and the members, particularly in the matter of negotiating with the jobbers.

"The Province of Quebec has recently passed into law a measure of obliging electrical contractors to take out licenses to do business, and also obliging wiremen to undergo an examination and take out licenses as journeymen electricians. It is expected that the law will be put into operation in the near future, and the Government is preparing offices for the staff of examiners. This is a step in the direction of stopping unauthorized and incompetent people from tampering with a business which may be considered hazardous when handled by those who are incompetent. Let me illustrate what I mean by hazardous. In the event of a fuse blowing out a copper is put in to replace it; as the cause has not been removed, another fuse nearer the transformer blows out, and this in turn is replaced by another copper. Eventually there is no protection, the wires become overheated, and a fire results. The legislation referred to will undoubtedly tend to minimize these risks, as it will oblige the work to be done by men with adequate knowledge. The safety entry switch will also go a long way to assist in the reduction of fire hazards. The co-operation between the Government, power companies, Underwriters' Association, and contractors is all in the direction of safeguarding life and property.

"During the last three or four months the meetings of the English Section have been much better attended. The members are taking a deeper interest in the work, and are appreciating the advantages of getting together and of endeavoring to place our business in that foremost position to which it is entitled. Many of the older members, who had experience of former efforts to get united action, had almost given up hope of obtaining a real effective organization, but late events have shown that something can be accomplished, and even the most pessimistic have to admit that our recent meetings have been very successful and that we have obtained results.

"As members of the Electrical Co-Operative Association, Province of Quebec, we have had an opportunity of interchanging opinions, of knowing members of other departments of the electrical industry, of getting to understand each other, and of obtaining such information as will lead to greater sympathy with the viewpoints of all branches interested in our industry.

"We hope as the result of this understanding between the jobbers and contractor-dealers to be able to report at the end of 1922 the most successful year in the electrical history of Montreal. The era of misunderstanding, so destructive to our business, is passing away; we are getting down to a basis of real co-operation which will make for the benefit of all sections of the industry."

# Your Share of $105,000,000!

## Canada Will Spend This Sum for Electrical Appliances. Are You Equipped to Do Your Share?

### By C. D. HENDERSON*

During the next three years Canadian people will spend at least one hundred and five million dollars ($105,000,000.00) in electrical appliances (clothes washers, ranges, vacuum cleaners, fixtures, ironers, irons, toasters, lamps, etc.) exclusive of wiring and repairs.

This figure is not a mere guess but rather a conservative estimate after a careful analysis of past and present conditions of the electrical industry in Canada.

Canada's per capita electrical development is larger than any other country in the world with the exception of Norway, while Canada's available horse power per capita is greater than any other country. There are more wired homes in Canada than in the United States, based on population and as "Wired Homes" is the most important element in the sale of appliances one can readily see that the sales possibilities offer wonderful opportunities.

Available records show that by the middle of 1922 there will be 1,000,000 homes in Canada using electricity, yet only about 5 per cent of these homes are enjoying the benefits of such labor-saving appliances as clothes washers, vacuum cleaners, electric ranges, etc., so that we need not worry about reaching the point of saturation for many years to come.

I suppose the reader wonders how I arrive at an estimate of $105,000,000 sales of appliances in three years. Well, let us do some figuring—every wired home must have tungsten lamps, and an average of say ten lamps per year, per home, gives us to begin with 30,000,000 lamps required at a sales value of about $15,000,000. This does not include the lamps required for stores, offices and factories.

Now, let us take ironers and toasters next. These appliances are generally the first to be purchased and during the next three years should amount to a couple of million dollars in sales value.

The next logical appliance is the vacuum cleaner. How many women will be satisfied to continue using the old broom and carpet sweeper when her neighbor is cleaning house with a vacuum cleaner?

Next let us consider the clothes washer. It is not necessary to prove to the woman in the home that an electric washer will wash clean without tearing the clothes. Almost every woman you meet, if she is not already using one, will tell you that she is going to invest in an electric clothes washer next.

Then comes the electric range. Show me the woman using an electric range who would go back to coal and gas, and with cheap electricity, such as we Canadians enjoy, I believe there will be 100 per cent more electric ranges sold in Canada during the next three years, based on population, than in any other country in the world.

Besides these main appliances that I have just mentioned there will also be a big sale for lighting fixtures, ironers, fans, vibrators, hot plates, heaters and numerous other appliances that are gradually coming into general use.

Considering these facts, would it be stretching our imagination to say that the average expenditure of each wired home would be $35.00 per year for the next three years? If this figure appears reasonable then simply mul-

tiply it by three (to cover the three year period) and then mutiply the balance by 1,000,000 wired homes and you get a total retail sales value of $105,000,000.00.

Let us visualize these figures by approaching the matter from another angle.

### Estimate of Three Years Sales

|  | Total |
|---|---|
| 100,000 homes will spend $300.00 each (electric ranges and sundries) | $30,000,000.00 |
| 100,000 homes will spend $200.00 each (electric washers, fixtures, etc.) | 20,000,000.00 |
| 300,000 homes will spend $100.00 each (vacuum cleaners, fixtures, etc.) | 30,000,000.00 |
| 200,000 homes will spend $80.00 each (fixtures and sundries) | 16,000,000.00 |
| 300,000 homes will spend $30.00 each (small items, lamps, etc.) | 9,000,000.00 |
|  | $105,000,000.00 |

These figures are exclusive of wiring and repair work and cover appliances only.

**Will this huge amount of money be spent through the electrical dealer who is the logical outlet or will the departmental and hardware stores received the largest portion of it?**

The answer is that the matter is entirely in the hands of the electrical dealers—if they make a determined effort to swing the business their way they will have everything in their favor, and it will be quite possible to keep it within the electrical family because, when people think of electric appliances their thoughts naturally turn to the electrical dealer as they realize he is more familiar with the different makes and is in a better position to advise them and give service, but the departmental store and the hardware man have heretofore proven themselves better merchandisers and you may rest assured, they will make a strong bid to swing the business their way.

The electrical dealer will not procure this appliance business by divine right—he must earn it. He should study carefully the different makes of appliances and make his choice from a standpoint of quality, service, prices and stability of the manufacturer. He must make his place of business look inviting to prospective purchasers and then go out and sell—instead of waiting for people to come and take the goods away from him.

In the next issue of the Electrical News I am contributing an article setting forth some ideas with a view to helping the electrical dealer capture and hold the appliance business. They are not original on my part but rather a summary of plans used by successful electrical dealers all over the Dominion.

It will do no harm to read the article. Even though you do not agree with the details you likely will agree with the principles and if any of the suggestions offered can be used to advantage by the readers of the Electrical News I shall feel amply repaid for the time and trouble spent in collecting and publishing them.

*President Henderson Business Service, Ltd., Canadian Manager A. B. C. Co., Brantford, Ont.

## and Sections in Touch with One Another's Operations

The Electrical Co-operative Association, Province of bec, have commenced the issue of regular bulletins to members. The first appeared on December 15 and con- much interesting information. It will keep the various bers and the various sections of the electrical industry uch with one another's operations.

One of the chief items of interest in this Bulletin No. 1 le announcement that the Association was negotiating a stmas Shoppers' Electrical Week, December 19 to 24. Association engaged window dressers to make window lays for that particular week in stores of members of

the Trade Relations Committee, which consists of represe tatives of the power companies, manufacturers, jobbers a both the English and French Sections of the contracto dealers association. This committee meets every Tuesd for a couple of hours, discussing matters of vital inter to the trade, representing, so to speak, the searchlight of t industry in the province.

For the benefit of the contractor members a job contra form, as used by the Southern Canada Power Company, attached to each copy of the bulletin and it is urged tl every contractor in the province should use a form of. t kind, as it protects him by covering minutely the work to done and the price to be paid for it. It leaves no room

Job Contract Form as used by Southern Canada Power Co.

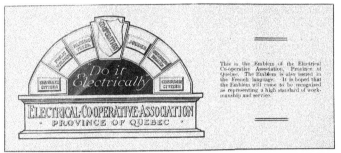

This is the Emblem of the Electrical Co-operative Association, Province of Quebec. The Emblem is also issued in the French language. It is hoped that the Emblem will come to be recognized as representing a high standard of workmanship and service.

## The New Secretary-Manager

The Electrical Co-Operative Association, Province of Quebec, began its second year's work with a new secretary-manager in charge. The Quebec association was looking for a secretary-manager with executive and organizing ability, of sound business principles and one that could impart the "do it electrically and co-operatively" spirit with equal ease to the English and French parts of the population of the Province of Quebec. They believe they have found such a man in Mr. Kon.

The new secretary-manager is a Westerner, and claims Winnipeg as his home where he lived for fourteen years, though for the past year and a half he was connected with the electrical appliance business in Eastern Canada. Mr. Kon was with the Grand Trunk Pacific Development in Winnipeg for some time and in 1910 represented the Grand Trunk

Mr. Louis Kon

and Grand Trunk Pacific Railways at the World's Fair in Brussels, Belgium, organizing and taking charge of their Immigration Department upon his return from Europe the following year. In 1916 he became Superintendent of Immigration and Colonization for the Province of Manitoba, entering in 1918 the services of the United Grain Growers' Securities, which work he left to assume the position of the secretary of the Canadian Economic Commission to Siberia at which work he spent nearly a year.

Mr. Kon is a university graduate in economics and having specialized in that subject is a firm believer in the theory

that the foundation of our industrial, commercial and social life must rest on co-operation.

## "Hoover" Improving Their Advertising

National advertising for the Hoover Suction Sweeper, always of a high order, has been materially improved for the coming year. The new design is not alone more pleasing in appearance but provides, as well, a more effective and more flexible vehicle for telling the interesting Hoover story. Fully twice as much copy can be used in the new advertisements without any sacrifice of good appearance. One of the most notable improvements is the substitution of exquisite line drawings for half-tones in illustrating the messages. Always one of the most extensively advertised of electrical products, the Hoover will continue, during 1922, to bid for the attention of millions of readers through the columns of practically all the larger national magazines. Response to Hoover advertising during 1922 should be even more general and generous than the response of other years. The artistic line drawings and correct typography of the newer messages should win instant attention from their many readers. The compelling messages of Hoover worth, emphasized in each advertisement, should sustain and quicken a desire for Hoover ownership; creating new business for this meritorious and aggressively advertised product.

## Electrical Co-operative Luncheon

At the weekly Electrical Co-operative Luncheon, held at Freeman's Hotel, Montreal, on December 7th, an interesting program was put on by the Entertainment Committee. Drawing for a handsome beaded necklace—the work of a disabled soldier at the hosital St. Anne de Bellevue—created considerable interest and amusement. Mr F. I. Spielman, of Spielman Agencies, added to the entertainment by the donation of one of his "Handilites"—a magneto flashlight which his company recently put on the market.

## Stratford Joins Electrical Page Boosters

The city of Stratford is the latest to add the Electrical Page to the activities of the electrical industry. A sample just received indicates that a good deal of thought and hard work have been put forth by the electrical men of that city. The complete list of cities now carrying electrical pages regularly, in Ontario, is as follows: Hamilton; Ottawa; Windsor; St. Catharines; Kitchener; Brantford and Stratford.

## Among Windsor District Contractors

The Windsor District Association of Electrical Contractors and Dealers held an important meeting on November 30th, 1921. After dinner at the Cadillac Cafe adjournment was made to the local Hydro office at the invitation of Mr. O. M. Perry, Windsor's hydro manager. Those present were Mr. V. B. Dickson, of the Barton-Netting Company, in the chair; Mr. F. D. Reaume; Mr. A. E. Roach, of the McNaughton-McKay Electric Co.; Mr. F. Garfat, of the Electric Supply Co.; Mr. W. A. Lefave; Mr. O. M. Perry, manager Windsor Hydro; Mr. McKellar, of the Northern Electric Co.; Mr. M. J. McHenry, manager Walkerville Hydro; Mr. Firth, of the Canadian General Electric Co. and A. H. Cook.

The chairman explained, for the benefit of the visitors, the purpose of the Association, emphasizing that "Goodwill to the industry" prompted the activities. He then requested that the floor be left to the visitors and asked them to voice their views of the situation and the local electric situation.

Mr. McHenry spoke in favor of the Association and an all-embracing central electrical organization. He said the Hydro had no desire to "hog" the market. To maintain the load was the main object—not competition, but the boosting of appliance sales.

Mr. Perry spoke on the magnitude of hydro sales and the servicing of large and small appliances. Mr. McKeller spoke of the new farmers', or rural, lines bringing in big business—estimated at $1,000 per farmer, spread over a period of five years.

Mr. Firth said that competition was the life of trade, and he found the meeting to be very valuable in an educational sense.

After they had all expressed themselves as favorable to the Association, and to Mr. McHenry and Mr. McKeller had been assigned the duties of providing the Electrical Food for next meeting and arranging for the advertising, respectively, the meeting adjourned until December 9th.

### December 9th Meeting

Owing to the nearness of Christmas, a full page advertising scheme in the local paper, the Border Cities Star, was the only business transacted. The advertising was arranged by Mr. McKeller, manager of the local branch of the Northern Electric Co. and makes a very effective showing.

## Successful Exhibit at Nicolet

"The Electric Service Corporation held an electrical exhibiton at Nicolet, Quebec, on December 1st, 2nd. and 3rd. The results of the exhibiton were very satisfactory, displays being made of electrically operated utility machines such as pumps, cream separators, meat cutting machines, utility motors, churning machines, etc, as well as the usual line of electrical appliances such as stoves, portable lamps and domestic appliances. The results were so satisfactory that further exhibitions will be held in the spring, the first one opening at Berthierville."

## Winnipeg Awards Meter Contracts

The city of Winnipeg recently purchased 3,000 meters, as follows: From Canadian Westinghouse Co. 1000, 10 ampere, 2-wire meters at $8.30, and 700, 20 ampere, 3-wire meters at $11.00; Packard Electric type R. 300, 25 ampere, 3-wire at $12.80; Sangamo Electric Co. 1000, 20 ampere, 3-wire at $11.00. It is worth noting that these prices are now as low, and possibly lower, than pre-war figures. It is true that in 1915, due to keen competitive conditions, prices went a couple of dollars lower, but during the next five years they almost doubled, so that the present prices indicate a reduction of from 30 to 35 per cent. from high.

## Miscellaneous Activities

The Toronto Transportation Commission recently purchased from the Packard Ontario Motor Company a 4-wheel drive tractor snow plow to be used in keeping the bus routes open. It has a powerful motor and a five-speed transmission with a low gear ratio of 80 to 1.

Graham Electric Co, 929 Pender St., Vancouver, are among the enterprising firms who believe in constant improvement to make attractive presentation of electrical displays. Quite recently they have completed remodelling of their shop and window fixtures on an extensive scale. Inlaid hardwood floors in the show windows are among other improvements. Both in their windows and their show rooms they are able to make effective showing of fixtures and appliances.

Messrs. Rankin & Cherrill, old-time electrical dealers and contractors of Vancouver, have added another store to their enterprise. They have opened salesrooms at 528 Hastings St., where a complete stock of fittings and appliances is being handled. They continue their shop at 55 Hastings St., W., also.

## Trade Publications

The Ward Leonard Electric Company, Mt. Vernon, N. Y., are distributing a booklet, Section D, describing and illustrating the use of Vitrohm (vitreous enamelled) Field Rheostats.

The Westinghouse Electric & Manufacturing Company, East Pittsburgh, Pa., are distributing catalogue 12-A, describing Safety Switches and Panel Boards. The catalogue is well illustrated.

The Wagner Electric Manufacturing Co., St. Louis, Mo., have issued a folder describing their "Pow-R-full," the latest addition to their line of motors. The new motor is a polyphase type. Bulletin No. 129 explains it more fully.

Conduits Company, Ltd., have issued catalogue No. 6, known as their Wire-Mold Catalogue. It deals with installation suggestions and is very fully illustrated, showing how the contractor may use this equipment in overcoming any difficulties he may meet with in his various installations.

Arc welding for repair and reclamation, general applications of arc welding, and arc welding for manufacturing processes are well described and illustrated in Leaflet 1825, just published by the Westinghouse Electric & Manufacturing Company, East Pittsburgh, Pa. A story is told of how costs are reduced by the use of arc welding.

The Robbins & Myers Company are distributing a folder describing their line of generators, motors and motor-generator sets for radio service. This line includes 500 volt and 1000 volt generators and motor-generator sets of 100, 200 and 500 watts capacity for use with vacuum tubes and for other special services; also synchronous motors of ⅛ and 1/6 h.p. ratings for operating the synchronous type of rotary spark gap.

The "EV" type of alternating current, vertical water wheel driven generators is described and illustrated in Circular 3439, just issued by the Westinghouse Electric & Manufacturing Company, East Pittsburgh, Pa. The vertical unit is being much used for small and moderate capacity low head hydro-electric plants and the type "EV" comprises a line of a.c. generators, to fit the modern hydraulic setting and the same factor of safety in design and reliability in operation required of larger units.

# What Is Newest in Electrical Equipment

### Lamp-Heated Warming Device

A new idea in a portable heater that can be used both as a bed warmer and for applying heat to bodily aches and pains is found in "Glo-Pax," made by the Twinplex Sales Company, 1627 Locust Street, St. Louis, Mo. This is a flat fan-shaped device, the warmth radiating from a small electric bulb held between two flat aluminum pieces, one face of which

is perforated. A fleece cover is also provided. A heat regulator dial on the cord has three degrees of heat indicated on the dial but any number of intermediate degrees can be obtained by adjusting the arrow point at the place desired. To open the device in case the bulb burns out, the knurled metal collar is turned slightly and pulled a half-inch down the handle, leaving the perforated face free to be lifted up with the finger.

### Hanging Lamp

A novelty hanging lamp device called the "Lampette," and made by the Colonial Lamp & Fixture Works, 5634 Lake Park Avenue, Chicago, introduces a new feature in the braid-

ed belt used to hang the lamp over any convenient point on chair or bed. The braids, shades, shields, and tassels may be had in a number of designs, colors and materials.

### Flush Door Receptacle

A new type of flush door receptacle designed to combine the "invisible outlet" features with safety and ease in use, has been put on the market by Harvey Hubbell, Inc. The flush plate contains double "in-folding" doors, through which a cap passes. The porcelain body, concealed behind these doors, is provided with narrow, beveled slots, set tandem.

These slots are just large enough to admit the brass blades of the cap, which, passing through them, are engaged by two double springs of phosphor-bronze. The springs are set within concealed contact chambers, and cannot be reached from the outside except by way of these slots. These receptacles are made with outlets for one or two caps. The receptacle body fits any standard outlet boxes.

### Mr. Leacock, Official Entertainer

Mr. W. A. W. Bundock, B. Sc., A. M. I. E. E., Commissioner of Public Works Department, of Sydney, Australia, has been touring Canada and the United States for some time, gathering ideas on high tension stations and equipment, in company with Mr. Wildridge, consulting engineer, of the same city. He spent a day in Toronto, under the guidance of Mr. Geo. D. Leacock, inspecting some of the principal sub-stations. Another recent visitor to Toronto, on a similar mission, was Lieut.-Col. B. C. Battye, Public Works Commissioner, Punjab, India, who was also Mr. Leacock's guest.

### Adds New Agency

The George C. Rough Sales Corporation, 134 Coristine Building, Montreal, announce that they have recently taken on the line of The States Company, of Hartford, Conn., which consists of apparatus for meter testing and laboratory specialties and accessories, including complete testing tables, phase sequence indicators, phasing transformers, phantom loads, potential phase shifters, etc. This line is of special interest to all companies who require extreme accuracy in the testing of meters. Descriptive bulletins will be mailed on request to those interested.

### Back to Pre-War Service

The Crouse-Hinds Co. of Canada, Ltd., and Harvey Hubbell Co. of Canada, Ltd., have issued an interesting folder entitled, "Back to pre-war service." The folder lists the various agents of these companies in fifteen principal Canadian cities, indicating their facilities for supplying customers with "24-hour" service every day. The folder also shows a number of interesting views of the factories and warehouses, exterior and interior, of Mr. Mack's two companies.

The Canadian Electrical Supply Co., Ltd., are now handling the "Brantford Locomotive" electric washer, manufactured by the Brantford Washing Machine Co. The Canadian Electrical Supply Company's territory includes eastern Ontario and the province of Quebec.

# Electric Railways

## Rehabilitation of Toronto's Street Railway System—II

### Details of Foundations, Rails, Intersections, Bonding and Surfacing

[This is the second of a series of articles covering the rehabilitation of the Toronto railway system. Our first article, which appeared in our issue of December 15, dealt only with the removal of old track and the excavation in preparation for the new track, and the foundation. The present article describes the foundation work, placing and bonding of the rails and the final surfacing.—Editor.]

#### Introduction

The track and road bed rehabilitation follows one or other of two general plans—(a) a concrete foundation; (b) crushed stone foundation. On foundations of the second type, granite block pavement only has been used, while on concrete slabs, either granite block paving or asphalt has been adopted, although the asphalt is in general merely a temporary substitute for granite blocks. Inability to secure sufficient material, on time, for all the track reconstruction under way compelled the commission to adopt asphalt. This will be taken up as it wears out and replaced by a more durable type of pavement, provision being made for this contingency by special construction, as described below. Plans are reproduced in this article illustrating the various types of construction.

The concreting of the base slab follows immediately upon the excavation, the sub-soil being first drained with a line of tile pipe, imbedded in gravel, along the centre line of the track allowance. The mixers are of the Foote type, producing a 21 cu. ft. batch. They are on caterpillar treads. The dry materials are pre-mixed at a central yard and conveyed in batch boxes to the mixer. Each batch box contains 33 cu. ft. of loose material, which is dumped into the loading skip of the mixer through bottom opening gates which are opened after the box is spotted over the skip by means of a derrick on the mixer body. Two batch boxes are carried on a truck and the trucks are so timed that no delay is occasioned through failure of supplies. Concreting operations are, therefore, continuous

and very rapid progress is made, so much so that it has been found necessary to keep the steam shovel very much in advance of the mixer. The foundation slab is of 1:2½:5 mix, trap rock running from 2-in. to ½-in. in size, as the coarse aggregate. Nine inches is the standard depth of slab. When this foundation has taken its set, ties and rails are laid and roughly aligned. Oak ties have been used wherever possible, but owing to their scarcity, softer ties have had to be used on some of the more unimportant streets. With soft ties, the rails are spiked through tie plates which are deemed unnecessary where oak ties are employed. On top of the foundation, a mixture of 60 per cent. ⅝-in. stone and 40 per cent. dust is laid and tamped under and around the ties to bring the tracks up to proper grade, and to give resilience to the track, thus greatly improving running qualities. The second course of concrete is then laid thoroughly embedding the ties and base of rails. This course is of 1:3:6 mix with a screened gravel as the coarse aggregate. The track is then ready for the pavement, which is mainly granite block laid in standard fashion on a sand base with cement grouting.

In cases where asphalt paving is adopted, the paving base is poured in two courses, the bottom course being allowed to stand for a day before the top lift is poured. This will make it possible to chip off the top course more easily when the asphalt wears out and it is desired to replace it with granite blocks. The granite blocks are laid on a cushion of cement and sand and are grouted with cement grout.

#### Practice in Special Cases

This general procedure has, naturally, been altered in certain details in order to suit the particular requirements of the street on which work is proceeding. For example, on a certain portion of Bathurst St., it was found impossible to divert traffic for any time longer than necessary to pour the foundation. The new tracks were, therefore, laid and carried on blocks resting on the sub-grade, only every alternate tie being used. The concrete foundation slab was then poured while traffic was using the tracks. As the load was carried to the sub-grade through the blocks, there was no interference with the setting of the concrete. When the concrete had sufficient time to set, the wedge blocks between the ties and the supporting timbers were removed, allowing the ties and rails to rest on the con-

These illustrations show three stages in the track laying operations on Bathurst Street hill, Toronto and indicate the rapidity with which the work is carried out. In this instance the line is entirely new and the same speed was not called for as on rehabilitation work, where traffic is waiting to make use of the rebuilt lines. However, only a month was necessary for the track laying from the time of breaking pavement to the completion of the road, ready for traffic. The illustration on the left shows the road bed as it appeared on September 16, while three days later, as shown in the centre picture, the foundation course of concrete had been poured. Ties, rails and granite block pavement were then laid, the finished appearance of the line being shown in the right hand illustration, which was taken on October 18.

crete. The alternate ties were then inserted and the track construction completed in accordance with the method detailed above.

### Pre-mixed Concrete Ingredients

A reference has been made to the use of the batch transfer method of handling concrete material in all of the Toronto Transportation Commission's work. By this system properly proportioned batches of dry ingredients were hauled on motor trucks to the mixer. The advantages are very obvious. A large labor force at the mixer is dispensed with; the streets are kept entirely clear of materials, and batches are in readiness as often as the mixer can handle them; at the yards, where the ingredients are proportioned, a large labor force is unnecessary,

loading machines. These machines are of the bucket type, which lift the materials into a measuring hopper or chute, according to whether a measured quantity is desired or not. They are mounted on caterpillar treads and are entirely self-contained, so that they may be readily shifted from one part of the yard to another. The various piles of materials are as follows: 1st, crusher run (for special foundation work); 2nd, tamping material (60 per cent. ¾-in. stone and 40 per cent. dust); 3rd, sand; 4th, screened gravel (for top lift of concrete); 5th, trap rock (for foundation slab). Each truck leaving the yard for a job carries two batch boxes, each containing 33 cu. ft. of loose material. They are filled first with the proper amount of sand and then receive the correct proportion of

## Types of Construction for Track Rehabilitation in Toronto

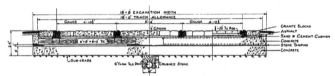

Concrete Base with Granite Block Paving

Concrete Base with Asphalt Paving

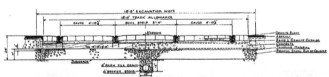

Crushed Stone Base with Granite Block Paving

as the materials are handled almost entirely by mechanical equipment. The commission has two yards where work of this character is carried out, namely, at Coxwell Ave., in the east end of the city, and at Bathurst St., in the central north part of the city. In addition there is also a large storage yard on Merton St., where rails, ties and other track supplies also come in direct by train. At the Bathurst St. or Hillcrest Yard, which has been selected as typical, there are two heavy locomotive cranes continually unloading stone and sand to the extent of 1,500 to 2,500 tons a day, which figures also represent the amount of material going out of the yards. The various materials are arranged in stock piles from which the trucks obtain their quota by means of a large fleet of Barber-Greene

gravel or trap rock, as the case may be. The proper number of bags of cement are then placed on top, these being opened and dumped en route by a man travelling with the truck. In this way much time is saved and any congestion at the cement platform is avoided. The cement platform is at such a height that very little lifting of the bags is necessary in loading up the trucks. When the job is in progress, there is a continuous stream of trucks coming to the yards, loading up and leaving for the site of the work. In this way a supply of materials is made constantly available and the continuous progress of the work is not held up.

The rails which have been adopted are 122 lb. grooved girder A. E. R. A. sections, although in certain instances, 100

The concrete materials are mixed in their proper proportion at a central yard and hauled in bottom dumping batch boxes to the mixer. This ensures accuracy of mix, saves time, dispenses with a large laboring force and keeps the street clear of stone and sand piles. Each batch box contains 33 cu. ft. of material, making 21 cu. ft. of mixed concrete.

In some cases, where traffic conditions made it impossible to wait until the concrete foundation had properly set before opening the street, the rails and ties were carried on blocks resting on the sub-grade. This enabled the street cars to proceed without interfering with the hardening of the concrete foundation. The blocks are kept well outside of the line of the rails, so that there will be solid concrete beneath the points of load.

Concrete materials are pre-mixed at a central yard and distributed in batch boxes by means of motor trucks. The boxes are loaded at yard by means of bucket loaders, which accurately proportion the amount of material necessary for each batch. This illustration shows three such loaders, one for sand and two for stone, in process of filling a truck load of two batch boxes.

Pneumatic tools are made extensive use of by the Toronto Transportation Commission's construction forces. Pneumatic drills break up old concrete and pavement, while pneumatic tamping tools are employed to bring up the new tracks to grade.

New rails are thoroughly bonded by being welded together solidly. The illustration shows the welding operation in progress, the plate being bonded to the rail along its full length at both top and bottom.

lb. A. R. A. "T" sections have been employed. These rails are carried on oak or jack pine ties, 8 ft. long, 6 in. thick and 8 in. wide, and spaced at 2 ft centres, standard tie plates being used if the ties are of soft wood. The whole track is then carried on a concrete or crushed stone foundation.

The gauge of street railway tracks in Toronto is 4 ft. 10 7/8 in. and on double track construction the commission is laying the tracks to 10 ft. 2 7/8 in. centres, giving a 5 ft. 4 in. devil strip. The rails are laid with joints opposite,

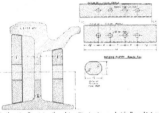

Section of rail and bonding plates. The bonds are electrically welded to the rails as explained in the article.

whether single or double track. They are tight-butted and the joint plates are welded to the rails along the top and bottom. Tie rods are inserted every six feet.

### Electric Welding the Bonds

The bond plates used were 24 in. long by ⅜ in. thick, with four holes; see plan herewith. By means of bolts the plates were drawn tightly into position, as the first essential is a good mechanical joint. Both the plates were welded top and bottom by the Lincoln process method, which consists of laying a 9/32 special low carbon rod at the joint between the plate and rail. A dynamotor, taking current from the trolley, furnishes 400 amperes at 35 volts at the arc; a 5/16 carbon pencil is used. The arc penetrates half into the rail and half into the plate, driving in the filler rod, making a mechanical and electrical joint at the same time. This process, which is faster than the metallic welding, is made possible by the use of a copper defining bar. This bar steadies and confines the arc and makes it possible to direct the arc along the welding

The Lincoln welding outfit supplying 400 amps. at 30 volts

line. It also holds the filler rod and flux in place. By conducting heat from the weld, it prevents excessive heating of the rail and by concentrating and confining the heat at the welding line, it assures ample crater depth. After the weld

is finished, the bolts may be removed. The Toronto Commission are only removing two of the four.

As the rail has to be aligned and the foundation built up to the ties before the welding is done, it has been found necessary in some cases to use two, or in special cases, all four, of the dynamotors on a section of track so as to keep ahead of the concreting gang. This was the case in doing the section of Yonge Street, from College to Queen, which section was put through in schedule time.

### Flexible Bonding

In addition to the joint welds described above, No. 0000 arc weld cross bonds are installed every 500 ft. between the two rails of each tangent track.

The thoroughness with which intersection special track-work layouts are bonded is of particular interest. Rail joints

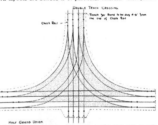

Illustrating the thoroughness with which special track-work lay-outs are bonded.

within the special work layout are not steel welded at the fish plates but are each bonded with No. 0000, 7 inch "U" arc weld copper bonds welded to the head of the rail. In addition to this complete bonding of the rail joints within the special work, the entire layout is shunted by a 1,000,000 c.m. bare stranded copper conductor laid around and outside of all tracks four feet outside of guard rails at all points of crossing. At the points of crossing the tracks, this bond, which is locally called the "ring bond" is attached to each rail through two 72 in., copper, No. 0000, arc weld bonds having terminals at both ends. Both ends of the 72 inch bonds are welded to the base of the rails and the mid point of the bonds is thor-

Wilson arc welding machines at work.

oughly soldered to the "ring bond." All soldered connections and welds are painted.

At outlying points the same system of special work bonding is used except that in these cases the "ring bond" is 500,000 c.m.

The sketch attached gives a clear idea of how the special work bonding is laid out.

The "ring bond" allows special work to be removed for reconstruction purposes without disturbing the return circuit. Where the track return circuit has to be supplemented by special negative feeders this "ring bond" system makes an easily obtained point of connection as the track foundation is not disturbed as is the case with bonds buried under the tracks.

The copper arc weld bonds used are of Ohio Brass Co.'s manufacture. Wilson arc welding machines also supplied by the O. B. Company are used to attach the bonds.

### An Effective Holiday Message

A particularly attractive Christmas message, in the form of the Christmas number of the Winnipeg Public Service News, was issued by the Winnipeg Electric Railway Company to its patrons on December 15. The spirit of Christmas at its best is the spirit of "service", and no more opportune time could have been found to tie in these two ideas—and no more effective message—than this expresion of goodwill. We reproduce herewith, in reduced form, the message, which was signed by Mr. A. W. McLimont, vice-president of the company. The reproduction necessarily weakens its effectiveness as the original was in two colors, but our readers will catch the underlying idea.

A report states that a $1,750,000.00 issue of 5 per cent bonds of the Montreal Tramways Company offered in New York recently was over-subscribed within 48 hours and that the issue had excited more than ordinary interest.

### M. C. Gilman New Sales-manager

Mr. M. C. Gilman has been appointed sales manager of Electric Utility, of the Winnipeg Electric Railway Company. Mr. Gilman has recently been connected with the Montreal, Light, Heat and Power Consolidated and previously was with the Toronto Electric Light Company, the Public Service Corporation, Newark, N. J., and the Brooklyn Edison Company, New York, in executive positions. He has always been a prominent member of the Canadian Electrical Association, where he was for several years chairman of the commercial section. He also was Canadian member of the Commercial Section of the National Electric Light Association, New York. The office of sales manager, Electric Utility, is established because of the greatly increased opportunities there will be for the disposal of power due to the greatly increased supply which will be received by the company from the Manitoba Power Company's development at Great Falls.

---

## A. M. E. U. Annual Meeting

*The regular annual meeting of the Association of Municipal Electrical Utilities will be held on Jan. 26-27, Thursday and Friday. The sessions, as usual, will be in the Chemistry and Mining Bldg., University of Toronto.*

---

# In the Transport Service of a Bustling Christmas Army!

IN THAT joyous army of Winnipeg citizens who will be happily celebrating Christmas, the Winnipeg Electric Railway Company will be occupying a place of great importance.

We will be transporting nearly one and a half millions of members of this army in Christmas week!

May we express the hope that, in the discharge of our duties there will be no untoward incident occur that might in any way detract from the happiness of each one of you. But rather through the courtesy of each member of the Company may the Yuletide spirit of "good-will" be engendered.

In addition to being assigned to this important branch of the "service" we are also connected with the "engineering staff" and the "commissariat department" of this army which is under the banner of "Santa Claus."

Approximately 20,000 Christmas puddings, piping hot, and as many turkeys (we hope), done to an extra turn of deliciousness will be taken

from the ovens of gas and electric cookers dependent on this Company for the supply of fuel.

And from even more homes in Winnipeg and vicinity bright lights, connected with our power service, will shine so the holly berries red may glow, the tinsel sparkle and the festoon glitter—and the solitary bunch of mistletoe be easily seen of men!

There are in the Company about 2,000 special agents of the genial old "Santa" whose absorbing task is now to see that so far as lies within their power nothing goes wrong to interfere with the bright plans you are making for a happy Christmas. The thoughts and actions of these 2,000 officers are directed to keeping alive the "spirit of Christmas."

And may the Christmas spirit never utterly depart from your heart and ours and leave us cold.

For so long as the mellow, kindly spirit of Christmas endures we will never despair of our efforts to win your friendship and good-will through "Service."

With Most Sincere Wishes, on Behalf of The Winnipeg Electric Railway Company and its Staff,

*A. W. McLimont*
Vice-President

# Current News and Notes

**Aylmer, Que.**

Mr. W. Lapensee, Hull, Que., has been awarded the contract for electrical work on three stores and four residences recently erected at Main & Bancroft Sts., Aylmer, Que., for Dr. F. W. Church, Aylmer.

**Havelock, Ont.**

A report states that because the electric light system of Havelock was not meeting expenses it has been found necessary to increase both the commercial and domestic rates one cent, bringing it up to six cents per kw. hr.

**Meaford, Ont.**

, The rate-payers of Meaford will vote on a by-law, at the municipal election in January, to bring hydro-electric power into the town.

**Montreal, Que.**

Messrs. J. A. Anderson & Co., 205 Mansfield St., Montreal, have been awarded the electrical contract on the gatehouse and stores of the Canadian Vickers, Ltd., Notre Dame Street E., Montreal, which are undergoing alterations.

**North Toronto, Ont.**

Messrs. Harris & Marson, 81A Parkway Ave., Toronto, have been awarded the contract for electrical wiring on a High School to be erected at North Toronto at an estimated cost of $160,000.

**Ottawa, Ont.**

Mr. S. J. Davies, 83 LeBreton St., Ottawa, has been awarded the contract for electrical work on a store and apartment building recently erected on Wellington St., Ottawa, for Mr. W. Burke, 871 Somerset St.

**Outremont, Que.**

The Sayer Electric Co., 87 Bleury St., Montreal, has been awarded the contract for electrical work on an apartment house on Girouard Avenue, Outremont, recently erected for Mr. E. Thompson, 1105 Bernard Ave., at an approximate cost of $35,000.

**St. Marys, Ont.**

Mr. A. Willard, St. Marys, Ont., has been awarded the contract for electrical work on the chopping mills of the St. Marys Milling Company, St. Marys.

**St. John, N. B.**

When the New Brunswick Power Company, St. John, N. B., failed to comply with the demands of its striking street car employees last summer, the strikers formed a company, known as the Union Bus Company, to operate a bus service that would give employment to several of the men on strike. Some 13 buses were finally put in operation at a cash outlay in vehicles alone, of about $15,000. The company has, however, ceased operations because of operating losses. Reports state that the public had ceased to patronize the buses to a considerable extent; also that some of the strikers have gone back to the employ of the power company.

**St. Catharines, Ont.**

Mr. H. C. Holmes, 30 Ann St., St. Catharines, Ont., has been awarded the electrical contract on a store and apartment building recently erected at Lake & Albert Sts., St. Catharines, for Messrs. Vine & Swan, 50 King St.

**North Bay, Ont.**

The Timiskaming & Northern Ontario Railway Commission has appointed a committee of engineers to make a report in three months' time as to the feasibility of electrification of that line.

**Toronto, Ont.**

Messrs. Bennett & Wright, 72 Queen St. E., Toronto, have secured the electrical contract on a Club House to be erected on College St., near Shaw St., Toronto, for the Parkdale & West Toronto G. W. V. Association, at an estimated cost of $55,000.

Mr. B. C. Taylor, 25 Marchmont Road, Toronto, has been awarded the contract for electrical work on an apartment building recently erected on St. Clair Ave., near Vaughan Rd., for Mr. W. B. Isbester, 209 Westmoreland Ave., Toronto.

Mr. W. F. Clarke, 62 Dundas St. E., Toronto, has secured the electrical contract on three stores being erected at Bloor St. & Beresford Ave., Toronto, for Mr. S. Winesanker, 3040 Dundas St. W., at an estimated cost of $36,000.

**Vancouver, B. C.**

Officials of the British Columbia Electric Railway, who have been busy altering their system to conform to the new rules of the road which comes into effect New Year's day, announce tat they have everything in readiness and that the change over will be made at 6 a. m. Sunday morning. Owing to the New Year's Eve traffic the change could not be made at midnight. The new rule means that all vehicles meeting will pass on the right instead of on the left as heretofore.

**Verdun, Que.**

Messrs. J. A. Anderson & Co., 205 Mansfield St., Montreal, have secured the electrical contract on a ten-room school being erected at Melrose & Verdun Aves., Verdun, Que. at an estimated cost of $100,000.

**Victoria, B. C.**

A delegation from Vancouver, New Westminster and Victoria, B. C., recently waited on the provincial authorities at Victoria to show why one-man cars should not be operated on the lines of the British Columbia Electric Railway. The result was that such cars may be given a fair trial and if, at the expiration of that time, they are deemed unsatisfactory or unsafe the question could be re-opened.

**Westmount, Que.**

Mr. H. R. Cassidy, 225 Regent Avenue, Montreal, has secured the contract for electrical work on a new bank building recently erected at Victoria & Sherbrooke Sts., Westmount, for the Canadian Bank of Commerce.

**Winnipeg, Man.**

The monthly returns of the Winnipeg Hydro-electric System shows that for the month of November the receipts for electric light and power were $166,996.75, an increase of $464.67 over November, 1920.

**Winnipeg, Man.**

Gross earnings of the Winnipeg Electric Railway Company for October, amounted to $454,224, an increase of $1,123 compared with the corresponding period last year. Operating expenses and taxes amounted to $324,263, against $337,881 last year. Gross earnings up to Oct. 31, 1921, total $4,550,250, against $4,285,817, in the preceding year.

### COMPLETION OF KANSAS CITY, MEXICO & ORIENT RAILROAD

Early completion of the gaps in the Kansas City, Mexico & Orient Railroad in Mexico is probable. At the present time, though service over this line—which has its terminals at Kansas City, in the United States, and Topolobampo, Mexico, on the Gulf of California—is limited to the lines in this country. Two unfinished gaps remain in Mexico—one between Fuerte and Sanchez, a distance of 155 miles, the other between Alpine and Falomir, approximately 158 miles.

The British interests controlling the line are ready to begin construction work with the object of putting the property in a revenue-producing condition. Labor is cheap and plentiful, and the movement would have the approval of the Mexican authorities as a source of employment for a considerable number of men. Furthermore, the Government is said to desire the completion of the road as a means of ready access to its north-west territory, and being considered a railroad of the first importance, the line will be entitled to a Federal subsidy.

The work, including necessary grading and tunneling will take about three years.

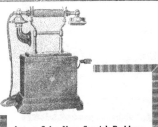

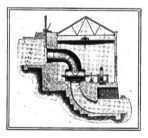

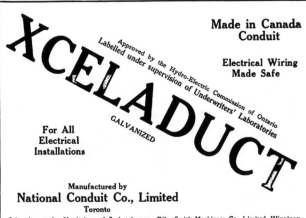

# CLASSIFIED INDEX TO ADVERTISEMENTS

The following regulations apply to all advertisers:—Eighth page, every issue, three headings; quarter page, six headings; half page, twelve headings; full page, twenty-four headings.

**AIR BRAKES**
Canadian Westinghouse Company

**AIR PRESSURE RELAYS**
Electrical Development and Machine Co.

**ALTERNATORS**
Canadian General Electric Co., Ltd.
Canadian Westinghouse Company
Electric Motor & Machinery Co., Ltd.
Ferranti Meter & Transformer Mfg. Co.
Lincoln Electric Company of Canada.
MacGovern & Company
Northern Electric Company
Wagner Electric Mfg. Company of Canada

**ALUMINUM**
British Aluminium Company
Spielman Agencies, Registered

**AMMETERS & VOLTMETERS**
Canadian National Carbon Company
Canadian Westinghouse Company
Ferranti Meter & Transformer Mfg. Co.
Hatheway & Knott, Inc.
Monarch Electric Company
Northern Electric Company
Wagner Electric Mfg. Company of Canada

**ARMOURED CABLE**
Canadian Triangle Conduit Company

**ATTACHMENT PLUGS**
Canadian General Electric Co., Ltd.
Canadian Westinghouse Electric Company

**BATTERIES**
Canada Dry Cells Ltd.
Canadian General Electric Co., Ltd.
Canadian National Carbon Company
Electric Storage Battery Co.
Exide Batteries of Canada, Ltd.
Hart Battery Co.
Hatheway & Knott, Inc.
McDonald & Willson, Ltd., Toronto
Northern Electric Company

**BOLTS**
McGill Manufacturing Company

**BONDS (Rail)**
Canadian General Electric Co., Ltd.
Ohio Brass Company

**BOXES**
Canadian General Electric Co., Ltd.
Canadian Westinghouse Company
G. & W. Electric Specialty Company
Hatheway & Knott, Inc.
International Machinery & Supply Company

**BOXES (Manhole)**
Standard Underground Cable Company of Canada, Limited

**BRACKETS**
Slater N.

**BRUSHES, CARBON**
Dominion Carbon Brush Company
Canadian National Carbon Company

**BUS BAR SUPPORTS**
Moloney Electric Company of Canada
Monarch Electric Company
Electrical Development & Machine Co.

**BUSHINGS**
Diamond State Fibre Co. of Canada, Ltd.
Electrical Development & Machine Co.
Steel City Electric Company

**CABLES**
Boston Insulated Wire & Cable Company, Ltd.
British Aluminium Company
Canadian General Electric Co., Ltd.
Phillips Electrical Works, Eugene F.
Standard Underground Cable Company of Canada, Limited

**CABLE ACCESSORIES**
Northern Electric Company
Standard Underground Cable Company of Canada, Limited
Steel City Electric Company

**CABLE END BELLS**
Electrical Development & Machine Co.

**CARBON BRUSHES**
Calebaugh Self-Lubricating Carbon Co.
Dominion Carbon Brush Company
Canadian National Carbon Company

**CARBONS**
Canadian National Carbon Company
Canadian Westinghouse Company

**CAR EQUIPMENT**
Canadian Westinghouse Company
Ohio Brass Company

**CENTRIFUGAL PUMPS**
Boving Hydraulic & Engineering Company

**CHAIN (Driving)**
Jones & Glassco

**CHARGING OUTFITS**
Canadian Crocker-Wheeler Company
Canadian General Electric Company
Canadian Allis-Chalmers Company
Lincoln Electric Co. of Canada

**CHRISTMAS TREE OUTFITS**
Canadian General Electric Co., Ltd.
Masco Co., Ltd.
Northern Electric Company

**CIRCUIT BREAKERS**
Canadian General Electric Co., Ltd.
Canadian Westinghouse Company
Cutter Electric & Manufacturing Company
Ferranti Meter & Transformer Mfg. Co.
Monarch Electric Company
Northern Electric Company

**CLEAT TYPE SWITCHBOARD INSULATORS**
Electrical Development & Machine Co.

**CLAMPS, INSULATOR**
Electrical Development & Machine Co.

**CONDENSERS**
Boving Hydraulic & Engineering Company
Canadian Westinghouse Company
MacGovern & Company

**CONDUCTORS**
British Aluminium Company

**CONDUIT (Underground Fibre)**
American Fibre Conduit Co.
Canadian Johns-Manville Co.

**CONDUITS**
Conduits Company
National Conduit Company
Northern Electric Company

**CONVERTING MACHINERY**
Ferranti Meter & Transformer Mfg. Co.
Lincoln Electric Company of Canada

**CONDUIT BOX FITTINGS**
Hatheway & Knott, Inc.
Killark Electric Manufacturing Company
Northern Electric Company

**CONDULETS**
International Machinery & Supply Co.

**CONTROLLERS**
Canadian Crocker-Wheeler Company
Canadian General Electric Company
Canadian Westinghouse Company
Electrical Maintenance & Repairs Company
Northern Electric Company

**COOKING DEVICES**
Canadian General Electric Company
Canadian Westinghouse Company
National Electric Heating Company
Northern Electric Company
Spielman Agencies, Registered

**CORDS**
Northern Electric Company
Phillips Electric Works, Eugene F.

**CROSS ARMS**
Northern Electric Company

**CRUDE OIL ENGINES**
Boving Hydraulic & Engineering Company

**CURLING IRONS**
Northern Electric Co. (Chicago)

**CUTOUTS**
Canadian General Electric Co., Ltd.
G. & W. Electric Specialty Company
Hatheway & Knott, Inc.
International Machinery & Supply Co.

**DETECTORS (Voltage)**
G. & W. Electric Specialty Company

**DISCONNECTORS**
G. & W. Electric Specialty Company

**DISCONNECTING SWITCHES**
Electrical Development & Machine Co.
Winter Joyner, A. H.

**DOORS FOR BUS AND SWITCH COMPARTMENTS**
Electrical Development & Machine Co.

**DREDGING PUMPS**
Boving Hydraulic & Engineering Company

**ELECTRICAL ENGINEERS**
Gest, Limited, G. M.
See Directory of Engineers

**ELECTRIC HEATERS**
Canadian General Electric Co., Ltd.
Canadian Westinghouse Company
Equator Mfg. Co.
McDonald & Willson, Ltd., Toronto
National Electric Heating Company

**ELECTRIC SHADES**
Jefferson Glass Company

**ELECTRIC RANGES**
Canadian General Electric Co., Ltd.
Canadian Westinghouse Company
National Electric Heating Company
Northern Electric Company
Renfrew Electric Company

**ELECTRIC RAILWAY EQUIPMENT**
Canadian General Electric Co., Ltd.
Canadian Johns-Manville Co.
Electric Motor & Machinery Co., Ltd.
Northern Electric Company
Ohio Brass Company
T. C. White Electric Supply Co.

**ELECTRIC SWITCH BOXES**
Canadian General Electric Co., Ltd.
Canadian Drill & Electric Box Company
Dominion Electric Box Co.

**ELECTRICAL TESTING**
Electrical Testing Laboratories, Inc.

**ELEVATOR CONTRACTS**
Dominion Carbon Brush Co.

**ENGINES**
Boving Hydraulic & Engineering Co. of Canada
Canadian Allis-Chalmers Company

**FANS**
Canadian General Electric Co., Ltd.
Canadian Westinghouse Company
Century Electric Company
Great West Electric Company
McDonald & Willson
Northern Electric Company
Renfrew Electric Company
Robbins & Myers

**FIBRE (Hard)**
Canadian Johns-Manville Co.
Diamond State Fibre Co. of Canada, Ltd.

**FIBRE (Vulcanized)**
Diamond State Fibre Co. of Canada, Ltd.

**FIRE ALARM EQUIPMENT**
Hart Battery Co.
Hatheway & Knott, Inc.
Northern Electric Company
Slater N

**FIXTURES**
Benson-Wilcox Electric Co.
Crown Electrical Mfg. Co., Ltd.
Canadian General Electric Co., Ltd.
Duguesne Electric & Mfg. Co.
Jefferson Glass Company
McDonald & Willson
Northern Electric Company
Tallman Brass & Metal Company

**FLASHLIGHTS**
Canadian National Carbon Company
McDonald & Willson, Ltd.
Northern Electric Company
Spielman Agencies, Registered

**FLEXIBLE CONDUIT**
Canadian Triangle Conduit Company
Hatheway & Knott, Inc.
Slater, N.

**FUSES**
Canadian General Electric Co., Ltd.
Canadian Johns-Manville Co.
Canadian Westinghouse Company
G. & W. Electric Specialty Company
Hatheway & Knott, Inc.
International Machinery & Supply Company
Moloney Electric Company of Canada
Northern Electric Company

## CLASSIFIED INDEX TO ADVERTISEMENTS—CONTINUED

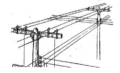

## CLASSIFIED INDEX TO ADVERTISEMENTS—CONTINUED

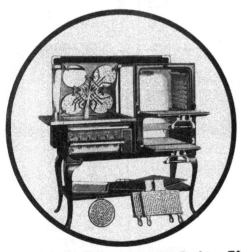

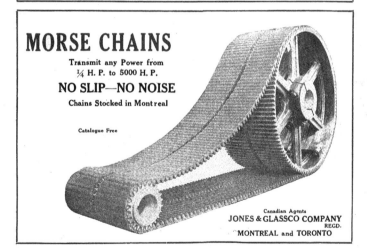

Vol. XXXI No. 2                                    Toronto, January 15, 1922

# Electrical News

**Engineering Contracting**    **Merchandising Transportation**

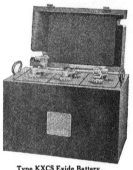

# You Never Heard of Property Partly
# Surveyed and Partly Estimated

There is no more use of partly surveying and partly estimating a piece of property than there is of measuring the actual kw. hour consumption of a customer and estimating his maximum demand. In each case the **whole** is really an estimation because **part** is estimated.

You bill your customers entirely on a measured basis only if you meter both their demand **and** consumption. A Lincoln Maximum Demand Meter will give you as accurate readings of demand as a watt-hour meter gives actual kw. hour consumption. It is these two meters working together which insures accurate billing in many of the Dominion's leading central stations today.

Don't delay an investigation—you will be under no obligation whatever.

Write for the section (or sections) of the Lincoln Meter catalog which interests you. SECTION I is a non-technical treatise on why demand should be measured. SECTION II, the technical treatise on demand measurement; SECTION III, Installation; SECTION IV, the Lincoln Graphic Demand Meter; and SECTION V, the Lincoln method of Volt-Ampere Demand Measurement (Power Factor recognition).

**The Lincoln Meter Co., Limited**
243 College Street, Toronto, Canada
Cable "Meters" Toronto    Phone College 1374

Hand now on left varies with *load and shows this load at any time of day.*

Hand now on right meter only to right and indicates maximum load since last resetting.

# Lincoln
## MAXIMUM DEMAND
# meter

*We wish to take this opportunity of thanking our
patrons for their favors during the past
year and extend to them the*

## Compliments
## of the Season

# THE DEVOE ELECTRIC SWITCH COMPANY
414 Notre Dame West, Montreal, and 105 Victoria St., Toronto, Ont.

# Can We Help You ?

There may be certain articles which you cannot find in these advertising pages that you would like to have information on. Do not hesitate to use this form, no matter what it is. If we can be of any service to you in supplying that information, it will be a pleasure to do so. We want you to feel that the Electrical News is published in your interests, and we want to help you whenever we can.

## INFORMATION WANTED

Electrical News,
    347 Adelaide St. West,          Date.................19
       TORONTO.

Please tell me ........................................................

...........................................................................

...........................................................................

Name ....................................................................

Address .................................................................

**ABC Alco Washers**
Single and twin tub models; power and electric types.

**ABC Super Electric**
Semi-cabinet washers; copper or galvanized iron tubs.

**ABC Electric Ironer**
Can be run by any ABC electric washer; also separately motored.

**ABC Super Electric**
The "Packard" of Electric washers; full cabinet.

# Why Not Sell *One* Complete Line?

Why tie up your capital in stocks of three or four makes of washers and ironers?

Why talk, advertise and sell three or four makes?

Why order separate shipments, enter into separate correspondence, and endure all the extra trouble and work that dealing with three or four sources entails?

Why not concentrate on one make that covers all popular types and meets every price demand?

You can also secure better discounts by that policy. Take on the famous A B C Line and make up combination orders. You can offer a greater variety of models to your trade, on a smaller investment, and make more profit per sale besides!

There are twelve A B C models from which you can select your assortments. No other line is as complete. Power and electric washers, and electric ironers, are included. The retail prices range from $210 to $295 on electric models. Power-driven washers are correspondingly less in price. All prices are at rock-bottom; greatly reduced from previous levels.

The manufacturers of the A B C Line, Altorfer Bros. Company, are pioneers in the industry. They are a big, progressive, reliable $2,000,000 organization, of broad and liberal policies. A B C Washers have been on the market for 14 years—their success is a matter of record. Every model is a high-grade one.

Exclusive dealerships for this famous A B C Line are assets. Real co-operation with the dealer goes along with them. They are worth getting. A few more are to be allotted at the present time. Immediate investigation is therefore advisable. Your inquiry does not place you under any obligation. Wire or write at once.

# A B C *Electric Laundress*

## WASHES    WRINGS    IRONS

### C. D. Henderson - Canadian Representative
Box No. 123,    Brantford, Ont.

WHOLESALE DISTRIBUTORS:

| | | |
|---|---|---|
| **MARITIME PROVINCES**<br>Blackadar & Stevens,<br>Roy Building,<br>Halifax - - N.S. | **ALBERTA**<br>Cunningham Electric Co., Ltd.,<br>Calgary | **SASKATCHEWAN**<br>Sun Electrical Supply, Ltd.,<br>Regina |
| **ONTARIO**<br>Masco Co., Ltd.,<br>78 Richmond St. - Toronto | **QUEBEC**<br>Dawson & Co., Ltd.,<br>148 McGill St. - - Montreal | **BRITISH COLUMBIA**<br>Rankin & Cherrell,<br>Vancouver |

## ALPHABETICAL LIST OF ADVERTISERS

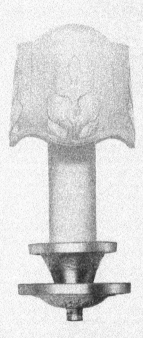

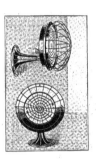

# What is "The Hoover Resale Plan"?

The Hoover Resale Plan was years ago originated by the Hoover organization. Many of the leading stores in this country and the United States have long operated under it. It has no parallel for achieving a large volume of sales.

Under this plan the Hoover organization is invited into a store where it conducts itself, under store rules, as a part of the store organization.

Trained, however, in the art of securing leads, demonstration, sale and servicing of Hoovers, the Hoover organization seeks and obtains business as only an organization thoroughly schooled in the advantages and features of a product can do.

The plan provides the dealer with such a specialty organization, relieves him of the responsibility for its selection, training and supervision, and, without risk, nets him a greater annual profit on his investment.

Experience has demonstrated that the plan is without an equal for success. The Hoover itself is an integral part of the plan. For there is no other electrical appliance which gives so convincing a demonstration, or serves its users in a more satisfactory way.

The Hoover and the Hoover Resale Plan are gaining every day. Are you willing to consider them now?

THE HOOVER SUCTION SWEEPER COMPANY OF CANADA, LIMITED
Factory and General Offices: Hamilton, Ontario

# The HOOVER

It *Beats . . . as it Sweeps as it Cleans*
MADE IN CANADA—BY CANADIANS—FOR CANADIANS

MONTREAL - - 119 Board of Trade Bldg.
WINNIPEG - - - Electric Ry. Chambers
VANCOUVER - - - - Winch Building
NEW YORK - - - - - 296 Broadway
CHICAGO - - Room 803, 63 E. Adams St.
LONDON, ENG. - - 16 Regent Street S. W.

ADVERTISEMENTS

Orders for advertising should reach the office of publication not later than the 5th and 20th of the month. Changes in advertisements will be made whenever desired, without cost to the advertiser.

SUBSCRIBERS

The "Electrical News" will be mailed to subscribers in Canada and Great Britain, post free, for $2.00 per annum. United States and foreign, $2.50. Remit by currency, registered letter, or postal order payable to Hugh C. MacLean Publications Limited.

Subscribers are requested to promptly notify the publishers of failure or delay in delivery of paper.

Authorized by the Postmaster General for Canada, for transmission as second class matter.

Vol. 32          Toronto, January 15, 1922          No. 2

# The Chippawa Development— First Unit of 650,000 H.P.

It is a human characteristic to follow the line of least resistance. That explains why so many people in this world are "knockers." It is easier to criticize a condition than to take the trouble to investigate the cause underlying that condition.

This explains in a measure, no doubt, some of the things that are being said about the Chippawa development, the first unit of which was officially turned over during Christmas week. For example, one of our foreign contempories waxes almost poetical in declaring that "the Hydro Commission of Ontario, lured on by ambition, has fallen into a pit of its own digging." This "pit" neither refers to the canal nor the forebay, the items that naturally occur to one in this connection. The writer is referring, figuratively, to the costs of the Chippawa development.

Now, there is no one in Ontario—or for that matter in Canada—who does not regret that the cost of the Chippawa development is as great as it is. Cheap power is a wonderful asset to any district and if, as seems probable, power rates in Ontario have to be revised upwards, it will be a matter of regret. It is neither fair nor reasonable, however, to charge this condition to the Commission's ambition—that is a superficial viewpoint that savors just a little bit, we think, of satisfaction at the consoling thought that things have turned out rather worse than they might have done.

The Chippawa development is about the biggest thing of its kind in the world. Perhaps we could defend the province with, at that time, her greatest need. In considering the question of the high cost of the Chippawa development, we must never forget, therefore, that two factors entered very prominently into these figures. First, the scheme was conceived and initiated at a time when construction costs were at their peak and, second, speed was the most essential element entering into the construction. Looking back on the whole situation through a distance of five or six years it doesn't take a particularly clever man to see that if the start on this construction work had been delayed until the present time, the cost would have been very much less.

It is not the province, nor the desire, of the Electrical News to defend any unnecessary cost in connection with this work or any engineering mistakes, where it is shown that these exist. We do not even raise the question as to whether this work might have been carried on more efficiently or with better results under private than under municipal management. The basic fact is: we needed power—it was a matter of life or death to us at a time when the country to which our contemporary belongs had not yet decided whether or not they were interested in the war. Further, it is an open question whether private capital could have been induced to undertake a work of this kind under the unsettled conditions prevailing at that time.

● ● ●

It may, of course, be argued again that it was highly improbable that a work of this nature could be completed before the ending of the war. This again is speaking after the fact. In 1915 none of us knew whether the war would last three years or thirty years and, in any case, this further fact must be recognized: that the moral effect of an enterprise of this nature, indicating the determination of Canada to continue the struggle as long as might be necessary, with every available resource, cannot be over-estimated.

The cost of Chippawa power is higher than we should have liked. Let us admit it. It is highly improbable that the first unit will take care of the fixed charges without an increase in rates. It would be foolish to expect it. It is even improbable that the completion of the first set of five units will do much more than carry the enterprise, even if the rates are slightly increased. We believe, however, that the Commission have stated their confidence in their ability to complete the total development of 650,000 h.p. at a cost that will enable them to establish at that time rates equal to those at present existing in Ontario, and probably reduce them. This, as we understand it, is the actual "ambition" of the Ontario Commission which, when it is considered that Ontario power rates are among the lowest in the world, must be admitted as a worthy one.

## The Contribution of Electricity to Our General Industrial Development

A significant statement in the recent annual report of the Southern Canada Power Company, Ltd., is to the effect that in the past three years seventeen new industries brought into the Eastern Townships of Quebec by that company have invested $10,000,000 of capital.

Do we, even in the electrical field itself, sufficiently appreciate our possibilities in attracting industries to Canada? How often do we hear it said "Or after all, the cost of the power in most manufacturing operations is too small to be a real factor in determining location?" But, in saying that, don't we overlook the inestimable value of the class of service rendered by electric power—which is out of all proportion to its cost?

The actual fact seems to be that we have all fallen into the error of emphasizing the cheapness of electric current instead of its dependability and utilitarianism. We have tried to sell on a basis of "cost" rather than "quality". Let us forget all about cost and talk "indispensibility." If the business man wants electricity in his factory, if he is convinced that he can't get along without it, he won't waste his own time or that of anybody else talking cost. He wants service—good dependable service—a power agent that will undertake to do a job and do it every day of the 365: cost is a detail. The Southern Canada Power Company has brought ten millions of capital into the district they serve, during the past three years—not because power was cheaper there than elsewhere but because they had much more convincing arguments than low cost. They talked the convenience, the advisability, the necessity of electric service, its dependability and finally, its ample supply with a guarantee for the extension of future years. On these arguments we have no serious competitors.

The electrical industry as a whole shows a tendency to forget that the modern conception of merchandising is "service first." Our energies should be concentrated on bringing the man and woman on the street into that attitude of mind where they "want" electric service because they are convinced they "need" it. The sale of electrical appliances and equipment would follow automatically and the price would scarcely enter into the matter at all.

### Our Wonderful Natural Resources

Attention is now being directed toward the country's natural resources as never before, since it is generally recognized that only by a more wide-spread utilization of Canada's undeveloped lands, mines, forests, water-powers and fisheries can present day economic problems be solved.

The Natural Resources Intelligence Branch of the Department of the Interior has published a map showing the leading natural resources of each province. In Nova Scotia mixed farming, mining and fishing predominate; in Prince Edward Island fur-farming and agriculture. New Brunswick has large areas of timber, while mixed farming and fruit growing are outstanding interests. In Quebec may be found a wealth of timber for pulp-wood, also minerals such as asbestos, graphite and molybdenite, while in Ontario somewhat similar opportunities exist.

In addition to information on natural resources, the map shows all railways and trade routes. An interesting and valuable feature is a series of comparative diagrams illustrating the production and exports of the various provinces. A copy of the map may be obtained free of charge upon application to the Natural Resources Intelligence Branch, Department of the Interior, Ottawa.

## A Diesel-Electric Plant That Has Made Good in the West

By S. H. EXCELL*

Just west of the Rockies and standing guard at the north end of the Okanagan Valley, is the town of Vernon, B. C., the snug centre of the famous apple growing district.

In these times when municipal ownership is so much to the fore it will not be amiss to recognize what has been a most successful municipal undertaking and to describe how thoroughly Diesel engines have proven their outstanding economy and ability to serve some 4,000 of a population, continuously and well.

From 1902 to 1908 a steam unit gave faithful service and in the latter year, the advertised economy of the Diesel engine was so marked, the Municipal Council decided to install a 4 cylinder, 200 h. p. unit to take care of a constantly increasing load. This machine was purchased from the old established firm of Mirrilees, Bickerton & Day, Ltd., Stockfort, England, and is of the vertical or marine type direct connected by flexible coupling to a Westinghouse generator and exciter of equivalent capacity. To indicate the serviceable qualities of this machine, which has been in constant service ever since, it is interesting to note that quite recently during a thorough examination all the major parts were found to be practically without wear, while the wear on minor parts amounted to only a few thousandths of an inch. This machine today is as capable of carrying continuous full load as it was when first installed. Load conditions became such that in 1912 a further extension to the plant became necessary, so the steam plant was shut down for good and a 500 h. p. Diesel unit installed in the new wing of the power house.

This machine is also of the 4 cylinder, vertical or marine type and was built in Sweden by the Alstiebolaget Diesel Moterer Co. The alternator and exciter were furnished by the Canadian General Electric Company and are coupled direct to the engine shaft.

The switchboard equipment throughout was supplied by the Canadian Westinghouse Company and consists of generator panels, light and power panel, meter panel and street lighting control panel. Here again the installation of the larger unit was so good that recently, under special micrometer test taken between the crank webs, the limit of wear showed a discrepancy of only some .003". As presently operated, the larger unit carries the load for, from 18 hours to 24 hours daily, according to season. while the smaller takes the night and week-end load and is kept available for the peak load.

The vertical type Diesel has proved to be highly satisfactory for continuous operation, the almost entire lack of wear and vibration showing the degree of balance attained.

The electrical equipment has been very satisfactory and, like the prime movers, is still good for many years of active service and "juice" extraction. What probably has added to the success of this plant is the routine adopted for plant operation and maintenance, and the justifiable pride of the operators for their charges. A definite schedule of inspection and repairs is followed out and each part in turn is examined and, where necessary, adjustments made. Nothing is overlooked; from the roof down to the pipe tunnel below the floor level a constant search to eliminate trouble is carried out. In the machine shop and spare room, the same thoroughness prevails. All supplies are in bins or cupboards available for instant use.

The entire plant is very efficiently lighted, while for artificial lighting high candle power nitro lamps in semi-in-

* City Engineer, Vernon, B.C.

direct fittings are used, giving well diffused illumination. Ventilation is amply taken care of, under all seasonal conditions. The power house is of ferro concrete construction and of a pleasing design.

### Engine Room details

Fuel oil is pumped by hand from storage tanks convenient to the nearby railway track to service tanks in the engine room, and so placed as to give a gravity head to the fuel supply to each engine. Cooling water is supplied from

# Aerial Construction for Carrying Heavy Cable

The somewhat unusual aerial construction shown in the two cuts was recently installed for the Ottawa Electric Company to carry a heavy power cable across the headrace; the previous cables had been laid as subaqueous cables in the bed of the race. Owing to local conditions, it is necessary to

200 h.p. unit, Vernon, B.C.

the city water system or from an emergency connection to a never-failing creek contiguous to the plant. Lubricating oil after passing through the engines is filtered through two Burt oil filter units. These filters are kept at good temperature by steam obtained from specially designed generators on the exhaust lines from the engines.

It is of interest to note these generators supply steam and hot water, not only to the filters but also to the fuel oil heaters and the jackets of the smaller engine to facilitate starting in below zero weather. The engines are arranged so that air could be supplied in an emergency from the bottles of either engine, but this has never been called into use, each engine starting and picking up promptly.

Cleanliness is paramount and to give the electrical equipment the best possible care, a motor driven "Curtis" air compressor is installed. This machine, besides blowing out the alternators and exciters, is put to many uses, such as clearing lubricating pipes, fuel pulverizers, blowing down switchboard, etc.

The operation of the plant is carefully logged daily, and besides the reports necessary to record output, expenditures and revenue, etc., close check is kept on all such terms as taking up of bearings, matters pertaining to lubrication, valve settings, indicator cards, etc.

For the year ending December, 1921, the operating expenditures and overhead were as follows:

| | |
|---|---|
| Interest and sinking fund and insurance | $17,755.67 |
| Fuel oil | 13,515.00 |
| Salaries and day labor | 7,800.00 |
| Lubrication and sundries | 1400.00 |
| Plant maintenance | 1500.00 |
| Output | 1,068,283 kw. hrs. |
| Load factor | 22.9 % |

Cost of fuel per kw.h. at average cost of 14.64 cents per gallon would be .......... 1.24 cents
Total annual cost of 1 kw.h. at switchboard ...... 4.11 cents

empty this each time a cable is laid; this means the closing down of the power station temporarily, and the load has to be carried by reserve steam plant, which is not only very costly, but inadequate.

To avoid closing down, it was decided to run the cable aerially over the head race, a distance of approximately 150 ft. The cable is carried over on a form of catenary suspension in practically a straight line. A main messenger

Carrying heavy power cable across headrace of Ottawa Electric Company

cable of ¾ in. diameter steel rope supports six "T" hangers, which in turn support two cable messengers of ⅜ in. strand and a third messenger which is supported on inverted trolley ears and is used for the cableman's aerial car, so that inspection and repair work may be easily carried out.

Each "T" hanger is tied to a ⅞ in. diam. steel rope wind or sway brace on each side, so that the cable is prac-

tically rigidly supported in all conditions of wind and weather.
The cable itself is a four conductor No. 00 B & S gauge, lead covered and paper insulated, for 13,200 volts, overall dia. 2.79 in.; it weighs 11.1 lbs. per ft., and is supported from the cable messengers by aerial cable rings placed 12 inches

Close-up view of heavy power cable

apart. There is provision for another similar cable on the same structure.

The cable was installed by the Northern Electric Company, and is part of a large improvement to the substation and transmission equipment recently completed under the direction of Prof. Herdt.

## Public Utility Outlook Is Encouraging

It is generally conceded that the year now opening will bring in its course the beginning of the long-heralded boom in electric light and power plant construction. The water power act is beginning to function, money is once more coming to a reasonable level, the power load is commencing to creep up, new customers are taking service, and costs of power equipment and building materials are being reduced. As an indication of this approaching activity, a prominent firm, Engineers and Constructors of New York and Chicago, advise that they have secured important contracts from the Duquesne Light Company of Pittsburgh and the New Orleans Railway and Light Company.

The work of the Duquesne Light Company includes the installation of the second unit of 60,000 kw. at the Colfax Power Station, together with three additional substations along the company's lines in the Pittsburgh district. The Colfax Station is designed ultimately to be one of the largest steam stations in the country, containing six 60,000 kw. units. The first unit was put in operation last year and the installation of the second unit now authorized is a part of the company's plan to provide additional facilities as the demand for power increases. The New Orleans Railway and Light Company is proceeding with extensive additions to its power plant and distributing systems, including the installation of a 20,000 kw. turbine and auxiliaries.

## Street Lighting Developments

The December meeting of the Toronto Chapter of the Illuminating Engineering Society was held on Monday evening, December 19th, in the Chemistry & Mining Building. Mr. M. B. Hastings, who has been in close touch with street lighting developments for many years, gave an illustrated address on that subject.

According to a statement of one of the provisional directors of the Prince Edward Light and Power Company, incorporated at the last session of the Legislature with a $2,000,000 capital, the development work proposed may not go forward unless an extension of time can be obtained.

# A. M. E. U. Convention Program

**Thursday, January 26**

Morning.
Registration at the office of the Hydro Electric Power Commission of Ontario, 190 University Avenue.

Afternoon. 2.30 o'clock, in Room C26, Chemistry and Mining Building, University of Toronto (College Street at McCaul Street).
President's Address.
Reports.
Election of Officers for 1922. Ballots will be distributed at the time of registration, and will be deposited at the beginning of this session. The scrutineers will announce the results of the election before adjournment of this session.
Illustrated Talk, 'The Nipigon Development," by T. C. James, Municipal Engineer, Hydro Electric Power Commission of Ontario.

Evening. 6.00 o'clock, at Bingham's, 84 Yonge Street (just north of King Street).

Address, "Co-operation," by F. A. Gaby, Chief Engineer, Hydro Electric Power Commission of Ontario.

Tickets may be obtained when registering. Subscription $2.00.

**Friday, January 27**

Morning. 9.30 o'clock.
Paper, "The Effect of Underloaded Transformers on System Power Factor," by G. F. Drewry, Municipal Engineer, Hydro Electric Power Commission of Ontario.
Paper, "The Measurement of Kilovolt Amperes for Power Billing," by S. L. B. Lines, President, Chamberlin and Hookham Meter Co., Limited. Toronto.
Discussion.

Afternoon. 2.30 o'clock.
Paper, "Further Developments in Rural Power Distribution," by J. W. Purcell, Farms Engineer, Hydro Electric Power Commission of Ontario.
Discussion.

# The Mushamush Hydro-Electric Plant

### A Scheme Undertaken by the Nova Scotia Power Commission Which, While Small, Conforms to Present-Day Standards of Efficiency, Simplicity and Economy.

By K. G. CHISHOLM
Assistant Engineer, Dominion Water Powers Branch.

On October 1st the Nova Scotia Power Commission put in operation a reconstructed development on the Mushamush river near Mahone, which by contrast with the original development strikingly illustrates the present practice of simple, sturdy, hydro-electric units of the vertical, direct-connected type with high efficiency, even at comparatively low heads. The new development, although small, is considered to conform to the present-day standards of efficiency simplicity and economy.

Previous to the present development a plant at the same site owned and operated by the Lunenburg Gas Company, supplied the lighting requirements of Lunenburg with night service only. In 1920, a request for current was received from the municipality of Riverport. In addition, West La-Have, Mahone, Blockhouse and Bridgewater, were in communication with the Commission and it became apparent that there were existing and future markets for power in the district considerably in excess of the available sources of supply, a condition which, under the Provincial Water Power Act, the Power Commission was authorized to meet in the way it deemed most advisable.

After careful study of several alternative propositions, it was decided to buy out the property and rights of the Lunenburg Gas Company on the Mushamush river, including among other items their small hydro development near Mahone with the storage dams and flowage rights pertaining thereto, as well as their transmission line to the outskirts of Lunenburg, and to utilize as much as possible these rights and properties in the construction of a new plant more liberally designed to meet the present and future demands of the district.

The company's development near Mahone consisted of a head dam with flume built into the dam, and wooden power-house immediately below. The dam about 400 feet long at highwater and 24 feet in maximum height consisted of a central timber spillway section with an earth embankment at both ends. The turbine installation consisted of a pair of 27 inch, horizontal, S. Morgan Smith cylinder gate wheels mounted on the same shaft and discharging into a common steel plate cylindrical draft tube. They were set in the bottom of the timber flume in a timber draft chest made up at the site. The shaft projected through the end of the flume into the power-house and the generator on the upper floor of the power-house was belted to a 6-ft. diameter cast iron pulley on the turbine shaft. The wheels were rated at 386 h.p., 280 r.p.m. under 24-ft. head, and the Westinghouse 3 phase generator at 200 kv.a., 3500 volts and 600 r.p.m. The exciter was belted to the generator shaft. This small plant had been in operation for 13 years and had given reasonably good service.

In connection with the studies made by the Commission's engineers, a test of the old plant was made in February 1921. The results of this test may be interesting as being representative of small plants of its class. A close inspection after the test showed the wheel and bearings to be in good condition, although some play on the bearings was observed. As far as the wheel and mountings were concerned, it was apparent that the full original efficiencies of the wheel should

have been obtained, but in fairness to the manufacturers, and in explanation of the low efficiencies actually obtained, it should be stated that the tail-race conditions immediately under the turbine were not good and piling up of the water took place, with consequent loss of operating head. The extent of this piling up could not be ascertained as it occurred around the draft tube beneath the flume. The average results of the tests made are given in the following summary.

| Gate | Head | Speed | Discharge | Turbine Effcy. |
|---|---|---|---|---|
| Full | 21.8 | 275 | 159 | 63.0% |
| ¾ | 21.9 | 280 | 139 | 56.6% |
| ½ | 22.1 | 280 | 99 | 46.2% |

For the new development it was decided to use the same head of 22 feet, utilizing the existing dam, but instead of the

Exterior view of power house of the Mushamush hydro-electric development in Nova Scotia

old timber flume and power-house to build a new plant with better construction throughout and with a larger turbine in a more permanent and efficient setting. The turbine was selected for a head of 24 feet as it is the intention to increase the height of the dam by two feet when the load becomes large enough to warrant doing so.

In selecting equipment for the new plant, it was necessary to bear in mind that the new development must be begun and carried to completion as early as possible. Some consideration was given to utilizing a high speed runner of the Nagler type but for various reasons this was not entertained, particularly when it was found that the S. Morgan Smith Company had already constructed a wheel suitable in every respect, which could be shipped on short notice. It was also fortunately learned that the General Electric Company had under construction in their works at Erie, Pa., a vertical shaft water-wheel generator of suitable speed and capacity for direct connection to this turbine.

The general layout of the new plant is shown in the figures presented herewith. It includes a concrete flume with a quarter turn flaring draft tube moulded in the solid concrete of the foundation. The setting is of the vertical type

with ring, gates and cover plate supported on a concrete
pedestal in the flume bottom, and with runner, shaft and
generator rotor supported by a General Electric spring thrust
bearing supported on the generator stator. The circular pit
beneath the generator permits inspection of exciter, govern-
or pump, governor fly-ball and brake pulleys mounted also
on the turbine shaft. The braces from the turbine cover
plate to the flume walls, as well as the intermediate guide
bearing supports and the brake supporting frame, are bolt
fastened and removable, so that if necessity should arise the
turbine runner may be hauled up through the opening in the
power-house floor within the generator pedestal.

To close off the water from the flume a timber gate and
hoisting mechanism is provided in front of the power-house,

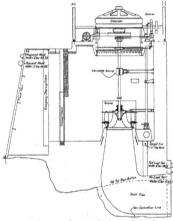

Section of power house showing the layout of the hydraulic
and electric units

and for safety's sake emergency stop-log checks are also
provided. A flap valve permits of unwatering the flume below the level of the turbine gates. Access to the interior of
the flume can be had through a trap-door in the power-house
floor and iron ladder rungs in the flume wall.

The flume is of reinforced concrete 12 feet 6 inches wide
by 33 feet long inside dimensions. A reinforced concrete deck
forms the power-house floor. The weight of the generator,
shaft and turbine runner is carried on an I beam resting on
the flume walls.

The rating of the new hydraulic unit for 24 feet head is
925 h.p. at full gate and 180 r.p.m. The generator rating is
750 kv.a. at 80% P. F., 6600 volt, 3 phase, 180 r.p.m.

Although no opportunity has yet arisen for testing the
new plant, the guaranteed turbine efficiency curve runs from
73% at half-gate to 88% at .95 gate, dropping off to about
80% at full gate. This is in sharp contrast to the efficiencies
obtained from the old plant. The generator efficiencies are
high, ranging from 91.5% at full load to 88.5% at half load,
at 80% P. F.

For reasons of economy and speed in delivery of equip-

ment, a standard horizontal shaft exciter is used, belt-driven
from the main shaft by means of an idler bolted to the under
side of the generator stator frame and a quarter turn in the
belt.    Undoubtedly from a purely mechanical standpoint,
either a belt-driven vertical exciter, or an exciter direct-connected to the main shaft would have been preferable, but economy and more particularly speed of delivery, decided
against them.    Not only would a direct-connected exciter
have been expensive but it would have held up delivery of
the generator which was already under construction. Furthermore, the Commission was already in possession of a
horizontal shaft exciter of the proper rating and this was
accordingly used. However, the change to a vertical shaft
exciter can be made if found desirable at any time. Some
trouble has been experienced with the exciter drive, first in
the idler pulley and then in the belt itself, due to stretching
of the new belt. This has now been overcome and the
drive is operating quite satisfactorily with a leather belt.

For the sake of economy, the power-house was made as
small as possible, consistent with properly housing the machinery. Its outside dimensions are 22 ft. by 24 ft. 2 ins. It
consists of a framework of 2 x 6 timber covered on the outside with cement stucco laid on Bishopric-board and on the
inside with gypsum fibre-board, the whole giving a building
cheap, fire-resistant and easily constructed. The roof covering is of red "Everlastic" asphalt slate shingles. The
building presents quite a creditable appearance as the illustration shows.

The use of the old site for the new development necessitated a shut down and a cessation in the delivery of power.
Every effort was made to reduce the period of shut down to a
minimum and speed of construction was an essential of the
work. Work was commenced on the morning of August 10th
and on October 1st the new plant was delivering energy to
the line. Furthermore, the two municipalities supplied were
without lights for only one night, on the 10th of August.

The old plant was shut down at midnight of August 9th
and immediately unwatering sluices were opened at the head
dam. At daybreak a gang had commenced to dismantle the
power-house. By seven p.m. exciter and generator had been
moved by hand to the site of a temporary plant 150 feet away
and connected to a steam engine and boiler already hauled
to the job and the generator rewired into the line. Except
for the fact that exciter trouble developed, the temporary
plant would have been in operation that same night. As it
was, current was turned into the line at dusk of the following evening and the people of Lunenburg, who had been prepared to go without lights for the whole period of reconstruction, actually were deprived for only one night. On a job
large enough to warrant the presence of adequate apparatus, this would be no feat worthy of mention, but considering
the manhandling of heavy machinery necessary under the
circumstances, it was considered very satisfactory. It is
further of interest to note that the initial schedule of 45
working days was realized.

The costs of the various phases of the work which were
very closely followed and carefully distributed, are unfortunately not available at the date of writing, but can be furnished later. The total cost is, however, within the estimate
of $48,000.00.

Both the design and execution of the work were carried
out by the Commission's own engineering and construction
organization, under the immediate direction of Mr. Harold S.
Johnston, hydraulic engineer of the Commission, with the
writer as engineering assistant and instrument man during the
period of reconstruction. K. H. Smith is chief engineer of
the Commission, which consists of Hon. E. H. Armstrong,
chairman, R. H. Mackay, secretary and F. C. Whitman, commissioner.

# Selecting and Training Salesmen

### The Most Important Factor in the Electrical Industry to-day—Health, Character
### Deportment, Reputation, Ambition, Self Confidence*

One of the major problems facing sales managers to-day is building up an efficient sales force to secure orders. Sales organizations are being subjected to the acid test; some of them will fail to survive that test. We have passed through four or five years recently when men who were called salesmen did not have to sell at all; they became very soft in their ideas of work, yet they made wonderful sales records. Now times have changed; order-takers are of no use now.

#### Get Down to Hard Work

It has occurred to me during the last six months to a year, carrying our share of worry as to business conditions, that there are many men who have got a lot of wisdom but do not know how to apply it. They do not realize the vital application of it, and that is in the hardest kind of work. Especially to salesmen who have had five years of easy going it is hard to realize that they have got to get down to hard work or they will not stay in a sales organization very long.

#### The New Conditions

The quicker we all realize that prosperity depends on the action of our own self-starter the better. A good many people are just waiting for something to turn up, even sales managers. To my mind there is nothing "going to happen" until you make it. Instead of the men going out on the 9 o'clock train they will have to go out on the 7 o'clock if they are to succeed, and instead of coming in Thursday night or Friday noon, they will come in Saturday morning—if they are lucky. Those are the new conditions that we are facing.

#### Memorize This

I want to convey this idea; it is not necessary to tell most men these things, it is necessary to "sell" them. In our organization we never tell a salesman to do something. We "sell" the proposition. That is the only way to get the proper co-operation and spirit. The art of selling is the art of buying. What do I mean?—well, the amount you sell depends on the amount of confidence you can buy, the respect you can buy from the prospective customer. I think that is an important principle to have the sales force memorize.

#### Obtaining Salesmen

Now where are we going to get our salesmen? Our experience is that we get our best men through our present sales force. Those men know the policy of the company, they have confidence in the company, they know the calibre of men the company want, and they have seen in action the men that they are going to send in to apply for the position. That is something you cannot do as a general proposition. They know whether he plays pool in the hotel until 2 a.m. or poker until daylight. They know his characteristics through working with him. I emphasize that, because looking back over my experience during the last thirty years I find that the best salesmen it has been my good fortune to secure came through the salesmen we already had.

#### The Man in Action

You will not buy an automobile unless you try it out, but there are many hundreds of salesmen employed by sales managers who have never seen them in action. A man comes to your office, naturally he is at his best, but that does not tell you what he will look like in Peterboro or Montreal or

Vancouver. His nails are clean and his shoes are shined. We did the same ourselves years ago. The important thing is to see the man in action in his regular working form, day in and day out, get a line on his characteristics, how he stands up under the test of the three a.m. bus, and the one old kind of pie and no cream.

#### The Other Fellow's Salesman

The next channel is from the salesmen that canvass you for orders; and that is a good channel. "Do you mean to tell me you would deliberately employ one of my salesmen?" someone may ask; I certainly do. If you say that is not a fair proposition I do not get your point of view. If my salesmen can make more money with you they will go with you. Therefore I am never too busy to see a canvassing salesman. The bank manager can wait, but this man may be the one I need as the corner-stone of a certain portion of our sales force.

I believe the successful sales manager is the one who is never too busy to see a canvassing salesman. If I am in a conference which would make it necessary to keep him too long, I go out and see him for a moment and apologize and make an appointment.

#### From a Great School

I spent twenty years with the National Cash Register Co.—Mr. Paterson, the president of that company, is known as a wonderful leader. We were in a convention of the sixteen sales managers of the North American continent. Mr. Paterson came in and remarked "these are the sales managers, yes; now just what do you gentlemen do for the high salaries you get?".

Well, it was kind of sudden. One chap stood up and said "I keep my eye on the territory, I see that the offices are kept nice and clean."

"And you?"

"I check up the accounts and payments and consignments—"

Well, I was thinking fast, it was coming close to me, I said to myself "these boys are not thinking right," so when he came to me I said "I employ and train salesmen." I was with the N. C. R. for twenty years.

#### Looking Them Over

I would have lost the opportunity to secure choice men had I been too busy to see salesmen. Often a man has come in and although I did not need him at all, he made an impression, I talked to him fifteen or twenty minutes, asked if he would mind coming in again day after tomorrow—next week,—I wanted to get another slant on him. They never act exactly the same twice in succession. I asked him,— "Now just tell me how much money you make and how much you would like to make. It may be that you are fitted for this business. I am going to find out in the next fifteen minutes if you are."

#### From the Retail Store

Another channel is the retail store. We have secured some excellent men who were retail store clerks. There is a field where you will find ambition; also, it is a field where men are not over-paid. An illustration: we had had difficulty in getting our devices in retail stores. Our time-recording division salesmen had been trained to interview manufacturers and factory men, but they fell down on the stores. I discov-

*By F. E. Mutton, vice-president and general manager International Business Machines Co. Ltd., Toronto, in "Business Methods."

ered that they did not know the language of the retail merchant, they had never been in that atmosphere. I looked about and found a young man who was head clerk in one of the large hardware stores in Hamilton, snappy, well-dressed, knew his job or he would not have been head clerk. I brought him to Toronto and employed him. We had to teach him the time-clock business as it related to the retail store; I said to him "Now you will have to go into the factory and put on a pair of blue jeans and work from 7 till 5.30 for two weeks, for which you will not get any pay."

"All right," he said, proving his mettle.

I was only testing him really, but that was his mechanical training. The first month that young man was out in the Hamilton territory he sold eighteen machines to eighteen stores. That was good work. He knew the atmosphere and the language.

### The Unemployed Man

The weakest source is advertising for salemen. I do not believe it is good business or sound policy to hire a man who is out of a job. That, perhaps, sounds severe, but there is generally some reason why he is out of a job, and it is difficult to get at the true reason. If you are going to put up a "Safety First" sign get a salesman who is in a position.

Now, I want to put before you in this graphic way, these principles:

**Get Them Clean—Mentally, Physically, Morally, Spiritually.** That to me is "The Bible" in hiring salesmen. There is no one of these things that you can afford to overlook. And you can test them all out in your interviewing.

**Mentally.**—The salesman who will consume time telling those stories that we sometimes hear is not clean mentally. The results show physically and morally.

**Spiritually.**—This is very important. Sometimes we meet a man who does not believe in such things as churches. I am not preaching a sermon, but I do know the value of that element; we all do.

Then I would like to add **Keep Them Clean** as to the environment they live in.

### Organization

In my opinion the following elements come under the head of organization in a sales force: **Employing Men; Training Men; Supervising Men; Promoting Men; Firing Men.** If we are capable of the first four, we will not have to worry about the last.

### Qualifications

As to qualifications here they are:—**Health, Character, Deportment, Reputation in Own Neighborhood, Guarantee Bond, Manner, Ambition, Confidence in Himself.**

I have a chart on my desk that is always ticked off 100 per cent before a man is taken into our organization. Health is most important. If a man has not got 100 per cent health he is anchored. A lot of talks have been made on salesmanship; to me salesmanship is the combination of a lot of qualities. Health is the foundation.

Our company spends a good many dollars a year taking care of the health of our organization. We talk to them of what to wear, what to eat, etc., etc. A lot is wasted but some gets home and it does a lot of good. I venture to say when you meet one of our salesmen you find he is "in the pink." If he is not, he will not go out, they are trained not to.

### Foundation of Everything

When a salesman approaches a prospect, one of the most important things in getting that man's order is the salesman's health. Selling is largely a matter of domination. You need all the health and energy at your command to dominate some people; if you are only feeling about 70 or 80 per cent fit the

buyer will do the dominating. Health is the foundation of everything. Take personality, the salesman's greatest asset. You cannot see personality, you just sense it. It is a thing you cannot define, and the greatest element of personality is health. His eyes are bright and his step springy, and he thinks just a little faster than his prospect; he cannot do it unless he has 100 per cent health.

### Supervision

Then **Supervision.** I think lack of supervision is one of the weak links in our work. Every man needs it. Let me illustrate. The chain is—**Salesmen, Sales Agents, District Managers, Sales Managers, General Managers, President.** Does it stop there? No, it goes on—**Executive Committee, Directors, Shareholders.**

I would like to make this subject clear.

Who supervises Salesmen? The Sales Agents.

Who supervises the Sales Agents? The District Managers.

Who supervises the District Managers? The Sales Manager.

Who supervises the Sales Manager? The General Manager.

Who supervises the General Manager? The President.

Who supervises the President? The Executive Committee.

Who supervises the Executive Committee? The Directors.

Then last, but not least, who supervises the Directors? The Shareholders.

### Enthusiasm Injections

One of the reasons salesmen fall down is because they are not supervised closely enough. I do not mean bossed. We do not know the meaning of "Boss" in our organization.

Supervision means for one thing that you get out with the salesmen occasionally, especially the men in the Provincial territory, who are meeting to-day probably the greatest resistance in the history of trade in this country. They are only human, they get lonesome, they have gone a week perhaps without an order. The greatest asset for a salesman to have is a big stout heart, not knowing the meaning of "Quit." Well, that man needs supervision it might be termed, or assistance, a new injection of enthusiasm, which he can only get from an executive higher up.

### The Heart Touch

I have recently returned from a trip to the Coast. I am running the manufacturing end as well as the selling end, but there is nothing else so important to my mind as going out and visiting these men, spending a day or two or three with them, getting that heart-to-heart touch that they appreciate so much—I know it because I have been there and received that treatment.

### Necessity for Training

As to Training. How many firms do not seem to realize its necessity? I have had men come to me who seem to have been given a price list and an order book, and left to grab any further knowledge they could out of the atmosphere. These salesmen are going to run up against other men in parallel lines who have been trained, and we have got to see to it that our men are equipped to meet these trained men. If in your own judgment the new salesman is the right material for your organization, then it is worth spending time and money to give him the proper ammunition.

### Peddling Troubles

It is important to train salesmen to know that it is no earthly use telling their troubles to fellow salesmen.

To salesmen I say, if you do not agree with the policy of the house don't go to Bill Jones about it, he cannot cure it; if the sales manager will not cure it, go to the general manager, and if he does not satisfy you, go to the president.

Get your men to have enough confidence in you and the house to now that they can not do that. Many good sales organizations are wrecked by "gossip" breeding disloyalty, so have them feel that if they think they have any complaint they can come to you, and either they will "sell" you that they are right or you will "sell" them that you are right. People are only suspicious of things they do not understand. It is of primary importance in training salesmen that they should know and thoroughly understand the policy of the house. Then it will have their support.

### Constructive Help

One of the chief duties of the sales manager is to give his men constructive help in their work. If a sales manager cannot help his salesmen make sales, what is he for? I have three sales managers in my business, one for each division. Each of them when he visits an agency helps the man there to make sales. If he could not he would not be sales manager. I do not mean for him to go out and canvass the stores, I mean he is everlastingly gathering material and thinking, so that he has something to take out with him and say "Put this down in your note book."

### A Talk on Cards

I have a little talk on cards for our salesmen. I am a great believer in teaching through the eye. Some of these cards are:—

Success is simple. You can talk five minutes on that to a convention.

Service. Action of one on behalf of another in the interest of both.

Add yourself up. Don't overlook the weak spot—we all have a lot of them.

Think in big figures. Teach the salesman who is selling two thousand to think of ten thousand.

We are only as big as we think. Let him look at that for a minute.

Increase health and knowledge. The salesman can never afford to stop doing that.

Men make the business. You can take the biggest concern in Canada and burn down the whole institution, and they could collect the insurance and build up again, but if you took my seventy-two salesmen I would have some trouble to get things going again. Trained men are of more value than machinery and plant. Let them know how much you value them.

Teach salesmen not to guess, to know what they are talking about.

Cut out gossip, fault-finding and destructive criticism. What is wanted is constructive criticism.

A man is known by the company he keeps, and a company is known by the men it keeps.

The open mind is the growing mind.

Men should not be employed at any task which a machine can perform.

Self Analysis. Teach the salesman to analyze himself, the most helpful thing he can do.

Better to aim at perfection and miss than aim at imperfection and hit it.

Organize the brains of your organization.

Don't make the same mistake twice! But if you never make any mistakes you are no good.

Co-operation.

Concentration. On the work you are doing.

Study, learn and teach.

Think.

Tell your griefs to the man who can help you.

### Through The Eye

The purpose of these cards that I use at conventions, and further that I distribute around our factory and office is to educate the organization through the eye, and at conventions you will find the salesmen copying them down in their note book, in other words, they are gathering ammunition that is going to help them at some time.

The sales manager who can select the proper men and train them properly will never have to look for a lucrative position. The position is looking for him.

# A Closer Relationship Between Central Station and Contractor Dealer

### By C. C. BAINES *

That a closer relationship should exist between the central station and electrical contractors and dealers is a fact which I believe will to-day be granted without discussion. The central station needs the co-operation of the electrical contractor as much as the electrical contractor needs the co-operation of the central station. Both have their place in the world to-day. Both must function with each other to work out problems which present themselves daily.

There has been in the past in a great many localities anything but a spirit of unity between the central station and the electrical contractor. The electrical contractor has been prone to knock the central station rather than boost it. If we will investigate carefully we will find that there is an underlying reason for this condition of affairs, and I will endeavor to point out as many as possible of the grievances which the electrical contractor has against the central station.

### Before the Electrical Contractor

Back in the early 90's when electricity was first being

*Before Indiana State Association

made use of for light and power purposes, we find that there existed no electrical contractor. What at that time took the place of the central station did not only the work of electrical construction but also managed the sales of what few electrical appliances then existed. There was no electrical contractor.

As time went on and the central station came into its own there arose the demand for the electrical contractor, and naturally the central station, which had controlled the electrical field prior to the advent of the electrical contractor felt somewhat as though the contractor was an intruder in its field. As the electrical industry was developed the contractor added to his line the sale of electrical appliances, and more and more the central station looked with anything but a co-operative feeling.

### First Evils

One of the first breaches that existed between the contractor and the central station was the practice of the central station replacing lamps without cost to its customers. With the advent of the mazda lamp the practice of replacing lamps free of charge was replaced, in a great many localities throughout the United States, by the practice of selling lamps at less than list price.

This practice was no doubt made in order to control the sale of mazda lamps. Perhaps the central station will say that the public demands this practice. But if the public demands this practice it seems rather strange that the public would purchase millions of lamps per year from the electrical stores and pay the standard list price established by the manufacturer even though it knows that it can purchase lamps at a discount from the central station.

Is it logical to think that the practice of replacing blown fuses is another demand of the public when we find on investigation that thousands of dollars worth of fuses are sold annually from places other than the central station?

### Practices Condemned

The practice of repairing appliances, sold by the central station without cost even though the guarantee has expired is another practice which seems to me is to be condemned. Likewise the practice of the central station of selling large motors and installing them at cost is most unfair to the electrical contractor. I have often come in contact with cases where appliances have been sold on terms which have extended over a period of two years or more. Is this fair?

This selling of lamps at less than list; this replacing of blown fuses without cost, and numerous other conditions just mentioned, certainly do not add to the income of the central station nor at the same ime could it offer a co-operative spirit between the central station and the electrical contractor.

In this day when central stations are asking for, and no doubt need, increased rates to cover the enormous advance in overhead expenses, it seems to me a rather ludicrous condition to allow this source of income to escape.

It is folly to believe that the public demands practices of this kind. For we believe that no sensible person would expect a central station or contractor to replace blown fuses or sell lamps at less than list. It is true that the public will accept these privileges if they are offered. But I believe you will agree with me that the public as a whole is not unreasonable in its demands.

### Public Unfair

The fact that lamps and appliances are oftentimes sold at less than list brings from the public the statement that the electrical contractor who does not cut the price is holding the prices unfairly. I have heard of appliances being guaranteed by central stations for an indefinite period of time. But I also think that no sane person today expects any kind of appliance to last forever. Of course as long as the public is furnished with service of this kind it would be foolhardy indeed for anyone to reject it, but I again ask does the public demand it?

It seems to me that the establishment of sub-collection departments in the store of the electragist is an excellent means of getting closer co-operation between the central station and the electragist. This would also afford a much better means of collection as far as the public is concerned and would tend to establish the fact that the central station was endeavoring to render better service at all times.

### Could Pay Bills Promptly

The housewife could pay her bills more promptly without a trip to the lighting company's office. Then some month when she did not understand why her electric bill was so high she could feel that by asking her electrical dealer she would get an answer which would be given in her best interest, rather than a stock reply from one of the clerks in the lighting company's office.

We all know how ready the general public is at all times to criticize a public utility and particularly a central station. It therefore seems only logical that the central station should do all in its power to avoid criticism. There is no question that the local electrical trade can exert a very powerful influence on public opinion.

Electrical contractors, dealers, and their employes are in constant contact with the lighting company's customers. Their opinions are therefore consulted quite frequently regarding service trouble, high bills and so forth. The kind of an answer will therefore depend largely upon the attitude of the electrical contractor based upon the treatment which he has had at the hands of the central station.

### Valuable Contact

To show the attitude of some central stations toward the electrical contractor permit me to quote the commercial manager of the Ft. Smith Light and Traction Company: "Have you realized that every contractor-dealer in your city is a personal salesman (without pay) for your central station, preaching the gospel of your business, working for your profit; that every electrician he employs can be made a good salesman for you; that their wives and their children, relatives and friends can all be made boosters that will build up your load, stop unfair criticism of the electric lighting company, and put electricity over?"

So in conclusion I desire to say it is my honest conviction that the central station will find the electrical contractor ready to meet it more than half way in working out a policy of better co-operation. The electrical contractor does not expect the unreasonable, as has been demonstrated in numerous cases in the immediate past where central station and electrical contractor have come together and worked out their individual problems as one.

Ever ready to boost one another with increasing confidence in each other, ready to co-operate to the limit, and always willing to serve each other, is the spirit which should exist between the central station and the electrical contractor.

## Trade Publications

The Mines Branch of the Department of Mines, Ottawa, has issued a booklet, No. 567, entitled "The production of coal and coke in Canada during the calendar year, 1920."

Marine equipment is the subject of Circular Reprint 106, issued recently by the Westinghouse company. The publication shows the diversity of marine products manufactured by this company through the reprint in the circular of advertisements that have recently appeared in leading marine journals.

Westinghouse turbine generator units are described and illustrated in Circular 1094-B, just issued by the Westinghouse company. This publication contains a discussion of the reactance and impulse types of turbines, both the semi-double flow type and the multiple cylinder type. Bleeder turbines and geared turbines are described, and each part of the turbine is elaborately treated. The generator is also discussed in detail, and illustrations are given to show the latest types of construction.

The Delta-Star Electric Co., Chicago, Ill., have issued a sixty-four page bulletin No. 37, devoted to outdoor substations of all voltages up to and including 66,000 volts. This bulletin contains substation equipment assemblies and is replete with illustrations of substations in commercial service. A copy will be sent upon request.

## A. M. S. Engineers Opinion of the Queenston Plant

An impartial viewpoint is always valuable on any question, and for that reason we publish a letter written by Mr. F. L. Stuart, a very well known United States engineer, to Sir Adam Beck, on the occasion of the official opening of the Chippawa canal scheme at Queenston on December 28. The letter was evidently written from entirely kindly and disinterested motives by a man who believed the statements he made were true. Mr. Stuart is vice-president of the American Society of Civil Engineers, and among other activities is consulting engineer to the Hudson River Bridge Corporation in charge of all foundation work and terminal facilities, involving an estimated expenditure of some half million dollars.

December 28th, 1921.

Sir Adam Beck,
    Chairman, Hydro-Electric Power Commission.

My Dear Sir Adam:

You are to-day formally opening an undertaking which is of epoch making importance and far-reaching in potentialities for the material progress of Canada as well as of the Niagara District of Ontario.

The vision, will and ability of the Commission and the people of the Niagara District to develop their natural resources in such an efficient degree will be appreciated more and more as the years go by.

The conception is simple—the water is taken by an open canal from a point at the head of the rapids above the Niagara Falls and delivered to a power house at the foot of the rapids below the whirlpool—thereby getting all of the energy possible from each cubic foot of water used in its passage from Lake Erie to Lake Ontario.

The magnitude and importance of the work in carrying out such a conception is one of the accomplished feats of our generation. It is the largest single hydro-electric development in the world and is the largest publicly owned undertaking of its kind ever attempted, and is a path-finder in such a field.

Its importance is not yet fully understood. By the magnitude of the cheap power produced and by the example of its usefulness it will directly and indirectly give an impulse to the upbuilding of the entire country, which will be second in importance only to the effects which the railroads have had on the country's growth. The difficulties of construction were many and at times seemingly unsurmountable—each foot of canal represented effort, each mile great difficulties overcome and each section of the work an established precedent in construction.

The personnel who carried out the work were faithful, hard-working, able engineers and superintendents of construction, meeting their problems daily with a resourcefulness, courage, and earnest effort which deserves your admiration. Their names Gaby, Acres, Angell, Goodwin, Brandon, Hogg, Hull, Hearn, Millar, Blanchard, Scriven, LeRoy, Anderson, Reid, Nablo and many others—when the list is published you should remember them as men who carry on and carry on well.

It should be a matter of congratulation and of gratitude to the directing force of and on the work that this large project has been carried through in such a workmanlike way with honest and able men, without discord, disaster, or scandal and now, without fuss or feathers, stands out as an accomplished fact, a splendidly effective addition to your resources which all Canada can be proud of.

Sincerely yours

(Sgd) Francis Lee Stuart

## Miscellaneous

The Western Electric Company announce that they have opened up a wholesale electrical supply house at 216 Bannatyne Avenue, Winnipeg.

———

The N. Slater Company, Ltd., Hamilton, Ont., have for distribution a comprehensive Hubbard catalogue, which is a complete list of Hubbard pole line hardware and Peirce construction specialties.

———

The contract has been awarded for the wiring and lighting of the new Paris arena to Mr. Percy Creen. Mr. Creen has just launched out as an electrical contractor and we wish him every success. The lighting will include 32 200-watt lamps in the main part of the building; 12 200-watt lamps over the curling rink and a number of 100-watt lamps in the dressing rooms. The circuits will all be controlled from the office at the main entrance.

———

Messrs. R. B. Turner and W. A. Johnson, under the name of Johnson-Turner Electric Repair & Engineering Company, have just established themselves at 24 Assumption St., Walkerville, Ont., where they will carry on an electrical business, including all kinds of repair work. Mr. Turner who is manager and electrical engineer, has had seven years' experience with the Westinghouse Electric & Manufacturing Company at East Pittsburgh. The firm will thus engage in both engineering practice and repair work.

### Radio Ramblings

"The melancholy days have come
The saddest of the year"
So sang our friend the poet once
But now he's wrong I fear.
The summer statics over and
The air is clear and bright
There's always something doing tho'
I listen every night.
On Saturdays a football game
Will find me full of pep
Tho' far away from scenes of strife
It seems but just a step:
When winter comes and piles of snow
Are heaped about my door
On Sunday morn why freeze my toes?
I'll venture out no more.
For in my humble cottage is
A simple wireless set
I'll hear the choir singing and
Perhaps a sermonette.
It has one great advantage if
The sermon fails to suit
I'll cut the preacher off real short
And practice on my flute.
Now after all is said and done
There's just this much about it
Take my advice and get a set
No home's complete without it.

—Peter Deets.

# The Development of Wireless—II

## A Series of Short Interesting Articles
## Covering Wireless Progress to Date

By F. K. D'ALTON

It was about the year 1864 that Clark Maxwell became convinced of the existence of the electromagnetic waves which are now used in wireless communication. He had completed some extensive calculations in which he discovered that the electrical energy entering a wire was greater than that which could be drawn out of it. This refers to a wire such as an aerial where the circuit is not complete, the wire being unattached electrically at the distant end. He then conceived the possibility of radiated energy but was not able to prove his theory for at that time there was not any known means of producing or detecting such waves. He claimed that the proportion of energy radiated increased with the frequency at which the direction of flow of current was reversed. He was satisfied that the radiated energy went out in the form of waves which had the same frequency as the current in the aerial but he did not know even how to produce these high frequency currents.

His theories were freely discussed in scientific circles and a few years later Prof. Fitzgerald suggested that possibly the electric spark might be a means of obtaining high frequency currents but he did not offer any suggestions as to a suitable detector.

In 1888, Prof. Hertz performed the first really satisfactory experiments in which he produced the waves by means of an electric spark and detected their presence by means of a ring in which there was a very short gap. The existence of waves was indicated by minute sparks appearing at the gap.

Having now proved that these waves did exist and thus confirming Maxwell's theory, he proceeded to study their laws. He found that they followed many of the laws of light but noted the remarkable peculiarity that they penetrated many substances which are opaque to light, and he advanced some further theories as to the nature of the radiations.

A year after Hertz had made his famous discovery, Prof. Elihu Thomson suggested that, with suitable apparatus, these waves might be used as a means of rapid communicaton. The distances across which Hertz could detect the waves were very short, only a few feet, and the next steps taken had the effect of increasing the distance considerably.

In 1894 Sir Oliver Lodge discovered that, in the presence of these waves, a group of metal filings would cohere and become conductive, allowing sufficient current to pass through to operate a telegraph relay. The chief drawback in this method of detection was that the filings would not fall apart and resume their original state when the waves ceased to come. The result of these discoveries was the invention by Lodge of a detecting device known as a "coherer," consisting of a small pile of filings between the ends of solid metal bars, the whole being enclosed in a glass tube, and a tapper, known as a "decoherer," was operated by the relay so that the filings were continually shaken while the waves existed but as soon as the wave ceased the next tap shook the filings apart, the relay then opened and remained in this position until the arrival of the next group of waves.

The relay could be arranged to operate a sounder or other device from which the signals could be read or the noise of the decoherer would suffice.

By means of this detector a distance of one half mile was immediately obtained and by gradual improvement and refinement in design it soon became possible to increase the range to several miles.

The first step in improvement was the placing of the elements of the coherer in a vacuum. This prevented oxidation of the small metal particles and made them more definite in action.

Up to this time wireless waves were only of interest to experimenters and physicists. Marconi now commenced to build and test out devices based on the then-known laws having in view a commercial application of the waves to communication. He made use of the appliances already developed by others, used vertical antennae for both sending and receiving and wireless communication became a fact.

Messages were sent in code with a spark transmitter and coherer detector. Owing to the time required for the relay and decoherer to complete their cycle of events, it was impossible to attain any appreciable speed.

This coherer simply indicated the presence of the waves but gave no character to the signals. The magnetic detector now made its appearance. In this device a band of soft iron was kept in continual motion through two concentric coils. The wireless signals, coming through one coil, had the power to magnetize the iron and this change in flux density caused currents in the other coil which was connected to a pair of telephone receivers. The signals were thus made audible, the stronger waves produced the louder sound and certain character in the received signals was noticeable. A decohering device was unnecessary, the receiver being always ready for the next wave. The moving iron band was kept in slow motion by a quiet clock mechanism which had to be wound once an hour. These detectors were found to be very reliable and in fact to-day are standard equipment for emergency on most of the ocean going vessels.

This magnetic detector was soon replaced by the much simpler, cheaper and still quite commonly used crystal detector consisting of a metal point resting lightly on certain crystals such as silicon, molybdenum, carborundum, galena, etc., which by virtue of their power to rectify an alternating current of small magnitude can so alter the incoming waves that a distorted wave of current is sent through the telephone receivers (hereafter to be called "phones") and sound waves are emitted by the diaphragms.

There is no mechanism connected with this detector, it derives the power it gives to the phones from the part of the radiated wave which it receives, and its only drawback is the fact that the whole surface of the crystal is not sensitive and good spots must be found in order to obtain fairly loud signals. The waves emitted from the sending equipment in the station where a crystal is installed, very frequently knock the metal contact off the sensitive spot which has been found and it becomes necessary to readjust the detector. The time lost in readjusting often results in the missing of part of the signals which are wanted, but this fault has not prevented the crystal from being used extensively by commercial and amateur stations throughout the world. The fact that there are no parts to wear out and no batteries to run down, make it very desirable for isolated land or ship stations.

Giving quantitative response, i. e., the intensity of sound varying with the strength of the wave received, it may be used for receiving wireless telephone stations as well as spark stations.

Probably the next step of importance was that of tuning both sending and receiving stations. Selection of desired stations without interference from those not wanted, and considerable increase in distances with given sending and receiving equipment were both obtained by tuning.

The fact that an electrical circuit is resonant at a certain frequency depending upon its electrical constants—capacity, inductance and resistance—is the fundamental principle in tuning, and by variation of one or both of the two former quantities, a range in resonant frequency is obtained.

**BETTER MERCHANDISING**

# Selling the "Idea"

The cities of Hamilton and Toronto are in the midst of an educational campaign to create in the home lovers of those cities the attitude of mind that says "I need electricity in my home."

Let us not lose sight of this basic idea. These campaigns are not interested in selling appliances or wiring—directly. They have only one object—to "create a desire." That desire is—to do things in the most modern, convenient, expeditious, sanitary, labor-saving way the world knows about today.

It is highly desirable, looking to the success of these campaigns, that indivdual interest should be kept as much in the background as possible. We must try our best not to confuse the mind of the man and woman we hope to reach. Above all we must try to keep him from thinking that back of it all, our chief aim is to sell him an appliance.

The publicity campaigns have been carefully thought out. The announcements are being so worded that the advantages of electricity are placed before the readers in a logical, readable, convincing manner. In the home itself no attempt will be made to sell anything but the value of electricity.

It will be patent to everybody that the ultimate result of such a campaign must mean an increased demand for electrical equipment of every sort. Let us in the meantime, however, keep our commercial desires in the back ground. Let us strain every nerve to sow the seed in the public mind that the best way and the only way to keep house is through the utilization of this great and wonderful power that Providence has supplied to us at Niagara Falls—the heritage of the people of the province.

# Hamilton's Electrical Campaign is Gaining Momentum

### The Essay Contest, Succesfully Applied to the Different Appliances, Arousing Wide Interest Among Householders

The Hamilton electrical campaign is going along persistently and effectively. Electric pages are appearing regularly and the essay contests, particularly, are arousing very keen interest in the homes. One by one the people of Hamilton are being led to investigate and analyze the merits of electrical appliances. An example was the essay contest on portable lamps. It is surprising how eagerly the young people can pick out strong points in these appliances and how eagerly they enter into the contests. It can easily be realized, too, how interested parents become in their essays and how valuable a form of advertising the Hamilton Electrical League has set in motion. The Hamilton Electrical League finds, also, that the interest increases with the various contests, i. e., the effect is cumulative. Samples of the essays are exceedingly interesting and are reproduced herewith:

### Essay on "Portable Electric Lamps"

"New lamps for old, new lamps for old!" was the cry of long ago, and no less is it the cry of the 20th century.

Who that remembers the drudgery of the daily "lamp cleaning" doesn't rejoice in the present day satisfaction of having no lamps to clean?

Just a turn of the switch, the light is there; no matches, no oil, no smell!

And what a variety of portable electric lamps there are!

The "standard" lamp for the parlor, so easily switched off when daughter's beau is calling, and on when father comes in.

The sewing lamp for mother, close to her work table or sewing machine.

A reading lamp for father by the side of his favorite arm chair.

If baby wakes at night, there is the lamp that clips on the side of the bed post, just where mother can reach it in a moment. Then the nervous old maid is catered for and she can hunt possible burglars by the aid of the 'electric torch," and many a district nurse has been glad of the same friend when going her rounds in dark alleys and stairways.

So what more acceptable gift at Christmas time than a portable electric lamp?

There are all kinds and all prices to suit all uses and all purses, and each will prove a real friend to its owner.

MRS. A. G. KEILLER,
Box 507,
Burlington, Ont.

### A Portable Lamp—A Ditto Ditty

Ho, people of Hamilton, ye who read
Here's news for you which you must heed.
A Portable Electric Lamp you need
For every room, as you must concede.
When through this list you all proceed,
You'll feel you must have one indeed.
You'll need it when you're a chaperone,
You'll need it when you're all alone.
You'll need it for the gramophone,
You'll need it at the telephone.
You'll need it at your office desk.
You'll need it when you when you take a rest,
You'll need it when you have a guest.
You'll need it when burglars molest,
You'll need it in the parlor,
You'll need it in the hall,
You'll need in the springtime,
You'll need it in the fall,
You'll need it when you're reading,
You'll need it when you write.
You'll need it when you're working,
'Twill save your good eyesight.
You'll need it when your mending,
You'll need it when you paint.
Buy your wife one for a gift,
She'll think you are a saint.
You'll need it for your parties,
It's fine for just two.
You'll need it when you're happy,
You'll need it when you're blue.
You'll need it when there's sickness,
You'll need it when there's health,
You'll need it in poverty,
You'll need it in wealth,
You'll need it when you're dining,
You'll need it when you lunch.
If you don't buy a couple
You surely are a dunce
So if you be a Gentile or if you be a Jew'
The price of a Portable Lamp you'll surely never rue.
HELEN PUGSLEY
Aldershot, Ont.
P. O. Box 63,

### The Portable Electric Lamp

My mother and I went shopping. We didn't know what to get daddy for Christmas. We always buy his present first. Mother suggested books, because daddy loves to read, but I reminded her that his eyes bothered him when he read at night. That gave mother an idea.

"I know," she said, "a portable reading lamp is what he needs. It will shade his eyes and throw the light on his book. He can read where he likes—in his chair or on the chesterfield if he is tired—and always have the lamp in proper position to take care of his eyes. Come, dear, we'll go to an electrical store."

Such an abundance of pretty library lamps were displayed! Mother selected one with an adjustable green shade. Green is the most restful color for the eyes.

While waiting for change, mother and I looked at boudoir lamps. There was one to suit every bedroom I could think of. One with a chintz shade looked as though it ought to belong to my sister's room. Mother agreed with me and thought the lamp would add much to the appearance of the room, and also be very useful, as it could be carried from the dresser to the secretary as needed. "You may send this lamp, too," said mother to the clerk.

"Let me show you the floor lamps, or piano lamps,

madame," said the clerk. "They are not a mere luxury in the home, but a necessity, if you want your home to have an attractive appearance with practical and useful furnishings. The floor lamp can be placed anywhere, for we all like a change in the arrangement of a room, and it is not only beautiful in itself, but the soft warm glow it reflects on the other furnishings just doubles the value of your efforts to make things look well."

Mother shook her head, although she looked longingly at one with a rose shade. But when I got home, I whispered to daddy about that lamp with the rose shade and I do believe mother will get it after all.

EDITH DARLING,
21 Stanley Avenue.

### The Comfort of Electric Lamps from Experience

It is a big pleasure for me to be able to let someone know outside of our house what extraordinary comfort can be gotten from a portable electric lamp. In our case it is a desk lamp.

On the anniversary of our moving into our new house, father bought a desk lamp for the library table. It is so convenient when studying lessons, writing letters, or, as father says, "When learning the daily news." The shade can be adjusted so that the light falls directly on the book, paper or whatever one is using, without hurting the eyes.

We are expecting some day to have a floor lamp, or at least a piano lamp; but now, while we have not either of these, sometimes when I am left alone to amuse myself, I take the desk lamp into the parlor and put it on a pedestal beside me at the piano, or on the top, adjusting the shade so that the light is shown on the music. In such conditions I could stay awake, I would not mind how long.

O boys! Who would be without an electric lamp? Not us, that is sure.

The new house which mother chose
For our new domicile,
Would be dark without that desk lamp,
Which gave us all the style.

MILDRED ELLIS,
37 East 25th Street,
Mount Hamilton.

### Annual Meeting of Hamilton District Electrical Association

A meeting of the Hamilton District Electrical Association was held in the Royal Connaught Hotel, January 4, 1922, at 8 p.m. The following officers for the year 1922 were elected: Fred Thornton, contractor-dealer, Ottawa St. N.; W. G. Jack of Jack Bros. Electrical Construction Co. 224 Main St. E.; V. K. Stalford, consulting electrical engineer, Sun Life Bldg. W. G. Jack was elected to represent the District in the Provincial Executive Committee.

An invitation was extended by the Canadian Westinghouse Co., Hamilton, to the members of the Hamilton District Electrical Association to visit their new lamp works. The members of the Association accepted this invitation and will visit the factory on Tuesday afternoon, January 24th. A very pleasant and beneficial trip to this plant is anticipated by a large number interested in the electrical industry in Hamilton. This is the first of a series of visits to be paid to various factories in Hamilton by the members.

At the next meeting of the Hamilton District Electrical Association, which will be held February 1, 1922, dinner will be served and the speaker of the evening will be Mr. E. H. Porte, general manager Renfrew Electric Products Co., Renfrew, Ont. Members of the Association are anticipating a very pleasant and instructive evening.

At a meeting of the Executive of the association held January 10th, Mr. Fred Thornton, contractor-dealer, was elected chairman of this committee and K. J. Donoghue of the Northern Electric Company was re-elected secretary-treasurer for the ensuing year.

### Chatham Radio Club Activities

The Chatham Radio Club which was formed in September, 1921, now boasts a total membership of sixteen members, eight of whom are licensed amateurs. Mr. Ivan Collins is president of the Club; Mr. Edward Davey, vice-president; Mr. C. Gammage, secretary-treasurer; Mr. A. Edwards, radio electrician; Mr. Harold Jackson, traffic manager. Equipment recently installed by the Club includes a one-half kw. spark transmitter and receiver and two-step amplifier. The meetings are held in the Chatham Industrial School. Both Senior and Junior classes are held.

### Electricity Wins First at Masquerade Ball

Harold A. Munnion of the Gate City Co., Ltd., Winnipeg, believes in boosting the electrical industry. On New Year's eve he attended a masquerade ball held in the Board of Trade building, attired to represent "Electricity." The head piece showed a model power house, transmission line, and street lighting lamps, which were illuminated at short intervals. The costume was a white duck suit with various electrical advertisements and slogans. Attached to the left shoulder was an old fashioned candle stick showing a lighted candle, representing 1821, and on the right shoulder a miniature floor lamp of the 1922 period, which flashed on and off periodically, the effect illustrating 100 years of progress. On the back of the coat was a card with the slogan "Say it Electrically, Flowers Die." Mr. Munnion was awarded first prize—a very fine electric reading lamp—for the best and most original costume exhibited, and was thoroughly deserving of it. Mrs. Munnion, who accompanied her husband, received the first prize for the ladies' most original costume. Dressed in black velvet she represented a very attractive pussy cat.

### Leaves Winnipeg for Toronto

Fred J. Kennedy, who for the past twelve years has been associated with Marshall Wells Co., Ltd., Winnipeg, in the capacity of manager of the electrical and builders hardware departments has resigned. Mr. Kennedy left Winnipeg on the 16th January for Toronto, to accept the position of sales manager with the Kennedy Hardware Co., Toronto, which business is controlled by his uncle. The electrical fraternity of Winnipeg will miss Mr. Kennedy, as he was a popular member and an energetic worker in the Manitoba Electrical Association.

### "Electrical Salesman's Handbook" Ready

Announcement is made by the "Electrical Salesman's Handbook" committee of the Commercial National Section of the N. E. L. A. that the first two units of the new handbook are ready. These sections are those entitled "Industrial Lighting" and "Lamp Equipment for Commercial and Industrial Lighting." They are punched to fit the old handbook covers and may be had from the committee at 50 cents each. As soon as practicable the remaining sections of the handbook will be revised and issued in the new form.

# Past, Present and Future of Merchandising Electrical Goods

### How Can We Create Electrical Merchants?   How Can We Sell
### More Goods, Give Better Service and Make a Sure Profit?

By GEORGE C GRAHAM

How can we, in the shortest possible time, create electrical merchants—men who understand the basic principles of retailing; know how to sell at a profit and when they are selling at a profit; know how to reach the public; to get them into their stores; to give them service when they get them. This is one of the problems of the electrical industry.

Mr. Graham read an interesting paper, full of useful hints, before the Rhode Island Electrical League, which is reported in the "National Electragist." He urged the necessity, not only of rendering service but of rendering it at a profit. Comparing the past with the present, in 1911 only 20,000 washing machines were manufactured while in 1921 1,026,000 were manufactured. Yet the industry is still in its infancy. There are 8,610,000 wired homes and less than half of them have vacuum cleaners and the number of wired homes is constantly increasing. The field cannot be saturated.

### Appliance Business Second

There is a great big future in the appliance business. Roger Babson, the statistician, has said that it is the second big best business. It may be larger in a few years than the automobile business. The stage is set. The people' want the appliances. The manufacturers have them for sale. The dealer must render the service of selling them at a profit.

To make a profit and to render service at the same time it is necesary to do four things:

1. Sell the best.
2. Maintain uniform price and policy.
3. Close sale as quickly as possible.
4. Maintain an efficient service department and know when to charge for service.

There are 187 washing machine manufacturers. They are not all sound. It is better for the dealer to sell only the well advertised standard appliances, those made by well known manufacturers who can be relied upon. It is bad for a dealer to begin selling a machine that will disappear from the market in a few years and for which when parts are needed it is hard if not impossible to get them.

Often a dealer is misled by discounts. The discounts may not be enough to make up for the increased cost of selling the machine. It certainly will not make up for the bad impression caused by carrying goods that are not standards and which cannot be replaced.

### Specialize But Don't Cut Prices

Specializing will increase sales and profits. It makes the rendering of service easier and makes it possible to give the service promptly when it is wanted.

Cutting prices means giving away the profits. There is not a large enough margin on any electrical appliance so that prices can be cut without giving away the profits. At the same time customers don't think much of a man who cuts down his price. Such practice may work all right in other countries but ours is a country of uniform prices and if the dealer is to make a good impression he must stick to the price that he has made.

Money is lost when the sale is not closed promptly. Don't leave a machine out too long before closing the sale. Turnover is necessary and this is reduced if machines are left out too long. A few days or a couple of weeks should be long enough for a person to learn whether or not he wants the appliance.

Service is a big item. We have found that service calls cost us on an average of about five dollars a call. By selling the best appliances you cut down the need of service. We used to give one year of free service. We have now cut it down to 60 days and have also found that many service calls can be handled in such a way over the telephone that it was not necessary to send out a man.

### Education Big Factor

It is largely a matter of educating the public. Before we cut down our free service, we gave the matter very careful attention. We feared that it would react, but it has not. The public is willing to pay for what it gets. It would rather have good service at a fair price than poor service free.

Appliances like refrigerating machines and washing machines fall into a different class from other appliances. These are new and require more free service. People have to be taught.

In selling we specialize. One set of salesmen sell washing machines and ironing machines. Some of them take the machines right along with them on a Ford truck and demonstrate as they go. Some of them are exceptionally successful in making sales in this way, a large proportion being closed on the first call.

The cleaners are handled by crews of five or six men working under a crew leader and all working on a commission basis, the leader getting a commission on all the cleaners sold by members of his crew. These men become very expert in their work and the best men close sales on the first call.

In the case of washing machines and cleaners we require 20 per cent of the purchase price for the first payment. In the case of dish washers the money is collected at the time of making the sale. It has been found that a person must use such a machine for two or three weeks before getting accustomed to it and that they are not likely to do this unless they have already paid down the full purchase price.

## Windsor District Activities

A regular meeting of the Windsor District Association of Electrical Contractors & Dealers was held on December 30. 1921, at the Barton-Netting Company's offices. The president, Mr. V. B. Dickeson, occupied the chair. There were also present Messrs. Roach, Reaume, Garfat. McKeller, Phelan and Cook. Mr. Dickeson read a paper on the subject of manufacturers fixing retail prices, which was followed by a useful discussion. Mr. Dickeson's paper will appear in the next issue of the Electrical News, as it represents a viewpoint that may very profitably be placed before the trade and discussed at greater length.

The January 13 meeting will be addressed by Mr. O. M. Perry, manager of the Hydro System in Windsor. The meeting will be held in the same place, at 7.30 p.m. Mr. A. H. Cook is secretary of the Windsor District Association.

# Selling Six Floors of Merchandise in Two Storey Building*

'.The chief reason why my store has been so successful in a merchandising way is that I sell six floors of merchandise in a two storey building," a small town merchant well known for his progressive merchandising methods told me the other day.

In other words, the merchant gets as many sales out of one square foot of space as some other dealer, less enterprising and less scientific in his merchandising methods, gets out of three square feet of space.

In these days of space economy space counts for a great deal. Eliminating the important item of rental, the merchant who can get more sales out of his store space shows himself a real up-to-date merchant.

Did you ever make it your business to try and find out what portion of your store space is the best selling space? Do you know what goods can most appropriately be displayed in a certain portion of the store?

I am not proposing an entirely new idea in merchandising. The sales per square foot idea has already been extensively used by most of the department stores in the country. With the department stores merchandising has been reduced to an exact science, and there is no earthly reason why the lead should not be followed by the specialty dealer.

### Floor Space Worth

Go to any merchandising man of a large department store. He can lay his finger on any square foot of space in his store and by referring to his card index and chart can tell exactly how much each unit of store space is worth to the store owners in dollars and cents. He can refer to two units of space in different parts of the store and he can tell exactly how these compare in point of selling efficiency.

What purpose do these figures serve the merchandising manager?

First of all, the merchandising manager knows exactly what part of the store is most valuable in sales returns. He wants to conduct a special sale, and he desires to display his goods where they will have the greatest attraction to the shoppers. He knows by referring to his records just what store space will serve the desired purpose.

He knows that a certain article will sell better on the second floor of the building than if it were displayed on the first floor, for he may have found by experience and due investigation that the second floor always attracts a better class of customers.

### Other Important Factors

There are numerous other factors involved, and the one I am going to mention is a very important one. Several years ago the directors of a certain department store found themselves pressed for space. As no additional space could be obtained the store directors decided that the only solution was to increase the sales per square foot of the store.

They took a rule and measured off the selling space in every department of the store. After totalling the number of square feet they took a total of the rent paid for the entire building. They then allotted percentages for each department of the store, based on the proportionate number of square feet in each department.

A conference of department heads was called, and this was the proposition put up to them: "We are pressed for space. The store is already small for the enormous business we are doing annually. In order to make the most of the situation we must increase our sales for each square foot of

ground. We must make the most out of every square foot of ground. Therefore, we are going to rent out a certain number of square feet of space to each department head. The store auditor will credit you with monthly rents on these space allotments."

### All Items Considered

Not only is the proportion of store rent credited to the department head, but the latter must also share in the store's expenditures. The store auditor credits each department head with so much for rent, so much for light and heat, so much for salaries of salespeople employed in the department, so much for depreciation of merchandise—and even advertising and delivery are apportioned among the various department heads.

The store directors get together at regular intervals and reapportion these items in accordance with new conditions. They go over the statistics of sales and profits in each department. They can tell whether the sales have increased and whether the overhead is eating into the profits.

If any of the actual expenditures in the department exceed the expenditures alloted by the store auditor the department head is credited with the difference, and he must watch his step thereafter to make up for the difference. This system has created a condition where there is continual competition among the department heads for best showing.

The net result has been a decided saving in overhead. Each department head feels his responsibility in the management of the store. Under ordinary conditions he wouldn't worry his head over delivery expenses and rentals.

### Quick Checking Easy

The store management has found it profitable in other ways. It is far easier to size up a situation when you have facts about a particular department at your finger tips. If something goes wrong in a certain department the store owners do not have to rummage all over the store to find the cause.

An investigator employed by a large research organization recently stated that it is very difficult to get a customer to buy on any floor above the fourth floor of any department store. At the same time, he stated, it is not always true that the best selling space is on the first floor of the building. He derived his conclusions from a study of sales per square feet records of a number of department and specialty stores.

The owner of a small retailing establishment need not bother getting up elaborate charts and tables of sales. He can keep close watch on the sales of his various articles. He can compare the sales of articles on various counters and sections of the store, and find out whether the respective sales could not be increased by a proper relocation of goods. If there are several floors of selling space, let the merchant compare the sales of the store, and find out why the sales of one floor exceed those of another.

---

Mr. G. Armstrong, 1217 College St., Toronto, has secured the contract for electrical work on a store and apartment building being erected at 691 St. Clair Ave., Toronto, for Mr. J. Becker of 265 Gladstone Ave., at an estimated cost of $50,000.

---

The incorporation of the Henderson Electric Company, Ltd., is announced in a recent issue of the Ontario Gazette. The new firm will carry on the business of electrical contractor-dealers, taking over the business formerly carried on under the name of Henderson Electric Company, Toronto. The firm is capitalized at $45,000.

*By Dr. Norris A Brisco, in "National Electragist"

# Heating the Domestic Water Supply the Modern Way—Electrically

### The Advantages are Many—The Cost is Not Prohibitive—Discussion of Proper Sized Unit

By J. A. MacDONALD*

At this time the problem of heating by electricity the domestic hot water supply is receiving considerable attention from central station engineers and those actively engaged in electrical merchandizing. It is generally admitted that the most desirable manner of obtaining results satisfactory to all parties concerned expresses itself in an extension of flat rate to low wattage heaters. These may be classed as the 660 and 1000 watt capacity, operating generally at 110 volts. For the former a rate of $1.50, and $3.00 for the latter, per 720 hour month, has been granted by leading central stations. In the past it has been customary to extend this rate for that period of the year known as the summer months, or that period of the year in which the operation of the furnace is unnecessary. This custom has now become prevalent because of the fact that the standard furnace coil has been used extensively during the winter months for heating the domestic supply of water, and until recent years electricity was not as economical for this purpose as coal. However, these conditions have become reversed by the steady advance in the price of coal, as well as a decrease in its efficiency, and on the other hand by a marked decrease in the cost of electrical energy. The following comments pertaining to the comparative costs of heating a thirty gallon domestic boiler by the standard furnace coil, with coal at $16.00 per ton, and heating a similar boiler electrically at a flat rate of $3.00 per kilowatt per 720 hour month, will be of interest to both central station and consumer.

Before considering this matter it is essential to emphasize the fact that no standard data is in existence that can be applied to the cost of heating the domestic boiler with a coil in the furnace. The cause for the absence of this data lies in the fact that important elements bearing on the exact solution are unknown. The more important of these unknown factors can be outlined as follows:—

(1) The number of gallons required per 24 hours, and the temperature at which it is to be delivered to the taps.

(2) The size of the boiler, the average temperature of the room in which installed, and its distance from the furnace.

(3) The exact position of the coil in the furnace, either in contact with or above the fire.

(4) The exposed area of the coil in square feet, and the number of pounds per square foot, of coal per hour, at which the furnace is fired.

(5) Whether or not the boiler is insulated and the class of insulation.

#### Cost of Water Heating With Coal

Considering the foregoing it is evident that the problem presents serious difficulties adverse to its exact solution. Actual experience has taught us that the standard furnace coil uses a great deal more energy than is usually required and it is not uncommon for those using this method, to be forced to waste large quantities of hot water in order to save their system from breakdown due to excessive pressure. This waste of energy cannot be overcome because in order to secure comfort in the home it is necessary to burn the furnace at a certain rate, independent of the fact that this rate may

be a great deal higher than that required to meet hot water demand.

Notwithstanding the fact that valuable data is absent, we can, by methods of comparison, arrive very closely at the cost of the furnace coil method. Therefore the following article will first consider the maximum demand that is likely to be possible, leaving it to the reader to judge the position of his particular case in relation to the demands suggested.

Theoretically, a pound of anthracite coal contains about 14,200 heat units per pound of combustible. Allowing a loss of 10 per cent for ash and non-combustibles, we have about 13,000 heat units to perform work, which would be possible if we could secure a coal burning device of an efficiency approximating 100 per cent. It is certain that such ideal conditions cannot be attained, and therefore it will be better to arrive at the actual heat units on a basis of the work they perform.

Experience and expert design have combined to make the modern high pressure boiler the most efficient fuel burner of the present day. No factor bearing on securing the maximum number of heat units per pound of coal has been neglected. Therefore an examination of standard data pertaining to these boilers, should furnish a basis for arriving at coal efficiency. A boiler horse-power is equivalent to 33,305 heat units per hour and allowing 3 pounds of combustible per horse-power we have 11,102 heat units or an efficiency of 80 per cent. We find that the same application to boilers of ordinary design, works out at 4 pounds per horse power, bringing the heat units down to 8326 or an efficiency of 60 per cent. Applying fuel efficiency to steam generation we find that a pound of coal with a noncombustible content of 16 per cent will evaporate 9 pounds of water. Therefore there would be an evaporation of 11 pounds of water per pound of combustible. The evaporation of 1 pound of water at 212 degrees into steam at the same pressure requires 964 heat units. Therefore the evaporation of 11 pounds of water by 1 pound of combustible requires 10,604 heat units or 76 per cent of the theoretical heat in the fuel.

#### Very Low Efficiency

The foregoing paragraph has been given to show that even with high grade apparatus the efficiency of coal as a fuel is surprisingly low. It is also apparent that the average householder does not burn his furnace at a higher rate of efficiency than the trained attendant operating high efficiency boilers. Therefore it should be fair to the coal to adopt an efficiency of 8000 heat units per pound as the value that the average household furnace operates on. At the present time heating engineers base their calculations on an efficiency of 55 per cent.

It has been demonstrated that the well known range water back will absorb 10,000 heat units hourly, with a bright fire burning, and in contact with the fuel. The position of furnace coils is well above the fire which naturally results in a lower consumption of heat units, and there is not a serious departure from actual facts when a value of 8000 heat units is adopted.

Dealing with an exposed 30 gallon boiler, and considering what can safely be defined as the maximum demand, it is

*Manager Thermo-electric Ltd., Brantford, Ont. before the Hamilton District Electrical Association

noted that the temperature of water in city mains varies between 45 and 60 degress Fah. according to the location and the time of the year. It is assumed that the water enters the boiler at 60 degrees and is delivered at the taps at 160 degrees. Therefore the coal will have to continually overcome a temperature of 100 degrees, supply all demand and take care of radiation losses. Taking the weight of water at 8.3 pounds per gallon we have 8.3 x 30 x 100 or 25,000 heat units per hour. For a 720-hour month there would be expended 18,000,000 heat units. Dividing this by 8000 we have 2250 pounds of coal, which at $16.00 per ton, represents a cost of $18.00 per month.

### Radiation Losses

Directing our attention to radiation losses we find that steam at 5 pounds pressure will radiate 250 heat units per square foot per hour. It is standard practice to rate hot water radiation at 60 per cent of this amount or 150 heat units per hour. The 30 gallon boiler represents 17.2 square feet of radiation surface, which with piping can be safely taken at 18 sq. ft. Then the radiation losses from a 30 gallon boiler for a 720 hour month would be 18 x 150 x 720 or 1,944,000. At the value adopted this means 243 pounds of coal, at a cost of $1.96, which brings the total cost to $19.96.

The foregoing analysis covers a maximum demand condition, which is not likely to be met with in actual practice, because common sense tells us that we do not use 30 gallons per hour. On the other hand we must not lose sight of the fact that even if we are not using the maximum demand that we have no way of proving that the coil is not furnishing heat units greatly in excess of our hot water requirements. In other words, the same coil may handle 100 or 250 gallon boilers just as well.

Turning our attention to a calculation based on the minimum demand, we assume that if sufficient energy is applied to a 30 gallon boiler, to raise the temperature at the rate of 30 degrees per hour, that it will take care of a reasonable demand. This is not unreasonable when you consider that it will take practically four hours to heat the boiler to 160 degrees, assuming a similar temperature in the main to that used in the coal calculation. The number of heat units required in this case would be 30 x 30 x 8.3 or 7,470 per hour or 5,378,400 heat units in a 720 hour month, dividing this by 8000 we have 672 pounds of coal. Adding radiation losses 915 pounds would be consumed, which would cost $7.32.

Somewhere between $7.32 and $19.96 per month lies the cost of heating the domestic supply of hot water with the furnace coil, with a tendency to the former estimate, and in isolated cases slightly below it, provided that the water is delivered at the temperature on which the calculation is based.

### Using Electric Heat

Considering the heating of the 30 gallon domestic boiler electrically, we are fortunate in the fact that we have at our disposal sufficient reliable data to proceed with confidence. We are aware that the supply of energy is constant and that electric heaters are now offered to the public of 100 per cent efficiency. Experience gained from actual installations has conclusively proven that a 1 kw. electric heater, operating continuously, will supply any reasonable demand that can be expected from a 30 gallon boiler. One kw.h. equals 3,412 heat units; therefore in a 720 hour month the number of heat units expended would be 2,456,640, or a saving over the minimum coal calculation of 4,856,760 heat units which were appropriated by the coil and in consequence lost to the home.

It is easily understood that coal cannot at present prices compete with electrical energy at $3.00 per kilowatt per month, and still less at $1.50 per month if a 660 watt heater is used. It is also noted that constant radiation losses amount to $2.00 per month; and that the $3.00 rate is only $1.00 in excess of this amount.

Having disposed of the matter of costs let us compare some of the factors related to the two different methods. We find that the objections to the furnace coil method can be summed up in the following:—

(1) The tendency to give to the coil an excess of heat units over that actually required.

(2) The waste in these units, which could otherwise be applied to heating the home.

(3) The inefficiency of coal as a fuel.

(4) The chilling effect on the fire caused by water at a lower temperature passing through the firebox.

(5) The detrimental action of excessive pressure on the system and valves, caused by high temperature when the hot water demand is low and the comfort demand high.

(6) The tendency to form leaks and the detrimental action on the fire by condensation at low water temperatures.

(7) The inconvenience caused by the presence of the coil in the firebox.

### Many Advantages

Regarding the application of electrical energy to the heating of water for the domestic supply, the following advantages are apparent:—

(1) The entire absence of fumes, dust, or noise.

(2) Constant supply of energy at a fixed value.

(3) The elimination of all danger from overheating.

(4) The entire absence of piping.

(5) Energy supplied 100 per cent efficient.

(6) Cost of heating a fixed sum.

(7) Higher furnace efficiency for home comfort.

(8) Lower costs of installation and upkeep.

(9) Independence and flexibility of control.

A point frequently in dispute is the temperature that water should be at for different purposes. It has been found advisable to keep the water in the boiler between 120 and 160 degrees, to allow for excess demands and radiation losses in transmission to the taps. The following temperatures have been adopted after considerable investigation:—

Hands, 103 to 113 degrees Fah.; face, 98; bath, 95; luke warm, 86.

The quantity of water heated is usually less than the quantity used, because water at 160 degrees will take up almost an equal quantity of cold water before it can be used with comfort. It is therefore apparent that the boiler is not required to deliver the total number of gallons used.

### Wattage for Various Sized Boilers

The question of wattage required for boilers of different capacity, can be judged very fairly by the use of the following data which has been secured from actual experiment. These values are applied to plain, galvanized, uninsulated boilers and piping. For boilers with 2 in. insulation and piping with 1 in. insulation, 50 per cent of the wattage given will be sufficient. It is assumed that the heater will operate continuously, that the water enters the boiler at 60 degrees and leaves it at 160, and that the hot water demand does not exceed 12 gallons per day.

For demands in excess of this amount add 10 additional watts, for each additional gallon used per 24 hours. In cases where the demand appears to be excessive the next larger size should be recommended:—

### Table I.

| | | | | |
|---|---|---|---|---|
| 12 | gallon | boiler | 600 | watts |
| 18 | " | " | 650 | " |
| 24 | " | " | 750 | " |
| 30 | " | " | 850 | " |
| 40 | " | " | 1000 | " |
| 60 | " | " | 1500 | " |

# Newest Developments in Electrical Equipment

### Motor Driven Electric Heater

The Carmean Electric Company, 2806-2808 East Eighteenth St., Kansas City, Mo., has placed on the market an electric motor-driven heater, as illustrated. The object of the device is to provide an arrangement whereby cold air is propelled through and over heat-generating coils by means of a motor-driven air impeller. The mechanism is entirely surrounded by a fool proof metal housing that keeps out the dirt and prevents bodily injury to persons.

The heater is controlled by a four-way switch, the first turn of which starts an electric fan only, for summer use; the second turn gives low heat of 660 watts; the third turn, cuts in an additional 1,320 watts. The fourth turn cuts off motor and all heating elements. The total weight of the

Combined
Fan and
Heater

heater is less than 12 lbs. It is claimed by the manufacturer that the important advantage of this type heater is, in the enforced circulation of the heat waves throughout the room almost immediately. Canadian patents on the Carmean Heater were allowed in December, 1921. Manufacturers interested in the Carmean line of heaters should communicate with Mr. Robert Yeomans, of the Department of Industries, 135 Bay Street, Toronto, where one of the heaters can be examined.

### High Voltage D. C. Generators for Wireless Telephony

The Robbins & Myers Company, Springfield, Ohio, has developed a new line of high voltage direct current generators and motor-generator sets for service with wireless telephone outfits. The generators are made in 500 and 1000 volt types in capacities of 100, 200 and 500 watts, for use with vacuum tubes and for other special services.

The motor-generator sets are furnished for operation on 110 or 220 volt, 25, 50 and 60 cycle, single phase, alternating current circuits, and on 32, 115 and 230 volt, direct current

Motor-Generator Set
Union Ring Type

500-1000 Volt Generator
Double Commutator Type

circuits. They are of compact construction. The 500 volt outfits are the two-bearing, union ring type, while the 1000 volt outfits are the four-bearing, sub-base type. The outfits are carefully balanced, insuring freedom from vibration and quiet operation. The large number of bars in the commutators of the generators does away with the objectionable

hum which is present when generators are used which have a smaller number of bars in the commutator.

All generators are shunt wound. In the 500-volt types they have a single commutator. The 1000-volt types have a commutator at each end of the generator, so arranged that the windings can be connected in series for 1000 volts or in parallel for 500 volts as desired.

The 500-6 volt and 500-12 volt type generators have two commutators, and other new features, and deliver 500 volts for charging the plate at one end and 6 or 12 volts for heating the filament at the other end, thus eliminating the necessity of providing a battery for heating the filament. The generators equipped with pulleys for belt drive are the same in construction as those used with the motor-generator sets, and the description preceding applies to them also.

### New Type Self-Cooled Transformers

On acount of self cooled. transformers being designed on an ambient temperature of 40 deg. C., advantage of the full rating of the transformer may not be obtained where temperatures are above this point. Such a case happened in a sub-station of the Southern California Edison Company, so that artificial cooling was necessary. To meet the conditions, air ducts were installed around the base of the 3000 kv.a., Westinghouse single phase, maximum rated, self cooled rad-

iator type transformer, so arranged with openings placed between the radiators to direct the flow of air supplied by a motor driven blower.

With this method of artificial cooling, these maximum rated transformers will carry 25 per cent overload continuously with the same winding temperatures as at full load without artificial circulation of air.

The transformer is provided with a hot spot temperature indicator, an oil expansion tank and a high voltage tap changer, the latter operated by a hand wheel in a convenient position above the transformer cover.

### Drying Machine for Hands and Face

The Airdry Corporation, of Groton, N. Y., is manufacturing a machine called "Airdry" for drying hands and face. The machine consists of a white enamelled iron standard, containing a motor, fan, heating element, and an adjustable

nozzle for directing the flow of air. Connected with the lighting circuit, the machine is set in operation by means of a foot-lever in the base of the standard. Pressing the lever downward causes warm air to be discharged through the nozzle. Releasing the lever stops the motor and the flow of air. This machine is adjustable to any position. It is said to dry thoroughly. It is safe and sanitary and eliminates towels. It is particularly adaptable to public washrooms, factories, etc.

### A Portable Wringer

Another labor-saving device under the trade name "Wring-O" is being placed on the market by the Petroleum Engine and Manufacturing Co., 120 Broadway, New York. This is an electrical wringer of the portable type, as shown.

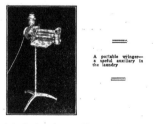

A portable wringer—a useful auxiliary in the laundry

It is designed to be used as an auxiliary to the electric washing machine or perhaps as a step toward the use of an electric washing machine by those who feel they are not

able at the moment to bear the full expense. It is claimed, also, that an efficient washing may be accomplished by thoroughly soaking the clothes in soapy water an then wringing and rinsing through several waters until clear water is wrung from them. It would seem to be specially useful in apartments and under similar conditions.

### Bank Hold-up Alarm System

For the protection of banks and other financial institutions against "hold-ups", a newly devised system has been brought out by Edwards & Company. The operation of the system is very simple. Tellers' foot switches, as shown in the illustration, are placed on the floor at tellers' windows and other places. Alarm bells or sirens are placed either inside or outside of the premises. The contact-maker can be

located where most convenient. In case of a hold-up, the bar is raised by the toe. This operates the contact-maker which immediately causes all bells or sirens to ring continuously until system is reset which can only be done by unlocking door of contact-maker and pressing button marked "Reset." It is important to note that releasing foot switch does not stop the alarm. G. L. MacGillivray & Company Ltd. of 35 St. Nicholas Street, Montreal, are the Eastern Canadian representatives for Edwards & Company.

The town of Sturgeon Falls, Ont., has bought out the property and franchise rights of the Northern Ontario Light & Power Company, within the town limits, and will operate it in future as a municipal enterprise under the supervision of Mr. L. E. Carter, town engineer.

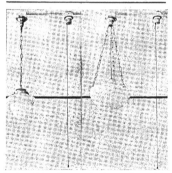

Two types of lighting unit used in the Imperial Office Building, Hamilton, described in our last issue

# Trolley Buses to Operate in Toronto on Mount Pleasant Road

A careful study of transportation problem convinces those who are familiar with all of its phases, that the bus will not take the place of vehicles operated on rails, but that there is a distinct field outside of that of the traction lines which the bus can fill advantageously.

A certain section of this outside field, where service is infrequent, can best be filled by the gasoline propelled bus. Another section, where service is more frequent, and is more closely allied to the traction lines, can be filled more advantageously by the electrically propelled vehicle or trolley bus, as it is called.

The St. Louis Car Company, recognizing the above facts, has recently completed a trolley bus which presents a number of attractive features. The bus was built for exhibition in connection with the proposed installation of trolley buses in Detroit. The electrical equipment was supplied by the Westinghouse Electric & Mfg. Company.

An important feature of the construction, is that the body is built directly on the chassis frame, thus avoiding the duplication of frame members, which occurs when a separate body frame is employed, and giving a very light and strong construction.

The wheel base is 194 inches, and the overall length of the bus is 26 feet. Nine cross and two longitudinal seats are provided, giving a seating capacity of twenty nine. The seats are deeply upholstered and provided with easy springs, which add greatly to the comfort of the passengers.

The driver is located at the front on the left hand side, and the entrance and exit door is directly at his right. From the driver's position, a clear unobstructed view ahead is obtained through a single window of the 'clear vision' windshield type, which extends across the entire front of the bus.

Foot operated band brakes are provided on all four wheels, and a separate pair of hand operated emergency band brakes are supplied on the rear wheels. The total weight of the bus equipped is approximately 10,500 lb.

Two Westinghouse type 506-AN-2, .25 h.p. ball bearing motors are used. The motors are mounted beneath the body and are connected in tandem with a short shaft and "splicer" universal joints. The connection is made from the rear motor through a propeller shaft and universal joint to a "Wisconsin" worm drive axle having a gear ratio of 6.5 to 1.

The control apparatus is mounted underneath the hood. This location of the apparatus is ideal in many ways as it provides for easy inspection, places the apparatus in a position where it is well above the mire of the road, simplifies the wiring, and adds a very appreciable factor of safety, as it places the current handling apparatus in a steel compartment entirely outside of, and not under the bus body.

The control is of the series parallel type. Eight magnetically operated switches give the necessary combinations for four steps and three in parallel. The control is operated by means of a foot pedal which is pivoted at a point under the center of the foot, so that the foot rests comfortably at all times on the pedal and a slight rocking motion forward and backward controls the various steps.

There are three operating positions to the foot pedal, which are indicated by "feel" to the operator by a star wheel provided on the foot operated controller. When the foot pedal is placed in the first position, the motors are connected in series and the total external resistance is thrown in circuit. When moved to the second position, the magnetic switches close automatically in proper sequence until all resistance is cut out and the motors are in "full series".

On the third position, the automatic step by step motion takes place up to the "full on" or "parallel" position. The pedal is returned to the "off" position by means of a spring. Both sides of the trolley circuit are opened in the "off" position, thus reducing the chance of leakage to the bus body to a minimum.

The automatic progressive action of the control is so arranged that it is slowed down as the load on the main motors is increased, but under no conditions is it entirely stopped, so the difficulty of "stalling" which may occur in other automatic types under heavy load, is obviated.

Reversing is accomplished by means of a simple knife switch type of reverser which is mounted on the front of the

The St. Louis type of Trolley Bus

dash, under the hood, with its handle projecting through the dash in a convenient position near the operator.

An overload trip opens all of the magnetic switches in case of excessive loads. The overload trip is mounted on the dash, under the hood and can be tripped or reset by the operator by means of buttons located on the dash, within his reach.

Two separate overhead collectors are provided which are similar in construction to the standard trolley used on light traction cars, except that they are provided with ball bearing swivel harps.

The collectors will permit a deviation of ten feet on either side of the overhead wires. The bus is equipped with electric heaters, electric lights and electric signals.

This bus makes a valuable addition to the growing fleet

of trolley buses; from which the electric traction companies can select a suitable type to supplement their "rail service" in localities where complete street car service is not at present warranted.

### The Trolley Bus in Toronto

The Toronto Transportation Commission has recently placed an order with the Packard Ontario Motor Company for four complete trolley buses for service in the city of Toronto. Bodies for these buses will be built by the Canadian Brill Co., and all electrical equipment will be supplied by the Canadian Westinghouse Co. The buses will be completely assembled in Canada by the Packard Ontario Motor Car Co. This is the first trolley bus installation sold for Canadian delivery.

The buses will be similar in construction to the demonstration vehicle completed by the Packard Motor Car Company on August 22, 1921, and already illustrated in the Electrical News. They will be used to furnish transportation service in the outlying districts of Toronto, which are at the present time inadequately supplied with service. The electrical equipment will be essentially the same as that used on the Packard demonstration bus mentioned above and on the bus recently completed by the St. Louis Car Co., and now m Detroit for demonstration purposes.

### Toronto's Radial Vote

On New Year's day the citizens of Toronto again had an opportunity to express their viewpoint regarding the purchase of outlying radials terminating in the city, and a study of the vote is unusually interesting. On former occasions the majority in favor of radials had always been so great that the council would almost have been justified in going ahead with them, even under the unfavorable trade conditions that existed. It is now plainly evident, however, that the electors have had time to think the matter over carefully. Their first enthusiasm has had time to cool and the latest vote, though giving a slight—a very small—majority in favor of radials, indicates that the pendulum is swinging back very rapidly and that the people are now in a much more thoughtful mood and are counting their pennies much more carefully. Under the circumstances, we do not believe the city council will be justified in continuing negotiations for the taking over of these radials at the present time.

In reaching this decision, it is doubtless the case that the citizens of Toronto have been impressed and influenced by the tremendous expenditures that have been made by the city in recent years. However it may be in theory, it works out in actual fact that the prosecution of the municipal ownership idea, and its application to the various utilities, piles up the debt of a city to an appalling extent and increases the obligation of every property owner in like proportion. It may be that, in many cases, individual citizens are not called upon to make good these obligations, but the fact remains that they must assume them and, apparently, the citizens of Toronto feel that for the time being they have assumed enough. We believe they are right and that the city council have a clear mandate, at least for the present year, to let this matter drop.

The situation with regard to the purchase of the N. S. & T. Railroad by the municipalities through which this road passes, is different. These municipalities are not already loaded with debt as is the city of Toronto, and they only are in a position to judge whether the road is being operated at the present time to meet the demands of the district. While Premier Drury was set his face against the province guaranteeing bonds to build or purchase radials at the present time, he does not object to the municipalities concerned

assuming these obligations if they are in a position to do so. The vote on January 1st indicated that the municipalities served by the N. S. & T. would prefer to control this road themselves rather than leave it with the Dominion Government.

Looking at it another way, the question was whether an electric road would be best operated by the Dominion Government or by the Hydro-electric Power Commission. When we consider that the management of the Canadian National Railway System is concerned chiefly with the operation of steam roads and that the Hydro-electric Power Commission has already a well established and equipped department for the operation of electric railroads, it would appear that the municipalities, in asking that these roads be turned over to the Hydro, have made a wise choice. The municipalities that have voted on the question appear to be of one mind on the matter and if the remaining municipalities, in which the vote will be taken, we understand, in the near future also endorse the scheme, the N. S. & T. will doubtless be handed over to the Hydro for operation in the near future.

### Employees Get 6 per cent on Savings

The Ottawa Electric Railway Company are trying out a generous plan with their employees' whereby savings accounts are accepted and interest paid on deposits at the rate of 6 per cent per annum. Mr. F. D. Burpee, manager of the company, has announced that they will accept loans from their employees at the same rate of interest that they have been paying the banks, namely six per cent, the only condition being that the employees must not withdraw their money frequently. In this way the company hopes to minimize the bookkeeping requirements connected with such a system. Up to the present time several thousand dollars of employees' savings have been deposited in this way and the arrangement will be continued if it meets with considerable response from the employees. The relations between the Ottawa Electric Railway Company and its employees have always been cordial, and generous treatment of this kind will naturally tend to maintain that condition.

### Renew Franchise by Heavy Vote

The electors of the city of Ottawa have voted by a large majority in favor of giving the Ottawa Electric Railway Company, whose present franchise expires next year, a new franchise for thirty years more. This company has given a good service but, more than that, has made a point of keeping the public informed on the problems that daily confront the operators of a street railway system. This policy has now proven its worth and will justify Manager F. D. Burpee in continuing his fixed policy that the public, when properly informed, and left alone, can generally be depended upon to make a sane judgment.

### Will Electrify M. C. R.

The vice-chairman of the London & Port Stanley Railway System, Mr. Phillip Pocock, stated recently that he had been informed that the Michigan Central Railway Company was considering the electrification of its Canadian line between Niagara Falls and Windsor. He considered that the starting up of the new power plant at Chippawa would have a considerable bearing on this project.

The Department of Mines, Ottawa, has issued a booklet, No. 564, entitled "The Preparation, Transportation and Combustion of Powdered Coal," by John Blizard.

# Current News and Notes

**Annapolis Royal, N. S.**

Mr. S. Rippey, Annapolis Royal, N. S., has been awarded the contract for electrical work on a skating rink and exhibition building being erected on St. Anthony St., Annapolis Royal, for Mr. F. J. Barnum.

**Brantford, Ont.**

A report states that when the Norfolk Telephone Company increased their rates last year, hundreds of their farmer subscribers refused to pay more and had the telephones removed from their homes. After being several months without the convenience of a telephone in the home, a number of farmers commenced negotiations with the company for the installation of 'phones, at lower rates. An agreement has finally been reached between the two parties, at the following rates:—

Simcoe Central, $17.50 per annum; Waterford Central, $16.50; Port Dover Central, $16; Delhi Central, $16; Scotland Central, $15.50; with a toll of five cents for each call from any one exchange to any other point on any other exchange in the system. The toll for the town of Simcoe remains the same.

The company agrees to replace all telephones that had been removed since July, 1921, free of charge, if notice to replace is received within 30 days from the date the board ratifies the agreement. After this a charge of $3 will be made.

**Cayuga, Ont.**

The town council of Cayuga, Ont., are contemplating the installation of a hydro system in the near future.

**Chicoutimi, Que.**

Messrs. Gilbert & Frere, Chicoutimi, Que., have been awarded the contract for electrical work on a $50,000 residence recently erected on Seminaire Road, Chicoutimi, for Mr. Louis Gagnon of that place.

**Conestogo, Ont.**

The Conestogo, Ont., town council are contemplating the installation of an electric lighting system for the town. The chairman of the council is Mr. H. W. Elber.

**Fredericton, N. B.**

The Province of New Brunswick is calling for tenders for bonds to the value of $1,800,000 for hydro-electric development and highway construction. These bonds will bear interest at the rate of 6 per cent, payable in Canada, and 5½ per cent, payable in the United States. They will run for a period of five years.

**Halifax, N. S.**

A draft of a contract has been prepared by a special committee appointed for the purpose whereby the city of Halifax will take over the St. Margarets Bay development, at cost, from the Provincial Water Power Commission.

The property of the Nova Scotia Tramways & Power Company, Halifax, N. S., suffered damage estimated at around $15,000 when a blizzard visited that city in the early part of January.

**Hamilton, Ont.**

The Ontario Gazette announces the incorporation of the S. U. & W. Electrical Manufacturing Co. Ltd., with a capitalization of $40,000. The new firm is authorized to manufacture and deal in all types of electrical machinery, appliances, etc. Head office: Hamilton, Ont.

**Kincardine, Ont.**

Mr. C. H. Wheeler, Kincardine, Ont., has been awarded the contract for electrical work on a $25,000 skating rink recently erected at that place.

**London, Ont.**

The Bowley Electric Co., London, Ont., has been awarded the electrical contract on a $10,000 gasoline station recently erected at Burwell & Dundas Sts., London, for the Canadian Oil Company, Ltd., London Junction.

It is stated that one of the first acts of the London city council for 1922 will be to revive the agitation for the electrification of the London, Huron & Bruce branch of the Grand Trunk Railway.

**Meaford, Ont.**

An enabling by-law was carried by the electors of Meaford, Ont., on New Years day. The electors also voted in favor of appointing a commission to handle the water works and electric power question.

**Milltown, N. B.**

The electrical contract on a new $90,000 school, recently erected at Milltown, N. B., has been awarded to Mr. H. D. Balkenay, Calais, Maine.

**Montreal, Que.**

Mr. A. Lafond, 470 Laval St., Montreal, has been awarded the contract for electrical work on a $25,000 residence recently erected at Sherbrooke & Addington Sts., Montreal, for Mr. A. Poirier, 347 St. Denis Street.

Mr. W. D. Michaud, 28 Lorne Ave., Montreal, has secured the contract for electrical work on a business block being erected at Bleury and St. Catherine Sts., Montreal, for Mr. A. H. Jassby, 4378 Western Avenue.

Messrs. Vallee & Hamelin, 1867 St. James St., Montreal, have been awarded the contract for electrical work on a chapel being erected on Ontario St. East for the Parish de la Nativite, 306 Desery St., Montreal.

Mr. J. A. St. Amour, 2171 St. Denis St., Montreal, has secured the contract for electrical work on an addition being built to the Ste. Justine Hospital for Children, St. Denis St., Montreal, at an estimated cost of $225,000.

Mr. A. Lafond, 470 Laval St., Montreal, has been awarded the contract for electrical work on a synagogue being erected at Duluth and City Hall Avenues, Montreal, for the Betha Juda Congregation at an approximate cost of $90,000.

**Niagara Falls, Ont.**

The Central Electric Co., Niagara Falls, Ont., has been awarded the electrical contract on a $70,000 theatre recently erected on Main Street, Niagara Falls, for the Drummond Hill Realty Co.

**Oshawa, Ont.**

Business was demoralized and the street railway system completely tied up for several hours in Oshawa, Ont., in the early part of December owing to an interruption in the hydro service, caused, reports state, by ice conditions in the Trent River.

**Ottawa, Ont.**

The Canada Gazette announces the incorporation of The Mis-Can-Ada Manufacturing Company, Ltd. The new firm will take over the business of the Mis-Can-Ada Manufacturing Company, Ottawa, who formerly manufactured electric washing machines and vacuum cleaners. The company is capitalized at $65,000. Head Office: Ottawa, Ont.

The Canada Gazette announces the incorporation of The James Weir Company, Ltd., with a capitalization of $12,500. The new firm will manufacture and deal in insulating varnishes, protective coatings, insulating compounds, etc. The head office of the company will be at Toronto, Ont.

### Palmerston, Ont.

The Palmerston, Ont., town council are planning improvements to their electric lighting system. A by-law was recently passed for the expenditure of $6,000 for this work. The council plans to rebuild part of line, change secondary system from 2 to 3 wires, using No. 2 copper conductors; change 30 services from 2 to 3 wires, including meters, and instal 20 new 3-wire and 30 2-wire services.

### St. Catharines, Ont.

Karl Wildern, district plant chief of the Bell Telephone Company, died on January 5. He came to St. Cathraines from Niagara Falls three years ago. Prior to that he had served the Bell Company in London, Windsor, Owen Sound and Woodstock, having been in their employ since 1891.

### Sarnia, Ont.

Mr. J. Filsinger, Christina St., Sarnia, Ont., has secured the contract for electrical work on a new store recently erected at Milton & McGibbon Sts., Sarnia, for Mr. Wm. Allaire.

### Simcoe, Ont.

The Utilities Board of Simcoe, Ont., of which Dr. A. T. Sibler is chairman, is considering the installation of additional equipment, their present system being badly overcrowded.

### Toronto, Ont.

At an "At Home" to celebrate the increase in users of their telephones from 40 to slightly over 100,000 in 42 years, the Bell Telephone Co. presented Mayor Church with a reproduction of the first telephone invented by Alexander Graham Bell, and a present day desk telephone, both done in silver.

The Toronto Electric Co., 101 Duke St., Toronto, has been awarded the contract for electrical work on an addition recently built to the arena and gallery at Roselawn Ave., near Avenue Road, of the Toronto Hunt Club, 542 Oriole Road, Toronto.

Messrs. Moss & Stocks, 14 Prince St., Toronto, have been awarded the contract for electrical work on a $35,000 Club House recently erected on Don Mills Road, near Fulton St., for the Finnish Society, 214 Adelaide St. W., Toronto.

Messrs. Smart & Walsh, 57 Kildonan Drive, Toronto, have secured the contract for electrical work on three stores being erected on Danforth Ave., near Broadview Ave., for Messrs. Coneybeare Bros., 108 Danforth Ave., Toronto, at an estimated cost of $105,000.

Messrs. Harris & Marson, 81A Parkway Ave., Toronto, have been awarded the electrical contract on a $150,000 school being erected by S. S. No. 26, York Township, Ont.

The Ontario Gazette announces the incorporation of the Battery Repair & Service Co., Ltd., with a capitalization of $40,000. The head office of the company will be at Toronto.

Messrs. Nichol & Fagan, 111 Logan Ave., Toronto, have secured the contract for electrical work on five stores recently erected on Danforth Ave., near Bastedo, for Mr.

W. W. Hiltz, 739 Broadview Ave., Toronto, at an approximate cost of $60,000. Also the electrical contract on six residences recently erected on Merrill Avenue E., near Moberly Ave., Toronto, at an approximate cost of $36,000.

### Waterloo, Ont.

The Electric Service Co., Waterloo, Ont., has been awarded the contract for electrical work on an addition being built to the Waterloo factory of the Forsyth Productions, Ltd., of Kitchener, Ont.

### Whitby, Ont.

It is reported that in an endeavor to make hydro power available for heating, at a figure that would be within the means of the average householder, the Whitby Public Utilities Commission has decided to reduce the power rates. At present the rate to small consumers is three cents per kw.h. and two cents per kw.h. to large consumers. The latter rate is to be reduced but it has not yet been announced to what extent.

### Winnipeg, Man.

The Star Electric Co., 185 Lombard St., Winnipeg, has secured the contract for electrical work on a Mission building being erected on Alexander Ave. at an approximate cost of $30,000.

The Swedish General Electric, Ltd., through their Western representatives, the Filer-Smith Machinery Company, Winnipeg, Man., have recently been awarded the contract by the City of Winnipeg Hydro-electric System for the supply and erection of five 45 kw. exciters for their Point du Bois generating plant. The machines will be built by the Swedish General Electric Company of Sweden.

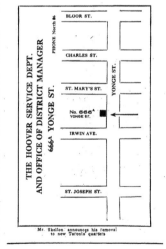

Mr. Skelton announces his removal to new Toronto quarters

# xclusive Designs in Table Lamps

The "Baetz" range includes many striking and usive designs—sure sellers. They are modely priced and we do not believe there is an trical dealer who cannot handle them without it. No matter where your business is located, what class of trade you cater to, the "Baetz" e will "fit in." Let us prove this to you.

## aetz Bros. Specialty
### Company Limited
### tchener :-: Ontario

### FOR SALE

Three Canadian General Electric Transformers, Size 40 K.V.A.
Dresden Flour Mills Ltd.,
Dresden, Ont.
24-8

### Do You Want a Good Man?

ACCOUNTANT, now with trust company, formerly in charge of consumers' accounts in large Hydro municipal plant, wishes to return to the electrical business, with power company or manufacturing or supply, concern in Toronto. 19 years accounting experience and practical electrical knowledge. War service with "Signals." Age 35. Present salary $2000. Box 741 Electrical News, Toronto.                    24-3

### BUILDING OPERATIONS IN NAPLES

Building activities are steadily increasing in Naples. At present five large apartment houses and one bank building are under construction. In the near future it is expected that there will be still further activities in this line, since the housing situation is acute and is becoming more serious every month.

From August, 1920, to August, 1921, prices paid for nearly all classes of labor have increased from 20 per cent. to 50 per cent. Practically all building operations are financed by building societies, which are backed by local banks. The banks lend usually from 25 per cent. to 35 per cent. of the total cost of the building at the beginning of construction and another 30 per cent. to 40 per cent. is advanced when the construction is half to three-quarters complete.

The tufa building stone, which enters into construction more than any other building material, comes from the local quarries. Cement, lime, and bricks are obtained from northern Italy, and from Spalato. Steel comes from Germany, and wood from Austria, Bohemia, Czecho-slovaki, and Calabria.

Up to the present time window and door sizes have not been standardized. The advantages of standardization are realized and as the matter is under discussion by the leading architects and construction engineers there may be a good opportunity for American sash, door, and blind manufacturers. Italian contractors are just beginning to put in parquet flooring in place of tile flooring that has been almost invariably used heretofore. This market may also offer possibilities to manufacturers of bathroom fixtures. The fixtures used at present are made in northern Italy.

### STEEL WORKS PLANNED FOR INDIA

According to information received from U. S. Department of Commerce, there has recently been registered in India the United Steel Corporation of India, Ltd., with a capital of approximately $60,000,000 to $100,000,000, depending upon the exchange calculated. The registration is in the name of the firms of Messrs. Bird & Co. of Calcutta and Messrs. Cammell, Laird & Co. The latter firm is acting as technical adviser to the corporation and is responsible for the design, erection, and staffing of the works. The plans of the new works are already well advanced. The plant is designed to produce 600,000 to 700,000 tons of pig iron, and the steel works and rolling mills will be capable of producing 450,000 tons of finished steel per year. As a first step, however, it is proposed to erect a unit capable of dealing with half these quantities, and the subsequent development will only be undertaken as occasion demands.

  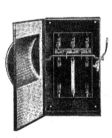

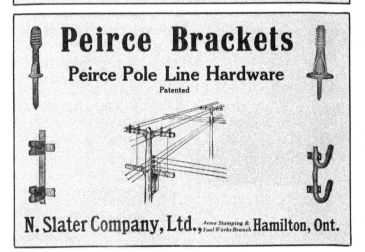

THE ELECTRICAL NEWS

# CLASSIFIED INDEX TO ADVERTISEMENTS

The following regulations apply to all advertisers:—Eighth page, every issue, three headings; quarter page, six headings; half page, twelve headings; full page, twenty-four headings.

**AIR BRAKES**
Canadian Westinghouse Company

**AIR PRESSURE RELAYS**
Electrical Development and Machine Co.

**ALTERNATORS**
Canadian General Electric Co., Ltd.
Canadian Westinghouse, Company
Electric Motor & Machinery Co., Ltd.
Ferranti Meter & Transformer Mfg. Co.
Lincoln Electric Company of Canada.
MacGovern & Company
Northern Electric Company
Wagner Electric Mfg. Company of Canada

**ALUMINUM**
British Aluminium Company
Spielman Agencies, Registered

**AMMETERS & VOLTMETERS**
Canadian National Carbon Company
Canadian Westinghouse Company
Ferranti Meter & Transformer Mfg. Co.
Hatheway & Knott, Inc.
Monarch Electric Company
Northern Electric Company
Wagner Electric Mfg. Company of Canada

**ARMOURED CABLE**
Canadian Triangle Conduit Company

**ATTACHMENT PLUGS**
Canadian General Electric Co., Ltd.
Canadian Westinghouse Electric Company

**BATTERIES**
Canada Dry Cells Ltd
Canadian General Electric Co., Ltd.
Canadian National Carbon Company
Electric Storage Battery Co.
Exide Batteries of Canada, Ltd.
Hart Battery Co.
Hatheway & Knott, Inc.
McDonald & Willson, Ltd., Toronto
Northern Electric Company

**BOLTS**
McGill Manufacturing Company

**BONDS (Rail)**
Canadian General Electric Company
Ohio Brass Company

**BOXES**
Canadian General Electric Co., Ltd.
Canadian Westinghouse, Company
G. & W. Electric Specialty Company
Hatheway & Knott, Inc.
International Machinery & Supply Company

**BOXES (Manhole)**
Standard Underground Cable Company of Canada, Limited

**BRACKETS**
Slater N.

**BRUSHES, CARBON**
Dominion Carbon, Brush Company
Canadian National Carbon Company

**BUS BAR SUPPORTS**
Moloney Electric Company of Canada
Monarch Electric Company
Electrical Development & Machine Co.

**BUSHINGS**
Diamond, State, Fibre Co. of Canada, Ltd.
Electrical Development & Machine Co.
Steel-City Electric Company

**CABLES**
Boston Insulated Wire & Cable Company, Ltd.
British Aluminium Company
Canadian General Electric Co., Ltd.
Phillips Electrical Works, Eugene F.
Standard Underground Cable Company of Canada, Limited

**CABLE ACCESSORIES**
Northern Electric Company
Standard Underground Cable Company of Canada, Limited
Steel City Electric Company

**CABLE END BELLS**
Electrical Development & Machine Co.

**CARBON BRUSHES**
Calebaugh Self-Lubricating Carbon Co.
Dominion Carbon, Brush Company
Canadian National Carbon Company

**CARBONS**
Canadian National Carbon Company
Canadian Westinghouse Company

**CAR EQUIPMENT**
Canadian Westinghouse Company
Ohio Brass Company

**CENTRIFUGAL PUMPS**
Boving Hydraulic & Engineering Company
Jones & Glassco

**CHAIN (Driving)**

**CHARGING OUTFITS**
Canadian Crocker-Wheeler Company
Canadian General Electric Company
Canadian Allis-Chalmers Company
Lincoln Electric Co. of Canada

**CHRISTMAS TREE OUTFITS**
Canadian General Electric Co., Ltd.
Masco Co., Ltd.
Northern Electric Company

**CIRCUIT BREAKERS**
Canadian General Electric Co., Ltd.
Canadian Westinghouse Company
Cutter Electric & Manufacturing Company
Ferranti Meter & Transformer Mfg. Co.
Monarch Electric Company
Northern Electric Company

**CLEAT TYPE SWITCHBOARD INSULATORS**
Electrical Development & Machine Co.

**CLAMPS, INSULATOR**
Electrical Development & Machine Co.

**CONDENSERS**
Boving Hydraulic & Engineering Company
Canadian Westinghouse Company
MacGovern & Company

**CONDUCTORS**
British Aluminium Company

**CONDUIT (Underground Fibre)**
American Fibre Conduit Co.
Canadian Johns-Manville Co.

**CONDUITS**
Conduits Company
National Conduit Company
Northern Electric Company

**CONVERTING MACHINERY**
Ferranti Meter & Transformer Mfg. Co.
Lincoln Electric Company of Canada

**CONDUIT BOX FITTINGS**
Hatheway & Knott, Inc.
Killark Electric Manufacturing Company
Northern Electric Company

**CONDULETS**
International Machinery & Supply Co.

**CONTROLLERS**
Canadian Crocker-Wheeler Company
Canadian General Electric Company
Canadian Westinghouse Company
Electrical Maintenance & Repairs Company
Northern Electric Company

**COOKING DEVICES**
Canadian General Electric Company
Canadian Westinghouse Company
National Electric Heating Company
Northern Electric Company
Spielman Agencies, Registered

**CORDS**
Northern Electric Company
Phillips Electric Works, Eugene F.

**CROSS ARMS**
Northern Electric Company

**CRUDE OIL ENGINES**
Boving Hydraulic & Engineering Company

**CURLING IRONS**
Northern Electric Co. (Chicago)

**CUTOUTS**
Canadian General Electric Co., Ltd.
G. & W. Electric Specialty Company
Hatheway & Knott, Inc.
International Machinery & Supply Co.

**DETECTORS (Voltage)**
G. & W. Electric Specialty Company

**DISCONNECTORS**
G. & W. Electric Specialty Company

**DISCONNECTING SWITCHES**
Electrical Development & Machine Co.
Winter Joyner, A. H.

**DOORS FOR BUS AND SWITCH COMPARTMENTS**
Electrical Development & Machine Co.

**DREDGING PUMPS**
Boving Hydraulic & Engineering Company

**ELECTRICAL ENGINEERS**
Gest, Limited, G. M.
See Directory of Engineers

**ELECTRIC HEATERS**
Canadian General Electric Co., Ltd.
Canadian Westinghouse Company
Equator Mfg. Co.
McDonald & Willson, Ltd., Toronto
National Electric Heating Company

**ELECTRIC SHADES**
Jefferson Glass Company

**ELECTRIC RANGES**
Canadian General Electric Co., Ltd.
Canadian Westinghouse, Company
National Electric Heating Company
Northern Electric Company
Renfrew Electric Company

**ELECTRIC RAILWAY EQUIPMENT**
Canadian General Electric Co., Ltd.
Canadian Johns-Manville Co.
Electric Motor & Machinery Co., Ltd.
Northern Electric Company
Ohio Brass Company
T. C. White Electric Supply Co.

**ELECTRIC SWITCH BOXES**
Canadian General Electric Co., Ltd.
Canadian Drill & Electric Box Company
Dominion Electric Box Co.

**ELECTRICAL TESTING**
Electrical Testing Laboratories, Inc.

**ELEVATOR CONTRACTS**
Dominion Carbon Brush Co.

**ENGINES**
Boving Hydraulic & Engineering Co. of Canada
Canadian Allis-Chalmers Company

**FANS**
Canadian General Electric Co., Ltd.
Canadian Westinghouse Company
Century Electric Company
Great West Electric Company
McDonald & Willson
Northern Electric Company
Renfrew Electric Company
Robbins & Myers

**FIBRE (Hard)**
Canadian Johns-Manville Co.
Diamond State Fibre Co. of Canada, Ltd.

**FIBRE (Vulcanized)**
Diamond State Fibre Co. of Canada, Ltd.

**FIRE ALARM EQUIPMENT**
Hart Battery Co.
Hatheway & Knott, Inc.
Northern Electric Company
Slater N.

**FIXTURES**
Benson-Wilcox Electric Co.
Crown Electrical Mfg. Co., Ltd.
Canadian General Electric Co., Ltd.
Duquesne Electric & Mfg. Co.
Jefferson Glass Company
McDonald & Willson
Northern Electric Company
Tallman Brass & Metal Company

**FLASHLIGHTS**
Canadian National Carbon Company
McDonald & Willson, Ltd.
Northern Electric Company
Spielman Agencies, Registered.

**FLEXIBLE CONDUIT**
Canadian Triangle Conduit Company
Hatheway & Knott, Inc.
Slater, N.

**FUSES**
Canadian General Electric Co., Ltd.
Canadian Johns-Manville Co.
Canadian Westinghouse Company
G. & W. Electric Specialty Company
Hatheway & Knott, Inc.
International Machinery & Supply Company
Moloney Electric Company of Canada
Northern Electric Company

## CLASSIFIED

**FROSTING**
Arts Electrical Company

**FUSE BOXES**
Canadian Drill & Electric Box Co.
Northern Electric Company

**GEARS AND PINIONS (Noiseless)**
Diamond State Fibre Co. of Canada, Ltd.

**GENERATORS**
Canadian Crocker-Wheeler Company
Canadian General Electric Co., Ltd.
Canadian Westinghouse Company
Electrical Maintenance & Repairs Company
Electric Motor & Machinery Co., Ltd.
Ferranti Meter & Transformer Mfg. Co.
Lincoln Electric Company of Canada
Northern Electric Company
Thomson Company, Limited
Toronto & Hamilton Electric Company

**GLASSWARE**
Consolidated Lamp & Glass Company
Jefferson Glass Company

**GUARDS (Lamp)**
International Machinery & Supply Co.
McGill Manufacturing Company

**HANGERS (Cable)**
Standard Underground Cable Company of Canada, Limited

**HEAD GATE HOISTS**
Hamilton Company, Wm.
Smith Company, S. Morgan

**HEATING DEVICES**
Benson-Wilcox Electric Co.
Canadian General Electric Co., Ltd.
Canadian Westinghouse Company
Electrical Appliances, Ltd.
Hatheway & Knott, Inc.
Hubbell Electric Products Co.
National Electric Heating Company
Northern Electric Company
Renfrew Electric Company
Spielman Agencies, Registered

**HIGH TENSION EQUIPMENT**
Electrical Development & Machine Co.
Electrical Engineers Equipt. Co.

**HYDRAULIC MACHINERY**
Boving Hydraulic & Engineering Company

**ILLUMINATING GLASSWARE**
Jefferson Glass Company

**INSPECTION ENGINEERS**
Allen Engineering Co.
Canadian Inspection and Testing Laboratories

**INSULATING VARNISHES AND MOULDING COMPOUNDS**
Canadian Johns-Manville Company
Continental Fibre Company
Diamond State Fibre Co. of Canada, Ltd.
Spielman Agencies, Registered

**INSTRUMENTS**
Canadian General Electric Co., Ltd.
Canadian Westinghouse Company
Hatheway & Knott, Inc.
Northern Electric Company
Sangamo Electric Company of Canada, Ltd.

**INSTRUMENT CUTOUT SWITCHES**
Electrical Development & Macnine Co.

**INSTRUMENTS—RECORDING**
Northern Electric Company

**INSULATION**
American Insulator Corporation
Canadian Johns-Manville Company
Continental Fibre Company
Diamond State Fibre Co. of Canada, Ltd.
International Machinery & Supply Co.

**INSULATION (Hard Fibre)**
Diamond State Fibre Co. of Canada, Ltd.

**INSULATORS**
American Porcelain Company
Canadian Westinghouse Company
Hatheway & Knott, Inc.
Illinois Electric Porcelain Company
Moloney Electric Company of Canada
Mica Company of Canada
Northern Electric Company
Ohio Brass Company
T. C. White Electric Supply Co.

**INSULATING CLOTH**
Packard Electric Company

**INSULATING VARNISHES**
Canadian Johns-Manville Co.

**IRONS (Electric)**
Canadian General Electric Co., Ltd.
Canadian Westinghouse Company
Electrical Development & Machine Co.
National Electric Heating Company
Northern Electric Company

**IRONING MACHINES**
Canadian Ironing Machine Co.
Mayer Bros.

**LAMP BULBS (Electrical)**
Jefferson Glass Company

**LAMP SHADES—SILK**
Manufacturers Trading & Holding Co.
Standard Silk Shades, Ltd.

**LAMPS**
Benson-Wilcox Electric Co.
Canadian General Electric Co., Ltd.
Canadian National Carbon Company
Canadian Westinghouse Company
Electrical Maintenance & Repairs Company
Hatheway & Knott, Inc.
McDonald & Willson, Ltd., Toronto
Northern Electric Company

**LANTERNS (Electric)**
Canadian National Carbon Company
Duncan Electrical Company
Spielman Agencies, Registered

**LIGHTNING ARRESTORS**
Moloney Electric Company of Canada
Northern Electric Company

**LIGHT-REGULATORS**
Dominion Battery Company

**LINE MATERIAL**
Canadian General Electric Co., Ltd.
Canadian Johns-Manville Co.
Canadian Line Materials Ltd.
Electric Service Supplies Company
Northern Electric Company
Ohio Brass Company
Slater, N.
Steel Company of Canada, Limited
Steel City Electric Company

**LIGHTING STANDARDS**
Canadian General Electric Co., Ltd.
Dominion Steel Products
Hamilton Co., Wm.
Northern Electric Company

**LOCK AERIAL RINGS**
Slater, N.

**MAGNETS**
Canadian Westinghouse Company

**MARBLE**
Davis Slate & Mfg. Company

**MEASURING INSTRUMENTS**
Northern Electric Company

**METERS**
Canadian General Electric Co., Ltd.
Canadian Westinghouse Company
Chamberlain & Hookham Meter Company
Ferranti Meter & Transformer Mfg. Co.
Great West Electric Company
Lincoln Electric Company
Northern Electric Company
Packard Electric Company
Sangamo Electric Company of Canada, Ltd.

**MICA**
Fillion, S. O.
Kent Brothers
Mica Company of Canada

**MOTORS**
Canadian Crocker-Wheeler Company
Canadian National Carbon Company
Canadian Westinghouse Company
Century Electric Company
Electrical Maintenance & Repairs Company
Electric Motor & Machinery Company
Ferranti Meter & Transformer Mfg. Co.
Gelinas & Pennock
Great West Electric Company
Lincoln Meter Company
MacGovern & Company
Northern Electric Company
Petrie, Limited, H. W.
Robbins & Myers
Thomson Company, Limited, Fred.
Toronto & Hamilton Electric Company
Wagner Electric Mfg. Company of Canada
Wilson-McGovern, Limited

**MOTOR AND GENERATOR BRUSHES**
Calebaugh Self Lubricating Carbon Co.
Dominion Carbon Brush Company

**MOTOR STARTING SWITCHES**
Canadian Drill & Electric Box Company

**OUTDOOR SUB-STATIONS**
Moloney Electric Company of Canada

**OVERHEAD MATERIAL**
Canadian General Electric Co., Ltd.
Northern Electric Company
Ohio Brass Company

**PANEL BOARDS**
Benjamin Electric Manufacturing Company
Canadian General Electric Co., Ltd.
Canadian Krantz Manufacturing Company
Devoe Electric Switch Company
Northern Electric Company

**PAINTS & COMPOUNDS**
G. & W. Electric Specialty Company
McGill Manufacturing Company
Spielman Agencies, Registered
Standard Underground Cable Co. of Canada

**PHOTOMETRICAL TESTING**
Electric Testing Laboratories, Inc.

**PATENT SOLICITORS & ATTORNEYS**
Dennison, H. J. S.
Fetherstonhaugh & Hammond
Ridout & Maybee

**PINS (Insulator and High Voltage)**
Ohio Brass Company
Slater, N.

**PIPE LINES**
Boving Hydraulic & Engineering Company
Canadian Line Materials Ltd.
Hamilton Company, Wm.

**PLUGS**
Benjamin Electric Manufacturing Company
Hatheway & Knott, Inc.

**POLES**
Lindsley Brothers Company
Northern Electric Company

**POLE LINE HARDWARE**
Canadian Johns-Manville Co.
Canadian Line Materials Ltd.
Northern Electric Company
Slater, N.

**PORCELAIN**
Canadian Porcelain Company.
Hatheway & Knott, Inc.
Illinois Electric Porcelain Company
Ohio Brass Company

**PORCELAIN BUSHINGS**
Electrical Development & Machine Co.

**POST TYPE INSULATORS**
Electrical Development & Machine Co.

**POTHEADS AND SLEEVES**
Electrical Development & Machine Co.
G. & W. Electric Specialty Company
Northern Electric Company
Standard Underground Cable Company of Canada, Limited

**POWER PLANT EQUIPMENT**
Boving Hydraulic & Engineering Company
Canadian General Electric Co., Ltd.
Electric Motor & Machinery Co., Ltd.
Hamilton Co., Wm.
MacGovern & Company

**PUMPS AND PUMPING MACHINERY**
Boving & Company of Canada
Canadian Allis-Chalmers Company
Hamilton Co., Wm.

**RECEPTACLES**
Benjamin Electric Manufacturing Company
Canadian General Electric Co., Ltd.
Diamond State Fibre Co. of Canada, Ltd.
Monarch Electric Company
Northern Electric Company
Paulding, John I.

**RECORDING INSTRUMENTS**
Canadian Westinghouse Company

**REFLECTORS**
Benjamin Electric Mfg. Co.
Pittsburgh Reflector & Illuminating Co.

**REMOTE CONTROL SWITCHES**
Electrical Maintenance & Repairs Company

**REPAIRS**
Electrical Maintenance & Repairs Company
Electric Motor & Machinery Co., Ltd.
Thomson Company, Fred.
Toronto & Hamilton Electric Co., Ltd.

**RHEOSTAT, BOXES AND ACCESSORIES**
Canadian Drill & Electric Box Company

**ROSETTES**
Benjamin Electric Manufacturing Company
Northern Electric Company

**SLATE**
Davis Slate & Mfg. Company
Bydeville Slate Works.

**SOCKETS**
Benjamin Electric Manufacturing Company
McDonald & Willson, Ltd., Toronto
McGill Mfg. Co.
Northern Electric Company

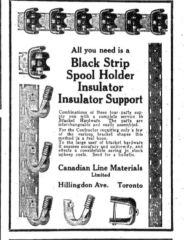

## CLASSIFIED INDEX TO ADVERTISEMENTS—CONTINUED

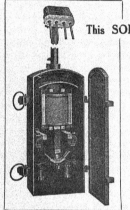
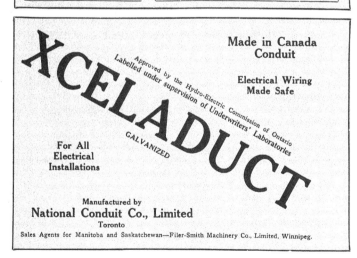

Vol. XXXI No. 3                                                Toronto, February 1, 1922

# Electrical News

**Engineering** **News** **Merchandising**
**Contracting** **Transportation**

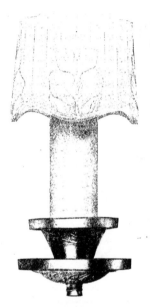

# Just the Shade for Candle Light

## No. 1032

Made to fit spherical bulbs without holder or fitter and finished in a variety of delicate tints of Rose, Amber, Blue, Old Ivory and Monolux. An assorted case will prove a good seller.

FACTORY & HEAD OFFICE
388 CARLAW AVE
TORONTO

DOWN-TOWN SHOW ROOMS
164 BAY ST. TORONTO

285 BEAVER HALL HILL
MONTREAL
272 MAIN STREET
WINNIPEG
510 HASTINGS ST.
VANCOUVER

Jefferson Glass
COMPANY LIMITED

## Four million megohms per inch cube

CONSIDER a switchboard material with 160,000 times the insulation resistance of slate, 16,000 times that of soapstone and 28 times that of blue Vermont marble.

Johns-Manville Ebony Asbestos Wood holds this superiority under ordinary dry conditions. It compares even more favorably under adverse conditions of oil, moisture and chemical fumes.

### Laboratory quality—practical price

Ebony Asbestos Wood is a laboratory quality insulating base made available for general use by low cost. It is made of Asbestos rock fibre cemented and impregnated with a special insulating compound.

Manufactured by formula, Ebony Asbestos Wood has none of the flaws of quarried materials—no metallic veins or cracks. While light in weight, it is rigid, permanent and tough.

### Many uses

Ebony Asbestos Wood is replacing slate, soapstone, marble and even porcelain for hundreds of different uses. It is comparatively easy to work, giving it a tremendous advantage where unusual construction is required. *Send for booklet.*

CANADIAN JOHNS-MANVILLE CO., Limited

Toronto    Montreal    Winnipeg    Vancouver
Windsor    Hamilton    Ottawa

Through
ASBESTOS
and its allied products
Electrical Materials
Brake Linings
Insulations
Roofings
Packings
Cements

Fire Prevention Products

# JOHNS-MANVILLE
## Ebony Asbestos Wood

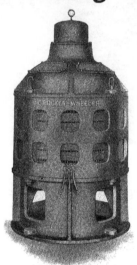

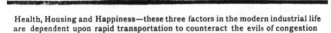

Health, Housing and Happiness—these three factors in the modern industrial life are dependent upon rapid transportation to counteract the evils of congestion

# C-G-E
## Electrical Equipment
## on Toronto Transportation
## Commission's New Street Cars

A BOVE is the new type of Motor Car used by the Toronto Transportation Commission and shown herewith is the CGE-241-B Railway Motor, 55 HP. rating—built by us in Canada,—4 of which will drive the motor car, this motor capacity providing for trail car operation.

A novel feature in connection with the control equipment, we are supplying, is the use of a line breaker.

This new line breaker equipment consists essentially of a ratchet switch in the controller, combined switch and fuse for the protection of the line breaker circuit, and a box (mounted under the car body) containing the line breaker, and overload relay. This equipment eliminates severe arcing in the controller as the line breaker breaks the main circuit, and also does away with the hand operated circuit breaker usually mounted in the motor car vestibule.

The control circuit for operating the line breaker is interlocked with the door opening and closing mechanism.

# Canadian General Electric Co., Limited
## HEAD OFFICE (C.G.E.) TORONTO

Branch Offices: Halifax, Sydney, St. John, Montreal, Quebec, Sherbrooke, Ottawa, Hamilton, London, Windsor, Cobalt, South Porcupine, Winnipeg, Calgary, Edmonton, Vancouver, Nelson and Victoria.

# ANOTHER BEAVERDUCT INSTALLATION

B EAVERDUCT has been adopted by the Toronto Transportation Commission for use on their new street cars. Once again, quality has fittingly demonstrated its ability to withstand competition.

Beaverduct is "Made in Canada" in a modern plant under the supervision of the best engineers available. Beaverduct is produced under the underwriters' inspection. Its use on such an installation as the new Toronto Street Cars is a fitting testimonial of its superiority.

Good conduit can only be produced in a plant where the equipment permits only absolutely clean material being used. Electrical apparatus and plating solutions must be kept up to the mark under scientific management. The proper application of the enamel, after the galvanizing of the outside, is of vital importance. And, lastly, rigid inspection.

*Always Specify*
*"Beaverduct"*

## Canadian General Electric Co., Limited

### HEAD OFFICE TORONTO

Branch Offices: Halifax, Sydney, St. John, Montreal, Quebec, Sherbrooke, Ottawa, Hamilton, London, Windsor, Cobalt, South Porcupine, Winnipeg, Calgary, Edmonton, Vancouver, Nelson and Victoria.

# ALPHABETICAL LIST OF ADVERTISERS

A great many merchants are. They are at a loss to know how to mark their merchandise, or what to charge for the supplies being sent out on the jobs

And they are losing money through lack of this knowledge. What they need is the

### Henderson Business System

It gives the correct retail price for every article of merchandise in the electrical dealer's store. It also contains other valuable information such as suggested charges for small repairs and house wiring rates per outlet, etc.

No matter how often cost prices change, your book is always kept

### Right-Up-To-Date

You are absolutely certain of making a profit on every article you sell. We guarantee this.

These figures are compiled from information received from reliable sources. The Henderson Business Service retail price book is, therefore, the STANDARD PRICE LIST.

Determine today that you will make 1922 a profitable year. Join the hundreds of other electrical dealers and use this system.

It will only cost you ten cents a day! Our booklet "Ten Reasons Why" is yours for the asking.

# The Henderson Business Service, Limited

Box 123                    Brantford, Ontario                    Bank of Toronto Building

# Westinghouse

## Motor and Air Brake Equipments
### For Toronto Transportation Commission

The illustration shows one of one hundred cars furnished the Toronto Transportation Commission, equipped throughout with Westinghouse motors and controllers and Westinghouse Air Brakes.

## Canadian Westinghouse Company

**Sales Offices:**

TORONTO, Bank of Hamilton Bldg.    MONTREAL, 285 Beaver Hall Hill.    OTTAWA, Ahearn & Soper, Ltd.
HALIFAX, 105 Hollis St.    FT. WILLIAM, Cuthbertson Block.    WINNIPEG, 156 Portage Ave., E.
CALGARY, Canada Life Bldg.    VANCOUVER, Bank Nova Scotia Bldg.    EDMONTON, 211 McLeod Bldg.

# What Will You Pay for Protection?
## Of Human Life! Of Valuable Property! Of Leaking Dollars!

THE late Peter Cooper Hewitt said that he would meet a problem in theory with a pad of paper and a pencil. But it took cold, hard experience to meet a problem in *fact.*

Dynamite as a toy for your child would be safe compared with handling high tension power equipment designed and installed on a basis of drafting room theory only.

Practical experience over a period of years with every phase of high tension control and transmission equipment fits Electric Power Equipment Corporation and Central Stations to qualify as high authority when it comes to these matters.

The Chief Engineers and Operating Executives of many of America's leading industrial plants have gone on record as to the concern's right to qualify as experts in this field.

Quoting from a letter written by a great manufacturer of talking machines. "This firm has made some of the best installations we have ever had put in at our plant. The largest piece of work they did for us was to furnish and install switchboard and switching apparatus, and cables for our transformer sub-station at our cabinet factory. All of this work was exceptionally well done and if we had any new work under consideration would not hesitate in awarding them the contract." From a plant manufacturing locomotives. "They have installed considerable electrical equipment in our various buildings at E—, the

total amount approximately $100,000: The character of the work and the service has been satisfactory in all respects."

From an internationally known ship-building concern. "The orders we have placed with them have been handled in all respects to our entire satisfaction both as to *quality of equipment furnished* and service."

The point we wish to bring to the electrical industry is this:

The equipment we offer is the result of unique and practical experience and without exception every piece of it has been designed to meet practical conditions.

These conditions had to do in every instance with safety to life and property and economy in the handling of nature's most elusive force.

Every piece of apparatus marketed under our name from switch to insulator has been subjected to Elpeco's high tension test before leaving the factory.

The cost of protection may be a little in excess of the price of ordinary material but we believe that every humane and thrifty engineer and contractor in America will agree with us that the pennies saved in the first place are not to be weighed against human life, and valuable property, and the dollars salvaged from leaky equipment over a period of years.

When you send for our catalog, bear in mind what the Elpeco crusader means—a crusade against danger and waste in the handling of high tension power.

ELECTRIC POWER EQUIPMENT CORPORATION

ELPECo

PHILADELPHIA PA.

SAFETY ECONOMY

Protection in the future from DANGER and LOSSES

ELPECO THE CRUSADER IN ADVANCED ELECTRICAL EQUIPMENT

Disconnecting Switches    Bus Supports    Pole Top Switches    Switch Boards

# WARD LEONARD DIMMERS

VITROHM DIMMER. SLOW-MOTION, CROSS-CONTROL, INTERLOCKING TYPE

*Built
for
continuous
duty*

*Built
for
continuous
duty*

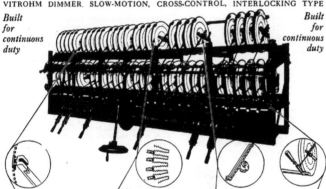

**Movable Contact** (Skate Shoe) The movable shoe is of the self-aligning, self-adjusting type. Springs hold the shoe firmly on the buttons assuring proper electrical contact and permitting the shoe to ride over projections of as much as ⅛" above contact buttons. This form of construction is simple and needs no attention or adjustment. It does not get "out of order" from usage. Moreover, a minimum of mechanical effort is necessary to operate the movable skate shoe over the stationary contacts and at the same time perfect electrical contact is assured. This has been a Ward Leonard standard construction for 20 years.

**Stationary Contact Buttons** are brass stampings, joined to the resistance wire under extremely high pressure while both metals are bright and dry. The joints are then imbedded in and covered by vitreous enamel, which holds the two parts firmly against any possible movement and protects the joint against oxidation or other chemical action. This also affords protection against mechanical injury. No solder, tin, or cement is used. The buttons project well above the enamel, so that dirt cannot readily become lodged between them and cause short circuit or poor contact.

**Space-Saving**
Vitrohm Dimmers occupy the smallest space and weigh less

for a given rating than any other dimmer.

### Continuous Duty
The foundation of all Ward Leonard Theatre Dimmers is a substantial iron plate of proper size, and great strength. It affords complete, permanent and rigid support of the resistance element, contact buttons, terminals and the *vitreous enamel* insulation. Vitreous enamel insulation is a glassy substance, non-porous and impervious to moisture. It does not contain cement, asbestos chips, or similar material, and it therefore does not have a spongy surface which might permit the atmosphere (always moist) to cause oxidation of the resistance element. When the wire is protected by vitreous enamel against oxidation and disintegration it cannot change in resistance value and therefore cannot have "opens" or "burn-outs". It is therefore obvious that Vitrohm Dimmers are designed and constructed for

### Continuous Duty
—to stand full rated load continuously without overheating or burning out. The homogeneous amalgamation of insulation, resistance wire and iron plate insure that the entire surface of the plates radiates the heat which is generated in the dimmer plate when in use.

### Delivery
In our 25 years' experience we have not held up or delayed the

opening of any theatre. We are always stocked to meet the demand.

### Positive Chain Drive
of the contactor arm is another desirable feature of Vitrohm Dimmer construction. A flexible brass link drive is used to operate the movable contact. There are no gears and pinions to become loosened, or "jam". Neither does this mechanism require lubrication which often dries out or "gums" and causes trouble. Adjustment of this chain drive is a simple matter, readily and quickly accomplished by turning the wing nut on a threaded stud, no tools required.

### Operating Levers
can be moved with much less effort and with much greater uniformity of motion than in other dimmers, this being due to the movable contact (skate shoe), stationary contact button and positive chain drive.

### Price
Because of our long experience and completed manufacturing facilities, we are able to quote prices which compare favorably with those of other dimmers which do not have our superior features. Moreover, our continuous duty dimmers can even successfully compete with intermittent duty dimmers in price.

**NOTE:** Vitrohm Dimmers installed in such houses as Hippodrome and Belasco Theatre, New York, etc., have remained unchanged since original installation over twelve years ago.

## WARD LEONARD ELECTRIC COMPANY,    Mount Vernon,   New York

THE beautiful design and finish of the R & M Fan appeal strongly to every fan purchaser. The gracefully formed, smooth drawn steel motor frame and base, the polished brass blades, the felt padded base, and the flawless enamel finish, reflect the quality built into the fan and make easy sales for the dealer.

And in every respect—performance, durability and efficiency— the R & M Fan measures up to its appearance. When the dealer sells R & M Fans, he knows that his customers will be as highly pleased with the service they give as they are with the attractive appearance of the fans.

## THE ROBBINS & MYERS COMPANY OF CANADA, LIMITED
### BRANTFORD, ONTARIO

# Robbins & Myers Fans
## And Motors

## The Power to Lift this Ponderous Drawbridge

Where demands are most exacting—where the safety of human lives rests upon their unfailing dependability—Exide Batteries will be found controlling mechanisms that *dare* not fail.

It is this recognized reliability that is responsible for their widespread use in railway signaling and interlocking work, in the driving of submerged submarines, operation of marine wireless and many other services of an equally important nature.

Exide Batteries are made for *every* purpose in which a storage battery can be used. They are the product of 34 years battery building experience—made by the oldest and largest manufacturers of storage batteries in the world.

> Information will cheerfully be furnished you regarding that phase of storage battery work in which you are most interested.

### EXIDE BATTERIES OF CANADA, LIMITED
153 Dufferin Street, TORONTO

*Exide Batteries are made in Canada, in England and in the United States.*

Generation, Transmission and Application of Electricity

For nearly thirty years the recognized journal for the
Electrical Interests of Canada.

Published Semi-Monthly By

## HUGH C. MACLEAN PUBLICATIONS
LIMITED

THOMAS S. YOUNG, Toronto, Managing Director
W. R. CARR, Ph.D., Toronto, Managing Editor
HEAD OFFICE - 347 Adelaide Street West, TORONTO
Telephone A. 2700

MONTREAL - - 119 Board of Trade Bldg.
WINNIPEG - - - 302 Travellers' Bldg.
VANCOUVER - - - Winch Building
NEW YORK - - - - - 296 Broadway
CHICAGO - - Room 803, 63 E. Adams St.
LONDON, ENG. - - 16 Regent Street S. W.

### ADVERTISEMENTS
Orders for advertising should reach the office of publication not later
than the 5th and 20th of the month. Changes in advertisements will be
made whenever desired, without cost to the advertiser.

### SUBSCRIBERS
The "Electrical News" will be mailed to subscribers in Canada and
Great Britain, post free, for $2.00 per annum. United States and foreign,
$2.50. Remit by currency, registered letter, or postal order payable to
Hugh C. MacLean Publications Limited.
Subscribers are requested to promptly notify the publishers of failure
or delay in delivery of paper.

Authorized by the Postmaster-General for Canada, for transmission
as second class matter.

Vol. 32          Toronto, February 1, 1922          No. 3

## Toronto Electric Home Arousing Tremendous Interest

How's the Electric Home going?

So great is the interest, in Toronto, in the Electric Home
campaign at present under way, that the inevitable query,
when members of the fraternity meet, is "How's the Electric
Home going?"

And the invariable answer is—"Tremendous success."

To begin with, the attendance is much greater than the
committee in charge had dared to hope it would be. Eighteen hundred and fifty visitors were shown through the home
after three o'clock on the opening day, and during the first
five days the house was open more than 6,000 people were
shown through it.

As a matter of fact, during the afternoon and evening
the number of visitors is regulated largely by the capacity
of the Home. It is an 8-room house of moderate size— just
about the size and price the committee considered would appeal to the greatest number of visitors. During the evening
some 18 or 20 demonstrators are constantly in attendance
and it is a rare circumstance that these demonstrators are
not fully occupied.

It is really too early to expect tangible results, but some
of the electrical merchants are already reporting inquiries and
requests from customers to come up to their homes and see
if they cannot "fix up the wiring" something like the Electric Home they were down looking at. The various men's
and women's organizations are being interested through

a special publicity committee. The architects' association
and the contractors, both general and sub-contractors, are
being told about the Home and a special day is being set
aside for them, when the general public will not be admitted.

One of the most satisfactory features of the whole campaign has been the support accorded the movement by the
daily papers. Representatives of the different papers have
been shown through the Home and given a thorough explanation of the underlying ideas. As a result, interesting articles,
written by the young lady representatives themselves, are
appearing from day to day and are doing much to explain
the whole campaign to the general public. These articles
are pointing out, for example, that while the average householder may not feel disposed to purchase at one time all the
electrical equipment shown in this model home, he will
nevertheless, doubtless expect to do so as the years go by,
adding a washing machine one year, an ironing machine the
next, etc. In this connection they point out the absolute
necessity of laying the proper foundation for the use of this
equipment when the house is built, by having the home properly wired, at a comparatively small additional expense.
These articles emphasize, also, the necessity of having this
work done by reliable contractors, using the most approved
methods, for example, a ceiling outlet in the laundry; a fixed
heater in the bathroom; wired furniture such as dining room
table and tea wagon; light, as one of them aptly puts it, "in
the darkest holes in the house—the clothes closets and cup-
hoards; the proper placing of switches to save steps; an
ample supply of outlets so that appliances may be used where
they are needed, etc.

At the formal opening of the Home on Saturday, January 21, Mr. P. W. Ellis, chairman of the Toronto Hydro-
electric Commission, officiated as master of ceremonies. He
briefly, but very enthusiastically, outlined the advantages of
the use of electricity in the home and its great possibilities
for placing the standards of living of the average man and
woman on a higher plane. Mayor McGuire also spoke
briefly, and Mrs. John Bruce, president of the Women's
Canadian Club, officially opened the Home with a golden
key provided for the purpose. Mrs. Bruce aptly remarked
that she felt, "like a good fairy opening the door of Fairyland
where electricity was the housekeeper, and household worries
were forgotten." The exhibit will remain open until the
evening of February 4th. After an interval of approximately
a month two other Electric Homes, which will be fitted up
in the meantime, will be opened to the public, one in the
east end and the other in the west end of the city.

## New Zealand Electrical Merchandizer Visiting Canada on Buying Trip

Mr. C. A. Seager, manager of the Auckland branch of
Messrs. Turnbull and Jones Ltd., electrical engineers and
contractors, who are represented throughout New Zealand
by five branches, is visiting Canadian cities. The object of
his trip to Canada, is to get in touch with Canadian manufacturers of electrical appliances.

Messrs. Turnbull and Jones Ltd., have in the past been
buying, in large quantities, goods manufactured in the United
States, the only reason for this being that they were not
aware of the fact that similar lines were obtainable in Canada, as the Canadian manufacturer has never had a representative in New Zealand, nor has he made any apparent
attempt to introduce his goods through the mails.

During the past three months, the government of New
Zealand has introduced a new custom tariff, which includes
a preferential rate of 15 per cent on electrical goods manufactured within the British Empire. When this announce-

ment was made. Mr. Seager received instructions from the directors of his company, to proceed to Canada at once, to ascertain what lines of electrical goods were being manufactured in Canada, and if prices and quality are right, to place orders in the Dominion.

## Now Keeping to the Right in Coast Province

After two weeks of the new "Rule of the Road" —"Keep to the Right" —British Columbia cities served by electric railways are beginning to settle down to the new order as a matter of course. It has been amazing, in such centres as Vancouver and Victoria, to see how well the change has been carried out. Not a little of the credit for this success is due directly to the organized efforts of the B. C. Electric Railway Company, which has for six months past been busy on the problems arising from the change. It has taken much thought and an enormous amount of labor to make advance preparations and be ready to instal the change-over at practically a moment's notice. The rule went into effect at midnight Dec. 31, and the great task was to complete the turning of all the switches on the system in readiness for the starting of traffic at 6 a.m. Jan. 1st. It was fortunate that the New Year came on Sunday, bringing with it a holiday on Monday. That made two days light traffic, which gave the workmen a little better show.

The casual observer, who might deem that it was only necessary to "turn the cars around and run them on the opposite track," would have been amazed to inspect the work which has been carried on at the B. C. Electric car barns for some months, when doors and platforms, on every car were being altered as the keeping up of the service permitted. With but few spare cars, much of the work had to be done at night. Gradually, as some cars were changed they were put on the road and others taken in. By that means most of the rolling stock had the preliminary work of right hand platforms and doors ready. But it was not possible to put on steps, hand-rails and other projecting appliances while still running to the left. That work, and the closing temporarily of the left hand openings and the removal of steps from left side, had all to be done during the last night of the old year.

A large extra force completed the work on a sufficient number of cars to keep up the schedule next morning.

It took a street force of fifteen extra gangs making the track changes on the last night of 1921. Eight tower wagons, some of them temporarily erected on motor trucks were put in commission, and an extra wiring gang with each, to make the changes in trolley overhead switches at the same time. In the car barns a large force of carpenters and workmen was busy making the changes from the moment the first car came in off the run, until the next morning when 160 were ready in Vancouver alone. One of the operations which was well advanced was that of changing the cross-over switches to the opposite direction. All but the electric wiring control of these was finished some time before the change.

Indicating the cost of the extensive changes required, the provincial government a year ago, in enacting the legislation for the change, decided to advance to the B. C. Electric Company a sum not to exceed $400,000, computed as half the cost of the necessary work.

## Prince Rupert May Build Plant

The city of Prince Rupert, in 1914, installed a development plant with a capacity of 1,650 h.p., on the Shawatlans River. During the war this supply satisfied the demand, but within the last couple of years it has become evident that a further development, either by an extension of the present plant, or by the installation of an entirely new plant, will be essential.

The Shawatlans power site is capable of developing another 1,000 h. p., but a second development on the Thulme River, distant only some thirty miles from the city, is estimated to have an ultimate capacity of 20,000 h.p. Figures have been prepared on the cost of this latter development by T. C. Duncan, superintendent of utilities, Prince Rupert. This estimate provides for an initial development of 2,000 h. p. with dam and transmission line capacity of 8,000 h. p., for $550,000. At last reports, the council was considering which of the two developments would be in the best interests of the city, but it is likely that the larger one can be shown to promise much better ultimate results.

## Canada's Favorable Position

Canada has her troubles. They are the lesser problems of resumed growth, not of reconstruction. She has no war currency to deflate. Her budget practically balances. In foreign trade her cash position is stronger than a year ago. Her production increases. No other country in the world can point to the combination of all these factors in January, 1922.

If the course of exchange be the true augury of 1922 eventualities, the Canadian outlook is 100% better than a year ago. The new year opened with Canadian dollars at a 5% discount. The old year began business with Canadian dollars worth 85 cents apiece in New York.

In the particulars which receive first consideration when methods of actual reconstruction are authoritatively discussed, currency, budget and foreign trade, the Dominion satisfies the requirements of international credit.

But for railway investments, public revenue for two-thirds of the current fiscal year would exceed expenditure by a surplus almost again as large as the deficit of $26,000,000 disclosed. Revenue is larger than in 1920.

Originating a generation ago, the Canadian railway problem outranks all others in fundamental importance. But it has always been and remains a detail of growth, huge as it bulks today.

Actual gain in cash position of Canadian foreign trade is stronger by $150,000,000 than 12 months ago. Both exports and bank clearings have declined less than our own.

Land values have happily escaped most of the inflation which carried wheat on the ground from 70 and 80 cents a bushel to $2 and $3. Pessimism, faithfully portrayed by commercial reviews, in their New Year's greetings, has, as was human, run riot in western mercantile centers. The fundamentals, calling for more industry in compilation, have been neglected.

Grain production is of record size. At the lowest prices since 1915 the farmer receives more real value of all kinds out of those prices than wheat boards could ever bring to him. Land values at least have not to undergo the sharper wrenches of deflation.

Immigration sets in. Building revives and building costs decline. Production increases. Where is the outlook better or as good? —Barron's Financial Weekly, New York.

# Kilovolt-Ampere Measurement and its Effect on Power-Factor

By S. L. B. LINES

General Manager Lincoln Meter Company before the A.M.E.U. Convention Jan. 26, 27

Broadly speaking the power factor at which a municipality buys power, is caused by two conditions.

(1) The exciting current of the distribution transformers.

(2) The loading conditions of the various customers.

It is the purpose of this paper to deal with some of the factors connected with (2).

For the purpose of simplifying the analysis we will divide the customers into three very general classes: Residential, Commercial, Power and Industrial.

The residential customers' load is almost entirely of the non-inductive type, i.e. cooking, heating and illumination, all of which give good power factors. Fans and small utility motors form an exceedingly small portion of the total load. Therefore this type of customer in general tends to improve the power factor, rather than create bad conditions.

The commercial customer is primarily a lighting and heating customer but may have motors of a sufficient total to adversely affect the power factor. To determine whether this be the case or not, it is necessary to study the individual load.

The power or industrial customer is the one usually responsible for bad power factor. The extent to which this is the case is not generally realized. An analysis recently made in a 60-cycle district showed the power factor of the average machine shop to be between 60 and 70 per cent at maximum load.

It is the practice of the Hydro-electric Power Commission to bill municipalities on 90 per cent of the maximum demand in watts if the actual power factor is below 90; if above 90, on the maximum demand in watts.

Some actual bills to municipalities within the past six months show what an effect this has on their monthly statement.

Muncipality A.

Bill without power-factor correction ..............$413.00

Bill with power-factor correction ..................$477.50

Increase ..................: ................. :............$64.50

Municipality B.

Bill without power-factor correction ............ $3,860.00

Bill with power-factor correction ................ $4,760.00

Increase ............... .......... ............. ... .900.00

Municipality C.

Bill without power-factor correction '............ $18,930.00

Bill with power-factor correction ....:.. ........ $21,300.00

Increase ...... ........ ......... ..,... ........ $2,370.00

Municipality D.

Bill without power-factor correction ............$652.00

Bill with power-factor correction .....................$766.00

Increase ......., ...... ......... ..... ...... $114.00

In analyzing the charts of "Municipality A" on the same day and time as that of maximum demand upon which their bill was based, we find that they have an industrial load of 95.5 horse power at 80 per cent. power factor. If, however, the industrial load has been taken at 90 per cent. power factor this would have resulted in the total being taken at 87 per cent. power factor, or a saving of $51.00 per month. The con-

ditions which increase the bill to this extent should either be corrected or passed on to the customer or customers responsible. Since, however, it is not usually possible by peace-persuasion to get the individual customer to amend his ways, even if he can, the second alternative should be adopted, i.e., pass the cost on to the customer responsible. It will, I believe, be readily admitted that this should be done, as the only alternative is to charge all customers, good, bad and indifferent alike, a higher rate to cover up the sins of the few.

### Some Municipalities Have Tried Methods

In some municipalities an endeavor has been made to do this in one or other of two ways; one, to have testers visit the premises of the various customers from time to time and take tests; the other, by means of two graphic meters—one reading watts and the other reactive kv.a.

The first method is better than nothing, but the cost is considerable and it has the objection that one cannot be sure that the power factor at time of test is the same as power factor at maximum demand. Thus errors in billing may creep in either for or against the customer.

The following examples were taken on six customers in one of our large cities, chosen at random,—

### Toy Manufacturer

Demand ........................................ 19 h.p.

Power factor by periodic test ...................... 80 %

Power factor by measurement ..................... 69 %

Error ............................................ 11 %

### 2. Leather Manufacturer

Demand ........................................ 25 h.p.

Power factor by periodic test ...................... 45 %

Power factor by measurement ..................... 45 %

Error .......................... None

### 3. Tannery

Demand ........................................ 35 h.p.

Power factor by periodic test ...................... 71 %

Power factor by measurement ..................... 86.6%

Error ............................................ 15.6%

### 4. Chocolate Manufacturer

Demand ........................................ 45 h.p.

Power factor by periodic test ..:................... 85 %

Power factor by measurement ..................... 74 %

Error ............................................ 11 %

### 5. Embroidery Manufacturer

Demand ........................................ 25 h.p.

Power factor by periodic test ...................... 85 %

Power factor by measurement ..................... 91 %

Error ............................................ 6 %

### 6. Textile & Knitting Mills

Demand ........................................ 550 h.p.

Power factor by periodic test ...................... 85 %

Power factor by measurement ..................... 78 %

Error ............................................ 7 %

Other municipalities, however, have neglected the question of power factor measurement and some results obtained in cases of this sort may be of interest,—

1. 16 kilowatts demand; power factor taken as ..... 85 %

Test showed ................................ 47 %

2. 190 kilowatts demand; power factor taken as .. 90 %

Test showed ................................ 66 %

3. 90 kilowatts demand; power factor taken as ...: 90 %

Test showed ................................. 71.5%
4. 26 h.p. demand; power factor taken as .......... 83 %
   Test showed ................................. 54 %
5. 22 h.p. demand; power factor taken as .......... 85 %
   Test showed ................................. 66.0%
6. 15.8 kilowatts demand; power factor taken as .... 85 %
   Test showed ................................. 57 %

The second method, i.e., using the two graphic meters, is so costly that it can only be applied to very large customers. The question therefore arises:—are there not some other methods by which one can make a just charge to each industrial customer? Let us compare the costs of the methods available at present.

### Method 1.

Two graphic meters as mentioned above, one to read watts and the other reactive kv.a. From these two readings a power factor can be accurately determined. The cost varies from $385.00 to $780.00 according to the type of graphic used, plus transformer cost.

### Method 2.

The use of two indicating demand meters connected as in method 1. This arrangement will give fairly accurate results, but since the maximum watts and r.kv.a. may not coincide, due to fluctuations in voltage, an error may be introduced which would be unfavorable to the customer. The amount of the bill would no doubt be questioned. The cost of this method varies from $117.00 to $126.50.

### Method 3.

The use of a graphic kv.a. meter. This arrangement gives accurate results. The cost varies from $275.00 to $280.00.

### Method 4.

An indicating kv.a. meter giving accurate results and costing $72.00 to $79.00.

From the point of view of the cost and accuracy, the choice, therefore, rests between 3 and 4.

Method 3 should be chosen where a graphic record is required or justified. Generally speaking, on 100 horse power, or over, the expense is warranted. Method 4 is preferable where a graphic record is not justified.

We have arrived at the point where the instruments available make it highly desirable that both large and small customers be measured in kv.a. ft is now well, perhaps, to run over rapidly what is meant by kv.a.

In a direct current circuit, kilowatts are the same as kilovolt amperes. In an alternating current circuit, this usually is not the case, because the question of power factor comes

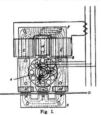

Fig. 1.

in. In this case watts are equal to volts multiplied by power factor (which is always less than one) whereas the volt amperes are product of the volts by amperes, neglecting the power factor. Therefore, a volt ampere reading is always higher than a watt reading.

There are two types of graphic meters for measuring kv.a. on the market.

### The Instantaneously Responding Meter

"Instrument A" is of the instantaneously responding type. It has the advantage of reading kv.a. on balanced loads at any power factor but penalizes the customer on unbalanced loads. It is a question whether this is a desirable feature or not. The Standard Interpretation of Rates and Contracts by the H. E. P. C. at present existing, do not permit of a penalty for unbalance. Therefore this instrument should not be used unless it is known that the load is balanced.

This instrument has the standard form of metering discs such as are found in an integrating watthour meter. An integrating wattmeter measures kilowatts because at the time the power factor departs from unity, the phase relation of the magnetic fields, (produced by the current and voltage cutting the meter disc) is charged. If the relation between these magnetic fields is maintained on varying power factors, then the instrument will read kv.a.

Figure 1 is an illustration of the magnetic circuit, shown in the single phase form for simplicity. The frame F is magnetized by the potential coils C and C'. The potential element A is mounted in the space of the frame, so as to be free to rotate. It does not, however, revolve like an armature of an ordinary meter, but changes its position only with the changes of power factor. The potential element A carries a polyphase winding which when energized, produces a rotating field about its periphery. The flux of the frame divides, the major portion passing across the gaps between the frame and the yoke S, cutting the disc at P and P'. The remainder of the flux crosses the path formed by the potential element at N. There is a reaction between this portion of the current field and the potential field. Therefore, the element A, which is free to rotate, will assume a position whereby the potential field is in phase with the current field.

### A Rotating Element Maintains the Phase Relationship

If the power factor of the circuit changes, causing one of the fields to lead or lag, the potential element rotates enough to re-establish this phase relationship. The element A does not move except to shift in either direction with changes in power factor. However, as flux is produced by the polyphase potential winding on it which rotates about its periphery at synchronous speed, the potential flux through the disc between K and N is at right angles to the potential flux path, from pole to pole through A.

The element A, always shifts so that the potential field is in time phase with the current field across the polar space, but with the element A, in this position the potential field across the disc between N. and K. will always be in time quadrature with the current field. This being the case, the torque on the disc is the same as if the power factor were maintained at unity. Therefore, the instrument measures volt amperes.

Upon the spindle of the meter disc is mounted a pen which writes on the chart. This is similar to the arrangement in standard graphic instruments.

### Description of the Demand Type of Meter

"Instrument B." is of the demand type and does not penalize the customer for unbalanced loads. It can therefore be used under any conditions, with the present wording of the contracts. It is limited however, in the range of power factor under which it will operate accurately. In its standard form, it covers power factor from 43 to 90 per cent. lagging by the use of two sets of taps, but by the means of special taps it can be made to cover any desired power factor.

### Method 4.

As far as the writer is aware, there is at present only one type of indicating kv.a. outfit available. It employs the same principle as the second type of graphic kv.a. meter. The principle of operation is to apply to demand wattmeter a small external transformer. The purpose of this transformer is to bring the voltage and current in the circuit into the correct phase relation and to do away with the phase differ-ence existing when the load is operating at power factors other than unity. It is immaterial whether the angular posi-tion of the current or voltage is shifted, as long as the two bear their correct relationship.

An inductive load causes the current to lag behind the voltage. If therefore we can cause the voltage to lag also, and by the same amount, we have the desired results and any standard wattmeter is caused to read kv.a. instead of watts. By means of the small transformer already referred to, the voltage is given the required movement backwards. The line voltage is applied to A B and C. (Fig. 2). Instead of C B being applied to the potential coil of the meter, 2-7 are connected for the one phase. Instead of A C being connect-ed to the other potential coil of the meter, 7-5 is used. It will readily be seen from the diagram that this has shifted the voltage 53.13 degrees away from its original position. This angle is the same as the lag in the current with a load at 60 per cent. power factor and the meter will now read the product of the volts by amperes, providing the power factor under measurement is 60 per cent. Furthermore, the power factor of the load can vary very materially with-out introducing an error of more than 2 per cent. as shown in Fig. 3.

Taps in the transformer, of which Fig. 2 is a diagram-atic sketch, can be taken out at various points and the volt-age caused to be in operative harmony at any desired power factor.

### Comparative Methods

When the contract is based on a charge per horse power plus an adjustment for power factor, the procedure would be as follows,—

Either measure the demand in watts obtaining the pow-er factor, from the tester who has visited the factory, and work out the formula as follows,—

Kw. demand reading multiplied by .746 to convert to horse power. Then multiply by the allowable power factor over the actual power factor.

It will be noted that both these multipliers increase the readings of the meter. To measure the customer in kilo-volts amperes on demand, cut out the testing on the custom-er's premises, with its uncertainties, and the formula then becomes as follows.—

$$kv.a. \; x \; \frac{1000}{746} \; x \; \frac{allowable \; p.f.}{100}$$

It will be noted that in this case that only one of the multipliers increases the readings. The other decreases it. Although the results of both formulas are the same; the cus-tomers are not as apt to feel quite so badly dealt with as where both multipliers increase the meter reading. An act-ual example of each method shows that the results obtained are exactly the same. A customer with 60 kilowatt on his

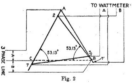

Fig. 2

meter, whose power factor is found to be 50 per cent. on tests, would have his bill worked out as follows:—

$$60 \; x \; \frac{1000}{746} \; x \; \frac{85}{50} \; = \; 137 \; h.p.$$

A kilovolt-ampere reading on the same customer would be worked out as follows;—

$$120 \; x \; \frac{1000}{746} \; x \; \frac{85}{100} \; = \; 137 \; h.p.$$

### Power Should be Sold on kv.a. Basis

I would at this juncture put in a plea for the sale of pow-er on a kilovolt-ampere basis in place of horsepower plus a power factor penalty. There are no instruments measuring horsepower. They all measure in watts or volt-amperes. This fact means that the meter reading must be multiplied by a factor which increases the figure appearing on the bill. The power factor penalty means, first, that the power factor must be measured and, secondly, that a multiplier has to be used which again increases the reading. By the time one has finished with the average layman, he is properly fogged. I am sure he will go away with the impression that most electrical engineers ought to be in jail. Whereas if power was sold at so many dollars per kilovolt-amperes (instruments now being available to measure this quantity) the reading would be taken from the meter, multiplied by dollars and cents and the total bill arrived at without any further fig-ures or multipliers. The objection may be raised that the customer will not understand what a kilovolt-ampere is. This same objection was raised to the change which took

Fig.3

place on lamps. Carbon lamps were all sold by the candle power whereas lamps to-day are sold on the watt basis and our better halves talk just as glibly about 25 and 50 watts to-day as they did in the old days about 16 and 32 candle power. It would appear reasonable to assume that the in-dividual customer would learn just as rapidly what a kilovolt ampere is as compared to the somewhat ambiguous term "electrical horse power."

May I point out that when a farmer, who has one horse doing a given time job, comes to install electrical power, he finds he has to have two of the electrical variety to do the same work. Electricity is somewhat belittled in his mind right at the start. Whereas if he was told that he required two kilovolt-amperes to do the work of one horse, he would be entirely satisfied. When his bill came in, if he were able to read the meter, his reading and the bill agree.

I will conclude by giving two sample bills together with the necessary calculations in each case;—

Load 100 h.p. at 50% p.f.

| Watts Demand Corrected for p.f. | Kv.a. Demand |
|---|---|
| Price per h.p.: $1.25 | Price per Kv.a. $1.51 |
| Meter reading, 74.6 kw. | Meter Reading 149 kv.a. |
| P. F. found by test, 50% | 149 x $1.51 = $225.00 |

$$74.6 \; x \; \frac{1000}{746} \; = \; h.p. \; = \; 100$$

$$100 \; x \; \frac{90}{50} \; (to \; correct \; for \; p.f.) \; = \; 180$$

180 x price per h.p. = Bill

or 180 x 1.25 = $225.00

# Further Progress of Rural Distribution

## By J. W. PURCELL
### Assistant Engineer Hydro-electric Power Commission of Ontario*

The tendency of the present times in rural districts is toward the use of power in different forms for field work, in the barn, and to a lesser extent in the house.

The western provinces of Canada and the United States have for a long time used the tractor for field work, but have found the large tractor too heavy, too unwieldly, too cumbersome for their needs and for their soil; the natural result has been the discontinuance of the use of it to a great extent; the small tractor coming during the war has been adopted quite generally in the West and East. There have been, and are extremes in the applications—in the East we find farmers purchasing tractors and not reducing the number of horses on the farm—in the West there are men like Schaal, twenty-five miles, southwest of Winnipeg, who operate 10,300 acres with no horses and no stock, but forty tractors; each one hauls two fourteen inch plows, and they plow an acre every 2 minutes; 12 foot drills seed 640 acres a day. In harvesting they cut and stook a section (640 acres) per day. Mr. Schaal, Sr., claims power cheapens production and increases profits by horseless farming.

We, however, are not interested in the sale of tractors excepting in the general application of power to farm work but we do not think it is unfortunate that millions of dollars should be expended on this form of power, when by doing so they feel they are not justified in making a smaller expenditure to equip their places with hydro-electric power, which would give them a greater service both in the house and barn, at only a fraction of the cost per annum.

Our Ontario farmers, however, in most cases have not considered the needs in the house as serious until recently. The wives and women of the country have been granted the franchise, and apparently this has affected their status to the extent that they now enter into discussions in many of the meetings,and are not backward in stating their views. The following table prepared by the U. S. Dept. of Agriculture shows the distribution of the work of the farmer's wife and the equipment she has to do the work. The notation at the bottom, by way of advice to manufacturers and distributors is a timely suggestion to the manufacturers, distributors and merchants in Ontario.

Shortly before the June convention of this Association, the Legislative Assembly of the Province of Ontario passed legislation authorizing the bonusing of rural power distribution primary lines whereby these lines would be bonused up to 50% of their cost, at the time they were constructed.

The Commission was advised that for the present the Province would bonus the primary lines to the full amount of 50%, so that all estimates sent out and rates given are based on this assumption.

Late last year the Commission found it necessary to classify rural service, using as a basis for each class the demand of such class of service, based on the average demand from our experience in service to a large number in each class. The following excerpts from the letters to townships giving rates, gives the classifications and the description of the service each may be applied to as well as the estimated service charge, the estimated annual cost, the class demand in kw., and in h.p. the average monthly kw.h. and an estimated total cost based on power at 5 cents per kw.h. for the first 14 hours use of the class demand, with a follow up rate of 2½ cents per kw.h. The Service Charge and total cost are given for both frequencies

*Before A.M.E.U Convention Jan. 26-7.

as there are some representatives here from districts using each, and it may be well to have these figures handy when your farmer friends come in to inquire about probable costs for service in your district.

Class 1—Hamlet service includes service in hamlets, where four or more customers are served from one transformer. This class excludes farmers and power users. Service is given under three sub-classes as follows:—

Class 1-A—Service to residences where the installation does not exceed six lighting outlets or twelve sockets. Use of appliances over 600 watts is not permitted under this class.

Class 1-B—Service to residences with more than six lighting outlets or twelve sockets, and stores. Use of appliances over 750 watts permanently installed is not permitted under this class.

Class 1-C—Service to residences with electric range or permanently installed appliances greater than 750 watts. Special or unusual loads will be treated specially.

Class II—House Lighting—Includes all contracts where residences cannot be grouped as in Class I. ' This class excludes farmers and power users.

Class III—Light Farm Service—Includes lighting of farm buildings, power for miscellaneous small equipment, power for single phase motors, not to exceed three horsepower demand, or electric range. Range and motors are not to be used simultaneously.

Class IV—Medium Single Phase Farm Service—Includes lighting of farm buildings and power for miscellaneous small equipment, power for single phase motors, up to 5 horsepower demand, or electric range. Range and motor are not to be used simultaneously.

Class V—Medium 3-Phase Farm Service—Includes lighting of farm buildings and power for miscellaneous small equipment, power for 3 phase motors, up to 5 horsepower demand, or electric range. Range and motor are not to be used simultaneously.

Class VI—Heavy Farm Service—Includes lighting of farm buildings and power for miscellaneous small equipment, power for motors up to 5 horsepower demand, and electric range, or 10 horsepower demand without electric range.

Class VII—Special Farm Service—Includes lighting of farm buildings, power for miscellaneous small equipment, power for 3-phase motors from 10-20 horsepower demand, and electric range.

Class VIII—Syndicate Outfits—Includes any of the foregoing classes which may join in the use of a syndicate outfit, provided the summation of their relative class demand ratings is equal to the kilowatt capacity of the syndicate.

The estimates on the cost of power delivered to users as herein set out have been based upon certain assumptions. The construction of the lines shall be undertaken and paid for by the Commission. The farmers in the vicinity of the roads along which the lines pass will assist in the construction and assistance will be paid for at a suitable wage. Lines constructed from the line on the highway to customers' premises will be paid for by the customer. The Commission proposes to supply the necessary expert labor to direct the construction of the lines and the installation of the equipment. It has been assumed that three farmers per mile of line, or the equivalent, are obtainable as an average for the entire district to be served. The supply of poles at low prices in the district or the vicinity of the district by efforts on the part of those desiring service will result in the reduction of the cost of construction and, corresponding reduction in the cost of service. Co-operation resulting in the reduction of the cost of construction is desired. The rates herein set out are also based upon a government bonus of 50 per cent of the cost of primary lines constructed on the highway or along the right-of-way.

Charges for power delivered shall consist of two parts, namely, the service charge and the consumption charge. The service charge which constitutes the greater portion of the

of the Power Commission Act as applied to individuals, therefore the liability in a district is assumed by this group of individuals, and any benefits accrue to them as individuals. The rural power contracts are covered by a guaranteeing agreement between the townships and the Commission.

Under legislation as revised in 1920, all new systems will be operated by the Commission. Distribution in districts such as yours, from existing or provided distribution centers, make this desirable. At the end of each year a statement will be rendered to each township in each district. Construction in rural districts was discontinued in 1917

TABLE NO. 1. WHAT THE FARMER'S WIFE DOES—AND WHAT SHE HAS TO DO IT WITH

|  | Rooms to care for | Stoves to care for | Kerosene Lamps | Water to Carry | | Wash and Iron | Sewing | Daily Mending Hours | Bread Making |
|  |  |  |  | Percentage | Distance Feet |  |  |  |  |
|  | Average | Average | Percent | Percent | Average | Percent | Percent | Average | Percent |
| Eastern States | 9.7 | 1.35 | 79 | 54 | 23 | 94 | 86 | 5 | 89 |
| Central " | 7.7 | 1.3 | 79 | 68 | 41 | 97 | 94 | 6 | 78 |
| Western " | 5.3 | 2.5 | 77 | 57 | 65 | 97 | 95 | 5 | 97 |
| Average | 7.8 | 1.6 | 79 | 61 | 39 | 96 | 92 | 6 | 94 |
| No Records | 9871 | 9210 | 9830 | 6611 | 6708 | 9767 | 9734 | 8001 | 9614 |

|  | Running Water | Power Machinery | Water in Kitchen | Sink and Drain | Washing Machine | Carpet Sweepers | Sewing Machines | Screened Windows & Doors | Out-Door Toilet | Bath Tub |
|  | Percent | Percent | Percent | Percent | Percent | Percent | Percent | Percent | Percent | Percent |
| Eastern States | 39 | 12 | 67 | 80 | 52 | 58 | 94 | 95 | 87 | 21 |
| Central " | 24 | 29 | 47 | 52 | 67 | 46 | 95 | 98 | 93 | 18 |
| Western " | 36 | 22 | 18 | 44 | 49 | 29 | 95 | 91 | 86 | 23 |
| Average | 32 | 22 | 48 | 60 | 57 | 47 | 95 | 96 | 90 | 20 |
| No Records | 9320 | 9080 | 6949 | 9334 | 9472 | 9513 | 9560 | 9667 | 9580 | 9784 |

What the farm woman needs, says the Department of Agriculture, and what she can be taught to buy, is more labor-saving equipment. In equipment the farm home is far behind the field or even the barn, with the result that the farm woman's task is no easy one. The Department looks to manufacturers and advertisers to do their share in improving conditions in the farm home; just as they have in other parts of the farm.

total cost of power delivered, consists of the operating, maintaining and fixed charges of the lines and equipment required to deliver the power to the users in the district. Consumption charges will be determined by a meter at each customer's premises, which will measure the quantity of power used to which a suitable rate will be applied. This cost can only be arrived at when the amount used has been determined. The rate used in the district will be determined by the cost of power at the transformer station supplying the district. The power supplied to the district will be metered at the transformer station.

The meter rates for users in that part of your township will be supplied from..........................................
are estimated as follows:—

5 cents per kilowatt hour for the first 14 hours use per month of customer's class demand rating.

2½ cents per kilowatt hour for all remaining uses.

Less 10 per cent for prompt payment.

The following table gives class demand rating, average monthly kilowatt hours, estimated consumption charge, estimated service charge, and total estimated annual cost for each class:—

Rural contracts have been developed to a form that agrees in principle and application to that which is in use in urban municipalities, excepting that in the former the liability is assumed collectively and benefits distributed individually through the municipality as a whole. Most service in rural districts is being supplied under Part II and Part IIB

on account of the war, and by reason of high costs and unsettled conditions was not undertaken again until this year.

At the end of the fiscal year 1920 the Commission had
Farm services ..................... 1136
Hamlet services .................. 857
Suburban services ............... 4103
Suburban service on local records,
  about ......2000
                                    ————
    Total ................. 8086
Miles of line ............. 502.74
Townships in which service is given..81

During 1920, however, the legislation regarding rural distribution was revised with a view to more extensive construction than had taken place before. During 1919 and 1920 preparatory surveys were made over a large area in 68 townships, divided as follows:
Niagara system ................. 30
St. Lawrence system ............ 14
Eugenia system ................. 3
Severn " .................. 3
Wasdell system ................ 7
Rideau " .................. 3
C. O. S. ....................... 6
Ottawa system .................. 2

Based on these surveys districts were laid out and submitted to the Government for approval.

Rates were approved for these districts on the assumption that a Government bonus of 50 per cent would be obtained and that the number of farmers signing contracts would not be less than three per mile.

The total number of townships included by these districts, and the number of townships adjacent to load centres, which might be served from H, E. P. C. lines, is 341 and the area covered is 17,365 acres.

tion it is possible to develop a system, where it was not possible as a township system.

The results of the meetings and canvasses by committees with the assistance of district engineers has been about 3500 rural contracts since June 16th, 1921, a large percentage being contracts for farm services, the balance hamlet and suburban services. Below is a list of rural construction approved.

**TABLE NO. II.**

| Class | Demand Rating K.W. | H.P. | Average Monthly K.W.H. | Est. Annual Consumption Charge | 25 Cycle Systems Est. Annual Service Charge | Total Est. Annual Cost | 60 Cycle Systems Est. Annual Service Charge | Total Est. Annual Cost |
|---|---|---|---|---|---|---|---|---|
| I Hamlet Service. .... | (a) 1/2 | 2/3 | 10 | 4.68 | 17.59 | 22.27 | 17.08 | 21.76 |
|  | (b) 3/4 | 1 | 15 | 6.84 | 20.50 | 27.34 | 19.74 | 26.56 |
|  | (c) 2 | 2-3/4 | 150 | 48.12 | 36.44 | 84.56 | 34.42 | 82.54 |
| II. House Lighting ... | 1 | 1-1/4 | 15 | 7.92 | 30.05 | 37.97 | 25.40 | 33.32 |
| III. Light Farm Service | 3 | 4 | 40 | 21.60 | 60.82 | 82.42 | 55.76 | 77.36 |
| IV. Medium single phase farm service | 5 | 6-2/3 | 70 | 37.80 | 66.94 | 104.74 | 60.00 | 97.80 |
| V. Medium 3 phase farm service.... | 5 | 6-2/3 | 70 | 37.80 | 84.50 | 122.30 | 76.04 | 113.84 |
| VI. Heavy farm service | 9 | 12 | 150 | 74.52 | 130.97 | 205.49 | 115.79 | 190.31 |
| VII. Special " " | 15 | 20 | 300 | 187.76 | 188.90 | 326.66 | 170.08 | 307.84 |

The above costs are calculated from our knowledge of the use of electric power in rural districts under average conditions. They have been adjusted by applying the rates as set out herein.

The rates will be re-adjusted by the Commission from time to time in your district to cover cost. Increase in the average number of farmers per mile or lower cost of power will reduce the annual costs to all.

Authorization having been received with regard to these districts, meetings were advertised and held in all prominent centres, in order that the new procedure and rates might be explained to the residents of these townships. The first meeting was held on June 16, 1921 and to date the following meetings have taken place—

Niagara system .................. 155
Eastern system ....................14
C. O. S. ......... ............. ....20
Northern systems ................ 100
                                   ———
                                   289

At about 50 per cent of these meetings committees were elected to carry on the work in the section covered by the meeting, including the making of the canvass for contracts; this has been the practice of the commission for some time, the endeavors of the committees being directed by the engineer for the district, keeping their efforts concentrated to a definite route, to a distribution centre, which in most cases is an urban municipality now being served.

The basis of former settlements submitted on the receipt of a petition was the petition, with an application of ordinary horse sense. This was in many cases found to be a wrong assumption—I do not mean the application of the horse sense —but the petitions did not represent the amount of business that could be secured when a canvass was made for contracts it was found necessary to establish a minimum amount of business as a basis for rates, to work from or to. It was also found that the 3 h.p. contract (Class 3) was more popular than the 5 h.p. contract (Class 4 or 5), as the total annual cost was less, therefore for the primary portion of this annual first charge they were put on the same basis and the minimum business referred to was placed at 3 of either of these classes per mile, or their equivalent, as an average for the district.

Under the legislation of 1920, distribution is made in districts or zones regardless of the geographical lines of townships. Under this more economical method of distribu-

for which the bonus has been approved. Canvasses in many prospective and existing systems are not yet complete, and are being carrried on by the local committees.

Construction is under way in quite a number of the districts approved, favorable fall weather permitting this. (It is the policy of the Commission to avoid as far as possible winter construction.) Expert labor is supplied to direct the construction and install the equipment.

Following is a list of rural systems and extensions under construction, or approved to contruct.

Shortage of poles in Old Ontario has forced the Commission to underground construction for rural service. In 1921 less than half of the poles required were secured in Ontario,

**TABLE NO. III.**

Rural Systems and Extensions under Construction, or Approved to Construct.

| Systems | Rural Power District | Consumers | Miles of Line | Load-K.W. |
|---|---|---|---|---|
| Niagara...................... | 13 | 1198 | 168—35 | 2050.5 |
| Eugenia...................... | 2 | 22 | 3.16 | 23.5 |
| Severn...................... | 1 | 60 | 7.6 | 85.5 |
| St. Lawrence................ | 3 | 88 | 17.75 | 131.5 |
| Wasdell..................... | 3 | 4 | 1.6 | 14. |
| Ottawa...................... | 1 | 76 | 18.61 | 194.5 |
| | 23 | 1448 | 214.57 | 2499.5 |

Rural Power Systems and Extensions Approved for Construction and Bonus Approved.

| System | Townships | Consumers | Miles of Primary | Trans. Capacity K.W. | Total Estimated Cost | Bonus |
|---|---|---|---|---|---|---|
| Niagara.... | 13 | 1094 | 173 | 1995 | | |
| Ottawa.... | 1 | 78 | 18.61 | 204 | | |
| Eugenia.... | 4 | 46 | 8.85 | 83 | | |
| Severn..... | 1 | 99 | 17.75 | 138 | | |
| St. Lawrence | 4 | 80 | 15.25 | 123 | | |
| Wasdell.... | 1 | 4 | 1.58 | 14 | | |
| | 24 | 1401 | 235.04 | 2577 | $506,020.06 | $146,867.16 |
| Additional Capital applied for— | | | | | $ 75,479.44 | $ 22,473.98 |

the balance coming from Quebec, New Brunswick and British Columbia. This is largely the cause of a great difference in cost of material for a standard rural 3 phase mile of line, on poles, the range being from $383.52, for material only, in 1911, to $843.94 in 1921 approximately 50 per cent of the cost of material per mile this year is for poles laid down.

is made, using plows, with a minimum depth of 12 inches. The cable is laid in without a duct, the fill being made by using a scraper. The Commission, appreciating that load density is low, and construction must be under way in many districts at the same time, has endeavored as far as possible to use equipment which can be secured locally thus decreas-

### TABLE NO. IV.
### TABULATION OF SOME ACTUAL FIGURES ON RURAL SERVICE.

The following series of tables are compiled from data taken in Brock Township, Ontario County, and are for the purpose of illustrating comparatively what can be accomplished in any rural community.

### Table Showing Comparative Data on Installation For Rural Electric Service

| Farm No. | I | II | III | IV | V | VI | VII | VIII | IX |
|---|---|---|---|---|---|---|---|---|---|
| **House Wiring** | | | | | | | | | |
| No. of Outlets.................. | 18 | 23 | 21 | 29 | 12 | 41 | 26 | 28 | 20 |
| Cost of Wiring................. | $200.00 | $190.00 | $100.00 | $115.00 | $100.00 | $130.00 | $ 75.00 | $100.00 | $ 75.00 |
| Cost of Fixtures............... | $ 10.00 | $ 75.00 | $ 80.00 | $ 18.00 | — | — | — | — | — |
| **Barn Wiring** | | | | | | | | | |
| No. of Outlets.................. | 17 | 22 | 16 | 17 | 13 | 20 | 16 | 15 | 11 |
| Motor........................... | 1 | 1 | 1 | 1 | 1 | 1 | 1 | 1 | 1 |
| Cost of Wiring................. | $151.50 | $150.00 | $170.00 | $158.73 | $138.00 | $200.00 | $175.00 | $166.00 | $145.00 |
| **Motor Installation** | | | | | | | | | |
| Sizes in H.P.................... | 5 | 8 | 5 | 5 | 5 | 5½ | 5 | 5 | 5½ |
| Cost........................... | $148.50 | $378.00 | $165.00 | $149.50 | $176.00 | $220.00 | $165.00 | $155.00 | $235.00 |
| **Line to Road** | | | | | | | | | |
| Length in Rods................. | 18 | 20 | 255 | 36 | 60 | 24 | 60 | 49 | 24 |
| Poles included................. | — | — | — | — | — | — | 2 | — | — |
| Cost of line................... | $ 30.00 | $ 50.00 | $119.34 | $ 63.45 | $265.00 | $ 49.56 | $100.07 | $ 53.87 | $ 54.34 |
| Total Cost..................... | $440.00 | $843.00 | $634.34 | $504.76 | $679.00 | $599.56 | $515.07 | $474.87 | $509.34 |

The Commission have laid out and have under way underground construction in the following township—

Salt Fleet ....................55 miles
Beverley .................. ........3½ miles
Ancaster ...................... 7 miles
Niagara .................... ... 3½ miles
Willoughby and Bertie ........8 miles

ing the cost. Farmers in the vicinity of the roads along which lines will be built, are assisting in the work, and are being paid at the current wage of the district; foremen, linemen and wiremen are brought in. It is possible methods of installation may be changed as the work progresses. The obstructions that are met with in districts are tree roots, boulders, and soil conditions. It would seem, from progress

### TABLE NO. V.
### TABLE SHOWING COMPARATIVE OPERATING DATA ON RURAL ELECTRIC SERVICE

| Farm No. | I | II | III | IV | V | VI | VII | VIII | IX |
|---|---|---|---|---|---|---|---|---|---|
| **Rates** | | | | | | | | | |
| Annual Service...... ... ...... | $60 | $60 | $36 | $36 | $60 | $60 | $60 | $60 | $36 |
| 1st Meter Rate............. | 7c. | 7c. | 7c. | 7c. | 7c. | 7c. | 7c. | 7c. | 7c. |
| **Annual K.W.H.** | | | | | ¾ yr. | | | | |
| In lighting .................. | 130 | 946 | 295 | 387 | 88 | 1138 | 434 | 604 | 276 |
| In power.................... | 284 | 1219 | 702 | 347 | 119 | 681 | 398 | 514 | 278 |
| **Work Done** | | | | | | | | | |
| Bu. chopping................. | 1000 | 5000 | 3000 | 1500 | 175 | 4000 | 3000 | 2400 | 1500 |
| Root pulping................. | 1200 | 35 Hrs. | — | 3000 | 700 | 3000 | 500 | 3000 | 4500 |
| Hrs. Milking.............. | 90 | 150 | — | — | — | — | — | — | — |
| Separating.........'...... | yes | yes | yes | — | — | — | — | — | — |
| Pumping........ .......... | yes | yes | — | yes | — | yes | — | 175 hrs. | 400 hrs. |
| Feed Cutt'g............ | — | yes | yes | — | 15 hrs. | 60 hrs. | 30 hrs. | 20 hrs. | — |
| Total Cost..... .............. | $85.12 | $196.10 | $95.15 | $79.45 | $26.48 | $151.07 | $110.57 | $126.49 | $73.78 |

part of this being three phase and part single phase, taking in all about 115 miles of cable. The cable is lead encased rubber insulated with stranded copper conductor. A trench

reports, and estimates, that 3 phase line underground will cost more than a similar line overhead, that single phase line underground will cost less than a similar line overhead, and that for a system with one-third three phase and two-thirds single phase underground, the cost of underground distribution would probably be lower than overhead. Most services from this system are overhead, transformers erected on a short pole at a suitable location for the service, or services and "run-offs" made to point of delivery to the premises.

This paper would not be complete without some reference to electric service in rural districts from other sources. The individual plant driven by gasoline engine with or without storage battery, is commonly known in Ontario, some having been installed for 9 years. These serve for lighting only; other uses, as a rule are a serious tax on most of these plants, and only result in impairing it for its most important use—lighting. By reason of the high depreciation of plant and battery, and the maintenance, the cost of service per unit of work or of electricity is quite high.   Recently a pamphlet on wind-produced electric power—not the "hot-air" kind commonly used—but a plant driven by a windmill, came to our notice.   It is manufactured in one of the Western states, costs $850.00 f.o.b. factory, has a capacity of 1.5 kw., and they claim in that State would generate and store 1013 kw.h. in the year. From our experience, the Class 3 consumer uses an average of less than 500 kw.h. per year. This amount of electricity in the average district as served by us, including all charges, would cost an average of 17.1 cents per kw.h., and if supplied by wind service, would cost 55.7 cents per kw.h.  Rankin and others give data which show costs per h.p.h., 3½ to 15 cents, the latter cost being for 8½ ft. mills, the size commonly used.  Individual plant users in each district are almost invariably the first to sign for Hydro-electric power.

The farmer as a user is usually more liberal in his uses, has a larger installation and more possible uses than urban customers.  The appended list of installations, cost of installation, annual uses, annual cost and work done are included as a part of the paper for your use as a reference if interested.

Syndicates have been formed in different parts of the province, whereby a number of farmers form to acquire, own and operate a large motor and outfit for doing their heavier work.  These have been a success from their point of view, and when operated on a system which is not too small, would be a benefit to the system by reason of the increased uses, if there be a sufficient number served by the syndicate.  In this we again come to the principle which must be kept in mind when considering rural services, viz.:—"Make the maximum use of the service when it is installed, and keep the peak low.—Good load factor."

From the manufacturers and distributors' viewpoint the rural business is no less a problem than from ours.  The residents in rural districts from time immemorial have been the butt of the "gold brick salesman".  So be careful at the start, and we believe the field now open for business will enlarge greatly.  The business is there but the rural resident is rather "gun-shy" of the city and large town salesmen; gain their confidence by sensible dealings, "don't advise them to use vacuum cleaners for the cows in cases where the stables need attention."  Local men in most cases can secure the business better than outsiders, and have the advantage of knowing the people, their faults and their virtues.

In conclusion, as the preacher says, the slogan "Do it Electrically" can now be applied in the country as well as in town, and the Hydro principle of "Co-operation" should be specially emphasized in approaching and dealing with farmers as a class. . The Commission will be pleased if you can help us in getting information to people in your several districts adjacent to the urban municipalities you represent, we for our part will endeavor to supply you from time to time with such information as we can.

## Dominion Engineering Agency, Ltd.

The Dominion Engineering Agency. Ltd., has just been formed with head office at 24 Adelaide St. E., Toronto, with Mr. Daniel M. Fraser as president.  In the meantime they are operating as manufacturers' agents but eventually, we are advised, they will handle anything of an engineering nature.  The firms represented include the Cutter Electrical & Mfg. Company, Philadelphia, (Eastern Canada); Schweitzer & Conrad, Inc., Chicago, (Ontario) and the Ferguson Manufacturing Co., London. Ont., (Eastern Canada). The first named company manufactures the well known type of circuit breakers; the second company specializes in high tension equipment, and the London firm manufacture Conduit Pipe fittings.  A further line soon to be added is the "Fraser"

Mr. D. M. Fraser

patent attachment plug cap, an invention of the president of the company, which will be placed on the market in the near future.

Mr. Fraser is very well known in electrical circles as estimating engineer and superintendent of construction for a number of years with the Canadian General Electric Company.  He was born and educated in Scotland and came to Canada in 1910.  For a time he was with the Montreal Light, Heat & Power Company and later with Mines Power, Ltd., Cobalt, (now the Northern Ontario Power Company).  Before joining the C. G. E. staff he also spent some 18 months on general electrical work.  He is a member of the Engineering Institute of Canada and a fellow of the American Institute of Electrical Engineers.  During the 1919-20 session Mr. Fraser was secretary of the Toronto Section of the A. I. E. E.

---

The Delta-Star Electric Co., Chicago, Ill., are distributing bulletin 32, describing a complete line of outdoor high tension switching and protective equipment. of Unit Type design. A copy will be sent on request.

---

Descriptive leaflet 3443 issued by the Westinghouse Electric & Manufacturing Company describes its new luminous top holder socket reflector.  These reflectors have several advantages, such as simplicity, easy to wire, mechanical strength and complete interchangeability of parts.

---

The Wheaton Electric Co., 318 Donald St., Winnipeg. has been awarded the electrical contract on a store building recently erected at Rosedale & Cockburn Sts., Winnipeg, for Mr. A. T. Menzies, 381 Roseberry St., St. James, Man.

# The Electrical Contractor

## Is The Contractor a Merchant ?

### Occupations of Contractor and Dealer as Distinct as Those of Blacksmith and Doctor

In our Jan. 1 issue appeared an interesting discussion on the question of whether a contractor-dealer is a possible combination or whether it is not better for the contractor to stick to his business and the merchandiser the same. The idea is gaining ground that it requires a different type of man to make a success of these two activities. A contractor reader has just taken the trouble to call up the editor to emphasize his own viewpoint in the matter. This contractor at one time had the idea of building up a merchandising business, but soon found that he was losing his grip on contracting without getting a corresponding hold on merchandising. He was wise enough to give up the idea of being a merchant and has concentrated on contracting, not forgetting, of course, to place an occasional appliance with a customer whose premises he may have wired; but he makes no attempt to "keep store" and has no sales organization.

A thorough discussion of these matters seems to be the only way to get at the facts. Below we reproduce another excellent viewpoint—an article published in the "Electrical Record" for December. We should be glad to have our readers' add their own viewpoint to the discussion:

A good deal has been said and written as to whether the electrical contractor should or should not also be a dealer in electrical appliances.

This question is easily answered in the case of the contractor-dealer who is successful; of course this type of contractor should also be a dealer, because he has proved that he is on the right track. And it is just as easily answered in the case of the contractor-dealer who has not met with success; there can be no question but what he should not continue in the joint business of contracting and merchandising, for the opposite reason, namely, that he is not making money.

All this would be quite simple were it not for the fact that it is also necessary to know in determining this question, whether the success or failure of the contractor-dealer is due to his aptitude or inaptitude in handling the contracting or the merchandising part of his business.

An investigation of such business, whether successful or unsuccessful, would soon indicate which part of the business was receiving the most expert attention. If the contracting end of the business was being better looked after due to the proprietor's knowledge of contracting, and the merchandising end was left to more or less shift for itself, and the business was successful, then one of two things is evident; either some of the good profits of the contracting business are making up for the unprofitable merchandising or there are sufficient profits in the merchandising end of the business so that it pays without close and expert supervision. In either event there isn't much to worry about as the business is successful and the creditors are being paid and the bank balance is growing—and everybody is happy.

#### The Unsuccessful Business

In the case of the unsuccessful business, however, such an investigation, while revealing the fitness of the proprietor for

contracting or merchandising, tells a different story; the business is not successful. Is it because of the contracting or because of the merchandising or because of both not being properly handled? If the last named is the case then there is no hope; the contractor-dealer should connect with someone else in business to whom he might render successful service and receive satisfactory compensation until he can find his weaknesses and overcome them either as a contractor or as a dealer rather than buck the tide of business before he is ready. If his failure is due to lack of knowledge of merchandising it is plain that he should drop this part of his business and stick to contracting only, or if his knowledge runs to merchandising and he knows little or nothing of the practical side of contracting then he had better let that part of his business go and attend to merchandising exclusively, because therein lies his best opportunity.

The human mind runs along certain channels; it has a tendency to perform a certain kind of work and to shirk another. That is why we have good doctors who like their work and are fitted for it and meet with success, and poor doctors who might have made good lawyers or carpenters.

The occupations as represented by the contractor and by the dealer are about as distinct and as far apart as those of the blacksmith and the doctor. There is no saying that a man can't be a blacksmith first and then decide to be a doctor and after going through the prescribed course not be a good doctor; but it will readily follow that if he is a good doctor he will no longer be a blacksmith, also that if it turns out that he is not a good doctor that he may return to his smithy at which he was proficient enough to earn a livelihood. It is safe to say, however, that he will not practice his trade as smithy and profession as doctor at the same time.

While the comparison may be far-fetched it is used to illustrate the point that the mind which runs to contracting runs away from merchandising and the merchandising mind acts likewise.

#### Exceptions Only Prove the Rule

It is true that cases will be found where one takes to both, but such cases, either mediocre successes or the rare geniuses, are so few that they are but the exceptions which prove the rule.

Therefore, it would seem that from the mental angle a clear distinction exists between the contractor type of mind and that of the dealer, and the contractor-dealer can determine by self analysis which direction he should take.

The mental angle, however, is not the only one to consider. We have also the financial burden to think of.

The contractor has his investment in material and labor and expense, and as a rule this outlay is heavy enough to keep him "humping" to meet his obligations as they mature. This is enough to keep him from adding on side lines that call for more money, the providing of which keeps him constantly in hot water and prevents him from functioning as he should in the conduct of his original business.

The same applies to the dealer. He has all he can do to have enough money on hand to meet his obligations for merchandise and selling and administrative expenses without taking on the additional burden of employing a construction crew on work with which he is not thoroughly familiar.

Granted, however, that the financial condition of either a contractor or dealer is such as to permit of expansion and it appears in the case of the contractor that the taking on of merchandising or in the case of the dealer the adding of a contracting business would utilize this idle capital to advantage, in most cases a thorough investigation will undoubtedly reveal it will be more sound to extend the business already operating than to branch out into another more or less unknown field.

### The Only Way to do Both

The exception to this rule will be found in that type of individual, who, having made a success in one line of business, organizes his business, places a capable individual at the head of his original business as well as another at the head of the new department, and then devotes his own time to supervising both, to the end that each pays its way.

When a man gets beyond the one-man stage it does not go into a two-man class, but rather skips that and takes the three-man organization. It begins to pyramid, and as in the case of the geometric figure of a pyramid the base is wider than the top; so in a business organization when pyramiding begins the base must widen to withstand the super-structure's weight.

In addition to the executive type of mind there is also required a mind that lends itself to the field of accounting, because while a one-man business can be kept 'under one's hat," a business organization must have records that always completely tell the story of "whither art thou going" and at what pace, or the rocks may loom ahead.

### The Public Point of View

So much for the contractor-dealer question from the inside angle. Now, from the outside point of view—from that source which controls business, which really determines whether it shall be a success or failure: The buying public.

If in a given community there is a contractor who has added a small display of merchandise to help meet the expense of operating his contracting business, and in another community there is a dealer who has a larger display of merchandise and does a little contracting on the side: enter the buying public. Which can give the better service. The contractor-dealer who is long on contracting and short on merchandise, or the dealer-contractor who is long on merchandise, but short on construction work? The answer is evident: If the public wants construction work it will be better served by patronizing the contractor-dealer, and if it wants merchandise then the dealer-contractor can render better service. The contractor-dealer has forgotten more about the requirements of construction work than the dealer-contractor will ever learn—it's his business to know it; and the dealer-contractor having a larger and more complete line of merchandise, more up-to-date, paying more attention to the business can render better service in that direction than the other.

Therefore, if the public wants the best service in both requirements it had better patronize the contractor for its construction work and the dealer for its merchandise.

It follows then, that if the public is to best served, the contractor, in adding a merchandising business to which he can't give the same expert attention as he does to the contracting, of which he has made a success, can't properly serve the public and has only added an overhead burden for his pains. It is the same in the case of the dealer, who, if he maintains the proper records, will soon discover that in lace of the revenue from contracting which he expected he

will be taking some of the merchandising profit to offset the new departure's loss.

### The Conclusion

And so the answer follows: where the community can support both contractor and dealer, let each confine his entire efforts to his own business which he knows well and which, by concentration, he can learn to know better.

Thoughtless and destructive competition which is bound to result where everybody is endeavoring to do somebody else's business, thereby neglecting his own, never served anyone well. Thoughtful and constructive co-operation, on the other hand, has given us all we have to-day and shall continue to be of the greatest service and benefit to all mankind in the time to come.

### The Wiggett Electric Co.

The Wiggett Electric Co., Ltd., of Sherbrooke, Que., incorporated under the laws of the Province of Quebec, have taken over the business of the Electric Repair & Supply Co. They have erected a new building, thoroughly up-to-date, at a cost of $20,000. This is a 3-storey building which will enable the company to carry on, in all its phases, a thoroughly modern electrical contracting and repair business. One section of the building is set aside for motor repairing and another for general electrical repairs, both of which are well equipped with all the latest machinery. The president of the company is Mr. W. J. Wiggett; vice-president, George Pearson; director, N. L. Wiggett. The company reports that business has been good and is to-day surprisingly good for this time of the year.

### Have Opened Electrical Department

The Cross & Sutherland Hardware Company, Ltd., Hanover, Ont., with a branch store at Durham, have recently established an electrical contracting and electrical merchandising department and are carrying an up-to-date line of electrical appliances and fixtures. Their contracting department has just been awarded the contract for reconstructing the electrical service and adding an additional 40 h.p. motor to J. Metzgar's grist mill at Neustadt.

### Short Course in Testing

"A short course in electrical testing," in its fourth edition, is published by the D. Van Nostrand Company of New York. It is written by J. H. Morecroft, E.E., B.S., and F. W. Hehre, E.E. This book covers the course at Columbia University designed for students in mining, mechanical, metallurgical, chemical and civil engineering whose courses do not include the theory of electrical machines but who, of necessity, come in contact with them and utilize them. It has thus been the intention of the authors to present the subject matter in such a manner that the student not well versed in theory can get the most out of his course in the shortest possible time. In the fourth edition certain experiments have been added on batteries, illumination, measurement of electrical energy, etc., with the idea that the non-electrical engineer who frequently has to pass judgment on these phases of electrical installation should have a general knowledge of these features of electrical engineering. 220 pages, well illustrated; green cloth cover; size, 6 in. x 9 in.; price $3.00.

The Swedish General Electric Co., Toronto has been awarded the contract for supplying the city of Winnipeg with exciter generators. The contract amounts to approximately $23,000.

# Hamilton Memorial Arts Building Equipped Electrically Throughout

The Hamilton Memorial Arts Building has just recently been completed, and contains much that is of interest to the electrical contractor. The building is situated in the school grounds adjoining the large 28 roomed Memorial Public School.

Deputations from all over Canada, and from the United States and Australia have visited these buildings and have pronounced them to be the most modern and complete of their type ever inspected.

The buildings are entirely fireproof construction throughout, with fireproof stairways towers. Large window area is provided to obtain the maximum amount of daylight in the classrooms.

A complete system of artificial lighting is provided as shown in the photographs. The wiring throughout is installed in concealed conduit work. The lighting and heating (cooking) circuits are entirely separate. The service supplying the building is installed underground. This service is connected to the switchboard in the sub-station of the main building.

The electric heating system is controlled by a special Krantz panel located in the household science director's private office. Thus, no person is permitted to use any of the electric heating or household science apparatus without permission from the director. Further, it prevents any of the apparatus being left connected, as the director can pull the main switch on this panel and disconnect all outlets.

This panel is provided with double doors to make it dead front and prevent unauthorized persons from tampering with the fuses. The doors are provided with locks and keys. All conduits leading to this panel are concealed.

The household Science Department consists of a bedroom, dining-room, kitchen, laundry and pantry fully equipped. The kitchen is equipped with an electric range. The laundry is equipped with electric irons, and ironing boards, washing machine and a clothes dryer. Stationary tubs complete the equipment of this very important department.

The Domestic Science and Cooking Department is on the second floor. By referring to the illustration, you will note how completely this room is equipped. Each student

has a combination gas and electric hotplate with stool and cupboard. These hotplates were manufactured specially for this installation. Each student is provided with a towel rack at the rear of the table, as shown in the picture.

The students are assembled in a square. At each corner of the square is a special table, mounted on casters, which is used similar to a serving table or dinner wagon. A large size range is shown in the corner of the room. With the combination gas and electric stoves the students can be instructed in the use of the two types of equipment.

It is needless to say that the students have shown a strong preference of the electric appliances owing to their cleanliness and convenience and there are no fumes.

The wiring for the hotplates is all mounted under the tables connected to waterproof floor boxes, as the floor is terrazzo in order that it can be washed. Artificial lighting has been provided for night classes, if required.

The Manual Training Department is very complete for the teaching of this art. Wiring has been provided for motor driven wood working tools which are to be installed later. This wiring is all concealed in the floor, with waterproof boxes, and will provide means of connecting to each motor on each machine.

The control for these motors is provided by enclosed motor switch mounted concealed in the wall. The tables are constructed of steel and each student has a very complete supply of tools. The lighting is provided by Benjamin R. L. M. Industial type units. Each student is provided with a steel locker for his apparatus. A storeroom for lumber, paints and varnishes is provided. All painting is done in a separate room. With the equipment used in this school the children should receive a very complete training before passing on to the collegiate and university. The building, is located in the section of the city which is inhabited by the middle class of people who are employed in the large number of industries in this section of the city.

Culley & Breay installed the electrical equipment in this building. Mr. Gordon Hutton, Bank of Hamilton Building, was the architect. The complete electrical installation, was designed and supervised by V. K. Stafford, consulting electrical engineer, Sun Life Building, Hamilton.

Hamilton Memorial Arts Building—View of Domestic Science and Cooking room to left

# A Golden Opportunity

Study the picture on the opposite page.

A wonderful airplane view of the Chippawa-Queenston development, you say.

Yes, it is, but it also means much more than that.

It represents the Ontario electrical industry's **opportunity**—the finest opportunity the province has ever had—the finest opportunity any province ever could have—ample electric power for years to some.

What opportunity and whose, you ask?

1.  To the citizens at large, comfort, convenience, leisure, health—a distinctly higher plane of living.
2.  To the contractor who equips the home—profitable business.
3.  To the merchant who sells appliances—profitable business.
4.  To the wholesaler who stands behind the contractor and merchant—profitable business.
5.  To the manufacturer behind the wholesaler—profitable business.

Think of the magnitude of the opportunity—50,000 h.p. available today; 500,000 h.p. just as soon as we can take it!

Our duty today and tomorrow, and for many days to come, seems perfectly clear. Everything depends on the attitude of mind of the men and women in our homes. If the viewpoint is "We want electric homes" there is no force that can stand against it. They will come and demand electrical appliances just as they now demand shoes and hats—but with greater enthusiasm.

Isn't it our plain duty to concentrate our efforts, to co-ordinate our efforts, to the sole end that men and women throughout the length and breadth of Canada shall rise in a body and

## DEMAND ELECTRIC HOMES

# Airplane View of World's Largest Power Scheme

A wonderfully realistic Bird's-eye view of the Chippawa-Queenston development of the Hydro-Electric Power Commission of Ontario showing Canal, diffuser, forebay, gatehouse, penstocks, powerhouse and Niagara River.

# Electrical Leagues and Associations

## Organizations Throughout Canada Doing Useful Work in Endeavor to Revive Trade and Improve Merchandizing Conditions

Mr. Rey E. Chatfield, secretary-manager of the Electrical Service League of British Columbia, is starting the year right with a vigorous campaign among the contractors and dealers of the province to improve their methods of doing business and increase their turnover by following the lead of older established lines of trade who, through many years of experience, have become more expert and exact in the science of merchandising than has been possible in the few short years during which the electrical industry has been a factor in the industrial world. Mr. Chatfield has sent out some very helpful letters to the League membership. Two of these have come to our attention and we believe that in passing them along to the contractor-dealers in other parts of Canada we shall be doing a real service to our readers. Here they are:—

### I—VISUALIZE YOUR JOB

The story of the three stone cutters leaves nothing of wisdom to be said. They were working on a stone. A stranger asked the first what he was doing. "I'm working for $7.50 a day," he replied. "And you?" the stranger asked the second. "I'm cutting this stone", growled the laborer. Then the question was put to the third stone cutter. He answered, "I'm building a cathedral."----The Christian Register.

At the end of the old year, Mr. Contractor-dealer, it is necessary to summarize the effort of your business for the past year. The facts may not be pleasant as shown by the yearly statement but how do you visualize your job? Are you working for seven dollars and a half a day, are you making wiring installations or are you building a business on a sound foundation?

If you are working for a daily wage; resolve to take more pride in the growth of your business: if you have faith in your business the public will have faith in you.

If you are merely selling wiring installations decide now to build a better business by rendering a real service to your customers.

If your are building a business, face the facts; during the year you have done a certain volume of business; the expense of operation has been a certain amount, which bears a definite relation to the volume of business which can readily be translated into a percentage figure. Use this figure in pricing jobs or merchandise and resolve to take no work at less; then adhere to your decision.

If you can do no better than make wages; why continue to tie up your capital in order that you make wages, far better would it be that you discontinue business and work for some one else.

Do your books show you an exact cross section of the condition of your business? If you are not sure of this, let the Service League help you decide this matter by assisting you to read the facts.

These are straight unvarnished facts that must be faced at this time of the year. Let's face them and decide upon a business policy that will bring renewed faith in our business. If we face the actual conditions and all work together for the betterment of the condition of the business 1922 will be a banner year for the electrical industry of British Columbia.

### II—GOGETIT

A country simpleton—not so simple as he looked—was sent out to hunt for a lost cow that had been unavailingly searched for by the entire community. To the amazement of all a few minutes after he started out he returned, driving the cow. The astonished bystanders immediately wanted to know how he had found the "critter."

"Well, I just set down and thought of the place I would go if I was a cow", he explained, "and then I went there— and there she was!"

The way to find business to-day is to figure out where business is likely to be--- and Gogetit (Forbes).

The above story opens great possibilities for the Electrical Industry in British Columbia for the New Year. The field for use of labor-saving devices, better lighting, and electrical industrial devices has been scarcely touched in British Columbia. Do not sit in your office and wait for business to come to you. Look the situation over and sell yourself the idea that a particular device will save your "prospect" money, by cutting production costs or by speeding up work; then tackle him with the proposition that by the purchase and installation of a particular device he will save money. In times like the present a man looks twice at a dollar before he spends it. He must be convinced that he will get value received for his expenditure.

Sit down and think what you might sell your customers or others, which will be a profitable investment for them. Then go get the business.

The Service League can be of great assistance in helping you gather ammunition for the attack on the reluctant prospect.

The Service League can be of assistance through the field man in helping you with your accounting and merchandising problems.

Use the League.

---

## Niagara Peninsula Contractors

The first regular meeting for 1922 of the Niagara Peninsula Branch of the Ontario Association of Electrical Contractors & Dealers held in the Chamber of Commerce rooms, St. Catharines, Ont., on Monday, January 9, was well attended by both members and outsiders. After the regular business had been disposed of, the chairman called on Mr. C. D. Henderson, of the Henderson Business Service, Brantford, Ont., to address the meeting.

Mr. Henderson chose as his subject, 'Making the Business Pay;" prefacing his address by congratulating the contractor-dealers on the co-operative spirit prevailing among members of the Association and their very evident desire to boost the electrical industry. He emphasized that making the business pay was the first fundamental for which we are all striving, some with success because they know what it costs to do business, others practically working for nothing because they guess at the costs and fail to correctly analyze them. The speaker outlined the items which constitute "Overhead" by means of five graphic charts showing the various estimated expenses for contractors with a $5,000 to $100,000 turnover per year. Members present were asked to criticize the various items constituting this expense but all were of the opinion that Mr. Henderson's figures were very conservative.

The Henderson Business Service, which has been developed to a remarkable degree by the energy and ability of Mr. C. D. Henderson himself, is one of the most valuable assets a contractor-dealer can have. The cost of this service may appear considerable to the man who does not realize its worth or who doesn't realize his own necessity, but we have yet to hear of the contractor-dealer who has made a study of Mr. Henderson's system and has not become a convert to its value. Contractor-dealers all over Canada are working hard, long hours, enthusiastically, and yet many of them are not making progress, simply because they are working in the dark, using wrong methods and overlooking essentials. To all such we urge a trial of the Henderson Business Service.

## Kitchener Exhibit May 1—6

The cities of Kitchener and Waterloo, Ont., will stage an exhibit in Kitchener during the week of May 1 to 6, to be known as the Twin City Electrical Exhibition. At the last regular meeting of the municipal council of Kitchener a resolution was passed heartily endorsing the exhibit and adding the rider that "Owing to this city enjoying the distinction of being the birth place of the present vast Hydro-electric Commission operations, it seems that the choice of Kitchener for this exhibition will be most fitting and popular and receive the co-operation and support of our citizens and neighbors throughout Western Ontario."

## Thermo Electric Ltd.

Thermo Electric Limited is the newest electrical manufacturing concern to locate in Brantford, Ontario. It is now manufacturing a number of water heating devices, notably the Thermo Electric single vent water heater. This heater is of the immersion type, requires no extra plumbing, and, it is claimed, no expensive wiring service. The heater is fully covered by patents and is approved by the H. E. P. C. It is being sold by the Northern Electric Company, Limited.

## A Unique Invitation Card

The Manitoba Electrical Association has hit upon a unique idea for sending out the invitations to their regular weekly luncheons. The cut herewith illustrates the idea. The actual invitation form is 7 in. long. The announcement itself is made on the reverse side of the card.

## Appointed Eastern Canada Representative

Mr. R. A. L. Gray has just received word from the secretary of the National Association of Electrical Contractors that he has been appointed Eastern Canada representative on the executive committee of the association. This position was formerly held by Mr. Kenneth A. McIntyre, who resigned on the occasion of his associating himself with the

Mr. R. A. L. Gray

Society for Electrical Development, in the capacity of Canadian representative. Mr. Gray is one of the best known contractor-dealers in the Dominion and is a representative of which the Canadian electrical industry has every reason to be proud.

## The Russell-Fowler Co.

A new firm of manufacturers' agents have just opened up an office, at 306 Notre Dame Ave., Winnipeg, under the name of The Russell-Fowler Co. Both Mr. Harold Russell and Mr. Ray Fowler have had considerable experience in the electrical business, Mr. Russell having until recently been on the purchasing staff of the Great West Electric Co., Ltd., Winnipeg, and Mr. Fowler on the sales force of the same company. They have some good lines at the present time but are in the market for a few more. Mr. Fowler will be calling on the Western trade as far as the Pacific Coast, while Mr. Russell will take care of Winnipeg and east to Port Arthur.

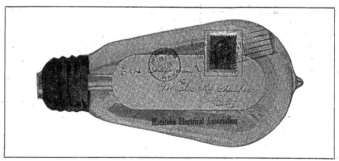

Manitoba Electrical Association

# Some Problems in the Industry

## We May Not All Think Alike But We Are All Working Toward the Same Goal—Confidence in One Another the First Step

### By W. B. DICKESON*

Our organization was instituted with certain ends in view and we, as members, are all aiming at the same mark. We have paid our fees and are all devoting more or less time to this association, so it is up to us to get our money's worth from it.

We are like ships on an open sea, bound for the same port and, as is often the case, steering our boats on somewhat different courses. It is only natural that each one of us thinks our particular course is the best one, but in so thinking it is better that we use our efforts to prove this than to criticise the other fellow for not aligning his boat along side ours for the entire course.

The idea of finding fault with the other fellow and forgetting our own interests in doing so seems to prevail to too great an extent among our members. Instead of using our club to bring the different dealers in our locality closer together, it seems to be, up to this time, driving us farther apart.

It is absolutely impossible for two individuals to think exactly the same regarding a common subject. We know that mechanics have their little hobbies as to their methods of finishing a piece of work which may be given to them. If one had to use another's tools, he would perhaps find fault as his own tools would be more familiar and easier for him to work with and therefore it would be quite natural for him to think that his were the best. At the same time another mechanic would be thinking exactly the same if he had to use someone else's tools.

It therefore goes to show that the same thing might prevail among merchants, or professional men for that matter. Now it has been proven in many instances through the formation of co-operative societies that these conditions can be overcome and the entire efforts centered on producing results which far exceed those of persons who find time to watch their competitors. Let us instead profit by what our friends are doing and work in harmony with them. If they should beat us in fair play, give them credit for it and take the incident as a lesson to ourselves.

You will notice, if you are a follower of athletic sports, that the coaches of different teams profit by the success of each other. Also the captain, if beaten, immediately collects the players and first of all they give three rousing cheers for the winners. What a fine spirit to exhibit among competitors in business.

It may be that your humble servant has not been making himself as plain as he would like to, but you will perhaps see from these comparisons what he is aiming at.

### Work in Harmony

Now we have launched out on a co-operative advertising scheme which is undoubtedly a step in the right direction. However, as it appears to me, the most important thing for us to accomplish at this time is a more united effort, working in closer harmony to the end that our line of business be accepted by the public on a higher plane than it has been in the past.

While it is farthest from our thoughts to establish any definite scale of prices for our merchandise, it would be remembered that in making a sale, we offer not only our goods,

but also the best possible service that we can give, charging a fair margin of profit for ourselves, at the same time giving our customers one hundred cents worth of value for every dollar they spend.

It does not require any sales ability to merely sell goods, but it does require considerable of it to make a sale and have yourself and your customer thoroughly satisfied in every way with the transaction.

### Pooling the Problems

If we as members of this Association can form a pool, and bring the different questions that confront us before our meetings for discussion, they will eventually be of profit to us inasmuch as we will know just how to grapple with them the next time they appear. You may have one that baffles you and another member may have already solved it. Why not thresh our problems out together? All the time you are getting better acquainted and to your surprise you find that your competitors are not such bad fellows after all. You know we all have some good points to our character, but as long as we stay apart and judge at long distance, we fail to discover any redeeming features. We think that because one of us has occasion to attempt anything other than the regular that he is trying to take some advantage and we do not allow ourselves to look upon it in any other light. It is just possible that if any other member found it convenient to attempt something of the same character that it would cause him to view the situation from an entirely different angle and form different conclusions.

I would therefore suggest that broad consideration be given these difficulties before any action is taken. Snap judgment on any matter is a poor policy and often leads to complications that otherwise could be avoided.

There is one condition however, that confronts us all and one which is worthy of generous consideration, and a recommendation on our part, backed by decisive action on our part, might work out for good.

### Resale Prices

This is the policy of many manufacturers of setting a resale price on their product which in some cases does not allow a merchant a reasonable profit. They get their own price for what they make and then advertise their article to the public at a certain figure, forcing the dealer to handle the goods without any say as to either cost or selling price. They do not know anything about the cost of the dealer's operating expense nor how much it is necessary for him to add to the cost of an article to insure him a profit, yet they take it upon themselves to set that sale price for him. They wouldn't allow anyone not connected with their own enterprise to figure out these details for them, so why should we be obliged to let them do it for us.

The method of marketing electric fans is a glaring instance of the above. There isn't any doubt in the minds of the average dealer or consumer that the price of fans is entirely adequate to take care of the expense of manufacture and profits for those who handle them, but I am sure that there is not a dealer in the country who has ever made a real profit from the sale of fans.

There are other devices and appliances which might be mentioned here if time and space would permit, but it is the

*President Windsor District Ontario Association Electrical Contractors and Dealers.

principle which is bad and I think we should aim to have it rectified.

### One Thing at a Time

I have endeavored in a crude manner perhaps to bring before you one or two ideas that are worth the time it will take to discuss them and try and have concerted action taken that may lead to a betterment of conditions. There are others but it would be well, I think, to apply ourselves to certain definite aims until they are disposed of.

In conclusion I would like to suggest the inadvisability of attempting anything in the way of reform until we have arrived at that stage where we can depend upon each other. If something occurs that makes it appear that one member is not inclined to play fair with the rest, the proper place to discuss it is at a meeting of the Association and not take the matter in hand as an individual. We might just as well not be organized if we are going to deal with problems singly. The whole scheme thus becomes a loss—lost effort, lost time and lost money. We are trying to overcome all these and make our vocation a profitable one for ourselves and our patrons.

Let us therefore unite as far as is possible and work to that end in a fair and honorable way, laying our cards upon the table so that each may see the other's hand. We will thus be off to a start to accomplish things. Otherwise our efforts are sure to fail.

# Some Suggestions for Better Lighting

By J. T. SCOTT
Sunbeam Lamp Division, C. G. E. Co

Try to imagine the old time office and work shop lighted by candle and oil lamps. That poor condition exists to-day in many offices and shops by reason of the improper use of quantity, quality and distribution of light.

It seems as if the eyes, because they never complain by an epidemic form of sickness, are to be unremittingly abused. A wise health department sees that conditions are proper for the maintenance of general health. In schools the eyes of children are sometimes examined and partially cared for. In industrial plants the optometrist visits and prescribes the glasses required where necessary. The evil is acknowledged. But what about the cause? No public department watches over man's need here. Indeed it would be difficult to expect anything to be enacted, beyond the setting of a standard minimum of lighting to be maintained where operatives are engaged. Much more than this is needed however, varying according to the particular demand. What this actual demand is, will be better understood in the course of time, and after sufficient educational work has been done to assist people towards making an intelligent interpretation of their needs. At the moment there are very few who know whether they require an intensity of light equalling 5 or 10 foot-candles, and many, struggle to perform their work by an insufficiency of light that does not measure up to one or two foot candles

It is the purpose here to indicate briefly the importance of an adequacy of light and to aid a recognition of plants which fall short of requirements in illumination.

The aim is more work, of better quality, and with less effort.

Poor vision, made worse by poor lighting, must mean either spoiled work or a longer time to do the job, both of which signify loss of time and money. The efficiency of expensive machinery is impaired also because of poorly placed light or not enough of it.

Good lighting is rather a matter of common sense than special engineering knowledge. Personal judgment is sufficient in most cases, but it must be good judgment based upon some knowledge of a common measure.

### Not a matter of cost

The purchases of goods is not made without reference to a measure of quantity. The stores are issued in the plant by similar standards. Then why equip the lamp sockets without a full idea of the quantity of light to issue? The cost does not end with the price of the lamps and the current consumed. It is not a matter of watts; they are cheap enough relatively; it is a question of sufficient foot-candles, lumens or intensity of light, having regard to the work to be performed. The illuminating engineer would here determine the amount of light distributed by use of the foot-candle meter, but where that handy instrument is not available to you, use the sense of sight as a guide to what is necessary. Cut out as much shadow as possible, reduce any glare by proper shading or by using frosted lamps, and avoid reflection from bright surfaces. Guard against the dim, eye-straining light that is usually mistaken for "soft light". Remember,—do not cut down the light, but shade it and keep the direct rays from the eyes. Unshaded light sources are far too prevalent.

A worker after gazing into an unshielded light may be injured by moving machinery as a result of temporary blindness due to after-images; deep shadows resulting from the same cause increase the possibilities of danger. These few facts have been often pointed out. They are worth repeating because a walk through some plants, and by some stores, shows the necessity for repetition. Dim lights are covered by dust laden shades, bright glaring lights are left unshaded —some hung so low as to be right in the line of vision, others hung so high as to give faulty distribution, and casting weird shadows. If the owner of these various sorry schemes could be compelled to a ten minute silence while they considered the violation to the intelligence of it all,—something might be done towards a betterment of lighting.

### Plays the More Important Part

Do not forget that light plays a far more important part than it usually gets credit for.

The correct lamp with its output of lumens equalling the required intensity of light distribution, is as necessary and important as is the correct motor for a particular load in the mill.

First determine how much illumination will be required. This of course will depend upon the purpose for which the light is used. Work requiring close attention will need a higher intensity of illumination than work of a coarser grade For instance, the machine shop will require good illumination of from 10 to 16 foot-candles as against the 5 to 6 foot-candles sufficient for the foundry, and a matter of 2 foot-candles for stairways and passageways.

The fundamental requirements of any installation are:— sufficient light on all parts of the working area and freedom from glare and reflection.

Help to make the store more pleasing, and get the factory made safe, cheerful and more productive by better lighting.

This also has its commercial side. Let the electrical dealer and salesman bear in mind the several phases outlined above, and use them in rendering service to the customer, and by suggestion, in trying to effect sales. Overcome the indifference and negligence and keep up with the wheels of progress.

## Organized Advertising Brought Results

Under the auspices of the Electrical Service League, the electrical industry of British Columbia conducted an intensive campaign for Christmas trade which lasted about six weeks. Beginning the first week in November, cards showing the cost of the principle electrical appliances, in common usage, were displayed in the shop-windows of all contractor-dealers' stores. This was followed by a so-called "advertising merchandising schedule" for a period from Nov. 30 to Dec. 15, when the window displays and newspaper advertising of the dealers was correlated. Beginning Nov. 17 and extending to Dec. 23, newspaper advertisements appeared in the British Columbia daily newspapers. These appeared for three days in the Vancouver papers, for four days in the New Westminster daily, and for three days in the papers in Victoria.

The cost of this campaign was borne jointly by the central station, the manufacturers, jobbers and the electrical contractor-dealers, so that the cost to the individual firm was small. Excellent results are reported, however, as the Christmas trade among the electrical stores was in excess of the anticipated business and even exceeded the volume of the Christmas business a year ago.

## Visited New Lamp Works

The Hamilton District Electrical Association and the Hamilton Electrical Development League were the guests of the Canadian Westinghouse Co. at their new lamp works in Hamilton on Tuesday, January 24. Mr Geo. Foot, assistant to the manager of sales, together with Mr. Stewart, superintendent, Mr. Kintner, engineer, and Mr. Orr of the sales staff, conducted the visitors through the works and explained the many different processes in modern lamp manufacturing. A great deal of interest was taken by the visitors in noting the care and selection of materials and the very close inspection of each process to prevent defective lamps being sold to the trade.

After the inspection of the plant, the visitors were conducted to the plant cafeteria, where supper was served—and enjoyed by all. Mr. V. K. Stalford acted as toastmaster. A toast to the Westinghouse Company was proposed by Mr. L. W. Pratt and very ably responded to by Mr. Geo. Foot. Mr. Kintner and Mr. Stewart. The visitors were presented with a souvenir pamphlet and lighting calculator which was very much appreciated. At the conclusion of the meal three cheers for Geo. Foot and the Westinghouse company were given in a very hearty manner. All who took advantage of this trip expressed their appreciation and agreed that it was a wonderful success.

The executive committee of the association are planning several trips to be taken in the future to some of the local plants of interest to the industry at large.

## "Felix Penne" at Vancouver Electric Club

A resolution congratulating and commending the management of the B. C. Electric Railway for the able manner in which the change in the rule of the road was carried out, insofar as it concerned that company, was passed at the luncheon meeting of the Vancouver Electric Club held at the Hotel Vancouver on Jan. 20th, and the resolution was ordered to be embodied in a suitable letter to the company.

The club listened to an interesting address from J. Francis Bursill,—"Felix Penne"—veteran journalist of London, England, and Vancouver. The theme developed was the importance of relaxation, of an interest in art, music, literature and the drama as a part of the broadening necessary

for a well-balanced life for any business man. A further suggestion was that the lighter side of literature and the drama was quite as necessary as the more serious and reflective side of the liberal arts, such as reading of standard literature. Mr. Bursill's address was full of practical suggestions most entertainingly presented. Vice-president H. Pim who was in the chair, seconded from his own life habits and experience the value of good reading and similar interests to the business man in keeping himself fit for his own immediate duties as well as those he owed as a citizen.

## Westinghouse Radio Broadcasting Newspaper

"Radio Broadcasting News", a weekly newspaper, has been established to mark the first anniversary of KDKA, the radio telephone broadcasting station of the Westinghouse Electric & Manufacturing Company at East Pittsburgh, Pa.

The newspaper is believed to be the first of its kind in the United States and the only one devoted solely to the publication of news concerning the activities at one broadcasting station.

About one year ago, the Westinghouse Electric & Manufacturing Company broadcasted its first program from KDKA, which was the first station in the world to give nightly broadcasting programs. Interest in the programs became so great that in the latter months of 1921, there came to the company an insistent demand on the part of "Listeners in" that they be informed "in advance" of the programs to be broadcasted from KDKA.

With this demand—good naturedly given, yet insistent—"Radio Broadcasting News" was born. Today, with only a few issues off the press, it is a fixture. It has come to stay because public opinion has demanded it. The birth of this newspaper marks one of the great forward steps in the marvelous history of the advancement of radio broadcasting.

Radio developments are chief items published in "Radio Broadcasting News," which derived its first circulation list from those friends of radio broadcasting, who after "listening in" on the KDKA programs, wrote to the Westinghouse Company expressing appreciation of the broadcasting service.

The publication gives in word and picture news concerning various broadcasting programs and pictures of artists who entertained radio enthusiasts. A feature of each issue is the program to be broadcasted nightly during the week following the date of issue of the newspaper.

It is estimated that more than 60,000,000 persons are within the range of four Westinghouse broadcasting stations, the calls, wave lengths and locations of which are as follows:—KDKA, 360 meters, East Pittsburgh, Pa.; KYW, 360 meters, Chicago, Ill.; WJZ, 360 meters, Newark, N. J.; and WBZ, 360 meters, Springfield, Mass.

The program broadcasted nightly by the Westinghouse broadcasting stations include concerts, church services, results of various games of sport, market reports, stories for children and news bulletins.

Copies of "Radio Broadcasting News" will be sent to all persons desiring to receive the newspaper who send their names and addresses to the Editor, "Radio Broadcasting News", Department of Publicity, Westinghouse Electric & Manufacturing Company, East Pittsburgh, Pa.

---

Mr. A. B. Cooper, formerly with the Canadian General Electric Company and for the past eight years transformer sales engineer with that company, has been appointed general manager of the Ferranti Meter & Transformer Manufacturing Co., Ltd. Mr. Cooper assumed his new duties on January 16.

We show herewith the Ha'lowe'en window display of the Acme Electric Company, of Moose Jaw, Sask. Mr. Burleson, proprietor of the Acme Electric Company, is to be congratulated on the windows, which indicate that this company is following correct lines in leading the Moose Jaw citizens to do more things electrically.

## Making the Most of Your Windows

Have you ever spent even a few precious moments looking over your daily paper—only to find that it was yesterday's?

What did you say?

— — — ! !

Then how do you think the regular passerby regards your windows, if you only provide him with something new to look at every two or three weeks?

Does the electrical merchant make the most of his windows?

Very few of them.

It was recently stated by a successful retailer that he could trace 25% of his sales to his window dressing. Think what that means! All the difference between success and failure—a loss at the end of the year or a fair profit— an A1 credit standing or a sheriff at your door.

Whatever you neglect in your business (try not to neglect anything), don't let it be your window. Close it in attractively; light it properly; dress it tastefully and often—every day is better than every two days, but never less than twice a week—keep it scrupulously clean and introduce action, movement, in some way.

A Brantford window. T. A. Cowan tells the women how they can improve their standard of living. This display was awarded first prize in a recent contest among some twenty merchants in Brantford.

# Newest Developments in Electrical Equipment

### Electric-Driven Cream Separators

D. M. Burrell & Co., Inc., of Little Falls, N. Y., have placed on the market two new cream separators, of the link blade type. Both separators are of high grade construction, with strength and simplicity of design. Due to the present day use of electricity on the farm, the machines have been designed specially for motor drive. The hand machines also

have been designed so that they can easily be converted to electric drive within a very few minutes. Three extra parts only are needed for converting the drive, these being the motor, drive belt and pulley. The total height of the new machines is less than 47 inches and the crank is at the standard height of 32 inches from the floor. On account of the simple gearing and the use of a high-grade annular ball bearing to support the bowl, the machine is unusually light-running and quiet. All high speed parts are in an oil-tight case. Westinghouse motors are used on all motor-drive equipped separators manufactured by the Burrell Company.

### New Outlet Box

The Canadian Drill & Electric Box Co., Ltd., 1402 Queen St. E., Toronto, are now making the outlet box shown herewith. The advantage of this box is that you can use a 3¼

in. or a 4 in. cover. Another feature is that the cross arm that carries the box will fit either a 12 in. or 16 in. joist. The manufacture of this equipment is in line with a recent regulation of the Hydro-electric Power Commission of Ontario.

The Metal Specialties Mfg. Co., 338-352 N. Kedzie Avenue, Chicago, Ill., announce that they have added the well-known Jorgensen Vapor Primer to their line of Presto products. This Vapor Primer was formerly manufactured and marketed by the Jorgensen Mfg. Co., Waupaca, Wis.

### "Thesco" Vertical Clamps

Among the products handled for The States Company by the George C. Rough Sales Corporation, are the "Thesco" vertical clamps, as illustrated. "Thesco" clamps simplify many of the awkward problems met with in line construction, as they make it possible to attach the vertical wire to

a common line insulator, keeping the insulator in its correct position where its surface will remain dry and maintain its insulating properties. The George C. Rough Sales Corporation announce that they have recently added two other agencies: Campbell & Isherwood, of Liverpool, England, manufacturers of electric drills; and Schweitzer & Conrad, manufacturers of high voltage protective and switching equipment.

### Hatheway & Knott Adapters

Hatheway and Knott, Inc., general sales agents, with offices at 117 West Street, New York City, have recently added Outlet Boxes to their line as well as several new devices manufactured by the Rattan Manufacturing Company.

Above—Adapter
Upper right—Connector
Right—Angle adapter

One of the new appliances is a galvanized Panel Box Connector Adapter, as illustrated herewith. Another is the 90 degree Angle Adapter for Squeeze Connectors, as shown. These two fittings, we believe, are entirely new, nothing of the kind at present being on the market.

The Cutter Company, Philadelphia, Pa., are distributing a booklet entitled "U-Re-Lite, the Circuit Breaker in the Steel Box;" well illustrated.

### Levolier Pull Socket

W. H. Banfield & Sons, Ltd., are advertising in this present issue, the Banfield Levolier pull socket. This socket was formerly manufactured by the McGill Manufacturing Co., Valparaiso, Ind., but is now being manufactured in its entirety at the Banfield factory in Toronto. For heavy duty 660 watt service, as required by the Hydro-electric

Catalog No. 65250
1/8 inch Cap

Power Commission, the Banfield Levolier is said to be specially adapted. The method of construction of the make and break mechanism inside the porcelain of the socket is on the same principle as the old type of knife switch. The break is positive, being made with a very fine coiled spring.

### New Toggle Flush Switch

Harvey Hubbell, Inc., Bridgeport, Conn., has recently placed on the market the neat toggle flush switch illustrated herewith. Its make-and-break action is almost instantaneous, the toggle compressing and guiding a spring which instantly

operates the switch. Entire freedom from flashing or binding is claimed by the manufacturer. This switch is provided with adjustable aligning lugs which fasten the switch body flush with the plaster, regardless of how carelessly the outlet box itself may have been installed. The toggle operates at the same tension without regard to the length of time the current may be on.

### Electric Door-opener

The Electrical Products Manufacturing Co., 69 Sprague St., Providence, Rhode Island, has recently placed on the market an electrical door opener, of sturdy design. The castings are die-cast manganese bronze, which gives in-

creased protection against forcing the door or breaking the lock and at the same time prevents excessive wear. The magnets are wound on automatic machines and claimed to be so produced that neither weather conditions nor abuse can affect the efficiency of the device. This equipment is very compact, the dimensions being: face plate, 1¼in. x 3½in.; mortise, 3-8/16 in. x 3 in. x 1½ in.

### New Fuse Plug on Canadian Market

A new manufacturing concern, The "Repeater 6" Fuse Plug Co. of Canada, Ltd., situated at Burlington, Ont., has purchased the Canadian manufacturing and selling rights of the "Repeater 6" Fuse Plug and will have its plant in operation by February 1st; its capacity will be 10,000 plugs per day.

In the Repeater 6, illustrated herewith, there are six fuses; when one blows out, by simply turning the knob to the right contact is instantly restored, leaving five more fuses for further use. This plug, it is claimed, thus removes many of the annoyances caused by the sudden blowing out of ordinary fuse plugs.

### Illuminating Engineering Meeting

The January meeting of the Illuminating Engineering Society, Toronto Chapter, was held on Monday evening, January 16th, in the Chemistry and Mining Building, University of Toronto. The subject of the lecture was "Gas-filled lamps," by Mr. Watson Kintner, lamp engineer of the Canadian Westinghouse Company. The lecture was illustrated by lantern slides. Mr. Kintner also brought several parts of lamps with which to demonstrate the process of manufacture.

The Sangamo Electric Co., Springfield, Ill., are distributing Bulletin No. 59, describing, with illustrations, Brooks two-stage current transformers.

## Rehabilitation of Toronto's Electric Railway System—III

**The new cars are Equal to the best on the Continent—and Then Some Improvements**

The type of street car as used by the Toronto Transportation Commission on the Yonge, Broadview and College routes had its first development in Cleveland, in 1914. After several months' trial, under close observation in that city, a standard design was adopted which is now used not only in Cleveland, but also in Buffalo, Toledo, St. Louis and other U. S. cities. For example, Detroit has just within the last few days put a number of these cars in commission.

This type of car was first tried on routes having a heavy transfer service and it was found to handle larger numbers of passengers in a minimum space of time.

### The Motor Cars

The majority of the new motor cars purchased by the Toronto Transportation Commission were made by the Canadian Car & Foundry Co. of Montreal. The Montreal order consisted of 140 motor cars and 60 trailers. A second order was later placed with the Canadian Brill Company for 50 of the same type of motor car. All these cars are much larger than the former cars used by the Toronto Railway Company. The motor cars are 51 ft. 8 in. long over buffers and 8 ft. 4-3/16 in. wide over side sheathing, the side sheathing being a special rolled plate for car purposes. The trailers are 49 feet long.

The information that follows immediately hereafter refers particularly to the Montreal cars. The Brill cars are referred to later in the article.

One hundred of these motor cars are equipped with Canadian Westinghouse 533-T-4 motors and K-35 controllers, while the remaining 40 are equipped with English Electric Company controllers. Gears and pinions are all helical.

The trucks are all 5 ft. 10 in. wheel base and have 30 in. wheels. They are inside suspension type with roller side bearings. The trucks are specially designed to make the car smooth riding; a system of spiral springs which provides easy action when the car is lightly loaded is a great advantage over former types of construction. When the car is fully loaded a semi-elliptic spring takes care of the load. The wheels have a 2¼ in. tread, though the intention is to increase this eventually to 2¾ in.; this will give much better traction and better wheel and rail wear conditions.

Westinghouse semi-automatic air brakes are used; this equipment includes a compressor having a capacity of 16 cu. ft. per minute. Slack adjusters on the brake cylinders provide proper adjustment for brake wear. Conductor's emergency valves to operate the brakes are located at the conductor's stand and at the rear of the car. The emergency valves supply full air to the brakes and sand to the wheels in one application. All cars are also furnished with an auxiliary hand brake.

The Peter Smith system of forced air distribution is used throughout. A very uniform distribution of fresh, warm air is provided by a fan which draws air from the outside, forces it through a heated chamber and discharges it evenly into the car through ducts under the seats. By this system the car is ventilated as well as heated, tests having shown that the air is completely changed approximately every five minutes. Ventilators placed in the roof are so designed as to be proof against rain or snow entering the car.

All the wiring in the cars is in rigid metal conduit. Illumination is supplied by fifteen 40 watt mill type lamps. Each lamp is equipped with an opalescent Moonstone reflector, supplied by the Jefferson Glass Co. There is also a circuit of five 23 watt mill type lamps for the signs and headlight. The headlight unit is Crouse-Hinds type. Indicating lamps on the rear of the car show green when power is on and red when power is off.

The body of the motor car is practically divided into two parts; these are of approximately the same area. Entrance is by the front door and the forward portion provides an ample loading area. The conductor is placed near the centre of the car and the passenger may either ride in the front part without having yet paid his fare or he may pass the conductor, depositing his fare as he does so, and take a seat in the rear half of the car. Both the entrance and the exit are provided with double doors, width 5 ft. 6 in., which provides sufficient room for two rows of passengers. The entrance steps drop as the doors are opened and are raised with the opening of the doors. All steps are fitted with safety treads. The doors are inter-locked with the control so that the power is shut off until the doors are closed.

The seats in the front half of the car are longitudinal, as shown in the figure. Cross seats are provided in the rear half. An overhead hand rail takes the place of the familiar strap. Both entrance and exit steps are low, being only about 14 in. from the ground.

For the convenience of passengers, a push button which sounds a signal buzzer in the motorman's cab is installed on every side post. In addition to this there is a signal system for the use of the conductor and motorman—a single stroke bell at each end of the car, which is operated by push buttons in the motorman's cab, at the conductors' stand and at the rear of the car.

In spite of their length, these cars can operate on a curve of 34 ft. 4½ in. centre radius. In fact, on some of the Toronto streets they are at present operating on curves of even smaller radius than this.

The cars are painted outside with a red enamel finish; the interior trim is of birch, stained cherry color. Double floors are used throughout and there are no trap doors or openings in these floors. Window guards are fitted on both sides of the car. The seats and backs are hardwood and there is practically no place in the car that can catch dirt or dust.

### The Trailers

The trailers, which have been purchased to the number of 60, are very similar in design to the motor cars. They are 49 ft. long, 8 ft. 4½ in. outside width; weight, 28,000 lbs. as compared with 49,000 for the motor cars. They will seat 61 passengers as compared with 57 for the motor car, the extra seating capacity being gained, in the trailer by the space taken up by the motorman's cab. In the trailer, entrance and exit doors are in the centre of the car, with the conductor's stand between them, as shown in the figure.

Motor Car—Seating capacity 57—Front entrance, centre exit—Toronto Transportation Commission

Here, also, the car is divided into approximately two equal spaces, the forward half of which provides a very ample loading area. The passengers may either remain in this forward portion without having yet paid their fares or they may move around the railing, shown in the reproduction of the trailer herewith, and pass the conductor, depositing their fares as they do so, and then seat themselves in the rear half of the car.

The air brakes and lighting circuits on the trailer are connected through couplers to the motor car. As will be noted in the picture, the entrance and exit doors are single in the trailer.

The experience in Toronto bears out the reports from other cities that have used this type of car regarding the facilities for loading and unloading. Anywhere from 20 to 40 passengers can be taken on or discharged about 30 per cent

Trailer—Seating capacity 61—Centre entrance and exit—Toronto Transportation Commission

Toronto Transportation train with seating capacity for 118 persons. These are all-steel bodies, birch interior trim, red enamel exterior, presenting a very handsome appearance. Motor cars are front entrance, centre exit; trailers, centre entrance and exit. In both cars forward half constitutes commodious loading platform.

quicker than in the older type of car with rear entrance and front exit. With larger cars, the advantage is still more in favor of the new type.

## Canadian Brill Company

An order for 50 of these cars was subsequently placed by the Toronto Transportation Commission with the Canadian Brill Company of Preston, Ont. There is practically no difference in the design of the cars except that they are somewhat lighter in weight and the roof design of the Brill car follows a somewhat sharper curve, i.e., it represents pretty accurately an arc of a circle, as compared with a flatter curve for the Montreal type.

In the following table there is outlined a list of the equipment and the names of firms that supplied it for both the Montreal and the Preston cars:

### Montreal Cars.

| Equipment | By whom supplied |
| --- | --- |
| Car bodies | Canadian Car & Foundry Co. |
| Air Brakes | Canadian Westinghouse Co. |
| Axles | Dominion Wheel & Foundries Co., Ltd. |
| Wheels | Dominion Wheel & Foundries Co., Ltd. |
| Bumpers | Hedley Company |
| Buzzers | Electric Service Supplies Co. |
| | (Lyman Tube & Supply Co.) |
| Conduits & condulets | Canadian General Electric & |
| | Crouse-Hinds |
| Couplers | Ohio Brass Company |
| Destination signs | Electric Service Supplies Co. |
| Door operating mechanism | National Pneumatic Co. |
| | (Dominion Wheel & Foundries Co.) |
| Fare boxes | Canadian Cleveland Fare Box Co. |
| Fenders & guards | Canadian Car & Foundry Co. |
| Gears & pinions | Allen General Supply Co. & |
| | R. D. Nuttall (Westinghouse) |
| Hand brakes | National Brake Co. |
| | (Lyman Tube & Supply Co.) |
| Heaters | Peter Smith Heater Co. |
| Headlights | Crouse-Hinds Co. |
| Headlining | Pantasote Company |
| Lightning arresters | Canadian Westinghouse & |
| | Dick Kerr |
| Motors | Canadian Westinghouse & Dick Kerr |
| Enamel paint | Pratt & Lambert |
| Sanders | Ohio Brass Co. |
| Sash fixtures | O. M. Edwards & Co. |
| Seats | Hale & Kilburn Corporation (C. S. Wright) |
| Treads | American Mason Safety Tread Co. |
| Tail lights & ventilators | Nicholls-Lintern Co. |
| | (Railway & Power Engineering) |
| Trolley Catchers | Earll Company |
| | (Railway & Power Engineering) |
| Trolley poles | National Tube Co. |
| Trolley wheels | Star Brass Co. |
| Controllers | K-35 and English Electric Q2-D |
| Trucks | Canadian Car & Foundry (cast steel frame) |

### Brill Cars

Same as Montreal cars, with following exceptions:—

| | |
| --- | --- |
| Car bodies | Canadian Brill Co. |
| Conduit & condulets | Conduits Co. Ltd. & Crouse-Hinds |
| Fenders & guards | Canadian Brill Co. (HB type) |
| Headlining | Nevasplit |
| Lightning arresters | Canadian Westinghouse |
| Motors | Canadian General Electric |
| Seats | Canadian Brill Company |
| Treads | Feralun |
| Controllers | K-35 |
| Trucks | 77-E-1 Brill (forged frame) |

# Current News and Notes

**Antigonish, N. S.**

The telephone office of the Maritime Telegraph & Telephone Co., Antigonish, N. S., was considerably damaged by fire recently.

**Calgary, Alta.**

The Crane-Cassidy Electric Co., 129 Fifth Ave. E., Calgary, Alta., has been awarded the electrical contract on a building to be erected at 8th Avenue & Centre St., Calgary, for the Royal Bank of Canada, at an estimated cost of $35,000.

**Guelph, Ont.**

The G. E. B. Grinyer Electric Co., Guelph, Ont., has secured the electrical contract on a dairy building being erected for the Ontario Agricultural College, Guelph, at an estimated cost of $60,000.

The Canada Electric Co., 175 King St. E., Toronto, has been awarded the electrical contract on a veterinary building recently erected by the Ontario Agricultural College, Guelph, Ont., at an approximate cost of $60,000.

A report states that the electors of Guelph, at the municipal elections, voted almost unanimously in favor of Sunday street cars.

**Hull, Que.**

In accordance with its previous notice, the Hull Electric Railway Company recently reduced the wages of its employees ten per cent. This action was disputed by the men with the result that a Conciliation Board was appointed. The Board's majority report recommends that the rate of wages in effect since August, 1920, continue until at least July 1, 1922, and that the wages deducted by the 10 per cent. cut be paid back to the men.

**Keremeos, B. C.**

Messrs. Farr, Robinson & Bird, 546 Howe St., Vancouver, have been awarded the contract for electric fixtures for the bank building recently erected at Keremeos, B. C., by the Canadian Bank of Commerce.

**Kitchener, Ont.**

The Kitchener Light Commission, Kitchener, Ont., are receiving tenders on 220 gross tons of steel rails and 1,000 steel ties. Work on street railway extension will be started in the spring.

**Lévis, Que.**

Messrs. Napoleon & Lepage, Bienville, Que., have secured the contract for electrical work on a Monastery to be erected at Levis, Que., for the R. R. Sisters Visitandines, at an estimated cost of $100,000.

**Moncton, N. B.**

The T. Johnston Co., Ltd., 895 Main St., Moncton, N. B., has secured the electrical contract on a $75,000 skating rink recently erected at Sunny Brae, Moncton.

**Montreal, Que.**

Messrs. Demers & Prevost, 189 Bernard St., Montreal, have secured the contract for electrical work on the St. Aloysius School, recently erected on Adam St., Montreal, at an approximate cost of $210,000.

Mr. W. Rochon, 454 Lafontaine Park, Montreal, has secured the electrical contract on a $90,000 warehouse recently erected for Mr. S. L. Content, 590 Marie Anne St., Montreal.

**Ottawa, Ont.**

Mr. Stan Lewis, 63 Metcalfe St., Ottawa, has secured the contract for electric wiring on a toboggan slide at the Ice Palace in Lansdowne Park, Ottawa.

**Paynton, Sask.**

The contract for a telephone system for the town of Paynton, Sask., was recently awarded by the Saskatchewan Department of Telephones, to Mr. F. A. Kelly, a contractor of Regina, Sask. It is understood that work on this system will commence at once.

**Penticton, B. C.**

The Penticton Electric Company, Penticton, B. C., has been awarded the contract for installing electrical equipment in the new Canadian Bank of Commerce, Keremeos, B. C.

**Port Moody, B. C.**

A report states that Messrs. Thurston & Flavelle, Ltd., of Port Moody, B. C. will electrify their sawmill, which is located on the water front at Port Moody.

**St. Thomas, Ont.**

The Hydro-electric Commission, of St. Thomas, Ont., announces that at the end of 1921 their books show a substantial balance on hand; this despite the fact that more than $10,000 was refunded to consumers, in rebates, during the year: Mr. H. K. Sanderson was elected chairman for his ninth consecutive term.

**Three Rivers, Que.**

Mr. J. B. Badeux, Three Rivers, Que., has been awarded the contract for electrical work on an extension being built to the St. Philippe Ward school, Three Rivers, at an estimated cost of $120,000.

**Toronto, Ont.**

Messrs. Bates & MacPherson, 33 Richmond St. W., Toronto, have been awarded the electrical contract on a church and Sunday School building recently erected on Belsize Drive, Toronto, for the Belsize Avenue Presbyterian Church.

Mr. J. Deemert, 34 Pauline St., Toronto, has been awarded the contract for electrical work on an apartment house recently erected at Ossington Ave. & Bloor St., Toronto, for Mr. Fred Reilly, 743 Dovercourt Road.

Mr J. B. Forsey, 159 Wychwood Ave., Toronto, has been awarded the contract for electrical work on two stores being erected at Roncevalles & Garden Avenues, Toronto, for Mr. W. R. MacDonald, 177 Roncevalles Ave.

**Victoria, B. C.**

Messrs. Fox & Mainwaring, Victoria, B. C., have been awarded the electrical contract on a building being erected on Fort Street, Victoria, for the Victoria Baggage Co., 506 Fort St.

**Walkerville, Ont.**

The Public Utilities Commission of Walkerville, Ont., received tenders up to January 31 for the erection of a building for offices for the commission and a sales room for electrical appliances. The building will probably cost in the neighborhood of $75,000.

**Welland, Ont.**

Mr. F. Sage, Welland, Ont., has been awarded the contract for electrical work on a fertilizer plant to be erected at Dain City, Humberstone township, at an approximate cost of $85,000 for the Cross Fertilizer Co., Ltd., Sydney, N. S.

**Westmount, Que.**

The Vincent & Say Electric Co., 344 Union Avenue,

Westmount, Que., has been awarded the contract for electrical work on an addition being built to the Dent-Harrison bakery, Prince Albert Ave., Westmount, at an estimated cost of $50,000.

Mr. H. R. Cassidy, 255 Regent Avenue, Westmount, Que., has secured the contract for electrical work on an addition being built to the Royal Bank of Canada building at Victoria Ave. & Sherbrooke St., Westmount, at a cost of from $15,000 and $20,000.

**Winnipeg, Man.**

The McDonald & Willson Lighting Company, 309 Fort St., Winnipeg, has been awarded the contract for electrical work on a ski jumping hill being erected at Broadway & Osborne Sts., Winnipeg, for the Winnipeg Winter Carnival, 203 Tribune Building, Winnipeg.

Revenue obtained by the Winnipeg Hydro-electric system for 1921 was $67,586 more than for 1920, it is reported. The total revenue from this source in 1920 was $127,254, and for 1921, $194,840.

Tenders were received by the Chairman, Public Utilities Committee, City Hall, Winnipeg, Man., up to Monday, January 23, 1922, for a supply of pole type transformers and approximately 53,000 ft. of high tension, lead covered cable. He will also receive tenders up to 3:00 p.m., Monday, February 6, 1922, for automatic induction regulators.

The Quality Electric Co., Simcoe St., Winnipeg, has secured the contract for electrical work on a building being erected at Redwood & Main Sts., Winnipeg, for Northern Motors, Ltd., 601 Union Trust Building, at an estimated cost of $32,000.

**Woodstock, Ont.**

Prices on a number of electric fire places are being asked for by Mr. R. E. Butler, Windsor, Ont. Mr. Butler is building an addition to his business block in Woodstock, Ont., and plans to install fire-places in a number of rooms.

---

The Electrical Co-operative Association, Province of Quebec, will hold a rousing "Get-together" Smoker on the evening of February 8, at the Windsor Hotel, in Montreal. Those who have experienced Montreal's hospitality will anticipate a thoroughly enjoyable evening. The prime idea underlying this meeting it to get better acquainted so that the co-operative movement may be more effectively and efficiently carried through. If any electrical man, within reach of Montreal, feels that he is not well acquainted with every other man in the industry, he should be found, on the evening of February 8, at the Windsor Hotel smoker.

---

**A pair of Wires and a Dial**

The Northern Electric Company are sending out an interesting booklet, entitled "A Pair of Wires and a Dial." This booklet is written with the idea of popularizing the Strowger automatic telephone equipment. The automatic telephone is being used more extensively all the time and should have the attention of the contractor-dealer to a greater extent, we think, than it has in the past. Every building from the size of the average home to the largest factory and warehouse is a prospect for an automatic telephone system. The Toronto office of the Northern Electric Co. have just advised their customers that they have on their staff an automatic telephone engineer who will be pleased to consult with the

contractor-dealer and co-operate with him in making the sale to his prospect.

---

**Know a Good Man**

At the recent municipal elections Mr. E. W. Sayer, of the Sayer Electric Co., electrical contractors and dealers, 85-87 Bleury St., Montreal, was re-elected alderman for the city of Outremont. The electors of that district evidently know enough to hold on to a good man when they once get him. This is Mr. Sayer's third term in office.

---

**Tenth Midwinter Convention of the A.I.E.E.**

The Tenth Annual Midwinter Convention of the American Institute of Electrical Engineers will be held February 15-17, 1922, in the Engineering Societies Building, 33 West 39th Street, New York, N. Y. This will be the first convention to be held under the new ruling of the Board of Directors which prescribes but four general technical meetings of the Institute to be held each year.

---

**More Leisure Hours**

To sell the idea of more time for the enjoyable things of life is the primary object of an attractive little booklet which The Society for Electrical Development is ready to distribute. "More Leisure Hours" is the title, and in 24 entertainingly written and illustrated pages the booklet tells how hours of drudgery may be eliminated by the use of electric service and appliances.

---

**The Telephone Echo**

The latest issue of the "Telephone Echo," published by the Manitoba Government Telephones, is a very creditable and interesting little magazine. An interesting item notes there are four and a half million independent telephones in North America—that is, outside of the Bell system, which comprises about nine and a half million—a total of fourteen million for the continent.

---

**Packard in new offices**

The Packard Electric Co, St. Catharines, announce that business has been increasing so rapidly that they have found it necessary to vacate their present Toronto office in the Temple Building, and are moving about the 15th of February into new offices at No. 308 & 309, third floor, of the Brass Building, at the corner of Adelaide & Yonge Sts. These offices are very commodious and we believe their customers will appreciate the change, as the Packard Electric Co. state they will now be able to take still better care of them than in the past.

---

The Robbins & Myers Company, Springfield, Ohio, are distributing their "Domestic Catalogue," No. 1177, dealing with electric fans for alternating and direct current circuits, non-oscillating, oscillating, ceiling and ventilating. The catalogue is well illustrated with an attractive color design, and very complete descriptions of the construction details of the various equipment.

---

Immediate consideration by the 1922 Winnipeg city council of the Winnipeg electric railway problem, and a prompt decision whether or not the city shall buy the property in 1927 as a public utility, was urged by Mayor Parnell in his address at the inauguration meeting of the new city council.

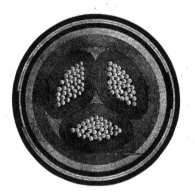

## THE ELECTRICAL NEWS

### HOUSE OF EARTH COST HALF PRICE OF BRICK

On the Yarmouth road near North Walsham, England, there is a new house which is being variously referred to as "homegrown" and "the most natural house in England." It is creating a great deal of interest, as it was built in the most primitive manner of domestic construction by a firm which specializes in cathedral work and dwellings of the most luxurious type.

The owner bought a field and made his house of the soil within it. It is an application by Messrs. Cornish and Gaymer, the builders, of Sunderland House, Mayfair, of the rammed earth principle to a country residence of fair size.

The walls are 17 inches thick, made of soil rammed between boards as hard as stone. The rest is timber work, with a roof of reed thatch 15 inches thick, laid on

### FOR SALE

### Do You Want a Good Man?

in the fancy style beloved of the old Norfolk thatchers.

Mr. John Gaymer, an authority on building construction old and new, describes it as being as "strong as a castle," and as proof of his contention cites a cottage he has just built in the same way a few miles distant on the cliff at Patson, which has stood unharmed against a 70-mile gale.

The walls of the new house have been washed a warm brown tint, the thatch lies like velvet over the dormers, and all the timbering is stained a rich dark brown. But a phase of this "natural house" more important at present than its pleasing aspect is its relative cheapness. Mr. Gaymer declares that the walls have cost just half the price of brick.

### A PIPE-ARCH BRIDGE
#### Engineering Curiosity, Which Carries Boston's Water Supply

An engineering curiosity, said to be unique in America and to have only one parallel in Europe, is the pipe-arch bridge over the Sudbury River, which carries Boston's water supply. The span is eighty feet, and the steel pipe, seven and one-half feet in diameter, rises five and a half feet above the horizontal at the centre.

The pressure on the abutments when the pipe is filled with water is very great, and is resisted by a mass of concrete forty feet thick behind each abutment. Across the curved top runs a hand-railed foot bridge. The steel of the pipe in the arched portion is five-eighths of an inch in thickness.

On November 15, 1921, there were 25 blast furnaces in operation in Lorraine (France), this number comparing with 18 active at the end of August and 27 on January 1, 1921.

### ELECTRIC RAILWAY CONSTRUCTION IN NORWAY

Information from Christiania states that the Norwegian Storthing has granted a concession to A/S Akershanerne for the construction of an electric railway from the centre of Christiania to Ostenajo, a distance of about 8 kilometers. Work on this line will be commenced simultaneously with the construction of the Majorsteun-Sognavandet railway, a concession for which was also granted recently.

A/S Kristiania Elektriske Sporvei (Electric Tramway Co.) will soon begin construction on the extension of its Lilleaker line to Stabaek. This work will cover a distance of about 4 kilometers.

### GREEK COAL IMPORTS

At present from 250,000 to 300,000 tons of coal are annual imported into Greece. In 1920 about 212,000 tons were imported, of which 50 per cent. came from the United States, 45 per cent. from England, and 5 per cent. from other countries. According to statistics received from official sources, up to August 31 of this year England sent 80,615 tons and the United States 41,890.

### NEW AND USED

# MOTORS

# Ideas

Study the ideas and methods in your trade paper. Find out how the other fellow does it and apply the principle—it may need some changes, but if it worked for him it can be made to work for you.

Keep a close watch on the advertisements. The manufacturers and jobbers are using their space to give you information on the goods you need. The latest styles, market conditions, prices, etc., are big factors in the success of your business.

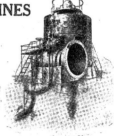
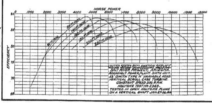

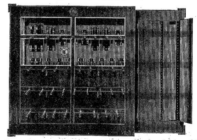

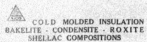

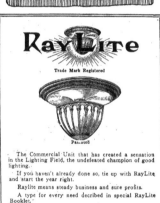

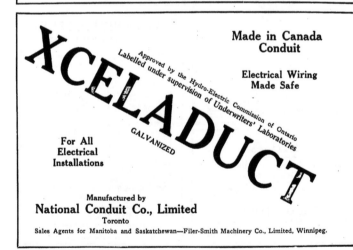

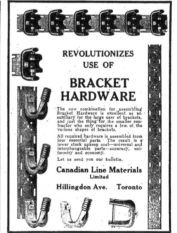

# CLASSIFIED INDEX TO ADVERTISEMENTS

The following regulations apply to all advertisers:—Eighth page, every issue, three headings; quarter page, six headings; half page, twelve headings; full page, twenty-four headings.

# CLASSIFIED INDEX TO ADVERTISEMENTS—CONTINUED

**FROSTING**
Arts Electrical Company

**FUSE BOXES**
Canadian Drill & Electric Box Co.
Northern Electric Company

**GEARS AND PINIONS (Noiseless)**
Diamond State Fibre Co. of Canada, Ltd.

**GENERATORS**
Canadian Crocker-Wheeler Company
Canadian General Electric Co., Ltd.
Canadian Westinghouse Company
Electrical Maintenance & Repairs Company
Electric Motor & Machinery Co., Ltd.
Ferranti Meter & Transformer Mfg. Co.
Northern Electric Company
Thomson Company, Limited
Toronto & Hamilton Electric Company

**GLASSWARE**
Consolidated Lamp & Glass Company
Jefferson Glass Company

**GUARDS (Lamp)**
International Machinery & Supply Co.
McGill Manufacturing Company

**HANGERS (Cable)**
Standard Underground Cable Company of Canada, Limited

**HEAD GATE HOISTS**
Hamilton Company, Wm.
Smith Company, S. Morgan

**HEATING DEVICES**
Benson-Wilcox Electric Co.
Canadian General Electric Co., Ltd.
Canadian Westinghouse Company
Electrical Appliances, Ltd.
Hatheway & Knott, Inc.
Hubbell Electric Products Co.
National Electric Heating Company
Northern Electric Company
Spielman Agencies, Registered

**HIGH TENSION EQUIPMENT**
Electrical Development & Machine Co.
Electrical Engineers Equipt. Co.

**HYDRAULIC MACHINERY**
Boving Hydraulic & Engineering Company

**ILLUMINATING GLASSWARE**
Jefferson Glass Company

**INSPECTION ENGINEERS**
Allen Engineering Co.
Canadian Inspection and Testing Laboratories

**INSULATING VARNISHES AND MOULDING COMPOUNDS**
Canadian Johns-Manville Company
Continental Fibre Company
Diamond State Fibre Co. of Canada, Ltd.
Spielman Agencies, Registered

**INSTRUMENTS**
Canadian General Electric Co., Ltd.
Canadian Westinghouse Company
Hatheway & Knott, Inc.
Northern Electric Company
Sangamo Electric Company of Canada, Ltd.

**INSTRUMENT CUTOUT SWITCHES**
Electrical Development & Machine Co.

**INSTRUMENTS—RECORDING**
Northern Electric Company

**INSULATION**
American Insulator Corporation
Canadian Johns-Manville Company
Continental Fibre Company
Diamond State Fibre Co. of Canada, Ltd.
International Machinery & Supply Co.

**INSULATION (Hard Fibre)**
Diamond State Fibre Co. of Canada, Ltd.

**INSULATORS**
American Insulator Corporation
Canadian Porcelain Company
Canadian Westinghouse Company
Hatheway & Knott, Inc.
Illinois Electric Porcelain Company
Moloney Electric Company of Canada
Mica Company of Canada
Northern Electric Company
Ohio Brass Company
T. C. White Electric Supply Co.

**INSULATING CLOTH**
Packard Electric Company

**INSULATING VARNISHES**
Canadian Johns-Manville Co.

**IRONS (Electric)**
Canadian General Electric Co., Ltd.
Canadian Westinghouse Company
Electrical Development & Machine Co.
National Electric Heating Company
Northern Electric Company

**IRONING MACHINES**
Canadian Ironing Machine Co.
Meyer Bros.

**LAMP BULBS (Electrical)**
Jefferson Glass Company

**LAMP SHADES—SILK**
Manufacturers Trading & Holding Co.
Standard Silk Shades, Ltd.

**LAMPS**
Benson-Wilcox Electric Co.
Canadian General Electric Co., Ltd.
Canadian National Carbon Company
Canadian Westinghouse Company
Electrical Maintenance & Repairs Company
Hatheway & Knott, Inc.
McDonald & Willson, Ltd., Toronto
Northern Electric Company

**LANTERNS (Electric)**
Canadian National Carbon Company
Duncan Electrical Company
Spielman Agencies, Registered

**LIGHTNING ARRESTORS**
Moloney Electric Company of Canada
Northern Electric Company

**LIGHT-REGULATORS**
Dominion Battery Company

**LINE MATERIAL**
Canadian General Electric Co., Ltd.
Canadian Johns-Manville Co.
Canadian Line Materials Ltd.
Electric Service Supplies Company
Northern Electric Company
Ohio Brass Company
Slater, N.
Steel Company of Canada, Limited
Steel City Electric Company

**LIGHTING STANDARDS**
Canadian General Electric Co., Ltd.
Dominion Steel Products
Hamilton Co., Wm.
Northern Electric Company

**LOCK AERIAL RINGS**
Slater, N.

**MAGNETS**
Canadian Westinghouse Company

**MARBLE**
Davis Slate & Mfg. Company

**MEASURING INSTRUMENTS**
Northern Electric Company

**METERS**
Canadian General Electric Co., Ltd.
Canadian National Carbon Company
Canadian Westinghouse Company
Chamberlain & Hookham Meter Company
Ferranti Meter & Transformer Mfg. Co.
Great West Electric Company
Northern Electric Company
Packard Electric Company
Sangamo Electric Company of Canada, Ltd.

**MICA**
Fillion, S. O.
Kent Brothers
Mica Company of Canada

**MOTORS**
Canadian Crocker-Wheeler Company
Canadian National Carbon Company
Canadian Westinghouse Company
Century Electric Company
Electrical Maintenance & Repairs Company
Electric Motor & Machinery Company
Ferranti Meter & Transformer Mfg. Co.
Gelinas & Pennock.
Great West Electric Company
Lincoln Meter Company
MacGovern & Company
Northern Electric Company
Petrie, Limited, H. W.
Robbins & Myers
Thomson Company, Limited, Fred.
Toronto & Hamilton Electric Company
Wagner Electric Mfg. Company of Canada
Wilson-McGovern Limited

**MOTOR AND GENERATOR BRUSHES**
Calebaugh Self-Lubricating Carbon Co.
Dominion Carbon Brush Company

**MOTOR STARTING SWITCHES**
Canadian Drill & Electric Box Company

**OUTDOOR SUB-STATIONS**
Moloney Electric Company of Canada

**OVERHEAD MATERIAL**
Canadian General Electric Co., Ltd.
Northern Electric Company
Ohio Brass Company

**PANEL BOARDS**
Benjamin Electric Manufacturing Company
Canadian General Electric Co., Ltd.
Devoe Electric Switch Company
Northern Electric Company

**PAINTS & COMPOUNDS**
G. & W. Electric Specialty Company
McGill Manufacturing Company
Spielman Agencies, Registered
Standard Underground Cable Co. of Canada

**PHOTOMETRICAL TESTING**
Electric Testing Laboratories, Inc.

**PATENT SOLICITORS & ATTORNEYS**
Dennison, H. J. S.
Fetherstonhaugh & Hammond
Ridout & Maybee

**PINS (Insulator and High Voltage)**
Ohio Brass Company
Slater, N.

**PIPE LINES**
Boving Hydraulic & Engineering Company
Canadian Line Materials Ltd.
Hamilton Company, Wm.

**PLUGS**
Benjamin Electric Manufacturing Company
Hatheway & Knott, Inc.

**POLES**
Lindsley Brothers Company
Northern Electric Company

**POLE LINE HARDWARE**
Canadian Johns-Manville Co.
Canadian Line Materials Ltd.
Northern Electric Company
Slater, N.

**PORCELAIN**
Canadian Porcelain Company
Hatheway & Knott, Inc.
Illinois Electric Porcelain Company
Ohio Brass Company

**PORCELAIN BUSHINGS**
Electrical Development & Machine Co.

**POST TYPE INSULATORS**
Electrical Development & Machine Co.

**POTHEADS AND SLEEVES**
Electrical Development & Machine Co.
G. & W. Electric Specialty Company
Northern Electric Company
Standard Underground Cable Company of Canada, Limited

**POWER PLANT EQUIPMENT**
Boving Hydraulic & Engineering Company
Canadian General Electric Co., Ltd.
Electric Motor & Machinery Co., Ltd.
Hamilton Co., Wm.
MacGovern & Company

**PUMPS AND PUMPING MACHINERY**
Boving & Company of Canada
Canadian Allis-Chalmers Company
Hamilton Co., Wm.

**RECEPTACLES**
Benjamin Electric Manufacturing Company
Canadian General Electric Co., Ltd.
Diamond State Fibre Co. of Canada, Ltd.
Monarch Electric Company
Northern Electric Company
Paulding, John I.

**RECORDING INSTRUMENTS**
Canadian Westinghouse Company

**REFLECTORS**
Benjamin Electric Mfg. Co.
Pittsburgh Reflector & Illuminating Co.

**REMOTE CONTROL SWITCHES**
Electrical Maintenance & Repairs Company

**REPAIRS**
Electrical Maintenance & Repairs Company
Electric Motor & Machinery Co., Ltd.
Thomson Company, Fred.
Toronto & Hamilton Electric Co., Ltd.

**RHEOSTAT BOXES AND ACCESSORIES**
Canadian Drill & Electric Box Company

**ROSETTES**
Benjamin Electric Manufacturing Company
Northern Electric Company

**SLATE**
Davis Slate & Mfg. Company
Hydeville Slate Works.

**SOCKETS**
Benjamin Electric Manufacturing Company
McDonald & Willson, Ltd., Toronto.
McGill Mfg. Co.
Northern Electric Company

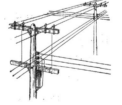

## CLASSIFIED INDEX TO ADVERTISEMENTS—CONTINUED

**SPECIAL HEATING APPLIANCES**
National Electric Heating Company
**STORAGE BATTERIES**
Canadian General Electric Co., Ltd.
Hart Battery Co.
Canadian National Carbon Company
Exide Batteries of Canada, Ltd.
Northern Electric Company
**STREET LIGHTING FIXTURES**
Benjamin Electric Manufacturing Company
Canadian General Electric Co., Ltd.
Northern Electric Company
Winter Joyner, A. H.
**SWITCHBOARD SLATE AND MARBLE**
Davis Slate & Manufacturing Company
**SWITCHES**
Benjamin Electric Manufacturing Company
Canadian Drill & Electric Box Company
Canadian Johns-Manville Co.
Canadian Westinghouse Company
Chamberlain & Hookham Meter Company
Devoe Electric Switch Company
Electrical Maintenance & Repairs Company
G. & W. Electric Specialty Company
Hatheway & Knott, Inc.
McDonald & Willson, Ltd., Toronto.
Moloney Electric Company of Canada
Northern Electric Company
**SWITCHBOARDS**
Canadian General Electric Co., Ltd.
Canadian Johns-Manville Co.
Canadian Westinghouse Company
Devoe Electric Switch Company
Ferranti Meter & Transformer Mfg. Co.
Moloney Electric Company
Monarch Electric Company
Northern Electric Company
**SWITCH GEAR**
Ferranti Meter & Transformer Mfg. Co.
Monarch Electric Company
**TAPE**
Canadian Johns-Manville Company
Packard Electric Company
Northern Electric Company

**TELEPHONE EQUIPMENT AND SUP-
PLIES**
Canadian General Electric Co., Ltd.
Century Telephone Construction Company
Northern Electric Company
**THRUST BEARINGS**
Kingsbury Machine Works
**TOASTERS**
Canadian General Electric Co., Ltd.
Canadian Westinghouse Company
Equator Manufacturing Company
Hubbell Electric Products Co.
National Electric Heating Company
Northern Electric Company
**TOOLS**
Klein, Mathias & Son
Northern Electric Company
**TOY TRANSFORMERS**
Canadian General Electric Co., Ltd.
Killark Electric Manufacturing Company
**TRANSFORMERS**
Canadian Crocker-Wheeler Company
Canadian Westinghouse Company
Electric Motor & Machinery Co., Ltd.
Ferranti Meter & Transformer Mfg. Co.
Kuhlman Electric Company
MacGovern & Company
Moloney Electric Company of Canada
Monarch Electric Company
Northern Electric Company
Packard Electric Company
**TRANSMISSION TOWERS**
Canadian Bridge Company
**TROLLEY GUARDS**
Ohio Brass Company
**TURBINES**
Canadian Westinghouse Company
Boving Company of Canada
Canadian Allis-Chalmers Company
Hamilton Company, Wm.
MacGovern & Company
Smith Company, S. Morgan

**UNDERGROUND INSTALLATIONS**
Gest, Limited, G. M.
Northern Electric Company
Standard Underground Cable Co. of Canada, Ltd.
**VACUUM CLEANERS**
Canadian General Electric Company
Franke Levasseur & Co.
Hoover Suction Sweeper Co. of Canada, Ltd.
McDonald & Willson, Ltd., Toronto
Northern Electric Company
**VARNISHES**
Spielman Agencies, Registered
**WASHING MACHINES**
Altorfer Bros.
Benson-Wilcox Electric Co.
Blackstone Mfg. Co.
Canadian General Electric Co., Ltd.
Coffield Washer Co. of Canada, Ltd.
Henderson Business Machines, Ltd.
Hurley Machine Company
McDonald & Willson
Meyer Bros.
Northern Electric Company
Slade Mfg. Co.
**WATER WHEELS**
Hamilton Co., Wm.
Smith Company, S. Morgan
**WIRE**
Boston Insulated Wire & Cable Company, Ltd.
International Machinery & Supply Co.
McDonald & Willson, Ltd., Toronto
Northern Electric Company
Phillips Electrical Works, Ltd., Eugene F.
Standard Underground Cable Company of Can-
ada, Limited
**WIRING DEVICES**
Canadian General Electric Co., Ltd.
Hatheway & Knott, Inc.
Northern Electric Company
**WOOD SCREWS**
Robertson Mfg. Company, P. L.
**WRINGERS (Power)**
BlueBird Corporation, Limited
Meyer Bros.

---

Vol. XXXI No. 4                                                    Toronto. February 15, 1922

# Electrical News

### Engineering
### Contracting

### Merchandising
### Transportation

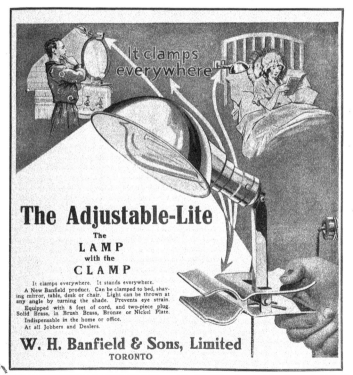
Alphabetical Index to Advertisers, Page 10          Classified Directory to Advertisements, Page 68

# KLEIN PLIERS

Pliers
Splicing Clamps
Sleeve Twisters
Climbers
Tool Belts
Safety Straps
Lag Screw Wrenches
Wire Grips
Tree Trimmers
Tool Bags
Charcoal Furnaces
Staysalite Torch

## "Man, What a Plier!"

That's the tribute an "old-timer" paid to a plier stamped with the Klein mark—that's the kind of reputation 65 years of quality tool-making has built.

Hammer-forged from the finest of special bar steel—carefully tempered knives—handles with just the proper "spring" —rigid factory inspection—no wonder Klein Pliers are the standard!

Equip your gang with sturdy, dependable tools. Buy Klein Pliers.

## Mathias KLEIN & Sons
Established 1857     Chicago. Ill. U.S.A.

# ALPHABETICAL LIST OF ADVERTISERS

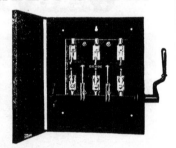
# Can We Help You ?

There may be certain articles which you cannot find in these advertising pages that you would like to have information on. Do not hesitate to use this form, no matter what it is. If we can be of any service to you in supplying that information, it will be a pleasure to do so. We want you to feel that the Electrical News is published in your interests, and we want to help you whenever we can.

- - - - - - - - - - - - - - - - - - - - - - - - - - - - - - - - - - - -

## INFORMATION WANTED

Electrical News,
   347 Adelaide St. West,                Date.................19
     TORONTO.

Please tell me ...........................................................

.........................................................................

.........................................................................

Name...................................................................

Address................................................................

# Sunnysuds
## Electric Washer & Wringer

## *How Many Families Have Waited Just for This?*

How often have people come into your store—sold on an electric washer and wanting one—but unwilling to pay the price that even the best manufacturers were compelled to ask? Those were sales that price prevented.

Not only for these people was the Sunnysuds designed. It was designed for any home that needs a standard size, all metal washer. But it retails for $150.00

Don't trust your imagination for the details of the Sunnysuds. Nothing essential has been omitted; no part has been "shaved." The cabinet is built of pressed steel, folded and braced to insure absolute rigidity. The six sheet tub is corrugated copper with sediment zone and copper baffle plates. The 4-position reversible wringer is aluminum.

A large number of department stores, appliance dealers, household shops, etc., read the foregoing Sunnysuds advertisements. They sent for literature and ordered samples, unwilling to believe that a satisfactory family-size, all metal washer could be made to sell for $150.00. All of them were convinced; many made application for territory; some are now Sunnysuds dealers—and making money! Write for complete information.

**ONWARD MANUFACTURING CO., KITCHENER, Ont.**

*Retail price* $150

*Winnipeg and West $165.00*

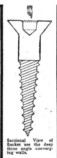

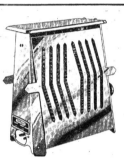

# Picking Brushes for the job

THE Columbia Data Sheet Service will assure correct brushes for each of your motors and generators. It is a service of investigation, analysis, and specification by Columbia Brush Specialists. The Columbia Pyramid Brushes they specify are guaranteed to be satisfactory.

Avail yourself of this service; fit each machine with the brush best suited for the operating conditions. The service is at your command whether you have one unit or a thousand.

Columbia Data Sheet Service
Solves Your Brush Problems
Write the address below

412144

# Canadian National Carbon Company Limited

Montreal        Toronto        Winnipeg        Vancouver

*This is one of the series of striking illustrations appearing in Hoover national advertising. Over six hundred thousand of Hoover full pages are circulated monthly through leading magazines*

# Become an "Authorized Hoover Dealer" in 1922!

*Convenient
Improved Handle Control
Exclusively Hoover—*

UNTIL the new "tilting bar" was brought out by The Hoover no satisfactory means of handle control on an electric cleaner existed. Now complete control of the machine is assured without stopping or stretching, without tightening bolts or screws. The user of The Hoover may easily tilt it backward or forward in order to pass any obstruction—may leave the machine to answer doorbell or telephone and return to find the handle held in a convenient operating position—may lower the handle for use under low furniture or lock it in an upright position for storage. All this is done quickly and easily by a slight movement of the foot. Protected by patents granted May 1, 1917, February 15, 1921, and pending. A total of 38 valuable patents are now the exclusive property of The Hoover Suction Sweeper Company of Canada, Limited. Still others pending.

This year every Hoover Dealer is to enjoy, more than ever, an enviable position among all dealers in household appliances.

Each dealer will be granted the Authorized Hoover Dealer's License, which publicly places the endorsement, confidence and moral backing of The Hoover Suction Sweeper Company of Canada, Limited, behind the store that secures an Authorized Hoover Dealership. Inquiries received as a result of our large national advertising campaign will naturally be referred only to such dealers.

Furthermore, the company will enter into a contract with each dealer which will be to his benefit.

The sale of Hoovers is to be limited to those dealers who are thus licensed to demonstrate, sell and service Hoovers. The license, framed and displayed in each store, informs the public that the dealer has been chosen to represent Hoover interests in his locality.

An "Authorized Dealer" window transfer is also furnished. It states that the dealer is licensed "to sell and service Hoovers bearing the factory guarantee."

Any Hoover purchased from other than an Authorized Dealer will carry no factory guarantee. Every legitimate effort will be made to protect licensed dealers.

The confidence of the buying public in the Authorized Hoover Dealer will be thus forcibly strengthened through this official authority granted by the factory.

Become an "Authorized Hoover Dealer" in 1922! The franchise will prove an asset of ever-growing value to you.

THE HOOVER SUCTION SWEEPER COMPANY OF CANADA, LIMITED
Factory and General Offices: Hamilton, Ontario

# The HOOVER
## It Beats . . . as it Sweeps as it Cleans
### MADE IN CANADA—BY CANADIANS—FOR CANADIANS

Generation, Transmission and Application of Electricity

For nearly thirty years the recognized journal for the
Electrical Interests of Canada.

Published Semi-Monthly By

# HUGH C. MACLEAN PUBLICATIONS
## LIMITED
THOMAS S. YOUNG, Toronto, Managing Director
W. R. CARR, Ph.D., Toronto, Managing Editor
HEAD OFFICE - 347 Adelaide Street West, TORONTO
Telephone A. 2700

| | |
|---|---|
| MONTREAL - - | 119 Board of Trade Bldg. |
| WINNIPEG - | 302 Travellers' Bldg. |
| VANCOUVER - - - - | Winch Building |
| NEW YORK - - - - - | 296 Broadway |
| CHICAGO - - | Room 803, 63 E. Adams St. |
| LONDON, ENG. - - | 16 Regent Street S. W. |

### ADVERTISEMENTS
Orders for advertising should reach the office of publication not later
than the 5th and 20th of the month. Changes in advertisements will be
made whenever desired, without cost to the advertiser.

### SUBSCRIBERS
The "Electrical News" will be mailed to subscribers in Canada and
Great Britain, post free, for $2.00 per annum. United States and foreign,
$2.50. Remit by currency, registered letter, or postal order payable to
Hugh C. MacLean Publications Limited.
Subscribers are requested to promptly notify the publishers of failure
or delay in delivery of paper.

Authorized by the Postmaster General for Canada, for transmission
as second class matter.

Vol. 31          Toronto, February 15, 1922          No. 4

## Canadian Electrical Council
## Will Co-ordinate Various Organizations

Action taken by a group of representative electrical
men, meeting recently in Montreal, looks·like the beginning
of a much closer co-operation of the various sections of the
electrical industry throughout Canada.

Our readers may remember a suggestion contained in
the issue of the Electrical News of July 15 last in which it
was pointed out that a considerable duplication of effort
existed in connection with the two central station organizations—the Canadian Electrical Association and the Association of Municipal Electrical Utilities— and the question was
raised as to whether these two organizations might not in
some way get together to prevent, or at least greatly, reduce this overlapping. In our August 15th, and following
issues, letters expressing the opinions of electrical managers,
superintendents and others holding prominent positions in
the industry, were reproduced in extract and the outcome
was that a meeting was held in Toronto, in October last,
composed of representatives of the C. E. A., the A. M. E. U.,
the A. I. E. E, and the Electrical Supply Manufacturers',
Jobbers', and Contractor-Dealers' Associations. At the conclusion of this conference a resolution was adopted favoring
the convening of a still more representative committee, at a
later date, to be composed of representatives of every electrical organizations within reach of the two provinces.

This second meeting was held in Montreal, on February
6, and the interest taken in the matter of closer co-operation

is, perhaps, best indicated by the personnel of the representatives of the various organizations. For the Canadian Electrical Association there were present Julian C. Smith, vice-
president and general manager of the Shawinigan Water &
Power Co., and president of the Association; A. A. Dion,
superintendent of the Ottawa Electric Company; J. B.
Woodyatt, general manager of the Southern Canada Power
Company, and M. K. Pike, sales manager, Northern Electric Company. For the Association of Municipal Electrical
Utilities of the province of Ontario, M. J. McHenry, general
manager Municipal Hydro System, Walkerville, and president of the Association; O. H. Scott, manager of the Municipal Hydro System, Belleville, Ont., and P. B. Yates, manager of Public Utilities, St. Catharines. The American
Institute of Electrical Engineers (Canadian Division) was
represented by W. A. Bucke, Canadian General Electric
Company and Wills Maclachlan; the Ontario Association of
Electrical Contractors and Dealers by Kenneth A. McIntyre,
president; R. A. L. Gray, Toronto, and W. McKenzie, St.
Catharines; Canadian Electrical Supply Manufacturers' As-
sociation, by M. K. Pike and Mr. Sorber; Canadian Electrical
Supply Jobbers' Association. by S. W. Smith; the Canadian
Electric Railway Association, by E. Blair, Montreal Tram-
ways Company; the Electrical Co-operative Association, by
K. B. Thornton, managing director Montreal Public Service
Corporation; J. A. Anderson, president English speaking
section Quebec Association of Electrical Contractors &
Dealers, and De Gaspe Beaubien. Secretaries J. A. McKay,
Eugene Vinet and Louis Kon were also in attendance.

### The Meeting Fully Representative

The whole industry may thus be said to have been well
and fairly represented. Every phase of the question of co-
operation, with a view to greater efficiency in the industry,
was discussed, and it can well be said that practically every
member of the committee took an absolutely broadminded
view of the whole situation, apparently having no thought
in mind beyond arriving at a conclusion that would have
the most far-reaching benefits for the electrical industry as
a whole throughout the Dominion of Canada.

### Results Represent Progress

The discussion was of passing interest but the final action
of the committee represents permanent progress. Before
the meeting adjourned it was moved by Julian C. Smith,
seconded by M. J. McHenry and agreed to unanimously,
that an advisory body be formed, to be known as the Canadian Electrical Council, to be composed of four representatives each from the Canadian Electrical Association, the
Association of Municipal Electrical Utilities and the Canadian Electrical Supply Manufacturers' Association. In the
meantime the duty of this Council shall be to receive or
initiate suggestions, and make recommendations thereon,
regarding the closer co-operation of the different elements
in the industry, i.e., the Council shall be, in a sense, a clearing house for suggestions regarding co-operation among
its three constituent members. A sub-committee is to be
appointed by the Council that will deal more particularly with
questions of common interest between the C. E. A. and the
A. M. E. U., and a second sub-committee, also appointed
by the Council, will give its attention to the merchandising
phase of the industry with a view to better and more general organization. It was also the sentiment of the Committee that, should the Canadian Electrical Council demonstrate
its ability to function usefully, its activities may be extended
by the addition of members from other all-Canadian associations, such as the jobbers', contractors', dealers' or electrical
engineers', and the discussion of the new questions that would
be involved.

First Meeting Toronto, March 13

The newly formed Canadian Electrical Council will hold its first meeting in Toronto on March 13. In the meantime the three organizations concerned will appoint their representatives.

## Railway Signalling

The regular meeting of the Toronto Section of the American Institute of Electrical Engineers was held in the Electrical Building of the University of Toronto on the evening of January 27, W. P. Dobson presiding. Mr. C. W. Parker of the Canadian Pacific Railway, addressed the meeting on the subject "Railway Signalling."

Mr. Parker first pointed out the inherent differences between double and single track signalling, showing how in the one case it was necessary only to protect trains from the rear, while in the other it was necessary to protect them both before and behind. He then called attention to the low value of the voltages and currents used in this work and to the various sources from which electrical energy was obtained for signalling purposes.

The speaker gave an outline of the development of interlocking systems, pointing out that the first interlocking system in Canada was installed in Toronto in 1894. He then gave a description of the working of the electrical staff systems in use, and showed how they were combined with the signalling systems along the right of way. A large number of other features of electrical, mechanical, and electro-pneumatic signalling systems were described, and some valuable data as to the costs of installation given.

## East Angus, Que, Installs Fire Alarm System

A most complete and up-to-date municipal fire alarm signal system was put into service at East Angus, January 13th. The apparatus is of the most modern type, the signal boxes being of the same design as those used in the larger cities of Canada and the United States. There are 21 street signal stations, an automatic whistle blowing machine, two electro-mechanical gongs, an automatic punch register and central office storage battery equipment. The boxes are of the positive non-interfering type,—that is the mechanism is so arranged that no matter whether one box or ten or more are "pulled" at the same time there will be no interference between signals and all boxes "pulled" (or operated) will sound their number, one after the other. For example, when the system was officially tested by the members of the council three boxes were "pulled" or started simultaneously by three different persons. All the boxes started running and struck their respective numbers one after the other. When a box is operated for fire it sounds its particular number on the special steam whistle at the mills of the Brompton Pulp and Paper Co., on a gong at the fire chief's residence and on a gong at the fire hall, at the same time operating a punch register which records the box number by punching the code in a paper tape at the fire hall for future reference and also as an aid to the fire department in checking their audible count of the alarm. The complete system was furnished and installed by the Century Electric Company, 619 St Paul St. W. Montreal, who have recently entered the municipal fire alarm field, representing the Foote Pierson Co., of New York and The Holtzer-Cabot Electric Co., of Boston.

## Mr. J. J. Wright Lives to see Phenomenal Development in the Industry He Pioneered

Mr. J. J. Wright, the pioneer of pioneers of the Canadian electrical industry, is dead, at the age of 72. In noting this fact, with deepest regret, may we simply recall to our readers some interesting facts in the life of the late Mr. Wright recounted in the Electrical News of July, 1911, when he was the subject of one of our series, then running, "The Makers of Electrical Canada." We quote:—

"In the year 1881, in a little back room loaned for the purpose, by the Firstbrook Box Factory, on King Street, Toronto, the first Canadian-made electric generator was in progress of construction. It was a big generator in those days, though only a twenty-five horsepower. The generator completed, a set of arc lamps of a primeval type were constructed and this apparatus was installed near the northeast corner of King and Yonge Streets. The twenty lamps were placed in various stores on Yonge Street, the late Timothy Eaton being one of the first customers.

"The building of the apparatus, its installation and the operation of this, the first electric lighting system in Canada, was the sole work of one man, a true pioneer in Canada's electrical development, Mr. J. J. Wright."

Further on, the article has this to say of Mr. Wright's pioneering: "It was in the same year that Mr. Wright showed his undoubting confidence in another form of electrical activity. In 1883 he and Mr. Van Depoele had fitted up and exhibited during the annual Toronto Exhibition the first Canadian electric

railway system. It was not an entire success the first year but the following season, 1884, Mr. Wright secured an old Grand Trunk flat car, placed a 25 h.p. motor on it, belted the motor to the car axle and successfully operated a train of three coaches from the foot of Strachan Avenue to the Exhibition grounds. For three or four years, at the same season, this road continued to demonstrate the feasibility of electric railway operation. This was only three years after Siemens, in Germany, had experimented on the first public electric railway the world had ever seen."

And again, "In still one other notable respect Mr. Wright has shown his interest in Canadian electrical matters. He was one of the small group which in 1891 was instrumental in organizing the Canadian Electrical Association. His helpful assistance was shown by the fact that he was the association's first president, which office he held for several consecutive years."

In recent years, Mr. Wright had spent most of his time within sight and sound of Lake Ontario, at Niagara-on-the-Lake, where to the last he maintained his love of the open air and the freedom of country life. One cannot but wonder with what satisfaction he must have followed the development of the industry from his own 25 h.p. generator, in 1881, to the 50,000 h.p. generator so recently installed almost on his very door-step—the yield of his own sowing—not the proverbial twenty or sixty or hundred fold but a full two thousand fold in his own day and generation.

# First 50,000 h.p. Unit of Chippawa-Queenston Power Scheme—I.

### A Series of Articles Describing the Preliminary Investigations, the Design and Construction of the Greatest Undertaking of its kind in the World. Some Wonderful Photographs. Whole work carried to Successful Issue by Canadian Engineers. Practically all equipment Canadian Made

From many view points the official opening of the Chippawa—Queenston hydro-electric generating plant on December 28th last, marks an epoch in Canadian history.

To begin with, it is a colossal enterprise, the ultimate success of which must have caused the engineers, who were charged with the responsibility, many anxious months.

As finally planned this development is the largest hydro-electric scheme in the world, utilizes the largest capacity units, and represents more pioneer work of a kind for which there was no precedent to act as a guidance.

Of even more significance, however, than the size of development adopted, the full natural force of Niagara's power is brought under control for the use of man. Other and earlier developments at Niagara had not attempted this ideal, and Canada's engineers must be given all credit for the courage that enabled them to grapple with and solve a problem upon the solution of which so much depended.

To the citizens of the province, it is most important to know that this natural resource is being utilized to the fullest extent and that all the benefits of this great natural heritage will accrue, either directly or indirectly, to every man, woman and child within electrical transmission distance of this great power centre.

Inasmuch as the official opening marks the completion of the civil and mechanical sections of the work, a general review of the whole project seems to be justified. If in the pages that follow we repeat certain information that has already been published in the Electrical News, it is because we desire that the present series should constitute a complete history of this greatest of hydro-electric projects.

## Selecting the Proper Location*

From the combined viewpoint of conservation and ultimate economy, the ideal Niagara development would be one that would utilize the whole of the future available water under the gross head of 327 ft. existing between Lake Erie and Lake Ontario. Several schemes, approximating this ideal in varying degree had been advanced during the last 20 years, the most practicable and promising of which was known as the Jordan-Erie scheme. This involved the intaking of water near Morgan's Point on Lake Erie, the building of an open waterway across the Niagara Peninsula to the brink of the escarpment above Jordan Harbour, thence carrying the water to the power-house at Lake Ontario level through a mile of pipe. Studied from an engineering standpoint, this scheme was open to serious objection for three main reasons, first, the unavoidable intake conditions at Lake Erie; second, the structural difficulties and unavoidable head loss in connection with the 24-mile canal;

*Harry G. Acres, Chief Hydraulic Engineer, Hydro-Electric Power Commission of Ontario.

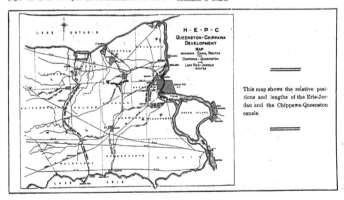

H - E - P - C
QUEENSTON-CHIPPAWA
DEVELOPMENT
MAP
SHOWING CANAL ROUTES
FOR
CHIPPAWA - QUEENSTON
AND
LAKE ERIE-JORDAN
ROUTES

This map shows the relative positions and lengths of the Erie-Jordan and the Chippawa-Queenston canals.

and the third the regulation difficulties attendant upon the control of a mile long water column in the penstock connection between the head of the canal and the power-house, where something over 16 feet of penstock would be necessary for each foot of effective head.. The economic effective head for this scheme worked out slightly less than 300 feet, the bulk of the losses being, of course, taken up in the long canal.

The problem was,. therefore, to find some feasible loc-

lems, it remained to determine whether or not it was feasible to construct a suitable waterway between the Chipawa intake and the power-house location above Queenston. An exhaustive series of surveys and core-drill borings established the fact that it would be entirely feasible to connect these two points by either the open canal or the pressure tunnel type of waterway. The next problem was to determine which of these two types of waterway would be the more suitable from the combined viewpoint of pure hydraulics,

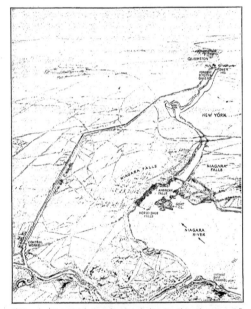

Bird's eye view of Queenston-Chippawa development showing intake from Niagara River above the Falls with Welland River section in foreground leading water to control works at upper end of canal which stretches to power house at Queenston where water is returned to the lower Niagara River.

ation that would overcome objections to the Jordan-Erie scheme. During the course of the subsequent investigations, it developed that by far the best intake conditions would be obtained at the mouth of the Welland River at Chippawa; also that suitable power-house locations were obtainable in the gorge between Foster's Flats and Queenston, which would require only about 18 inches of penstock connection for each foot of effective head, thus reducing the regulation problem to one of minor importance.

### Pressure Canal versus Open Canal

Having tentatively solved the intake and regulation prob-

structural difficulties and hazards, and comparative cost.

In the matter of comparative cost, carefully compiled estimates indicated that throughout the full range of assumed carrying capacities, the open canal had a decided advantage over the pressure tunnel.

### Pressure Tunnel Disadvantages

In the matter of structural difficulties and hazards the following main points were given consideration in the case of the pressure tunnel:—

(a) The necessity of driving the headings at an acute angle through the various limestone, shale and sandstone formations, involving the certainty of a heavy overbreak and expensive timbering and lining.

. . (b) The unfavorable conditions as regards the disposal of excavated material.

(c) The unknown water hazard and the impossibility of predicting the cost of unwatering within reasonable limits, of accuracy.

(d) The difficulty and hazard attending the driving of, and maintaining a pressure tunnel of unprecedently large diameter in the clay formation of the Whirlpool Ravine.

(e) The difficulty in connection with the construction of a distributing chamber in the shale and sandstone at Queenston.

As against the above, the difficulties and hazards in the case of the open canal were limited to two main points; first, the removal of the earth overburden in the canal prism, and second, the permanent holding of the slopes subsequent to such removal. While it may never be possible to establish finally the comparative importance of the above points on the basis of actual construction, the fact remains that the work on the open canal has demonstrated beyond doubt that the overburden can be removed with no more difficulty than was an-

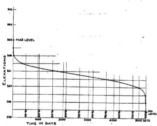

Fig. 1. Duration curve of water levels at Chippawa, 1902-16

ticipated and that the means originally devised will hold the banks safely within the limits of the predetermined slopes.

### Hydraulic Comparison

In the matter of purely hydraulic comparisons, the first point to consider is that both types of waterway would, of necessity, have the same point of intake at Chippawa and the same point of discharge at Queenston, so that they are exactly on a par as regards the utilization of available gross head, neither having any primary advantage over the other in this regard.

Since 1902 the water level at Chippawa has been observed and recorded twice daily. Fig. 1 shows the mean daily elevations for the ensuing period compiled in the form of a duration curve. The following facts are deducible from this curve.

(a) The mean level for the entire period is about elevation 560.8.

(b) A level of elevation 559.5 or higher is obtained for nearly 99 per cent. of the entire period.

(c) A level of elevation 561 or higher is obtained for a little more than one-third of the above period.

(d) It is reasonable to assume that the effective operating range of levels lies between elevations 559.5 and 561.

As to the possibility of the carrying capacity of either type of waterway being seriously affected by a permanent

lowering of the natural levels of the Chippawa-Grass Island pool, due to present and future diversions of water therefrom, it is essential to consider two facts; first that any diversion for power purposes from the pool itself will be largely compensated for by the intercepting effect of the diversion works, and second, that the level of the pool can be controlled independently to compensate for any diversion

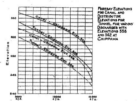

Fig. 2. Comparison of head losses chargeable to tunnel and canal waterways

whatever, whether from the pool itself, from the upper reaches of the river, or from Lake Erie direct.

Fig. 2 shows in graphic form a comparison of the head losses chargeable to each type of waterway under discussion. In making this comparison, a possible extreme low elevation of 558 has been assumed for head-water, and the open canal losses calculated on this basis for a carrying capacity of 15,000 sec. ft. On the basis of this loss a tunnel was designed of the requisite diameter for the same capacity of 15,000 sec. ft.

These curves have been computed for the extreme range of possible operating levels, elevations 558 minimum and 561 maximum. The shape of these two pairs of curves illustrates clearly the basic difference between the two types of waterway. Under the assumed conditions the tunnel and canal curves for the head-water elevation 558 and 15,000 sec.ft. discharge have a common point of origin. As the discharge drops off, however, it is seen that the canal delivers any fixed discharge to the forebay at a consistently higher elevation than in the case of the tunnel. This is simply due to the inherent characteristics of the two types of waterway. In the case of the pressure tunnel, the discharge area is necessarily constant and any gain in head is due to decreased friction only. In the case of the canal the reduction in vel-

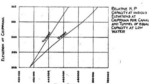

Fig. 3. Relative h.p. capacity at various elevations at Chippawa for canal and tunnel

ocity not only reduces the friction losses, but the retardation of flow increases the effective discharge area of the canal section. By reason of this extra factor, the open canal has an advantage over the tunnel ranging as high as 5 feet of head loss. When the high discharges involved in the problem are considered, it is evident that this difference in head loss is a very important factor.

The curves shown on Fig. 3 have been plotted on a different basis, but with the same factors involved. In Fig. 2 head-water level and carrying capacity have been assumed constant and forebay level the variable. In Fig. 3 head-water and forebay level are the constants and carrying capacity expressed in horsepower is the variable. In this latter curve forebay level is assumed constant at the fixed minimum elevation for peak load capacity and from this common point the comparative carrying capacities for the two types of waterway have been calculated for specified levels of head water in the Chippawa-Grass Island Pool.

The conclusions which may be drawn from a full discussion of all the factors involved are; first, that starting from the common basis of equal loss and carrying capacity at extreme minimum head-water level, the open canal will deliver the required quantity of water to the forebay with a materially less loss of head than the pressure tunnel, for any head-

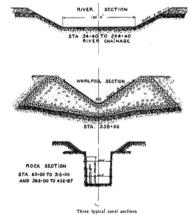

RIVER SECTION
STA 34-40 TO 244-40
RIVER CHAINAGE

WHIRLPOOL SECTION
STA. 336+00

ROCK SECTION
STA. 65+00 TO 315+00
AND 365+00 TO 452+87

Three typical canal sections

water level above the assumed absolute minimum; second, that starting from the common basis of a fixed minimum forebay level, the canal will deliver a constantly increasing quantity of power in proportion as the level of the head-water rises above the assumed extreme minimum level, which the pressure tunnel cannot do to any appreciable extent by reason of its inherent hydraulic characteristics; and finally, that the open canal is the only agency, under the above conditions, which can make automatically available the large quantities of excess power resulting from any temporary or permanent increase in the level of the Chippawa-Grass Island Pool, above the extreme minimum level which has been used as a basis of comparison.

The above were the primary reasons which led to the final choice of the open canal for the connecting waterway between Queenston and Chippawa. This canal, consists of 4¼ miles of the improved natural channel of the Welland River near Montrose to the forebay site above Queenston.

Besides being only about half the length of the alternative Jordan-Erie Canal, the average operating head for equal qualities of water carried is about six feet greater at Queenston than it would have been at Jordan Harbour, despite the fact that the elevation at the point of intake is about 9 feet lower and the elevation at the point of discharge about 2 feet higher than would have been the case with the Jordan project.

Another distinct advantage of the open canal is the fact it can effectively and inexpensively take advantage of any water which might now or in the future be available from the Welland Canal system. This open waterway would furthermore furnish the only means, in connection with the upper reaches of the Welland River of reclaiming the unused 9 feet of head in the Niagara River above Chippawa.

The Queenston-Chippawa development in its final form is a much larger installation than that contemplated in 1915, when under the exigencies of war conditions, it was proposed to construct a power development and canal with an initial capacity of only 100,000 horse power, and an ultimate capacity of 190,000 horse power at a cost for the initial development of $10,500,000—a figure, which it has been computed would have been raised to approximately $29,000,000, due to war prices prevailing during the years 1918, 1919, 1920 and 1921.

# The Design of the Canal [*]

The canal is divided into four sections. The first of these is the Welland River section 21,000 feet in length, with a bottom slope 0.000119 and side slopes of 2 to 1. This was excavated by means of dipper dredge and cableway. The earth section which follows the river section is 6,250 feet long with a bottom slope of 0.0001208 and side slope 2 to 1. For each of these sections a roughness factor of .035 in Kutter's formula was used. The earth section of the canal was originally designed as a concrete lined section of much smaller cross-sectional area but a study of the economic, constructional and operating conditions indicated that the advantages of the larger section would be sufficient to compensate for the cost of the extra excavation. This portion of the canal has a capacity of over 15,000 c.f.s. with uniform flow at the assumed roughness factor of 0.035, and extreme low water in the Niagara River at Chippawa.

At the end of the earth section is located a transition, 300 feet long, in which the trapezoidal cross-section is changed to the rectangular rock section of 48 feet finished width with concrete sides and bottom. Beyond this are the control works, which are described later

The rock section proper is 36,252 feet long and is divided into two parts by the Whirlpool section which has a length including transition, of 2,450 feet. The water section of the rock portion of the canal has concrete lined sides and bottom with a finished width of 48 feet. The bottom slope is 0.0002113 and the roughness factor used in Kutter's formula 0.014. This value is conservative in view of the method of placing the concrete lining. With the steel forms that were used and the special provisions made for alignment of the forms, a smooth plane surface was obtained on the concrete facing.

For 27,000 feet the concrete lining is carried up 33.5 feet above the finished grade of the canal; for the next 9,000 feet the lining is 31 feet high and for the remainder 30 feet high, except in the Whirlpool Section where it is carried up to elevation 563.0 For the greater part of the time the water surface will be above the top of this concrete lining but the

[*] T. H. Hogg, Assistant Hydraulic Engineer, Hydro-Electric Power Commission of Ontario.

friction loss will be reduced by the lower velocities that will then exist in spite of the greater roughness of the unlined rock. Numerous hydraulic studies have been made to determine the surface slopes in the canal for various discharges and for various water levels in the Niagara River. In cases where the water surface was above the top of the concrete lining a composite roughness factor was used in which the proportions of the wetted perimeter on the lining and on the rock surface were taken into account. Roughness factors as high as 0.019 resulted in some of these instances.

### Determination of Depth and Slope

The depth and slope of the rock section were fixed by an economy study and the decision to use a concrete lining throughout its length was also reached in the same way. The method of arriving at the economic section of the canal will be explained later.

An examination of the profile of the canal indicates that the rock surface falls far below the grade of the canal about Sta. 333, rising again to grade about Sta. 849. This occurs in the Whirlpool Section, which is located at Bowman's Ravine or the Whirlpool Gully. Here it was necessary to carry the canal partly on fill and to use a trapezoidal cross-section on account of the foundation upon which the canal is carried. A concrete lining was essential on account of the high velocities. The bottom width is 10 feet and the side slopes 1½ to 1. The slope of the bottom is the same as that of the rock section, viz., 0.0003113.

The Whirlpool section of the canal was designed to have the same cross-sectional area at the lowest possible operating water level as that of the rectangular rock section. This minimum water level would be, somewhat above elevation 542, which is the elevation of the curtain wall at the screen house. The area of the cross-section below elevation 542 is the same for both and for greater elevations the Whirlpool section has the greater area so that there is no danger of the canal capacity being "choked off" at this point.

In locating the Welland River section the river course was followed closely so as to take advantage of the area of the natural channel. This necessitated leaving in all the bends that occurred in the unimproved stream. As the deflection of these curves is not great they will not produce any appreciable loss.

The first important change in direction occurs at the beginning of the earth section at Montrose and is followed by a second bend at the Michigan Central Railroad crossing at Montrose. In addition to these there are only five changes of direction in the rock section of the canal, the deflections of which are 51, 27, 31, 33 and 46 degrees. The radius of curvature in every case is 300 feet and this radius is used for the inside and outside of the bend as well as for the centre line. That is, the curves of the two sides and the centre line of the canal are not concentric, resulting in a greater width of canal at the middle of the bend than at either end, the expectation being that the energy losses will be less than in a bend with concentric curves. It is probable that a shorter radius than 300 feet would give even better results, but this minimum was fixed by the size of the electric shovels that are being used for the excavation work.

The question of surges of the water surface in the canal, due to changes of load on the plant, is of great importance. This problem received an amount of study proportionate to its importance. The sides of the canal and the floor of the screen house are built to such an elevation that with the worst combination of conditions the water will always be contained within the sides of the canal.

Observations of river stage at Chippawa have been available since 1902 and show a minimum W. S. elevation of 558.5, which low stage was reached only on two days. An examination of the past records of Lake Erie stage indicate that as low a stage as 558.0 may be possible at Chippawa. This latter water level is therefore treated as extreme low water and the canal is designed to carry full load at this stage of the river.

While the low water conditions control the size and slopes of the canal, on the other hand the mean water conditions were assumed to be those on which the economic proportions should be based.

### Design of Canal Section

Certain limitations were met with at the outset. The Welland River section of the canal had to be maintained as a navigable stream, and as the excavation is in earth this portion of the canal was therefore designed for a low non-scouring velocity. The minimum width of the rock section was fixed by the type of electric shovel used for excavating this portion of the canal and was placed at 48 feet.

The problem thus resolved itself into selecting the best proportions for the trapezoidal earth section and the best depth and slope for the 48 foot rock section. The procedure in the latter case is the one that will be described.

It is, of course, possible to design any number of canals 48 feet wide, but with depth and slope varying so that all will give the same discharge at low water. For a low velocity the wetted cross-section must be deep but its slope may be moderate. For a high velocity the depth of the wetted cross-section will be small but the slope may be so great that the depth of the cut at the down-stream end may be greater and the total cost of excavating greater than for the low velocity design.

The procedure in determining the economic depth was as follows:

First, the design of a number of cross sections for vel-

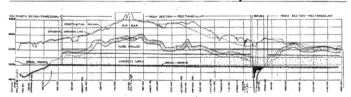

Profile of Queenston-Chippawa power canal, exclusive of river section

ocities of 3, 4, 5, etc., feet per second, the determination of the requisite slope of the bed in each case to give the full load discharge with uniform flow, and the determination of the variation in cost of these canals with low water velocity.

Second, the determination of the friction loss in each of these canals with the river stage at its mean value. This friction loss represented so much lost power, which was, of course, small in amount for the lesser low water velocities, and greater as the low water velocity increased up to a certain point.

Third, the plotting of the differential curves for items 1 and 2 thus showing the variation in delta cost with low water velocity and the variation in delta power with low water velocity. From these two differential curves a third curve can thus be obtained giving the value of delta-cost by delta-power plotted against low-water velocity. Thus there is obtained what will be called an economy curve, showing for any given low water velocity, the rate at which further gains in power may be made at any low water velocity by enlarging the canal slightly and so cutting down velocity and friction loss.

Fourth, the selection of the best low water velocity from this economy curve.

### Economic Water Velocity

The gain in power which results from a slight enlarge-

ment of the canal comes as a result of the reduction in friction loss. The additional cost for power house equipment is so small, within the limits in which we are working that they can be neglected. It is reasonable then to continue the enlargement of the canal until the interest charges on the cost of excavation for the last horse power gained are equal to the average value of the power from the whole plant, including interest, depreciation, operation and maintenance. By stopping short of this point we would be in a position to gain more power at a cost less than the average that we were willing to pay for the power from the whole plant. It is interesting to note that the economic velocity determined in this way is but slightly larger than the minimum cost at which the canal could be built to get the required discharge at low water. The minimum cost occurs for a velocity somewhat greater than the economic velocity.

The advantage for this method of attack is that it permits an economic size to be selected for the canal without the inclusion in the estimate of the cost of anything that does not vary with the low water velocity. In this case, the width of the rock cut being fixed, the earth excavation does not vary with the various designs for the rock section and as a matter of fact in computing the cost of rock excavation only that below some assumed horizontal plane at a lower elevation than natural rock surface but above canal grade was considered.

# Annual Toronto Convention A. M. E. U.

### Many Important Matters Discussed and Reports Presented—Extension of Meter Seal Period to Ten Years, being Considered—Attendance almost 100 per cent of Membership

The Association of Municipal Electrical Utilities of Ontario held the regular annual conventon in Toronto, Chemistry & Mining Building, U. of T., on Thursday and Friday, January 26-27. Mr. M. J. McHenry, the president, was in the chair. The attendance, as is usual at this convention, was very complete throughout and the attention paid to the papers as read, and the discussion that followed, indicate the keen interest shown by the various municipal managers in the matters that come before these conventions.

The chairman's address was brief but much to the point. In effect, he spoke as follows:

Before we proceed with the routine business, gentlemen, there are one or two things which I should like to bring to your notice. Matters of great importance have arisen and have been dealt with by your executive, and I trust that the action taken will receive your approval.

#### Merchandising Electrical Appliances

The first of these matters is the merchandising of electrical appliances by the Hydro municipalities. At the suggestion of the Hydro Electric Power Commission, through the chief engineer, Mr. F. A. Gaby, a committee was appointed by the executive of this association to co-operate with the Hydro Electric Power Commission and consider the problems of merchandising. At the present time a certain number of our larger municipalities are merchandising electrical appliances, but the number doing so on any scale worth considering is very small. The problems confronting the smaller municipalities are very difficult to cope with, and it has been felt that closer co-operation in this respect would be a distinct advantage. Consequently this committee has been formed and already a considerable amount of work has been accomplished. We are not yet in a position to give you any details, because as a matter of fact the situation so far has been discussed only in a general way, but I hope that in the very near future every municipality will receive definite and detailed information.

#### Extension of Seal Period

Another matter taken up with which you are practically all acquainted by now was that of having the seal period of inspection of electrical meters extended from five to ten years. This matter has been taken up in conjunction with the Canadian Electrical Association and memorials have been prepared endorsing the extension. I think practically every municipality received a memorial for signature and return, and these memorials will be presented to the Department of Trade and Commerce at Ottawa.

In this connection memorials were received from all over the Dominion covering more than 75 per cent. of the meters now in use in this country. These memorials should have great weight at Ottawa and we are looking forward to having some action taken by that Department as the result of our efforts in this direction.

#### Central Electrical Organization

The third matter I can give you no definite information on as yet. Possibly many of you remember some months ago an editorial in the Electrical News with reference to the overlapping and lack of efficiency in our electrical organizations on account of each organization carrying on the same or very similar work in different sections, and therefore doing a large amount of work for itself which might very well be combined. The editorial suggested that something might be done to form a central electrical organization for the Dominion.

This matter has reached the stage where our executive has been asked to appoint a committee to meet with committees from the various other electrical organizations to consider whether such a central body could be formed and on what basis. A committee has been appointed by your executive, and I may tell you that a meeting is called for February 6th, in Montreal at which your committee and the committees from the Canadian Electrical Association and from all the other electrical organizations will meet to discuss this matter and report back to the executive and to you. So you may expect later on to have a definite report as to the possibilities of forming a central electrical organization and the basis of its formation.

### Closer Co-Operation

Let me add this thought, gentlemen. What we need at the present time more than anything else is closer co-operation among all the municipalities forming this organization. A short time ago it was brought to my attention that one of the small municipalities did not care to join us because its representatives felt that it would get nothing out of membership. I think that was a mistake on the part of the municipality and was due to its people not understanding all of the conditions, for I think most of you will agree with me that everything has been done for the benefit of all municipalities and not for a certain group.

Nevertheless this incident shows that it is incumbent upon us to see that in future no municipality shall be able to say that it does not receive any benefit from membership in this association. The municipality I referred to claimed that the conventions were solely for the benefit of the larger municipalities in the association. We must always keep before our minds that we have all classes of municipalities in our membership, and therefore we must carry on our work for the benefit of all our members generally, and not for the benefit of any one group in particular. That means that we must co-operate to the fullest possible extent.

### Committee on Regulations and Standards

#### By R. H. STARR

I received a complaint from one member relative to the expense of installing a separate switch on range installation. I communicated his complaint to Mr. Hall, chief inspector, and in his reply to me he stated that although the installation was recommended it was not insisted upon. This information I forwarded to the gentleman who had registered the complaint.

The executive meeting held last October authorized me to write Mr. Hall that owing to the different interpretation of rules and regulations by various inspectors it would be well to have a general meeting at least once a year for the discussion of the rules to insure their uniform application in the province.

Mr. Hall in reply stated that it was questionable whether the benefits to be derived would warrant the expense of bringing inspectors in for a general meeting, as a great many of them carry on their work single handed and their distance from Toronto would mean a day coming here and another day returning to their districts. However, he stated that before the next issue of the rules and regulations is ready for distribution it is just possible that a general meeting will be called. If this cannot be arranged, meetings will be held at the principal centres in various parts of the province.

### Standardization of Oils

There is another matter I should like to bring before the convention. Along about Christmas the Electrical World had a very good editorial on the Standardization of Oils. This struck me as being such an excellent editorial that I wrote to the different manufacturers throughout Canada asking them for expressions of opinion in regard to it.

Where five or six different oils are handled there is liability of the lineman or foreman getting hold of coal oil or some other oil rather than the proper standard oil, and the consensus of opinion of the various manufacturers was that something should be done towards standardization, one manufacturer suggesting that the whole thing should be referred to Mr. R. J. Durley, secretary of the Engineering Standards Association at Ottawa.

In view of the importance of the subject I think this association ought to take action on the lines of this suggestion, and I should like to have an expression of opinion from the members present.

It was moved by Mr. Shearer, seconded by Mr. Yates, that the report of the Regulations and Standards Committee be adopted, and that the executive of the association be empowered to take up with the department at Ottawa the question of standardization of transformer oils.—Carried.

### Twin City Electrical Exhibition

The secretary, Mr. S. R. A. Clement, reported that he received a letter from Mr. George O. Philip, manager of the Twin City Electrical Exhibition as follows:

"During the week of May 1st to 6th next, there will be an electrical exhibition held in Kitchener. This exhibition is being put on under the auspices of the local Electrical Contractors and Dealers Association and the Waterloo and Kitchener Light Commission. The exhibition promises to be a great success and will contain some new and unique features illustrating what can be done with electrical current.

[Concluded on page 49]

M. J. McHenry, re-elected President

A. T. Hicks, Vice-president

S. R. A. Clement, Secretary

## Sociability as an Aid to Co-operation—Little Speech-making and Much Fun Brings Montreal Fraternity Together

The use of the social element is a potent method of propaganda. Many men have little or no use for a lecture or an address, but a smoking concert, luncheon, or dinner is another matter. The prospect of a "good time" in this work-a-day world is not to be resisted. This was shown by the second smoker of the Electrical Co-operative Association of the Province of Quebec, held in the Rose Room, Windsor Hotel, on February 8, when about 500 attended, representing every branch of the industry. The arrangement of the hall was admirable. It consisted of small tables, each accommodating five or six people, with a head table, and a large table running down the centre of the room.

The speech making was brief and to the point. The key-note was co-operation. The chairman, Mr. L. I. Mc-Mahon of the Bell Telephone Company, introduced the speakers—K. B. Thornton, president of the Association, and K. A. McIntyre, Canadian representative of the Society for Electrical Development, New York. After welcoming the visitors, Mr. Thornton remarked upon the harmonizing of all sections of the industry, as shown that night—a thing which would have been considered impossible a few years ago. Now was the time to get together and boost the industry. He was of opinion that trade prospects had improved, but it was essential that they should co-operate in order to decide the maximum results. They must get together. Mr. Thornton's speech was very cordially received.

Mr. McIntyre pointed out that in co-operating they must have a definite object—something that was calculated to benefit everyone. In this connection he spoke of the Electrical Home to be shown in Montreal, with the principal idea of educating the public on better wiring. In Toronto an exhibition of this character had been a great success. In 13 days 19,555 had visited the home, each receiving a brief talk on electrical wiring. In order to attain success, it was necessary to have volunteers to do this work, and he appealed for help in this derection when the Montreal Home was opened. Every section of the industry would benefit—and here was an opportunity for practical co-operation.

In the course of the evening Mr. J. W. Pilcher spoke briefly on the subject of securing greater business by concerted action.

The arrangements were made by the following committee: Messrs. N. Richards, Northern Electric; Geo. Atchison, Southern Canada Power; A. L. Jones, Canadian General Electric; L. I. McMahon, Bell Telephone; E. S. M. McNab, C. P. R.; Jack Kavanagh, Northern Electric; W. McCarthy, Northern Electric; T. Obennell, Bell Telephone; W. Gibson, Canadian General Electric; J. A. Dick, Canadian General Electric; C. B. Sifton, Northern Electric.

The programme was varied and included songs, dramatic recitals, card tricks, and boxing, the latter by boys from the University Settlement. The following also contributed: Miss Thevenard, Miss Ryan, Joe Bochon, Jock Hunter and C. Cooper. Some impromptu fun was provided by Messrs Pilcher and Hiller. Mr. Bochon acted as leader in some community singing.

Electrical and other firms donated some eleven prizes, mainly domestic electrical appliances, which were drawn for.

The concert was a complete success, the attendance being even larger than was anticipated. There was apparent the spirit of co-operation which was so strongly emphasized by the speakers and which, in fact, was one of the main objects of the gathering.

## Mr. E. H. Porte tells Hamilton Contractor-dealers about "Better Merchandising."

A meeting of the Hamilton District Electrical Association was held on Wednesday evening, February 1st, in the Board Room of the Y. M. C. A. Dinner was served at 6.30 p.m. to about thirty members. A report was received from the Legislation Committee on the Licensing Act for electrical contractors and dealers, which was adopted and carried unanimously. Formal approval of this Act has been forwarded to the secretary of the Ontario Association of Electrical Contractors and Dealers. Mr. James A. Daly, manager of the Northern Electric Co. at Hamilton, was called upon to introduce the speaker of the evening, Mr. E. H. Porte, manager Renfrew Electric Products Ltd. Mr. Porte gave a very interesting talk on merchandising of electrical equipment. He urged the necessity for more "system" in the average electrical contractor's business; better accounting; proper determination of selling prices; better merchandising methods, and everyone present benefitted to a large extent from the speaker's remarks. Mr. Norman Crawford of the Crawford Electric Co. moved a vote of thanks to the speaker for his very inspiring address; this motion was seconded by Mr. William Thornton. The members were also entertained by Mr. William Doerr of the Repeater Six Fuse Co., Burlington, who told many jovial stories of travel, and rendered songs that brought a hearty laugh from all present. The new executive have decided that in future the monthly meetings will be held in the Board Room of the Y. M. C. A., commencing with a dinner at 6.30p.m. At each meeting there will be a speaker who will address the members on various subjects of interest to all connected with the electrical industry. Mr. K. J. Donoghue is secretary-treasurer of the Hamilton Association.

## Nesbitt Electric Manufacturing Co.

The Nesbitt Electric Manufacturing Co., Ltd., has been formed, with head office at 95 King Street East, Toronto, as successors to The Canadian Krantz Co., to manufacture and sell in Canada the well-known Krantz equipment, such as safety and live face switchboards, panel boards, switches, motor starters, etc.—in fact, the complete line manufactured by the Krantz Works, Brooklyn, N.Y. The officers of the company are particularly well-known men in the trade. Mr. R. H. Nesbitt is president; Mr. A. Ross Oborne, vice-president, and Walter Warren, late manager of the Central Electric Supply Company, treasurer. In the hands of these men the trade may be assured of quality products in keeping with the Krantz reputation, backed by an excellent service.

## Twenty Thousand Interested People
## Visited Toronto's Model Electric Home

# Campaign for Educating the Public

### In the Necessity for Adequately Wired Homes meets with Tremendous Success — Industry has Already Experienced a Decided Impetus Locally — Results will be Cumulative

It is too early to attempt to relate the whole story of the successs of the Toronto Electric Home campaign; but enough has been demonstrated in the first exhibit to indicate that the results will be very far reaching and will fully justify the trouble and expense in which the electrical men of Toronto have been involved. Though we cannot as yet speak entirely definitely of results, the story of the campaign is full of interest and to other cities that may be contemplating a similar movement, prove of benefit, in the way of a precedent—for which there is altogether too little in our industry.

Like many other movements that have started from small beginnings, and developed as the objects became clearer and the obstacles were removed one by one, the Electric Home idea had its birth at a meeting of a few representative electrical men in Toronto who lunched together in the Board of Trade Building and discussed the necessity of doing something to arouse interest in the minds of the general public in electric wiring and electrical appliances and, at the same time, stimulate the flagging enthusiasm of the members of the industry themselves. At this little meeting it was decided to form an Electric Home League, and before the meeting broke up ways and means were discussed of financing the undertaking, the necessary committees that should be involved, the personnel of these committees and their duties throughout the campaign. Fortunately, through the helpful co-operation of the Society for Electrical Development, Mr. K. A. McIntyre, the Canadian representative of the Society, was placed at the disposal of the Electric Home League and matters were left very largely in his hands to arrange and develop. Mr. McIntyre has throughout practically acted as general manager, and the results obtained are the best testimonial of his untiring efficiency.

The committees formed included a Central Committee, which also functioned as a Finance Committee; a Publicity Committee; a House Committee; an Appliance Committee; a Wiring and Illumination Committee and last, but by no means least, a Results Committee.

Generally speaking, the Central Committee's duties were to co-ordinate the plans of the various other committees; receive and pass up on reports of these committees from time to time; offer suggestions and assist in every possible way in carrying on the work. The Finance Committee was delegated with the important duty of raising funds and keeping a watchful eye on the expenditures. The Publicity Committee had charge of advertising in the newspapers; publicity posters on the street cars; indicating cards on the distribution poles; electric signs; window cards for contractors and others; stickers for letters; literature to be handed to the visitors as they passed out; the selection of dignitaries to officiate at the official opening; keeping in touch with the publicity side of the daily press; providing speakers before this and that organization of men and women, etc.—anything that would bring the idea in a favorable light before a large number of citizens.

The House Committee had charge of the Home itself, after all the work of installation had been completed and the public began to arrive. No visitor was allowed to pass through the Home without carrying away a definite idea of the purpose of the campaign, every room having been visited and the cause and effect of the outlets having been explained.

The Appliance Committee secured the particular appliances that were shown in the Home and it was their business, also, to see that these appliances were on hand as required.

The Wiring & Illumination Committee, as its name implies, had charge of the rehabilitation of the wiring to bring it up to a standard in keeping with the demands of a real Electric Home. They also were responsible for the illumination of the various rooms.

The work of the Results Committee has just begun. It is their duty to see to it that the contractor and the merchant follow up the advantage provided by the exhibition in placing

---

The plans on the following page indicate the location and the number of the outlets in the Electric Home; the photos also do this to some extent although it will be noted that only corners of the rooms are shown and, in consequence, only a percentage of the outlets. In room "A", the living room, arrows indicate the approximate locations of a number of outlets, including two under the mantel-shelf, a telephone outlet and the floor outlet for lighting. Room "B" shows two baseboard outlets, a telephone outlet, a reading bed lamp, etc. Room "C" is a corner in the kitchen, with the refrigerator showing through the open door in the hallway. Note the switch controlling the heater down m the furnace room. "D" is a corner of the dining room. Note the decorative wall fixture, the baseboard outlet and the outlet in the table. This table is equipped with a number of outlets. "E" shows the use of a wired tea table, where the ladies are using a toaster and tea urn. In "F" note the ceiling outlet, also the outlet for the iron and the arm controlling the cord. The ironing machine does not show in this picture. In "G" note the lighted closet; two outlets are indicated, through at least two more were supplied in this room. The central picture is a typical view, taken by flashlight as an auxiliary to the flood light. This indicates how the people streamed into this Home, hour after hour, throughout practically every afternoon and evening of the demonstration. For a number of the interior views we are indebted to G. R. Anderson, Professor of Illumination, University of Toronto.

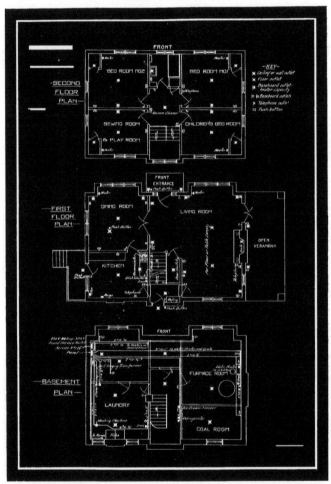

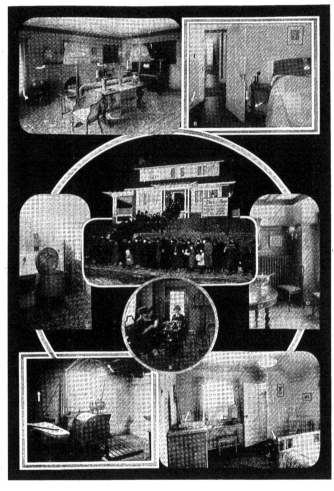

the public in the frame of mind where they are interested in having more outlets and more appliances.

The actual operation of the various committees is covered in greater detail below:

#### The Finance Committee

A campaign of this kind cannot be run on sentiment or "best wishes" and the Electric Home League had no false ideas on this point. The very first activity of the Central Committee was to prepare a schedule of costs, considering in detail the various items that must enter into the expense, allowing a fair amount for contingencies. In a city the size of Toronto it seemed necessary that more than one Home should be shown, and so it was decided, in a general way, that the first Home should be centrally located in a section of the city which represented approximately one-quarter of the population. Within the next few months three other Homes will be shown at intervals of one month or six weeks, or thereabouts, where the other sections of the city will be brought more closely in touch with the idea. So far could be judged by a careful study of the knowns and unknowns in this campaign, the sum of $13,000 was the amount necessary, and this money it was proposed to raise approximately the following way: the Toronto Hydro-Electric System was asked for—and contributed—$5,000; the manufacturers and jobbers together are contributing $3,000; the contractor-dealers have been asked for $3,000 and the real estate firms, of which four will be interested before the campaign is completed, are expected to contribute at least $2,000. While the whole of this sum has not yet been assured, splendid progress is being made and there is every indication that each section of the industry will do the part that is required of it.

#### The Publicity Committee

It goes without saying that a very large percentage of the total expenditure is incurred in newspaper advertising, and of the total sum used for this purpose the proportion required to stimulate interest in the first Home is necessarily much greater than will be required for the others. A scheme was very carefully worked out, by which the interest was stimulated before the Home was opened, but, after the public were fully advised, the expenditure was curtailed—the announcements being much smaller. This has been found to work out very satisfactorily. For the second Home it will not be necessary to spend so much money to get the public interested and by the time the last Home is put on exhibition it is expected the public will be looking for it, rather than having to be coaxed to read about it. Mere announcement will thus serve the purpose.

Street car publicity, in the way of posters on the front of the cars, proved very effective. "This car to the Electric

Home, Oakwood & Regal Road" was shown daily on all cars passing in the vicinity of the Home. Stiff cardboard signs reading "To the Electric Home," with finger pointing, were installed at a height of about eight feet on all main streets leading to the Home. Hundreds of these signs were used, the Hydro poles being utilized for the purpose. A limited number of electric signs were also used.

The Electric Home was quite prominently located on a corner lot, the house itself being placed well back from the street, with a high terrace intervening. For this reason it lent itself admirably to flood lighting, and one of the illustrations shows what a striking effect caught the eye of the visitor as the Home first came into view. The roof of a house on the opposite side of the street served to carry two powerful flood lights, which illuminated the Home, the sign and the long string of visitors who patiently waited—some of them for hours—to get a peep at this fairy home.

Of very great importance, in the way of popularizing and advertising the Electric Home, were the addresses given by various members of the Publicity Committee, and others, before organizations of every kind. For example, the matter was explained by five or ten minute talks at regular meetings of the Electric Club of Toronto; the Rotary Club; the Canadian Club;

### THIS CAR
### TO THE
# ELECTRIC
# HOME
### OAKWOOD & REGAL RD.

the Kiwanis Club; the Women's Canadian Club and business men's and women's organizations throughout the city. As an example of the care taken to reach the right kind of people, one may mention the choice of Mrs. Bruce, president of the Women's Canadian Club, to officially open the Electric Home. Mrs. Bruce was so pleased and enthused that she carried this enthusiasm back to her own organization, thus reaching thousands of the best homes in Toronto by a direct appeal. This is merely an example of the efficient way in which the Publicity Committee functioned to popularize the Electric Home in the minds of the people who were best suited to avail themselves of its advantages.

Nor was it necessary only to enthuse the general public. As with all new movements, electrical men themselves were somewhat slow about throwing up their hats, and it was thought advisable to hold a general meeting to which all members of the electrical industry were invited. This brought together some 200 or 250 of the livest men in the trade, where they were fully advised of the activities of the League and the possibilities underlying them. The program at this gathering was most unique, one of the items, for example, being a dialogue between a householder—a sharp business man—and two contractors, one of whom represented the ideal of a "cheap" job, the other the ideal of a "good" job. The amusing result was that a contract, previously let for some $80 after the "cheap" contractor had urged his customer to cut out a number of necessary switches and outlets, was cancelled and given to the second contractor, who sold his customer a $700 job. Following this meeting, interest in the campaign was much more general and enthusiastic, and the Central Committee received a much wider co-operation.

These are typical instances of the work of the Publicity Committee, to whom due credit must be given for the popularizing of the whole movement.

#### The Wiring Committee

When the League took over the control of the Home its

construction had just been completed and it was still on the market. As might be expected, it was not properly wired, and the first duty of this committee was to attend to this matter. The wiring of the Home, in the first instance, must have represented about 1 per cent. of its total cost; when the committee had finished with it, the wiring would represent about 5 per cent. of the total.

Plans and specifications were prepared by a firm that specializes in that work, and the wiring of the Home was let to a reliable contractor. This particular contractor was determined by lot, the members in the district all having an equal chance. On account of the short time at the disposal of the League, this work was done on a cost plus basis, the real estate dealer being assured that it would not exceed a certain amount. The owner, of course, paid for this work. During the installation the Wiring Committee constantly inspected the work and, by the evening of the day before the official opening, everything was complete; the house was well equipped with single and duplex outlets, switches and lights and the capacity of the service in every case was sufficient to take care of any demand that would be made upon it.

### The Plans

The plan of the Home is shown herewith and the outlets are indicated. It will be seen that there was nothing faddish about the installation—nothing unnecessary. In every room one outlet was wired heavy enough for a heater service and the word "Heater" was printed in prominent letters across the plate. A number of the outlets were duplex. It will be seen that the living room has no ceiling outlets. The wall lights were intended for decorative effect chiefly, the real lighting being furnished by semi-indirect floor lamps fed from a floor outlet suitably placed. The house was wired for telephone service, two outlets being supplied—one upstairs and one down—so that the same telephone could be plugged in and used either in the living room or at the bedside in the main bedroom. This service might have been extended, but the committee considered in all points the advisability of not overdoing anything—not letting the average man or woman go away with the idea that this was something beyond their reach.

The appliances in the living room included a wired tea wagon, upon which was demonstrated the use of two appliances such as a toaster and a tea urn; an electrically operated phonograph; a fan; an electric grate; an electrically operated player piano, etc. Another feature of this room was the light in the coat closet operated by an automatic switch as the door opened and closed.

### Followed a System

Nothing helped so much in making the exhibition a success as the system followed in keeping the different parties, or groups, together and in a certain fixed order. With people clamoring for admittance the greater part of the day, it was found necessary to limit the time taken by each demonstrator to three minutes, and groups of ten were thus admitted and expelled just that often. As a group of ten entered it was directed according to a predetermined plan. The first demonstrator explained in a few words the general object of the exhibit, emphasizing, not the appliances, but the wiring of this Home and illustrating his remarks later by a display of the various outlets in the room. This consumed from nine to twelve minutes and in the meantime two or three other groups of ten had been admitted and placed in charge of other demonstrators. It was found necessary sometimes to keep as many as five or six demonstrators in the living room at the same time.

### 100 Per Cent. Load-Factor

In each room in the basement, in the kitchen, the dining room and in each of the four rooms upstairs it was found necessary the greater part of the day to keep an attendant in charge. At the busiest times it was also necessary to have three or four "traffic" men whose duty it was to see that the various parties kept together and followed their proper course. This was absolutely necessary for, difficult as it may be to one who was not present to realize the extent of the crowding, it was a fact that from early afternoon each day the Electric Home was working absolutely on a 100 per cent. load-factor basis and the only reason more people were not shown through was that the capacity of the house was not greater.

The plan of any Home must regulate the proper course for the visitors to take, but in this case, after leaving the living room they moved downstairs past the electric refrigerator into the furnace room where the electric water heater was operating, then into the switch room where the advantages of a properly equipped safety board were shown; from there into the laundry with its washing machine, fed from an outlet conveniently and safely placed in the ceiling, its ample wiring capacity—in conduit—for heating the electric ironing machine and operating the motor, the auxiliary sad electric iron with indicating lamp attachment, all properly illuminated. From here the course led to the kitchen, the centre of interest to most of the ladies, wired for a range; an outlet for buffer or polisher, coffee grinder, etc.; dish washing machine, all connected up to the hot water system and sewer, etc. The kitchen was also supplied with a wall telephone. In the kitchen, also, was a switch controlling the water heater down in the furnace room; some specially effective lighting and, of course, the centre of the annunciator system throughout the house.

### The Wired Dining Room Table

In the dining room, amply supplied with baseboard outlets, the feature of greatest interest was the wired dining room table. A floor outlet of ample capacity supplied four outlets in the table. The energy was led by a flexible cord up the middle leg of the table where, through a suitable junction box, it was distributed through flexible conduit to different points of the table. The visitors were particularly interested in the utility and luxury of this feature and were convinced of its practicability by the demonstrators themselves, many of whom had experienced the value of wired tables in their homes over a period of years. From the dining room, visitors were shown upstairs, convenient switching arrangements being explained as they passed along first to the play room, which was also called a sewing room.

And then on to three bedrooms; there was of necessity a certain amount of sameness in these four rooms, but the main point was kept in mind throughout—the necessity for having a sufficient number of outlets for illumination, heaters, warming pads, reading lamps, telephones, hair dryers, vibrators, milk warmers, etc. etc. In this Home the bathroom was, unfortunately, so small that it was not possible to allow people to inspect it.

The stream of visitors was finally directed downstairs and out of the front door, an endeavor being made in every case to answer any questions by the way or pass a pleasant remark along the line such as—"Be sure to have wires heavy enough, and plenty of outlets," and the answer, invariably, was "We surely shall"; "We'll do our best," etc.

As a final send-off each visitor was presented with a descriptive booklet showing the plan of the house and the various outlets, giving a quantity of general information and on the back page a list of all the contractor-dealers who had shown an interest in the enterprise by becoming members of the Electric Home League. During the period some 15,000 of these booklets were carried away.

### The Signal System

Among the items of special interest may be mentioned the signal system for control of the groups, devised by Mr. McIntyre. Buzzers were installed throughout the Home with a control in the lower hall. At this point a man was constantly stationed, watch in hand. At the end of two and a half minutes a preliminary signal indicated to the demonstrators that they had one-half minute longer to finish their story and start the people on their way. A second signal at the end of the full three minutes saw each group pass in an orderly manner to its next position, to be followed automatically by the next group. As a result of this systematic control it is probably safe to say that twice as many people saw the Home, and those who did see it obtained many times more information than they would have obtained, had they been allowed to pass through the building in less systematic manner.

### The Number Who Attended

As has been already intimated, the house was full to capacity during a large part of the afternoon and the whole of the evening on each of the thirteen days the Home was open. Beginning at 10.00 o'clock in the morning, the house remained open until 10.00 o'clock at night, a period of twelve hours. Giving each group three minutes, it is evident that twenty groups could be passed through in an hour, i.e., 200 people; in 12 hours, 2,400 people. As a matter of actual fact, the number of counted visitors each day averaged 1,508, so that the load-factor for the whole twelve hours of the thirteen days averaged well over 60 per cent.

### Results

It is too early yet to expect definite evidence of results, but one contractor-dealer in the vicinity, before the end of the first week of the campaign, volunteered the information that fourteen different people had made a definite request to him to come up to their homes and advise them regarding the installation of more outlets, like some they had "seen in the Electric Home." Evidence of so definite a nature is surely proof positive of the value of the scheme. Another feature that was quite noticeable throughout the campaign was the sympathetic attitude of the daily press. That, of course, might be expected in view of the fact that a considerable amount of money was being paid them for advertising, but, quite aside from that, the nature of the exhibit caught the fancy and appealed to the intelligence of the women on the reporter staff, with the result that the objects of the exhibit were explained in a very logical way. This, by the way, is another activity of the Publicity Committee, of which mention was omitted in its proper place, namely, arranging for a visit of the lady reporters of the daily papers to the Home.

Space does not permit naming the demonstrators, who gave liberally of their valuable time to conduct the people through this Home. Men high up in the industry considered it a pleasure to take their share of this work, and consulting engineers, manufacturers, wholesalers, contractors and dealers alike lent their co-operation by being present in person and experienced the exhilaration of listening to the expressions of pleasure and appreciation of the general public as the various matters were brought to their attention.

### The Appliances

The Home was completely decorated throughout and furnished by the T. Eaton Company. This furnishing did not include any electrical equipment whatever. All apparatus and appliances were supplied by the manufacturers or wholesalers. The particular firms asked to supply these were chosen by lot. For example, if there were a dozen firms manufacturing electric irons, these names were first given numbers; the numbers were then written on cards that were later shuffled and placed face downward. The order in which the cards were drawn then indicated the order in which the manufacturers would be asked to supply irons. If the first one were unwilling or for any reason unable to do so, No. 2 was given the choice, and so on.

A particularly important feature in connection with the display of appliances was the removal of all marks of identification. Name plates were either removed or covered up in such a way that the general public could not read them, and the demonstrators were asked to explain, in case of inquiry, that

the pieces of apparatus simply represented types and that the various makes could be seen at any of the stores of the contractor-dealers whose names appeared on the back of the folder distributed to them as they would pass out. This procedure was followed so that no one manufacturer would gain any undue advantage over another. No one was expected to advertise any particular type of equipment, any advertising or publicity carried on during the period being concentrated on selling the people the "idea" of electric outlets in ample quantity and capacity.

The selection of these appliances, as a further safeguard, was placed entirely in the hands of a consulting engineer, in whom the industry recognized an absolutely impartial arbiter.

### "Cashing In"

This, briefly, concludes the story of Toronto's first Electric Home. Whether or not the full value of the effort is realized depends very largely on the contractor-dealers themselves. This exhibit has created in some 20,000 people a sympathetic, inquiring attitude of mind towards the better wiring of their homes. Without doubt, hundreds of these will be aggressive enough to approach their contractor-dealer, but there are hundreds and, doubtless, thousands, of others who will wait to be approached. This, then, is the contractor-dealer's opportunity. He should see that every home in his district is personally canvassed by a salesman who knows his business—either by telephone or by a personal call at the door. A cheerful "Good morning, Mrs. Brown; I expect you were one of the 20,000 that visited our Electric Home" will be sufficient introduction in numberless homes to gain ready access and the privilege of suggesting an outlet here and there, and later the supply of equipment for those outlets. It is a golden opportunity, indeed, for the contractor and the dealer which, indirectly, will be reflected in the added orders that reach the central station, the manufacturer and the wholesaler.

## Work of Wiring Committee

The Wiring Committee of the N. E. L. A. recently made the following report:—

The standardizing of attachment plugs and receptacles having been completed the Committee has been working on advertising this fact.

The Committee is working on standardizing medium size plugs and receptacles for large heater ranges, small motors, etc.

Also on the standardization of the appliance plug, where the cord connects to the flat-iron or other device.

The Committee is pushing the idea that there is a great difference in durability between the various heater and other cords on the market and that the average quality can be brought up by tests of different brands.

The Committee is endeavoring to get changes in the code to prevent the solid neutral. i. e., the omission of fuses on all grounded wire. Also to simplify and improve the so-called 660 watt rule.

Also to discourage the requirement of larger mains inside the buildings than the sizes which the central station experience with diversity factors shows to be safe.

Also to provide machinery to relieve the present situation whereby nothing is put in the code until tried in the field and whereby nothing may be tried in the field until after it has been put in the code.

The Committee is studying the best means for selling wire to the extent that this differs from selling electricity.

The Committee is developing the policy that the so-called electric fire loss can be dealt with by devoting special attention to each particular cause as flat-irons. over-fusing, poor portable cord. etc., rather than lumping all electrical fires including light. static. electricity, etc., together.

# Your Credit Standing Five Years Hence?

## A Few Outspoken Suggestions for the Electrical Merchant—If You Don't Cash in, Your Opponent Will

### By C. D. HENDERSON*

Will your business, five years from to-day, be referred to as a sound, successful institution, or will you be struggling along working night and day in order to keep a few paces ahead of the sheriff!

It is entirely up to you!

To my mind the electrical business offers better opportunities right to-day than any other line of business. Every place you go you hear people discussing electricity. Governments and financiers are banking water power projects; railway companies are electrifying their lines; house owners are realizing the comfort and convenience electricity will bring them, and almost daily some new way is found to utilize this mysterious power.

The electrical contractor-dealer is the connecting link between the user and the producer, and his success from every angle will be in proportion to the manner in which he conducts his business. He is the logical outlet for the distribution of electrical goods, but it will not come to him by divine right. He must put forth every effort at his command or be swept aside by those in other lines of business eager to cash in on the millions of dollars that will be spent in this potential market.

"Now" is the time to act. Not "When business comes back"—but right now when you have a little time to study the situation and put your house in order. Otherwise you will wake up some morning to find that the electrical business is gradually slipping away into other channels.

It is not necessary to learn everything by bitter experience. You know, without trying it, that if you jump into the river you'll get wet; or if you have a dollar and spend 20 cents you will have 80, not 90, cents left. These things have been proven conclusively; and so have certain fundamental principles of business which should form the foundation on which your business and mine may prosper.

Let us review some of the important features of business success applied to the electrical contractor-dealer:

### Profit

This should be the first and last word in your business

Just a few paces ahead of the sheriff

vocabulary. It stands out head and shoulders above everything else; it is the exact measure of business success, or you might say, it is your score in the game of business. Therefore, start now with this motto: "No article shall leave my place of business unless I know for a fact that it will produce a reasonable profit above overhead expenses."

### Overhead Expense

This is too big a subject and too important, to discuss fully in this article, and I have covered it fairly well in previous articles. Therefore, I will simply recall that from all facts, figures and experiences during the past few years overhead expense in the electrical retail business is not less than 25 per

cent. on sales, which is the same as 33 1/3 per cent. on cost. Keep an accurate list of all items of overhead expense so that you will know just what it is costing you every week to conduct your business. Don't estimate these figures—get the facts. Facts are the pulse of your business.

### Place of Business

In choosing a place of business remember—where there are crowds there is business—so locate as near the shopping centre as your capital will permit. Of course, you cannot compete with Woolworths for the choice sites; your class of business will not permit it, but don't be too saving on this point. It is quite easy to make up an extra $20 or $30 for rent in increased profits with a good location.

Make your place of business attractive to the feminé taste

### Appearance of Store

During the next few years most of your dealings will be with the womenfolk because the greatest advances in electricity will be confined to the home. This being the case, it behooves you to make your place of business attractive to the feminine taste. Always keep your store clean and neat, with your goods tastefully arranged so that the best class of women in your city will not hesitate to come in and look around. Learn from the experience of the big department stores that spend thousands of dollars to induce the women to feel at home in every part of the establishment.

### Merchandise

You cannot give too much thought and attention to choosing the different lines you are to handle. Don't allow price to be the deciding factor, bear in mind that most people are willing to pay a little more for articles of merit. This is particularly true regarding electrical goods, in view of the fact that they are not being replaced every year as is the case with other classes of merchandise.

Confine yourself to a few well-known, firmly-established makes, rather than attempt to carry a little bit of everything—which is not only confusing to the customer, but to the salesmen as well. Furthermore, the fewer the lines, the less investment, overhead expense and service trouble.

### Employees

You and your business will be judged by the class of people you employ. Impress this on the minds of each and every one of them from the apprentice up to the manager. Make them feel they are part and parcel of your business and that every act of theirs will re-act either favorably or unfavorably. Develop a spirit of co-operation and responsibility, and even a little competition within the organization will do no harm. Eliminate the words "boss," "weekly wage" and other relics of old-fashioned days, and in its place try to develop a spirit of family compact.

### Advertising

A lot of money can be wasted in a haphazard method of advertising. Yet a little thought and attention to this im-

*By President Henderson, Business Service, Ltd., and Canadian Manager A. B. C. Co., Brantford, Ont.

portant matter will pay big dividends. To my mind your show window can be made your best paying advertisement at very little expense. Place some one in your establishment in charge of this work and insist that your window be changed once or twice a week. As often as possible get something in motion in your window. This attracts attention, and when you get attention you get the opportunity to tell your story, and this is half the sale.

### Bookkeeping

No business should attempt to operate 24 hours without an accurate, simple set of books that will tell the owner each week or month just how he stands. A couple of hundred dollars spent for a good system at the outset will save its cost many times in succeeding years. A good bookkeeping system is the business man's compass; it leads him through the rocks and over the stormy seas into the calm waters of success.

### Service

Remember that "service" is a very important feature in connection with 90 per cent. of your transactions, and it can be made a curse or a blessing. People buy electrical goods with the idea that they will give years of lasting satisfaction, and so they should if properly sold and explained at the time of the sale; but repairs are necessary at times and by giving your customers quick and efficient service, you cement the bond of friendship and satisfaction. After selling a vacuum cleaner and seeing to it that the customer receives the proper service, your sales resistance is lessened for the sale of a washing machine, an electric range and other appliances. But don't overdo this thing called "service," or it will become a burden. I believe that during the first year after a sale is made

But—give them service

every reasonable service call should be given free; after that make your service department stand on its own feet and produce a profit and you will find most people quite willing to pay. But give them service!

### Records of Sales

By all means keep a record of the sale of each appliance sold, the make, style, serial number and other information about the article itself; also the name and address of purchaser and date of purchase. Of course, this need not apply to the very small appliances. These records will be found a wonderful help from many standpoints. They will enable you to settle disputes; will be found helpful for references to satisfied customers and also act as a sales record of any particular line.

### Sales Plans

Prepare a card index system of prospective purchasers of electrical goods, but prepare it—go through your books—the telephone directory—the lighting company's lists—and make a record of the people whom you know should buy and are able to do so. With this list made up and kept corrected you have something definite to work on. The manufacturers that you buy from will gladly supply you with circulars, letters, etc., to create the desire in these prospects' minds.

But this alone will not sell goods. You must follow it up with personal calls and arrange for demonstrations. Remember we are in a buyers' market and goods must be sold. The order-taker is a thing of the past.

Procure the services of a neat-appearing, well-mannered young man with an inclination towards selling. Teach him all

you can about your different appliances and, if possible, send him to the different factories for further training—then turn him loose on your prospect list. Do not make the fatal mistake of using this man for different odd jobs, or he will never make a successful salesman. There is plenty of work for him 10 hours a day on sales work that will produce big profits for you if properly handled.

This covers a few important features relative to the successful operation of an electrical contracting business. Analyze them—study them and talk to some of your competitors who are making headway. Then convince yourself you are in the best business in the world and resolve here and now that you are going to make a success of it.

---

### Mr. Pope, Manufacturers' Agent

Mr. Macauley Pope, son of Major W. W. Pope, secretary of the Hydro-electric Power Commission of Ontario, has established himself as manufacturers' agent at 166 Bay St., Toronto. At the present time Mr. Pope is representing Campbell & Isherwood, of Liverpool, Eng., manufacturers of electric drills; Fuller's United Electric Works, Ltd., London, Eng., manufacturers of batteries, cables, etc., and the Engineering & Lighting Equipment Co., Ltd., of London, Eng., manufacturers of au extensive line of lighting fixtures and fittings. A novel feature being shown by Mr. Pope is the "Anti-break" lamp economizer, designed to overcome the problem of vibration and shock. This economizer consists of a plurality of strips of phosphor bronze gauze cut and assembled in such a way as to damp out the waves of vibration before they reach the fragile filament. This economizer has been thoroughly tested out and is said to have proven efficient in engine rooms, over propellers on board ships, as well as under vibration set up by a pneumatic hammer on steel plates to which the fittings were attached. The device has even been subjected to tests on board naval vessels, where all the lamps not fitted were broken, the only lamps left to light the compartments being the ones fitted with "Anti-break."

---

### Result of Hoover Contests

The Hoover Suction Sweeper Company of Canada, through their general offices at Hamilton, Ont., announce the results of the two contests in the Fall of 1921. The prize to the leading district manager in Canada, was awarded on points, given for sales to dealer, newspaper advertising (depending on the circulation of the paper in which the advertisment appeared), and for sales to users, and was awarded to H. M. Potticary, district manager at Montreal. There were 6 prizes offered to resalemen, based on sales of sweepers and complete sets of attachments, and the prizes were awarded as follows:— 1st-R. N. E. Connor, Robert Simpson Co., Toronto; 2nd J. M. Murray, Public Utilities Commission, London, Ont; 3rd-H. L. Husel, Hudson's Bay Company, Edmonton, Alta; 4th-C. R. Robinson, Toronto Hydro Elec. System, Toronto; 5th-P. T. Brinsmead, R. H. Williams & Sons, Regina, Sask; 6th C. E. Lillie, Toronto Hydro-electric System, Toronto.

---

Mr. Herbert Smith, secretary to Commissioner Lowry of the Manitoba Telephone System, was married on the fourth of January to Miss A. K. Cleverly, late of Sussex, England, the ceremony taking place at St. Lukes Church, Winnipeg. To show the high esteem in which Mr. Smith is held by his associates, the heads of departments and chief clerks presented him with a complete set of silverware; he was also the recipient of a cut glass water set from the staff.

# The Merchandising of Electrical Appliances

### The First Step is to Create, in the Public Mind, the Desire for Electric Goods — Just as Possible, Just as Easy, to sell at a Price that Will Yield a Proper Profit

#### By W. W. FREEMAN

President Society for Electrical Development, before Electrical Supply Jobbers Association at Cleveland

I have heard it said that there are central station people who seem to think that the jobber is a peculiar species, one that must be regarded tolerantly, but not taken too seriously, and I have heard it said by representatives of the jobbers that the central station people might be similarly described. In other words, there seems to be an impression in some quarters that we are two different breeds of cats. I haven't felt that way myself, and I feel that way less and less the more I mingle with the other elements in our industry. I have given up, if I ever held it—I don't think I did—the thought that we can expect to agree, to have our minds meet absolutely with regard to the policies and issues that are of importance to each of us in our own operations. I don't believe we should agree if we have a diversity of interests, but there are many things upon which we can agree as to what is for the best interests of all. I believe that the successful jobber is successful because he practices the same intelligence and the same perseverance in his business as is exercised by the successful central station man. He differs from the central station man in some of his views, not because he is more intelligent or less intelligent, or more fair or less fair, but just because he is a jobber and the central station man differs from him just because he is a central station man. We must, necessarily, continue to disagree in regard to some things. The interests are not identical all along the line, and the man who is successful will sincerely follow the line of his own interest as he sees it.

In the large proportion of activities in which we are engaged we can agree, and we can work in the closest harmony and with the greatest enthusiasm, and the reason why I believe in the Society for Electrical Development, if I may refer to that for a few minutes, is because I know of no other basis that is at the present time available for a combination of actions and a combination of effort that should serve in equal degree to benefit the interest of each of the elements that can combine. You work together upon your problems, as you should; we work together upon our problems, as we should. But there are other problems which are of a mutual character, and other work to be performed in which we must join hands if we are to get the best results, and the purpose of this organization is to afford the vehicle for such action.

I submit to you that, irrespective of anything that this Society has undertaken to do in the past, or may undertake to do, the fact that it brings together a board of directors of twenty-five men, representing equally the diversified interests in the electrical industry, should, in itself, prove the value of the organization, because the opportunity, in itself, for you to send into such a Board of Directors four representatives of your branch of the industry who have free and full opportunity to present your problems as you see them, as affecting the problems of others, and to work in co-operation with the representatives of the other interests in formulating a program of general benefit, that I think in itself should carry convictions as to the value and the possibilities of such an organization.

One of the things that I like to think about, as opportunity affords, which is an inspiration to me, and I believe to all, is the progress that has been made in this industry of ours in its relatively brief period. The whole progress of the elec-

trical industry is within the memory of many who are within this room. I don't wish to mention names, but if you doubt that I will tell you some of them, proving my statement. The pioneers in each of the branches of the electrical industry, electric lighting and power, and telephone, the street railway, are in strong and vigorous health, and in active work to-day. It is very new, and yet notwithstanding that fact it ranks to-day high up in the matter of accomplishment, investment, and in earnings with the very biggest industries of the country. Compare it with the gas industry, one hundred or more years old, and yet there are only three or four thousand gas companies in the United States, compared with fourteen thousand central stations or electrical companies manufacturing electric current.

I have some figures furnished by the Society just bearing on the relative importance of our industry in relationship to others which I think are startling to those who have not realized this relationship.

#### A Very Heavy Investment

In the first place, the total investment in the electrical industry to-day—and in that I include manufacturing companies, the central stations, the telephone service, street railway service—is over seventeen billion dollars, and the revenue, in one year, is over four billion dollars.

Now that compares with the other large industries of the country—steam railways rank first, according to statistical information, and yet the total earnings and value of the product supplied during the year is only six billion, two hundred million dollars.

Second in volume is the slaughtering and animal products, which is five billion dollars.

The third, and tied with the electrical industry, is the petroleum industry, which has an annual income of four billion dollars, so, as a matter of fact, in this brief experience the electrical industry ranks third in all the industries of this country in the matter of the annual sale, the value of the products of the industry.

In presenting that to you I have another thought in mind, not one of mere congratulation. It is a record of which we all have a right to be proud. It should be regarded, however, principally in the light of pointing out the possibilities of the future rather than of singing our song as to what has been done in the past. The greater proportion of the results of the forty years of the industry have been realized in the last decade. What has been done in forty years can be duplicated in the next fifteen or twenty years, readily, if we fulfill all the possibilities. In these forty years of experience there have been wired about seven million houses for electric service. That is a very large figure. But in half that time Henry Ford designed a flivver, and, through the operation of one company, has sold five million autos. I can't conceive that, in contrast with the relative importance of having electricity in the home, and of having an auto, even if we can afford autos, that the auto would win out, yet here is one organization that in half the space of time has done more in the way of relative accomplishment than the entire electrical industry with regard to getting electric service into homes throughout the country. I cite that because that is a comparison that I think justifies the reflection and thought of you men.

### Creating the Desire

What was done to sell the auto, what counted for success, ultimate success, in selling the auto, was the **creation of the desire in the public mind** for the auto. It wasn't done through individual salesmanship as much as it was done through creating a wave of thought throughout the country, deep down into the consciousness of every individual, and the demand for these autos flowed in upon the agency and the autos flowed out, and were sold. The electrical industry, in my judgment, can be advanced more rapidly than in any other way by the creation of this wave of thought, and it has begun, and it has reacted very encouragingly up to date, but by no means to the extent that is possible.

Now all this is preliminary to telling you what, in my judgment, although that perhaps isn't shared entirely by the other directors of the Society, is the one thing, the one big thing, the one thing that is going to do more than any other one thing, or one act on the part of the industry to sell electricity throughout this country to every citizen, and that **is to have everybody throughout the country reading about electricity and electrical service, having it hit him between the eyes** when he picks up his magazine, paper or what not.

One of the things that this Society has been doing in past years has been to get into trade papers and the monthly national magazines and every conceivable form of literature the story to the average person as to what electricity means to him in his home, and in his factory, and in every application and walk of life.

Now that, as I said before, seems to me to be one thing alone, irrespective of everything else, that may be done by this united effort which will count in tremendous results, in very many times over the money spent, and furthermore, it is one thing that is absolutely impossible to do in any other way than through a national movement that fits in with, and in no sense contradicts, all of the possible effort that can be put into effect in each locality and in each place by the local company and by local activity.

### Something About Merchandising

I am asked to say something about merchandising. I am not a merchant of electrical appliances, and therefore I am not speaking from my own experience or outlining any policy which I am carrying out with the purpose of selling electrical appliances direct. I am tremendously interested in the accomplishment of the merchandising of electrical appliances. I believe this affords a tremendous opportunity in the immediate future. I have in mind, as one of the old men in the electrical industry, that it is not so very long ago when the electric motor was a novelty. The central station companies were organized primarily to do a lighting business, the power business was very, very incidental, but the power business is now seventy-five per cent of the volume of business of the larger companies. So that this motor business has been developed within a much shorter space of time than the lighting business. There are now twelve million, nine hundred and thirty thousand h.p. of electric motors on the lines of the central station companies throughout the country.

The next step in the enormous development along electric lines, paralleling to a considerable extent, if not entirely, the experience of the electric motors, will be electric appliances. Already central station companies find that in residential districts during at least the period of the year for which statistics are available, they sell as much electrical energy for other than lighting purposes as they do for lighting purposes. Our general manager in Cincinnati figured out for me the day before yesterday that there were potential sales with electric consumers now on our lines in the form of electric appliances for domestic use only to the amount of at least ten million dollars. Maybe I shouldn't tell that to these boys of Cincinnati, but I suppose I can't get them out of the room. There is ten million dollars in electric appliances that can be sold in Cincinnati before we connect another electric consumer to our lines, and I think I am entitled to credit, perhaps, for saying that I am not going to get into the business of selling any of those appliances myself. And furthermore, I am absolutely certain in my own mind that there is no necessity and if no necessity, no justification, for those appliances being sold on any other than a sound merchandising basis yielding the seller a proper profit on his work.

### Unlimited Possibilities

I believe with the right amount of effort it will be just as possible, perhaps just as easy, to sell those appliances, at reasonable rates—and so far as my influence goes with my own company or elsewhere, that is the policy I very heartily and cordially recommend. This merchandising business, therefore, of appliances, if that is a fair statement of the prospects, in a relatively small city, if that is the case in one city, multiplied by the number of cities in the country shows the unlimited possibilities.

Sometimes we have spent a lot of time discussing policies. Some of us are of the opinion that it is the policy that produces the results rather than the amount of work that is put behind the policy. I will tell you a secret, that in past years we have spent considerable time in our Electric Light Association Conventions discussing policy, and some companies have had one policy, some companies have had a policy totally different, and I noticed on more than one occasion that the representatives of two companies have given almost identical results as to percentage in increase of business over a corresponding period of time, and each one has said that he believed it was the policy that they adopted that led principally to the result. Then when he found out that another man in another city claimed the same thing for a policy diametrically opposed, he had to fall back upon the suggestion that perhaps a policy good for one city was not equally good for another city. It merely proved to me that it wasn't the policy at all that got the results, but it was the amount of work that was put back of it. I am rather growing to the philosophy in my own mind, resulting perhaps from age, that economists are disposed to place too much stress upon the value of certain policies and too little stress upon undertaking to get results, even though we are not entirely agreed to the policy that should be pursued.

I want to say to you men that I have only one sure rule for getting results, and that is WORK. I know that the right amount of intensive work will produce results. In our own territory in a city that has shown very little growth, where we have had to depend upon population that has been there for years, we have in five years multiplied our business and our number of electrical consumers practically by three. We have shown for five years past an average gain in the number of consumers and in the output of electrical energy of forty per cent per year, and we are showing in this month, as in each preceding month of this year, a record fully up to any part of that five years' record.

### Use Intensive Methods

When there was an inclination in the early part of the year to a falling off in results, when there was an inclination to say "industrial conditions are not going to be so good, we cannot expect such good results," we tried the plan of putting on simply more men to intensify our effort and increase our results, and we have proven to our satisfaction that it works.

What I had in mind in mentioning that, and of course what applies generally can't be applied too literally to each individual case, I recognize that, as you do, but what I have in mind is this: We are all of us thinking more or less that times are not as good as they were, we are thinking of curtailing and curtailment in expenditures along some lines is absolutely necessary. I want to merely suggest to you that when business is poorest, when business is not coming in, or the volume of business that comes in without very much effort is at the lowest point, that generally is the time for the most intensive effort on our own part to make the period of depression as short as it can possibly be. We can to a considerable extent, I know we can in the central station industry, and therefore I have the temerity to suppose that perhaps you can in your line, shorten the period; we can control these periods of depression by a greater activity on our own part. We can very often spend more money, even when every dollar seems like a drop of blood that we are spilling, we can spend more money to better advantage than would result from saving money, if the money is spent along productive lines.

My message to you men, insofar as I have a message, coming from our own industry, is a very firm belief, based on my own experience, that there is every occasion for encouragement, there is every occasion for the spending of money, real money, along lines of constructive and intensive effort, and I believe the more money we spend wisely, the more effort we put forth wisely, even in times when business seems to be poor, the sooner we will ride over that period and get back into signs of real, enthusiastic prosperity. I am sure that you are meeting here with that enthusiastic point of view. I have seen nothing on the faces of anyone here to indicate otherwise, and I am sure that in voicing these sentiments I am not saying anything that is contrary to your own point of view, but it may be encouraging to you and re-enforce your own ideas to have that message come from another industry.

## The Basic Fundamentals of Water Power

On Thursday the 19th January, Mr. C. H. Attwood, District Chief Engineer, Dominion Water Power Branch, gave a very interesting and instructive address, before the members of the Manitoba Electrical Association.

In the course of his address Mr. Attwood said:-

I think we all realize the extent and the richness of the agricultural lands of this Province, but very few people realize and appreciate Manitoba's bountiful inheritance of varied natural resources among which the water powers are of paramount importance.

The total power possibilities of Manitoba at known power sites or falls as at present estimated is 3,200,000 h.p. at ordinary minimum flow and 5,200,000 h.p. for maximum development. While many of these water powers are at present somewhat remote from the more thickly settled parts of the province they are for that reason more particularly important for the exploitation of the natural resources of the hinterland. Of the larger powers such as those on the Nelson River which amount to some two and a half million horse power we can confidently expect that with the great advances in the art of electrical development and transmission that most of these powers will in time prove to be important factors in the solution of fuel-power and industrial problems of the province.

Practically every great industrial centre in Canada is now served with hydro-electric energy and has within easy transmission distance ample reserves of water power. In those localities where water-power is not available nature has bountifully supplied fuel reserves of coal, gas or oil.

According to a recent computation, the water power resources of the British Empire have been placed at from 50 to 70 million horse power. To this total Canada contrib-

utes some 20 million horse power. From statistics just compiled there is installed throughout the Dominion some 2,430,000 turbine or water wheel h.p. and the ultimate capacity of the present plants and those now under construction totals some 9,400,000 h.p.

Over 90% of the primary power used in central station's throughout the Dominion is derived from water power, and of the total water power installed about 75% is installed in central electric stations, i.e., stations which are engaged in the development of electrical energy for sale and distribution for lighting, mining, electro chemical and electro metallurgical industry, milling and general manufacturing. The pulp and paper industry utilizes some 475,000 h.p. of which 385,000 h.p is generated directly in the pulp and paper plants, and some 90,000 h.p. purchased from hydro-electric central stations. Other industries may be listed as follows: For lighting purposes, 435,000 h.p.; in the mining industry, 180,000 h.p.; in flour and grist mills, 48,000 h.p.; in timber and saw mills, 38,000 h.p.; in other manufacturing industries, 175,000 h. p. These figures are evidence of the widespread manner in which the Dominion's water power resources are being applied to the furtherance of its industrial development. In addition to the above some 230,000 h.p. is exported to U.S.

With a water power development of 274 h.p. per 1000 of population, Canada stands well in the forefront in respect to the utilization of hydro power resources. The enormous water power reserves still untouched form a substantial basis for the progressive exploitation and dvelopment of other natural resources and if properly co-ordinated with the development and utilization of the enormous fuel resources of the Dominion, are an assurance of continued industrial expansion and prosperity.

## U. of T. Research Bulletin

A valuable bulletin has just been issued by the University of Toronto, Faculty of Applied Science and Engineering, School of Engineering Research. The bulletin contains five papers on Current Transformers, as follows:

Effects of Magnetic Leakage in Current Transformers.

An investigation of flux distribution in the cores of current transformers of different types and related effects. Some remarkable conditions are found in certain standard types.

Method of Measuring Ratio and Phase Angle.

A description of the method employed, and some discussion of errors in measurement from changing turns-compensation, due to change in wave form. Oscillagrams showing wave forms.

Through-Type Portable Current Transformers.

A quantitative investigation of the variations in ratio and phase-angle errors with changes in position and arrangement of primary conductor.

Determination of Turn-Ratio of Current Transformers.

A method too inconvenient for commercial use, but having some features of interest.

Minimizing the Errors of Current Transformers by Means of Shunts.

Effects on ratio and phase-angle errors of turns-compensation, and secondary and primary shunts of resistance or condensance or both.

The authors of these papers are H. W. Price, professor of Electrical Engineering, and C. Kent Duff, research assistant, either individually or in collaboration.

Mr. V. Judge, 51 Gladstone St., Montreal, has been awarded the contract for electrical work on five residences recently erected on Kensington Avenue, Montreal, at a cost of approximately $40,000, for Mr. G. E. Blackwell, 4184 St. Catherine St. W.

## A Tribute to the Work of the Water Powers Branch at Ottawa

Public recognition in the form of a glowing tribute to the work of the Water Powers Branch of the Dominion Government, was voiced by Edward Anderson, K.C., at a largely attended meeting of the Manitoba Electrical Association held recently, over which M. E. Deering presided.

Mr. Anderson said that due to the ability and vigilance of Mr. Challies and Mr. Attwood, and their colleagues in the Water Powers Branch, the maximum power development of the Winnipeg River was now assured for all time, and added "If the people of Winnipeg knew how much they owed to the officers of the Water Powers Branch they would be grateful indeed, for their work has been of inestimable benefit."

Mr. Anderson's remarks followed an address by Mr. C. H. Attwood, District Chief Engineer of the Dominion Water Powers Branch, on the "basic fundamentals of water powers" in which he described the organization which existed for measuring water, collecting data and putting same in shape so that it might be used when it came to the point of designing for water power developments.

Mr. Anderson said:—"I am indeed happy to have an opportunity in a public way of expressing my great appreciation of the efforts of the Water Powers Branch of the Dominion Government, which is represented in Winnipeg in the person of Mr. Attwood.

"Perhaps I am in a better position than anyone else to offer a tribute of appreciation of the work of the Water Powers Branch because of certain connection I have had with that Department and with the Dominion Government in connection with the Lake of the Woods levels investigation before the International Joint Commission. It was my privilege to act as Counsel for the Dominion Government in that investigation and in that capacity I had first-hand opportunity of observing the work which had been done by the Water Powers Branch of the Dominion.   I can tell you it was a real eye-opener to me, and if the people of Winnipeg realized just how much they owed to that Water Powers Branch they would be grateful indeed to that department.

"I have come into considerable contact with Mr. Challies, Chief of the Department, and never in my experience have I found public servants so zealous and energetic in the discharge of their duties as the gentlemen of the Water Powers Branch are—and always have been. They did not strike one at all as public servants in the ordinary sense of the word but rather people who have the interests of their particular department at heart, and who are ever on the alert to serve, not themselves, but the public.

'Let me illustrate in one or two instances how much the gentlemen of this department have done for the city of Winnipeg.   The Lake of the Woods is a great reservoir where the water that flows down the Winnipeg River is collected. Were it not for the Lake of the Woods, the Winnipeg River would not exist as a potential source of power. It is extremely important, therefore, that all data in connection with the lake and the flow of the river, should be known and the Water Powers Branch has gathered all possible data and have intimately studied the situation.   It is also exceedingly important that the water of the Lake of the Woods should be so regulated that there will be an even and continuous supply so that the maximum benefits of the Winnipeg River will be obtained.  The question of these levels was referred to the Joint Commission and the work done in this connection by the Water Powers Branch of the Dominion was the most valuable work that was done and presented before that Commission, and Mr. Challies and his associates were always on the job and alert, and kept in the forefront the

possibilities of the Winnipeg River and water powers to be developed there. Due to the ability and energies displayed by Mr. Challies and his assistants, and their ceaseless watchfulness, a good deal of good was accomplished, with the result that when the recommendations of the International Joint Commission come to be carried out as they undoubtedly will, you will have the maximum development that is possible from the water powers on the Winnipeg River.

"You probably all remember that not very long ago there was a good deal of discussion in the papers as to certain concessions obtained by Mr. Backus, and their effect on the flow of the Winnipeg River.   The Dominion Water Powers Branch has only jurisdiction over rivers where the title has not passed to the Province. In the case of Manitoba the title is still vested in the Dominion—though in most Provinces the title is vested in the Provinces.  The Winnipeg River runs through Ontario, as well as Manitoba.   Mr. Backus has great developments at International Falls, at the head of the Lake of the Woods.  Now there is a dam, called the Norman Dam, which is a strategic dam in connection with the outflow from the Lake of the Woods.  The Water Powers Branch has exercised certain jurisdiction over that dam but has never possessed complete legal authority to control it absolutely.  Mr. Backus got possession of the Norman Dam and acquired the Kenora Municipal Hydro-electric plant there and also got a concession known as White Dog Falls.  It appeared to the Province of Manitoba, the City of Winnipeg, and the Winnipeg Electric Railway Company, that by virtue of the concessions he had obtained, Mr. Backus might be able to so control the outflow of the Lake of the Woods and the water in the Winnipeg River that it would act to the disadvantage of the users of the water power, developed in Manitoba on the Winnipeg River.  And it was a serious, a proper and well-grounded fear that the people interested had in that connection.  But Mr. Challies and his department, were right on the job, and it was largely through the instrumentality of that department—Mr. Challies. Mr. Attwood, and other colleagues—that matters were so arranged that the Dominion Government has now, beyond all question of doubt, the jurisdiction of regulating the outflow of the Lake of the Woods into the Winnipeg River, and we may be assured that for all time to come we will get the maximum development of the Winnipeg River.

"I am just giving these as illustrations which came to my attention whereby the services of this department, under Mr. Challies, have been of inestimable value to the citizens of Winnipeg and it is only proper the people of Winnipeg should know it.  These gentlemen work in an unostentatious and quiet way.  They do not advertise and the public does not know just how much, in their interests, they accomplish. But they are doing a great public service, and it gives me great pleasure to pay this small meed of tribute to them. No occasion has ever arisen in my experience in which I have been so glad of the opportunity of saying a few words in public in appreciation of the work of public servants.

---

Mr. G. A. Magowan. 94 Glendale Ave., has secured the contract for electrical work on a two storey building, being erected for the Goldstein Manufacturing Co. at 12-18 Beverley St., Toronto, at an estimated cost of $75,000.

## Visiting Western Branches on Power Business

A recent visitor in Vancouver was Mr. W. O. Taylor, sales manager of the Northern Electric Company's power department.  He arrived from Montreal on Jan. 23rd and left for the east on Jan. 31st, going via C.P.R. and is visiting all of the company's western agencies en route.

## Contractors Meet in Montreal to hear Discussion on Merchandising Topics

The French & English sections of the Electrical Contractor-Dealers' Association of Montreal held a meeting at the Windsor Hotel, Montreal, on February 9th, at which Mr. C. D. Henderson, president of the Henderson Business Service, of Brantford, Ont., and Mr. Kenneth A. McIntyre, the Canadian representative of the Society for Electrical Development, New York, were the chief speakers. Those present also included some jobbers. Mr. Gaspe de Beaubien presided.

Mr. McIntyre's speech dealt principally with the electrical home shown in Toronto, with a view to giving some pointers in connection with the electrical home to be shown in Montreal. He referred to the advantages which would accrue to all sections of the industry from such an exhibition, remarking that it had an educational value to men in the industry as well as to the general public. He read a letter just received from Toronto outlining the results of the electrical home in that city. These included numerous sales of appliances, increased inquiries directed to controctor-dealers, inquiries for wired furniture, and also for information as how to lay out houses on the lines of the electrical home. He declared that the contractor-dealers were themselves partly to blame for the comparatively small amount spent on wiring, as in order to secure contracts, they had often suggested the elimination of a number of outlets.

Mr. Henderson described his business service and its operation. He also referred to the loose commercial methods of many electrical contractor-dealers. Of the four links in the electrical chain, that of the contractor-dealer was the weakest. Apparently the aim of many contractor-dealers was to get business, irrespective of profit. Some jobbers and manufacturers were also affected in a like manner. Volume was no doubt to be desired, but the paramount end in view was to secure a profit on the business done. Many contractor-dealers declared that if they added overhead to their charges they would lose jobs, which was a poor system of conducting business. Mr. Henderson said it would pay the large companies to undertake a campaign for educating contractor-dealers. During last year nearly three hundred thousand dollars were lost by the failure of contractor-dealers in Canada, a sum which could be cut at least in half if contractor-dealers would use good business sense, cease competition which was neither profitable to themselves nor to their neighbor, and would take the trouble to know the amount of profit they were making. He believed in making a legitimate profit, which in the end, was better for all concerned in the industry.

Mr. J. A. Anderson, president of the Association, spoke of the progress which the Association had made in securing the co-operation of the contractor-dealers and jobbers. Let them, he said, charge prices which were fair and reasonable and which would allow decent wages to the workmen. Many contractor-dealers were working at rates which did not allow them a fair profit, and it was one of the objects of the Association to educate such men on the necessity of adopting a higher standard in order that it would benefit not only themselves but the industry generally. Mr. Anderson added that the English section now numbered 35 members, nine new members being added within a week.

Mr. S. W. Smith believed that the efforts of the Association would gradually result in stronger co-operation between contractor-dealers and the jobbers. They must get together in order to put their business on a better basis.

In the course of further discussion it was stated that the contractor-dealers and jobbers had agreed on a code of ethics.

## Annual Convention A.M.E.U.

[Concluded from page 35]

"It is proposed to have on May 4th, a Hydro day and I hereby wish to extend to the members of your association a hearty invitation to visit this exhibition on the above date. If you will be kind enough to send me the names of members of your association, I shall be very pleased to mail them complimentary admission tickets to the exhibition.

"I shall also be pleased to send you more information regarding same from time to time."

Mr. V. S. McIntyre, Hydro Manager at Kitchener, supplemented this letter by a very cordial invitation to the members of the electrical fraternity.

He pointed out that this scheme is not primarily for the purpose of advertising certain lines of electrical goods, but rather to demonstrate the adaptability of electricity to our everyday needs and that there will be about 52 exhibits of the uses of electrical current. He mentioned certain unique applications of electricity. For instance, the Rennie Seed Company are going to show a lawn grown by electricity without ground for soil. They are going to clip the lawn inside of 24 hours and mow it inside of five days. The grass seeds are put on a piece of wet flannel, and electric current passed through this flannel germinates the seeds. Another novelty will be a wireless concert played by an orchestra in Pittsburg.

### Election of Officers for 1922

It is always at this meeting that the officers are elected for the year. The balloting resulted in the election of the following gentlemen:

President, M. J. McHenry; vice-president, A. T. Hicks; secretary, S. R. A. Clement; treasurer, G. J. Mickler; directors at large, P. B. Yates, O. H. Scott, H. H. Couzens.

District Directors—Niagara, J. J. Heeg; Central, W. E. Reesor; Georgian Bay, E. J. Stapleton; Eastern, H. F. Shearer; Northern, R. H. Staford.

### The "Get-Together" Dinner

The usual "Get Together" Dinner was held on Thursday evening, and every delegate must have been present, for the dining room was crowded to its limit. The speaker of the evening, as announced beforehand, was Mr. F. A. Gaby, chief engineer of the Hydro-electric Power Commission of Ontario, but in Mr. Gaby's enforced absence Major Hume Cronyn, of London, called upon at the last moment, very kindly consented to take Mr. Gaby's place. His address dealt with Canada's necessity for an industrial research laboratory. Major Cronyn was Member of Parliament at the time this matter was brought up in Ottawa by the Advisory Council for Scientific and Industrial Research. Indeed, he was one of the strongest advocates in the House of the formation of an Industrial Research Department at some strategic point or points in Canada, where the problems of the Canadian manufacturer could be studied and solved. He outlined the fight put up by the Government for the establishment of this laboratory and the Bill's final defeat in the Senate. The address was a very masterly presentation of the needs and claims of Canadian industries for governmental assistance of a kind such a laboratory could offer and there was no question but that the sympathies of the audience were entirely with the movement. Indeed, this sentiment was expressed in every quarter, with the hope that the present Government would see its way to introduce and carry through a similar Bill in the near future.

The meeting adjourned early so that the members might have the evening to themselves to attend theatres. A number of the delegates took advantage of this lull in the proceedings to visit the Electric Home being shown in Toronto at that time by the Electric Home League.

# What is Newest in Electrical Equipment

### Convenience Arrow Plug

The Arrow Electric Company are making the "Convenience" plug illustrated below, by means of which it is more easily possible to attach the appliance to a lamp socket with a shade on it. The use of a lamp socket for electrical

appliances is, of course, not the ideal method or a method that can be encouraged indefinitely, but, in the meantime, it is much better that householders should utilize devices of this sort than be without electrical appliances.

### "3-in-1" Searchlight

A searchlight equipped with three lamps has been placed on the market by the Franco Electric Corporation, Pearl and Tillary Sts., Brooklyn, N. Y. Features of this device include a scientifically developed reflector throwing a sharp, concentrated beam of light; a newly developed lens ring with a nickel finish; and an improved contact box which permits each

lamp or all three to flash intermittently or light continuously. The device is equipped with spring lamp sockets so as to hold reflector in place when lens and lens ring are removed. For signalling purposes red, green and clear lamps may be inserted. The shock absorber is independent of the reflector and lamp base and this will protect the lamps against any undue jars. Equipped with 3.8 volt Mazda lamps and "Franco Monocells."

### Electric Vaporizer for Motor Cars

The device illustrated herewith, an Electric Vaporizer manufactured by the Kase Electric Co., Sherwood Building, Duluth, Minn., will interest motor car owners. It consists of a set of four metal grids in a heat insulating case, made in the form of a gasket, which is placed between the carburetor and

the intake manifold flanges. One terminal of the heating grid is connected to the battery, through a snap switch placed in any convenient location, and the other terminal is grounded. The object of the heater is to dry and warm the gasoline vapor as it passes into the manifold so that ignition may take place promptly.

### New Street Lighting Unit Gives Good Distribution

The Westinghouse company has recently developed a highly efficient post top of novel design for use with Mazda "C" lamps, in which upper and lower parabolic reflectors are used to direct the light on to the plane of illumination. The quality of the light emitted by the Reflecto-Lux units is brilliant and sparkling, and they have been designed to distribute a flood of light on the streets, with a small amount upwards to light the fronts of the adjacent buildings.

The maximum light is emitted at approximately 20 degrees below the horizontal and the distribution is ideal for mounting heights and spacings customary with ornamental street lighting. The distribution is obtained by upper and

lower parabolic reflectors, which direct the light outwards and, in addition, a portion of the light in the upper hemisphere is re-directed by an opal glass band around the upper hemisphere of the lamp, or by a band of enamel on the lamp itself. The construction of the Reflecto-Lux units is rugged, the frame is of galvanized cast iron, and the glass panels are set in felt gaskets, making the whole thoroughly dustproof. In the post top, the lamp burns in a "tip up" position and is readily accessible for cleaning and replacement through the hinged top cover. In the pendant unit, the lamp burns "tip down" and access is obtained to the interior of the lantern through the bottom casting, which is hinged.

### Babbitt Metal and Babbitting Motor Bearings

Babbitting motor bearings is described in detail in Circular Reprint No. 104, published by the Westinghouse Electric & Manufacturing Company, East Pittsburgh, Pa. The publication is a discussion of the production of babbitt metal by J. S. Dean, of the Railway Engineering Department, Westinghouse Electric Company, and it contains a number of photographs of equipment used in the manufacture of babbitt metals as well as the results of various tests of samples of alloys.

A general discussion of the cost of lead base babbitt metal is contained in Folder No. 4476, just published by the same company. The publication describes the properties and applications of Westinghouse lead base babbitt metal, known as Westinghouse Alloy No. 23, and genuine babbitt, known as Westinghouse Alloy No. 14.

The Crouse-Hinds Company of Canada, Ltd., are distributing a folder describing ZY Condulets for control of small motors. This folder is well illustrated, both in the detail of these condulets and in their actual installation:

## New Sign Receptacle

The Canadian General Electric Company have a new sign receptacle in which the screw shell can be removed and replaced by a new one just as readily and easily as a burned out lamp can be replaced. Those who have had experience in keeping electric signs in service will realize the importance of this removable shell without the necessity of dismantling the fixture. In the ordinary way two men are required for the opera-

tion, but with the new receptacle one man can correct the trouble without difficulty. Another feature is the bushing and nut in the lugs of the receptacle. The manufacturers have spun a bushing and nut combined, which is made fast into the porcelain, so that the screw hole is easily found when fastening the receptacle to the back of the sign.

## Ferranti's New General Manager

As mentioned in our last issue, Mr. A. B. Cooper has been appointed general manager of the Ferranti Meter & Transformer Manufacturing Company. Mr. Cooper's many friends will recognize in his photograph, reproduced herewith, a particularly pleasing likeness.

Mr. Cooper brings to his new position wide experience in the electrical field. He has been with the Canadian General Electric Company for the past eight years, as transformer sales engineer. Immediately previous to that period he was in the Transformer Sales Department of the General Electric Company at Schnectady and Pittsfield. At a still earlier date he was on the staff of the Rio de Janeiro Tramway, Light &

Mr. A. B. Cooper

Power Company as inspector at the Westinghouse plant in East Pittsburg, Pa., and was for a time stationed in Brazil as assistant engineer.

Mr. Cooper is a Canadian, of United Empire Loyalist stock, but by force of circumstances received his early training in the United States. He graduated from Tufts College in 1903 with the degree of B.A.Sc. He is also a graduate of the General Electric Test Course at Schnectady. He has always taken a keen interest in both engineering and social electrical activities. He is a member of the Engineering Institute of Canada, the American Institute of Electrical Engineers, and is a past chair-

man of the Toronto Section of the A.I.E.E. He is also a director of the Engineers' Club of Toronto. In addition, he is known as an enthusiastic golfer.

It is certain that Mr. Cooper's wide experience will be of very material assistance to his company. He is planning to spend a few months in England in the near future so as to thoroughly acquaint himself with the practice and policy of the parent company, Ferranti Limited, of Hollinwood.

## Mr. Daly Goes to Regina

The Northern Electric Company has just announced that Mr. J. A. Daly, manager of the Hamilton office, has been transferred to Regina, where, beginning March 1, he will have charge of the Saskatchewan province territory. Mr. Daly has been associated with the Northern Electric Company since 1908, first in Montreal and later in Toronto. From 1910 he represented his company in Northern and Eastern Ontario, and likes to recall the old corduroy roads of the early days of the Golden City and Porcupine mining camps, the big bush fire of 1911, the booming days of Cobalt, and the National Transcontinental Railway construction, when $15,000 orders were common affairs in which "Jim" rather more than held his own. During these years Mr. Daly organized and assisted in the establishment of many rural, municipal, co-operative and private telephone systems, with the result that the north

Mr. J. A. Daly

country to-day, even though miles from a railway, has as good a telephone service as Central Ontario. During the past three years he has been in charge of the Hamilton office, which is responsible for the Niagara Peninsula district. He has been active always in association work and one of the pillars of such organizations as the Hamilton Contractor-Dealers' Association, and the Hamilton District Electrical Development League.

Mr. Daly, during his comparatively short stay in Hamilton, has made many friends among the electrical fraternity who are exceedingly sorry to lose him. A very handsome reminder of this fact in the shape of a silver cigarette case, was presented to him by the Executive Committee of the Electrical Development League of Hamilton on Friday evening, February 3, on the occasion of this Committee's visit to Toronto to inspect the Electric Home. Mr. W. H. Childs, chairman of the Hamilton committee, made the presentation, and Mr. Daly, replying, expressed his appreciation of the gift and his regret at leaving so many good friends.

The A. C. Gilbert-Menzies Company had an attractive exhibit of their full line of equipment at the Made-in-Canada Toys exhibit, recently shown at the Queen's Hotel, Toronto.

# Electric Railways

## Rehabilitation of Toronto's Electric Railway System—IV

### Keeping in touch with the work from day to day—Speeding up the Service

In articles 1 and 2 of this series, which appeared in our issues of January 1 and January 15, the methods adopted in preparing and placing the track base, laying the rails and surfacing the roadway were described in detail. One of the most interesting features in connection with this work was the method adopted to keep the management in daily touch with the progress. For this purpose a chart was used, of which the accompanying illustration is an attempted reproduction. The original chart, however, told the whole story at a glance, as it was drawn in colors, thus making each day's progress stand out more prominently.

Each particular piece of work, i.e., each street, was taken care of by a separate chart. The reproduction refers to Yonge

each day's work was written in the chart, but these have been omitted so as not to complicate the figures. The distances may readily be judged, however, as the chart is divided into vertical sections 500 ft. in width, as indicated on the top of the chart.

As will be seen from this reproduction, work was started on October 1. On that date 700 sq. yds. of wearing surface was removed, the concrete broken and the grading completed. On October 2 the slab was poured, the rails placed and the track aligned and surfaced. This was a special piece of construction work where it was necessary to have the cars operating again as soon as possible, and a special method of placing the rails and pouring the concrete, as described in our previous article, was used.

The chart then shows that no further work was done on this line until October 8, when the joints were completed on the section between King and Wellington Sts.; the wearing surface removed, concrete broken and the grading completed from Wellington to Front Sts.; also a certain amount of work on the first two operations, between Queen and King. On

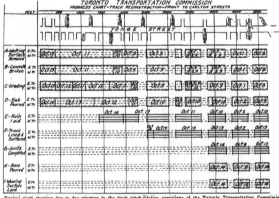

Typical chart showing day to day progress in the track rehabilitation operations of the Toronto Transportation Commission

Street between Front and Carlton, where the track was entirely replaced (with the exception of certain intersections) as indicated on the left by A, B, C, D, E, F, G, H and I. Daily tabulation on the chart indicated the progress on these various items—wearing surface removed; concrete broken; grading; base slab poured; rails laid; track aligned and surfaced; joints completed; base poured and wearing surface relaid.

The dates written on the chart indicate the particular operation carried out on any particular day. In actual practice, the amount of work complete (square yards or lineal feet) in

October 9, all the wearing surface had been removed as far as Alice St. The concrete was also broken to this point, grading was completed on a section between Albert and Queen and between Adelaide and King.

On October 10, the chart shows a small amount of wearing surface was removed between Alice and Dundas, and the concrete broken; two short pieces of grading; a considerable amount of concrete poured—at two different locations; and two sections of rail laid.

Thus progress of the work was indicated from day to day

in accurate detail. The chart finally indicated the conditions of the work on the evening of October 14.

Similar charts, covering the other streets where rehabilitation work was going forward, afforded, each day, a condensed but absolutely accurate view of the whole situation—a very important consideration, in view of the large-scale upon which the work was carried out.

## Some Comparative Loading Figures

Interesting figures have been prepared by J. F. Layng, electric traction engineer for the General Electric Company, covering the average loading and unloading time of the rear entrance, front exit P.A.Y.E. type of car, and the front entrance, centre exit type, the former used on Toronto streets up to September 1 last, and the latter since that date, in considerable numbers. These figures have also been plotted and are represented by the curves A and B in the accompanying illustration. The horizontal line measures passengers, and the vertical line time in seconds. Curve A represents the average loading and

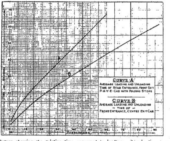

Curves showing the relative time consumed in loading and unloading two standard types of street car

unloading time using the rear entrance, front exit P.A.Y.E. type of car. Curve B represents the average loading and unloading time of the front entrance, centre exit type, where practically half of the car is used as a loading platform. These curves indicate that a very appreciable saving in time is made with the newer cars and that the gain improves as the number of passengers increases. The feature in the new Toronto cars of an ample loading platform which is responsible in a considerable degree for this gain, is working out extremely satisfactory.

---

A report of the London Hydro-electric System states that there are now more than 2,000 electric ranges used in that city. The activities of the past year are shown in the statement that 800 of these were added during 1921. The general manager, Mr. E. V. Buchanan, has just placed in operation a new distributing station, which is said to be one of the finest in the whole Hydro system, both in the matter of construction and design.

---

The Swedish General Electric Ltd., Toronto, have been awarded the contract for supplying the city of Winnipeg with two 310 kv.a. alternators, with accessories, for local service in the municipal power house at Point du Bois.

## Hydro Development in the Maritime Provinces

### By our Eastern Correspondent

The province of Nova Scotia, some time ago, created what is known as the Nova Scotia Power Commission, with power, among other things, to investigate and, if thought wise, to develop water powers at various points of the province. After careful investigation a hydro development has been carried out at St. Margarets Bay, situated some 15 miles out of Halifax, this power being primarily developed for use in that city. A power company, called the Halifax Power Company, had done some preliminary work on this same development but finally the province took it over and completed it. The expectation was that the city of Halifax would buy the entire output but, after long negotiations, the city council decided against this course. The power was then offered to the Nova Scotia Tramways & Power Company, Ltd., with whom a contract has just been signed. In brief, the contract remains in force for 30 years and by it the Nova Scotia Tramways & Power Company purchases power up to 18,000,000 kw. hrs. per year, at actual cost, and distributes the energy to the citizens of Halifax. The cost price of the current to be paid by the company is determined by the Nova Scotia Power Commission, but the selling price at which the energy is to be retailed to the ultimate consumer is determined by the Provincial Board of Public Utilities. A clause in the contract provides that the Commission may sell direct to the city of Halifax as a municipality, for the municipality's use, but not for resale.

It will be noticed by the above that a Board of Public Utilities is mentioned. This Board controls the rates in the province of Nova Scotia for electricity and gas, as well as street railway fares. The situation in this province, therefore, is that a provincial body (the Nova Scotia Power Commission) is selling power to a private company (the Nova Scotia Tramways & Power Company) and the final rates paid by consumers is being adjusted by another provincial body (the Board of Public Utilities). This certainly seems to protect the public adequately but, in addition to this, there is a clause inserted in the contract providing for one representative of the city of Halifax on the Board of Directors of the Tramways & Power Company, this representative to be named by the Governor-General-in-Council.

It would appear that the Nova Scotia provincial government, and the Nova Scotia Power Commission, are to be complimented on the outcome of these negotiations. The consumer seems to be adequately protected and there has been no duplication of equipment or drastic action that might be construed as equivalent to confiscation.

Mention should be made of the activities of Mr. K. H. Smith, chief engineer of the Nova Scotia Power Commission, who has been actively interested in the negotiations and largely instrumental in bringing them to such a happy issue.

---

The Harvey Hubbell Company of Canada, Ltd., are distributing a folder describing the Hubbell Te-Tap Ten, with illustrations. This is an assortment of ten standard Hubbell electrical specialties such as might be sold to any one customer as easily as any one of them might be sold singly. This assortment, therefore, means, ten sales instead of one.

---

Mr. Arthur E. Wilson, for some 26 years with the Canadian General Electric Co., and recently sales manager of Earle Electric, Ltd., has been appointed sales manager of Factory Products, Ltd., 220 King St. W., Toronto. Mr. Wilson succeeds Mr. Langmuir, who is severing his connection with Factory Products, Ltd., to engage in another line of business.

# Current News and Notes

**Arnprior, Ont.**

The Modern Electric Co., Arnprior, Ont., has succeeded Mr. D. O. Booth, who formerly carried on an electrical contractor-dealer business in that place.

**Calgary, Alta.**

The Cunningham Electric Co., Lesson & Lineham Block, Calgary, Alta., has secured the electrical contract on the building of the Royal Bank of Canada, 3rd St. and 8th Ave. W., Calgary, which is undergoing alterations.

**Fredericton, N.B.**

A report states that the city of Fredericton, N.B., has been temporarily deprived of electric power because of a mishap in the plant of the Maritime Electric Company.

**Hamilton, Ont.**

Tenders for the electric wiring of a new school to be erected at Hamilton costing in the neighborhood of $425,000, will be called for by the architect, F. W. Warren, Hamilton, on March 1.

**Kingsville, Ont.**

Messrs. Harris & Russell, Kingville, Ont., have been awarded the electrical contract on a building being erected by Messrs. Birch & Serigley, Kingsville, for a garage and sales rooms.

**Kitchener, Ont.**

The Doerr Electric Co., Kitchener, Ont., has secured the electrical contract on the Evangelical Church, Kitchener, recently erected at an approximate cost of $40,000.

The Electric Light Commission of Kitchener, Ont., is having plans prepared for a street lighting system costing in the neighborhood of $60,000. The city clerk will receive tenders between May 1 and May 15 for about 50 iron standards and 50 200-watt electric lights.

**London, Ont.**

The Lavender Electric Co., 7 MacDonald Ave., London, Ont., has secured the electrical contract on a store recently erected on Horton St., London, for Mr. R. H. Smith of that place.

**Montreal, Que.**

Messrs. Vallee & Hamelin, 1867 St. James St., Montreal, have secured the contract for electrical work on an addition recently built to a store at 414 St. Lawrence St., Montreal, at an approximate cost of $18,000.

Mr. J. J. Valois, 444 Durocher St., Montreal, has been awarded contracts as follows: Electrical work on a $28,000 apartment house on Champagneur St., Outremont, Que.; wiring and electric fixtures on an apartment house to be erected at St. Denis and Demontigny Sts., Montreal, at an estimated cost of $80,000; electric fixtures, wiring and ranges for a $28,000 residence recently erected at 54 Victoria Ave., Montreal; electric fixtures on a $32,000 apartment house recently erected on Girouard Ave., Outremont.

Messrs. E. Marcou & Frere, 450 Roy St. E., Montreal, have been awarded the contract for electrical work on two stores and four residences recently erected on Amherst St., Montreal, for Mr. Murphy, 29 Drummond St.

The Croteau & Labelle Electric Co., 233 St. James St., Montreal, has secured the contract for electrical work on a building at 241 Beaver Hall Hill that is being altered for offices at an estimated cost of $15,000.

The annual meeting of the Laurentide Power Company is set for February 21. Figures available for 1921 operations indicate that this company had a very prosperous year, and, for the second time in its history, gross earnings have passed the million dollar mark, being $1,237,561, as compared with $1,040,887 for the year 1920. Net earnings amounted to $570,740, as against $432,225 for the previous year. The president of the company, in his report, states that the earnings for the past year are based upon the sale of 112,500 average horse power for the year.

**Outremont, Que.**

Messrs. Crane Ltd., St. Patrick St., Montreal, has secured the contract for electrical fixtures for an apartment house recently erected on North Bloomfield Ave., Outremont, at an approximate cost of $53,000, for Mr. E. Hogue, 365 De Lanaudiere St.

**Quebec, Que.**

The newly formed Quebec and District Amateur Radio Association decided, at a recent meeting, to meet every Wednesday evening. Club rooms are being arranged for. The secretary of the association is Mr. L. P. Souey, 82 Aberdeen St., Quebec.

**St. Thomas, Ont.**

Radio enthusiasts of St. Thomas, Ont., are endeavoring to form a Radio Club in that city and have called a meeting to discuss the matter.

**Saskatoon, Sask.**

As a result of the introduction of one-man cars on the Saskatoon street railway system a deficit of $8,996, as compared with a deficit of $58,844 for 1920, is shown for the year 1921, a reduction of approximately $50,000.

**Simcoe, Ont.**

The Norfolk county Chamber of Commerce will urge that the steam road from Simcoe to Port Rowan be electrified and extended to Port Burwell.

**Three Rivers, Que.**

The North Shore Power Company, Three Rivers, Que., has secured the contract for wiring, poles, brackets and outside work on a $10,000 fire alarm system being installed by the Northern Electric Co., Montreal, at Cap de la Madeleine, Que.

**Toronto, Ont.**

Messrs. Beattie-McIntyre, Ltd., 72 Victoria St., Toronto, have been awarded the electrical contract on an addition recently built to the factory of George Weston, Ltd., 134 Peter St., Toronto.

Mr. B. C. Taylor, 25 Marchmont Road, Toronto, has secured the contract for electrical work on two stores recently erected on St. Clair Ave., near Lauder Ave., Toronto, for Mr. O. Extence, 103 Westmount Ave.

Mr. R. M. Mitchell, 140 Leslie St., Toronto, has been awarded the contract for electrical work on four stores recently erected at Greenwood and Danforth Avenues, Toronto, for Mr. John MacLean, 1204 Danforth Ave.

Messrs. Smart & Walsh, 57 Kildonan Drive, Toronto, have secured the contract for electrical work on a store and apartment building being erected on Danforth Avenue, near Glebemount Ave., Toronto.

Mr. A. C. Plumb, 33½ Burgess Ave., Toronto, has secured the contract for electrical work on a store building recently erected at Danforth and Glebemount Avenues, Toronto.

Wanted University graduate Electrical Engineering, single, 27, wants position with electrical manufacturer or allied industry, any department. Salary to start minor question, but something more than a job required. Location no object. Present salary $2100. References. Box 810 Electrical News. Toronto. Ontario.    4-7

Wanted by electrical engineer position as manager or superintendent of Hydro Town or private power company, wide experience in power plant and line construction, twenty years of practical experience, at present managing civic utilities for Eastern town. Can give excellent references. Wish to settle further West. Married. Box 776, Electrical News, Toronto.    2-4

POSITION WANTED—Electrical Engineer, technical and commercial training, over twenty years' practical experience with large power and industrial companies in operation, construction and maintenance of hydro-electric plants, sub-stations, transmission lines, distributing systems and motor installations, desires position with power or industrial company. Present position electrical engineer-manager of Street Railway, and Light and Power System. Excellent reasons for desiring change. Box 652, Electrical News, Toronto.    17-tf

Manager Public Utilities, with technical education and sound business training, desires position with municipality of from five to ten thousand population, experienced in the construction, operation and successful management of water works plants, hydro electric, steam and gas generating stations, and distribution systems; can furnish splendid reference and am an active member of American Association of Engineers. Box 765 Electrical News, Toronto. Canada.    2-5

## FOR SALE

Motor repair business, best district in Ontario, owner sick. Write Jarvis, 20 Grosvenor St., Toronto.    4

## MANAGER WANTED

Written applications will be received for position of Manager for a waterpower driven electric plant located in Western Ontario. State qualifications and salary expected. Box 795 Electrical News, Toronto.    4-5

## Corporation of Penticton

## MACHINERY FOR SALE

The Corporation has for disposal certain Electrical Generating Equipment, including
1—200 B.H.P. Diesel Engine
1—200-250 B.H.P. Semi-Diesel Engine
1—125 K.V.A. Generator
1—215 K.V.A. Generator
Detailed information will be supplied upon request.

4-5     B. C. Bracewell,
    Municipal Clerk.

## FOR SALE

Motor-Generator set in first-class condition. Capable of handling elevators and lighting system Equipment as follows:—
Motor—30 H.P. 200 volts, 3 phase, 66-2/3 cycles complete with common bedplate and starter.
Switchboard—Slate panel 24" x 48" x ¼" mounted on standard 1¼" W. Iron pipe frame complete with 300 volts voltmeter 150 ampere Ammeter. Double pole 100 ampere Knife Switch and Main enclosed fuses.
Generator—20 K.W. 200 volts, 100 amperes, 2 wire complete with Rheostat.
Will sacrifice for quick sale.
Apply
    The G. W. Robinson Company, Limited,
4     Hamilton, Ontario.

## ROAD CONTRACT LET

The contract was recently awarded for the construction of 7½ miles of the Duane-Lake Titus road, linking up with the Mountain Pond road constructed last year, and completing the route from Malone to the Adirondacks via Paul Smith's. W. J. Senter, of Waterdown, N.Y., was the successful bidder out of eighteen, the amount of his tender being $232,617.71. Work will be started as early as possible in the spring and will be expedient in order to make this road available for motor tourists before the season closes.

## PUBLIC WORKS IN PERU

Mr. Robert Dunsmuir, formerly of Victoria, B.C., but now of London, Eng., has secured large concessions for railway and harbor works in Peru. The contract is for some £13,500,000 and will take seven years to complete. It is stated that the concessions carry with them land grants amounting to more than 50,000,-000 acres, the exclusive rights to the oil, coal and all other minerals found within this area for a term of 33 years, and to a company organized by the concessionaire, the tobacco monopoly for the whole of the republic for 33 years.

Excerpts from a booklet by
The H. K. Ferguson Co., Engineers and Builders

# Advertising
# and
# Construction

An actual and definite relation between the two

The past twenty-five years, marked by the ever-increasing application of the force called advertising to a wider and wider range of business problems, have, of course, built up a growing fund of general information with regard to the effects of advertising.

We have all recently seen the publication of figures indicating that 84% of all business failures in 1920 were of concerns which failed to advertise. An analysis of the factory-construction records of my own company shows that 83¾% of our customers are advertisers.

Stated differently, the figures indicate that the manufacturer who advertises is five times as likely to stay in business as his non-advertising competitor, and is five times as likely to need more factory space which he will probably buy in two and one-half times the quantity required by the non-advertiser.

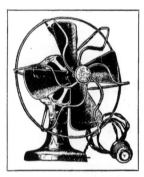

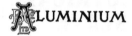

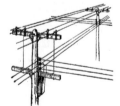

# CLASSIFIED INDEX TO ADVERTISEMENTS

The following regulations apply to all advertisers:—Eighth page, every issue, three headings; quarter page, six headings; half page, twelve headings; full page, twenty-four headings.

# CLASSIFIED INDEX TO ADVERTISEMENTS—CONTINUED

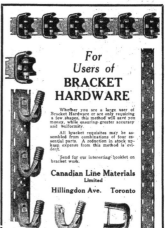

## CLASSIFIED INDEX TO ADVERTISEMENTS—CONTINUED

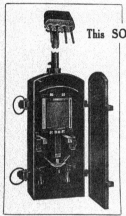
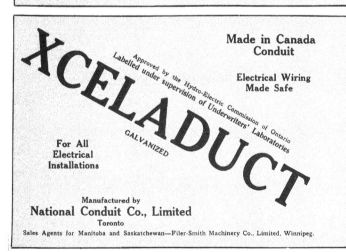

Vol. XXXI No. 5                                                    Toronto. March 1, 1922

# Electrical News

Engineering Contracting · Merchandising Transportation

---

## STANDARD
## Wires and Cables

Bare Copper, Brass and Bronze Wires
Colonial Copper Clad Steel Wire
TRADE **C.C.C.** MARK
Weatherproof and Magnet Wires
Rubber Insulated Wires
Varnished Cambric Insulated Cables
Paper Insulated, Lead Covered Cables
Rubber Insulated, Lead Covered Cables
Steel-Wire and Steel-Tape-Armored Cables
Cable Terminals and Junction Boxes
"Ozite" Insulating Compounds, etc.

Catalogues and prices on request to our nearest office.

## Standard Underground Cable Co.
## of Canada, Limited

Sales Offices and Warehouses
Montreal, Toronto, Hamilton, Winnipeg, Seattle
Factory: Hamilton, Ont.

---

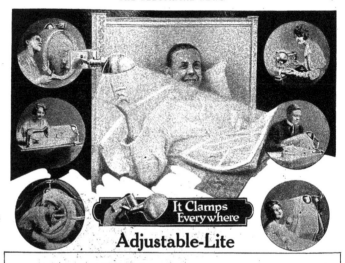

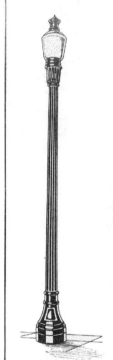

# How about new duct lines?

## Can you afford to wait till 1923?

Nineteen twenty-two will be a big year for underground construction because:

1. The Bond Market is eagerly absorbing Public Utility securities as rapidly as they are offered. Money is no longer difficult to secure for capital expenditures by Central Stations.

2. Many cities are conducting campaigns of public improvement in order to assist in reducing the number of unemployed workers. This municipal work means, in part, new pavements. It is always wise, as a matter of good financial judgment and public policy, to lay extra duct lines before new pavements are placed, so that it will not be necessary to disturb the new street surface for several years.

3. Labor of the type used in underground construction is available at satisfactory rates.

4. Materials used in duct lines are at very low prices. Fibre Conduit is back to its 1914 price. Portland Cement is approximately at its 1916 price.

This summary of fundamental conditions represents our judgment of the outlook in the underground construction field in 1922. We further believe we should honestly counsel buyers of Fibre Conduit to get their inquiries into the hands of responsible producers as early as possible.

The present low prices of Fibre Conduit and the unusual spring demand now expected make early action a wise course for buyers who will require prompt service.

**CANADIAN JOHNS-MANVILLE CO. LIMITED**

Toronto    Montreal    Winnipeg    Vancouver
Windsor    Hamilton    Ottawa

# "Leadership"

No one in the Electrical Industry was ever born to leadership—nor had it thrust upon them.

No Sir—

Leadership in this business predicates hard-won recognition of superiority earned only in the face of years of keen, brainful competition.

Therefore the admitted leadership attained by the Canadian Crocker-Wheeler Company in the design and manufacture of motors, generators, transformer, etc., is full of meaning to you, Sir, who buy these products.

In the transformer illustrated note the large surface of the windings in contact with the oil and the extra staunch method of bracing the coils.

Bore and coils of 1500 K.V.A. 45700/26400/13200 Volts to 4000/2300/575 Volts, ·3 phase, 25 cycles, oil insulated, water cooled Canadian Crocker-Wheeler Transformer.

## THE CANADIAN
# CROCKER-WHEELER CO.
### LIMITED
Manufacturers and Electrical Engineers

Head Office and Works:
## ST. CATHARINES. Ont.

District Offices:
## TORONTO   MONTREAL

Also all
Branches of

*Northern Electric Company*
LIMITED

Montreal, Halifax, Ottawa,
Toronto, London, Winnipeg,
Regina, Calgary, Vancouver.

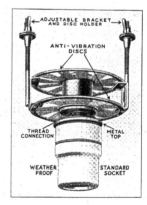

# ALPHABETICAL LIST OF ADVERTISERS

# A Remarkable Washer
# at a Remarkable Price!

The interest created by the announcement of the $150 Sunnysuds proves one thing: Both washing machine dealers and the public were actually waiting for a dependable, standard-size, sure-action, all-metal electric washer retailing at a figure that made selling and buying easy.

Detailed comparison with *any* washer at *any* price is the most convincing proof that the $150 Sunnysuds is a remarkable washer at a remarkable price.

> **TUB:** Heavy copper. Rolled and soldered seams. Six sheet capacity. Corrugated agitators. Sediment Zone. Oscillates silently. No springs.
>
> **FRAME:** Heavy pressed steel. Turned in stamping to insure absolute rigidity. No angle irons. Heavily enameled. Open. Attractive. Easy to clean. Crown casers.
>
> **WRINGER:** Aluminum. Snap locks into four positions. Patented single turn-screw controlling pressure and safety release. 12 in. rolls. Metal waterboard.
>
> **DRIVE:** Gears and shafts entirely enclosed in grease. Housings finished to prevent grease seepage.
>
> **MOTOR:** ¼ h.p. Ball Bearing Domestic. Splashproof. Direct shaft connection. Mounted on wood to insure silence.

Numerous applications for dealerships have been received. Appointments are being carefully made to insure a solid, able, and permanent dealer organization. Correspondence is still solicited from domestic appliance shops, hardware merchants, electrical dealers and department stores sufficiently experienced, aggressive and financed to sell a remarkable washer at a remarkable price.

ONWARD MANUFACTURING COMPANY, KITCHENER, ONTARIO

## Retail price $150

*Winnipeg and West $165.00*

**Sunnysuds**
Electric Washer & Wringer

# *Announcing –*

## The Fuse That Can be Refilled
### For Economy With Ease and Safety

Heavy well constructed molded-insulation parts insure permanence—these are kept, used over and over. The small part that holds the fuse element which is discarded when blown, is very inexpensive, ranking in cost with a postage stamp.

### COTE BROS.
## $IMPLICITY
#### REFILLABLE FUSES

### Approved in All Capacities by the National Board of Fire Underwriters

Provides genuine protection for homes, schools, hospitals, offices, factories and other buildings in which safety, economy, and ease of operation are essential.

**SIMPLICITY**
Simplicity Fuses consist of only two parts and the tiny refill—no little parts to juggle and lose—not a tool needed when refilling.

**SAFETY**
The heavy pressed-insulation parts insure permanence and safety —this can be refilled even with wet hands without danger of electrical shock.

**PROTECTION**
The tiny refill has its ratings stamped on both ends, one of which shows through the aperture in the cap. You know the rating of SIMPLICITY. Protected circuits.

**ECONOMY**
Simplicity Refills rank in cost with a postage stamp—protection with out expense.

To people interested in either the use or sale of this new fuse, we will gladly send a sample and a descriptive booklet. Drop a line on your letterhead!

## COTE BROS. MANUFACTURING CO.
### 401 Somerset Block
### WINNIPEG   -   -   -   MANITOBA
#### Branch Office: Montreal

Side captions:
Out of the panel without a shock.
Apart with a slight twist.
Out with the blown cartridge.
Refilled at the cost of a postage stamp.
Together with the rating sure to show.
Back at work in 20 seconds

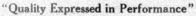

# Protect the "Devil Dogs"!

### What is it worth to safeguard the lives of the men who work around high tension power transmission?

BRAVE as the Marines are the men who keep the switchboards, substations and transmission of America's high power lines in operation and free from trouble.

So valuable are these men that every accident is an economic loss hard to replace.

That is a forceful reason why Plant Engineers, Contractors, Executives and Purchasing Agents are specifying equipment and apparatus designed by specialists in high tension practice.

Electric Power Equipment Corporation qualify as authorities in such equipment because every piece of material they put out has been designed from practical experience in building and installing high voltage generating and sub-stations, switchboards and line construction.

The ELPECO trademark—the Crusader with two shields—stands for Protection to Life and Property and Prevention of Loss Thru Leaky Equipment.

It costs little to specify ELPECO.
It is worth much!

**Indoor Bus Supports**

All material passes special high tension test. Tongues centered and offset to equalise stresses. Metal parts cemented to porcelain. No low temperature flowing metals. Interchangeable mounting base-pipe or flat—adjustable for any bus position.

**Indoor Disconnecting Switches**

Hinge and contact specially ground-in for perfect contact surface. All double blade construction.

**Indoor Switchboard and Structural Fittings**

Designed for interchangeability and clean construction.

**Outdoor Disconnecting Switches**

All truss-type blades. Soft brass head clamps shaping to insulator head, non-crushing. 45-degree switches equipped with heavy H section pins. Upright and inverted switches with X section pins. Special direct-acting patented latch for outdoor lock. No spiral springs. No hinges. Non-rusting. All outdoor apparatus mounted on hot-dipped galvanized channel iron sections. All outdoor steel equipment hot-dipped galvanized.

**Choke Coils**

All 25,000 volt service and over fitted with longitudinal braces. No distortion under short circuit.

**Copper Fittings**

High grade copper castings micrometer machined for shrink fit.

**Control Switches**

Specially designed for positive action preventing accidental throwing. Both lamp and mechanical indication. Protection from all hazards to operator.

**Pole Top Switches**

Circuit opened positively with no destructive effect upon the system. All phases broken simultaneously. Applies greatest pressure at point of contact with smallest stress at insulator.

Protection in the future from DANGER and LOSSES

ELPECO—THE CRUSADER IN ADVANCED ELECTRICAL EQUIPMENT

Disconnecting Switches    Bus Supports    Pole Top Switches    Switch Boards

ELECTRIC POWER EQUIPMENT CORPORATION, Philadelphia, Pa.
Canadian Representatives: FERRANTI METER AND TRANSFORMER MFG. CO., Limited, Toronto & Montreal

Generation, Transmission and Application of Electricity

For nearly thirty years the recognized journal for the
Electrical Interests of Canada.

Published Semi-Monthly By

## HUGH C. MACLEAN PUBLICATIONS
LIMITED

THOMAS S. YOUNG, Toronto, Managing Director
W. R. CARR, Ph.D., Toronto, Managing Editor
HEAD OFFICE - 347 Adelaide Street West, TORONTO
Telephone A. 2700

| | |
|---|---|
| MONTREAL - - | 119 Board of Trade Bldg. |
| WINNIPEG - - - | 302 Travellers' Bldg. |
| VANCOUVER - - - - - | Winch Building |
| NEW YORK - - - - - | 296 Broadway |
| CHICAGO - | Room 803, 63 E. Adams St. |
| LONDON, ENG. - - | 16 Regent Street S. W. |

### ADVERTISEMENTS
Orders for advertising should reach the office of publication not later
than the 5th and 20th of the month. Changes in advertisements will be
made whenever desired, without cost to the advertiser.

### SUBSCRIBERS
The "Electrical News" will be mailed to subscribers in Canada and
Great Britain, post free, for $2.00 per annum. United States and foreign,
$2.50. Remit by currency, registered letter, or postal order payable to
Hugh C. MacLean Publications Limited.
Subscribers are requested to promptly notify the publishers of failure
or delay in delivery of paper.

Authorized by the Postmaster General for Canada, for transmission
as second class matter.

Vol. 31       *Toronto, March 1, 1922*       No. 5

## Sir Adam Beck replies
## to Sutherland Commission Report

Sir Adam Beck has published an interesting reply to
the majority report of the Sutherland Commission, appointed
some time ago by the Drury Government to inquire into the
subject of hydro-electric radials in the province of Ontario.
Sir Adam reviews the situation from the inception of the hydro radial idea up to the present time, stating that the result
of the controversy has been to unsettle the public's mind and
leave it in doubt, first, as to whether the radials could be
constructed and operated so as to be self-sustaining and, second, as to whether this could be attained without a duplication of railway lines that would be injurious to the country as
a whole. He charges that the findings in the majority report show that the commissioners were "apparently unable to
rightly weigh the great assemblage of material which they
brought together, or rightly interpret its import."

Sir Adam makes some interesting remarks regarding the
value that may be attached to the evidence of certain of the
experts who gave evidence before the Commission, and points
out, for example, that a number of these had had little or no
experience except in steam railway work. He states that
Bion J. Arnold, a witness who had expert knowledge of interurban electric railway operation, had reported favorably on
hydro radials. Sir Adam deals specifically with the evidence
of some of these witnesses. For example, the opinion of F.
P. Gutelius, whose estimates of traffic possibilities are claimed
to be ten times too small in view of actual figures available

of interurban electric traffic in both Canada and the United
States. For instance, Mr. Gutelius had said that nineteen
trips per capita per year of population in Oshawa on a To-
ronto-Oshawa line would be "about ten times too much." Sir
Adam answers this by quoting the figures between Toronto
and Aurora, which represents 38 trips per capita per annum;
Detroit & Northville, 30 trips, and a number of other similar
instances. Sir Adam also points out that according to their
own admission, the witnesses before the Sutherland Commission who strongly condemned the proposition, had made only
a cursory study of conditions.

Sir Adam's reply, finally, deals specifically with the
Sutherland Commission's three special objections—high
construction costs; too low estimate of operating costs; too
high an estimate of revenues.

### High Construction Costs

He argues that the Sutherland Commission did not take
into consideration the fact that the construction and equipment of the railways with which they compared the proposed
hydro radials were not of a sufficiently high standard, and
quotes the cost figures of a number of the better class inter-
urbans in the United States to show that the Commission's
estimates of costs were entirely unreasonable. He quotes examples of capital costs very much higher than the Commission's estimate, where a surplus is being shown year by year
and increasing as the efficiency of these systems becomes
greater. Sir Adam states that the Sutherland Commission
failed to emphasize the fact that the most notable exceptions
to the railways, which are in trouble are the very railways
whose construction and equipment are of the highest class.
The average cost of the hydro radial system was placed at
$140,000 per mile by the Sutherland Commission itself, and
Sir Adam's reply quotes costs up to $225,000 per mile for lines
in the United States operating successfully and doing better
than paying charges.

### Operating Costs

In the matter of operating costs, Sir Adam charges that
the Sutherland Commission chose to deal in generalities. The
operating ratio of the hydro radials had been placed at 55.7
per cent. The Sutherland report contended that this ratio
was too small and published a table of fifteen radials in Canada and the United States where the operating percentages
varied all the way up to 79.3 per cent. In his reply, Sir Adam
charges that the figures quoted by the Commission apply
only to the years 1919 and 1920, when operating costs were
abnormally high. He also charges that the Commission failed
to include in its list such figures as that of the Texas Electric
Railway which, even in 1920, had an operating ratio of 50.9
per cent. Sir Adam submits a table of operating ratios for
the years 1914, 1915, 1916 and 1917, for the most part, of six
interurban railways in the United States where the percentages were as low as 46.49 per cent and the highest was 57.2
per cent.

### Revenue

In the matter of revenue Sir Adam is even more critical
of the Sutherland report, and states "Probably no other feature of the majority report discloses the inconsistent, inac-
curate and inadequate character of the reasons given by the
Sutherland Commission in support of its conclusions than the
manner in which this subject of passenger revenues is dealt
with." He states that the Sutherland Commission compared
passenger revenue per capita upon a wrong basis; that they
made poor comparisons; that they employed population data
improperly—in one case including the population of the city
of Toledo twice, thus reducing the per capita revenue unduly.

Sir Adam's argument closes with the following:

### Conclusion

In dealing with public problems such as the Hydro-electric Power Commission has dealt with, the Commission has regarded that adequate knowledge of any problem in hand, coupled with the zeal and enterprise for its successful solution, have really been more important factors than the mere acquirement of Capital. Once the essential factors are determined to be sound and favorable, they will irresistibly carry Capital with them. The Hydro-electric Power Commission investigated this subject of Hydro-Radials with a staff of expert investigators—and these cannot justly be discredited by the slighting personal remarks contained in the Majority Report of the Sutherland Commission. The Hydro-electric Power Commission's investigators have proved themselves competent to appraise essential data.

The Hydro-electric Power Commission asserted its own confidence in Hydro-Radials before inviting the confidence of others. The Commission's confidence was in the particular project it has recommended. It called for high-speed transportation between certain terminals traversing certain territory. The railway itself was to be of the highest standard in order to meet the demands which were to be placed upon it. The whole proposition was unique and of exceptional business promise. It is of no real pertinence to compare such a hydro-radial project with other railways which should not have been built. A rapid-transit electric road through a populous and prosperous territory and adjacent to thriving industries, is in a different category from a steam railway trying to tap the Arctic zone.

The very fact that hydro-electrical energy for driving the railway would be available at low cost, is of itself a factor of great significance and one which, had it been possessed by many electric roads in the United States, would have permitted them to operate in a manner they were unable to do with expensive steam-generated electric power. There is a wide difference between a cheaply constructed electric railway system, over-capitalized, using expensive power and operating on an ordinary street railway basis; and a strictly modern, high-class, rapid-transit electric railway, operating with cheap power. The latter is the Hydro-Radial proposition submitted for the acceptance of the public of Ontario.

Now, instead of appraising this proposition upon its merits, the Sutherland Commission has treated it in certain important respects as though it were one of the cheaply constructed electric railway systems, over-capitalized, operating more or less on the highways and streets, and depending upon expensive power. This statement is warranted because even though clearly recognizing that the Hydro-Radial project was unique, the Sutherland Commission, nevertheless, employed data relating to cost, operation, revenue and other features, of inferior electric railways as the criterion by which to judge the merits of the Hydro-Electric Power Commission's proposed radials. The STATEMENT here presented will, it is believed, sufficiently demonstrate the unjust character of the conclusions submitted by the Sutherland Commission in its Majority Report. Moreover, it should be fully recognized, not only that the data of revenues and other factors germane to the subject of Hydro-Radials have been incorrectly and unjustly applied, but also that if these data had been rightly applied by the Sutherland Commission there would have resulted a clear and outstanding demonstration of the soundness of the estimates of the Hydro-Electric Power Commission's experts; in other words, the Hydro-Radial project could not, even with the semblance of justification, have been discredited, but on the contrary would have had to be confirmed by sheer weight of evidence on record before the Sutherland Commission.

It is believed that the Hydro-Radial project may be advanced with caution and yet with great benefit to the Province as a whole. Later, after the success of the first Hydro-Radial installation of, say, 325 miles is assured, these lines may

be extended and other new lines may be constructed as circumstances warrant. The whole radial scheme will thus gradually acquire a province-wide character, just as has been the case with the transmission and distribution of Hydro-Electric power, which has developed from its limited initial installation to its present province-wide proportions. It remains, therefore, for the people of the Province of Ontario to decide whether they will be guided by the conclusions of the Majority Report of the Sutherland Commission or by the representations made by the Hydro-Electric Power Commission based as they are upon the extensive and detailed research of experts in whose judgment the Commission has full confidence—a confidence, in fact, which has been increased rather than diminished by the criticisms which have been directed against them.

It is confidently believed that the Hydro-Radial Project which is recommended by the Hydro-electric Power Commission of Ontario will assuredly be of very great social, commercial and financial benefit not only to the municipalities directly concerned, but also to the Province and, indeed, to the Dominion as a whole. It remains, therefore, for the Public, in whose general interest the Hydro-Radial Railway Project has been conceived, to decide whether or not this Hydro-Radial policy of the Municipalities shall be consummated.

## Symposium on Hydraulics

The interesting announcement has just been made that a Symposium on water power development will be held at Toronto University, in Room 22, of the Mining Building, from February 27 to March 6. This Symposium has been arranged by the Hydraulic Section of the Department of Mechanical Engineering of the University of Toronto, of which R. W. Angus is the professor.

Lectures will be delivered on various subjects by engineer graduates of the University of Toronto, and others. The following programme has been arranged: February 27, 4.30 p.m., "Some of the principles controlling the use and design of hydraulic turbines," by Lewis F. Moody, who also speaks at 8.00 p.m., the same date, on "Hydraulic turbines and installations"; February 28, 4.30 p.m., "Hydraulic equipment as affecting power house design", by Max V. Sauer; March 2 4.30 p.m., "Design of power canals, headworks and surge tanks," by Thomas H. Hogg; March 3, 4.30 p.m., "Power house machinery," by W. M. White; March 6, 4.30 p.m., "Regulations and testing of hydro-electric machinery," by Norman R. Gibson; March 6, 8.00 p.m., "Some economic aspects of hydro-electric power development," also by Mr. Gibson.

Speaking of the purposes of the Symposium, the announcement states that from the standpoint of the student, the purpose is more especially to bring him into direct contact with the engineers responsible for the great hydraulic developments in America. It is intended to supplement the courses in water power engineering and hydraulics now given, and to give students a somewhat broader vision.

To the practicing engineer it is believed the lectures will also be of value on account of the eminence of the lecturers, and it is also hoped that it may bring these engineers into a more vital relationship with the University. The hours set for the lectures are specially arranged for the benefit of engineers.

---

The Auto and Electric Service Co. have opened a workshop and garage at 391½ Richmond Street, London, for motor and generator repairs, also re-winding. They will maintain a Gould battery service station and specialize on automobile electric service.

## City Superintendent of Vernon, B. C., Issues Challenge to Contemporaries

Mr. S. H. Excell, city superintendent of Vernon, B. C., has sent us further interesting figures in connection with the cost of operation and generation of power. For a plant of this size, the figures of Mr. Excell are unusually satisfactory, but perhaps some western superintendent can go him one better:

### City of Vernon B. C. Power Plant

Report on operations for the period, (Dec. 1920-Dec. 1921.)

Capacity of plant 532 kw. (Size of Units 157 & 375 kw.)

| | |
|---|---|
| Output | 1,068,284 kw. hrs. |
| Distribution of load.—Light & Utilities 61%, Power 18% | |
| Heating 22.6%. | ¾ |
| No. of services, all metered. | 1180 |
| Load factor | 22.9% |
| Maximum peak | 360 kw. |
| Average peak | 320 kw. |
| Kw. hrs. output per capita of population | 267 |
| Total expenditure on service | $54,847.59 |
| Gross revenue | $64,152.73 |
| Surplus over expenditure | $9,305.14 |
| Fuel used | 87,917 gal. |
| Cost of fuel | $12,515.00 |
| Average cost of fuel per gallon | 14.64 cents |
| Average cost of fuel per kw. hr. | 1.24 cents |
| Average overall cost per kw. hr. | 5.13 cents |
| Average return per kw. hr. | 6.05 cents |
| Cost of generation per kw. hr. | 3.596 cents |
| Operating cost of dist. system per kw. hr. | 64 cents |
| Capital value of plant, (1921) | $145,741.72 |
| Power plant, $80,741.72, ' Dist. system, $65,000.00 | |

Note:— Average cost of fuel was higher than in 1900, owing to unusual high cost for the months of December to April, inclusive.

### Analysis of Cost of Generation.—Year 1921

| Item | Cost | Average Cost per kw. hr. |
|---|---|---|
| Fuel | $12,515.00 | 1.24 cents |
| Wages | $ 7,800.00 | .730 cents |
| Interest, sinking fund & insurance. | $12,250.00 | 1.146 cents |
| Mg'nt & office | $ 1,900.00 | .181 cents |
| Lub. oil & supplies | $ 1,400.00 | .131 cents |
| Repairs; govt. insp. chgs. | $ 1,550.00 | .168 cents |
| Total | $38,415.00 | 3.596 cents |

## "The Installation of Power Plant Equipment" Before Toronto A. I. E. E.

On Feb. 10 the Toronto Section of the A. I. E. E. were favored by Mr. F. H. Farmer, of the Canadian Westinghouse Compny, who presented a paper dealing with "The Installation of Power Plant Equipment". He opened his talk by stating that while standard equipment such as motors and small transformers might be installed by the purchaser, it was wise in the case of large generating and transforming apparatus, particularly when this was of such magnitude that complete shop tests could not be made upon it, to have the installation made by the manufacturer. He pointed out the necessity of the closest co-operation between the manufacturer and the purchaser in such cases.

Attention was called to the rapid growth of the dimensions of equipment. But a few years ago, 5,000 kilowatts was considered an immense machine, and the sizes had quickly jumped to that of the present Queenston generators which were 45,000 kv.a., and were likely soon to be surpassed by 66,-

000 kv.a. machines in the plant of the Niagara Falls Power Company.

Mr. Farmer then gave considerable details of the dimensions of the Queenston generators, comparing the magnitudes of the various parts from small items such as the spiders weighing a mere trifle of 28 tons to the completed rotors or stators weighing over ten times that amount. It was evident that as sizes increased a greater percentage of the work usually done in the factory would have to be carried out at the plant, under conditions not as favorable as in the shop, and special methods would have to be developed to meet each individual case.

The speaker then gave a general outline of the methods followed in the setting up of large machines, starting from the turbine which, being less flexible in its placing, determines the final location of the generator. He showed how alignment was carried out, and pointed out the sources of error which might be looked for, and the care that must be exercised to avoid these. The placing of rotor shafts, the assembling of rotors, and the winding and drying out of the armatures, was explained in detail.

The talk was concluded by some observations on the handling and installation of large transformers and switching equipment, after which a most interesting set of slides was shown, illustrating experiences which had been met with by the speaker in installation work in various parts of the Dominion.

## Converting Winnipeg's Telephone Service to Automatic

The Manitoba Telephone System have contracted for additional automatic telephone equipment necessary to convert two of the Winnipeg exchanges to automatic service. The Winnipeg Office of the Northern Electric Company, were awarded the contract for the port, Rouge exchange; the contract price being $423,254.00, and will supply their well known Strowger equipment. This type is now in use in the business or, down town section of the City. This equipment will be a Canadian product, being manufactured in the company's Montreal factory. It is expected the first load of equipment will leave the factory during July, and the work will be completed and automatic service inaugurated in Fort Rouge area January 15th, 1923.

The Siemens Electrical Works of Woolwich, England, obtained the contract for St. John's exchange, which differs in type somewhat from the Strowger or Northern Electric. The English firm is expected to complete its work in April 1923. The price of the Siemens contract amounted to $483,438.00.

The Manitoba Telephone System will undertake to convert telephones and outside plant to automatic conditions with their present staff.

Some 13,000 lines are involved in the two exchanges, and the work to convert telephones, adding dials, etc., will begin about April 1st., so as to be completed before freeze up next winter.

The Winnipeg public are well satisfied with the service rendered by the automatic system, in downtown offices, and the telephone officials anticipate like satisfaction will follow a similar service in the residential areas. It is understood that the automatic telephones make possible a substantial reduction in the annual charges by the elimination of the operator.

When these contracts are completed, two thirds of the city of Winnipeg will have the automatic system.

The Commercial Electric Company, London, Ontario, have been awarded the contract for the supply of new Raylite, 300 watt units for the new rotunda of the Belvedere Hotel, London, Ont.

# Water Powers of the World

### Figures Prepared by the Dominion Water Powers Branch, Department of the Interior

Many estimates of the world's water power have been published from time to time* but of late the available information has increased rapidly. The great increase in the cost of coal and the need of more power for manufacturing due to the war turned the attention of all civilized countries to the subject. Most of them increased or instituted active investigation of their power resources and the majority have since passed legislation nationalizing all power resources or provided for state bonuses or other forms of financial assistance to power development. As many of these investigations started four or five years ago, considerable information is now coming to hand and it is of interest to consider the latest estimates of the total water power of the world.

In addition to the new information available from Europe and the British Dominions, much matter of great interest concerning countries of which but little was known as to their water power resources, such as Africa, South America, and parts of Asia and Oceanica is now available in a recent publication of the United States Geological Survey.† This is an elaborate production containing 10 maps covering all parts of the world and the prologue states that every available source of information open to the U. S. Government has been made use of in compiling the particulars. Extremely valuable and interesting as this publication is, the time necessarily required to compile, print,. engrave and issue so complete a paper prevents some of the figures from being the latest available, as in the case of Canada and a few other countries.

The following are a few points of especial interest from the above mentioned United States Report.

North America contains less than 15 per cent of the world's available water power but has developed more than all the rest of the world. The United States has 41 per cent of the developed power of the world, Canada over 10 per cent (See below for comparison of development in U. S. and Canada).

The largest water power development in the world is still that at Niagara Falls, where the total capacity in operation is 870,000 h.p. and this is being increased by 114,500 h.p. in the U. S., and 300,000 h.p. in Canada, which will bring the total up to 1,284,000 h.p.

The U. S. estimates are based on 75 per cent of the theoretical power from flow available at least 75 per cent of the time and may be summarized thus:—

| | Horse-Power | | |
|---|---|---|---|
| | Developed | Per Cent | Potential Per Cent |
| North America .....| 12,210,000 | 53.5 | 62,000,000 | 14.1 |
| South America .... | 424,000 | 1.9 | 54,000,000 | 12.3 |
| Europe ...... ..... | 8,877,000 | 38.8 | 45,000,000 | 10.2 |
| Asia ...... ...... | 1,160,000 | 5.1 | 71,000,000 | 16.2 |
| Africa ........ .... | 11,000 | .1 | 190,000,000 | 43.3 |
| Oceanica (Including Australia & New Zealand) | 147,000 | .6 | 17,000,000 | 3.9 |
| | 22,829,000 | 100.0 | 439,000,000 | 100.0 |

It will be noted that Africa is credited with over 43% of the total potential water power of the world. According to the above quoted report this power lies mainly in the tropical portion, the north and south portions having small rainfall. In this area the Kongo river and its tributaries have a drainage basin of 1,500,000 square miles, nearly all of which is well watered. The Kongo affords greater water-power

*See issues of Electrical News Dec. 15/21, Jan. 1/22.
†World Atlas of Commercial Geology: Part II. Water Power of the World, U.S. Geological Survey, Washington, D.C.

resources than any other river system in the world and more than one-fourth of the potential water power of the world is in this one river basin—more than 100,000,000 h.p. in two stretches alone.

A point of special interest to our readers is, naturally, how Canada stands in relation to other countries in water power development. It will have been noted above that the statment is made that the United States has 41% of the developed water-power of the world and Canada 10%. It has to be remembered, however, that the United States has a population of 100,000,000 and Canada of about 9,000,000 and to truly exhibit the comparative position the figures require to be put on a per capita basis. The comparison then stands thus:-

| | Horse-power installed per 1000 of population. | | |
|---|---|---|---|
| | 1902 | 1912 | 1920 |
| Canada ............ | 47 | 198 | 280 |
| United States ...... | 26 | 51 | 93 |

It will be seen that the horsepower installed per head of population is three times as great in Canada as in the United States.

Particulars of the water power resources are now available for over 100 separate nations or countries and therefore form too extensive a list for space available, but the figures may be given for a few of the countries in which the greatest development has taken place.

| | Population. | Water Power | | | |
|---|---|---|---|---|---|
| | | Developed. | | Available. | |
| | | :Per 1000: | | | : Per |
| | | H. P. | : popula- | H. P. | : cent |
| | | | tion. | (minimum) | : dev. |
| Sweden | 5,814,000 | 1,460,000 | 251 | 4,500,000 | 32.4 |
| Norway | 2,700,000 | 1,350,000 | 500 | 5,500,000 | 24.5 |
| Italy | 40,000,000 | 1,150,000 | 287 | 3,800,000 | 30.2 |
| Switzerland | 4,000,000 | 1,070,000 | 267 | 1,500,000 | 71.4 |
| France | 41,500,000 | 1,400,000 | 34 | 4,700,000 | 29.8 |
| United States | 105,683,108 | 9,823,540 | 93 | 28,000,000 | 33.0 |
| Canada | 8,769,489 | 2,755,980 | 314 | 18,255,316 | 15.2 |

It will be seen from the above table that only one country, Norway, exceeds Canada in water power development per capita. In ratio of developed to available power it will be remembered that Canada is incomparably the less densely populated country and in this respect is over 100 years behind the next newest on the list, the United States.

---

## Would It Not be Advisable to Use Larger Size Meters?

In the letter reproduced below, a reader draws attention to an apparent error of judgment by someone in our industry—the result of the use of too small a meter on the average household load. He points out that the loss in revenue, due to meter inaccuracy on over-load, is much greater than the loss due to inaccuracy at part-load, and raises the question whether it would not be advisable, therefore, to install larger meters. The point, we believe, is well taken and we should be very glad to have the opinion of any readers on the point raised in this letter:

Editor Electrical News:—

There has come to be, in recent months, a quite general demand that house service meters should have an overload capacity of 100%. There are, of course, three functions of overload capacity:

1—The capability of withstanding short circuits without mechanical or electrical damage.

2—The current carrying capacity without undue heating.

3—The accuracy of the meter's measurement on overload.

It is with No. 3 that I propose to deal in this letter. There is no great difficulty in fulfilling the demand for a meter which shall register within 3% of accuracy on double load, but in the light of full knowledge of the induction meter's capabilities, it seems impossible to avoid a falling accuracy curve from full load to double load.

In explanation of the supply companies' requirements of 3% accuracy (which always means 3% slow) at double load, I am told that:

(a) The 10 amp. meter is enough for ordinary lighting purposes, but consumers instal additional apparatus without advising the supply authority, so that their consumption is liable to get as high as 40 amp. or more.

(b) The percentage of consumers who do this is nearly 100, and moreover the use of such additional apparatus is very great as it includes generally a cooking outfit.

(c) The meters usually installed are within 3 to 4% at double load (always slow).

I am unable to understand why a 20 amp. meter is not used as the standard instead of the 10 amp. meter when it is known that 20 amps. will be the usual load on the meter.

According to information (b), the use of apparatus such as stoves, irons, radiators and the like, is equal at the least to two hours per day, thus:—

$$\frac{20 \text{ amps. x } 110 \text{ volts x } 2 \text{ hours}}{1000} = 4.4 \text{ kw.h.}$$

The meter which measures this is, as shown, generally about 3% slow at 20 amps.

The use of a single lamp is not common, but 1/20th load of the 10 amp. 110 volt meter = 55 watts, and this may be taken as the lowest likely lamp load.

55 watts for say 12 hours = 0.66 kw.h.

It will be seen that the extremes of a meter's accuracy curve are very different in their degrees of importance. As that part of the curve represented by 1/10th load to full load is within say 1% of accuracy in nearly all meters, we can ignore it for comparative purposes.

Therefore, in one day's consumption of power, we get:

0.55 kw. for 12 hours=0.66 kw.h.

2.20 kw. for two hours=4.40 kw.h.

Total consumption 5.06 kw.h.

3% error at 200% load, i.e., 3% of 4.40 kw.h. = 0.132 kw.h.; that is, nearly 3 % of the total bill.

3% error on 1/10 of full load, i.e., 3% of 0.66 kw.h. = 0.0198 kw.h.; that is, 0.39% of the total bill.

It is thus clearly established that a 3% error at double load is 9 times more important than the same error at 1/10th load. If the 55 watt load were not registered at all, the omission would probably be unnoticed on the total bill of the average consumer.

Will any engineer say why the main source of revenue, i.e., the 20 amp. load, should be registered by a 10 amp. meter which is known to be 3% slow on 20 amperes.

Why not use a meter big enough for its job and increase the revenue by approximately 3 per cent?

Yours truly
"Retem."

Geo. R. Wright, Winnipeg, district manager of the Canadian General Electric Co. Ltd., was married on the 18th of February to Miss Vera Whitmore. The staff presented him with a very fine case of Community silver. Mr. and Mrs. Wright are spending their honeymoon at the Pacific Coast. They expect to return to Winnipeg about the middle of March.

# Pointers on Electric Power Rates for Industrial Engineers

### By GEORGE MACLEAN*

Three subjects that are frequently little understood among men who are responsible for the purchase of power for industrial plants are the flat rates, kilowatt-hours and maximum demand. A flat rate is based on the connected load. Assume an installation of several motors, amounting in all to 200 h.p., and a rate of $40 per horsepower year is charged. Then the cost of power per month is 40 x 200÷12=$666.66 for the total connected load of 200 h.p. The advantage or disadvantage of a rate of this kind depends on the load-factor, or the ratio of the horsepower employed to the motor horsepower. From this it can be readily seen that if only 100 h.p. of the total connected load of 200 h.p. is being utilized, then the connected load-factor is 50 per cent., or 50 per cent of the connected load is being paid for and produces no return. Thus if the contract is on a flat-rate basis, watch the motors and keep them loaded up to full load, if possible. Also, the processes should be arranged so as to keep all motors in service as much as possible, for when they are stopped the power expense is still going on.

The kilowatt-hour is the equivalent of 1,000 watts for one hour, or 1,000 watt-hours. This method of buying power has the advantage over the flat-rate system, in that only the power used is paid for. If on 200 h.p. connected load only 100 kilowatt, or 134 h.p., is used continuously for one hour, the power consumed is 100 kw.h. and this is what is paid for. The total 200 h.p. is not paid for, as on the flat rate, but only for the actual power consumed by the motors.

Maximum demand presents a problem different from either the flat-rate or kw.h. rate. A special meter is connected in the circuit, which automatically registers the highest amount of power used over a certain interval of time. If you have contracted for 200 h.p., and the maximum demand does not exceed this, then only 200 h.p. is paid for. However, if the maximum demand is 250 h.p., say for 15 minutes, for any one day during the month, then 250 h.p. is charged for. The reason for this is that the power company is supposed to have to hold in reserve sufficient capacity to meet the needs of the customer. If it is evident that much can be done to keep the peak loads down by the proper manipulation and arrangement of machinery. For example, if there are a number of heavy processes that operate only intermittently, then, if it is possible to have different processes going on at different times, the load will be much lower than if all were in operation at once.

The maximum demand rate can also be used in conjunction with a kilowatt-hour rate; that is, so much per kilowatt-hour is charged and in addition to this, so much per kilowatt maximum demand for a given interval during the month is also charged.

The McDonald & Willson Lighting Company, 309 Fort St., Winnipeg, has secured the electrical contract on an office building being erected at 300 Main St., Winnipeg, at an estimated cost of $55,000, for the Northern Life Assurance Co.

The Manitoba Gazette announces that the Gamble & Willis, Electric Co., Ltd., has been formed with headquarters at Winnipeg. The new concern will carry on the business of electrical contractor-dealers. Capitalized at $4,500.

*Electrical Engineer, Canadian Connecticut Cotton Mills, Sherbrooke.

# First 50,000 h.p. Unit of Chippawa-Queenston Power Scheme—II.

### Our First Article Described the Work in Connection with Location of Intake, Canal and Power House; also the General Design. Details of Hydraulic Installation Herewith.

## Details of Hydraulic Installation*

Before dealing with the constructional details of the canal and power house, we shall treat briefly a few of the outstanding features in the hydraulic installation. The features to be touched upon cover the canal control gate, mentioned already as being situated within a few hundred feet of the point where the canal proper joins with the Chippawa River; the intake; ice chutes; screens; removable gates; penstocks; Johnson valves; turbines; governors, etc.

### The Intake

The Commission has found it necessary to delay work on the intake to enable them to concentrate on other portions of the construction work. This was possible on account of the limited demand that will be made upon the water supply from Niagara River during the present winter. When completed, this intake will be of special design to take care of ice trouble. The design has been determined by a long series of tests and experiments with models. It is probably sufficient to say that at the moment the engineers believe the design they have adopted, will permanently remove the ice

*M. V. Sauer, Assistant Hydraulic Engineer Hydro-Electric Power Commission of Ontario

troubles which have caused a number of the power houses at Niagara Falls so much trouble in the past. In the meantime, water to operate the plant enters the Welland River through a natural channel which will eventually be closed when the permanent intake works have been installed.

### Canal Control Gate

A single motor-operated vertical lift roller gate for the purpose of controlling or entirely shutting off the flow is installed at the upper end of the canal near Montrose where the earth section of the canal merges into the rock section. The combination of span and head make this gate the largest ever built and particular attention was paid to the roller design in order to secure bearing loads well within safe, structural and operating limits. The clear width is 48 feet and the height of the gate is 42½ feet.

The general arrangement of the control gate is shown. It will be noted that the lift is extended to a point 14 feet above the water level and this has been done in order to permit a patrol tug to pass freely up and down the canal when the gate is in normal position, i. e., wide open.

The gate is counterweighted and operated by a motor connected through a worm drive to the two main hoisting gears, and the motor is provided with distant as well as local control so that the gate can be operated from the power house if required.

The gate itself is made up with horizontal trusses span-

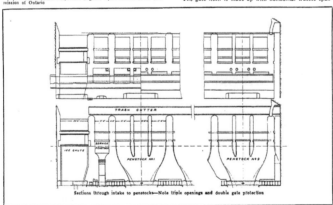

Sections through intake to penstocks—Note triple openings and double gate protection

# Two Views of Forebay from Opposite Points

Close-up view of diffuser where canal widens into the forebay

Screen house at end of forebay showing entrance to pipe-lines. The racks for screening the water and the gates for closing off the entrance of the pipe-lines are supported between the piers. There are three openings to each pipe-line.

ning the opening and vertical beams to which the skin plate
will be rivetted. It is not important that the gate be absolutely watertight when closed although "stanching" bars
are provided to insert between the skin plate and the end
guides which make it practically tight.

A comparative analysis of costs and operating conditions was made on the single gate as against two gates with
an intermediate pier, and it was found that the cost of a
single gate and superstructure did not exceed the cost of the

View of penstock mouth from one of the three openings

twin gates, on account of the additional construction required
by the intermediate pier and the necessary widening of the
canal to keep down the velocity to normal. When the added
advantage of having a clear unobstructed waterway was
taken into account as well as the simplified operation of a
single gate in place of two, the decision was entirely favorable to the single large gate.

### Forebay Ice Chutes

It is expected that the intake as designed will take in
water from the Niagara River, absolutely free from floating
ice, but to take care of any ice that may form on the surface
of the Canal or the Chippawa River channel, a small ice
chute is provided at the lower end of the forebay. It consists simply of an opening through the screen house and
down the cliff and under the power house to the lower river
in a reinforced concrete pipe 10 feet in diameter.

### Screens

The only feature of the screens worthy of comment is
the wide bar spacing of 4½ inches in the clear and the layout of the bars and frames which are so designed that the
whole frame with the bars attached is removable, thus leaving a completely unobstructed passage when they are removed. There are three bays of screens for each penstock
and two frames in each bay. The tops of the screens are
eight feet below the normal surface of the water and the
maximum velocity of water through the screens is 2.35 feet
per second. With these provisions it is not anticipated that
anchor ice will cause much trouble.

### Removable Gates for Penstocks

In view of the fact that a Johnson valve is being installed at the lower end of each penstock, adjacent to the
turbine, it was decided to omit permanent gates in the screenhouse at the penstock entrances. To take care of any failure in the valves, removable structural gates made up in
sections are provided which can be lowered into any penstock entrance by means of an electric travelling crane in
the screenhouse.

The quantity of water used by each turbine at full load
and under normal head is approximately 1800 cu. ft. per sec.,
and in the design of the penstocks, the diameter was fixed
by plotting up various curves showing the value of power
lost due to varying velocities and their consequent friction
losses as against the carrying charges on the corresponding
penstocks. By this means a diameter of approximately
fifteen feet was found to give the best value, but so great a
diameter at the lower end required a plate thickness of over
1½ inches and this was considered beyond the limit for safe
field rivetting. On this account the diameter of the upper
two-thirds of the pipes was made 16 feet and the bottom
third 14 feet which made the construction work feasible and
at the same time gave the desired economical results. The
loss is considerably reduced by the use of butt girth joints
with an outside cover plate as against the customary practice
of using inside and outside courses.

### Johnson Valves

One of the figures shows a longitudinal section through
the Johnson hydraulic operated valve located at the lower
end of the penstock. The operation of these valves is very

Close-up view of penstock mouth

simple, no outside power being required, the valve being
opened or closed by means of the penstock pressure. The
valve plunger is of the differential type and seats against a
ground fit ring in the neck of the body. The annular chamber A and the central chamber B are connected through a
control valve and piping either to the penstock pressure or
to the atmosphere. Admitting penstock pressure to A and
atmosphere to B opens the valve while the reverse operation
closes it.

The advantage of a valve of this type is its simplicity of
operation. Furthermore, because of its circular section it
can be built for any head and thus located at the lower end
of a penstock, obviating by this arrangement the necessity of
emptying and filling the penstock for each shut down.

### Turbines

Five turbines are at present under contract, two
of which are completely erected. They are each of 50,

## Two Views of Finished Queenston-Chippawa Canal

Curve on canal showing retaining walls above concrete lining where natural rock surface lies below water level

Power Canal showing change in shape of the canal as it passes from vertical faces of rock cut to sloping sides of Whirlpool section

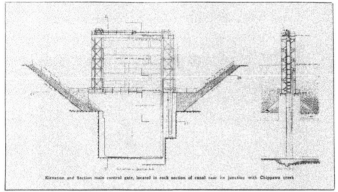

Elevation and Section main control gate, located in rock section of canal near its junction with Chippawa creek

000 h.p. rated capacity and of the vertical spiral case, single runner Francis type and will operate at a speed of 187½ r.p.m. This gives a specific speed of 36. The maximum guaranteed efficiency is 90 per cent, although in view of recent practice it is expected that this efficiency will be exceeded. On model runners of homologous design tested at Holyoke 91 per cent was obtained. The inlet diameter of the scroll case is 10 feet and the diameter of the runner is 10 ft. 5 in. at the inlet. An open space has been left in the power house foundations below the runner, so that by removing a section of the draft tube the runner can be taken out from below, thus obviating the necessity of dismantling

the generator when a renewal of the runner is necessary. The runner is designed for a capacity of 61,000 h.p. and is "gated back" to a maximum capacity of 55,000 h.p. The reason for this is that the turbines, which will normally operate at or near full rated load, will also therefore operate at their maximum efficiency.

### Governor System

The centrifugal head, relay valves, and hand control for each governor are located on the generator floor while the main automatic valve control is located directly under the governor stand at the level of the turbine regulating cylinders. The advantages of this arrangement are the short piping between the main valve and the regulating cylinders and the separation of the two main parts of the governors, giving freer access for repairs and maintenance. The pressure fluid is water, treated with soluble oil, which will pre-

Photo of canal gate, open, shown in plan above

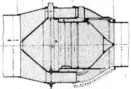

Longitudinal section through Johnson valve

vent rusting of the wearing parts and at the same time give a lubricating value to the water. A central pumping system is used, with duplicate motor driven multi-stage centrifugal pumps, either one of which has sufficient capacity for all the governors. The pressure fluid will be piped to all the governors through accumulator tanks, one located near each governor so as to eliminate any inertia effects through the piping system. The pump motors are automatically controlled by relay switches which are controlled by pressure variation in the system. As a further safeguard for preserving continuous operation, in the event of failure of the pumps or motors, penstock pressure can be turned into the governor system. When the plant is finally extended to its full capacity a complete duplicate pumping system, similar to the one above described, will be installed and interconnected with the present system.

## Control Pedestals

A control pedestal will be set up adjacent to each generator, and on this will be mounted the various indicating instruments and control handles. The principal use for such an arrangement is that the communicating devices between the floor operator and the chief operator in the control room, together with the local control and indicator, will be located in such a way that the floor operator can handle the machine while in touch with the chief operator.

## Service Units

For furnishing heat, lights, and power service to the plant two service units, each of 2500 h.p. capacity are being installed. Each of these consists of a vertical turbine running at 300 r.p.m. direct-connected to a generator. The turbines are supplied by a single 5 foot diameter penstock branching into two pipes at the turbines, each branch being provided with a Johnson valve. It is expected that the service plant will be duplicated when the power house is completed to its full capacity.

# The Coal Equivalent of Water Power

### By H. E. M. KENSIT, M.E.I.C., M.A.I.E.E.

In relation to the value and benefits of water power development and the tendency to ever rising cost of coal, the question of the average saving of coal by water power development has recently received considerable attention, and it is of interest to review recent studies of this matter.

In the case of an individual plant the conditions are known and the comparison can usually be worked out fairly accurately, so that no special remarks are called for.

What is of special interest is the result that can be anticipated from a general policy of water power development over a large industrial area, and it is in this connection that the following particulars are given. In such a case the study must include all uses of power and such may be broadly divided into public utility purposes, manufacturing industries and railroad electrification.

One of the most complete studies that has been made is that of the United States Superpower Survey of the region between Boston and Washington.[*] This region, approximately 450 miles long by 100 to 150 miles broad, with an area of 45,000 square miles and a population of about 25,000,000, is less than 5 per cent. of the total area of the United States, but contains 70 per cent. of the total industrial development and uses 40 per cent. of the total coal consumption, and is therefore an intensive industrial district. Within the zone there are 559 electric utilities, 18 railroads and 96,000 industrial plants, 76,000 of which use power and average 350 horsepower each. The 18 railroads represent 36,000 miles of single track, of which 19,000 miles can be profitably electrified. Practically all conditions obtaining in a manufacturing country are therefore represented. The power survey is a broad one and it appears to have received very detailed examination.

The horsepower in use is as follows:

|  | Total | Per Cent. |
|---|---|---|
| Factories | 5,850,000 | 34.4 |
| Public Utilities | 4,150,000 | 24.4 |
| Heavy Railroads | 7,000,000 | 41.2 |
|  | 17,000,000 | 100.0 |

It was found that the average coal consumption in electric power stations in the area is 4.7 lbs. per kw.h. (3.5 lbs. per h.p. hour), in steam locomotives, 9 lbs. per kw.h. (6.72 lbs. per h.p. hour), and that the average factory uses more, some of them more than 20 lbs. per kw.h: (15 lbs. per h.p. hour).

If we take the average consumption in factories at 12.5 lbs. per h.p. hour, the mean of the above results is 7.93 lbs. per h.p. hour, and this at a 33 per cent. load factor is equivalent to 11.5 tons per h.p. year.

In Great Britain the subject was examined by the Coal

[*] Professional Paper 123, United States Geological Survey, Washington, 1921.

Conservation Committee of the Ministry of Reconstruction, and its final report (Cd 9084, 1918) stated that figures based on the last census showed that if 75 per cent. of the total coal used in industry was used for power production it gave a figure of 8.03 lbs. of coal per h.p. hour, that since that time greater efficiency had been obtained, that the records of the large power companies in the North East Coast Industrial Area showed that the average coal consumption of present power users before the adoption of electric driving had been 7 lbs. per h.p. hour, that these power companies themselves used but 1.54 lbs. per h.p. hour, and that "accordingly the present consumption for the country as a whole is taken at 5 lbs. of coal per h.p. hour instead of the actual figure of 7 lbs." This, it will be seen, is on account of the results obtained by large power companies operating over extensive areas.

The Hydro-electric Power Commission of Ontario examined the matter of average coal consumption with some care, but on a less extensive scale. Its "Report on the Rate of Coal Consumption" published in 1918 gives a careful digest of the results obtained from 73 central stations of all sizes in Canada and the United States, and from 135 industrial plants in the Niagara district; special care is being taken to include only those plants that used coal exclusively and used it for power only. The average result obtained from both central stations and industrial plants was 6.85 lbs. per horsepower hour, which on a 33 per cent. load factor is equivalent to 10 tons per h.p. year.

These three extensive investigations may be summed up thus as regards general results in industrial districts before extensive systems of central station distribution over large areas are installed:

| United States | 11.5 tons per h.p. year |
|---|---|
| Great Britain | 7.0 tons per h.p. year |
| Canada (mainly) | 10.0 tons per h.p. year |
| Average | 9.5 tons per h.p. year |

so that 10 tons may be taken as a safe round figure for estimating purposes.

## The Saving of Coal

It is now to be considered how this figure should be applied in estimating the saving of coal due to water power development.

With respect to coal consumption per h.p. it is only in the case of public utilities that the horsepower actually used on the load is closely known—in the case of factories and steam railroads the coal used per horsepower is usually based on the installed or rated h.p. of the plant, which usually has but little margin over requirements, and from the table given above of the distribution of load in the United States it will be seen that public utilities represent less than one-quarter of the total horsepower used. Therefore, for approximate estimating purposes it may be taken that coal consumption per horsepower represents the consumption per installed horsepower.

In estimating the coal equivalent of water power, therefore, the calculation should not be based on the average power the stream will develop, but on the horsepower of plant that will be installed to meet the maximum requirements, as this will more nearly correspond to the conditions obtaining in the steam power plants over a large area. Investigation will show that in water power plants the installed horsepower is on the average at least 50 per cent. greater than the minimum capacity of the stream, this being due to the fact that the pondage (reserve of water created by the dam) and load factor enable temporary peak loads to be carried that are far above the nominal capacity of the stream. It is not considered necessary for the present purpose to treat these well-known facts in technical detail.

### 10 Tons per h.p. Year

In calculating the coal equivalent of water power over a large area or over the country as a whole, therefore, the 10 tons of coal per h.p. year should be multiplied by the horsepower installed or estimated to be installed at the water power plants.

To illustrate this, the figures given for Canada in evidence by the writer before the Special Committee of the House of Commons on the future fuel supply of Canada during the session of 1921 may be quoted.

It is there shown that the total coal consumption in Canada in 1920 was 35,227,000 tons, that the total developed water power actually in use at that time was 2,214,721 h.p., equivalent to 22,147,000 tons of coal, most of which would otherwise have had to be imported, and that at the average price of $8.32 per ton this represents over $183,000,000 per annum.

It is also shown that, looked at from another point of view, the consumption of coal per capita is 50 per cent. less in Canada than in the United States, although the colder climate might be expected to lead to an opposite result; that this is mainly due to the water power development per capita being 194 per cent. greater in Canada than in the United States, that the resultant saving in coal is now equal to $146,500,000 per annum, and that from the national point of view this is equivalent to a return of 27.5 per cent. per annum on the capital invested in water power development.

---

## The Complex Nature of a Modern Telephone Plant

At a recent meeting of the Montreal Electric Luncheon, Mr. N. M. Lash, chief engineer of the Bell Telephone Company, delivered an address on "The complex nature of a modern telephone plant." At the outset Mr. Lash dealt a body blow to that well-known fallacy that, as a telephone system grows, the cost, per unit, of giving service should go down. As more subscribers are added and more expensive and intricate equipment is added, costs invariably go up and not down. Increased traffic, with greater first cost, power consumption and maintenance charges, all contribute to this result. As towns grow, Mr. Lash showed, bigger switchboards, more expensive land and buildings are needed, with bigger power plant, more complicated apparatus, more extensive underground and cable systems and heavier traffic loads requiring more operators. Lines to subscribers are longer, fewer lines must be assigned to each operator, and cost of installation and maintenance becomes greater. When a city requires more than one central office, trunking facilities between them add greatly to cost. Toronto, for example, with its eleven central offices, requires wire, for trunks alone, sufficient to girdle the earth 1.6 times. All these factors of ever-increasing intricacy involve increased first cost per line, increased operating costs and increased maintenance and annual charges.

Mr. Lash showed how in his department engineers plan 15 to 20 years in advance of a city's growth so that telephone facilities, especially underground, and cable distribution, may keep pace with demand. The remarkable accuracy of these estimates was shown from the fact that it is never necessary to dig up a conduit to increase its capacity. The average efficiency of these plans was said to be 92 per cent correct, really absolutely correct for all practical purposes. To keep ahead of demand, there must be a vast investment in spare plant—more for fast-growing than for slow communities! When it is necessary, for lack of capital, to postpone the opening of a new central office, there are always big expenditures necessary for temporary construction. If the work cannot be laid out 3 and 4 years ahead and orders placed for equipment and initial steps taken, costly duplication of plant and expensive construction are inevitable.

### Long Distance Service

Mr. Lash next dealt with the wonderful development of the long distance system, citing cases in which open wire must soon give way to wires in cable to care for heavy demand. The actual productive talking time on long lines is constantly being increased as a measure of economy and operating efficiency improved, but each advance along this line called for bigger expenditures in new equipment. The demand for longer and ever longer talks has meant greater refinements in methods and required apparatus costing much money. It is only the cumulative effect of scientific devices that makes such long talks possible, and to apply new devices and inventions to the long distance network invariably means extensive rebuilding of both inside and outside plant. The problem is one of infinite delicacy—not of power! Maintenance and repairs call for the highest quality of engineering, supervisory, mechanical and technical skill. The standard must be kept high, else long talks would be impossible.

The speaker emphasized the need for adequate rates for service if the public demands for extensions were to be met. It is only in the public interest that earnings should be such as always to attract new investors to put their money into the telephone business. The millions needed for extensions to meet demand could be secured in no other way! Not less than 13 to 14 millions of new money is needed every year to meet demands for telephones and provide an adequate margin of spare plant. If the public once realized that the cost of giving service really does increase as the plant grows and understood some of the complications involved in a large modern telephone plant designed to maintain a high grade of local and long distance service, universal in its scope, they are fair enough, Mr. Lash believed, not to oppose a fair schedule of rates.

Following Mr. Lash's talk some splendid moving pictures were shown that gave his audience a better appreciation of a telephone system than perhaps could be gained in any other way.

Mr. Lash also addressed The Electric Club of Toronto on February 22. His subject on that occasion was slightly different—"A Growing Telephone System"—but the general data was along similar lines. The members were highly pleased with Mr. Lash's address.

---

L. C. Barbeau & Company, Limited, have just taken over the top floor of 320 St. James Street, Montreal, for their offices and rooms.

---

The Easy Washer Company are opening a store at 409 Yonge Street, Toronto, where they will demonstrate and retail their electric washers.

# The Development of Wireless-III

## A Series of Short Interesting Articles Covering Wireless Progress to Date

### By F. K. D'ALTON

Capacity and inductance are the important properties in an electrical circuit which determine its resonant frequency. These terms may be defined as follows:—

"Capacity" (C) is the ability of all conductors or metal parts to hold a charge of electricity: the quantity in the charge is proportional to the voltage applied and a slight change in voltage is accompanied by a considerable current flowing into or out of the parts being charged, hence a tendency to establish a current instantly upon a variation in voltage.

In the wireless apparatus capacity is added in the form of condensers, being plates or sheets of some metal, which are insulated from those adjacent but alternately connected to one or other electrical terminal. Thus large surfaces are presented, and this is one essential in obtaining capacity.

"Inductance" (L) is that property of a coil of covered wire whereby it tends to prevent any change of current through it. A change in voltage is followed by a comparatively slow change in current. This effect is the opposite of that found in a condenser.

The inductance increases roughly as the square of the number of turns, i. e., doubling the number of turns gives four times the effect.

"Resonant Frequency" is that frequency of reversal of current at which the retarding effect of the coil equals the assisting effect of the condenser; then the current that will be built up is limited only by the resistance, (a property of the materials of which the conductors, wire and plates are made).

The formula for this relation is:—

$$frequency = \frac{0.159}{\sqrt{LC}}$$

where L and C are expressed in their usual units, i. e., henrys and farads, respectively.

Now the length of the wave is equal to the velocity of its propagation through space divided by the frequency or

$$wave\ length = \frac{velocity}{frequency} = velocity \times \sqrt{LC} = K\sqrt{LC}$$
$$\frac{0.159}{}$$

where K is a constant including velocity which we know is fixed at 188000 miles per second.

Since resonant frequency, and wave length, depend upon the product of the coil effect (inductance) and the capacity, there is an unlimited variety of values of L with corresponding value of C to give a particular wave length, also, either L or C may be variable, (the other fixed) or both variable if preferred, to give us a range.

In a tuned transmitter, the spark occurs in a circuit containing both inductance (a coil) and capacity (a condenser). The coil is placed in proximity with another coil which is connected between the aerial and the ground wires, the aerial forming capacity, and thus a tuned circuit. Then the "spark circuit" and aerial circuit are tuned, adjusted so as to have the same resonant frequency. This results in a pure wave with maximum power being radiated, and reduces the interference of this station with others to a minimum.

In the receiver, a circuit containing a coil and condenser is tuned so as to be in resonance with the transmitter of the station which it is desired to "read" and the voltage built up in the coil and across the condenser is carried to the detecting device.

The receiving aerial obviously receives simultaneously the waves from many stations but it is only the waves at a particular frequency which build up an appreciable current in the tuned circuit. All other waves, unless very strong, pass through to ground unnoticed, causing absolutely no interference. By tuning alone, the workable distances have been very much increased.

It must be understood that the writer has attempted to describe only the simple systems, showing the fundamental principles. In long-distance commercial stations much more elaborate schemes are used for both transmitting and receiving:

In the case of a string giving forth sound of a certain pitch, it was stated that this pitch depends upon the constants of the string such as material, tension, length, etc., and that each time the string was plucked the same note was heard but that the sound gradually diminished in intensity and finally died out altogether. In just the same way a spark starts a current in a tuned circuit and this current gradually diminishes although keeping the same frequency. This is termed a "damped" wave, the rate of dying down depending upon and varying with the resistance of the circuit.

Obviously the most energy, equivalent to the loudest sound, is in the first cycle. The damped wave attempts to get the current in the tuned circuit of the receiver built up immediately and if the damping be excessive its effect will be much the same whether this receiving circuit be tuned exactly to the transmitted wave length or to some other wave of a different length.

A station radiating a highly damped wave will then interfere over a considerable range in wave length and is very objectionable for this reason. Lightning causes a highly damped wave of great strength and is therefore the worst offender.

The importance of wireless communication for ships at sea with the necessity of keeping interference at a minimum has been the cause of the making of a number of laws regarding "damping" and a limit has been set. A station is breaking these laws if its wave dies out in less than a given time.

Spark stations at first used the well known spark coil, but in order to get higher power and greater distances, transformers were employed. These were designed with high reactance in order not to draw too great a current when the spark occurred, short-circuiting the high voltage winding. The spark gap at first was a simple one consisting of two rods with a small air space between. The received note from such a gap is high in pitch and hard to read, and as a result a rotary gap was employed. In this device an arm revolves between stationary points and a spark occurs whenever an end of the arm passes a point. The sparks and wave groups are thus broken up and a lower note obtained; this note being adjustable at the transmitting end only.

Let us now consider the manner in which the crystal acting as a rectifier of current becomes a detector of wireless waves.

The damped waves from a spark station are picked up by the aerial and a current in the tuned circuit results. This current is alternating, and its average value is at all times zero. If it were turned directly into a "phone", no sound would result for several reasons—(a) the frequency is above the range of sound; (b) the diaphragm cannot respond to such high frequencies; (c) the coil in the phone acts as a choke and would only allow very little current to pass.

By rectifying the current we do not change its frequency but we change its average value, and make the half wave consist of two parts, one the normal high frequency, the other a frequency equal to the number of wave trains or groups of waves. If the rotary gap of the transmitter gives us one

(Concluded on page 45 )

# Ten Years From Now?

The merchandising of electrical equipment is in the balance.

Ten years from now—through what agency will electrical appliances be reaching the public?

It may be the hardware man—or the department store—or the druggist—or the furniture dealer—or all of these.

**Or it may be the Electrical Store**—almost exclusively.

That's why our business is in the balance. It may develop along many different lines—there are arguments in favor of each. If it is profitable, each trade will make a bid for it.

It's little use to argue that the electric store is the **logical** channel and that it's bound to work out this way. Isn't it logical that the furniture store might sell floor lamps; that the seller of carpets might sell vacuum cleaners; that the merchant who retails coal stoves might also sell electric stoves?

The only **really logical** phase of the whole situation is that this business should, and will, gravitate in the direction of the trade that is most aggressive in its development.

And in this one respect the electrical merchant has the lead—he, alone, realizes the possibilities. To that extent his competitors are handicapped. For that reason he has more than an even chance to hold the lead.

Don't let us make the mistake, however, of thinking this appliance business is ours by divine right. It's **anybody's business**. If the electric merchant as a class makes a determined fight he can establish himself, within the next ten years, in the minds of the public, as the one and only reliable and desirable source of supply.

But, if he fails to make that fight—?

# Following up the Advantages Created by Lively Electric Home Campaigns

### The "Electric Home" Exhibit Leaves the Housewife's Mind in a Receptive Mood—She Wants to Learn More—Contractors and Dealers are Following up The Lead

At various points in Canada, at the moment, there is an epidemic of "Electric Home" campaigns. Toronto has it; so has Montreal; Hamilton is getting it; so, they say, is Sarnia; Kitchener, likely, too. East and West, it's spreading. Why all this sudden interest and enthusiasm?

The fact is that the industry has hit upon something that catches the popular fancy. Every woman is interested in her home—so are most men. The Electric Home, tastefully decorated, neatly finished, wired for the use of every convenience and comfort, touches a responsive chord in the home-lovers' hearts and sends them away feeling that there are many places in their own homes where another outlet should be installed.

The big thing is that the men and women go away from these exhibits in a mood that is entirely sympathetic to electric outlets and appliances and to anyone who may approach them regarding these matters. The people may not, in the majority of cases, be sufficiently enthusiastic to approach the contractor or merchant to buy these things—though a percentage will even do this—but they are all ready to talk about electric outlets and appliances, if the matter is broached to them in a diplomatic way.

For example, the Electric Home makes it possible for the contractor or dealer to call any of his old prospects on the telephone and be assured of a considerate hearing—or to get an easy entrance into the home of a prospective customer for a vacuum cleaner, washing machine or whatnot.

The people are ready to discuss, ask questions and benefit by instruction where formerly they were merely "not interested."

The ultimate value of these exhibitions will thus depend on the policy of the contractors and dealers—on the extent to which they follow up their advantage. A number of contractor-dealers have put salesmen out whose business it is to locate prospects and follow them systematically. One often hears it said that "you can't get salesmen," but in one case brought to the writer's attention a newspaper advertisement brought more than half a hundred replies from which eight or ten really live young men were chosen that are since "making good." For the selling of appliances, it doesn't require a man highly versed or experienced in electrical matters. He needs, rather, a good address and the ability to inspire confidence. It goes without saying that he must be an enthusiastic believer in the articles he is selling.

# Canada Due For Burst of Home Building

### Accumulated Shortage Will Force Construction Activity in Immediate Future—Every Home Should be Fully Wired and Equipped Electrically

The Society for Electrical Developement has been gathering statistics in the United States regarding the number of houses wired, and places the total at 7,636,409. On the supposition that the population of the United States is 110,000,000, and the average family consists of four persons, there should be 27,500,000 homes. Thus, there are only 25 per cent. of the homes in the United States wired and supposing, again, that only 50 per cent. of the total are within reach of electrical supply (that is a very low estimate, of course) the conclusion is that less than half the houses that ought to be wired have actually been equipped for even electric lighting. This point again emphasizes the great possibility in old house wiring.

The Society figures that the number of houses wired during the year 1921 was 1,001,700; that of this number, 700,000 were old and 301,700 were new. They also estimate on an increase of 25 per cent. during 1922. These figures are given to show, not only the very large number of wiring contracts available for the electrical contractor, but also to indicate that electric merchants were provided with a very fertile field for the sale of their appliances. Their final figure is that 8,500,000 homes in the United States need better fixtures and more appliances.

Figures in Canada do not check up exactly in proportion to the population, but the comparison is interesting, nevertheless. According to the most authentic reports, 16,000 new homes were built in Canada during 1921. There

is every indication, however, that this number will be tremendously increased during 1922 for statistics also show that there were 110,000 marriages. 94,000 extra families must be taken care of in some way, and while the apartment house will do its share, the inevitable result must be that home building in Canada will see a very marked increase during the next few years.

Just for the sake of argument let us suppose that in Canada 110,000 new families were in a position to wire their homes and buy electrical equipment. Putting the wiring at $100, and the electrical appliances at the same amount, the purchasing power of these new families is $22,000,000.

Do we realize yet the possibilities in this electrical business? Do we realize, it may also be asked, how closely this development is tied in with ample home accommodation? In the construction industry the prospects for great activity are very bright. Money is cheaper and will, naturally, be looking towards mortgages at 7½ per cent rather than 5 per cent. bond yields. Rents are higher today than they have been at any time between now and 1914. The costs of building material, in general, have been reduced, perhaps, 25 per cent.

Don't let us be caught napping. Every new house that is built should be wired to the extent of from 3 to 5 per cent. of the total cost of the home—and that just paves the way for equipment sales to begin.

# Bonspiels—Yesterday and To-day

### Electric Illumination Has Magnified, a Hundred Fold, the Glory of Canada's Unique Outdoor Sport— Winnipeg Carnival a Blaze of Glory

Winnipeg, the Gateway to the West, held its first winter carnival at the same time as the famous annual Bonspiel—during the week, February 6th-11th. Electricity played a most important part in making the event a success, and was the subject of much favorable comment by the thousands that visited the grounds. The Carnival program consisted of many winter sports, pageants, parades, crowning of the Carnival Queen, mocassin dancing and stunts by officials of Hickville, whose band, police force, and fire brigade provided plenty of amusement. The program continued until 11 p.m. daily and therefore the matter of lighting had to be given careful consideration to ensure the public being able to properly view all the events. It was also thought desirable to make the grounds attractive by the display of plenty of light.

#### Flood Lighting on Parliament Buildings

The Carnival was held on grounds adjoining the new

Parliament Building. The dome on this building is about 240 feet high, and is permanently lighted by flood lights mounted at intervals on the main roofs.. They throw light up to and all around the dome.

The steps at the west entrance to the Parliament Buildings were used for the coronation of the "Queen", and her reception each evening. Her court and retinue consisted of hundreds of attendants, horses, etc., which occupied a space about 100 feet wide, the steps forming the throne.

For lighting this space, 4-1000 watt floods were erected on the roof of a bandstand about 200 feet distant; here also was a large movable search light loaned by the "Free Press"; this was used to light the throne or scenes of interest on the grounds.

#### Illuminated Ice Wall

The entire grounds were surrounded by an ice wall, con-

Upper Left—Illuminated Band-stand, Winnipeg Winter Carnival; note ice-block foundation; Manitoba capitol buildings in background.
Lower Left—Illuminated ski slide with flood light at top; building in the foreground erected and used as one of the fun making features of the Carnival, and labelled "Hickville."
Upper Right—An inner gateway of the Carnival grounds, constructed entirely of ice and illuminated by the Winnipeg Hydro-electric System.
Lower Right—Electrically illuminated transmission tower erected by the Manitoba Power Co.

sisting of ice blocks about 6 feet high, standing on end; at intervals of 33 feet were ice pillars about 4 feet by 8 feet high.

Evergreen trees 5 feet high were placed at intervals of 3 feet on top of the wall and pillars. For lighting the wall, festoons were used—about 1500 feet in all, 25 watt lamps being placed at about 2 feet centres. These were in the Carnival colors of red, green, and white and looked very pretty mingling among the trees.

### Lighting of Band Stand and Dressing Rooms

The main entrance was the old Fort Osborne gateway, which was decorated by festoons across the top and 6 streamers down the front, of colored lamps.

The bandstand was in the centre of a huge rink used for moccasin dancing. This building was octagonal in shape and the exterior was decorated by 8 streamers from a central flagpole down to the roof corners, these being illuminated by colored lamps, 25 watt, 15 inch centres. For exterior lighting 8 brackets were projected about 4 feet on which were supported No. 402 "Lumo" balls and 300 watt nitro lamps. The interior was lighted by 100 watt lamps suspended from the ceiling.

A building 280 feet long used for dressing and check rooms, concessions, etc., was lighted by lamps projecting over the cornice, these being set at 2 foot centres. The tower on this building was lighted around the top and on each corner by lamps at 1 foot centres. About 300-25 watt colored lamps were used on the entire building.

### Ski Slide—General Illumination and Toboggan Slides

The ski slide was lighted by 4-1000 watt flood lights, 2 at the top and 2 midway down the slide. For decoration, 2 streamers were run up the 120 foot tower over the top, and held by flagpoles. At intervals above the handrail—each side of the slide—25 watt colored lamps were used at 15 inch centres.

The toboggan slide and towers were a brilliant spectacle. On each tower was a 1000 watt flood, lighting the slide and the entire 900 feet runway. 15 watt colored lamps were placed at 15 inch centres, suspended about 9 feet from the ground.

The general lighting of the grounds was accomplished by means of flood lights, about 30 being used, mostly 1000 watt; these were mounted on poles or tops of buildings; and, being adjustable, could be swung to any desired position.

### Manitoba Power Company's Display

An original idea was carried out by the Manitoba Power Co., who erected one of their 45 ft. steel transmission towers, on top of which were placed 6-1000 watt flood lights. The tower was outlined with solid rows of lights, and about half way up were bands around the tower consisting of many rows of red, white and green lamps; over 1200 60 watt lamps were used. The Winnipeg Electric Railway Co. supplied this current. The idea was to typify the coming to Winnipeg of more power, which work the company is now engaged in.

Advertising which was carried out by the company in a plan which linked up with the tower notified all carnival visitors that power from the company's 170,000 h.p. plant on Winnipeg River would be available for industries in Winnipeg in 1923.

The tower was erected under the supervision of D. K. Lewis and M. C. Gilman, of the Winnipeg Electric Railway Company.

### Winnipeg Hydro Display

Winnipeg Hydro-electric erected a huge arch and 2 pillars of ice. Surmounting the top were festoons of colored lamps. On either side of the arch was a sign announcing the name, the letters being outlined with 50 watt nitro, blue sign lamps, which showed up to great advantage. Current generally was supplied by the Winnipeg Hydro who charged for service only, contributing the current.

The Queen's ice palace was heated by a 3 k.w. Moffatt grate and twenty three kilowatt air heaters, while the lighting consisted of beautiful silver plated candle fixtures and brackets, the product of the Standard Bronze Company, Toronto. J. R. Aikman, J. Swan and W. T. King spent a lot of time to make this display most attractive.

### Let by Contract

The electrical work was let by tenders, the contract, amounting to approximately seven thousand dollars, being awarded to McDonald & Willson Lighting Co., Limited, Winnipeg. The contract included maintenance and removal of the system. The same company bought the salvage by bid.

There was a remarkable absence of trouble, one fuse being blown by a wagon backing into and breaking a festoon on the ice wall. Fred Shipman had charge of the contract work and is to be congratulated upon the splendid display.

### Materials Used

Several services were brought into the grounds; these were controlled by Langley externally operated switches. Mains were 3 wire, 110-22 volts a.c., and all branch circuits 110 volts, 2 wire, not exceeding 1500 watts.

Wiring was open work on knobs or cleats; Norbitt 118 decorative sockets were installed with Laco lamps. Davis flood lights were used.

The entire load amounted to about 250 kw. The contract provided work for 15 men for about three weeks, and was completed 2 days ahead of schedule and dismantled in 1 day.

## Canadian National Carbon Co. Appoint New General Sales Manager

Mr. Alexander MacKenzie was recently appointed general sales manager Canadian National Carbon Company, Limited, and Prest-O-Lite Company of Canada, Limited. Mr. MacKenzie joined the staff of the Prest-O-Lite company in 1917.

Mr. Alex. MacKenzie

After some months in the head office in Toronto, he took charge of the Montreal office and service station of the Prest-O-Lite company and built up large and profitable branch business in Quebec. On August 1st, 1920, he was made sales manager of the Canadian National Carbon Company, Ltd. His work for this company earned for him the promotion to general sales manager of both companies.

# There's Big Business for the Contractor in Wiring Old Houses

### Most Old Homes are Wired for "Light" Only—It is as Important that they be Entirely Remodelled as that New Homes Should be Fully Wired

With so much attention being paid to the business of wiring new homes, the average contractor has lost sight, perhaps, of the fact that there is much to be done in the way of rewiring old houses. For one reason or another, the contractor fights shy of the old house jobs. There are so many uncertainties connected with it, he will tell you, that it is impossible to form an estimate before-hand of the actual cost of the work, and he runs a big chance of losing money.

A contractor, who has specialized in old house work, recently stated that there was practically no excuse for this uncertainty. He pointed out that too many contractors take merely a casual glance at the situation and "guess" its cost. Possibly they take the precaution of counting the outlets.

Light in an old bathroom "where you want it"

If they find, as the work nears completion, that they are losing money, they prevail on the owner to cut out one or two or more outlets—"because they are unnecessary."

#### Requires Skill but Can be Done

This contractor points out that determining the cost of wiring an old house requires, of course, a little greater skill, experience and engineering ability than the new job requires. If the wiring is to be concealed, it is as necessary to make investigations as an engineer would find it necessary to take borings before fixing the cost of the foundation of a dam or large building. With the necessary care, however, the ele-

ment of uncertainty can be removed in practically every case.

Then there is the alternative method of cost plus percentage arrangement which many householders will readily agree to when the difficulty of forming an exact estimate is explained to them.

#### Old Houses in Big Majority

This contractor, however, was emphatic in urging that the "old house" business should receive more attention from the electrical contractor. In the smaller towns, a new house is an event, and even in some of our smaller cities the number of new homes in a year may be counted on the fingers of our two hands. How diminutive the new business looks when compared with the thousands of old homes that are waiting to be rewired.

#### Don't Destroy the Furnishings

A very real factor entering into the problem of convincing the owner of an old house that he ought to have it rewired is the feeling that the walls have to be torn to pieces, practically, and the whole place badly messed up. To the housewife, comfortably settled, her walls well decorated, floors and furniture well polished and everything placed "just so", the thought of all these things being upset is too repugnant to be easily overcome. In this case it is plainly the first duty of the contractor to convince the housewife that he can do the necessary work without interfering in any considerable degree with the comfort and appearance of her home. In a case of this kind much depends upon the contractor. Some workmen are naturally untidy, careless of the surroundings, careless of personal property; such as these have no business to be engaged in the wiring of occupied houses—or, for that matter, of new houses, either. It is quite possible, however, to do this work practically without injury to the decorations and furniture.

#### Why not Exposed Work?

There was, perhaps, a prejudice in the early years of electrical wiring against exposed work. However, there seems very little justification for it to-day. A modern exposed conduit is very inconspicuous and has the added advantage of an installation that costs less, takes less time to install, upsets the household less and is capable of adding a wonderful flexibility to the wiring arrangements. With exposed wiring one can lead the wires anywhere irrespective of joists or stone walls, doorways, chimneys or whatnot that might interfere with the work of concealing the wires in many an old home.

In Canada we don't seem to have realized the possibilities of exposed wiring. To the contractor, it opens a wonderful field. A few feet of molding run along the corner of a room, decorated the same color as the walls, substitutes a baseboard outlet for an inconvenient ceiling outlet. Molding for this purpose, used nowadays, is of most inconspicuous dimensions, the whole molding of one manufactured product, with its three wires, measuring less than ¾ of an inch in width and less than half an inch in thickness. Painted or decorated the same color as the ceiling or walls, or covered with paper at the time the rest of the room is papered, all possible objection to its appearance is overcome.

## *Typical Examples of Exposed Wiring*

Upper left—Table reading lamp from wall outlet.

Upper right—A convenience outlet in the workshop.

Left—The best of light at the base of operation.

Right—Extra wall outlet for reading and work table.

Lower left—Ironing outlet conveniently placed.

Lower right—Ceiling outlet for washing machine; keeps cord off damp floor.

These installations made with "Wiremold."

## Fine Sample of All Steel Safety Switchboard Installation in Hamilton

### By V. K. STALFORD

The regulations which are becoming effective in different municipalities throughout Canada and the United States for dead front switchboards have been the means of introducing a new type of switching equipment for low voltage up to 600 volts a.c. and d.c. designed to meet these requirements. The first condition to be met by the designers was that it had to be sold for the same if not a lower first cost than slate boards had been in the past. They had to be suitable for the same class of mounting such as angle iron frame and wrought iron pipe frame. The panels had to be such a size that they would make up the same as the standard widths and heights universally adopted and at present in use.

The manufacturers of steel safety switchboxes have recently placed on the market steel panels, punched, formed and enamelled, which meet all the requirements for this new type of switchboard. They have been tried out in several of the large industrial plants and several municipal buildings such as schools and colleges and have proven very satisfactory in every way. Some of the large electrical manufacturers are now using them in their plants in place of slate switchboards, as formerly.

The flexibility of the steel panel is a very important point in favor of this new type of switchboard. The panels, with

All-steel Safety Switchboard

the switchboxes, may be carried in stock and a board can be made up on very short notice to meet the demands in industrial plants where alterations and extensions are made on short notice. This condition is a very important one from the contractor-dealer's point of view. as he is able to take from his stock or obtain them on short notice from the jobber's stock. They can be shipped to the job and made up to suit any requirements which might arise and as is often required, as changes have to be made when the job is nearly completed due to the fact that some condition has arisen which had been overlooked previously. Several of the local contractors have expressed their approval of this type of equipment. The delays in shipment which have been customary in the past with

the slate board are entirely eliminated with this equipment.

The control for groups of small motors up to 5 h.p. may be grouped on boards such as this, making a very efficient and mechanical arrangement.

The accompanying cut portrays an installation in an industrial plant which is operated by a steam engine direct-connected to alternator of 220 volts, 60 cycles. The control for the generator with meters and rheostats is mounted on panels No. 1. The control with meters for the exciter is mounted on panel No. 2. The remaining switches are for feeder control. This type of switch is very desirable for feeder control as it is possible to lock the switches in the off position when it is desired to prevent any circuit from being made alive when work is being done on this particular circuit, thus preventing the accidental closing of switches which so often occurs with other types of switchboards. The question of grounding can be very easily taken care of as the several equipments are all mechanically connected, which makes ideal conditions for good grounding, and by grounding at one point you are sure to have all the equipment properly protected.

The panels are mounted on an angle iron frame similar to the types used for standard switchboard construction, with a channel iron base and channel iron wall supports. The panels are provided with flanges at each end for the purpose of connecting them together. Openings are provided in the flanges for bolting them together. These flanges can be used for bolting the panels to the vertical members of the frame. In the wrought iron pipe frame suitable cross angles can be used if the panels are to be used with this type of frame.

The mains are entirely enclosed in a gutter box, all conduits being connected to this box, which provides a very good bond for the grounding of circuits. It is very convenient to interchange the circuits to the extreme sections of the board through the gutter box.

The cases of meters are all well grounded, as they are bolted to the steel panels. The entire switchboard both front and rear is dead, so that it is unnecessary to provide guard rails or enclosures. Any arcing or heating is retained in the boxes and gutters, which prevents fires being started from the switchboard equipment as is the case with the open type.

---

## Recommendation of Wiring Committee

The Wiring Committee of the N. E. L. A. has for some years been recommending the "solid neutral from the generator or transformer to the lamp."

Originally, electric distribution circuits were insulated throughout. With this method, it was desirable that in case of trouble both the wires of the circuit should be opened by a fuse or switch or circuit breaker whenever the circuit was to be disconnected.

After a while, however, experience showed that the balance of advantage was in favor of a grounded system, and after long struggles permission was given for grounded systems rather than completely insulated systems. To-day, grounding is in many cases compulsory.

In connection with grounded systems it was soon found that if the grounded wire was opened, trouble resulted. This applied chiefly to the middle or neutral wire of a three-wire system. When the fuse on the middle or neutral wire of a three-wire system opens, there are two kinds of trouble. One is that unless the load is perfectly balanced, the voltage, instead of being say, 115 on each side, becomes say, 150 on one side and 80 on the other. the exact figures depending on the load. This, of course, causes lamp burnouts and other injuries.

The second trouble is that if there are two or more ground connections on the system, either intentional or accidental, then when the fuse on the neutral blows, the current, instead of flowing through the neutral wire as it should may go from one of the ground connections to the other through an improper path.

When these troubles appeared, permission was sought and finally granted to omit the fuse on the neutral wire, since if it did not blow it was no use, and if it did blow it caused trouble.

For some reason, however, the rules, while they permit, and sometimes require the omission of the neutral fuse on three-wire systems, still require two fuses, one on each wire of the two-wire branches from three-wire systems.

In this case, the trouble from unbalanced voltage does not arise. The other trouble, however, from stray current is just the same.

If the fuse on the grounded side of a two-wire branch should blow, then the following may happen.

If high tension current, as 2,300 volts, should in any way get on to the system, then with the fuse on the grounded side blown, the householder can get a 2,300 volt shock. The blowing of the fuse on the grounded side removes the protection of the grounding.

Also if there should be any accidental ground on the two-wire circuit (as there often is), then if the fuse on the grounded side is blown the current goes through the lamps to the accidental ground, and instead of returning through the wire intended for it, returns to the main ground through an improper path.

It is true that few troubles arise from blowing of the fuses on the grounded side, since usually the other fuse blows at the same time, but the fuse on the grounded side never does any good; it may do harm and what is also of importance, it costs money. Panel boards in particular can be much

smaller and simpler, if the fuses and switches are all single-pole on the potential wire and the grounded wire is made solid throughout.

This construction is required in England, and what means more to us has been for a long time required in Providence, (U. S.), where insurance is freely granted.

The Wiring Committee of the N. E. L. A. is asking that this construction be officially approved in the code, but pending such approval there would seem to be no reason why local inspectors should not allow it in their districts since it is perfectly safe.

The only point to be watched is that the neutral wire should be adequately identified. In the case of the two wire services, there is a possible danger of transposing the two wires, but whenever a three-wire service enters a house this danger practically disappears, and there would seem to be no reason for paying for two-pole cut outs and two fuses when single-pole cut outs and single-pole fuses would be both cheaper and safer.—The Bulletin.

## Canadian Electric Supply Co. Have Added New Fixture Department

A new lighting fixture department has been added to the premises of the Canadian Electrical Supply Company, Limited, 165 Craig St. W., Montreal. As seen from the illustration, the lines of fixtures carried are many and varied. The whole general lay-out of the light and fixture department is both pleasing to the eye and arranged in such a fashion that it is a very easy matter for the intending purchaser to see at a glance each individual fixture just as it would appear when installed. The side tables, which are used for glass ware, toasters, table lamps, hot plates, irons and portable heaters, etc., are so placed that a demonstration is a very simple matter. Mr. J. C. Sullivan recently appointed sales manager, is very proud of this addition to their showrooms.

The Canadian Electrical Supply Co. have added the attractive lighting fixture department shown herewith

## Northern Electric Company, Toronto District, Hold Three-Day Conference

The Toronto District of the Northern Electric Company held a very important conference in Toronto on Monday, Tuesday and Wednesday, February 20, 21 and 22. Meetings were held at the King Edward Hotel, at which were representatives from Toronto, London, Hamilton and Windsor. A number of heads of departments from Montreal were also present, including M. K. Pike, general sales manager; C. F. R. Jones, telephone and wire & cable sales manager; L. A. Johnson, general supply sales manager; William Carswell, chief accountant; W. H. Smedley, general stores manager; G. R. Allerton, general advertising manager, and A. J. Soper, Montreal district manager. Mr. W. R. Ostrom, manager Toronto district, acted as chairman.

The conference concluded with a Banquet, at the King Edward Hotel, on the evening of February 22. In addition to some fifty members of the "Northern" staff, there were also present a number of guests. The invited guests included Messrs. H. S. Balhatchet, vice-president Benjamin Electric Mfg. Company; H. A. Burson, sales manager Canadian Crocker-Wheeler Company; W. J. Campbell, manager Crown Electrical Mfg. Company; H. Eccles, general manager Easy Washer Company; W. N. Elliott, sales manager N. Slater Company; N. E. Geery, sales manager Jefferson Glass Company; J. Herbert Hall, president Conduits Company, Ltd.; T. Kaufman, general manager Square D Company; S. Lynn, factory manager Sangamo Electric Company; E. G. Mack, managing director Crouse-Hinds Company; Alex. MacKenzie, general sales manager Canadian Carbon Company; K. A. McIntyre, Canadian representative Society for Electrical Development; A. L. Page, president Frost Steel & Wire Company; E. H. Porte, general manager Renfrew Electric Products, Ltd.; K. J. Shirton, general manager William Shirton Company, and C. E. Walsh, general manager United Electric Company of Canada.

The evening was given over to community, quartet and individual singing and orchestral selections,—not forgetting the "bones," by Mr. Baulch—but was necessarily brought to a close at an early hour to enable the out-of-town visitors to catch their trains.

An interesting event was the presentation of gold "Eversharps" to Messrs. J. A. Daly of Hamilton, and A. L. Brown of Toronto, who are leaving immediately for Regina and Halifax, respectively. Mr. Pike and Mr. Campbell, in company with Mr. Daly, left the same evening for the West, where conferences will be continued at the headquarters of the company's different districts.

---

## Big Market for Electric Ranges

Electric ranges will be vigorously promoted in Vancouver and vicinity by the British Columbia Electric Railway Company, according to E. E. Walker, its sales engineer. Mr. Walker, in company with W. Saville and P. Gill, two other officials, recently made a circuit of Seattle, Tacoma, Portland and Spokane and were impressed with the growth of the electric range load in these cities.

"We have a total of 600 ranges on our lines," said Mr. Walker on his return, "whereas Spokane, a city with much smaller population, has 3500 ranges. Our cooking rate is 3 cents a kilowatt hour so that there is no reason why the electric range should not be adopted in a much wider manner.

"There is a market, I believe, for not less than 500 electric ranges a year in and around Vancouver and our company is therefore about to institute a vigorous campaign to endeavor to put that number on the line, not only this year but continuously and in increasing numbers.

"We are just completing a new 11,000 feeder line to West

Point Grey which will enable us to serve that rapidly growing community. There were 2000 new homes put up in Greater Vancouver last year and this year the number will be 3000. Many of these home builders will be in the market for electric ranges, especially those in the newer sections of Point Grey.

Part of the B. C. Electric Railway Company's plans take the form of a vigorous advertising campaign, while the territory will be divided up among a number of salesmen.

### A Good Outlet for Canadian Products

Mr. C. S. Seager, manager of the Auckland branch of Messrs. Turnbull & Jones, Ltd., electrical engineers and contractors of New Zealand, who, as noted in a recent issue, was spending some time in Canada for his firm, has just left for New York. Before leaving, Mr. Seager appointed The George C. Rough Sales Corporation, Coristine Building, Montreal, his Canadian purchasing agents.

### Moloney Electric in New Offices

The Moloney Electric Co. of Canada, Limited, are vacating their present office No. 1221-4 Bank of Hamilton Building and are moving into new offices at suite No. 501, York Building, which is located on the north west corner of King and York streets, Toronto.

Mr. A. Bauer of Mitchell, Ont., has taken an agency for the Canadian Fairbanks Morse Lighting Plants, for Mitchell and district, and has also gone into the electrical contracting business.

Mr. H. R. Cassidy, 255 Regent St., Montreal, has been awarded the contract for electrical work on two residences recently erected at Roslyn Ave. and Sherbrooke St., Westmount, at a cost of approximately $30,000, for Mr. Chas. J. Brown, 4263 Sherbrooke St. W.

Messrs. Fox & Mainwaring have been awarded the contract for lighting Cedar Hill School and Strawberry Vale School in the municipality of Saanich; they are also busy fitting up electrical homes for Mr. Ayton on Wilkinson Road, and Mr. Taylor on Newport Ave., Oak Bay.

The Star Electric Company, Kitchener, Ont., have moved from 22 Yonge Street to a new and up-to-date store at 91 King Street West, Kitchener, where they will carry a complete line of electric washers, ranges, vacuum cleaners, etc. Messrs. Bremner & Schnur are the partners of the Star Electric Company.

### Trade Publications

Meter service switches are described and illustrated in Circular No. 4484, recently issued by the Westinghouse Electric & Manufacturing Company. The switches are designed to meet every condition encountered by central stations in supplying service to homes and apartment houses and they can be used in connection with standard makes of meters on the market, or independently, if desired. The publication also contains a description and illustration of the Boston-type outfit. Other features discussed are meter trims, end walls and accessories, methods of banking, dimensions and wiring connections.

The Hazard Manufacturing Company, of New York, are distributing a little booklet, entitled "Keystone Wire." The booklet describes the construction and characteristics of this wire, with illustrations.

# News from the Maritimes

## Nova Scotia Electrical Association Showing Active Interest in Better Lighting

The Electrical Association of Nova Scotia held an important meeting on the evening of February 20. The discussion centred around street lighting, the chief speaker being Mr. C. H. Wright, eastern manager of the Canadian General Electric Company.

The situation in Halifax as regards street lighting is particularly favorable. The power being developed at St. Margarets Bay, which by agreement—as explained in our last issue—may be obtained by the city of Halifax, at cost, either from the Tramways Company or from the Nova Scotia Power Commission, leaves the city in the fortunate position of having ample supply at a minimum expense.

The speaker, Mr. Wright, introduced his lecture by tracing the development of street lighting from early times up to the present day. Lantern slides were used to illustrate the various phases in this development. Mr. Wright dealt more particularly, then, with the lighting conditions in Halifax, pointing out that the present system is necessarily out of date, owing to the fact that it was installed many years ago and to the further fact that tremendous development has taken place in illumination methods, construction of glassware, efficiency of lighting units, etc., during the last few years.

He also made a very strong point in connection with the safety features of good lighting, dwelling upon the fact that fatalities from accidents in the United States during the war were very much greater than among U. S. soldiers in Europe due to their participation in the war. Good illumination is the best form of police protection, as well.

The point was also made that the city of Halifax, even under the old system, was paying much less than the average city, for its street lighting. It was stated that in Springfield, Mass., the cost of street lighting was $1.20 per head while in Halifax it was only 40 cents per head. The city could, therefore, afford to spend more money than it was doing at the present time and, considering the greater efficiency of equipment that could now be obtained, the amount of illumination would also be in much greater proportion to the money the city did spend.

In the discussion that followed Mr. Wright's address, Prof. Knight of the local Technical College spoke briefly on the benefit of electricity in technical work. Other speakers included City Engineer Doane, and Mr. W. H. Hayes, assistant manager of the Maritime Telephone Company.

Mr. G. H. Durling, city electrician of Halifax, has drawn plans for the illumination of the city streets on a modern scale. These plans were discussed at some length and referred to in a complimentary manner as being very comprehensive and artistic, and quite in keeping with the many other improvements this city has inaugurated from time to time in the past.

The address was preceded by an informal dinner and a short musical programme. Mr. Wm. Murdock, the president, was in the chair. The secretary of the Association is Mr. E. A. Saunders.

---

The Electric Specialty Co., 781 Granville St., Vancouver, has secured the contract for electrical work on the Bank of Toronto building, Broad and Yates St., Victoria, which is undergoing alterations.

---

Mr. Fred Dubois, 394 Clarence St., Ottawa, has secured the contract for the electric wiring of a parish residence, recently erected at Anglesea Square, Ottawa, at a cost of approximately $30,000., by St. Anne's parish.

## New Manager Halifax District, Northern Electric Company

Mr. Arthur L. Brown, of the Toronto office of the Northern Electric Company, Ltd., has been appointed manager of the Halifax branch of the company. Mr. Brown is almost a newcomer in the electrical industry and the confidence being placed in him by the Northern Electric Company is all the more remarkable. During the war he was connected with the Imperial Munitions Board and was on the staff of the British Munitions Company, Ltd., under J. D. Hathaway, who was president of that company and is also vice-president of

Mr. Arthur L. Brown

the Northern Electric Company. At the end of the war Mr. Brown was engaged for some time in winding up the affairs of the British Munitions Company, after which he was taken on by the Northern Electric Company and sent to Toronto as assistant district stores manager. A little later he was made district stores manager, which position he held up to the present time, when he leaves for the larger sphere in the Maritime Provinces.

---

## The Development of Wireless—III

(Continued from page 35)

thousand wave trains per second, this second component will when turned into the phones give us a note, more or less pure, at one thousand sound waves per second. A variation in the speed of the rotary will cause a change in the pitch of the sound received.

The crystal is simply a rectifier, and in its stead have been used several types of electrolytic rectifiers and the original two electrode values which consist of one hot and one cold body both in a vacuum. The incandescent tungsten filament usually forms the hot body, the cold electrode being composed of any one of several metals. These electrodes, known as "filament" and "plate" are placed in a bulb which is later exhausted. With the filament heated and a potential applied between plate and filament a current is allowed to pass through the "value" in one direction only—namely from plate to filament. This device therefore acts as a rectifier, and when used in this capacity for wireless reception gives clear signals.

It is the improvement in this little device; the value that has made possible the present-day signalling across tremendous distances of thousands of miles and has also brought into existence the wireless telephone.

---

The contract for one 450 h.p. motor has been awarded to the Canadian Westinghouse Company, Toronto, by the Hull city council.

# The Latest Developments in Electrical Equipment

### Miniature Connectors

The Canadian General Electric Company have placed on the market the cord connector and series tap connector illustrated herewith. The miniature cord connector, on the right, is for use wherever a small separable connection is desired. It is only 7/8 in. wide by 2-9/16 in. long. The fingers are recessed to snap over raised points on the heavy phosphor-bronze spring contacts, making a firm connection. A black composition is used for the outer part of this con-

nector, which has a capacity of 600 watts, 250 volts. The series tap connector, shown on the left, is specially designed for attaching electrically operated sewing machines. Its material, finish, design and operation are the same as that of the cord connector, but, in addition, has the necessary connection to join it to the motor and treadle of a sewing machine.

### Adjustable Bed Lamp

The Faries Manufacturing Co., Decatur, Ill., has placed on the market the bed bracket and lamp here shown. This bracket is designed to fit any bed, either wood or metal, and can be used in any place where there is some means of attaching the device. Its usefulness as a bed lamp is at once apparent, serving to light the room until after one has retired, to

see the time during the night, etc. It is made of brass, simple and strong in construction, and the standard finish is brush brass, although it may be secured in any desired finish. By means of a ball joint the bracket can be adjusted to any angle and the light thus directed where needed. The device is attached by means of a brass strap, which fits over the post, rail or rod of the bed, and then tightened by a thumb-screw.

The application of the Bell Telephone Company for a further increase in rates has been refused by the Dominion Board of Railway Commissioners.

### Hotpoint Reversible Toaster

The Canadian Edison Appliance Co., Limited, Stratford, Ontario, are now offering to the trade an improved Hotpoint reversible toaster. The new toaster is simply constructed with a "turn-over" feature that always works, and the element

provides an even distribution of heat—thus insuring even toasting. An important feature is that the toast is kept hot while both sides of the bread are being toasted. The design of the toaster is particularly pleasing making the appliance an addition to any dining table.

### Electric Hair Cutter

The Continental Electric Company, 505-11 King Street E., Toronto have placed on the market the "Royal" electric hair cutter illustrated herewith. This appliance has three moving parts—the eccentric, oscillating arm, and blade. The

motor is equipped with a patented speed control, with which the cuts can be regulated from 400 to 1000 per minute. Freedom and facility in handling are obtained by the shaft being swiveled at both the motor and the cutter handle. All gears are located at the motor, resulting in a light weight, smooth-running and vibrationless cutter.

### For testing the Ignition

"Spark-C" Ignition Tester, recently introduced at various automobile shows by the Westinghouse Lamp Company, 165 Broadway, New York, is a device for testing the ignition of any internal combustion engine. With its aid it is possible to locate at once any spark plug which is not "firing" properly.

It not only tells which cylinder is giving the trouble but what the trouble is. Essentially "Spark-C" is a small Geissler tube filled with a rare gas. This gas has the property of glowing in the presence of an electric current. Its peculiar behavior is put to practical use in "Spark-C". By the intensity of the glow in the small tube, one can test with certainty the condition of his ignition system. This device also has many applications in other fields connected with high tension work. In power distribution it is used for a number of purposes. It is invaluable to the owner of automobile, motor boat, aeroplane, tractor or gasoline power plant.

### An Electric Tilting Kettle

An electric kettle designed so as to permit water to be poured without lifting it from the stand has been placed on the market by Landers, Frary & Clark, New Britain, Conn. The body is made of heavy spun copper, highly polished, with a nickel or silver-plated finish. The inside is coated with pure

tin, silver finished. Other features are its one-piece spout and its ebonized handles, feet and knobs. It is furnished complete with 6 ft. of mercerized cord and a lamp socket plug. The capacity of the kettle is 2½ pt., its height over all, 12 in., and its weight, when packed, 7 lbs. The device consumes 420 watts and is equipped with a patented safety fuse plug.

### Ace Bell Transformer

The Dongan Electric Manufacturing Company, Detroit, are placing upon the market the Dongan Ace porcelain clad bell ringing transformer of 8 volts secondary. This type is mounted in white glazed porcelain case, the name appearing in red. The "Ace" contains the same grade construction embodied in the steel clad transformers now being manufactured

by this company. The coils are designed to withstand a voltage of 2500 volts between primary and secondary windings, and will not be affected by an indefinite short circuit. The back cover consists of a heavy steel plate rigidly secured to the transformer unit, eliminating any possibility of same becoming loose or allowing the wax to flow out of the case should the bell wires become short circuited thus causing the transformer to overheat.

### Instantaneous Water-Heater

The "Hot Flo" Faucet has recently been put on the market by the Hot Flo Electric Company, 535 Seventh Ave., at 39th St., New York City. A heating element is inserted in the Hot Flo faucet, the handle of which throws the electric

current "on" or "off" as it turns on or shuts off the water. The faucet is furnished with a 30-inch No. 14 heater cord encased in waterproof flexible brass tubing, plug, receptacle and plate. It is nickel finished throughout.

### Toggle Appliance Plug

The Bryant Electric Company are introducing a new appliance switch plug which will be ready for the market March 1, 1922. It will fit practically all makes of heating appliances and its most important features are a toggle switch mechanism, self-adjusting springs, which make practically tight con-

nection to the pins of the appliance, and very hard, dense composition casings which will stand a lot of abuse. By removing two screws, the contacts can be renewed when they become pitted and worn. This is a very practical feature.

### Electric Cigar Lighter

The Cuno Engineering Corporation, Meriden, Conn., has placed on the market the cigar lighter here illustrated. The device is provided with an improved automatic switch which, it is claimed by the manufacturer, makes the lighter practically trouble-proof. It is attached to the battery and is equip-

ped with four feet of automatically rewinding cord. It is so designed that the heating unit, held in place by a bayonet lock, may be removed when necessary.

## A More Detailed Description of the Express and Local Passenger Locomotives for the Chili State Railways

A general description of the electrification of the first zone of the Chilean State Railways has already been given in these pages, when it was announced that the contract for this work was awarded to a Chilean firm, all electrical equipment to be furnished by the Westinghouse Electric International Company. This contract included 39 electric locomotives. A description of the two types of passenger locomotives is given in the following article which, in view of the ever recurring discussion on electrification of certain Canadian steam railways will have a special interest for our readers.

The six Baldwin-Westinghouse electric locomotives which are being built for express passenger service on the Chilean State Railways will be capable of hauling 300 ton trains in either direction between Valparaiso and Santiago without the aid of helper engines as is now necessary on the Tabon grade with steam operation. The locomotive will weigh 127 tons and will have 105 tons on the drivers. It will have a nominal rating of 2250 h.p. corresponding to a speed of 37 miles per hour at a tractive effort of 23,400 lb.

The wheel arrangement will be 2-6-0 + 0-6-2, consisting of two main trucks, each of which has three driving axles and a two-wheel guiding truck. The cab will be of the single box type and the motors will be geared direct to the driving axles. The general dimensions and weights will be as follows:

| | |
|---|---|
| Classification | 2-6-0 + 0-6-2 |
| Length over buffers | 57 ft. 4 in. |
| Length over cab | 38 ft. 0 in. |
| Total wheel base | 48 ft. 4 in. |
| Rigid wheel base | 14 ft. 5 in. |
| Diameter of driving wheel | 42 in. |
| Diameter of guiding wheel | 30 in. |
| Weight of complete locomotive | 253,500 lb. |
| Weight of mechanical parts | 160,000 lb. |
| Weight of electrical parts | 93,500 lb. |
| Weight per driving axle | 35,000 lb. |
| Weight per guiding axle | 21,800 lb. |

The trucks will be connected at the inner ends by a draw bar held in tension by spring buffers. The frames will be cast steel, bar type, located outside of the wheels, connected by cast steel bumpers and cross ties and carried on semi-elliptic springs over the journal driving boxes.

The cab underframe will be of rolled steel longitudinal members connected by cast steel and rolled steel cross members. M. C. B. couplers will be used with Continental spring buffers. Although eventually all drawbar equipment on the Chilean railways will be changed to M. C. B. standard, the Chilean freight cars now use the Continental type draw-hooks and for this reason the M. C. B. couplers will be provided with attachments for chain couplers.

The automatic air brake equipment will be the type 14-EL. a standard similar to that used on the present steam locomotives. With this equipment, straight air is available for handling the locomotive alone and the automatic feature for both locomotive and train.

### Motors Operate in Series

The express passenger locomotives will be equipped with six 275 h.p. driving motors provided with field control and geared direct to the axles by Nuttall flexible spur gears. These motors are designed for operation two in series on 3000 volts and will be grouped in three speed combinations, all six in series for low speeds, three in series with two groups in parallel for two-thirds speed and three groups each with two motors in series for full speed.

There will be six running positions, the change from one motor combination to another being made by the shunting method of transition.

The control equipment is designed for operation of the locomotive from either end and provision for regenerative electric braking is included. This enables the locomotive to return energy to the overhead system when descending grades. The main motor armatures will be connected in the same combinations when regenerating as when motoring, the excitation for the motor fields during regeneration being supplied by a constant voltage motor-generator set.

There will be two master controllers, one in each end of the cab, and the same controller will be used for both motoring and regulating. This controller will have four levers with a total of 51 notches available in the three combinations. Westinghouse type HLF control establishes the main circuit connections by the use of individual unit switches operated by compressed air controlled by electro-magnetic valves.

Motor-generator sets will supply low voltage current for the control equipment, blowers and compressors. This is a two-bearing type of machine with a common frame for both motor and generator. The normal rating of the set is 35 kw. at 95 volts with 3000 volts at the motor terminals.

The current collectors will be spring raised. air lowered and mechanically locked in the lowered position and controlled throughout by compressed air.

On level tangent track these locomotives will have a running speed of 61.5 miles per hour when hauling a 300-ton trailing load. On the Tabon grade which is 2.25 per cent, the average running speed will be 33.5 miles per hour. The maximum tractive effort based on 25 per cent adhesion will be 52,500 pounds, and the maximum speed 62.6 miles per hour. The range of speed in regenerative braking will be 12½ to 50 miles per hour.

### Local Passenger Equipment

In general appearance the eleven Baldwin-Westinghouse electric locomotives for local passenger service will be somewhat similar to the express passenger locomotive. This locomotive will weigh 80 tons and the wheel arrangement will be 0-4-0 + 0-4-0. It will be capable of hauling a trailing load of 350 tons from Puerto to Vina del Mar, 260 tons from Vina del Mar to Llai Llai and return and 300 tons from Las Vegas to Los Andes and return. These locomotives will have a rating of 1300 h.p. corresponding to a tractive effort of 15,600 pounds at a speed of 31 miles per hour. The maximum tractive effort under standing conditions will be 40,000 pounds and the maximum speed will be 56 miles per hour. The cab will be of the single box type and the motors will be geared direct to the axles. The general dimensions and weights will be as follows:

| | |
|---|---|
| Classification | 0-4-0+0-4-0 |
| Length over buffers | 40 ft. 9¾ in. |
| Length over cab | 31 ft. 0 in. |
| Total wheel base | 28 ft. 0 in. |
| Rigid wheel base | 9 ft. 0 in. |
| Diameter of driving wheel | 42 in. |
| Weight of complete locomotive | 160,000 lb. |
| Weight of mechanical parts | 96,000 lb. |
| Weight of electrical parts | 64,000 lb. |
| Weight per driving axle | 40,000 lb. |

The two trucks each having two driving axles will be connected at the inner ends by an articulated coupling in the form of a Mallet hinge. The frame and cab construction, the couplers and the brake equipment will be similar to the express passenger locomotives.

The local passenger locomotives will be equipped with four 275 h.p. driving motors, with field control and geared direct to the axle with Nuttall flexible spur gears. There will be two combinations by connecting the motors in series and in parallel and additional speed variations will be obtained by varying the fields of the motors.

The control equipment is designed for operation of the locomotive from either end but the grade conditions on the section of line where these locomotives will operate do not justify the use of the regenerative braking feature. There will be two master controllers, one in each end of the cab, each controller having two levers, namely speed and reverse, with a total of 28 notches available in the two combinations. The switching equipment duplicates that on the express locomotives.

The motor-generator set will be a double armature machine, each armature consisting of a motor and a generator. The normal rating of the set will be 22.5 kw. at 95 volts with 3000 volts at the motor terminals. The current collectors will be the same as on the express passenger locomotives.

On leven tangent track these locomotives will have a speed of 56 miles per hour when hauling a 220-ton trailing load. The maximum tractive effort based on 25 per cent adhesion will be 40,000 pounds.

A great many of the electrical and mechanical parts are interchangeable between the express and local passenger locomotives and this will both facilitate the maintenance of locomotives and reduce the maintenance cost.

## Major Burpee Becomes a Director o. His Company

Maj. F. D. Burpee, manager of the Ottawa Electric Railway Company, has been elected a director of the Ottawa Traction Company, Ltd., of which the Ottawa Electric Railway Company is a subsidiary. Maj. Burpee succeeded Col. J. E. Hutcheson in 1912, when the latter became manager of Montreal Tramways, and has maintained the splendid record of the Ottawa system for both service and efficiency. The life history of Maj. Burpee is a story of constant promotion. His first position with the Ottawa Electric Railway Company was as a stenographer; later he was secretary-treasurer, then acting superintendent, superintendent and, finally, general manager.

If one may point to any special characteristic of Maj. Burpee, it is his faculty of understanding the attitude of mind of the man on the street, and of being understood by him. Quite recently Ottawa has passed through the usual campaign to municipalize the street railway system, and it was largely due to the Major's logical and insistent presen-

tation of the facts, through the company's little house organ, that the people became convinced that their wisest course lay in leaving matters as they were. Consequently the franchise of the company is being renewed under conditions which appear to be entirely harmonious.

## Miscellaneous

Representatives from the municipalities, both urban and rural, along the G. T. R. line between Galt and Elmira, Ont., recently held a meeting in Galt at which a resolution was passed favoring the electrification of the Galt-Elmira line of the Grand Trunk.

At the annual meeting of the Dominion Power & Transmission Company, held recently in Hamilton, the following officers were re-elected:—President, Lieut.-Col. J. R. Moodie; Vice-president, Cyrus A. Birge; Treasurer, James Dixon; Managing Director and Secretary, W. C. Hawkins; General Manager, E. P. Coleman; Directors, Sir John Gibson, Lloyd Harris, C. E. Neill, W. E. Phin, Robert Robson and John Dickenson.

Employees of the Dominion Power & Transmission Company were notified recently by the Minister of Labor that he had appointed two Boards of Conciliation to deal with their wage disputes with the company.

Fred Bancroft and F. H. McGuigan of Toronto will inquire into the dispute as it affects the Hamilton street railway, radial lines and engineers; J. C. O'Donoghue of Toronto and Mr. McGuigan will act in the dispute as it affects the Canadian Federation of Electricians.

The employees of the London Street Railway Company, London, Ont., have forwarded to the company a formal notice that they refuse to accept a cut of three cents per hour which the company proposes to make on March 1. The men have requested a conference to discuss the matter.

The Toronto Transportation Commission are having plans prepared for machine shops, to cost in the neighborhood of $1,000,000. It is understood these shops are to be erected at Davenport Road & Bathurst St., and that tenders will be called during March.

The electrical equipment in the Waterloo factory of John Forsyth Ltd., is being installed by Mr. H. C. Kress, electrical contractor, Kitchener, Ont. In a recent issue we reported, in error, the awarding of this contract to another company.

The Bureau of Standards, Washington, D. C., has just published Scientific Paper No. 427, entitled "Some effects of the distributed capacity between inductance coils and the ground."

The action of the executive of the Hollinger Mines against the Northern Canada Power Company to collect alleged damages to the extent of nearly $2,000,000. has been dismissed by Justice Middleton, with costs.

Messrs. McKeown Bros., 1259 Lansdowne Ave., Toronto, have been awarded the contract for electrical work on residences recently erected at 703 and 705 Windermere Ave., Toronto, for Messrs. Argent & Banks, 29 St. Mark Street.

# Current News and Notes

**Brandon, Man.**

The council of the city of Brandon, Man., proposes opening negotiations with the provincial government with a view to having a power line built west from Brandon, to serve rural districts and points by the way.

**Courtright, Ont.**

The ratepayers of Courtright, Ont., recently carried a by-law authorizing the expenditure of $11,000 for the installation of Hydro power in that town.

**Galt, Ont.**

The report for the year 1921 of the municipal hydro-electric department of Galt, Ont., as submitted recently by manager W. H. Fairchild, up to December 31, 1921, shows a revenue for the year of $127,562.01, and an expenditure for operation and maintenance of $110,678.82, leaving a gross profit of $16,983.19. Domestic lighting revenue for 1921 was $44,879.01, an increase of $6,418.67 over last year; for commercial lighting, $19,055.01, an increase of $1,479.94; for power, $47,079.49, a decrease of $2,079.96; and for street lighting $16,548.50, an increase of $195.60.

**Halifax, N. S.**

Mr. W. W. Hoyt, Hollis St., Halifax, N. S., has secured the contract for electrical work on a Club House recently erected at Point Pleasant, Halifax, N. S., at an approximate cost of $15,000., for the Royal Nova Scotia Yacht Squadron.

**Montreal, Que.**

The Canadian Comstock, Ltd., 10 Cathcart St., Montreal, has secured the contract for electrical work on the $8,000,000. hotel being erected at Peel & St. Catherine Sts., Montreal, by the Mount Royal Hotel Company, 422 Drummond Bldg.

Mr. J. W. Tousignant, 637 Marie Anne St., Montreal, has been awarded the contract for electrical work on a store building being erected at Marquette & Marie Anne Sts., Montreal, at an estimated cost of $30,000.

**Niagara Falls, Ont.**

Mr. Arthur Green, Niagara Falls, Ont., has secured the contract for electrical work on a parish hall recently erected at Fourth & McRae Sts., Niagara Falls, at an approximate cost of $12,000.

**Ottawa, Ont.**

Plans for a dam for hydro-electric power development in the Calumet channel of the Ottawa River at Bryson, Quebec, have been deposited by the Ottawa & Hull Power and Manufacturing Company, with the Minister of Public works at Ottawa and the Land Registry Office at Quebec.

**Oxford, N. S.**

Mr. Geo. McIntosh, Oxford, N. S., has secured the contract for electrical work on a restaurant building recently erected on Water St., Oxford, N. S., for Mr. H. A. Amos of that place.

**Peterborough, Ont.**

Messrs. Miller, Powell & Watson, Peterborough, Ont., have been awarded the contract for electrical work on a building at 168 Brock St., Peterborough, that is being altered for apartments.

**Pt. Credit, Ont.**

Mr. Jas. McMahon, Port Credit, Ont., has been awarded the electrical contract on a residence, garage and swimming pool situated near the Mississauga Golf Course and owned by Mr. J. P. Bickell, Standard Bank Building, Toronto, on which work has recently been started.

**St. John, N. B.**

Messrs. Mackie & Driscolle, King Square, St. John, N. B., have been awarded the contract for electrical work on an addition to be built to the epidemic hospital on Delhi St., St. John; also the contract for electrical work on deposit vaults to be built for the Eastern Trusts Company, 111 Prince William St., St. John.

**Quebec, Que.**

Messrs. Jobin & Paquet, 78 Abraham Hill, have secured the electrical contract on a Girls Home, being erected at Darphine & St. Angele Sts., Quebec City, at an estimated cost of $125,000.

**Toronto, Ont.**

Mr Chas. Wilson, 167 Dovercourt Road, Toronto, has been awarded the contract for electrical work on two residences recently erected at Armadale Ave. & Colbeck Ave., Toronto, by the Godson Home Building Co., 1154 College St., at an approximate cost of $85, 000; also the contract for electrical work on an apartment building to be erected on Glenholme Ave., near St. Clair Ave., Toronto, at an estimated cost of $30,000.

Mr J. H. Harris, 683 St. Clair St. W., Toronto, has secured the contract for electrical work on a Club House recently erected for the Oakwood Lawn Bowling Club, 184 Oakwood Ave., Toronto.

Messrs. Taylor Bros., 25½ Norwood Ave., Toronto, have secured the contract for electrical work on nine residences in course of erection on Gates Avenue, between Westlake & Meagher Aves., Toronto, at an estimated cost of approximately $60,000., by Messrs. Crawford & McClintock, 189 Cedardale Ave.

Mr J. Ritchie, 720 Annette St., Toronto, has been awarded the contract for electrical work on two residences recently erected on Willard Ave., near Colbeck Ave., Toronto, at a cost of approximately $15,000., for Mr. F. Taylor, 180 Willard Ave.

Mr. F. Johnson, Curzon St., Toronto, has been awarded the contract for electrical work on a number of residences to be erected on Stephenson Ave., between Morton Road & Westlake Ave., Toronto, at an estimated cost of $60,000 for Messrs. Grant & Stephens, 4 Moscoe Ave.

---

An official of the wireless station, recently erected at High River, Alta., held a wireless telephonic conversation with the operator of a radio station at Denver, Col., a distance of approximately a thousand miles.

---

### Paving the Way for More Appliance Sales

J. W. Thomas, electrical contractor of Niagara Falls, Ont., is not suffering from any depression in his business, judging from the following contracts. At the present time he is booked to wire homes for Geo. Dingwall on Willmott Street; for Wm. Hodgkins, on Ryerson Avenue; John Lamb, Simcoe Street; Harry Ingham, Willmott Street; Arthur Ingham, Fourth Avenue; Geo. Dingwall, John Street, Highland Avenue, and Fourth Avenue, and a house for John Goring, on Willmott Street. All these residences are being wired for electric ranges, and several of them for electric fire-places.

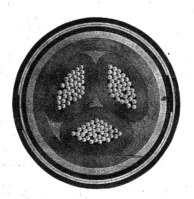

### TO SPAN THE DELAWARE

A salute of 17 guns from the Olympia, Admiral Dewey's old flagship, was the signal recently for the opening of the exercises which officially marked the beginning of work on the Delaware River bridge between Philadelphia and Camden, N.J.

The bridge will be one of the longest of the suspension type in the country. The main span will be 1,750 feet long and the total length of the structure will be 1.82 miles. Its cost is estimated at $26,000,000.

### FIRES DURING 1921 IN MONTREAL

There were 1,776 outbreaks of fire in the city of Montreal during the year ending December 31, 1921. Of this number Investigator Constantin of the Fire Commissioner's Court found that 568 merited inquiry. This necessitated 1,582 witnesses attending the Fire Commissioner's Court.

The causes of 809 fires were established, leaving 967 outbreaks, the origin of which was not determined. It was stated at the Fire Commissioner's Court that 50 per cent. of the fires were due to negligence.

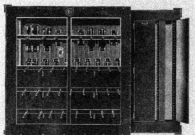

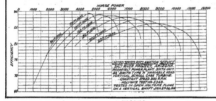
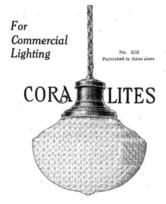

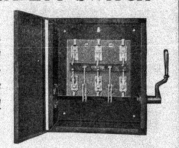
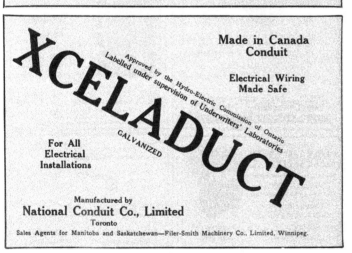

# CLASSIFIED INDEX TO ADVERTISEMENTS

The following regulations apply to all advertisers:—Eighth page, every issue, three headings; quarter page, six headings; half page, twelve headings; full page, twenty-four headings.

**AIR BRAKES**
Canadian Westinghouse Company

**ALTERNATORS**
Canadian General Electric Co., Ltd.
Canadian Westinghouse Company
Electric Motor & Machinery Co., Ltd.
Ferranti Meter & Transformer Mfg. Co.
Northern Electric Company
Wagner Electric Mfg. Company of Canada

**ALUMINUM**
British Aluminium Company
Spielman Agencies, Registered

**AMMETERS & VOLTMETERS**
Canadian National Carbon Company
Canadian Westinghouse Company
Ferranti Meter & Transformer Mfg. Co.
Monarch Electric Company
Northern Electric Company
Wagner Electric Mfg. Company of Canada

**ARMOURED CABLE**
Canadian Triangle Conduit Company

**ATTACHMENT PLUGS**
Canadian General Electric Co., Ltd.
Canadian Westinghouse Electric Company

**BATTERIES**
Canada Dry Cells Ltd.
Canadian General Electric Co., Ltd.
Canadian National Carbon Company
Electric Storage Battery Co.
Exide Batteries of Canada, Ltd.
McDonald & Willson, Ltd., Toronto
Northern Electric Company

**BEARINGS**
Kingsbury Machine Works

**BOLTS**
McGill Manufacturing Company

**BONDS (Rail)**
Canadian General Electric Company
Ohio Brass Company

**BOXES**
Canadian General Electric Co., Ltd.
Canadian Westinghouse Company
G. & W. Electric Specialty Company

**BOXES (Manhole)**
Standard Underground Cable, Company of Canada, Limited

**BRACKETS**
Slater N.

**BRUSHES, CARBON**
Dominion Carbon Brush Company
Canadian National Carbon Company

**BUS BAR SUPPORTS**
Ferranti Meter and Transformer Mfg. Co.
Moloney Electric Company of Canada
Monarch Electric Company
Electrical Development & Machine Co.

**BUSHINGS**
Diamond State Fibre Co. of Canada, Ltd.

**CABLES**
Boston Insulated Wire & Cable Company, Ltd.
British Aluminium Company
Canadian General Electric Company
Phillips Electrical Works, Eugene F.
Standard Underground Cable Company of Canada, Limited

**CABLE ACCESSORIES**
Northern Electric Company
Standard Underground Cable Company of Canada, Limited

**CARBON BRUSHES**
Calebaugh Self-Lubricating Carbon Co
Dominion Carbon Brush Company
Canadian National Carbon Company

**CARBONS**
Canadian National Carbon Company
Canadian Westinghouse Company

**CAR EQUIPMENT**
Canadian Westinghouse Company
Ohio Brass Company

**CENTRIFUGAL PUMPS**
Boving Hydraulic & Engineering Company

**CHAIN (Driving)**
Jones & Glassco

**CHARGING OUTFITS**
Canadian Crocker-Wheeler Company
Canadian General Electric Company
Canadian Allis-Chalmers Company

**CHRISTMAS TREE OUTFITS**
Canadian General Electric Co., Ltd.
Masco Co., Ltd.
Northern Electric Company

**CIRCUIT BREAKERS**
Canadian General Electric Co., Ltd.
Canadian Westinghouse Company
Cutter Electric & Manufacturing Company
Ferranti Meter & Transformer Mfg. Co.
Monarch Electric Company
Northern Electric Company

**CONDENSERS**
Boving Hydraulic & Engineering Company
Canadian Westinghouse Company

**CONDUCTORS**
British Aluminium Company

**CONDUIT (Underground Fibre)**
American Fibre Conduit Co.
Canadian Johns-Manville Co.

**CONDUITS**
Conduits Company
National Conduit Company
Northern Electric Company

**CONVERTING MACHINERY**
Ferranti Meter & Transformer Mfg. Co.

**CONDUIT BOX FITTINGS**
Northern Electric Company

**CONTROLLERS**
Canadian Crocker-Wheeler Company
Canadian General Electric Company
Canadian Westinghouse Company
Electrical Maintenance & Repairs Company
Northern Electric Company

**COOKING DEVICES**
Canadian General Electric Company
Canadian Westinghouse Company
National Electric Heating Company
Northern Electric Company
Spielman Agencies, Registered

**CORDS**
Northern Electric Company
Phillips Electric Works, Eugene F.

**CROSS ARMS**
Northern Electric Company

**CRUDE OIL ENGINES**
Boving Hydraulic & Engineering Company

**CURLING IRONS**
Northern Electric Co. (Chicago)

**CUTOUTS**
Canadian General Electric Co., Ltd.
G. & W. Electric Specialty Company

**DETECTORS (Voltage)**
G. & W. Electric Specialty Company

**DISCONNECTORS**
G. & W. Electric Specialty Company

**DISCONNECTING SWITCHES**
Ferranti Meter and Transformer Mfg. Co.
Winter Joyner, A. H.

**DREDGING PUMPS**
Boving Hydraulic & Engineering Company

**ELECTRICAL ENGINEERS**
Gest, Limited, G. M.
See Directory of Engineers

**ELECTRIC HAND DRILLS**
Rough Sales Corporation George C.

**ELECTRIC HEATERS**
Canadian General Electric Co., Ltd.
Canadian Westinghouse Company
Equator Mfg. Co.
McDonald & Willson, Ltd., Toronto
National Electric Heating Company

**ELECTRIC SHADES**
Jefferson Glass Company

**ELECTRIC RANGES**
Canadian General Electric Co., Ltd.
Canadian Westinghouse Company
National Electric Heating Company
Northern Electric Company
Rough Sales Corporation George C.

**ELECTRIC RAILWAY EQUIPMENT**
Canadian General Electric Co., Ltd.
Canadian Johns-Manville Co.
Electric Motor & Machinery Co., Ltd.
Northern Electric Company
Ohio Brass Company
T. C. White Electric Supply Co.

**ELECTRIC SWITCH BOXES**
Canadian General Electric Co., Ltd.
Canadian Drill & Electric Box Company
Dominion Electric Box Co.

**ELECTRICAL TESTING**
Electrical Testing Laboratories, Inc.

**ELEVATOR CONTRACTS**
Dominion Carbon Brush Co.

**ENGINES**
Boving Hydraulic & Engineering Co. of Canada
Canadian Allis-Chalmers Company

**FANS**
Canadian General Electric Co., Ltd.
Canadian Westinghouse Company
Century Electric Company
Great West Electric Company
McDonald & Willson
Northern Electric Company
Robbins & Myers

**FIBRE (Hard)**
Canadian Johns-Manville Co.
Diamond State Fibre Co. of Canada, Ltd.

**FIBRE (Vulcanized)**
Diamond State Fibre Co. of Canada, Ltd.

**FIRE ALARM EQUIPMENT**
Northern Electric Company
Slater N

**FIXTURES**
Benson-Wilcox Electric Co.
Crown Electrical Mfg. Co., Ltd.
Canadian General Electric Co., Ltd.
Jefferson Glass Company
McDonald & Willson
Northern Electric Company
Tallman Brass & Metal Company

**FLASHLIGHTS**
Canadian National Carbon Company
McDonald & Willson, Ltd.
Northern Electric Company
Spielman Agencies, Registered.

**FLEXIBLE CONDUIT**
Canadian Triangle Conduit Company
Slater, N.

**FUSES**
Canadian General Electric Co., Ltd.
Canadian Johns-Manville Company
Canadian Westinghouse Company
G. & W. Electric Specialty Company
Moloney Electric Company of Canada
Northern Electric Company
Rough Sales Corporation George C.

# CLASSIFIED INDEX TO ADVERTISEMENTS—CONTINUED

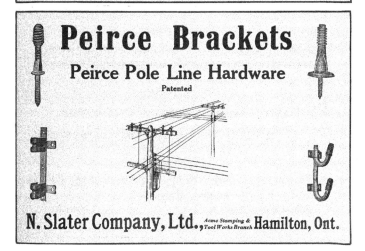

## CLASSIFIED INDEX TO ADVERTISEMENTS—CONTINUED

# Westinghouse
# 1922 Electric Fans

8-inch "Whirlwind" Fan

Gyrating Fan

12-inch Desk and Bracket Fan

10-inch Oscillating Fan

12-inch Exhaust Fan

16-inch Desk and Bracket Fan

Westinghouse 1922 Fans are designed to meet the demands of the most critical buyer.

The motor is made of the very best materials obtainable. The entire oiling system is automatic, and so arranged that no oil or grease can drip on the furniture. The felt base prevents scratching polished surfaces.

The attractive velvety-black finish of Westinghouse Fans is very practical. It allows the washing of the blades with ordinary soap and water. Shiny brass blades soon become tarnished after washing.

The Westinghouse line of fans is complete:
8-inch "Whirlwind" Desk and Bracket Fans.
10, 12 and 16-inch Oscillating and Non-Oscillating Desk and Bracket Fans.
Gyrating Fans. 56-inch Ceiling Fans, Floor and Counter Column Fans.
12 and 16-inch Exhaust Fans, etc., etc.

## Canadian Westinghouse Co., Limited, Hamilton, Ont.

TORONTO. Bank of Hamilton Bldg.    MONTREAL. 285 Beaver Hall Hill.    OTTAWA. Ahearn & Soper, Ltd.
HALIFAX. 105 Hollis St.    FT. WILLIAM. Cuthbertson Block    WINNIPEG. 158 Portage Ave. E.
CALGARY. Canada Life Bldg.    VANCOUVER, Bank of Nova Scotia Bldg.    EDMONTON. 211 McLeod Bldg.

REPAIR SHOPS

MONTREAL—113 Dagenais St.    VANCOUVER    WINNIPEG—158 Portage Ave. E.
TORONTO—663 Adelaide St. W.    1090 Mainland St.    CALGARY—316 Third Ave. E.

Vol. XXXI No. 6            Toronto, March 15, 1922

# Electrical News
### Engineering Contracting — Merchandising Transportation

# *Northern Electric*
# Box-Line Fixtures are—

**Individually Packed**
**Photo - Labelled**
**Easily Identified**
**Quality Products**

## Seventeen Styles to Select From

This new plan of packing each of these Quality Fixtures in an individual carton and labelling the carton with a photograph of the fixture itself and all descriptive detail, saves unnecessary handling of the fixture and helps you serve customers quickly and intelligently.

You don't have to unwrap the package to see what it contains.

The fixtures remain protected until you are ready to wire them. After wiring return them to their cartons for delivery to the job of installation.

The finish of every fixture is protected and your shelves are always neat and clean.

*Write our nearest house for catalog and details of this "Picture-on-Package" Plan.*

## *Northern Electric Company*
### LIMITED

MONTREAL   TORONTO   WINDSOR   CALGARY
HALIFAX   HAMILTON   WINNIPEG   EDMONTON
QUEBEC   LONDON   REGINA   VANCOUVER

### *"Makers of the Nation's Telephones"*

**MANUFACTURING**
Manual Telephones
Automatic Telephones
Wires & Cables
Fire Alarm Systems
Power Switchboards

**DISTRIBUTING**
Construction Material
Illuminating Material
Power Apparatus
Household Appliances
Electrical Supplies
Power & Light Plants
Marine Fittings

# Here It Is At Last!

*Ready*
*for*
*Delivery*
*Mar. 16*

Thoroughl
covered b
Patents ir
Canada.

*Made in Canada by*
## The Canadian Floor Wax
### Toroi

# ALPHABETICAL LIST OF ADVERTISERS

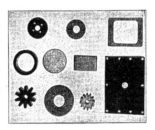

# "STAYSALITE" TORCH

*In the circle above is shown the torch and shield detached*

The "Staysalite" Linemen's Torch is nothing short of a boon to the telephone or telegraph lineman!

Giving a hot flame quickly, it may be used either as a small heater or soldering iron. It burns alcohol without odor or noise and may be lit or extinguished in a moment. Light and handy, it may be carried in the lineman's belt, and does away with a ground man on the emergency job.

With the wind shield, it "Staysalite," and for the man on the pole it is unusually handy. He simply suspends the torch on the wire under the joint with a short half turn of the upper rim. Finished, he simply drops it back in his belt by means of the hook.

Gives you an economical hot flame where you want it, when you want it. A Klein quality product that is a good buy—do you use them?

Pliers
Splicing Clamps
Sleeve Twisters
Climbers
Tool Belts
Safety Straps
Lag Screw Wrenches
Wire Grips
Tree Trimmers
Tool Bags
Charcoal Furnaces
Staysalite Torch

**Mathias KLEIN & Sons**
Established 1857     Chicago Ill USA

FOR
## STREET LIGHTING

The town of Weston, Ontario, has recently installed this new
street lighting system, using Sol-lux glass throughout. The
above illustration is an actual night photo not re-touched.

*Installation by A. H. Winter-Joyner Ltd., Toronto*

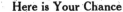

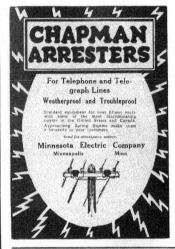

*This is one of the series of striking illustrations appearing in Hoover national advertising. Over six hundred thousand Hoover full pages are circulated monthly through leading magazines*

# We *Guarantee* the Sale of Hoovers on Our Co-operative Plan

**Lubrication Made Simple and Safe**

*(Another Patented Hoover Feature)*

COMPLICATED methods of lubrication on electric cleaners cause many troubles. Grease-cups puzzle women and are "messy" to use. Hidden oil-cups are usually neglected. The Hoover is therefore designed to be oiled in only one place, located for convenience and as a reminder, right on top of the motor. Three drops of oil from an oil-can are sufficient for an ordinary week's use. Any excess of oil that may be given drains off into the dust chamber instead of getting on the commutator to cause pitting, of sticking on the carbon brushes to cause sparking, or of soaking the electrical windings to cause shorting. This invention protects Hoover users from troubles occasioned by improper lubrication. Covered by a Hoover patent. Eighteen other patents protect the salient features of electric cleaner construction. Still other applications for patents are pending.

We do not ask the Authorized Hoover Dealer, who accepts our famous Co-operative Plan, to order a large stock of Hoovers.

On the contrary, we endeavor to restrict his first order—as well as the subsequent orders—to a month's supply. For we are strong advocates of the monthly stock turnover idea. We do all in our power to enable our dealers to make *twelve profits a year* on Hoovers.

Furthermore, we relieve the dealer of risk by unequivocally guaranteeing to sell his Hoovers for him.

We agree to secure leads, demonstrate, sell, deliver and service Hoovers for him. All we ask in return is a small space in his store for a demonstration

booth, and his hearty co-operation.

He receives the benefit of our years of experience in selling Hoovers for other merchants. We use the same methods, in his store and community, that we are successfully using in so many cities and towns.

*Now is an Excellent Time to Become an Authorized Hoover Dealer*

While Hoovers are year around sellers, the Spring Housecleaning period always brings many additional prospects into the market.

So you could not ask for a more favorable time to join forces with us.

Send immediately for a Hoover representative!

THE HOOVER SUCTION SWEEPER COMPANY OF CANADA, LIMITED
Factory and General Offices: Hamilton, Ontario

**MADE IN CANADA        BY CANADIANS        FOR CANADIANS**

# *The* HOOVER
## *It BEATS···· as it Sweeps    as it Cleans*

Generation, Transmission and Application of Electricity

For nearly thirty years the recognized journal for the
Electrical Interests of Canada.

Published Semi-Monthly By

# HUGH C. MACLEAN PUBLICATIONS
LIMITED

THOMAS S. YOUNG, Toronto, Managing Director
W. R. CARR, Ph.D., Toronto, Managing Editor
HEAD OFFICE - 347 Adelaide Street West, TORONTO
Telephone A. 2700

MONTREAL   -  -  119 Board of Trade Bldg.
WINNIPEG    -     -   302 Travellers' Bldg.
VANCOUVER   -   -   -   -   Winch Building
NEW YORK    -   -   -   -   -   296 Broadway
CHICAGO   -   -   `Room 803, 63 E. Adams St.
LONDON, ENG.   -   -   16 Regent Street S. W.

ADVERTISEMENTS
Orders for advertising should reach the office of publication not later
than the 5th and 20th of the month.  Changes in advertisements will be
made whenever desired, without cost to the advertiser.

SUBSCRIBERS
The "Electrical News" will be mailed to subscribers in Canada and
Great Britain, post free, for $2.00 per annum. United States and foreign,
$2.50. Remit by currency, registered letter, or postal order payable to
Hugh C. MacLean Publications Limited.
Subscribers are requested to promptly notify the publishers of failure
or delay in delivery of paper.

Authorized by the Postmaster General for Canada, for transmission
as second class matter.

Vol. 31          Toronto, March 15· 1922          No. 6

## Recognition of Public Utilities Coming with Trade Revival

Judging by the usual signs, not only are we on the eve
of a trade revival but this revival will bring in its wake· a
more universal recognition of the ·all-important part that
public utilities play ·in any community's industrial life.

During the period immediately following the war, when
the prices of practically every commodity that "retails"
into the daily life of our citizens sky-rocketed in price, public
services of one kind and another—electric light and
power, street railway fares and water rates—were
the only things that. stood by the citizens in their·
hour of need, giving the same value for the same
price as in former years. It may be quite true
that this service was given by force of necessity
as many public utilities were operating under hard and fast
franchise conditions—but the fact remains the same. As
a result, it has been brought home to our citizens that public
utilities are essentials and while the average man on the
street, pinched from every quarter as .he was during the
years 1919 and 1920, had not sufficient moral courage, at
that time, to admit the injustice of his treatment of these
utilities, he is plainly in a different mood today. With the
pressure easing up in other directions, he is now able to
pay more attention·to that  still small voice that told him
all along he was in the wrong, and he is daily becoming
more sympathetic to the needs of the central station, the
electric railway and the other utilities that served him so
faithfully through the crisis.

This change of heart has also been brought about, in

part, by the systematic and intelligent campaign of education that has been carried on by many of these utilities, in
which the customer has been given a more intimate knowledge
of the problems of the industry concerned and been led to
see that the maintenance of a good service ·is a business
necessity, requiring capital and permanent income just as
any other business does. The public always responds to
"education" and, in the main, can be trusted to make a correct diagnosis when possessed of the facts.

All this tends to place the public utility in a much more
enviable position than it occupied .in the recent past. There
are other encouraging factors, too. Material costs have
receded—and will go lower. Labor has given way a little
and must still recede further as wages in other lines, such
as construction, for example, gradually sag back. Finally,
"Capital" is gradually losing interest in Dominion bonds,
which held the stage up to the recent past, on account of
the unusually high return; from now on much more money
will be available 'for construction work of every kind—for
industrials and for properly conducted public utility
enterprises.

While it can probably be said that public utilities, as a
class, were more ·adversely affected by the war than any
other of our industrial activities, present indications point
to a steady revival, with indications of greater permanency owing to the general public's fairer understanding of
their problems and of their necessity as a foundation for
any substantial commercial prosperity.

## The Influence of Home Building on Electrical Activity

The whole electrical industry has much to expect from a
revival in the construction industry. This is true more
particularly as regards home building. A few years ago
the building of a new house costing. $10,000 meant a small
wiring job of fifty, seventy-five or a hundred dollars at
most, with the sale of a few lamps. Today, thanks to
"Electric Homes" and similar· propaganda, a fair amount
of people are awake ·to the fact that ample wiring is as
necessary as ample radiation in their heating systems. Nowadays one finds a goodly number of homeseekers who actually ask about· the wiring and, if building their own home,
think it· only reasonable that the electrical contract should
run into four or five per cent. of the total cost of the home..
This, therefore, brings the contract on the house mentioned above up to four or five hundred dollars.

And, logically enough, it follows that the properly wired
home· means electric furniture—floor lamps, table lamps,
appliances of every kind. These run into considerable expenditure—perhaps a couple of hundred dollars on the
average—bringing the total revenue to the contractor-dealer to six hundred or seven hundred dollars.

There were 16,300 homes built in Canada last year.
There were 110,000 marriages. Is it any wonder, with this
disproportion, that rents continue their upward trend and
that the clamor for more homes is more insistent than
ever? It is highly probable that 30,000 homes will be built
during 1922. This means 30,000 times (say) $350, for the
industry in w iring appliances and labor—ten and a half
million dollars in new houses alone.

So we see how intimately the electrical industry is tied
in with house building. There is also, of course, the office
and factory of every kind, but the demand is not so insistent
in this direction. If the contractor-dealer will work closely
with the architect during the present summer the end of
1922 will tell a very different tale, not only for them, but
for our manufacturers, central stations, and wholesalers as
well.

And, added to that, don't let us forget the wiring of
"old homes."

# The Booster Rotary Converter

### With Special Reference to the Installation at the British America Nickel Corporation's Refinery

By RUPERT F. HOWARD*

The synchronous booster rotary converter is not a piece of apparatus new to the electrical business, as it has been in use for several years. However, it is a type of apparatus not generally known. Indeed, very few, perhaps aside from those whose work brings them in contact with it are familiar with this form or rotary converter. The number of installations of these machines in Canada can be counted on the fingers of one hand.

The first rotary converters were used where a constant voltage supply was required, and since the ratio of the alternating and direct voltage is practically constant, the rotary converter is inherently incapable of voltage regulation except within very narrow limits. They were used extensively on 25 cycle circuits feeding Edison 3-wire systems, and to a limited extent in some industrial applications. In fields other than these the booster rotary converter has until recently been practically unknown.

Ten years ago the 60 cycle rotary was often shunned when it came to the choice of a good reliable operating unit. On 60 cycle systems in particular where companies have Edison 3-wire networks, the converting agent was the motor generator. With the improved 60 cycle rotary, reliability of operation is assured, and we are becoming more and more inclined toward this machine.

From the improved 60 cycle rotary it is but a step to the booster type.

A booster rotary converter is a combination of a specially designed shunt wound converter, and an alternating current generator or booster, having the same number of poles in its field as in that of the converter.

The most satisfactory type of booster rotary so far developed has revolving armatures and stationary fields for both the converter and booster. The booster armature is mounted on the shaft between the collector rings and the converter armature, and the booster and converter frames are mounted side by side on a common bedplate. In this

*Chief Engineer for the Corporation.

construction the windings on the booster armature are connected to the collector or rings on the one side, and to the converter armature taps on the other.

The purpose of the booster is to make it possible to obtain a variable continuous voltage from the converter. They are made shunt wound with commutating poles, the purpose of these additional poles and windings is to provide the proper commutating field at all loads with fixed brush position. The addition of commutating poles is one of the most important developments in these rotarys, and has been responsible for the manufacturers redesigning complete lines of machines,—speeds were increased, weights and floor space reduced, efficiencies increased, and commutation placed on a totally different plane.

Fig. 1—Complete machines as installed and shows the relative positions of the booster and the main rotary.

Fig. 2—Shows a cross section of a standard booster rotary converter.

Probably the most useful fact to remember in connection with rotary converter operation is that the ratio between the a.c. and d.c. voltage is fixed once for all by the design proportions of the converter.

When varying direct current voltage is obtained, as by compounding the converter, or by the addition of the booster, it is obtained only by changing the alternating current voltage delivered to the converter armature winding. This characteristic of the rotary converter, more than any other, differentiates it from the direct current generator, and causes many of the differences in behaviour between these two classes of machines under the various conditions that arise in operation.

Fig. 3—Shows the development of booster and converter armature windings.

With this arrangement any generated voltage in the armature of the alternator, or booster, will add to, or subtract from, the impressed voltage at the rings or a.c. end, depending upon the direction of this generated voltage, which in

Fig. 1—Left, Rotary sub-station. View looking towards local switchboard    Right, Rotary and d.c. Panel—British America Nickel Corporation

turn depends upon the direction in which its field is excited.

When the d.c. voltage of a booster rotary is to be raised, the field strength of the alternator, or booster, is increased in the proper direction, and its armature voltage adds to the impressed voltage at the rings, with a resulting increase in the d.c. volts. Conversely, if the d.c. voltage is

poor regulation where d.c. service may require close independent regulation.

(2) Where large corrective effect for low power factor is desired.

(3) Where regulation of frequency and voltage is so poor as to prohibit the use of synchronous apparatus, in

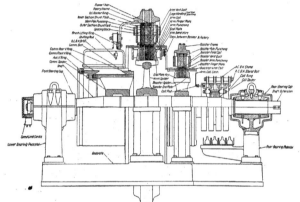

Fig. 2—Cross Section of a Standard Booster Rotary Converter

to be lowered, the field of the booster is reversed so that its armature volts oppose the impressed voltage at the rings, resulting in a lower d.c. voltage.

When the converter is operating at normal voltage with the booster idle, its action is no different from that of a straight rotary converter. When the booster is operating to raise the voltage, its action is positive, and it operates as a generator being driven by the converter, which acts as a synchronous motor. When the booster is operating to lower the voltage, its action is negative, and it operates as a synchronous motor driving the converter, which acts as a direct current generator.

The outstanding advantage of the booster rotary over the straight rotary is that it has a d.c. voltage range of approximately 12% above, and 12% below any neutral voltage. This feature makes it a desirable machine for lighting and railway work, as well as its application to electrolytic operations.

The straight rotary has an automatic compounding effect of about 5%, but where special design features are used in the transformers and supply circuits to the rotary, something higher than this can be obtained.

The booster rotary with transformers, from the standpoint of economy, is superior to the motor generator.

In the smaller capacities their first cost is in excess, and in the larger capacities is less than that of the motor generator. The greater economy usually makes the matter of first cost of small importance.

Consequently, the rotary converters can be recommended in preference to motor generators, except:—

(1) Where they are to be used on a system having

which case the induction motor generator is used.

The British American Nickel Corporation, near Ottawa, uses a large block of power for electrolytic refining of nickel and copper. Direct current being used exclusively in the deposition tanks.

The choice of generating equipment for this plant lay

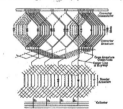

Fig. 3—Development of Booster and Converter Armature Windings and Connections

between motor generators and booster rotary converters.

The load factor of a plant of this kind, if properly operated, is in the vicinity of 95%. The greater economy of the rotary operating under the above conditions, coupled with the lower first cost, made it the logical choice over the motor generator. This electrolytic load with rotary converters and consequent high power factor, makes it a very desirable one from the central station standpoint.

# First 50,000 h.p. Unit of Chippawa-Queenston Power Scheme—III.

### The Story of the Methods of Construction One of the Most Interesting Features of This Big Development. Removing Seventeen Million Cubic Yards of Material and Cutting Through Rock and Earth One Hundred and Forty-five Feet Deep.

From the intake at Chippawa to the power house at Queenston, as already explained, the canal measures 12¾ miles in length, of which the first 4¼ miles follow the route of the Welland River, formerly a sluggish stream with a flat gradient, whose flow has been reversed. The remaining 8½ miles of canal is dry excavation in earth and rock and is the part of the work that has involved construction operation carried out on a scale and with a speed hitherto unprecedented. From Montrose, on the Welland River, where the canal proper begins, the first mile was dredged through earth and is similar in section to the Welland River cut; from this point the canal, which runs in a general northerly and easterly direction, is all in rock with the exception of the built-up, rock-filled section, 2,500 feet long, adjacent to the Niagara Whirlpool. The lower end of the canal opens up into a triangle-shaped forebay cut into the rock at the top of the cliff near Queenston. Where the canal is in rock, the sides and bottom are lined with concrete for the purpose of increasing its carrying capacity by virtue of the smooth surface thus obtained. It was estimated that the capacity of the canal would be increased 20 per cent. by means of the lining, but in view of the exceptionally fine quality of work as actually constructed, it is now evident that an even greater increase in flow will be obtained.

Among the special features of the canal may be mentioned the massive electrically-operated control gate located near Montrose, for controlling the flow in the canal; the deep excavation at Lundy's Lane crossing, where the bottom of the canal is 143 ft. below the level of the ground; the various railroad and highway crossings over the canal; and the "Whirlpool section." The last named is that part of the canal which crosses over an old gorge, the bottom of which was largely composed of quicksand. The construction of the canal across the valley was accomplished by first filling the whole gorge with rock excavated from other parts of the canal, and then allowing this rock-fill to come to a final settlement. After this it was excavated for the canal section and faced with a heavy reinforced concrete lining.

### Unusually Large Construction Machinery

The Hydro Electric Power Commission's engineers when investigating the economics of the development, made very exhaustive investigations into the construction methods that would have to be adopted, inasmuch as the type of construction plant and the manner of its operation would have very considerable bearing on the cost of the project. From the start it was obvious that something entirely unusual in the way of plant and methods would have to be adopted to handle a construction project of such unprecedented magnitude. In the determination of the exact type of plant to be employed in the construction work certain conditions expressed themselves. These were: (1) The availability of cheap electric power for operating construction plant, (2) large quantities of earth and rock to be removed—about 17,000,000 yds.—which made it possible to consider excavating equipment of the heaviest type and largest cap-

acity obtainable, and, (3) excellent facilities for the disposal of the spoil along the Niagara escarpment, within short hauling distance of the site of the work. With these conditions in mind, a very thorough study of various kinds of construction equipment was made by the Hydro engineers and a large amount of information was obtained concerning the output, operating costs, working conditions, etc., of all types of electric and steam operated excavating equipment.

### Largest Shovels in the World

The result of these investigations was that new construction appliances, such as large shovels and special concrete forms, were created to carry on operations of the stupendous magnitude involved in this project. The excavation was carried out with the aid of fourteen shovels, most of which were electrically operated, and five of which were larger than any heretofore built. The largest shovel handled an eight cubic yard bucket for earth and one of six cubic yards capacity for rock. They weigh over 400 tons each, are operated by motors aggregating 750 horsepower for each machine and have a capacity of 150,000 cubic yards of earth or 70,000 cubic yards of rock per month per shovel. These shovels could load cars standing 73 feet above their location. They were also very speedy and were capable of loading a car of 20 cubic yards capacity, standing 60 feet above the shovel in 1½ minutes. The booms of these shovels were about 90 feet long and the dipper sticks 58 feet. Sixteen wheels, eight on each side, comprised the mounting, the machines running on two tracks (four rails) spaced 30 feet apart on centre. Two of the large shovels were steam operated instead of electrically driven only because the manufacturers were able to place better delivery on steam equipment. Other shovels with capacities of 7/8 to 4½ cubic yards were also used, some of these having caterpillar traction and the majority being electrically driven.

Other equipment for excavation operations included 17 channelers, numerous rock drills, dredges and a cableway excavator, the latter being used for widening and deepening the Welland River section of the canal.

### Electric Construction Railway

In order to efficiently carry away the enormous amount of excavated material, a standard gauge, double track electric railway in which standard 80 pound main line rails have been used, was constructed along the entire route of the canal, with branches to the disposal areas. Eighty-two miles of track were laid altogether. The rolling stock consisted of 50 locomotives, 24 of which were electric, and over 300 air-dump cars of 20 cubic yards capacity each. The trolley wires on this railroad were offset on one side of the track to allow the dippers of the shovels to pass over the cars freely when loading them and to permit locomotive cranes to use the tracks. On temporary tracks, alongside the dump, for example, the trolley wires were carried on framed timber trestles which were mounted on wheels to facilitate removal when necessary to shift the track.

For the operation of the electric railway, electric shovels and other machinery and equipment, over 13,000 electrical horsepower were required.

In addition to the equipment above noted, other special machinery and devices were developed, which will be described in detail as this article proceeds.

## Excavation Methods

On the Welland River section of the canal, the excavation was carried out largely by a cableway excavator and by dredge. The former equipment was a Lidgerwood cableway of 800 ft. span on an 80 ft. head tower and 60 ft. tail

The plant for lining the side walls of the canal. Two form carriers and a mixer unit comprise the equipment.

tower, fitted with a 8 cu. yd. clam. The towers were mounted on railroad trucks running on parallel double tracks. The capacity of the plant was such that a trip of the bucket was possible every two minutes. For the first 4,400 ft. of the river a dipper dredge was used, as bridges and other impediments interfered with the operation of a cableway In the latter part of 1920, when it was decided to make strenuous efforts to complete the canal by the end of 1921, the suction dredge "Cyclone," the largest in the world, was transferred from the harbor development at Toronto to the Welland river, where it assisted in the excavation of both the river section and the earth section of the excavated canal. On account of the tremendous size of the dredge, considerable trouble was experienced in transferring it from Toronto to the Chipawa canal, particularly in navigating the Welland river, where bridges, power lines and shallow water offered impediments, which were successfully overcome, however, as described in the Contract Record of May 11, 1920, page 458. The work of the dredge has amply justified its use on the Chippawa work, as in 5½ months it removed 1,160,000 cubic yards of earth, an average of over 200,000 yards a month.

On the artificial section of the canal, excavation included removal of rather heavy overburden and deep rock cuts. The overburden was partly fine wet clay sand and

partly very stable red clay, all of which was removed by the revolving shovels mentioned previously.

On the rock cuts, the first ten feet of the sides was channelled, but below that, the rock was close drilled. The spoil was removed by the shovels referred to. The method adopted was to operate on a series of benches or cuts resembling stair steps, with one shovel following the other until the proper grade was obtained. The spoil was loaded directly by the shovels into the dump cars on the track situated anywhere from 45 to 75 feet above shovel grade.

## 17,000,000 Cubic Yards of Excavation

The amount of material excavated from the canal proper was over 17,000,000 cubic yards of earth and rock, the earth excavation amounting to 13,200,000 cubic yards and the rock to 4,182,000 cubic yards. The deepest cuts were 72 feet in earth and 85 feet in rock. At one point—Lundy's Lane crossing—the combined earth and rock cut was as much as 145 feet deep. From these figures it is evident what a tremendous undertaking this excavation was. The speed with which it was carried out was remarkable as the entire work of constructing the canal was accomplished since March, 1918.

Placing concrete lining on rock sides of canal. Completed sections, 40 feet in length, in foreground.

when active excavation with the large shovels began. Of course, a huge construction force was employed, at one time there being 8,100 men on the payroll, although the average force was much smaller than this.

This wonderful achievement was largely due to the type of excavating equipment adopted. The economy of the large shovels is illustrated by the fact that in 1917 when some preliminary work was commenced with railroad type shovels, direct labour cost comprised 29 per cent of the total unit cost of excavation, while in 1920, when labour was at its peak, costing 250 per cent more than in 1917, the labour cost per yard of excavation had only increased 4 per cent over the 1917 figure of 29 per cent. This would indicate that the saving of man power resulting from the use of the large excavating units practically offset the 250 per cent increase in labour expenditure.

The rapidity of operations possible with the plant at the

disposal of the construction organization is best realized by the fact that in the past year 3,710,000 cubic yards of earth and 2,410,000 cubic yards of rock were removed. Twelve months ago only 7,000 feet of the rock cut was completed to a maximum depth of 60 feet. Since that time, the whole canal has been completed, quite aside from work on the screen house and power house—surely a stupendous and wonderful achievement.

In excavating the canal, the general method was to start a pilot near one side of the canal prism with a railroad shovel loading cars on the ground surface. In this cut were run the loading tracks for the big shovels which followed with the main line at both ends. The shovels worked through in steps, as already explained, loading the spoil into dump cars, which were operated in 8 and 10 car trains. As many as 300 trains a day were often necessary to dispose of the excavation.

The rock was drilled, except for the first 10 feet, which was channelled. Rock drills of standard type were employed

construction was essential to the prosecution of the excavation and for that reason were constructed before any of the digging was carried out in their vicinity. This gave rise to the remarkable phenomenon of bridge structures erected below the ground surface, the shovels later removing the earth from around and beneath them. When bridges were built in this way, an enclosure of steel sheet piling was used as a cofferdam. This manner of constructing bridges made concreting operations very easy inasmuch as a mixer was placed on the edge of the excavation to spout the concrete directly into the forms.

### Disposal of Spoil

Practically all of the excavation had to be wasted. Some of the rock was used for concrete purposes and some for riprap, and at the whirlpool gulley about 1,500,000 yards of rock obtained nearby was dumped and allowed to consolidate but practically all the rest was conveyed to a 200 acre dump capable of holding 21,000,000 cubic yards. Some smaller dumps were used, but the one mentioned which was two

View of canal showing a typical railway crossing

as well as submarine drills mounted on high frames similar in design, to a pile driven rig. Each drill on equipment of this sort was carried in the 55 foot leads of the frame from the main line of a hoist mounted on a platform behind the leads, a spare line on the drum being used to elevate the drill steels when required. The frames were loaded on skids or rollers to facilitate shifting within the canal prism. The drills were of a maximum diameter of 4½ inches and minimum diameter of 2 inches.

The forebay excavation was begun by shooting out a 10 foot lift over the entire area, about 1,100 holes being fired at once. About one pound of dynamite to the yard, including that used for springing the holes, was used. Each hole was sprung with 5 or 6 sticks and loaded with 15 or 20 sticks, the spacing being 7 feet each way. As a result of this initial blast about 60,000 cubic yards was broken up fine enough to be handled by the railroad shovels.

### Bridges Built Before Excavation

About 17 bridges were required to carry highways and railways across the line of the canal. On some of these early

miles from the canal received the greater part of the spoil. This fill was reached by a double track spur of the construction railway. The trolley wires along this dump were carried, as explained before, on portable frames. These frames were mounted with outriggers for the trolley supports were mounted on flanged wheels which run on short sections of track set up on the filled side of the tracks on which the dump trains ran. As the fill progressed and the filling tracks needed to be shifted, the trolley towers were moved accordingly. When first dumping the fill, a trestle was built well out into the middle of the dump area and the fill started from it. In addition, the tracks fanned out over the side from the corner where they entered the dump. In this way sufficient places of dumping were maintained to take care of the spoil without any train interference.

### Crusher Plant

The crusher plant was located near the forebay and received material by rail in the 20 yard dump cars, which discharged direct into a large hopper lined with 2½ x 6 inch steel bars laid flat. The crusher fed from this hopper was a 60 x 84 inch jaw outfit operated by a 250 h.p. motor, reducing the stone to 8 inch size and delivering it to a belt which took it to the top of the secondary crusher house. In this were three gyratory crushers reducing the stone to 2 inch size. From these crushers the material passed through a screen which removed dust and oversize stone and then was carried on a suspended belt conveyor over the storage pile. At the end of the storage pile was the bin structure for the receipt of 1 inch stone used for reinforced concrete. The 1 inch stone was obtained by passing the oversize into small auxiliary gyratory crusher which delivered its product directly to the bin mentioned above.

The sand was brought in by car and dumped into a track hopper from which it was carried on a 14-inch conveyor to a bucket loader which lifted the sand to an overhead conveyor running over the storage pile which was adjacent to the stone pile. Reclaiming conveyors beneath the piles carried the aggregates to the concrete mixing plant at the top of the escarpment, from which plant the screen house and part of the power house were concreted.

### Scaling of Rock

All of the rock, except in the forebay, was scaled before the concrete lining was applied, in order to remove uneven projections. This work was done with movable towers which practically occupied the entire canal prism. These scaling towers were simple wooden frames mounted on trucks for movement on sectionalized track. At each side was a platform suspended by wire cables from sheaves on top of the frame and counterweighted for easy movement. From these platforms the scalers removed all pieces of rock projecting over the line of the clean-cut which was indicated by a plumb line hung from the top of the trestle. The scaling was done with picks and hand air tools, the air being supplied from the pipe lines parallelling the canal.

At the same time as the scaling was carried out, dowels of old steel rods were grouted into the sides of the cut at 4 ft. intervals in both directions, in order to bind the concrete lining to the rockface. These rods were about ¾ inch in diameter, 2 ft. 9 in. long, embedded 2 ft. in the rock and bent up vertically so as to hold the concrete lining.

### Lining of the Canal

After the rock sections of the canal were excavated and sealed, they were lined with concrete to reduce water friction. This lining was from 30 to 34 ft. high. The floor was at first paved with at least 6 in. of concrete laid in three longitudinal slabs—two slabs each 13 ft. wide, spaced 18 ft. apart, this 18 ft. space being later filled in. The typical plant for carrying out this work consisted of a 1-yd. paving mixer with chute mounted on a flat car, which also carried an overhead bin to hold sand and stone, and a cement platform. A chute fastened to the frame and extending to the construction railway on top of the cut, permitted sand, stone and cement to be passed to the mixing unit.

When the floor was laid, the sides were poured, special equipment being used to do the work. Nine of these plants were in use, each one consisting of one concreting unit and two form carrying units, the three comprising a "battery." Each section of the battery was mounted on trucks running on two parallel tracks (4 rails) and while each was capable of independent movement, the three were operated in conjunction as a self contained outfit, forming and lining the two sides of the canal at the same time.

The centre of the three sections of the battery formed the mixer unit and comprised hoppers for the storage of sand and stone, a large capacity steam mixer and a hoist tower with spouts for chuting the concrete to place in the forms. The other two sections were form carrying units, each consisting of a timber frame filling the canal prism and carrying the wall forms which were steel plates on a steel framework. Jacks were inserted between the frame and the forms, so that the form could be adjusted in proper alignment and so that it could be swung back after the concrete had set sufficiently to allow the unit to be shifted to the next position.

The walls were poured at least 6 in. thick in alternate 40 ft. sections, one form unit being used for the first pourings and the other for the intermediate pourings. The forward form unit was provided with side bulkheads to retain the concrete but the second unit, of course, did not require these and the wall slabs already poured acted as retainers. A fillet was formed at the junction of the wall and the floor, this being poured with the wall. For easy chuting and for spading, etc., the wall forms were built in removable sections. As soon as the concrete had set, finishers patched up any unevenness.

On the Whirlpool section, a different procedure was adopted. This portion of the canal was trapezoidal with side

One of the curves in the Chippawa canal. At this point the rock cut is 85 ft. and total cut 148 ft. in depth

A transition portion of the canal where the rectangular section merges into the trapezoidal section. Note the excellence of the concrete lining

slopes of 1 on 1½ and the lining was laid on the slope and heavily reinforced. Three-foot strips of concrete were first laid by ordinary methods up the slope at 18 ft. centers. The intervening spaces were then filled in. At the beginning wooden forms were used, these being in panels 3 ft. wide by 16 ft. long, attached to the concrete strips by anchor bolts. To get away from the carpenter work entailed by these forms, a steel screed or sliding form was later devised. This measured 16 ft. long by 17 ft. wide, and consisted of one thickness of ¼-in. steel plate with a 9 in. channel around the outside and reinforced with 9 in. channels placed horizontally across the plate at 9 in. centers. The drawbar was a continuation of side channels, brought together about 6 feet ahead of the form. A hand winch was placed at the top of the slope by which the form was moved upward.

The bottom 20 ft. of each panel was poured behind the regular forms and allowed to set. The steel form was then placed on this and kept moving slowly up the slope, the concrete placed being kept almost even with the top. By the time the plate had passed over the concrete it was set sufficiently to stay in place. The result of the use of the steel form was a much more uniform face and a better chance to finish, as finishers could get to work sooner and could work on green concrete. Only one finisher was allotted to each panel, he following the form as it went up the slope.

Some of the finest concrete on the canal is at the transition sections, where the canal changes from earth to rock and from rock to Whirlpool sections. At these points, the lining was applied in smooth curves to reduce friction to a minimum and special care was taken in pouring to avoid any irregularities.

In the forebay a smooth lining was not deemed essential, as the velocity of the water is very low. However, the rock wall was sealed by a coating of gunite to a depth of a few inches.

### Power House Construction

The power house, located at the foot of the cliff, near Queenston, involved very interesting construction methods. Inasmuch as the whole of the building is not required at present, the superstructure as constructed today is only one-fourth of the ultimate length. The substructure, however, is completed for the entire building.

The excavation for the power house substructure was entirely in rock on a comparatively flat site immediately adjoining the river. In order to obtain sufficient area under the turbines to permit the spent water to discharge into the river, it was necessary to excavate the rock for a depth of 25 feet below the surface of the river. To enable this work to be done in the dry, the rock shore adjacent to the river was left undisturbed so that it formed a dam between the river and the work. It proved to be very watertight, only one small pump, operating intermittently being required to maintain the site free from the usual water difficulties. This favourable circumstance enabled a high quality of foundation work to be done in this vital part of the plant and at the same time afforded an exceptional opportunity to make a thorough examination of the underlying rock strata, which proved to be dense and sound. At no point was any sign of settlement apparent.

The excavation was straight rock work, the spoil being removed by dump car over a siding which was built along the river edge. This railway also served to bring in material from the sorting yard located near Queenston.

For concreting, a large mixing plant was stationed on the face of the escarpment a short distance from the top. From this plant the concrete was chuted through suspended troughs to the work. Relay towers were erected at various points from which the concrete was re-directed as necessary.

The steel erection was carried out by means of a two-derrick wooden traveller on wheels.

Mr. John Wilson. Jr.. who for the past fifteen years has been associated with the sales organization of the Dominion Bridge Company and its subsidiary companies. the Dominion Engineering Works. and the Structural Steel Company. has accepted the position of sales-manager with the Walsh Plate & Structural Works Limited of Drummondville. Que. Mr. Wilson assumed his new position on March 1.

The Diamond State Fibre Company are distributing Catalogue No. 20, covering Condensite Celoron, which may be had upon request. Condensite Celoron is made especially for radio panel work and is furnished in three grades: Condensite Celoron Fibre Veneer; Shielded Condensite Celoron; and Condensite Celoron Grade 10. All these can be furnished in full sheets. or panels cut to size or machined to customers' specifications.

## Electrical Co-operative Association
## Province of Quebec, boosting business

# Montreal's Electric Home Display

### Arousing interest of thousands of well-to-do citizens who are eager to learn more of the use of electricity and to experience the comforts and luxuries it affords in well-wired homes

Apart from the purely organization aspects the most important work of the Electrical Co-Operative Association of the Province of Quebec since its formation is the Montreal Electrical Home movement. The Association's work may be broadly divided into two parts—that appertaining to questions affecting the industry itself and that relating to the promotion of sales of electrical appliances. The latter naturally brings the Association, through its members, in contact with the general public, which is for the most part lacking in appreciation of the value of the manifold services which electricity can render.

Experience in every line of business shows that if substantial results are to be secured something more that a mere casual attempt to secure trade must be made. A persistant follow-up system is imperative.

There is a wide field for the energies of the contractor-dealers, the supply houses, and the light and power companies. The Electrical Home may be described as a wedge which provides the opening for the contractor-dealer in particular, who will miss one of the best opportunities ever given him if he neglects to turn to account, and to translate into dollars and cents, the educational work begun at the Electrical Home. The success of such an exhibition must necessarily benefit all sections of the industry.

As has been frequently pointed out some contractor-dealers are in a sense responsible for the small amounts spent on wiring. In their eagerness to secure contracts they have of their own accord cut down the number of outlets and other work when tendering for jobs. The result is that building companies, general contractors, and private persons who construct houses spend as little as possible on electrical work. They give great attention to plumbing and heating and other essentials, but hardly spare a thought on wiring and convenience outlets with the result that the sums appropriated for this work are absolutely ridiculous. Builders and architects and even electrical contractor-dealers need a lot of education on such points, and the Electrical Home is one of the best means for driving home the lesson of better and more wiring.

#### Organization Plans

The great advantage of the Modern Electrical Home lies in the fact that it provides a medium for demonstration, and demonstration constitutes the most effective selling argument in the world. The public is shown the convenience of the various outlets and also the utility and economy of the appliances, a far more potent method than a mere general talk which often leads nowhere. The purchase of all the electrical appliances shown would involve a considerable sum and probably few of the visitors could make such an outlay at once, but most of them could buy one, two, three or more, and add to their stock from time to time. In this way the Home is educating the public to do things electrically to a far greater extent that at present, and is also educating the public on the advisibility of making provision in the way of additional outlets for present and future needs.

The organization of the Electrical Home involved an amount of work which only those who participated in the campaign can appreciate. As far back as December preliminary preparations were made, but certain obstacles had to be overcome before definite headway could be made. First a vacant house had to be secured, then arrangements made for its furnishing and equipping with appliances, and finally the active co-operation of all departments of the industry definitely obtained. Mr. Louis Kon, the secretary of the Electrical Co-operative Association, Province of Quebec, planned and superintended the general arrangements, receiving generous assistance from Mr. Kenneth A. McIntyre, the Canadian representative of the Society for Electrical Development. Mr. McIntyre made two visits to Montreal, on the first occassion addressing three meetings of the contractor-dealers and jobbers and the Electrical Co-operative Association, detailing his experience with the Electrical Home in Toronto and describing how matters were dealt with in that city. Mr. Kon was unceasing in his endeavors to make the campaign a success, and with the happiest results, particularly when it is remembered that he had no large fund of money to draw on.

Unlike Toronto, there was only one committee, the House Committee, which controlled the arrangements for taking care of the visitors. The committee consisted of Messrs. C. Slimpin, assistant manager, Montreal Light Heat & Power Consolidated, chairman; George Atcheson, superintendent of merchandising department, Southern Canada Power Co., vice-chairman; J. P. Sullivan, Canadian Electrical Supply Company; C. C. McGovern, sales-manager, Appliance Department, Northern Electric; J. A. Anderson, president of the Contractor-Dealers' Association (English Section); Lee Jones, superintendent of sales, Canadian General Electric; J. W. Tremblay, director, Contractor-Dealers' Association (French section); N. W. Fairley, sales-manager, Monarch Electric Co.; W. S. Parke, eastern representative Renfrew Electric Company; J. B. Woodford, Bell Telephone Company; Norman Clarke, eastern representative, National Electric Heating Co.; P. E. Ostiguy, Aqua Electric Heater Company; Fred. Baker, eastern rep. Canadian Edison Appliance Company; and J. A. St. Amour, director Contractor-Dealers' Association (French section). Messrs. E. W. Smith, Northern Electric Company, and C. B. Caulson, Canadian General Electric, acted as general assistants to the chairman of the committee. The committee found no difficulty in securing all the help they needed, a practical illustration of the co-operative spirit which is the basis of the associaton.

#### Publicity

Publicity in such a matter as the Electrical Home is of the first importance. The association received a certain amount of help from the newspapers, and in other directions were the recipients of hearty support. For instance, the Montreal Tramways Company allowed the Association to display posters, in French and English, on the cars and also gave permission for electrical signs to be displayed on their poles. The Montreal Light, Heat & Power Consolidated and Montreal Public Service Corporation used space in the daily press, referring to the Home. The association issued a large number of stickers which were freely used by depart-

# "Co-operatives" in Montreal Demonstra

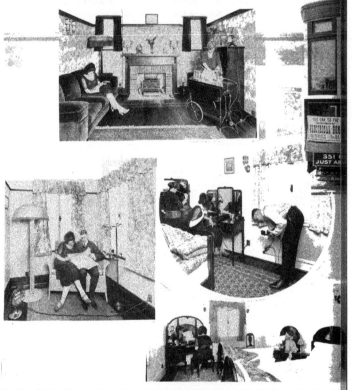

Living-Room—Besides what you see there are two more convenience outlets in this room for an electric gramophone, the use o
of light throughout, including one over the sink, a convenience outlet for a dishwasher and a duplex one on the kitchen cabine
Here is the best illustration of the advantages of a duplex convenience outlet.  Mark the heater cord stretched on the floo
plug telephones, self-lighting clothes closet and well distributed bracket lights controlled by a switch make the bedroom
besides those seen: one next to the serving table and one near the sideboard.  Basement—Convenience outlets in a basem
The 26 thousand subscribers of the Montreal Gazette might have missed the most important news but they could not have

# del Electric Home for Public Inspection

and placing of the wired tea-wagon or portable lamps where required. It saves steps and avoids crowding. Kitchen—Plenty uty motor or use of any useful appliance, and an electric range make a fully modern kitchen in a fully modern home. Sun Parlor— a single outlet instead of a duplex. The Bedrooms—Convenience outlets on the side of the beds and dressing tables, jack and Room—The old and the new ways of getting "juice" for the table appliances. There are two duplex convenience outlets ed in the ceiling. It prevents the dragging of cords on wet floors. A pilot light on the ironing board saves the iron and current. picture of the street car of the Montreal Tramways Company shown in the centre above.

mental stores, the Bell Telephone Company and the light and power companies. The members of the Electrical Co-operative Luncheon also gave considerable assistance. They were present at the official opening of the Home, being conveyed in special cars placed at their disposal by the Tramways Company.

The roads approaching the House, situated at 351 Cote des Neiges Road, were liberally dotted with signs so as to guide the visitors by day and night. At the commencement of the road leading to the Home there was a large sign in French and English inviting a "A Visit to the Montreal Electrical Home—Wired for Every Convenience." Then as the visitors turned in the path to the Home they received a "Welcome to the Modern Electrical Home—Bienvenue au Foyer Electrique Moderne." At night a large flood light mounted on a pole was directed onto the Home and in this way it was easy for visitors to find their way after leaving the tramcars.

### The Electrical Home

The House was one of a group centrally heated. It consisted of nine rooms and two bathrooms on two floors. It was furnished—and very tastefully furnished too— by N. G. Valiquette, Limited, Montreal, The electrical equipment was supplied by members of the association, but in every case the indentification marks were either removed or so covered up as to preclude possibility of the names of the manufacturers being known. In the event of the demonstrator being asked as to the make of any of the appliances, the visitor was assured that the object was not to sell any given type but that information regarding such goods could be obtained from any electrical firm whose name appeared on the printed list distributed. Emphasis was, indeed, laid on the fact that, in the ordinary sense of the term, the Campaign was not a commercial one—it was educational in its scope. The wiring was done by the owners of the house, at their expense, and by a contractor selected by them.

With a limited amount of accomodation it was of course necessary to adopt a plan by which overcrowding was avoided and the visitors given an opportunity of hearing and understanding the different demonstrators in charge of the rooms. At the periods of the largest attendances, the visitors were admitted in groups. The objects of the Home were first of all explained, special attention being directed to the wiring—the necessity and economy of good wiring and the importance of its being of sufficient capacity—followed by a reference to the convenient outlets in the room. The visitors then passed on to the other rooms in turn where those in charge explained and demonstrated briefly the objects of the various outlets and the convenience of the appliances. Other groups were admitted as the space became vacant, an attendant regulating the entrance of the groups in turn.

### General Wiring Plan

On the ground floor of the Home are four rooms. The sun-parlor calls for little comment except to say that there are two duplex convenience outlets for a heater, fan, portable lamp or other use. The living room has a large number of outlets for candelabra, portable lamps, electric fireplace and heater, fan, and tea wagon, the latter wired for connection to any outlet in the room. Then there are duplex outlets for the gramophone and the portable lamp. Owing to the height of the ceiling, the illumination scheme is by four double wall brackets.

In the dining room, adjoining, the wired dining table attracted the chief attention. The electricity is supplied through a floor outlet, and distributed by means of outlets in the table, for a toaster, chafing dish or any other appliance. Outlets are made in different parts of the room for grill and egg boiler, candlesticks, etc. The illumination consists of a central fixture for five candlesticks and two brack-

ets, giving abundant light without glare and harmonizing with the general decorative scheme. The butler's pantry is provided with a duplex outlet for a plate warmer, while the kitchen has a number of outlets for appliances designed to save labor and to insure efficiency in cooking and other domestic work, such as an electric range, a dish washer, a vacuum cleaner, a disc heater, and a general utility motor for whipping cream, grinding cutlery, and buffing the silverware. The maid's room contains a duplex outlet to connect with an adjustable lamp for the sewing machine and also for operating the machine.

The next course was to the basement where visitors were given details, and saw demonstrations, of the ironing machine for large pieces, with the sad electric iron and the attached pilot lamp; the instantaneous water heater enabling hot water to be supplied by merely turning the tap; and the electric washing machine. The outlets are placed in the ceiling in order to avoid cords dragging on the floor.

A special desk was provided in the basement for the purpose of securing the names and addresses of those interested in getting any special information as to the various electrical features.

The second floor consists mainly of bedrooms—the main bedroom, second bedroom and the child's room. In the first named, between the twin beds, there is a duplex outlet for a boudoir lamp, with provision for a jack and plug telephone connection. Near the dressing table, a duplex outlet is shown for a hair drier and massage vibrator and for other toilet appliances. In the next room are outlets for heating pads and dimmer light, violet rays and curling tongs. The lighting in both rooms is by brackets with silk and parchment shades. The nursery has outlets for a portable heater, bed-warmer, and a transformer for electric toys. In each bathroom lights are set on both sides of the washing basins and mirrors, while a single outlet gives a connection for an immersion water heater. A small den has outlets for reading lamp, cigar lighter, fan or percolator.

All cupboards are equipped with lights, controlled by switches, putting the lights on or off automatically by the opening and closing of the doors. An inter-communicating telephone system and jack and plug extensions are installed, enabling communication to be secured in all parts of the house.

### Improving on "Yesterday"

The visitors were not allowed to depart without literature in order to deepen the impression of the demonstration. This literature was in the form of a folder—"Improve on Yesterday—Home Comfort and Economy through the use of Electric Service". The folder gave a brief description of the advantages and conveniences of electricity as applied to illumination, domestic work, and the home. Visitors were invited to consider the house in the two-fold function of a home and domestic work-shop, and "See what the Service of Electricity can give you." One page, entitled "The Modern Electrical Home," read as follows:—

So rapidly has the Electrical industry, particularly in its application to the home, advanced in recent years, that it has been difficult for the average man or women to keep pace with its development.

"The chief aim of the Electrical Co-Operative Association, Province of Quebec, in having the "Modern Electrical Home" exhibit, is to impress the public that electricity is the most effective means available to-day for making the house a home.

"The plans for wiring are largely a matter of being sure that the outlets for lighting, fixtures and appliances are located in convenient and accessible places, and that enough of them are put in to take care of all possible needs.

"It is obviously cheaper to include a few extra outlets in the original wiring; however, it can be done later at a comparatively slight expense.

"A house to be fully modern must be properly wired for all the domestic uses of electricity.

"Get your suggestions for electrical improvements from

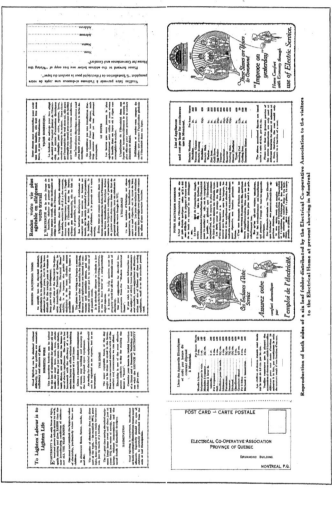

Reproduction of both sides of a six leaf folder distributed by the Electrical Co-operative Association to the Electrical Home at present showing in Montreal the "Modern Electrical Home."

"If you wish to get more information on the subject of wiring a home for convenience and comfort, more suggestions on illumination and use of labor and time saving appliances, mail to us the attached postal card."

A postcard, section of the folder, addressed to the Electrical Co-operative Association, Province of Quebec, requested a free copy of "Wiring the house for Convenience and Comfort." The last page of the folder gave a list of the appliances and the cost of electricity for continuous use, in Montreal—

| Electric Washing | Per hour | Watts |
|---|---|---|
| Machine | 2c. | 345 |
| Vacuum Cleaner | 1c. | 185 |
| Electric Ironing Machine | 2¼c. | 490 |
| Electric Dish Washer | 1c. | 200 |
| Flat Iron | 3c. | 575 |
| Grill Stove | 3c. | 600 |
| Egg Boiler | 2¼c. | 475 |
| Toaster | 3c. | 575 |
| Percolator | 2¼c. | 450 |
| Samovar | 2¼c. | 450 |
| Chafing Dish | 3½c. | 690 |
| Heater (Radiator Type) | 3c. | 600 |
| Heating Pad | 1/2c. | 80 |
| Curling Iron | 3c. | 550 |
| Immersion Heater | 2c. | 345 |

The above mentioned figures are based under a rate of 4.8c. kw.-hr.

Please note that the greatest portion of the above apparatuses are only used at different intervals for about 10 minutes at a time, therefore, the cost would only amount to a fraction of a cent.

Besides the folder each visitor was presented with a printed list of wiring contractors, appliance and fixture dealers in the Montreal district.

Mr. Kon has stated that the industry is very much gratified with results, particularly as regards attendance. Although this is not as large as it might be if they were carrying on an advertising campaign in the daily press, they consider they eliminate the merely curious type and that the visitors they do get are eager to be educated on properly wired homes. Regarding results, Mr. Kon adds:

"We average about 300 people a day, children, of course, not counted, and we are in a position to give them individual attention which is already to an extent, fruitful in results. Some very interesting things are transpiring in connection with the Home. I had a couple of builders who built some ninety houses in Notre Dame de Grace district, which is one of the good residential parts of our city, and they were perplexed to see the duplex convenience outlets, and their number in the various rooms, as well as the location of switches and distribution of light fixtures. They admitted that they never knew of these things before. During the conversation I asked them what is the approximate expenditure of plumbing in the houses they build, which I was informed by them would represent 14 to 15 per cent of the cost of the house, but when asked how much they spend on electric wiring per house they would not say any percentage but shamefacedly admitted, "we would not like to say, it is ridiculous, it may be between sixty and seventy-five dollars per house," and they added "we can see where we were wrong".

Then some of the men working for the contractor-dealers who are on duty as hosts in the home, mentioned to me, after the house had been opened for about five days, that they were called by people to bring them duplex convenience outlets and have them installed in the kitchen and sewing-room, the orders being given by men who added, "My wife went to that darn Modern Electrical Home and got the duplex convenience outlet bug in her head, insisting upon having it installed right away." This happened in three or four cases.

"We hope that by the first of May, the time everybody moves in and out in Montreal, our contractors will be very busy properly wiring some of the residences."

### The Effect on Members of the Industry

One of the results is not to be computed in dollars and cents. It is the effect upon the members of the industry. In educating the public, those engaged in the work have themselves received benefit, in the way of co-operative effort, and also in the way of enlarging their experience in dealing with electrical subjects. One member who acted as host, and who was dubious as to his ability to talk to the public on the need for better wiring, confessed that his experience had been of immense value to himself, and that it enabled him to deal in a much more capable manner with customers in the matter of wiring and appliances. Besides this, contractor-dealers who previously lacked the imagination—if we may put it that way—to suggest duplex convenience outlets to prospective customers, are now alive to the possibilities of increased business in this direction, and are putting this matter before their clients. The Home has also had the effect of cementing the industry, of promoting that get-together spirit which is so essential to the progress of any movement, and which makes for the better understanding between different sections.

### Inquiries Coming In

Although the full results cannot be tabulated at this stage sufficient data is available to be able to state that the industry will receive appreciable material benefit. Inquiries are daily coming in. Many replies have been received to the invitation to write for the pamphlet, "Wiring the Home For Convenience and Comfort". These are passed on to the Contractor Dealers' Association to be dealt with. As an example of direct results we may mention that one visitor wrote for information as to where he could get certain work done, the outcome being an order for wiring a house, for an electric pump, an electric washer, etc. Another result was the visit of an architect who is drawing plans for some houses. He went in response to the request of his client, who had previously inspected the house, and desired the architect to get some ideas on convenience outlets. In yet another instance a contractor who was doing work was instructed to put in some convenience outlets; he was told that the order was due to the wife of the client visiting the Home.

### Wiring Dressing Tables

Some of the C. P. R. officials connected with the company's hotels department were among the visitors. It has been decided to put in wired dressing tables for curling tongs and other appliances in the bedrooms of the Chateau Frontenac, Quebec City, and also to install special reading lamps over the beds.

An average attendance was 300 people per day. It was noticeable that husbands who inspected the Home made a second visit, bringing their wives, and vice versa.

The women's organizations—Montreal Housewives League, Local Council of Women, and Montreal Women's Club—have been lined up. At first there was a little difficulty in this matter, owing to an impression that the Home was a mere advertising exhibition, but after a visit of the president of the Montreal Housewives League, this idea was speedily removed, and a meeting of the organizations was held in the Home. The students of the domestic science classes of the High Schools, and the students of Macdonald College also visited the Home.

### Other Homes to Follow

Four other Homes in different parts of the city will be wired and furnished. These will be held in April, May, September and October. At the meeting of the Electrical Co-operative Association on March 7, the following committee were appointed to take charge of the different branches of these Homes—House (the old committee was made permanent), Wiring, Financial, Advertising & Publicity, Illumination, and Appliance.

# BETTER MERCHANDISING

## $10,500,000 for the Contractor and Dealer in 1922, in New Homes Alone

Great house building activity during the present season seems a foregone conclusion. Rents are to-day at the highest point in history; the house shortage continues to accumulate; material and labor costs are down, and money is more easily available for mortgages. Following an unusually mild winter contractors have already commenced active operations.

Canada's building program for 1922 includes 30,000 new homes.

These will range in value, mostly, from $3,000 to $20,000. The average will be around $5,000.

Thirty thousand homes at five thousand each is $150,000,000.

If the Electrical Industry does its part—"sells" the architect and owner—the wiring of these homes will run into five per cent.

There is no place in the electrical industry for the "two per cent." contractor. To accept a house contract for less than five per cent. of the total value admits lack of confidence in our own product.

Five per cent of one hundred and fifty millions is $7,500,000 for wiring alone.

Electrical equipment may well add another five per cent.—let's say two per cent. and be on the safe side.

Two per cent. of one hundred and fifty millions is $3,000,000.

$10,500,000, in all for Canadian contractors and dealers, in new homes.

In addition, there are all the unwired homes, the poorly, partly wired and the homes with little or no equipment.

**These are facts.** In face of them, the man who **thinks depression** had better look for the cause withni himself.

## Hamilton Will Open Electric Home in May—
## Plans Well Advanced

A meeting of the Hamilton District Electrical Association was held March 1, in the Board Room of the Y. M. C. A. where dinner was served at 6.30 p.m. A report was received from the Hamilton Electrical Development League on the Electric Home as follows:

"It is expected the Electric Home will be opened on or about May 15.

"Mr. Norman Ellis has been selected by the Realtors' Association to provide the Electric Home for exhibition. This home will be erected on Main-St. E. near Sherman Ave.

"The Arcade Ltd. have agreed to decorate and furnish the Electric Home throughout. Arrangements have been made for the installation of a wireless outfit and persons visiting the Home in the evenings will have an opportunity of hearing wireless concerts.

"Special electrical effects will be used each evening during the exhibition of the Electric Home. It was reported 20,000 people visited the Electric Home in Toronto which was on exhibition for 13 days.

"Nothing will be sold at the Electric Home during the exhibition as the object of this Home is to show the advantages in the increased use of labor-saving devices and better wiring for the home."

Mr. V. K. Stalford introduced the speaker of the evening Mr. Russell T. Kelly of the Hamilton Advertisers, Ltd. Mr. Kelly's subject was "Better Business and More Business" and the address was appreciated by all present.

Mr George Foot, on behalf of the Canadian Westinghouse Company, presented each person present with a beautifully mounted photograph of the largest generator in the world which is now set up in the plant of the Canadian Westinghouse Co. at Hamilton. Mr. J. Cutley, of Culley and Breay, in a few remarks in behalf of the Hamilton District Electrical Association thanked Mr. Foot and the Canadian Westinghouse Company for the great interest they have shown in the Hamilton District Electrical Association.

## The Lamp Business—Is There Anything Wrong

Vancouver, B. C.

Editor Electrical News:—

In the hope of getting expressions of opinion from other quarters, I am bringing the above matter to the notice of your readers after soliloquized over it for several months.

We sign a contract with a lamp manufacturer to take say, $15,000 worth of his lamps in a period of 12 months which entitles us to a discount of 33% with an extra cash discount of 5% and although many doubtless take advantage of the latter, yet I will show later why it should not enter into the discussion, and it would be fair to call our discount merely 33%—we being in a business on which our bread and butter depends on the number of lamps and other electrical material and appliances we sell.

We then turn 'round to the consumer to sell our lamps and what happens? What are we compelled to do? To any man, whether he be a lumberman, owner of an apartment house, butcher or second hand dealer, if he purchases $150. worth of lamps in a period of 12 months, being one one-hundredth part of the lamps we purchase, we have to give him 17% or a fraction over 50% of our discount. But there is worse to come. If the same type of consumer should purchase $2,500 worth, being one-sixth of the lamps we purchase, we have to give him 29%, or 88% of our discount—and I leave the cash discount we receive out of the question as it is more than offset by the two, three and sometimes four months that we have to wait for our money.

Let us examine for a moment the consumer, and let us

take, for example, the stores of P. Burns & Co. I quote this company solely because I know that they are among those who get these ridiculous discounts: they purchase lamps, not for resale but for the operation of their business, in the same way as they have to purchase wrapping paper, stationery and other things. But are the vendors of these other goods found giving such enormous discounts; or on the other hand, if one buys meat to the value of $150 worth in 12 months as many electrical men do, does one's butcher offer 17%?

This is merely an example of hundreds of similar cases of consumers all purchasing lamps which should be reckoned as part of their overhead expense and which I maintain should be entitled to a maximum of 15% no matter how large a quantity they buy—they are not earning a living by the electrical business.

Look for a moment at another side to this matter and let us ask ourselves "If the present consumers' discounts were cut in half, would the manufacturers suffer? They would only suffer if there were fewer lamps sold, and the possibility of this happening is so remote that it hardly merits discussion.

If my theory is wrong, I shall be very glad to be enlightened on the subject. If on the other hand a wrong exists, it means that real cash is now being handed to the consumer which legitimately belongs to the electrical dealer. If this is true. I believe we have only to point the matter out to the manufacturer in the proper manner, to get it remedied. What, I ask, is the opinion of your readers.

Yours very truly,

E. Brettell

## National Co-operation

March 8 was one of the dates set apart by the Montreal Electrical Co-operative Luncheon for a speaker secured by the Electrical Co-operative Association of the Province of Quebec. The address was by Mr. Walter J. Francis, Montreal, whose subject was the appropriate one of "National Co-operation." He referred to the wide-spread efforts to co-operate by associations throughout the Dominion and stated that co-operation, in its proper sense, was opposed to improper competition. It was the essence of service, helping those who gave as well as those who received. As illustrations of co-operative effort, Mr. Francis mentioned the churches, scientific institutions, and the Boy Scouts. Co-operation had accomplished much in the way of placing professions, on their present high planes.

Mr. Francis discussed, as examples, the co-operative aims of the Rotary and Kiwanis Clubs. He pointed out, as far as the first named is concerned, there are three elements which enter into the service. and co-operation—the individual element, that of service, and that of profit. They were all essential if there was to be that co-operation which would give results. Everyone must co-operate, and the organization must be on broad lines which would make for that end. He was satisfied that the electrical industry was working on those lines, and the results would justify the efforts that were being made.

Mr. Richards stated that the Trades Relationship Committee of the Co-operative Association suggested that there should be monthly social gatherings. These would take place in the evening. and would include a light meal. entertainment and dancing. The idea was approved and left to the Entertainment Committee to arrange details.

The Toronto Hydro-electric System has sent out cards announcing a demonstration of electric cooking. at 1151 St. Clair Avenue West, from 2:00 p.m. to 9:30 p.m. on the dates March 8 to March 25. inclusive.

# 43

# Selling Wiring — It Pays to Specialize

## An Interesting Story of a Contractor Who Systematically Canvasses His Territory— Entirely Separates the Two Operations of "Selling" and "Wiring"

An interesting story of selling wiring at the rate of $10,000, weekly, one salesman keeping ten journeymen electricians busy, is told by James H. Collins in " Electrical Merchandising." Mr. Collins states that a contractor's advertisement, reading "Houses wired for electricity on monthly payments"was the start of the whole scheme. This advertisement caught the eye of a house builder who at the moment was out of employment The idea of wiring houses on the instalment plan interested him, and he said to the wiring contractor "If you are ready to wire houses on that plan, I'll find customers for you."

Between them they worked out an idea. Most wiring contracts were made with people building new houses, but this instalment plan ought to put wiring within reach of thousands of people living in finished houses. Most wiring contract jobs are landed by salesmen for individual contractors working in competition with each other, and there isn't always enough new construction to go around. But suppose the whole community could be canvassed for wiring contracts on old construction—couldn't enough business be found to keep all the contractors busy?

This pair made a study of more than $350,000 worth of wiring installations to learn costs and strike an average, and tried out their plan. It was soon clear that they could save at least 5 per cent on selling expenses. With that, they organized a sales company, which is today operating as an independent selling organization for electrical contractors in the most populous section of New Jersey, and covering a number towns and cities. Since they began last summer more than $150,000 worth of wiring has been signed up in that territory, $50,000 worth of it in the six weeks before Christmas when householders met salesmen with the double-barreled objection of hard times and Christmas spending. Every journeyman electrician in that section is working, where before there was a considerable amount of unemployment.

### Wiring Jobs Without Expense

When this independent sales organization makes an agreement with an electrical contractor to secure for him at least $50,000 worth of wiring jobs without expense to himself, it is so attractive a proposal that he usually signs immediately. In a little while he is working full tilt at wiring, and nothing else. With his own sales force, at least half his attention was given to sales problems. He has to advertise for salesmen, interview applicants, try them out, train them one by one and replace them. If his business ran to even a moderate volume, he would need a sales manager. Under the new plan he concentrates on wiring, and all his wiring and training goes into building up and supervising a more efficient technical force.

This sales organization now has 150 canvassers at work, few of them with electrical training, or even selling experience. Almost any intelligent, willing fellow can be turned into a capable canvasser in ten days, because he is trained and supervised by men who devote all their time to sales work.

"Why hard times is one of our best selling arguments!" said one of the partners. "Last week some of our men began to slack up because people did not want to talk wiring until after Christmas—needed all their money for presents, were busy shopping, and so on. 'How many of you boys have a prospect like that—willing to talk wiring, but

not until after Christmas, I asked. Three salesmen said they each had such a prospect in mind. I went out with those three fellows and closed the first contract in twenty minutes, the second in fifteen minutes, and the third in an hour, In each case the argument was 'hard times.' 'You have the money to wire your home,' I told them. 'There are thousands of men out of work. If you order this job done now instead of waiting, you will have the comfort of electric light and convenience that much sooner, and you will do your share in making a Merry Christmas for other people by putting them to work.'"

### "Hard Times" A Good Sales Argument

"We have emphasized that so persistently that we would miss the hard times argument if there was a sudden prosperity boom. We are not only selling wiring, but trying to find work for men. One salesman can keep ten journeymen electricians at work. Besides 700 or 800 additional mechanics who have been given employment by our organization, many more have been taken on by the electrical equipment factories and the central stations. Hard times talk is a good sales argument for us, but apart from that I have no patience with it. Hard times is largely marking time—if people would stop talking about depression and get busy, they would find, as we have, that there is money to be spent and business to be got if they will go after it."

Under the agreement made with electrical contractors, this sales organization promises $50,000 worth of wiring and fixture sales within twelve months, $16,000 worth in four months. Moreover, this business is to be found entirely in finished houses, so it does not interfere with new construction jobs. The contractor pays $200 bonus when he signs, furnishes desk room for salesmen, and pays a sales commission weekly on all business turned in, settlement being made each Saturday for contracts closed up to the previous Wednesday. The sales organization operates independently of the contractor and also of central station, manufacturing and jobbing interests. It advertises in the daily papers, offering house wiring on the instalment plan. The contractor also agrees to put a certain amount of advertising on his motor trucks and get behind the general idea.

### Combing Neglected Territory

Probably the strongest point in this sales plan is that canvassers work in comparatively neglected territory, and cover it thoroughly. Though many households have been solicited by contractors' salesmen, most of their sales activity has been concentrated on new construction, and in seeking business among people living in unwired houses, there has been a tendency to pick and choose the most promising prospects. This organization pulls every door bell in the block, the only prospects passed by being people living in rented homes in sections where rents obviously will not permit an outlay for wiring.

An old sales story in nearly every line of business is that about the salesman who complained that his territory, a mile square, was "worked out." Whereupon his boss gave him a single block, saying, "Let that be your territory," with the outcome that the salesman had to visit every house and office in that block, and found more business than he had got from his square mile of territory. That this principle will work in wiring is shown in the fact that even a metropolitan section like Newark has three unwired houses for every one

that is wired. Estimating 100,000 homes in the city and that
a contract is secured in one of every ten, at the present rate
of working, it will take this sales force five years to cover
Newark alone, and when the job is done, it could start right
over again!

### Householder Invariably Interested

Calling at an unwired house, the salesman explains that
he represents a group of electrical contractors, and would
appreciate the opportunity of examining the house and sub-
mitting an estimate for wiring. The householder may not
be ready to have the work done then, but he is usually interested in knowing what it will cost. The salesman has
average figures, and also a schedule blank covering property
floor by floor and room by room. When this is filled out,
it gives a record of the number of outlets and fixtures, as well
as the character and condition of the building. A rough
estimate can be given on the spot, with the distinct understanding that it is only an estimate—a definite price will be
submitted if asked for, and it may be lower. As a matter
of policy, average wiring prices in a given community are
neither cut nor raised—they vary in different sections according to wages and other conditions. Generally, the
estimate is lower than the figure most people have in mind
for the cost of wiring their homes—so much so that one
difficulty is persuading them that a satisfactory and sightly
job can be done for the price. Objections are raised—the
householder has heard that fires are caused by rats gnaw-
ing exposed wires, and does not want his home burned up.
When he finds that all work is done with conduit and
first-class material, and that the job must conform to fire
insurance, municipal and central station requirements, he is
reassured.

### Sell Electric "Service"

One significant point is that salesmen do not talk or
sell electric lighting, but electric service. Householders almost invariably think of wiring in terms of lighting. Electric service with additional outlets for appliances is a viewpoint so new and interesting that they stop thinking of
cost for the time being and discuss possibilities with the
salesmen, deciding where outlets for fans, irons, heaters and
motors will be most convenient.

The salesman's distinction between lighting and "luxury
outlets" makes a decided impression on most prospects.
The idea of electrical luxury as well as light is novel, and
attracts by its suggestion of greater convenience and comfort
in the home. Moreover, it is a basis for selling more outlets where the home is already wired.

Asking for permission to look over a home that has
been wired for lighting, maybe years ago, the salesman almost invariably finds twin plugs in many lamp sockets, some
placed for use with fans, curling irons and the like, with
others supplying auxiliary lights that have been found necessary as lamps and candlepower costs have been cheapened.
It is often possible to show that such additions to the lighting equipment are hazardous and secure a contract for
wiring that will give additional lights with luxury outlets.

### Average Contract, $250

The average contract in territory thus far canvassed
comes to about $250, providing about twenty outlets with
fixtures. The average value of property is about $7,000, so
that credit and collections involve practically no problems.
Thus far, collections have been 100 per cent, and the very
few contracts rejected have been in cases where a householder's equity in his property was not large enough to give the
moderate security required.

### Financing

All jobs are financed through a finance company, the

Contract Purchase Corporation, so that contractors and the
selling organization have practically no capital tied up.
This corporation's method of financing the instalment wiring of homes, as well as consumer purchase of the more
costly appliances, is doubtless familiar now to electric men.

Briefly, six separate corporations cover the country, with
headquarters in New York, Philadelphia, Cleveland, Chicago,
Dallas and San Francisco. Having closed a wiring contract, the account is sold to the purchase corporation by the
contractor for cash, at the full price for the job, the householder paying down a first installment that covers moderate
interest. The contractor acts as collection agent for the
purchase corporation, receiving a commission for that service when the last payment has been made.

### Middle-Class Are Best Customers

The best customers for wiring are middle-class home
owners—the busy, foresighted, thrifty-spending and neighbor-
ly people, "just folks," who appear to have been overlooked
by the electrical contractor's salesmen because no way
had been contrived to canvass them collectively. Once signed
up, they are something better than good individual customers. The signature on the contract is hardly dry before
they want the wiring job begun. The contractor agrees to
start each job within ten days after credit has been approved,
and to complete it within reasonable time. The householder
is usually right on his heels, keenly interested in details.
One of the difficulties that must be overcome by salesmen
is the apprehension that wiring means several weeks' work and
a torn-up house.

When work is begun, and the householder sees with his
own eyes how conduit is run between partitions and floors,
he is more than delighted. He becomes a volunteer salesman for modern wiring. He explains to the neighbors just
how it is done, and talks about the "luxury outlets" and invites them in to see how convenient and up-to-date a wiring
job can be. Thus, one contract leads to others, and the salesman and contractor usually benefit because the householder
almost invariably advises neighbors to place their work with
the same people—the salesman took such pains to see that
the householder got the worth of his money in electrical
service, and the contractor was so prompt and careful in doing the work, and so forth.

### Demonstration of Possibilities

If it takes five years to cover a city like Newark, wiring
only one house out of twenty in the first canvass, new business ought to gravitate constantly and automatically to the
contractor who does satisfactory work. That one house in
twenty becomes a demonstration of electric possibilities, and
the householder is a self-appointed and well-disposed demonstrator working among his neighbors. As a matter of experience, even in the short time this campaign has been carried on, contractors have found enough people interested in
wiring jobs to make up a list of live prospects for the sales
force.

---

Mr. Harold B. Fisk, manager of the Walsh Plate &
Structural Works of Drummondville, Que., was also elected
secretary of the company for the ensuing year, at the Annual Meeting held on January 25th. Mr. Fisk was recently
admitted an associate member of the Engineering Institute
of Canada.

---

The contract for the erection of a building to be used
by the Walkerville Hydro-electric System as offices and
sales rooms, at Argyle Road and Wyandotte St., Walkerville, has been let to the Windsor Construction Co., Ltd.

Notice that the maximum of illumination is produced at the level of the desk surfaces. Within the private offices note the bright effect below with comparative darkness above. This all results from choosing proper units and spacing them properly at the correct heights.

## Two Illustrations of Good Lighting

### If We Want to Be Efficient in Our Business We Must Have Proper Working Conditions

On this page we show two illustrations of pleasing and effective illumination. One is the interior of a bond house, recently fitted up. The desks and all woodwork shown are of dark finish and the photograph is taken without any auxiliary lighting. It is particularly noticeable that the maximum of illumination has been produced in the neighborhood of the desk tops. For example, looking into the private offices, one notices at once the higher illumination on the walls just above the desk than is apparent on the walls higher up. This result has been produced by proper choice, spacing and adjustment of the lighting units.

The other picture represents the store front of George Wilkinson, electrical merchant, Toronto. Note the detail with which everything in the window is portrayed. Mr. Wilkinson states that this window is proving of double value to him commercially. His appliances are selling very much more rapidly because they are well shown, and he is also doing a nice business in putting in lighting installations, similar to his own, for other merchants.

Both of these pictures represent results from Holophane equipment, in the skilful hands of Frank T. Groome. Mr. Groome is now devoting his entire time to illumination. It is a big field with possibilities for almost unlimited development. In a country where electric current is cheaper than anywhere else in the world, we have some of the most lamentable exhibitions of inadequate illumination. Work such as Mr. Groome is engaged in has a moral as well as a commercial side.

The store front of George Wilkinson. It has sold both appliances and lighting contracts

## Suggestions for the Window Trimmer

Some time ago, there appeared in the Westinghouse "Contact" a number of suggestions for trimming windows. They seemed so complete that we felt our readers also would like to see them.

1. Keep your windows clean. Have them washed frequently. If your own employees haven't time, get outside help. It will pay.

2. Don't overcrowd your window.

3. Avoid the other extreme. Too little in a big window will cause the merchandise to be "lost".

4. Card holders are useful. They'll keep price cards from falling over on their faces.

5. Make your display attractive to the eye—and the purse —but don't make it so "pretty" the merchandise is forgotten in admiration of the "trimmings".

6. Make your store front reflect you. It is the exterior which most people see. Impressions are made by exteriors.

7. Put the emphasis on the goods, not on the decorations.

8. Use art only to create a desire to buy the goods displayed.

9. Be sure your window lighting is the best obtainable.

10. Have the backing of your window high enough to shut off the view of the store interior.

11. Use a dark color in the background when displaying light colored goods, and vice versa. Get contrast.

12. To express coolness in a winddow use gray, light green or light blue for the color scheme

13. To show warmth use reds, yellows, oranges—warm colors.

14. Dust out the window space frequently.

15. Never allow soiled or flyspecked cards or merchandise to remain on display.

16. To help the eye to travel quickly from a card to the object displayed, connect the two with white tape or ribbon. An arrow will have the same effect.

17. Invest a little money in stands on which to display your merchandise. It will pay.

18. Empty cigar boxes make good "building blocks" to erect most any size or shape foundation for a display.

Crepe paper, bunting and cheese cloth are inexpensive coverings and draperies.

20. Make your display fit the season.

21. Get ideas from merchants in other lines of business.

22. Plan your displays ahead—days and even weeks ahead.

23. Get all material ready for the new arrangement before the old display is taken out.

24. Keep a "window note book". Jot down in it ideas you see which you may use later.

### May Electrify Planing Mill

Thurston-Flavelle, Ltd., manufacturers of cedar lumber at Port Moody, B. C., recently purchased a Goldie-McCulloch cross compound engine and direct connected generator from the electric light department of the city of Edmonton. This will be used to supply additional power for the new gang saw being installed by the lumber company. About 200 h.p. will be developed in the meantime, which will probably be added to in years to come if the company decides to electrify their planing mill, which they have considered doing, using group rather than individual motor drive.

### New Books

Practical Electrical Engineering; Direct Currents; by Harry G Cisin, M. E., engineering editor "Electrical Record." D. Van Nostrand Company, New York, publishers. This book is described as "a manual for industrial and evening classes and for home study." The basic principles of direct-current electrical engineering are presented, simplicity being the keynote. The chapter headings will indicate the scope of the work:

1. Electricity and Magnetism; 2. Fundamental Laws of Electrical Engineering; 3. Electrical and Magnetic Circuits; 4. Electrical Instruments; 5. Direct Current Measurements; 6. Primary and Secondary (Storage) Cells; 7. The Electrical Generator; 8. The Direct-current Motor, with an appendix on laboratory experiments. 68 illustrations; 305 pages; size, 5 in. x 8 in.; stiff green cloth binding. Price $2.00.

Mechanical Trades, Ltd., 54 University Ave., Toronto. have secured the contract for electrical work on a warehouse to be erected for Barber-Ellis, Ltd., at 370-374 Adelaide St. W., Toronto. at an estimated cost of $120,000.

Northern Electric officials and members of Toronto staff in recent convention in Toronto. Note the determination in every face to make 1922 the "best yet."

# A "Pure Food" Cafeteria

## Hamilton Can Boast the best Equipped Restaurant—Electrical Throughout

A "Pure Food" cafeteria which was opened recently at 52½ King Street E., Hamilton, is equipped with most modern kitchen and restaurant equipment. It goes without saying, therefore, that the kitchen equipment is electrical throughout. It consists of: one electric bake oven; one electric hotel range; one electric steam table; one electric hotel toaster & griddle; three electric urn heaters; and two electric water heaters.

### The Oven

The electric bake oven, with three separate decks, each individually controlled by separate three-heat switches, has a capacity of twenty-five 1½ lb. loaves per bake. With this oven, one complete bake is possible every forty minutes when bread only is baked. When an assortment of bakery goods is handled, it is possible to show a daily output such as follows: 60 pies; 1 19-in. x 27-in. pudding; 40 meat pies; 150 rolls; 3 doz. tea biscuits; 6 doz. cookies; 1 pan macaroni; 1 pan spaghetti; 12 cakes and 50 lbs. meat.

As each compartment is individually controlled, it is possible to bake three different products in the different compartments at different temperatures, all at the one time.

### The Range

The heavy duty range is of a capacity to take care of 300 fancy or a la carte meals at a sitting, and from 350 to 400 table d'hote meals at a sitting. This range is equipped with the latest development sheath wire units, which are specially built to withstand the heavy service of hotel and restaurant work.

### Other Equipment

The electric steam table is equipped with units of 6 kw. capacity, which insures an even distribution of steam. The toaster is of 16-slice capacity and toasts both sides of the bread at one time. The top of the toaster is fitted with a special polished steel top, for the preparation of griddle cakes, etc. The electric urn heaters are equipped with 3-heat switches, and are used to heat the coffee urns installed. Hot water is supplied from a tank conveniently located in the main kitchen, the water being heated by two sheath wire water heaters, of a total capacity of 6 kw., which assures a continuous supply of hot water.

The total connected load of the electric kitchen equipment is 52½ kw. Power is supplied by the Cataract Power, Light and Traction, Ltd., Hamilton, who installed one 50 kw. transformer, and a 3-wire, 110-220 volt service of 4/0 cable.

All of the electric equipment was manufactured by the Canadian Edison Appliance Co., Ltd., Stratford, Ont. Mr. Sariotis, the owner of the cafeteria, is highly pleased and justly proud of his installation and the service he is now enabled to render. Cuts of the kitchen and self-serve dining room are shown herewith.

## Laboratory Lightning Storms

The General Electric Company have recently perfected, through the efforts of Dr. C. P. Steinmetz, an equipment for producing actual lightning in the laboratory. The practical value of this will be seen in the means that will thus be provided for testing lightning arresters, under what will amount to actual storm conditions. The voltages used do not constitute a record in any way—120,000 volts—, the necessary energy being produced by storing up capacity in a condenser consisting of 200 large glass plates. Dr. Steinmetz points out the essential difference between high voltage and lightning. Lightning is a high voltage coupled with a heavy amperage, the discharge lasting for a very short time and so giving explosive effect. In the lightning "generator" Dr. Steinmetz gets a discharge of 10,000 amperes at 120,000 volts, which means something over 1,000,000 h. p., but the length of time it lasts is only one hundred thousandth of a second. It is interesting to note, in this connection, that while we are far from producing the amount of power represented by a natural lightning flash, we are not so very far from equalling the voltage represented by it. Dr. Steinmetz represents the voltage of a lightning flash at around 50,000,000 volts. Recent laboratory experiments have developed apparatus that will produce 1,000,000.

## Radio is Popular

That radio telephony is growing in popularity has been quite apparent for some time past. This fact was emphasized by speakers before the annual convention of the Electrical Supply Jobbers' Association, held at Atlantic City, N.J., who stated that manufacturers are unable to keep pace with the demand for apparatus and that "radio telephony is sweeping the country with a popularity that is a revelation."

Kitchen with Electric Bake-oven, Heavy Duty Range, Water Heater and other equipment

The Electric Steam Table has units of 6 kw. capacity. The toaster does 16 slices at once. Electric Urns shown in backgrown.

# The Latest Developments in Electrical Equipment

The Killark Electric Manufacturing Co., 3940.48 Easton Ave., St. Louis, Mo., have recently put the rectangular push button here illustrated on the market. This push button is made in either solid brass or copper, with a brush satin finish;

contact parts are all spring brass. Damp or salty atmospheres do not corrode or damage it. The manufacturer states that while the design of this button is suitable for all types of houses, they find there is a particularly heavy demand for it by builders of the bungalow type of homes.

### All-porcelain Tumbler Switch

The Canadian General Electric Company are marketing a new all-porcelian tumbler switch, as shown. This switch is specially designed for use in damp places, such as laundries, basements, factories, etc. The insulation af-

forded by the porcelain cover is, of course, superior to that of brass. Another interesting feature claimed for this switch is that it is the only all-porcelain tumbler switch made in the Dominion.

### Short-handled Vacuum Brush

The O. K. Machine Co., Inc., Fairfield Ave. & Poplar St., Fort Wayne, Ind., have placed the "O. K." Vacuum Brush on the market. This is a complete machine in itself—not an attachment—and is specially designed for cleaning upholstering, draperies, clothing, billiard tables, etc. and has been found almost indispensable in dry cleaning

establishments. The brush is driven by a motor, controlled by a thumb operated switch on the motor case. The dust bag is placed inside the short handle and is quickly emptied by removing a metal cap on the rear of the handle and turning on the motor; a larger bag, which fastens over

the end of the handle, may also be used. The machine is equipped with a universal motor and operates on 110 volts a. c. or d. c. The brush can be used with one hand as it weighs less than three pounds.

### Tumbler Water Heater

The tumbler water heater, manufactured by the Westinghouse Company, is designed to heat a small quantity of water quickly and efficiently. The product is substantially and durably constructed, making it practical for heating any small quantity of liquid. The element is inserted in copper tubing bent to afford a large heating surface. It is nickel-plated and highly polished. The through-switch of black composition is extra large to provide a convenient rest as well as a practical switch for the heater. The voltage of the heaters ranges from 110 to 120 volts and the wattage is 350.

The net weight is one pound. The heater has three distinctive features; in that it will not tip the tumbler over; it will not break the glass, and it will not burn or scorch the article on which it is laid. When it is placed in a glass of water, it rests on the base of the switch, thus preventing the weight of the heater from tipping the glass. The switch also affords a rest for the heater when it has been removed from the liquid. It will not break the glass because it does not come in contact with the glass. The danger of burning or scorching is eliminated by laying the tumbler heater so that the tubing extends in the air, thus making it impossible for the table upon which it is lying to be scorched or burned.

### A Smaller Size Curler

The Northern Electric Company, 224 North Sheldon St., Chicago, Ill., manufacturers of electric heating appliances, are now manufacturing a smaller sized curling iron than their standard No. 55. This iron is 10½ inches by 5/16 inches and also has the removable clamp feature. The same company are manufacturing a 3-heat heating pad, 12

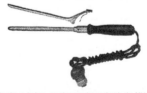

inches by 15 inches. Another popular product is the Midget toy flat iron, with a weight of 1½ lbs. and a consumption of only 27 watts. This is strictly a toy for children, but is nickel plated and well made; complete with cord and plug to use on standard current, "just like Mother's."

## New Cleaning, Waxing and Polishing Machine for the Home

An entirely new idea in electrical equipment is announced in this issue of the Electrical News. It is an electric floor cleaner, waxer and polisher combined in one. This is not the first time, it is true, that floor polishers and waxers have been placed on the market, but up to the present time, none of them have proven entirely satisfactory. The new machine is claimed to cover the faults of all previous designs, and to have many added features. In short, it is a machine which the manufacturer claims has been fully tested out and will do all that is claimed for it, namely, it

It Cleans Waxes and Polishes

will clean our hardwood floors, then it will wax them and, finally, it will polish them—the whole operation being so simple, and the machine itself being so light and well balanced, that the operation may be conducted by a child.

The main principle underlying the construction of this machine is the revolving disc, floating on its bearings in such a way that it adjusts itself to any little unevenness in the floor. The machine is supplied with two discs, (either pad or brush, depending on the operation under way) belt-driven from the motor. In the initial operation, one of these discs would apply a cleansing material; the other disc following it would rub this material into the floors, absorbing and drying at the same time. These discs would then be replaced—the work of a few seconds, only—by a waxing pad and a distributing brush or, under certain circumstances, a polishing brush, or this latter operation may be deferred until later, when a coarse and fine polishing brush may be used together so as to leave the floors in the finest possible condition.

Another interesting feature about this new equipment is that it is manufactured in Canada, by the Canadian Floor Waxer & Polisher Co., Ltd., Toronto. Deliveries will be made from the middle of March. The two principals in this firm are **Mr. J. Skelton** and **Mr. R. N. Connor**, both already well known to the trade as Hoover enthusiasts. Mr. Skelton has been Toronto representative of the Hoover Suction Sweeper Company for twelve years and leaves them with the greatest regret and only because he is convinced that his new waxer and polisher will do all that is claimed for it and that he will thus be in a position to supply the public with an electric appliance for which there is a very urgent demand. Mr. Connor has also been with the Hoover company for many years. He is sales manager of the new organization. The headquarters of this company is at 666A Yonge Street.

## Mr. Tufford Goes to Hamilton

Mr. A. A. Tufford succeeds Mr. J. A. Daly as manager of the Hamilton office of the Northern Electric Company, Mr. Daly, as recently announced, having been promoted to Regina. Mr. Tufford is a graduate in electrical engineering of the University of Toronto, obtaining his degree of Bachelor of Applied Science in 1917. Previous to gradua-

Mr. A. A. Tufford

tion he had spent some time with the Canadian Westinghouse Company, Hamilton, and in the engineering department of the Hamilton Hydro-electric System. Following graduation he joined the Northern Electric, but in 1918 received a commission in the Canadian Engineers and went overseas. On his return the Northern Electric Company sent him to Vancouver, where he spent two years, returning to Toronto office in November last. Mr. Tufford took charge of the Hamilton office on February 15.

Mr. J. Skelton    Mr. R. N. Connor

## Ventilation of Electric Sub-stations

By A.M. STEPHEN, A.S.H. & V.E.
Of Stephen & Boyle Ltd, Heating & Ventilating Engineers, Vancouver, B.C.

A peculiar and unique problem in ventilation is afforded by the electric power station. The excessive heat generated by the transformers, which in many cases are loaded to capacity or over, in combination with the temperature of the air on warm days, makes these buildings extremely unsanitary and oppressive for those working there. In

**Ventilating New Westminster Station**

some cases the machines are cooled by the use of blowers but there is always the objection that dust and dirt may thus be introduced. The more practical method would seem to involve the removal of the over-heated air at a sufficient rate, thus allowing for the supply of cooler, fresh air to take its place. Fans sufficiently powerful to handle this super-heated air at a rate sufficient to keep the air cool mean draughts dangerous to the health of the employees.

At the B. C. Electric sub-station at New Westminster, B. C., a cut of which appears on this page, the problem of ventilation is solved by means of the Autoforce air pump, four of which are installed on the roof of the sub-station. Each of these machines has a definite displacement in cubic feet of air per minute, making it possible to have a complete change of air in the building every fifteen minutes. Their construction is such that there is no possibility of down-draft and besides making conditions more comfortable for the staff in charge, the generators and transformers are kept from overheating since the air is kept in motion and the surplus heat removed as soon as generated.

### Another Pioneer Gone

Another pioneer in the electrical industry has passed away, in the person of Mr. A. A. Wright of Renfrew. Mr. Wright as president of the Renfrew Electric Company was also one of the earliest presidents of the Canadian Electrical Association. At one time he represented his constituency in Parliament. Locally, he was always spoken of as the town's most prominent citizen. Mr C. H. Wright, Halifax, Manager, of the Canadian General Electric Company is a son.

Electric alarms will be installed at railway crossings in Tavistock, Ont. early this spring, by the Grand Trunk. Mr. W. J. Piggot, Stratford, Ont., will supervise the work.

### Will Merchandise, Also

We are advised that The Western Quebec Power Co., Room 603, Drummond Building, Montreal, has the intention of adding merchandising to their activities as an accommodation to the many customers in the territory already covered by its system, which includes to-day about 15 municipalities. This latter figure is expected to be raised to 25 within a short time.

Mr F. D. Reaume, 1015 Wyandotte St. E., Windsor, has secured the contract for electrical work on a hotel building at Chatham & Ferry Sts., Windsor, that is being altered, to a newspaper plant, for the Windsor Telegram Company.

The Northern Electric Co., Ltd., 599 Henry Avenue, Winnipeg has been awarded the contract, by the city of Winnipeg, for a quantity of lead covered cable.

The Ontario Gazette announces that Letters Patent have been issued to a corporation to be known as The Monarch Battery Co., Ltd., capitalized at $40,000. Head office; Kingston, Ont.

Fire losses in Canada during the week ending February 8, are estimated by the Monetary Times at $736,500, compared with $1,163,000 the previous week.

It is stated that Ottawa labor in building trades will be asked to accept a general cut of 10 cents an hour this year, although no agreements have been reached as yet.

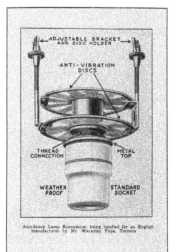

ADJUSTABLE BRACKET AND DISC HOLDER

ANTI-VIBRATION DISCS

THREAD CONNECTION

METAL TOP

WEATHER PROOF

STANDARD SOCKET

Anti-break Lamp Economizer being handled for an English manufacturer by Mr Macaulay Pope, Toronto

*Views during construction of Queenston-Chippawa Canal*

view of concreting plant showing how the walls were built up in alternate sections

Placing concrete lining in the Whirlpool section of the canal showing various stages of construction

# Current News and Notes

**Campbellton, N. B.**

The town council of Campbellton, N. B., are considering the placing of all electric light and telephone wires underground. Mr. H. G. V. Farrer is electrical engineer.

**Edmonton, Alta.**

By-laws will be submitted to the rate-payers of Edmonton, Alta., regarding the following expenditures: $250,000 for telephone exchange and equipment; electric light extensions $12,500; power plant extension $115,000.

**Ft. William, Ont.**

At a joint meeting of the Utilities Commissions of Pt. Arthur and Ft. William, Ont., recently, a resolution was passed favoring the use of one-man street cars in these cities. If the approval of the Railway Board is obtained it is understood about twenty cars, in each city, will be changed over for one-man operation.

**Kitchener, Ont.**

The Kitchener Board of Trade recently endorsed the proposal of the Waterloo & Wellington Street Railway Company to extend the company's line through to Guelph, via Bloomingdale and New Germany, providing such action is taken within two years' time. At present the line runs from Kitchener to Bridgeport.

**London, Ont.**

At the annual meeting of the London Street Railway Company, held recently in London, Ont., it was decided to approach the city council, at an early date, with a request for an increase in fares in order that needed improvements might be made. For the year 1921, the following figures apply: Gross earnings, $567,866.26, an increase of $43,137.96 over 1920; operating expenses,$487,344.13, an increase of $32,786.78 over 1920; wages, included in operating expenses for 1921, amounted to $351,476.24, an increase over the year 1920 of $22,542.93. The net earnings from operation amounted to $80,522.13, and after deducting fixed charges and depreciation, the net income for 1921 amounted to $11,512.90.

The Northern Electric Co., Ltd., at present located in the Home Bank Building, London, Ont., have leased larger quarters at 175 King St., London, which they are having remodelled and fitted up for offices and show rooms.

**Montreal, Que.**

Messrs. J. Ortiz & Co., 1395 Bordeaux St., Montreal, have been awarded the contract for electrical work on a store building recently erected at Masson & Chabot Sts., Montreal, for Mr. T. Lefebvre. 292 Clifton Avenue.

**Mount Royal, Que.**

Mr. H. H. Cassidy, 355 Regent St., Montreal, has been awarded the electrical contract on 24 residences to be erected early in the spring, at Mount Royal, by the Rockland Housing Co., at an estimated cost of $150,000.

**Oakville, Ont.**

Mr. F. Sullivan, Oakville, Ont., has been awarded the contract for electrical work on a service station recently erected on Dundas St., Oakville, Ont., by the White Rose Oil Company, of Toronto.

**Ottawa, Ont.**

The Canada Gazette announces the incorporation of the Brazilian Hydro-electric Company, Ltd., with a capitalization of $5,000,000; head office, Toronto. The new firm is authorized to carry on hydro-electric developments throughout the Dominion.

Officials of the Hydro-electric System, Ottawa, Ont.,

are having plans prepared for a sub-station, which they propose to build this spring in the south part of the city, the cost of which is estimated to be in the neighborhood of $200,000.

That the Ottawa Hydro-electric Commission passed through a most successful and profitable year in 1921 is indicated by the annual statement and balance sheet submitted by J. A. Ellis, chairman of the commission, which shows a gross surplus for the year of $861,791.96, after deducting all charges for maintenance and operation, power, interest, and sinking fund on $700,000 debentures. The operating report shows a revenue for the year of $328,108.97.

**Palmerston, Ont.**

The Library Board, Palmerston, Ont., are contemplating a complete new installation of electric wiring and lighting fixtures in the public library at that place and are asking for prices.

**Quebec City, Que.**

The Quebec Railway, Light, Heat & Power Company, Quebec City, recently received five of the ten new double truck cars ordered some time ago. These cars will be placed in commission in the very near future.

**St. Thomas, Ont.**

The city council of St. Thomas, Ont., propose an expenditure of about $25,000 in hydro improvements and have asked for the approval of the Railway Commission. It is planned to enlarge the Wilson Avenue sub-station and install a new transformer; also to install a number of transformers at other points.

**Stratford. Ont.**

The Stratford Public Utilities Commission have opened new showrooms and offices on Ontario St., Stratford, Ont. Mr. R. H. Myers is manager of the Public Utilities Commission, and Mr. E. W. Tobin is manager of the Hydro Shop, which also includes a staff of three salesmen.

**Toronto, Ont.**

The Canada Electric Co., 175 King St. E., Toronto, has secured the contract for electrical work on an office building to be erected at Bay & Temperance Sts., Toronto, for the General Accident Assurance Company of Canada, Ltd. Continental Life Building, Toronto. It is estimated that the cost of this building will be in the neighborhood of $500,000.

Messrs. Douglas Bros., 2137 Yonge St., Toronto, have secured the contract for electrical work on a tabernacle being erected on Christie St., near Bloor, Toronto, for the Christian & Missionary Alliance.

Messrs. Taylor Bros., 23½ Norwood Ave., Toronto, have been awarded the contract for electrical work on a store building being erected at Danforth and Westlake Avenues, Toronto, for the Wickham Hardware Co., 2424 Danforth Ave., at an estimated cost of $18,000.

Messrs. Nichol & Fagen, 411 Logan Avenue. Toronto, have secured the contract for electrical work on four residences being erected at Balfour Ave. near Dawes Road, and Barrington Ave. & Colemen St., Toronto, by Mr. T. A. Lewis, 77 Balfour Ave.

**Vancouver, B. C.**

Mr. J. C. Reston, 411 Howe St., Vancouver, has been awarded the contract for electrical work on an addition and alterations being made to the factory and store building of Jas. Reid, 559 Granville St., Vancouver, at an estimated cost of $25,000

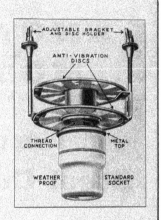

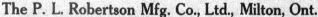

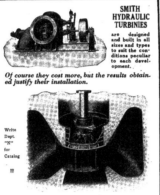
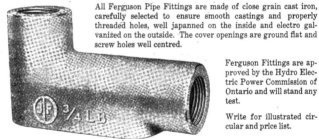

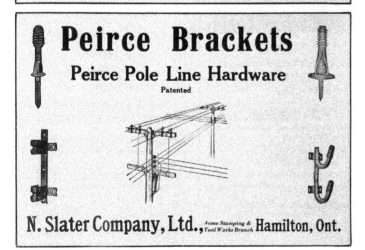

# CLASSIFIED INDEX TO ADVERTISEMENTS

The following regulations apply to all advertisers:—Eighth page, every issue, three headings; quarter page, six headings; half page, twelve headings; full page, twenty-four headings.

# CLASSIFIED INDEX TO ADVERTISEMENTS—CONTINUED

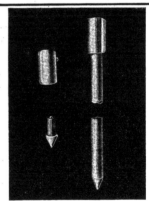

## CLASSIFIED INDEX TO ADVERTISEMENTS—CONTINUED

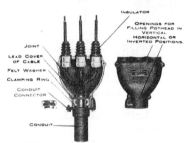

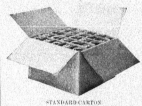

Vol. XXXI No. 7

Toronto, April 1, 1922

# "Canadian" Ironer
### Made in Canada

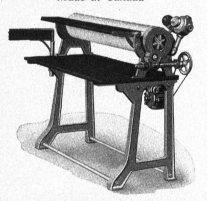

The "Canadian," Electrically Heated, Motor Driven
Also, Gas and Gasoline Heated
*Approved by the Hydro Electric Power Commission, June 14/21.*
L. A. No. 580

Write for Revised Dealers' Prices and Discounts on
Ironers, also Electric Washers and Electric Clothes Dryers

Manufactured only by

## MEYER BROS., 101 Queen St. E.
### TORONTO  -  CANADA
*Established 1884.*

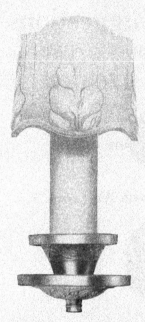

# Hundreds of Fuses
## —but only one
# "Noark"

E LECTRICAL men who measure fuse costs on circuits — instead of on paper— continue to demand *indicating* non-renewable fuses.

The indicator didn't get on the fuse by accident—experience with fuses demanded it years ago. To deny this is to slide backward twenty-five years.

It is more than convenience insisting that a "Noark" fuse carry an indicator, for a decade of records shows that it is an ultimate economy which more than offsets the slight additional first cost.

CANADIAN JOHNS-MANVILLE CO.,
Limited

Toronto   Montreal   Winnipeg   Vancouver
Windsor   Hamilton   Ottawa

*Through*
## ASBESTOS
and its allied produc'

Electrical Materials
Brake Linings
Insulations
Roofings
Packings
Cements

Fire
Prevention
Products

# JOHNS~MANVILLE
# ELECTRICAL MATERIALS

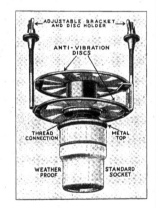

# How much shall I Charge?

O NCE—twice—twenty times a day this all important worrisome question comes up. A customer comes in the store for a repair part of an iron, or your man is installing a 60 amp. service box and necessary materials. You don't know what you ought to charge and no one else does without figuring it out.

## It's Some Task

You want to be fair to the customer and yet you must have a reasonable profit above overhead expenses. First you must procure the lists and discounts, and after getting your net price must then add a percentage for overhead, and another for profit.

## You Risk Your Profits

No matter how careful you are, or how much time you take, there is always the chance of the information in hand being out of date and very often you are too busy to take the time to figure out all those retail prices, with the result that they are often guessed at, and if you yourself are not in the office, there is no telling what prices will be used.

## Henderson's Price System

Gives you the correct retail price any time of the day on a second's notice, on Two Thousand electrical items, such as Conduits — Wire — Condulets—Sockets—Receptacles—Appliances—Repair parts, etc.

## The Cost of the Service

This wonderful service which includes a Leather-covered Binder, and is kept right up to date, week in and week out, costs only about 70c per week. Hundreds of the best contractors all over the country are using this service.

## Why Not Try It Out for a Year?

Even if it's only half as good as we claim it is easy worth the 70c.

## The Henderson Business Service Ltd.

Box 123          Farmers' Building
Brantford     -     Ontario

# ALPHABETICAL LIST OF ADVERTISERS

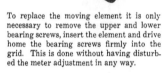

Front View
Male Type Outlet Box used in
straight installations with

# "TRICABLE"

### For

## Economy Wiring

Write for information.

*Selling Agents—*

## CONDUITS COMPANY LIMITED
### 33 Labatt Ave.   -   -   Toronto.

# Westinghouse
### DISCONNECTING     SWITCHES

**600 ampere, 135,000 volt Disconnecting Switches**

These 600 amp., 135,000 volt, 3 pole, single throw, disconnecting switches, embody a number of special features which are of particular interest.

The porcelains are the largest one-piece pillars of this type yet built, being 42½″ in height. Flexible multiple finger type contacts are employed and are enclosed in a corona shield.

They are designed for remote manual or electrical operation as a three pole unit and may be mounted in any desired position. The centre insulator of each pole rotates on double race ball bearings which insures very easy operation.

Fifty-eight of these three pole switches, comprising the total 135,000 V. disconnecting s w i t c h equipment for Queenston, are being furnished by the Canadian Westinghouse Company.

In addition to the large generators, transformers and oil circuit breakers illustrated in the preceding pages, a great deal of the auxiliary and detail apparatus for this installation was manufactured in the Canadian Westinghouse Works at Hamilton, Ont. This consists of the following:—

      wo 2200 kv.a. vertical water wheel service generators, having characteristics of 3 phase, 2200 volts, 25 cycles.
      All of the 2200 and 550 volt oil circuit breakers on the station service feeders.
      All of the large electrically operated field discharge switches.
      All of the 2200 volt and 550 volt disconnecting switches on the station service circuits.
      All of the control switches for the numerous electrically operated circuit breakers, field switches and rheostats.
      Also a large number of instrument transformers, meters and relays.

## Canadian Westinghouse Co., Limited, Hamilton, Ont.

TORONTO, Bank of Hamilton Bldg. MONTREAL, 285 Beaver Hall Hill    OTTAWA, Ahearn & Soper, Ltd.
HALIFAX, 105 Hollis St.       FT. WILLIAM, Cuthbertson Block     WINNIPEG, 158 Portage Ave. E.
CALGARY, Canada Life Bldg.      VANCOUVER, Bank of Nova Scotia Bldg. EDMONTON, 211 McLeod Bldg.

COMPLETE satisfaction is the rule wherever R & M Fans are used—whether in the office, home, store, theatre or hotel. And because of the satisfactory service they always give, the dealer benefits not only to the extent of his profits on fan sales, but to a considerable degree also from the good will they help build for his business in general.

This is the reason R & M Fans are sold so extensively in the stores of those dealers who are building their businesses for permanence, on a quality-service basis.

## The Robbins & Myers Company
## of Canada, Limited
Brantford        Ontario

# Robbins & Myers Fans

# Why Elpeco Indoor Bus Supports are Designed for Safety First

Elpeco
Indoor
Bus Support

YEARS of experience handling high tension transmission, designing and installing station and switchboard apparatus have, taught us that human life and thousands of dollars of property are dependent upon the right design and the careful building of each individual support.

These supports are too important to put through the ordinary processes of manufacture in quantity. We believe the Electrical Industry will always 'pay the slight extra cost necessary to insure safety.

Designing-board-theory is insufficient to measure the great electrical and physical stresses put upon Bus Supports in High Tension work. Elpeco apparatus is built from experience on the ground—from the facing of hard facts.

Every Elpeco Bus Support is subjected to special high tension test and passed before leaving the factory.

To balance physical stresses, center and offset tongues are provided to meet even and uneven bars. All metal parts are cemented to porcelain. There is no danger of the melting of low temperature metals causing short circuits. Each support has inter-changeable base for mounting—pipe or flat. Adjustable for any bus position.

The Elpeco trademark—the Crusader with two shields—stands for specialized knowledge of high tension apparatus and its installation.

We will be glad to discuss high tension matters with Engineers, Contractors or Operating Executives anywhere.

Protection in the future from DANGER and LOSSES

ELPECO-THE CRUSADER IN ADVANCED ELECTRICAL EQUIPMENT

Disconnecting Switches          Bus Supports     Pole Top Switches     Switch Boards

ELECTRIC POWER EQUIPMENT CORPORATION, Philadelphia, Pa.
Canadian Representives: FERRANTI METER AND TRANSFORMER MFG. CO., Limited, Toronto & Montreal

Generation, Transmission and Application of Electricity

For nearly thirty years the recognized journal for the
Electrical Interests of Canada.

Published Semi-Monthly By

## HUGH C. MACLEAN PUBLICATIONS
### LIMITED
THOMAS S. YOUNG, Toronto, Managing Director
W. R. CARR, Ph.D., Toronto, Managing Editor
HEAD OFFICE - 347 Adelaide Street West, TORONTO
Telephone A. 2700

MONTREAL - - - 119 Board of Trade Bldg.
WINNIPEG - - - 302 Travellers' Bldg.
VANCOUVER - - - - Winch Building
NEW YORK - - - - 296 Broadway
CHICAGO - - Room 803, 63 E. Adams St.
LONDON, ENG. - - 16 Regent Street S. W.

**ADVERTISEMENTS**
Orders for advertising should reach the office of publication not later
than the 5th and 20th of the month. Changes in advertisements will be
made whenever desired, without cost to the advertiser.
**SUBSCRIBERS**
The "Electrical News" will be mailed to subscribers in Canada and
Great Britain, post free, for $2.00 per annum. United States and foreign,
$2.50. Remit by currency, registered letter, or postal order payable to
Hugh C. MacLean Publications Limited.
Subscribers are requested to promptly notify the publishers of failure
or delay in delivery of paper.

Authorized by the Postmaster General for Canada, for transmission
as second class matter.

| Vol. 31 | Toronto, April 1, 1922 | No. 7 |
|---|---|---|

## Standard Rules and Regulations Governing All Canada

Little by little, with gradually improving organization, the electrical industry is functioning to remove obstacles that for a long time have barred the road to real economic progress. The lost motion due to overlapping, misunderstandings, and failure to grasp the situation as a whole, is being saved by closer working arrangements between central stations, manufacturers, contractors and merchants, who all now realize much more fully than formerly that their operations are interdependent and that success to any one section of the industry is impossible without the success of all.

There is one item, however, in which we are still carrying on in a woefully inefficient manner. It is in the matter of standard rules and regulations regarding electrical equipment and its installation. The province of Ontario is governed by a code drawn up and administered by the Hydro-electric Power Commission; the province of Quebec operates under a very similar code of the National Board of Fire Underwriters. We are frequently told, however, that outside of the larger cities this Code is more or less a dead letter on account of the lack of proper inspection. The western provinces, in many cases, have no inspection; in the larger centres the matter of specifications is left largely with individual inspectors, whose ideas vary, naturally, as to the essential standard. Some of them are unduly lax, others over-severe, with a large section of the country inadequately controlled. This works particular hardship to the

manufacturer and increases the cost of electrical equipment to the consumer. It also means that poorly constructed equipment, hazardous alike to property and life, is frequently distributed in these districts.

Why not have standard rules and regulations for all Canada? Generally speaking, a device that is suitable and safe for use in the province of Quebec is suitable and safe for use in the province of Manitoba or British Columbia. Minor changes might be necessary to suit very special local conditions, but the exceptions could be minimized. What we seem to need is a recognized authority of some kind that meets with the approval of the various provinces and has the Government sanction in each province.

This matter is brought to the attention of our readers with the object of creating discussion out of which a solution of this weakness in our economic machinery may be crystalized. We shall be glad to have comments for publication from men who can speak with knowledge of the situation in different districts throughout Canada. The question, of course, is not a new one. It has had the attention of the Canadian Engineering Standards Association, and has been discussed by the two central station organizations, as well as by the Manufacturers', Jobbers', Contractor-dealers' and other electrical associations, all over Canada. There doesn't seem to be any doubt that electrical men are uniformly aware of this problem and the need of its early solution. Let us discuss the best way to handle it and get the thing done promptly.

## First Canadian Electrical Council Meeting

In our issue of February 15 we outlined the formation, in Montreal, of the Canadian Electrical Council, composed of representatives of the Canadian Electrical Association, the Association of Municipal Electrical Utilities and the Canadian Electrical Supply Manufacturers' Association. The first meeting of this council was held as scheduled, in Toronto, on Monday, March 13. Messrs. A. A. Dion, J. B. Woodyatt, A. P. Doddridge and L. W. Pratt represented the private companies; M. J. McHenry, E. V. Buchanan, O. H. Scott and P. B. Yates represented the A.M.E.U.; F. Jno. Bell, J. Herbert Hall and Geo. F. Foote the manufacturers. Messrs. Eugene Vinet, J. A. McKay and Walter Carr were also present.

The proceedings consisted chiefly of discussions on various points of common interest that were brought up by one or another of the delegates. These discussions included such important matters as a standard Code, for all Canada, covering the manufacture and installation of electrical equipment. At the present time there exist the greatest possible irregularities in this connection, which works a very real hardship on the industries — more particularly to the manufacturers and jobbers; Ontario follows a Code of the Hydro-electric Power Commission; Quebec, the National Code; the other provinces leave the matter largely in the hands of individuals, who are vested with authority to use their own judgment in the matter of what they shall require locally. The matter of Dominion standardization was advanced a stage and will have the further attention of the Council, which will work to the end of bringing the different provinces into touch and agreement with one another.

Other matters touched upon included the question of a joint convention of the C.E.A. and the A.M.E.U.; standardizing of transformer oil, attachment plugs, range elements, etc., and on a number of these topics recommendations were made to the parent organizations.

Two permanent committees were appointed, the first

composed of Messrs. Julian C. Smith, M. J. McHenry, A. N. Dion and O. H. Scott, whose function it will be to consider matters of common interest between the C.E.A. and the A.M.E.U. A second committee, composed of L. W. Pratt, for the C.E.A., P. B. Yates, for the A.M.E.U., and Geo. F. Foote, for the manufacturers, will consider the question and advisability of adding to the membership of the Council any other organizations with Dominion-wide affiliation. It will also devolve upon this committee to encourage Dominion-wide organization of different groups, where such does not exist already.

J. A. McKay, the secretary of the Canadian Electrical Supply Manufacturers', the Canadian Electrical Supply Jobbers', and the Ontario Electrical Contractors' Associations, was appointed secretary of the Council. Walter Carr was appointed temporary chairman. The address of the secretary is 24 Adelaide Street West, Toronto.

## The Luminaire

The coming into vogue of appliances which can be removed from place to place and to which the misnomer "movable fixtures" has been attached has brought the matter particularly to the attention of the lighting people. The Illuminating Engineering Society referred the consideration of this question to its Committee on Nomenclature and Standards with the suggestion that a term be recommended. Requests for suggestions of suitable terms were sent out to the membership of the Society and a considerable number of such suggestions were received. It was found, however, that most of the terms suggested were manufactured or coined words which had no legitimate ancestry and were therefore objectionable. Amongst the terms proposed, however, was one which met with the approval of the Committee, namely the word "luminaire." This word is used in this connection in the French language. Its construction and ancestry are such that it can be adopted in the English language as readily as "garage," "hangar," et cetera, which have recently been taken in. The significance of the word is evident on the face of it. It is believed that this word could and should be introduced into the English language and that it would be a distinct advantage so to do.

The Committee on Nomenclature and Standards recommended the use of this generic term for "lighting unit" in its report as presented to the annual convention of the Society at Rochester last September.

The Council of the Illuminating Engineering Society at the March meeting formally approved and adopted the use of the word "luminaire."

An expression of opinion favorable to the adoption of this term has been received from Engineering Societies and other organizations, and the term has already been adopted by the National Council of Lighting Fixture Manufacturers.

## C. E. R. A. Convention in Quebec Jan 1, 2, 3

Preparations are in full swing for the annual meeting and convention of the Canadian Electric Railway Association, to be held in Quebec on June 1, 2 and 3. Most of the associate members have agreed to put on exhibits of specialties and arrangements have been completed for the drill hall for both the exhibits and meetings. The drill hall is about five minutes walk from the Chateau Frontenac, and there will be ample space for 25 or 30 first class exhibits. Every indication points to a convention which will exceed those of previous years in interest and attendance.

All the space for the exhibit, and the general decorations, are donated by the association. Exhibitors desiring to display a moving exhibit will also be supplied with alternating current in their booths. This convention inaugurates the exhibit idea and it is expected that with this added attraction the attendance will be the largest Canadian electric railway operators have had. The Exhibit Committee have made arrangements to take care of shipments arriving in advance of the convention, so that exhibitors will do well to lay their plans early in order to avoid possible delay.

We are advised that the membership of the Association, to which municipal enterprises are now eligible, has been greatly increased during the past year. This will add not only to the attendance at the convention but also to its value, in the wider exchange of ideas and experiences.

## Wireless and Carrier Wave Telephony

An open meeting of the Toronto Section of the American Institute of Electrical Engineers was held on March 10 in the assembly room of the Canadian General Electric building, on Wallace Avenue, Toronto. The ladies had been invited and the attendance was one of the largest in the history of the Section, approximately 500 being present. Mr. W. P. Dobson, chairman of the Section, presided, and Charles A. Culver, Ph.D., addressed the meeting on the subject "Wireless and Carrier Wave Telephony."

Dr. Culver outlined the development of communication without wires from its inception, and introduced present day wireless by giving a few of the fundamental definitions. He then showed the principal methods of generating the high frequency voltages used as carriers. The matter of resonance was demonstrated by means of pendulums of different lengths, and its application to the electric circuit shown by analogy. The necessity of close "tuning" for the reception of the carrier waves was thus brought out, and the principle of "beat notes" explained. The speaker then followed up the development of the audion in its different forms and explained the variety of uses to which it was applicable in wireless work.

During the lecture, a loud speaking set was connected in, and the speaker frequently interrupted his discourse while parts of the programmes from broadcasting stations were received. At the close of the talk, the 10 o'clock time signals were received from Washington. After the technical session, light refreshments were served, and those present given an opportunity to meet socially.

---

> The annual convention — the thirty-second — of the Canadian Electrical Association will be held this year in Ottawa, Chateau Laurier, June 15-16-17.

The Shawinigan Water & Power Company, Montreal, will increase their capital from $20,000,000 to $40,000,000. A bill authorizing this increase has been passed through the Legislative Council Committee of the Province of Quebec.

The New Brunswick Electric Power Commission have issued the second annual report, for the year ended October 31, 1921. In this report it is stated that New Brunswick possesses undeveloped water powers with a continuous capacity of over 200,000 h.p., provided storage facilities are developed. With the exception of Grand Falls, of approximately 90,000 h.p., the different water powers vary from 1,000 to 10,000 h.p., but they are well distributed over the province.

# First 50,000 h.p. Unit of Chippawa-Queenston Power Scheme—IV.

### The Design of the Power Plant and the Specifications and Installation of the Major Apparatus

The present article, which is the fourth of a series on the Chippawa-Queenston power development of the Hydroelectric Power Commission of Ontario, deals, in a general way, with the electrical design and equipment in the power house, which is located on the bank of the Niagara River, near Queenston. The first unit was placed in operation, officially, on December 28, 1921; a second unit commenced operation during the second week in March. Each of these units has a capacity of 45,000 kv.a. and generates power at 12,000 volts, 3 phase, 25 cycle, which, in turn, is transformed to 110,000 volts for transmission. During the present year it is expected that five units of this capacity will be installed and, ultimately, the plant will consist of nine or ten units. The subsequent units will in all probability have greater capacity than those being installed at present. The two units now in service are already developing in the neighborhood

the rotating part of the machine, plus the hydraulic thrust impressed by the turbine.

#### Generators Air-cooled

The generators are air-cooled, the quantity of air utilized being 120,000 cu ft. per minute. Air is drawn either from the outside of the building or from the generator room, or from both, and may be discharged either into the atmosphere or through ducts into the different sections of the building, for heating purposes. With five units operating, the heat available from this source will be the equivalent of 1.2 kw. per 1,000 cu. ft. of building content—an amount sufficient for heating the building in the coldest temperatures.

According to ratings, the maximum observable temperature which any portion of the units may attain, as indicated by thermo-couples, will not be in excess of 105 degrees Centigrade, with 40 degrees Centigrade ambient air. Each thermo-couple indicates the temperature both on the control pedestal near the generator and also in the control room. Generators are protected by differential relays which open the main and field circuits, shut down ventilator fans and close dampers in the air ducts.

A generator, complete, weighs 1,400,000 lbs., the rotor itself 615,000 lbs.

#### Transformers

There will be five banks of three transformers connected

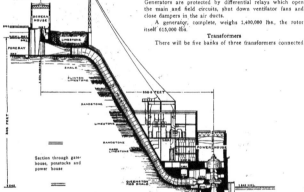

Section through gate-
house, penstocks and
power house

of 110,000 h.p., which is being delivered to the existing 110,000 volt system at Niagara Falls for transmission to various points throughout southwestern Ontario.

The generator type is vertical, with direct connected exciter. Each unit will be capable of operating continuously at 49,500 kv.a., with either voltage or current 10 per cent in excess of the rated value. At 80 per cent power factor the over-all efficiency of the generators will be slightly in excess of 97 per cent.

The thrust bearing is designed to support a load of 1,000,000 lbs. This is slightly in excess of the weight of

delta to "Y" to give 110,000 to 132,000 volts, at rated load. These are of the shell type, and individually rated at 15,000 kv.a., 25 cycle, 12,000/63,500/75,000. They are guaranteed 98.2 per cent efficient at 80 per cent power factor. The weight of a complete transformer is 100 tons; this includes 6,500 Imperial gallons of oil, which is required with each transformer for cooling purposes. The transformer dimensions are 28.5 ft. high, 9.5 ft. diameter. Each bank is equipped with differential protection.

#### Switching Apparatus

The switching apparatus consists of 12,000 and 110,000

## Airplane View of Queenston-Chippawa Plant

A wonderfully realistic Bird's-eye view of the Chippawa-Queenston development of the Hydro-Electric Power Commission of Ontario showing Canal, diffuser, forebay, gatehouse, penstocks, powerhouse and Niagara River.

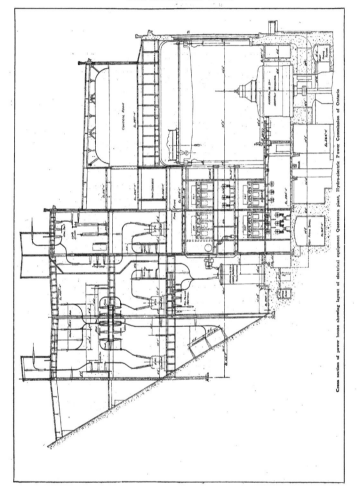

Cross section of power house showing layout of electrical equipment Queenston plant, Hydro-electric Power Commission of Ontario

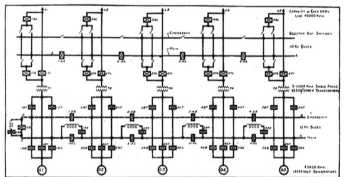

Wiring Diagram Queenston Power House—Hydro-electric Power Commission of Ontario

volt oil circuit breakers and disconnecting switches. Both high and low voltage breakers have rupturing capacities of the order of 1500 M.V.A.—considerably greater than anything of the kind previously produced in Canada. One of the 12,000 volt breakers already installed measures 36 inches in diameter and has been tested to 500 lbs. per square inch. The high voltage tanks are 6 feet in diameter, 14.5 ft. high and have been tested to 250 lbs. per square inch; this breaker is 25 ft. in height over all.

The 12,000 volt disconnecting switches are mechanically operated. They are rated at 3,000 amperes and are so arranged that the three switches forming a group operate simultaneously. The 110,000 volt disconnecting switches are of the three insulator type, with centre insulator rotating; they are rated at 600 amperes. The supports for the high voltage disconnecting switches are either standard suspension insulator or 42 inch corrugated posts.

### Service Equipment

Service power is obtained from two generators, having a capacity of 2200 kv.a. at 2300 volts, 25 cycles, 500 r.p.m. An auxiliary 12,000 volt line also connects the Ontario Power Company. Motors of 25 h.p. or larger operate at 2200 volts; smaller 3 phase motors are 550 volts. The power house is provided with two main cranes having a capacity of 150 tons each, so that the combined service of the two cranes is required to handle a main generator rotor.

The location of the main generator room floor is an innovation in this plant. This is placed just a little below the top of the generator frame. This construction provides space for the exhaust ducts, fans and field equipment adjacent to the unit, on the lower floor, and still provides a clear space around the generator on the main floor.

The main leads from the generator leave near the top and are carried to the switching equipment in the form of bus construction. The heavy capacity single unit bus supports are used for all leads, which consists of three 4 x 1/4 copper bars. The supports are spaced at a distance not exceeding 5 feet.

The main units are spaced 50 ft. centres, and all equipment involved in each unit—transformers, oil switches, reactors, fans, etc.—are located in the corresponding 50 ft. parallel space, each piece of equipment being symmetrically locat-

ed within this unit space in regard to corresponding equipment of other units.

Each 12,000 volt oil circuit breaker is located in a room by itself, as are also the disconnecting switches belonging to each oil switch. The main 12,000 volt bus is mounted on a floor, the phase spacing being 4 feet. In no case are the connections spaced closer than 3' 1" centres.

The 110,000 volt (arranged for 132,000 volts) wiring is spaced 8 ft. between phases and 4 ft. to ground.

The control room will be located over the centre section of the generator room. This portion of the building is not yet constructed, so it has been necessary to provide a temporary control room in the service section. The permanent equipment is now being developed. This will embody some improvements which have developed as the result of operating the temporary boards. The control room will be arranged in units also, so that bench, instrument, relay and recording boards for each unit will be symmetrically arranged.

All control instruments and relay circuits will run to the control room. For each unit there are 36 such cables each with an average of 6 conductors; total length of conductor per unit is about 11 miles.

These cables are all carried in conduit to the cable gallery within the unit space for the unit, thence are grouped and carried in specially designed pans longitudinally to control room. The location of the cable gallery is such that the run of each cable is direct, i.e. there is no doubling back on itself of any cable; consequently the total length of cables should be a minimum. As the total length of control relay and instrument conductor for the complete plant, without the emergency 12,000 volt bus, is in excess of 120 miles, one appreciates the necessity of selecting the proper location for this gallery and the establishment of direct runs.

### Arrangement of Apparatus

The main wiring diagram, herewith, shows the arrangement of the main electrical apparatus. A generator, bank of transformers and a line is considered a "unit," each with a normal capacity of 45,000 kv.a. As it will be impracticable to dispose of the power in blocks of this capacity, grouping of units will be essential. For this purpose both 12,000 and 110,000 volt buses are provided. The diagram shows a double 12,000 volt bus, only one of which, however—that with reactors—is being installed, but space is being provided for the second or emergency bus, which may be installed later if conditions warrant.

The remainder of this article will deal more specifically with the major apparatus included in the installation of Unit No. 1, which has been operating successfully for some months. The generator was supplied by the Canadian West-

is of the vertical type with direct connected exciter which is mounted above the generator, as will be seen from the illustration (Fig. 1) herewith. The driving turbine is directly underneath the generator. Between the generator and exciter is located the thrust bearing which carries not only the weight of the complete rotating element, but also that due to the water thrust on the turbine. The guide bearings at the upper and lower ends of the generator maintain a

Fig. 1—First 45,000 Kv.a. Unit (Westinghouse in Queensto n' Plant of Hydro-electric Power Commission of Ontario

inghouse Company, Ltd., of Hamilton, Ont., and has a capacity of 45,000 kv.a. (55,000 h.p.) Generated current is 3 phase, 25 cycle, 12,000 volts, alternating. The generator is coupled directly to a 55,000 h.p. Wellman-Seaver-Morgan water turbine. While this generator is only one of a number of units of equivalent output being installed in the same station, it has the distinction of being the largest of its kind in operation at the present time, not only in Canada but in the whole world.

The generator to which reference has been made above

uniform clearance (or air gap) between the rotating and stationary parts.

### Stationary Part

The stationary part, or stator, of the generator comprises a number of integral parts each bearing a definite relationship to the others. Of these parts the frame is one of the most important since it carries directly or indirectly, all the other parts.

The frame is in the form of a webbed annular ring and is made of cast iron. By reference to Fig. 1, it will be seen that in the outer part there are numerous openings which serve as outlets for the air forced through the generator for cooling purposes. Inside, there are heavy axial or cross ribs, the inner surfaces of which are machined and have dovetail slots for holding the armature punchings cut in them. The cross ribs are braced by circumferential ribs uniformly spaced along the cross ribs. Due to the large size of the frame for this machine, it was found advantageous to divide it into a number of parts arranged for bolting together to form the complete unit. By the first division, a

sulated from the punchings. Fig. 2 illustrates the assembly of the frame and armature core, one quarter section being removed.

The armature winding of this generator consists of a large number of coils formed to shape and thoroughly insulated for insertion into the slots in the core. Bare copper strap rectangular in section, is used in making the coils, there being a number of straps used to make up each conductor. After the coils are formed to shape, the individual straps and also the groups of straps forming the conductors

Fig. 2—Internal view of Armature Core of first 45.000 kv.a. generator with one quarter section removed— Queenston Plant Hydro-electric Power Commission of Ontario.

lower and an upper frame are provided. The lower frame rests on the bedplate and carries the lower bearing bracket with the lower guide bearing. This part of the frame is divided radially into two equal parts. The upper frame is supported by the lower frame and holds in it the armature core and windings while on top it carries the upper bearing bracket which in turn supports the exciter, the upper guide bearing and the thrust bearing. This part of the frame is divided radially into four equal parts, the joints being made in the centres of cross ribs to facilitate belting the sections together. Some conception as to the magnitude of this frame may be had from the fact that the approximate dimensions of each section of the upper frame are 10 feet in height and 19 feet in length, while the weight of each section is 42,000 pounds approximately.

The armature core is built up of thin sheet steel laminations which are punched to exact size. Due to the large diameter of this machine segmental punchings are used, having, on the outside, dovetails for holding them in place in the frame, while on the inside are the slot openings into which the armature coils are inserted. After being punched to size and before being assembled in the frame, these segmental punchings are annealed and enamelled, a coating of enamel being put on both sides of each punching. In building the armature core from these punchings, vent ducts spaced about 2 inches apart, are put in between the layers and extend all around the core. The punchings after being stacked in the frame, are drawn tightly together under pressure and held firmly by means of heavy cast iron end plates which are keyed to the frame, in addition to being braced by heavy bolts which extend from one end plate through the punchings to the other end plate. These through bolts are in-

are insulated from each other by taping overall with mica tape. In insulating the complete coils micarta folium insulation comprising a process patented by the Westinghouse company is used. This insulation consists almost entirely of mica and in applying it to the coils precautions are taken to see that the layers are drawn tightly together in order to eliminate air spaces between the layers. The result is an insulation which is not only very compact, but also one which has a high dielectric strength. It will readily conduct heat away from the coils and at the same time will withstand high temperatures without deterioration, thus giving an insulation which is practically indestructible under all conditions of operation. One of the armature coils completely insulated in this way is illustrated in Fig. 3.

Slot cells are used to protect the surface of the insulation from the iron and at the same time they serve as wedges to make the coils fit tightly in the slots. The coils are retained in the slots by fibre wedges fitting into grooves in the side of the slots. The ends of the armature coils are braced similarly to those of turbo-generators. There are cast iron coil supports bolted to the end plates and to these the coil ends are fastened by means of wooden blocks placed under, between and above the coils. Insulated bolts extending through the openings in the ends of the coils to the iron coil supports hold the wooden blocks rigidly in place.

The connections between the various parts of the winding are made of insulated copper straps carried on supports above the upper ends of the coils. The windings are arranged for star connection and both ends of each phase winding are brought out through the frame in order that protective relays may be inserted in each circuit, while between coils

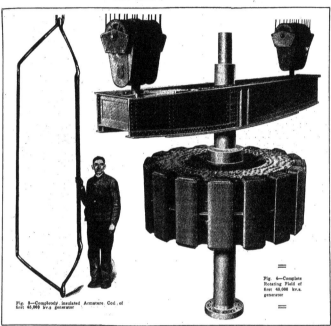

Fig. 3—Completely insulated Armature Coil of first 45,000 kv.a generator

Fig. 4—Complete Rotating Field of first 45,000 kv.a. generator

in the slots, a number of thermo-couples are distributed throughout the winding for checking the temperature of the various parts of the winding.

The upper and lower bearing brackets are each made in the form of a spider having the arms heavily webbed for strength and rigidity. Cast iron is used for these brackets since it is more rigid and subject to less deflection than cast steel for equivalent strength. The brackets are each made in two sections, being divided radially through the hub. The surfaces at the intersections are machined and the two sections are bolted rigidly together. The approximate weight of the lower bearing bracket is 45,000 pounds, while that of the upper bracket is 80,000 pounds.

### The Bearings

There are three bearings in connection with the generator, the main one of which is the 69 inch Kingsbury thrust bearing carried on the upper bearing bracket. This thrust bearing is designed to carry a load of 1,000,000 pounds including the complete rotating element together with the water thrust on the turbine. The rotating part of the bearing consists of a cast iron collar carefully surfaced on the bearing side and fastened to the shaft. This collar is supported by babbitt-faced segmental steel shoes which are pivoted so that they can rock in any direction a limited amount. The complete bearing is immersed in oil which is kept cool by means of copper cooling coils through which water is circulated to carry away the heat generated in the bearing. A means of circulating the oil is also provided. This is a well known type of thrust bearing having proven satisfactory in many large hydro-electric installations, and is highly favored on account of its low friction loss.

Two guide bearings are used in connection with the generator, one of which is placed above the rotating field and mounted on the upper bearing bracket below the thrust bearing, while the other is below the rotating field and is mounted on the lower bearing bracket. The bearing housings are made of cast iron arranged with water jackets through which cooling water is circulated. These bearings are lined with babbitt of a high grade having numerous grooves for carrying the oil to all parts of the bearing surface. The oil is forced through these bearings by means of pumps.

Underneath the lower frame and supporting the complete generator is a circular cast iron bedplate made in halves which are bolted together. The top of the bedplate is machined so that by means of adjusting screws the proper alignment of the generator can be obtained. Heavy foundation bolts are used to hold the bedplate firmly on its foundation.

### Rotating Part

The rotating field, illustrated in Fig. 4, consists of laminated field poles on which are mounted the field coils. The poles are attached to a spider mounted on the shaft.

The laminations for poles are punched to size from sheet steel and are held together by heavy plates at each end together with long rivets extending through the punchings between these end plates. The poles are dovetailed to the rim of the spider and are held in place by means of two tapered steel keys driven into the slots beside the dovetail of the pole. In making the field coils, heavy bare copper strap is used which is formed to shape and then insulated. Sheet asbestos and shellac are used in insulating between turns while on the inside of the coils, adjacent to the poles, and also on the upper and lower edges of the coils, insulation consisting of mica, asbestos and bakelite is built up and moulded to size and shape. The coils completely insulated in this way are baked to consolidate the insulation. Heavy insulating washers placed at the top and the bottom of the coils protect them from the iron of the supporting pole tips and also from that of the spider rim. They also act as wedges holding the coils tightly in place.

The yoke of the rotating field is formed by the rim of the spider on which the poles are carried. Although the peripheral speed of the spider is not excessive, a laminated steel rim is used, the uniformity of the material for which being much more certain than that for cast steel. The rim punchings have dovetails on the inside for mounting on the cast steel hub of the spider and dovetail slots on the outside for holding the poles. Cast steel end plates bolted through the punchings support the laminated rim. The hub of the spider is a steel casting made in the form of a webbed spider of comparatively small diameter. Dovetail slots are cut in the outer surface of the cross ribs to receive the dovetail of the rim punchings, while the inside is bored out to receive the shaft. Due to the length of this part, three machine fits are used on the inside to facilitate inserting the shaft. A heavy steel key prevents this hub from revolving on the shaft. On the lower side of this cast steel hub there is a brake ring to which air brakes mounted on the lower bearing bracket, can be applied in case of emergency. This cast steel hub weighs approximately 55,000 pounds.

The shaft used in this generator is hollow, there being a hole 8 inches in diameter throughout, which gives a member of maximum strength for minimum weight. Through the portion of this hole between the spider and the cast iron collector rings mounted on the shaft above the thrust bearing, the leads from the generator field to the collector rings are carried. This hole is also useful when it is desired to remove the runner of the water turbine without taking out the rotating part of the generator, since cables can be let down from the crane above the generator through this hole, to support the runner when the shaft is disconnected at the coupling between the generator and the turbine. A heavy cast iron collar keyed to the shaft at the upper end and held by means of a large steel nut, is used to support the shaft together with its impressed load, on the thrust bearing. The shaft of the generator alone is approximately 30 feet in length when measured from the face of the coupling to the top of the thrust bearing and is 32 inches in diameter inside the cast steel hub of the spider. Its approximate weight is 70,000 pounds.

### Ventilation

Forced ventilation is used to keep the machine cool. Free air is drawn up from underneath the generator, some of which passes through the spider to the upper side of the rotating field. Vane blowers attached to the spider distrib-

ute the air uniformly to various parts of the generator from which it passes through the various ducts in the iron and openings in the coils and the frame to the space between the generator and walls of the pit in which the generator is mounted, whence it is drawn by an exhaust fan. Openings in the generator leading to the generator room above, in which is located the upper portion of the machine, consisting of the thrust bearing and exciter, are closed by means of sheet steel covers so that the heated air does not pass up into the generator room. The quantity of free air passing through the generator is approximately 120,000 cubic feet per minute.

When it is realized that this generator as completely assembled is approximately 35 feet in height from the face of the coupling to the top of the exciter, and 24-1/2

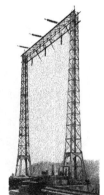

No. 1 Unit Escarpment towers, looking north

feet in diameter, some conception will be conveyed as to the enormity of the generating units being built and installed in Canada at the present time. While the manufacture of these large generating units embodies only features that have proven satisfactory in smaller sizes, some of the outstanding features worthy of special note which have made feasible the development of large generators of the type described are (a) the rugged mica insulation adopted for armature coils, (b) the laminated steel construction of the spider rim, (c) the low friction losses in the Kingsbury thrust bearings and (d) the rigidity of cast iron utilized in making brackets capable of supporting heavy loads with a minimum deflection.

### Other Equipment

As already noted, the first generator unit installed in this station was manufactured by the Canadian Westinghouse Company in their Hamilton plant. The same applies to the transformer, the breakers and the disconnecting switches. Unit No. 2, just installed, was also built by the Canadian Westinghouse Company, and unit No. 3, by the same company, is nearly completed.

Units 4 and 5, of the same capacity, are being manufactured by the Canadian General Electric Company, who also supplied certain auxiliary equipment in the earlier units. All the 110,000 volt lightning arresters, oxide film type, are of C. G. E. manufacture, as are also the reactors.

The 12,000 volt disconnecting switches for all five units are being supplied by the Electrical Development & Machine Company.

The 110,000 volt transformers for all five units are being supplied by Canadian Westinghouse; this company is also supplying the two 2200 kv.a. service generators.

# The History of the Development of Electric Welding

### By J. M. F. WILSON, B. Sc.*

When two pieces of metal are to be welded together we have two distinct methods of treatment. Where the metals are first heated to a point below fusion and then pressed or hammered firmly together, we have what is known as compression welding including the ordinary forge weld, the electric spot and butt welds and the most recent electro-percussion weld. Where the metals are raised to such a temperature that they will flow together without the use of force or pressure, we have the autogenous or fusion weld, including the oxy-acetylene, the thermit and the electric arc processes. It is my part of the program to deal with the electric arc process.

Two systems of electric arc welding are in use and get their names from the type of electrode used: (1) carbon electrode, (2) metal electrode.

### The Carbon Electrode

The carbon electrode process first used by Bernardos more than thirty years ago employed a direct current arc drawn between a carbon rod as positive and the work as negative, a metal filler rod being fed into the flame of the arc by hand. By reversing the polarity several advantages were obtained. The greater heat developed in the positive crater of the arc is transferred to the work where it is most needed, the arc is more stable and the carbon particles are prevented from flowing into the weld, thus preventing the weld from becoming too hard to machine. The carbon electrodes vary in diameter from 1/4" to 1½". Greater stability for the arc is obtained by tapering all electrodes to 1/8" and a lower rate of carbon consumption as well as a softer weld is found when graphite electrodes are used in place of carbon. Currents range from 100 amperes to 800, but in general 300 to 400 amperes with a 1" electrode is found sufficient.

As regards the filler material, commercially pure iron rods, low in carbon, from 3/8" to 1/2" diameter are found necessary for a strong, sound weld. In operating, the surfaces should be chipped clean and the arc formed by withdrawing the graphite electrode from a clean surface of solid metal or from the end of the filler rod held in contact with the parent metal. By this means particles of slag are prevented from adhering to the end of the electrode and deflecting the arc. An inclination of the electrode at 15° to the vertical makes the deflecting force of the air currents constant in direction and steadies the arc.

A surface may be built up either by puddling or by depositing in layers. In puddling the arc melts a small quantity of the parent metal; a small portion of filler rod is next melted off, the rod withdrawn and the added material fused with the parent metal, a rotary motion being imparted to the arc. The puddling drives the slag and oxidized material to the edge and the rotary motion heats a comparatively large area about the weld with consequently less danger of cracking the work or making a hard weld.

In depositing the filler material in layers, we have a

*Before Manitoba Branch E.I.C.

similar deposit to that of the metal electrode but wider and higher. The layers are then smoothed over with the arc. When a fine grain structure is required the expedient of hammering the surface as soon as the metal starts to cool is resorted to.

### The Length of Arc

As regards the length of the arc, even the lower temperature of the graphite rod as negative will vaporize the carbon in the rod and if a hard weld is to be prevented this must be oxidized. By increasing the arc length so that sufficient atmospheric oxygen will diffuse through the arc stream this may be accomplished. To obtain the same degree of oxidation for the different currents used, the length of the arc would have to be increased in proportion to the arc diameter. As this latter varies with the square root of the current (for the same current density, $c \div d^2 = K$, therefore $d$ varies as $\sqrt{c}$) we find 1" quoted as the maximum for a 250 ampere arc and 1½" for a 500 ampere (Viall). On the other hand, we have J. H. Bryan (see Lefax) quoting 3 to 4 inches as the usual length. This latter would necessitate a voltage bordering on the dangerous side. As to its uses, the carbon arc is successfully applied without the filler rod to melt two flanged edges together. With a bronze filler rod, low in tin and zinc, but with at least .25% phosphorus to act as a strong de-oxidizing agent, copper and bronzes have been successfully welded. The class of work to which the carbon process particularly applies is the cutting or melting of metals, repairing broken parts and building up materials, but where strength is a consideration it calls for great skill on the part of the operator. Overhead and vertical surface welds are ruled out but it has a field in the finishing of welds made by the metallic arc process. It is particularly adapted to welding cast steel and non-ferrous metals, cutting cast iron, melting and cutting scrap metal, hardening rails, frogs, and wheel treads. When the art of electric welding is better understood, it may be that the pendulum will swing back again to the carbon arc process from the more recent metal electrode process.

### The Metal Electrode

The metal electrode was first used by Slavianoff. Here the negative terminal of the arc is the filler rod itself and we find both bare and coated electrodes in common use. The voltage is low, about 22 volts for bare and 35 for coated electrodes. A highly luminous central core of iron vapour is formed as the arc is withdrawn from the work and this is surrounded by a flame of oxide vapours formed by union with the atmosphere. The intense heat sets up a draft which, combined with local air currents, tends to break through this protecting wall. Consequently it is only by the maintenance of a short arc (1/8") and proper inclination of the electrode that a successful weld can be produced. The amount of current and size of electrode has an important bearing on the weld. In a test carried out by Wagner for the Welding Research Sub-Committee a double V weld in 1/2" ship plate showed an increase in tensile strength as the current was increased from 100 to 200 amperes and he would advocate even 300 amperes for 3/4" plate. British practice at present uses 4000 to 6000 amperes per square inch of electrode, but the Welding Committee recommend 10,000 as more suitable for 1/8" and 5/32" diameter electrodes. That is, a 1/8" electrode could carry 120 amperes through a 5/32" is more generally used for this current.

The Westinghouse company publish a graph showing the relationship between size of electrode, current and thickness of plate, the latter being an important factor, as also would be the temperature of the plate and the type of joint. The values quoted are lower than the above:

| Thickness of Plate | Current in Amperes | Electrode Diameter |
|---|---|---|
| 1/16 | 20 to 50 | 1/16 |
| 1/8 | 50 to 85 | 3/32 |
| 3/16 | 75 to 110 | 1/8 |
| 1/4 | 90 to 125 | 1/8 |
| 3/8 | 110 to 150 | 5/32 |
| 1/2 | 125 to 170 | 5/32 |

This is for a butt weld.

A lap weld between two 1/2" steel plates has a heat storage equivalent to that of a butt weld of two 1 inch plates, hence makes a better weld with 225 amperes and a 3/16" electrode.

An absolutely steady current would seem to be an essential factor when the correct current is obtained. The Wilson Welder Co., who used their "plastic arc" process so successfully on the recovery of the German ships interned during the war, claim that much of the success was due to the automatic control of the current, that as there is a certain critical temperature at which steel can be worked to give the greatest tensile strength and ductility, a variation of 20 amperes above the proper value will cause collection of carbon and slag pockets in the metal, while the same value below will cause the metal to be deposited in globules with holes in the weld to prove that there has been insufficient heat.

## Constants Involved

In the types of electric welding machines on the market by far the greater number are d.c. At first, designed from the view of economy in operation, the inclusion of resistances on even a 120 volt d.c. circuit being very wasteful, so called constant potential types were used. This type necessitated a series resistance and in order to dispense with this resistance constant current types of generator were next developed. A third type where the product of volts by amperes is kept constant and termed the constant energy type, is also designed to deliver current to the arc without the use of resistors or other energy-consuming devices. In examining the various methods used for regulating the machine we have to bear in mind that the arc itself has some inherent regulating characteristics such that current and voltage are so inter-related that an increase in one causes a decrease in the other. A 1% decrease of voltage from that of the customary 20 volts will cause about 2% increase in current. An increase in the voltage across the arc, when other conditions remain the same, indicates an increase in length of the arc, thus giving the metal a greater opportunity to oxidize in the arc and cause a reduction in the quality of the weld. Hence a type of machine which would make it difficult to maintain a long arc would be an advantage. Now the rate of consumption of the electrode is proportional to the current and is independant of the voltage across the arc and the length of the arc. In the constant-energy system when the operator shortens the arc, the increase of current consumes more electrode, and brings back the length of the arc to its original length. Lengthening the arc decreases the current and at the same time the consumption of wire, again automatically bringing the length of arc to normal. Too great a length will reduce the current to such an ex-

tent that the arc will go out. With a constant current this regulating factor is absent and a long arc is made possible. The oxidation of a long arc is shown by the increase in red oxide ($Fe_2O_3$) covering the weld. The black oxide ($Fe_3O_4$) does not contain as much oxygen.

## Constant Potential Type

As examples of the constant potential type we have the Wilson Welder Company's plastic arc set, a flat compounded generator where the current is automatically controlled by a solenoid operating a carbon pile resistance for small variations of current. Large current variations are obtained by means of a variable grid resistance. The volt-amperes at the arc are constant for each setting, hence the set might also be classified under the constant energy type. It is usual to include a reactance in the welding circuit to smooth out variations of current caused by sputtering of the arc, but in this particular set the series winding on the solenoid acts in this capacity. The welding current passing through the carbon pile and the solenoid causes the plunger of the solenoid to slacken the springs compressing the pile when the arc is struck. When the current decreases, the pull of the plunger decreases, and the current is adjusted automatically. In the earlier models the operator is able to control the amount of current which is to be kept constant by means of a pilot motor which shifts the fulcrum of the pressure mechanism as desired.

## Constant Current Type

As an example of the constant current type we have the Lincoln Electric Company's set comprising a separately excited shunt field and a differentially wound series field changing the voltage from 65 on open circuit to 20 on load. The series field is varied by means of a diverter so that currents of 80 to 200 amperes may be obtained. An external reactance is used.

Another example of the constant current type is the Westinghouse company's latest model. The generator is a commutating pole machine, with a series winding, a separately excited shunt field and a second shunt winding connected between one exciter terminal and one generator terminal, the other exciter and generator terminals being connected through a resistance $R_1$. Under open circuit and normal welding conditions this second winding is excited by the generator voltage and assists the first shunt winding (separately excited) in maintaining the generator voltage through $R_2$. Under short-circuited conditions the second shunt winding is excited from the exciter through $R_1$ and opposes the field of the first shunt winding. As the series winding also opposes this first shunt winding all the time, both assist in reducing the short-circuit current to practically the same as the welding current. The variable rheostat in the generator separately excited field allows control of currents from 50 to 200 amperes. An external reactance in the welding circuit also keeps down the short circuit current at the moment that the arc is struck and stabilizes the arc. The exciter voltage is maintained by a compound winding at 125 volts and the generator pressure falls from 60 to 20 volts.

The latest constant energy type of the General Electric Co., uses no separate exciter but makes use of the third brush idea for controlling the current through the shunt winding. Time will not admit of full details.

I might mention here that very good results have been obtained in welding and in holding a steady arc by using an ordinary battery charging generator of 80 amperes capacity, using a very weak shunt field and a cumulative series winding.

## Selection of Electrode

The physical properties of a weld depend on the crystal structure, gas holes, slag inclusion, impurities and composition. What we require in arc welding is a fine grained casting, free from blowholes, slag inclusions and impurities. It is noticeable that the first part of a weld will show a fine grained structure merging gradually into a coarse grain as the plate heats up. These coarse grained crystals are what is known as columnar and wherever these occur we have brittleness. Varying methods have been tried to produce the finer grained structure, viz., hammering the weld, depositing the metal rapidly in thin layers and cooling by a stream of water. The structure in the interior of the weld, however, (see Rawson) is of fine grain and it is to be inferred from this that the subsequent heatings of the successive layers refine the first layers. It has also been found that the higher the current density the greater the collection of columnar crystals.

Gas holes found in all electric welds are a great source of weakness. It is not the electrode alone that can be blamed for this but also the plate. Any substance, such as carbon, having a great affinity for oxygen will produce gas holes. Carbon monoxide is formed and trapped in the weld by the rapid solidification. Usually any dissolved gas in the electrode is liberated as it passes through the arc stream causing certain electrodes to be unworkable through "sputtering," (the General Electric Company claim that by merely dipping electrodes inclined to sputter in milk of lime (white wash) sputtering may be avoided) but the carbon in the plate dissolved in the welding pool reacts with

the iron oxide in the pool and forms more CO with no chance to escape. Examination of welds with a carbon free electrode on a carbon free plate showed only an occasional small gas hole, while it is well known that steels containing only .3% carbon are often difficult to weld.

Certain electrodes are on the market which are annealed. As the effect of annealing is to decarburise the outside layers and hence reduce the amount of carbon, it may be for this reason. Slag inclusions are also a source of weakness, particularly where covered electrodes are used in inexperienced hands. Slag becomes entangled in the deposited metal and fails to rise to the surface. Wrought iron welding wire contains considerable quantities. Slag aids the weld, however, by preventing surface oxidation of the pool.

### Impurities

Under the term impurities the oxygen content which is characteristic of the electric weld is not alone sufficient to account for its want of ductility, but we will deal with the oxide first. In all specimens examined microscopically are found tiny globules of oxide having no definite arrangement and scattered indiscriminately throughout the crystals of iron, (Fig. 2). A film of oxide is also found surrounding the "metallic globule inclusions," Fig. 1. It is supposed that these globules are small metallic particles formed as a sort of spray at the top of the electrode and

have failed to fuse in with the metal subsequently deposited over them.

It is a curious fact, borne out by analysis and tests from the Bureau of Standards, Strauss and others, that electric welds show over a .12% nitrogen content, while acetylene welds give only .02%.

Nitrogen in steel is an element producing a great degree of brittleness, as small a quantity as .06% reducing the elongation on a .2% carbon steel from 28% to 5%.

Microscopic examination shows a large number of lines, needles or plates in the weld which resemble iron highly nitrogenized. These occur within the ferrite (or pure iron) crystals, and are more abundant in the columnar and coarse grain crystals. It might be assumed that rupture takes place along these nitride plates (see Fig. 3) but it is found that the effect of the oxide inclusions and films, the metallic globule inclusions, and the lack of adhesion between successive layers of the weld, makes the nitrogen feature of small importance as a cause of rupture. While these unsound spots are no doubt the cause of the low ductility of the weld, the metal retains its inherent ductility, as microstructure shows after bending tests on the weld.

Other impurities which must be kept to a minimum, both in plate and electrode, are sulphur and phosphorus, the latter causing a brittle envelope round the crystals. As regards the composition of the electrode, some tests which

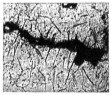

Fig. 1—Microscopic Evidence of Unsoundness. The Globule at the right is surrounded by a film of Oxide, and the fracture of the specimens when tested in tension originated in such unsound areas

Fig. 2—Characteristic Feature of Arc-Fused Iron. The "Needles" or "Plates" are due to the increase which occurs in the nitrogen content during fusion

Fig. 3—Relation of Path of Rupture of the Microstructure of Arc-Fused Iron. The course of the crack or tear in the metal which was produced by stressing in tension does not appear to have been influenced appreciably by the "Nitride Plates"

were carried out for the Welding Research Sub-Committee by the Bureau of Standards bring out some interesting points. Two electrodes, one (A) a "pure iron" electrode, and the other a low carbon steel (B), prepared to specifications, were tested for content before and after welding.

|  |  | C | Si | Mn | N | Tensile Strength |
|---|---|---|---|---|---|---|
| "A" Iron | Before | .058 | .33 | .042 | .003 | 60,000 |
|  | After | .031 | .007 | 0 | .143 | 50,000 |
| "B" Mild Steel | Before | .15 | .06 | .47 | .003 | 70,000 |
|  | After | .024 | .008 | 0 | .14 | 50,000 |

Note the heavy loss in C and Si and Mn where they occur in large quantities, and the gain in N.

(To be continued)

Mr. J. J. Seguin after 27 years of service in various capacities with the Bell Telephone Company, has resigned to take up commercial activity on his own behalf. It was in 1895 that he entered the employ of the company as fireman in the Eastern Division plant department and worked his way up to the positions of climber, sub-foreman, foreman, district plant chief, and, finally, plant chief of district No. 1, at Quebec. He has thus had a varied and valuable experience, and is widely known. The industry's good wishes are with him in his new undertaking.

# The Development of Wireless IV

### A Series of Short Interesting Articles
### Covering Wireless Progress to Date
By F. K. D'ALTON

The two element valve which is used as a rectifier was invented by Dr. J. A. Fleming. A remarkable improvement was made by Dr. Lee DeForest, who inserted a third element commonly called a grid between the plate and the filament.

Whereas, in the two element valve, the plate current increases with both the temperature of the filament and the potential of the plate with respect to the filament, it was found that with filament temperature and plate potential both fixed, the plate current will vary with variations in the voltage of this grid with respect to the filament. The current was so arranged that the grid, with practically no power supplied to it was able to control a measureable amount of power in the plate circuit. The valve is a power relay and what is more the moving parts are electrons, parts of molecules, without inertia, hence variations in plate current follow the same wave form as the grid voltage (voltage of grid with respect to filament). Thus the valve with the three elements may be used to amplify waves of almost any frequency. As a detector it is much better than the crystal, giving louder signals and not being thrown off adjustment by the transmitter in the same station.

The telephone receivers are placed in the plate circuit and across them is usually placed a small condenser to act as a by-pass for the high frequency (radio frequency) components in the current.

The received and tuned waves are not rectified, but are distorted so that a component is added to the otherwise steady plate current; this gives the necessary variation to enable the phones to give forth sound waves.

If a small transformer, known as an amplifying transformer, have its primary winding connected in the plate circuit instead of the phones, and also have its secondary winding connected between the grid and filament of a second valve in the plate circuit of which the phones are now placed, we have what is known as a cascade connection with one step of amplification. The signals will be amplified and be from six to ten times as strong as with one valve only.

Considering that each step of amplification multiplies the signal strength by ten, the effectiveness of amplifiers will be at once recognized, and their use in increasing the workable distances by leaps and bounds will also be realized.

The three electrode valve, as well as all the other detectors which use phones, gives quantitative response and can be used for detecting any waves where the variation in the amplitude is sufficiently fast to be within the range of sound.

We may mention that quantitative response in detectors was suggested by Prof. R. A. Fessenden. He also advocated the use of continued waves,—i.e., undamped waves,—and suggested both a method of producing them and also a scheme for detecting. His transmitter was to be a high frequency (100,000 cycles, or higher) alternator which would charge the aerial by direct connection between it and ground thus causing the radiation of long wireless waves. Messages could be sent by interrupting the current with a key. No spark gap was necessary and an audible note was not produceable with the detectors so far described.

The continuous wave when received simply displaced the diaphragm but did not cause vibrations and hence no sound, so different schemes had to be used. The ticker detector was the first step in reception of "C. W." (continuous waves). It consisted in a mechanical interrupter in the receiving circuit between the tuning device and the detector which chopped up the incoming signals and produced an audible signal in the phones while the waves existed. This, however, was con-

sidered an improvement, since the magnification by tuning was greater for a given strength of wave than with damped waves.

The high frequency alternator is used in many transoceanic stations and is considered best for this class of work.

Prof. Fessenden suggested a new and novel means of detection for continuous waves. He generated at the receiver a current of nearly the same frequency as the signals he was receiving but allowed the difference in frequency to be within the range of sound, thus if the incoming wave had a frequency of 100,000 cycles per second and the current generated at the receiver was 99,000 or 101,000 cycles a second, the "beat" note equalling the difference would have 1000 cycles per second and be audible, whereas neither received nor generated frequencies are within the audible limits. This method of detection is known as "heterodyning" and gives a greater volume of sound from a given signal than any other scheme. This again increased the workable distances and permitted the pitch of the signals to be adjusted at the receiving end, where they may be given a suitable note, enabling them to be detected through the interference. It allows a greater discrimination between stations than was ever obtainable with spark stations.

Poulsen then brought out his famous "arc" transmitter wherein the higher harmonics in an electric arc were made to build up high frequency currents in electrical circuits. The frequency was determined by the constants, capacity and inductance, and a fairly uniform continuous wave of radio frequency was obtained. Some attempts were made to use this wave for radio telephony but without great success, as no suitable transmitter could be made to carry the heavy currents and at the same time be actuated by the voice.

Arc sets are used for telegraphy across very long distances and are preferred in the navy stations of the United States.

There is still another, and very convenient method of generating continuous waves. It is by means of the three electrode vacuum tube already described as a relay detector.

It will be noted that the three element valve receives the signals from the tuning device in the form of voltage variations which, due to the action of the valve, appear as current variations in the plate circuit. Captain E. H. Armstrong inserted a coil in the plate circuit and placed this so close to the coil of the tuning circuit that it supplied part of the power of the plate circuit back to the grid circuit. This had the effect of still further increasing the intensity of the signals.

Since variations in grid potential cause corresponding variations in plate current due to the valve alone, and the changes in plate current cause variations in grid voltage through the coupling of coils alone, Armstrong found that an oscillation was set up in both circuits. The valve then became a generator of high frequency currents, the actual frequency of which depended solely upon the electrical constants of the circuits.

These quantities—capacity and inductance—were easily varied, thus causing changes in frequency of the generated currents. For heterodyne reception these valves are very convenient. With one "oscillating" valve a continuous wave station may, when tuned in, be detected and the note adjusted very readily by manipulation of the small variable condenser in the tuning circuit.

Furthermore by properly proportioning the coils and condensers in circuit—i.e. in the coupling—these high frequency oscillations may be made to cause currents of the same frequency in a tuned aerial circuit, thus becoming a transmitter and radiating a continuous wave of almost pure (sine wave) form.

A key may be inserted in the grid circuit, in the plate circuit or possibly in the ground or aerial leads and thus break up the radiations into code letters, permitting the sending of messages by dots and dashes. Since the same little

(Continued on page 40)

# Licensing Contractors and Journeymen

## An Ontario Act Respecting the Examining and Licensing of Electrical Contractors and Journeymen Electricians

Electrical contractors and dealers throughout the province of Ontario will be keenly interested in the Bill (No. 105, 1922) introduced on March 8 in the local Legislature by Mr. Swayze, labor member for St. Catharines. The Bill is entitled "An Act respecting the examining and licensing of electrical contractors and journeymen electricians", and is a revision of a similar Bill that was given its first reading in the dying hours of the 1921 session of the Legislature.

Reproduction of the Bill in full would occupy more space than we have at our disposal, but the following copious extracts contain the more important points of interest:—

1· This Act may be cited as "The Electricians' Licensing Act, 1922."

### Definitions

**2.** This section defines the various terms "contractor," "journeyman," "license," "inspector," "apprentice" and "board." This section differs from the corresponding section in the 1921 Bill in that the Act is to be administered by the Department of Labour, where formerly it was the Department of Public Works.

### Application

· **3.**—"(1) The provisions of this Act shall apply to all contractors, journeymen and inspectors engaged within the province in the business of placing, installing, maintaining, repairing, replacing or inspecting in or on any class of structure any conduits of any description, designed for the purpose of enclosing or carrying any electrical conductor upon which is impressed and E. M. F. equal to or higher than the voltage prescribed in the wiring regulations issued by the Ontario Hydro-Electric Power Commission, between any two conductors and ground independent of the characteristics of the current; of placing, installing, maintaining, repairing, replacing or inspecting in or on such structures, of any conductor switch, attachment, fitting or any element whatsoever of an equipment designed for the purpose of supplying such electrical service, or for any purpose in connection with such an electrical service.

"(2) The provisions of this Act shall not apply to such work within power houses, sub-stations or other places wherein the business of generating or distributing electrical power is carried on by public service corporations or by municipal departments, and where such work is installed by employees under the direction of officers of such public corporations or municipal departments, except in structures wherein the public, other than employees of such public service corporation or municipal department have free access on business.

"(3) The provisions of this Act shall not apply to such work on street railway cars or locomotives, or on railway cars or locomotives which are the property of municipal departments or of public service corporations, and where such work is installed by employees under the direction of officers of such municipal departments or public service corporations.

"(4) No apprentice or other person shall perform any electrical work or install any electrical material or appliances within the meaning of the Act, except as an assistant to, in the presence of, and under the direct personal supervision of a journeyman continuously employed on the same contract or job and licensed under this Act, and only one apprentice shall be allowed to each journeyman as assistant on any job.

"(5) Nothing in this section shall be taken to apply to the insertion of incandescent lamps in sockets or receptacles, or the replacement of such lamps, the carboning, trimming, or operation of arc lamps, the lawful connection of utilization equipment to supply by means of attachment plugs or the use or operation of the same, or the lawful replacement of fuses controlling circuits or equipment."

**4.** States that all electrical contractors and journeymen shall be subject to examination.

**5.** This section is a little indefinite and we reproduce it in full; it runs as follows:

### Board of Examiners

"The Lieutenant-Governor in Council, upon the recommendation of the Minister, may appoint a board of examiners, consisting of three members, who shall be qualified in practical electrical work, one of whom shall be an electrical contractor, and one who shall be a practical journeyman electrician, and who are conversant in a practical degree with the qualifications necessary to be held by a person described as an electrical contractor or an electrical journeyman, who shall hold office during pleasure and who, subject to the regulations mentioned in the next following section, and to the approval of the Lieutenant-Governor in Council, shall prescribe the subjects in which candidates for a contractor's or journeyman's license shall be examined, and conduct and provide for and supervise the examinations of candidates and report thereon to the department."

**6.** Outlines the duties of the Board of Examiners.

### License Forms

**7.** "Four license forms shall be issued, designated as follows:—

**License A,**—which may be issued to any person who has satisfactorily passed the examination prescribed for journeymen electricians, and has filed an application to be registered as a contractor in the examiner's office and paid the fee prescribed by subsection (2) of section 8 of this Act.·

**License B,**—which may be granted to any company, association, corporation, or firm doing or wishing to do business as contractor for electrical installation; provided one of the members of the said association, company, corporation, or firm, or at least one person in its employ, holds a certificate of journeyman electrician given by the examiners, and that the fee for the license has been paid.

**License C,**—which may be given to a journeyman electrician having at least four years' experience, and who, after passing his examination successfully and complying in every respect with the prescription contained in the forms prepared by the examiners, has paid the fee prescribed by subsection (2) of section 8 of this Act.

·**License D,**— which is the special license authorizing a person with a knowledge of electricity employed in a factory, warehouse or public building subject to exceptions specified in subsections 2 and 3 of section 3 of this Act, to do work in connection with the repair and maintenance of electrical installations in the said public buildings, and the person applying for such license must pass an examination before the board of examiners."

**8.** Has reference to the meeting and organization of the Board of Examiners, the issuing of licenses and the fees. Subsection 2 of this section differs from the 1921 Act in that it definitely appropriates all the fees to reimburse the examining board for its services. This subsection is as follows:

### Fees·

(2) "If satisfied as to the competency of the applicant, the

board shall thereupon issue to such applicant a license in accordance with section 7 of this Act, authorizing him to follow, engage in or work at the trade or occupation of electrical installation in the Province of Ontario as specified under
that section, and the examination fee shall be $5.00 for con-
tractors and journeymen electricians and shall be applied in
reimbursing the examining board for its services, and the license
fee, which shall be $35.00 for contractors and $2.00 for journeyman electricians, shall be renewed annually upon payment
of $5.00 by a contractor and $2.00 by a journeyman electrician, and shall be applied in reimbursing the examining board
for its services."

9. Outlines the duty of the secretary.

10. Notes that candidates failing in one examination may
take additional examinations.

**Bonding**

11. "Every contractor before obtaining a license shall file
a bond with the board of examiners in the penal sum of two
hundred dollars, conditioned for the faithful performance of
his duty as licensed contractor, and for his not permitting any
electrical installation that he is called upon to do, to be performed by any person in his employ, except by such persons as
are authorized to do electrical installation under this Act, and
for his not violating any of the terms and conditions thereof,
or any amendment from time to time made thereto."

12. License may be revoked.

13. Penalty for failure of contractors to obtain license,
a fine of not less than $50. nor more than $200.

14. Penalty for journeyman not obtaining license, a fine
of not less than $5 nor more than $50.

15. Special license to industries.

16. Penalty to industries for not obtaining license, not
less than $50 nor more than $200.

17. Deals with the appointment of inspectors.

18 & 19. Requires the contractor or journeyman to furnish satisfactory proofs that he is licensed, subject to a penalty of not less than $50. and not more than $200.

20. This Act shall come into force July 1, 1922.

This Bill will doubtless come up for its second reading
at an early date and, in view of the importance of the matter
of licensing contractors, in Ontario, it is very important that
this Bill should have their careful consideration immediately.
It would be advisable to bring any criticisms to the attention
of the secretary of the Ontario Association, Mr. J. A. McKay,
at 24 Adelaide St. West, Toronto.

---

## Toronto District Electrical Contractors and Dealers Have Bumper Meeting

The regular monthly meeting of the Toronto District
of the Ontario Association of Electrical Contractors &
Dealers was held on March 22. The Canadian General
Electric Company. very kindly placed the assembly room
in their Wallace Avenue plant at the disposal of the Association and, in addition, provided the entertainment portion
of the evening's proceedings.

Following the general business meeting, with Mr. Drury
in the chair, a varied program was provided, consisting of
songs, jazz band selections, short addresses, etc. A particularly interesting item was a series of sales dialogues—on
vacuum cleaners, electric washing machines and electric
ranges. Some very clever suggestions were passed along
to the dealers, which will doubtless be used in their daily
sales talk.

Another interesting feature of the entertainment was a
model electric store that had been set up for the special
benefit of the members of the Association. A number of
interesting points were incorporated in the plan of this store;
for example, the difference between dilapated looking shelves with boxes of different colors and sizes, in comparison
with neater shelves, containing boxes of uniform color and
size, were shown. One corner of the shop was given over
to the display of lighting fixtures, the idea underlying this
display being that the various fixtures for the same room
should go together. These fixtures were shown in groups,
i.e., the ceiling fixtures, wall fixtures, etc., were shown in
the same little compartment, with no other designs present
to attract the attention while consideration was being given
to that particular design. Another feature was the placing
of the lamp counter at the extreme rear of the store, the idea
here being that more people came into the store to buy
lamps and, therefore, are obliged to walk past the other
appliances in the store to make their purchases. A couple of
good model window displays were also shown and Mr. Rim-
mer, who had this matter in charge. should be congratulated
on the results of his labor.

Another item on the program was a short talk on artificial sunlight, by Mr Dan Logan, followed by a moving
picture story illustrating the development of illumination from the earliest days, when the caveman produced his
fire by striking two flints together, to the most highly developed incandescent lamp of the present day.

Mr. C. S. Mallett, superintendent of the Wallace Avenue plant, made a few remarks of appreciation to the men
who had assisted—both members of the company and others
—in ensuring the success of the evening's program.

---

## Electric Home No. 2, Toronto

The second Electric Home exhibit and demonstration campaign being put on in Toronto by
the Electric Home League will open on Saturday, April 1, and will continue for two weeks.
As the first Home was shown in the northwest
section of the city, the second Home has been
chosen in the east end. The house is located near
the end of Gerrard St., just beyond Coxwell
Avenue, in what is known as the Kelvin Park
Estate. Mr. Kenneth A. McIntyre, who acted
as manager of the first Home and has since
visited a number of cities in Canada and the
United States in connection with his work for
the Society for Electrical Development, is again
in charge. The various committees are working
hard and duplication of the success of the first
Home is fully anticipated.

---

## The Development of Wireless

(Continued from page 38)

vacuum tube may be used as detector, amplifier or oscillator
it becomes a very useful device. It occupies very little space
in comparison with other equipment giving the same strength
in radiation of continuous waves, and in the smaller sizes is
so low in cost that it is within the reach of most amateur
operators. By means of vacuum tubes, wireless waves are
generated, detected and amplified. The whole system then
depends upon them. They have revolutionized radio communication and made possible very simple radio telephone
transmission, where even after hundreds of miles have
been travelled, the receiver signals, voice or music may be
made louder than the original.

In the sending of continuous wave radiations, where a
valve is used as a generator. the transmitter is particularly
quiet. the tapping of the key being the only noise.

# Electrical Co-operation in Vancouver

### Pacific Coast Province Meeting Returning Activity More Than Half Way— A Visit From Special Representative of National Contractors' Association

An extremely interesting and, to the Vancouver contractor-dealers, very profitable two days' visit from Mr. Laurence W. Davis, special representative of the National Association of Electrical Contractors and Dealers ended at a largely attended dinner meeting of the Electric Service League on the evening of March 2nd. Mr. Davis was the chief guest and the speaker of the evening, and two other visitors prominent in electrical circles in Canada, Mr. Pike and Mr. Carswell of the Northern Electric Company, Montreal, were also welcomed. Mr. Pike, who is no stranger to the fraternity in Vancouver was asked to address the meeting, and he gave a sort of curtain raiser by describing the successful Electric Home staged at Toronto, which he had visited before coming west. He also spoke of the similar display at Hamilton and at Montreal, the latter being at the time on exhibit.

Mr. W. G. Murrin, assistant general manager of the B. C. Electric Railway Co., who occupied the chair, introduced Mr. Davis with the comment that the visit of that gentleman a year ago had shown prompt results in the new life taken on by the Vancouver Contractor-Dealers Association. His present visit showed his real enthusiasm for his chosen work, as he had been in the hospital for an operation in San Francisco on his way to Vancouver.

As he was addressing not only the contractor-dealers but the Service League, Mr. Davis in his opening remarks put the question of what the Service League stands for. Answering, he said it was the grouping of the entire industry, the common meeting ground for all branches of the industry. The electrical business is unique in being so closely tied together. No other industry has all its branches so closely related. Sometimes, it seemed as if they were prone to forget that it was a number of units after all, and that one cannot lean too much on another. The object of the Service League was, he said, "to sell to the public the idea of electric service. But," he warned "it is not going to take the place of individual effort. Each of you has to build his own business. The league is responsible to the public and each of you has a duty, through the league, to the public."

Touching on a comment made during the early part of the evening on the small attendance of the contractor-dealer class at meetings of the industry, Mr. Davis came to their defence. He pointed out that while he was very much in sympathy with making the contractor attend all meetings, yet as a class these men had to give up their own time for all meetings held in business hours, while nearly all of the other representatives were coming, and quite rightly too, on their company's or employer's time. Referring to the Service League he said, "You have in the league, what they aim at in the East."

Analyzing the electrical industry, Mr. Davis showed how it was primarily based on the distribution of electric energy— to get the people to use the product of the central station. The manufacturer is an outgrowth of this need, as he has been making the appliances required to use electric energy. The jobber came into the industry to meet the need of the manufacturer for a means of distributing his products. There is a definite function for the jobber—to deliver to the retail trade. Finally came the contractor-dealer. And he has a problem no other branch of the industry has, "though you will recognize it as your problem too.". The manufacturer can study his costs and fix the price of his product; the jobber can do the same; the central station can arrive at

the cost of production and distribution of electric energy. They can also have technical and expert men to work out their problems of cost.

### Contractor-Dealer Represents Whole Industry

Then comes the final "contact" with the public. All the effort, all the production, all the output of the electrical industry must pass to the public through the contractor-dealer. And each time he faces an unknown quantity in every job that comes up. "Do you realize what he faces?" queried the speaker. "He has a very serious problem. The central station, the manufacturer, the jobber, all depend on the kind of an outlet they have through the contractor-dealer."

Speaking for the contractor-dealer, he said: 'We are the representative of the central station, the manufacturer, the jobber, to the public. How much of an advance have we made since the day of the first incandescent electric globe being put before the public? How do we stand before the public?' What is the impression we make? The industry cannot afford to let the contractor-dealer struggle alone with his problem," was the emphatic assertion Mr. Davis made. He showed how good will depended on the type of representative of the industry meeting the public. The electrical industry had not yet attained to placing the contractor-dealer on the same plane with the other branches of the industry. They hoped to see the time when this would be so.

"Here is the one idea. Let us make these men competent to be successful. Then it is up to them."

"It is easy to be up-to-date when you are prosperous" said the speaker, adding that the contractor-dealer could not be prosperous unless he was doing business on a profitable basis. How to get at the problem was through the Electric Service League and the Contractor-Dealers Association. They needed all the men in the industry "sold" on the idea. Contracting on competitive basis was only as high as the lowest bidder. And that is the real source of weakness. Some firms might become successful by getting out of the competitive field and working up a business on time and material basis. Such a course would come as a firm built up confidence on the part of the public. So long as contracting depended on competition all must be in the league and association work.

### Success is Based on Knowing Costs

There can be no success till the cost of operation has been met. There, said Mr. Davis, addressing himself to the contractor-dealers, is the whole secret of why the contractor-dealer has not advanced. Each job has a definite relation to the whole year's business, or to that of, say, a five year period. But this was no exception to the rule, for that fundamental principle applied to every part of the electrical industry and to every line of business in the world. Can the jobber say today that he will sell for 10 per cent less because he happens to want the business? He has to apply the same rule.

Following his theme Mr. Davis undertook to explain what is overhead—the cost of giving service to the public. When a contractor-dealer figured on a job, the cost of material and labor was only the beginning of his cost. When, he asked, are you going to pay the jobber? The first of the month. But at the first of the month the landlord had to be paid, and every one of the other costs of operating a business. While acknowledging that the contractor did "not always forget to add the overhead" he asked "Why don't you give the customer the material sometimes? Or the labor? It would be just as reasonable." On this question of figuring cost and profit he asserted that the average man actually does try to deceive himself into the belief that he has figured himself a profit when he submits a bid.

In a study of 2300 contractor-dealers made by the National Association, in 1920, the figures showed that the average cost of operation—the item of overhead—was 23.63 per cent. That was on a gross business of $87,000,000. The 1921 figures were not yet complete but they were going to be about 1½ per cent higher. But that is much lower than the average contractor can operate under. Of the contractor-dealers written to for the purpose of obtaining these figures, only half replied. Very naturally, the most satisfactory returns were reported by the largest concerns. Of these, there were five whose business ran from $800,000 to $1,000,000 a year. Their overhead was low. In a personal examination of the books of over 100 contractor-dealers, Mr. Davis said, there were only one or two who had got down to the average of 23 per cent overhead. Some were as high as 38 per cent and the average was nearer 29 than 28 per cent.

### The Man the World Seeks

"If you can succeed in making your overhead 18 per cent, then you are the man the world is looking for," said Mr. Davis. "But you cannot succeed if you are fooling yourself into thinking it is."

One of the greatest difficulties under which the contractor-dealer class labors is the fluctuating, varying membership. In Los Angeles a year ago, stated the speaker, there were 262 contractor-dealers. Today there are 498. But that is not the worst of the record. Every week last year in Los Angeles 17 men entered the business of contractor-dealer and 11 withdrew. They have a way of checking it up there, because each man engaging in the business is required to put up a cash deposit of $50. which is returned to him when he gives up business in the city.

"Does that make you realize what the industry is up against?" queried the speaker. "---A constant succession of failures."

Falling back on the trusty blackboard, with a piece of chalk Mr. Davis rapidly and graphically illustrated his discussion of the true basis of working out cost, the rock on which so many had struck and gone down. With a gross business of $10,000 per year, which was, he emphasized, the average for a large number of the smaller contractors, it required of that sum, $2,300 as a mimimum amount for the mere cost of doing business. "And of that $2,300 you can only pay yourself an average of $848, wages for the year. The rest goes in upkeep of "fliver" with which tools, men and materials are taken to the work, taxes, the tools required, rents, insurance, leakage on materials, and the hundred and one large and small items paid out in operating. Eighty per cent of the contractor-dealers are not doing a ten thousand dollar gross business yearly," was his assertion.

### Wrong and Right Estimating

Turning to details, Mr. Davis told of a case which he had gone into with a firm in a U. S. city, which he visited. They were doing a $200,000 business, and were considered successful and enterprising. Their claim was to a 20½ per cent overhead item. Testing their methods of figuring, he asked for an example: They took 23 per cent as an overhead, and gave him the following:

Cost of materials ........................... $100.00
Overhead .................................. 23.00
Profit .................................... 10.00
　Selling price quoted .................. 135.00

To show the error in this method of figuring, he pointed

out that 23 per cent overhead is of all the business you do, or of the selling price of all jobs. Then he gave a blackboard example of the correct method of working out the problem. "The selling price is your final figure. Therefore it is 100 per cent. With overhead assumed at 23 per cent, and profit say 10 per cent you have:

Overhead .................... 23 per cent
Profit ........................ 10 per cent
Cost of Material .............. 67 per cent
　　　　　　　　　　　　　　　　　100 per cent

"I do not want you to have any rule," urged Mr. Davis with great earnestness, "I want you to know why." Applying the figures of percentage to the job above referred to he worked it out as follows:

Cost of material, $100.00, is 67 per cent of selling price.
67 per cent=$100.
　1 per cent= 100÷67=$1.4925.
100 per cent=$1.4925 x 100=$149.25
23 per cent, or overhead=$1.4925 x 23=$34.23.

Subtracting this corrected amount for overhead from the quotation above, you have:

Selling price quoted .............. $135.00
Cost plus overhead ................134.23
　　　　　　　　　　　　　　　　　　　　　　——————
Actual profit ........................... .67.

Mr. Davis said the struggle of destructive competition is killing the business of the contractor-dealer. No fixed price basis is possible, and most of them were not working on the correct method of cost, as illustrated. The small contractor had no chance in competition, so he goes out to get the cheapest material possible. "What does inspection mean?" he asked. "You know it simply means that the city inspector says:'This job is just as poor as we will permit to pass'." That, he showed, was the wrong basis. He illustrated, by the plumbing industry, how the public were sold to the idea of comfort. They spend their money on it, on getting something that will benefit them. "But because we are always telling the public that their work ought not to cost so much, we cannot sell them service as it should be sold. Competition is the cause of all the evil."

It was a condition, he said, which cannot go on. The manufacturer wants to sell his products, the central station is looking for a load to even up the low places in the 24 hours, and thus cheapen the cost. "We have this job to do —it is nothing more nor less than putting the industry on a business basis. We do not want cost any lower than necessary to make the men in the industry successful. There would not be the constant influx of cheap men if we convinced the public we were trying to give them better service, that we were banded together to do it.

On the subject of residence 'wiring' he warned his hearers not to mix up the labor question with the educative campaign. Vancouver had led the way in securing a permanent secretary for the Electric Service League. Mr. Chatfield had a splendid record and was doing good work since coming to Vancouver.

### Business Best Game in the World

"Business is one of the best games in the world if you know the rules and can stop worrying over how to play," was one of the typical remarks with which Mr. Davis' address was freely punctuated. and it seemed to touch a responsive chord. On the question of cutting costs, speeding up collections, and thus increasing net profits, Mr. Davis' blackboard again came into play. He asked a number of questions, such as: "When do you send out your bills?" to which the answers varied widely. Some sent them out when the job was done, some when the time slips came in, some at the first of the month. He said the average was found to be five months. The man who shortened his collections to thirty days was turning his money over oftener than the man who waited five months. The same applied to buying materials. It was false economy to buy materials for five or six months ahead. Even a discount for quantity did not cover the loss in having the capital locked up. Often too the man who bought a thirty day stock took his discounts where the other could not, gaining an advantage there. Another saving in short date buying was avoidance of loading up with stock that became unsaleable. These points taken into consideration soon cut the overhead—it was the basis for doing good business.

### Function of Association

"While your association gives you ideas, helps you with your problems, there never will come a time," said Mr. Davis, "when your success will not depend on your own desires to use what your association gives you. You can sit down and ride on a street car, but in business you have always to keep on running. There was a danger in success, that a man might ask himself why keep on belonging to the Electric Service League. But this League is never going to put you in such a position. It is going to need you and you are going to need it." Of the 43 founders of the National association, 38 were still in it, and every man was successful. In their own League in Vancouver he said "You will get your reward in proportion to the time and interest you put into the League. It does not want your money—it wants your time, thought and energy—you.

Capital is now becoming available for building and there will be more and better business in the future. It was up to the contractor-dealer to make use of the opportunity. In no place in all his extensive travels in U. S. and Canada has Mr. Davis found the contractors as a class more awake to their problem, and more anxious to find the way out. There was a great improvement in merchandising methods in the stores, he had observed. Turning to the general subject of the central station and merchandising, he said that when the power load and the distribution of appliances was satisfactory there would be no need to ask them to drop out of merchandising. The central station is only "in" merchandising to see that the public gets adequate distribution of electric appliances and not to see what they themselves can sell. The central stations were educative pioneers in telling the public the advantages of electric appliances. Now it is different; there are many educative agencies. Concluding, Mr. Davis said if the League succeeds in achieving co-operation in the ranks of the electrical industry in Vancouver it will have proven its worth. The contractor-dealer must come into the League. The others should know his problems, know him, and be with him in his efforts to extend electric service.

### Mr. Pike Praised the "Electrons"

Mr. M. K. Pike, sales manager and Mr. Wm. Carswell, chief accountant, Northern Electric Company, who spent a week in Vancouver recently, on an annual visit to their western office, were guests on Friday, March 3, at the Vancouver Electric Club luncheon. Mr. Pike told the "electrons" that from what he knew of Vancouver they need not be afraid to tackle the enterprise of an electrical home exhibit. He told them how well the Toronto Electric Home had been handled in January, and what a success it had been in getting the public interest. He warned them that they need not spend too much money to get results. They could find some enterprising real estate man or housebuilder who would provide the building and pay for the wiring. The rest was comparatively easy. He detailed how the electrical industry in Toronto supplied the staff of demonstrators and salesmen each afternoon and evening, to explain and "sell the idea" to the public. Their greatest difficulty was in the limited space they had to handle the crowds to advantage.

### Vancouver Electric Club

If Thos. Pearson, member of the provincial legislature of British Columbia, were a poet, he would surely inherit the mantle of Dr. Drummond, for he has all that talented writer's knowledge of and love for the sturdy French-Canadian habitant. Like Dr. Drummond, Mr. Pearson was born in the province of Quebec, and made his home there until ten years ago, when he removed to the Coast. In an address before the Vancouver Electric Club at the weekly luncheon on Friday, 10th March, Mr. Pearson paid a tribute to the worthy character—and characteristics—of the French-Canadian. To live among whom was to respect and love them, he said. From a life-time spent among them, he had observed their thrift, their willingness to work, their kindly and hospitable natures, their loyalty to their leaders. Mr. Pearson gave some amusing reminiscences of political meetings and typical election speeches and election "dodges."

### Remodelled Electric Shop in Royal City

Chas. Moulton, New Westminster, has entirely remodelled the interior of his electric shop and has altered his show windows in unique manner. The hardwood floor of the window is removable, and below it at street level a cement floor and walls are exposed, giving the appearance of a basement for the display of electric washing machines, the result being very effective. The interior of the shop has been panelled throughout in laminated wood (a local product) and the panels have been stained in pleasing shades.

Vancouver Electric Service Leaguers have been busy welcoming their secretary-manager, Rey E. Chatfield, and his bride on their return from San Francisco. Mrs. Chatfield was Miss Olive Mills of Alameda, Cal., and the wedding took place at the home of her parents in that city on Feb. 21st.

The month of April is expected to see a new wireless plant in operation at Estevan, on the west coast of Vancouver Island. It is said this will be the most powerful apparatus on the British Columbia coast. At present there is a 5 kw. set, with a normal sending range of 400 to 500 miles, but the new equipment is a 25 kw. set, which will have a day range of 1500 to 2,000 miles and a night range of twice or three times that distance. The new station will be used entirely for ship work, handling messages to and from ships at sea.

A simple description of controllers for electric motors and definitions of the terms used in that connection have been compiled by the Electric Power Club and published in handbook form. Words which do not appear in the regular dictionaries are here explained in simple language and in addition to being a handbook for users of electrical control apparatus, it gives the meaning of terms with which all users of electrical apparatus should be familiar. The handbooks may be obtained from leading U. S. manufacturers of electric power and control apparatus, or from S. N. Clarkson, executive secretary, 1017 Olive St., St. Louis, Mo.

---

### Manufacturers and Jobbers

The annual meeting of the Electrical Supply Manufacturers' Association will be held in Toronto on April 6.

The annual meeting of the Electrical Supply Jobbers' Association will be held in Toronto, April 7.

## The Latest Developments in Electrical Equipment

### New Apparatus for Radio Reception

Another step in the advancement in radio reception has been made through the invention and manufacture of new apparatus by the Westinghouse company, whose nightly programs from its radio telephone broadcasting station, KDKA, are heard by thousands of radio enthusiasts. The new products are the Aeriola, Sr., and the Vocarola. The former is a regenerative receiver and the Vocarola is a reproducing apparatus which serves the purpose of a sound chamber. Through the use of the new apparatus, reception of programs broadcasted by radio telephone will be made comparatively easy. The Aeriola, Sr., is a single circuit tube regenerative receiver. It is contained in a nicely finished wood box with a cover and.

The Aeriola, Sr.

as supplied, includes a Brandes head set and special type WD-11 Vacuum Tube. The oscillating circuit of the receiver of the Aeriola, Sr., is identical with that of the Aeriola, Jr. It consists of a mica condenser of two steps in series with a variometer inductance. The steps of condenser are brought out to two binding posts, the lower capacity being used for wave

lengths up to 350 meters, and the higher capacity for wave lengths ranging from 300 to 500 meters. Tuning in between these steps is accomplished by the variometer inductance.

The Vocarola consists of a specially designed metal horn mechanically attached to the mechanism of a single Baldwin

The Vocarola

telephone receiver. The standard Baldwin mica diaphragm has been replaced by a special metal diaphragm which will stand practically any amount of abuse without damage. A large amount of experimental work was carried on by the radio experts before final decision was made on the horn and it is believed that the design furnishes as fine a quality of reproduction as can be obtained except through the use of a very elaborate sound chamber such as is found in high-priced talking machines.

The National Association of Cost Accountants, Bush Terminal Building, 130 West 42nd St., New York, has issued a bulletin entitled, "The Scrap Problem." The chief purpose of this booklet is to discuss the methods that may be best used in reducing losses from scrap, with particular reference to plants of some size, in the electrical industry.

## Striking Example of Flood Lighting

The illuminated tower shown herewith is that of Emmanuel Church, Baltimore, Md., the four sides of which were illuminated to a bright intensity on Christmas Eve and Christmas night, last year. This effect was produced by the use of 96 250-watt X-Ray floodlighting lamps, placed

in suitably designed reflectors. The floodlights were placed on the roofs of adjacent buildings. This is a new use for floodlights, and suggests a field where the sale of this class of equipment might be largely increased.

## Bell Ringing Transformer

The Canadian General Electric Co. report that they have recently made improvements in the design of their 5 watt, 25 and 60 cycle, household type of bell ringing transformer. In general their steel encased form of transformer is retained but the terminals, both primary and secondary, are made more sturdy and reliable by use of japan impregnated fibre insulating washers securely riveted to the frame with brass eyelets, thus overcoming any possibility of chipping or breaking of insulation under most severe usage. Improvements have been made in insulating the coils by the introduction of horn fibre anchor plates protecting the outside windings on the ends of the coils and also affording a secure anchorage for the primary leads. It is slightly smaller in the new design and with its baked japan finish has a very pleasing appearance. The transformer is designed to withstand an

insulation voltage of 2500 volts between primary and secondary windings and also between windings and the case, and each transformer is given this test before leaving the factory as well as a ratio checking test and a capacity test of ringing three bells in multiple. The coils are designed to stand a short circuit test at rated voltage with secondaries

closed for 3 hours with a temperature on the casing not to exceed 50 degrees Centigrade rise which means the transformer is not affected by an indefinite short circuit. Comparative tests on these transformers has brought out the fact that in ordinary household service the transformer only draws one third of a watt when contact is made.

## An Economee Switch

The Economee Rheostatic Switch Co., 3551 North Fifth St., Philadelphia, Pa., has recently placed on the market the "Economee" Switch here illustrated. In appearance, it is very similar to the ordinary snap switch; its operation is also the same. However, it differs greatly in operation, i.e., it is a liquid rheostatic switch and the current is applied

gradually, the resistance being controlled by a simple timing device. The cutting off of the current is accomplished instantly, as in the usual snap switch. The manufacturer claims that this gradual application of current prevents "starting" troubles in motors, lengthens the life of electric lamps, etc.

## Reversible Toaster and Grill

A reversible toaster stove and grill is now on the market, the product of the Triangle Appliance Mfg. Co., 160 N. Wells St., Chicago. Two slices of toast can be made at the

same time, then released from the holder and deftly turned onto a plate, without either touching the toast or burning the fingers. The same appliance becomes a grill by simply removing the toaster part of the device.

# Electric Railways

## Rehabilitation of Toronto's Electric Railway System—V

### Buses and Trackless Trolleys—Serving the Outlying Districts at Minimum Capital Cost and Maintenance

The adoption by the Toronto Transportation Commission of gasoline propelled buses and trolley buses for use in certain districts, may be considered as added proof of the fact that there is a distinct field for these types of vehicle. The Commission has been faced with the problem of providing transportation in the less thickly settled areas, and the choice of providing rail or bus facilities,—"mass transportation", as commonly defined, has not entered into the question and the decision of the Commission to use buses and eliminate the heavy cost of track construction is in keeping with a sound policy. That a decision has not been made as to the relative merits of gas bus, gas-electric bus and trackless trolley bus may be adduced from the fact that they are at present operating one or more of each type.

It would be unwise to assume that exact conclusions as to economy and usefulness of bus operation may be drawn from the results of such service in England or foreign countries. The Canadian field has peculiarities of its own; indeed, every district presents a special problem that can not be solved in any "slap dash" fashion, and a cautious policy is to be commended.

Most street railway men will agree that there are certain districts where service is essential, but where the demand is not sufficiently great to warrant the laying of tracks; admitting this as a basic fact, the installation of a service with the least possible investment is desirable and is answered by the use of some form of trackless service. This narrows the question down to the choice of gas, gas-electric or trolley bus, and it is proposed to mention some of the characteristics of each, classing the gas-electric type with the gas bus.

#### Comparison of Types

Winter weather has always been a bugbear to Canadian operators but in regard to this, one type of vehicle requires about as much attention as another. In the recent heavy weather in Toronto, all bus trips were made on schedule, thanks to the fact that there was an organization to take care of the job, the apparatus was got out early and the routes kept clear. The Commission possesses a specially designed 4-wheel drive tractor, with two ploughs attached, which did great work in snow clearing.

The special advantages of the gas bus include:

(1) Cost of power plant and overhead trolley eliminated.

(2) Greater freedom of mobility. They may operate anywhere that pavements permit.

The trolley bus also eliminates the cost of track and gives freedom of mobility as compared to rail cars. Its use of centralized power generation and its general similarity to rail cars are points that appeal especially to railway men.

The personal equation enters into the question to a certain extent. Some people like to ride on a gas bus simply

because it is a gas bus; whether they think they are getting back at the railway for its past sins, or that the upholstery makes them believe they are on a pleasure vehicle, it would be hard to say; the buses of the Toronto Transportation Commission always have a fair percentage on the top deck, smokers of course filling their reserved accommodation.

#### Maintenance and Depreciation

Every street railway has to carry a large investment in paving and the bus saves this expenditure. Obsolescence is also an extremely high figure in the accounts of a street railway company and it should be considered that a bus service cuts down this item, which is entirely apart from the cost of maintenance and depreciation of tracks, overhead line and power station facilities. When motor buses are used, they may be written off as they become worn out, and the company is not faced with the problem of wiping out a large portion of its investment. For an accurate comparison of cost figures, in the case of the rail cars, a proper allowance must be made for depreciation, followed by a correct estimation of obsolescence of track, overhead and power station facilities as well as cars, and the figures thus obtained compared with those relating to the cost of bus operation, which as is apparent, embodies a self-contained power unit and no investment in tracks and overhead line.

In many cases franchise taxes and other imposts for paving, sprinkling, sweeping, etc., now expected of electric railways, are not imposed on bus service. It also should be borne in mind that the idea of a 5-cent fare is not fixed in the mind of the public with regard to bus services, and for this and other reasons previously mentioned, bus service appears to be superior from a merchandising standpoint.

#### Cost of Operation

No absolutely reliable figures as to cost of operation seem to be available. It is too soon as yet to expect data of that nature from the Toronto Transportation Commission, and the table which is given below is from a United States source;—it relates to trolley bus operation on a typical service involving a fifteen-minute headway with buses weighing approximately 10,000 lbs. Total operating and fixed costs per bus mile are given as 24 cents.

Cost Per Bus Mile of Trolley Bus*

| | |
|---|---|
| Cost of conducting transportation | 7.3 cents |
| Cost of power | 2.5 cents |
| Maintenance of vehicle | 5.0 cents |
| Maintenance of way and structures | 1.05 cents |
| General and miscellaneous expense | 4.0 cents |
| Road tax | 1.0 cent |

*Sept. 10, 1921, Electric Railway Journal.

It is a very risky business to go far on such figures as the above for though it is a trite saying that "conditions are different", the fact remains.

The statement that power generated at central stations and transmitted over wires is cheaper than that generated in a self-contained unit must be borne in mind along with the consideration of the cost of feeder lines, crosstown lines and country lines where infrequent headway obtains and with the cost of investment in transmission lines, overhead lines, etc., not lost sight of. Also the cost of generating power in a self-contained unit would be considerably reduced by the introduction of a more economical fuel, of which several varieties are being experimented with at the present time.

### Will the Bus force Development?

It would seem that any company operating gas buses or trolley buses must take into consideration the ultimate replacement of these by rail facilities. Canadian cities spread, the average native-born citizen abhors the congestion typical of European cities and gets out somewhere, but the congestion keeps right at his heels. The buses serve him for a while, but experience shows that they tend to force the development of a district; just how far this item should be considered in the installation of buses is a problem that will only be settled by very careful consideration. It might be economy to install a bus service that would very shortly have to be replaced by rail service to secure adequate facilities.

### Personnel and Equipment

It is hardly necessary to point out that a great deal depends on the personnel secured, and that thorough training is necessary. The Toronto Transportation Commission has been fortunate in securing some men as bus operators who have had similar experience in England.

The accompanying photographs illustrate the various types of vehicles used by the Commission. The differences are interesting and so far, performance has not been remarkably dissimilar.

It is unfortunate that there is not at present available any entirely Canadian-made bus of proven service. The Veteran is built by a Canadian company but it is the first of its kind. It will be seen that the policy of the Commission has been to buy, if not Canadian, at least, as far as possible, British-made buses.

### The London Bus (A. E. C.)

An example of the type of bus which has proved useful abroad is illustrated in the machine acquired by the Commission from the Associated Equipment Company, of London, England. This company build the well known London buses, and the one shown is of the same general construction, with certain modifications necessary to meet the peculiarities of Canadian service. It is motored with a 4-cylinder

Associated Equipment Co's Bus

engine of 35 h.p., which is equipped with Zenith carburetor and the usual pump cooling facilities. The mechanical features include full floating rear axle, 4-speed selective gear box and dual ignition. A 50-gallon gasoline tank is carried. Standard practice of solid tires and cast wheels is followed,

as is also the feature of left hand drive which, though necessary to Canadian conditions, is a complete reversal of the English type.

Lighting is looked after by a 12-volt Delco system, and heating is from the exhaust. The weight of the bus is 13,000 lbs. Its dimensions are: length, 24 ft. 3½ in.; height, 12 ft. 4 in. As will be noted from the cut, it is of the double deck type (top deck removable) and has an inside front stairway.

### The Fifth Avenue Type

At this date the Fifth Avenue Coach Co., New York. N. Y., through the Packard Ontario Motor Co., have supplied the Toronto Transportation Commission with four gas buses, two more being on order. These machines include the single

Fifth Avenue N.Y. type, single deck

and double deck type, the single deck carrying 29 and the double deck 51 passengers and involving the use of single and two-man crews, depending on the type. Some of the later buses to be supplied are to have the convertible features. As may be seen from the illustration, the buses are designed on the front entrance, pay-as-you-enter idea. The double deck bus is exceptionally low, the total height being 12 ft., "from the ground to the top of a derby hat worn by a 6-ft. man on the top deck."

The motor is a 4 cylinder special silent sleeve valve type, of 40 h.p., lubricated by means of a force pump. Ignition is from a high tension magneto; the thermo-syphon cooling system is employed, and a special feature is the radiator, so designed that its freezing up would not result in damage. Selective sliding transmission is employed, with four gear changes in forward motion. The drive is of the external gear and plain bevel type. The rear axle is of heat treated forged steel, equipped with steel low level housing, and the front axle is of the I beam type, of heat treated forged steel. The wheels are of the steel disc type, the front ones being equipped with 34 x 5 in. single solid tires, and the rear with 34 x 5 dual solid tires. The service brakes are of the duplex air cooled type, and the bus is also equipped with external contracting cable emergency brakes. Worm and nut steering gear is employed, with the steering wheel located at the left. The gasoline is conveyed to the carburetor by gravity; gasoline tank is of 40 gall. capacity, and is provided with a large demountable strainer and a hand hole for cleaning. The interior is heated by utilizing the hot gases from the motor, these passing through protected radiator coils on their way to the exhaust opening. The interior is provided with artificial illumination by 12 lights, current being supplied by the generator and also by storage batteries.

### A Gasoline-Electric Unit

Another bus, which has been acquired by the Commission from Tilling-Stevens Motors, Ltd., Maidstone, Eng., is a gasoline-electric powered unit. A gasoline engine is used to drive a generator which supplies current to an electric motor used for the propulsion of the vehicle, there being no mechanical connection between the engine and the driving mechanism. The design follows automobile practice insofar as that the engine is located in front; it is rated at 45 h.p. and embodies the well recognized principles of water cooling and high tension ignition. Lubrication is automatic, the oil being circulated by a gear pump. The dynamo and motor are carried on two pressed steel frame members. They are built with yokes of high permeability magnet steel and main poles of the customary annealed iron laminations; the armature shafts are of especially sturdy construction and carbon brushes are used. Armature and field coils are tested at a pressure of 2,000 volts a. c. between windings and frame.

The controller and speed regulator are carried in separate aluminum cases. The controller is of the tramway type, with screw adjustment to the contact fingers. The speed regulator is of the multiple contact type, and operates by varying the resistance in the shunt field of the generator, and by shunting the series field of the motor.

The chassis is designed for a maximum load, including

A gasoline-electric Bus

body, of 4 tons 10 cwt. for passenger work, or a gross weight of 8 tons 8 cwt. The accompanying illustration shows the bus body construction. The passenger entrance

nance cost is reduced. Control being obtained by the throttle pedal and electrical field regulation lever, it is said that very easy starting and handling is obtained.

### Another English Bus

A gas bus supplied by the Leyland Motors, Ltd., differs essentially from the usual type built by that company (for

Leyland Bus to special specifications

use in England) in that it was designed to the specifications of the Transportation Commission and represents pioneer work in an attempt to overcome the more arduous Canadian conditions. Its high engine rating of 55 h.p. and feature of single or double convertible deck are innovations in Leyland design. The machine has a capacity of 59 passengers and will carry a substantial overload. It is equipped with special auxiliary shock absorbing springs, dual ignition and has a high ground clearance. The convertible feature allows of its use as either a one-man or two-man bus, in regard to which it will be seen that the stairs are just behind the driver.

The engine is, perhaps, the most interesting feature of this bus; it is the same type as used in the standard Leyland motor fire engine. Some features of its construction are: aluminum crank case; extra long water jacket; aluminum pistons; composite valves; centrifugal water pump; main gear oil pump; auxiliary pump and Zenith carburetor. An aluminum cone clutch, fabric faced, is used and the gears permit of four changes of speed.

The Leyland bus is being operated at present on the Mount Pleasant Road run, in the northern part of the city.

A Pierce-Arrow Bus to be used in Rosedale section, in N.E. Toronto

is at the rear, from the right side. The total seating capacity is 48, viz., 22 in the interior, and 26 on top.

It is claimed by the Tilling-Stevens Company that as there is no positive connection between the engine and back axle there is less strain on the working parts and mainte-

### The Pierce-Arrow Product

The Pierce-Arrow Company have been turning out buses for some time and one of these is included among the recent orders of the Commission.

The machine is equipped with a dual-valved, 4 cylinder

engine which will develop 46 h.p., a Stromberg carburetor is used and ignition is by the Delco system, with 2 sets of spark plugs. The length inside the floor is 17 ft. 19½ in. and height from floor to ceiling is 6 ft. 2½ in. Twenty-five passengers are carried.

Several design innovations are present; these include the use of a cushion wheel with semi-pneumatic tire; a special steel window casing which does not wear loose, and an ingenious seating arrangement. The illustrated machine differs slightly in wheels, body, steps and painting from the one ordered by the Commission. It is claimed by the Pierce-Arrow Company that one of these buses has averaged better than 11 miles per gallon.

The body of the machine is being built in Canada and it is expected to be in operation by April 1st. It will be used in the Rosedale district of Toronto, from the end of the Church Street car line.

### One of them Made-in-Canada

One of the buses purchased by the Transportation Commission has been made in Canada. This order was given to the Eastern Canada Motor Truck Company, Ltd., of Hull, Que., and the bus itself has been named "The Veteran," doubtless having in mind that Messrs. Bell & MacDowell are

The "Veteran"—Made in Hull, Que.

veterans of the late war, the former, Capt. G. G. Bell, D.F.C., Legion of Honor, and Croix de Guerre, and the latter, Maj. T. W. MacDowell, V.C., D.S.O.

The chassis of this bus is specially designed for bus service; it is under slung, with drop frame—floor level 27 inches from ground; wheel base 174 inches. This combination of low centre of gravity and long wheel base is a valuable feature in double deck buses.

The motor is a Buda transmission, constant mesh type, which facilitates the changing of gears; radiator cooling

Interior of "Veteran."

system, thermostatic. The wheels are cushion type, with demountable rims. Body, wood and steel construction; standard red color, gold leaf striping; interior, mahogany, with rattan seats. Ventilators draw air through grids in aga-

Rear entrance of "Veteran". Inside passengers leave by front door

sote ceiling. "Perfection" exhaust heaters are used; the body being lined throughout and specially constructed for Canadian winters. The windows are drop type and provided with curtains. All fittings, such as seat handles and railings are brass.

A unique feature of this bus is the front door exit, enabling full load of 51 passsengers—30 upstairs and 21 inside—to be unloaded in half the regular time; the operators report that a full load has been discharged in 30 seconds. There are two steps at the entrance, (all enter at rear) one to platform and the second to enter the lower section of the car.

### Trolley Buses

Trolley buses are also being given a thorough try-out by the Toronto Transportation Commission. The four thus far ordered have a chassis of standard Packard design, with such changes as are necessary to accommodate electrical instead of gasoline equipment. The body is of metal frame construction, the lower side and rear of car of ply-metal and the roof of water-proof veneer, sufficiently strong to carry a man. At the bottom of body at chassis frame the overall length is 20 ft. 7 in.; the width over posts outside is 7 ft. 4 in., and the height from top of floor to under side of ceiling is 6 ft. 2 in. Complete with motors, control apparatus and resistor, the chassis weighs 7,190 lb., there being an almost equal distribution of weight on the front and rear axles. The wheels are equipped with single solid tires in front and double solid tires at the rear. The buses are built with one double outward folding door on the right, in front, with wire plate glass in the lower section so that the driver can see that the step is clear; it is operated pneumatically, by a hand lever at the left of the driver. An emergency door is provided at the left side, near the rear, and is controlled by an electric lock in such a way that the breaking of a glass will automatically cause the lock to operate.

A stanchion, extending from floor to ceiling, is located at the rear end, and a similar one just inside the door at the front, to assist passengers in entering and leaving. A mirror is placed in the interior so that the driver can see the rear. Provision is made for route signs and advertising space. Adequate heating arrangements are made by the installation of six Cutler-Hammer, 500 watt, 286 volt, single unit truss plank heaters, two units in series, the wiring of which is installed in conduit, as is also that of the lighting circuits.

The two motors which are used are interchangeable and are duplicates of those in the safety cars at present operating in Toronto; they are connected through a universal joint to the automobile type of differential in the rear axle. They are rated at 25 h.p., 37 amps., at 600 volts.

The control is to be of the automatic series parallel type, having four notches in the series position, and three in the parallel. The first notch in the series position is independent of the accelerating device, so that very slow running is had for emergency use in slow traffic. All other notches are under the control of a time element current limit. The action

control resistor, all but two of the control circuits are at low potential, tending to provide more reliable operation of the control apparatus on account of the less liability of insulation breakdown. Overload protection is obtained by a two pole, single magnet, coil-operated line switch, mounted with an overload trip. This switch is also used to close or open the main circuits each time the master controller is moved to "on" or 'off" positions. This switch is mounted adjacent to the contactors on the central carrier, and an operating lever is brought through the dash for closing or tripping the switch.

The proper sequence of switches is obtained by a single drum master controller which is mounted under the hood and operated by a foot lever. A two-pole snap-switch, complete with fuses, is supplied for opening the control circuits for

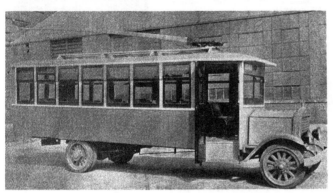

On the Mount Pleasant line Toronto will have the first Trolley Buses on this Continent

of this acceleration limiting apparatus provides a minimum time element for complete operation of the sequence switch when no current is flowing through the main motors. With current in the main motors, the speed of operation is approximately inversely proportioned to this current. Three operating positions of the master controller are provided for "starting," "full series," and "full parallel." The last two notches are running ones. Operation of the control apparatus is by a foot pedal for normal running. Reversal of running direction is by a switch type reverser, controlled by a lever which interlocks with the main operating pedal.

The control apparatus includes one set of six magnetically operated switches with control resister and sequence relay, one foot operated master control, one motor field reverser, one double pole, magnetically operated line switch with overload relay, one automatic time element sequence switch, one motor cutout switch and one set of grid resistance. All of the control apparatus is mounted beneath the usual engine hood, the resistors, control resistance, reverser and master controller being mounted on one side of the panel.

The power supply for the control circuits is obtained from the main supply circuit, and is reduced to the proper operating voltage by a control resistor. By the use of this

inspecting or repairing, and is mounted on the dash board.

A motor operated drum switch is used to obtain a time element current limit acceleration of the control equipment. This switch constitutes a secondary master switch for limiting acceleration of the vehicle, and is in turn controlled by the master controller. The speed of the drum of this sequence switch is limited by the current in the main motor circuits. However, the drum continues to move, although at slow speed until the full parallel position is reached, if the main motors do not accelerate. The sequence switch, therefore, provides for acceleration on heavy grades or with heavy loads.

A three pole double throw switch, similar in construction to the reverser, is furnished to cut out either of the two main motors. For collection of current, one No. 13 U.S. trolley base complete, with 16 ft. pole, swivel harp and 4 in. wheels is supplied. This single pole is equipped with two shoes, and an equalizing spring serves to maintain contact with both the positive and negative wires despite irregularities and differences in elevation as between the two wires.

---

W. M. Preston of Simcoe, Ont., has taken over the electrical contracting, plumbing & heating business of J. H. Madden, Norfolk Street, Simcoe, Ont.

# Current News and Notes

**Brantford, Ont.**

Officials of the General Hospital, Brantford, Ont., are considering the installation of an auxiliary lighting plant. The recent sleet storm, which damaged transmission lines and left that city in darkness, brought out the necessity of some plan whereby electric light, at least, might be obtained in cases of this kind.

**Brockville, Ont.**

Messrs. A. G. Dobbie & Co., King St. W., Brockville, Ont., have been awarded the electrical contract by Mr. C. S. Cossitt, Brockville, on his stores on King St., now undergoing alterations.

**Copper Cliff, Ont.**

Mr. Percy Morrison, Sudbury, Ont., has been awarded the contract for electrical work on a new school to be erected at Copper Cliff, Ont., at an estimated cost $110,000.

**Edmonton, Alta.**

The Hillas Electric Company, 10041 Jasper Avenue, Edmonton, Alta., has been awarded the contract for electrical work on a Home for the feeble minded to be erected at Oliver station at an approximate cost of $120,000.

**Forestburg, Alta.**

Mr. N. K. Lund, Forestburg, Alta., contemplates the installation of an electric lighting plant at Forestburg, in the near future.

**Goderich, Ont.**

The Water & Light Commission, Goderich, Ont., contemplates replacing the electric transmission lines on West and Waterloo Streets with heavier wires on account of overload.

**Hamilton, Ont.**

The Hamilton Street Railway Co., Hamilton, Ont., are now charging a straight five cent fare, the result of an agreement with the Hamilton city council. This provision may be withdrawn at any time by the council.

**Hull, Que.**

Mr. Ephrem Labelle, 9 Labelle St., Hull, Que., has been awarded the contract for electrical work on a store and residence recently erected at 13 and 15 Youville St., Hull, by Mr. J. Baillot, 32 Amherst St.

**Kitchener, Ont.**

The announcement is made that preparations for the Electric Show, to be held in Kitchener during the first week of May, are already well advanced and that considerable space has been subscribed for by manufacturers of electrical equipment.

**Meaford, Ont.**

The town council of Meaford, Ont., recently passed a resolution authorizing the town clerk to submit an offer of $12,000 to the Georgian Bay Power Company, Meaford, for its power plant, equipment, distribution lines, etc.

**Moncton, N. B.**

The Canada Gazette announces the incorporation of the Johnston Company, Ltd., with a capitlization of $49,000. The new organization will take over the electrical contracting and plumbing business formerly carried on by the Thomas Johnston Company, Moncton, N. B. The head office of the company will be at Moncton.

**Montreal, Que.**

Mr. J. J. Valois 444 Durocher St., Montreal, has secured the contract for electrical fixtures and wiring on the Harvard Apartments being erected at 5628 Sherbrooke Ave., Montreal, at an estimated cost of $80,000, for Dr. G. A. Belanger, 21 St. Famille St.

Messrs. J. A. Anderson & Co., 205 Mansfield St., Montreal, has secured the contract for electrical work on a warehouse at 30 St. Helen St., Montreal, that is undergoing alterations.

Mr. H. R. Cassiday, 255 Regent St., Montreal, has secured the contract for electrical work on stores recently erected at Park & Pine Avenues, Montreal, at an approximate cost of $86,000, the property of the J. H. Peck Estate.

**Niagara Falls, Ont.**

Work has been commenced on a transformer station at South St. & Victoria Avenue, Niagara Falls, Ont., being erected by the Niagara Falls Hydro-electric System, at an estimated cost of $125,000.

**Penticton, B. C.**

The Penticton Electric Company, Penticton, B. C., has been awarded the contract for a motor pump and equipment, for use in irrigation work at Ellis Creek & Skaka Lake Flats, British Columbia, by the city of Penticton.

**Seaforth, Ont.**

Reid Brothers, of Seaforth, Ont., have recently installed an Electrion isolated lighting outfit on the farm of Robert Campbell, who is one of the most up-to-date farmers in South Western Ontario. Mr. Campbell is reported to be very enthusiastic over his new installation. This is the second plant Reid Brothers have recently installed.

**St. John, N. B.**

The Webb Electric Company, 91 Germain St., St. John, has been awarded the electrical contract on a recreation hall to be erected on Prince Street, St. John, by the New Brunswick Division of the Red Cross Society.

**Toronto, Ont.**

Mr. J. E. Harman, 42 Arundel Ave., Toronto, has been awarded the contract for electrical work on four store buildings recently erected on Danforth Ave., for J. H. Rooke, 4 Playter Crescent, Toronto.

The Toronto Transportation Commission recently purchased the office building of the Security Life Insurance Co., at 37 Yonge St., Toronto, and are having it renovated for head office purposes.

Mr. W. J. Crone, 27 Sparkhall Ave., Toronto, has been awarded the electrical contract on a store and apartments building being erected at 1999 Danforth Ave., Toronto, by Mr. E. Dodson, 684 Woodbine Ave.

Mr. Geo. O. Lee, 6 Howard St., Toronto, has been awarded the contract for electrical work on two stores recently erected on St. Clair Ave., near Northcliffe, Toronto, for David Lavine, 71 Gloucester St.

**Vancouver, B. C.**

The Jenkins Electric Company, 539 Main St., Vancou-

ver, has secured the contract for electrical work on a warehouse to be erected at 168 Water St., Vancouver, by the J.
Leckie Co., Ltd., of that city.

The Jarvis Electric Co., 570 Richards St., Vancouver,
have secured the contract for electrical work on a building,
located at 445 Richards St., Vancouver, owned by Mr. J. H.
Roaf, 1285 Harwood St., that is undergoing repairs.

**Verdun, Que.**

Mr. E. Robillard, 1346 St. Joseph St., Verdun, Que.,
has been awarded the contract for electrical work on a store
building recently erected on Wellington St., Verdon, for
Mr. A. Cuerrier, 1342 Wellington St.

**Vernon, B. C.**

At a meeting of the local civic power committee, held
on March 8, it was decided that a preliminary report be obtained on the cost of installing a power plant at Shuswap
Falls, and building a transmission line into the city. Messrs.
Yuill & Knight will make the report. This power plant is
one of a number of alternatives discussed at the meeting,
which included a development on the Adams River; an offer
from the Kootenay Light & Power Company, and also the
advisability of adding another unit to the present plant. Mr.
Excell, superintendent, stated that the present equipment is
operating to capacity, with no opportunity for overhauling,
and urged the necessity for early action in relieving this
situation.

**Wellington, Ont**

The Canada Electric Co., 175 King St. E., Toronto, has
been awarded the contract for electrical wiring on a consolidated school being erected at Wellington, Ont., at an
estimated cost of $115,000.

**Windsor, Ont.**

A report states that the city council of Windsor, Ont.,
is considering a plan whereby householders may be assisted
in the purchase of electric ranges, etc.

**Winnipeg, Man.**

The Canadian General Electric Company, Winnipeg,
have been awarded the contract, by the city of Winnipeg,
for a number of automatic induction regulators.

The Canadian Westinghouse Company, Ltd., Winnipeg,
have been awarded the contract, by the city of Winnipeg,
for the supply of a number of pole type transformers.

The Moloney Electric Company of Canada, Ltd,
Toronto, have secured the contract for thirty 37½ kw., one
100 kw., and three 75 kw. pole type transformers, for the
city of Winnipeg.

Mr. S. H. Wilson, Curry Building, Winnipeg, Man., has
secured the contract for electrical work on a building, owned
by Messrs. Peace & Co., at 260 Portage Ave., Winnipeg,
that is undergoing repairs.

### District Branch in Guelph

A District Branch of the Ontario Association of Electrical Contractors & Dealers has been formed at Guelph. At
a meeting held on Friday, March 10, Mr. V. K. Stalford, of
Hamilton, was present and assisted in the formation of the
Branch. The following officers were elected: Mr. W. W.
Stuart, chairman; Mr. R. Christie, Mr. F. Martin, Executive Committeemen; Mr. George B. Grinyer, secretary-treasurer; Mr. W. W. Stuart, representative on the provincial
executive.

It is reported that the membership of this district is 100
per cent. At the inaugural meeting all themembers were
present and at the close of the business proceedings, which
were marked by great enthusiasm, the members dined to·
gether.

---

### Montreal "Home" a Great Success

The results of the Montreal Modern Electrical Home are regarded as satisfactory by the
organizers of the campaign. The results of the
other houses which will be opened are expected
to exceed those already obtained, owing to the
cumulative effects of the movement. One building company, which will construct fifteen houses
this season, has agreed to wire them according
to specifications supplied by the Electrical Co-
operative Association of the Province of Quebec.
Signs to this effect will be placed on the houses
during construction. A committee, consisting
of two contractor-dealers, an electrical engineer,
and an architect, will prepare plans and specifications for wiring houses.

---

A folder, entitled "Simplicity Fuses" has just been issued
by the Cote Bros. Mfg. Corporation, Chicago, describing
the new refillable fuse of that name, whose outstanding
characteristic is its simplicity. This fuse consists of only
two parts and the tiny refill cartridge. The salient features
are safety from shock while refilling, great economy, and
consistent operation over a long life.

---

The Electrical Equipment Company, of Johnstown, Pa.,
announce that S. R. Burd has been appointed acting sales
manager of the company, to succeed H. A. Selah.

---

### Curling Electrically

On March 9th & 10th, the Manitoba Electrical Association put on a very successful curling tournament, which
was held at the Granite Rink, Winnipeg, fourteen rinks taking part in this annual affair. Winnipeg Hydro presented a
fine Silver Cup to be held for a period of one year by the
winners, who are as follows:—Messrs. G. White, J. Stein-
hoff, G. Newman and J. C. Munro, skip. The runners up were
Messrs. H. C. Howard, R. D. Smith, F. J. Malby, and J. R.
Aikman, skip. After a very exciting finish for the consola-
tion event, J. M. Russell's rink, which was made up of
Messrs. A. Esling, J. Swan, E. V. Caton, and J. M. Russell,
skip, won out with a score of eight to seven over J. E.
Lowry's rink, which was composed of H. Smith, H. Allan,
F. Filer and J. E. Lowry, skip. The prizes were presented to
the winners by L. M. Cochrane of Cochrane, Stephenson &
Co., Ltd., at the fortnightly luncheon, held on the 23rd of
March. The prizes were generously donated by various mem-
bers of the Manitoba Electrical Association.

### Electrical Sales or Office Man

### CANADA'S MINERAL PRODUCTION

The Dominion Bureau of Statistics has published a preliminary report on the mineral production of Canada, which shows that the economic minerals produced during the calendar year 1921 reached a total value of $172,327,580, as compared with $237,422,357 for the preceding year. The report was prepared under the direction of Mr. S. J. Cook.

chief of mining, metallurgical and chemical branch of the bureau.

By classes the value of the mineral production during the year comprised metallics, $52,580,000; non-metallics, $89,405,000, and structural materials and clay products, $30,342,000.

The principal mineral producing province according to the returns for 1921, was Ontario, the mineral output from this province being valued at $54,505,770. British Columbia came second with a mineral production worth nearly $35,-000,000. Nova Scotia was a close third with $33,500,000; Alberta ranked fourth with $29,000,000; Quebec was fifth with $14,600,000, and Manitoba, Yukon Territory, New Brunswick and Saskatchewan followed in the order named with productions between one and two million dollars each.

The ten principal products of the mineral industries of Canada in 1921, arranged in order of value were, coal, $74,273,000; gold, $21,327,000; silver, $9,185,000; copper, $7,459,000; nickel, $6,752,000; natural gas, $4,902,000; asbestos, $4,807,000; lead, $3,855,000; zinc, $2,758,000, and gypsum, $1,726,000.

### LARGE HOTEL PLANNED

Plans for the world's largest hotel, to contain 3,000 guest rooms and to be built on Michigan Boulevard, Chicago, at a cost of more than $12,000,000, were announced recently. It will be 25 storeys high.

The announcement followed the sale of a block of land for $2,500,000, on which the new hotel, to be known as the Stevens, will be built. Construction, it was said, will start soon after May 1. The announcement was made by James W. Stevens, president of the LaSalle Hotel Company.

# 178,000,000 Acres
## of Agricultural Land

are located within the Canadian Provinces of Manitoba, Saskatchewan and Alberta.

In 1921, with less than ten per cent. of the arable land under cultivation, the three Prairie Provinces produced 260,000,000 bushels of wheat.

Agricultural wealth increased from $4,761,000 in 1880 to $636,501,000 in 1920.

These figures emphasize the enormous possibilities of the Canadian West.

To-day an important market, it promises to become of much greater importance within the next few years.

# Electrical
Engineering **News** Merchandising
Contracting Transportation

## PRAIRIE NUMBER
### To be published April 15th

offers the manufacturer of electrical equipment and appliances an opportunity to lay the foundation for the bigger business which he hopes to develop from this field. His message will be carried to the Central Station man and the electrical dealer in the most direct way—and by a publication devoted primarily to the Prairie Provinces. Saturation will be complete—every known executive in the Central Station and electrical merchandising business will be reached with this number.

**Guaranteed circulation 3,500 copies**          **Forms close April 12th**

*Send advertising copy to "ELECTRICAL NEWS" at any address below*

| Travellers Building, | 347 Adelaide Street West | Board of Trade Bldg. |
| Winnipeg, Manitoba | Toronto, Ontario | Montreal, Quebec |

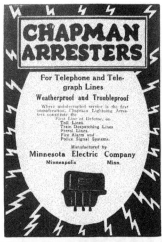
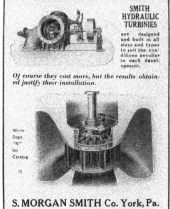

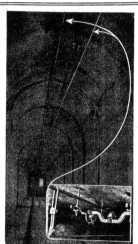

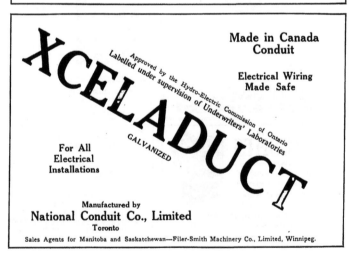

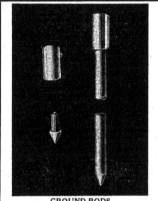

# CLASSIFIED INDEX TO ADVERTISEMENTS

The following regulations apply to all advertisers:—Eighth page, every issue, three headings; quarter page, six headings; half page, twelve headings; full page, twenty-four headings.

## CLASSIFIED INDEX TO ADVERTISEMENTS—CONTINUED

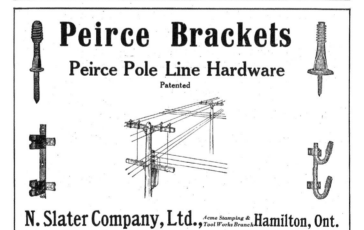

---

## CLASSIFIED INDEX TO ADVERTISEMENTS—CONTINUED

---

---

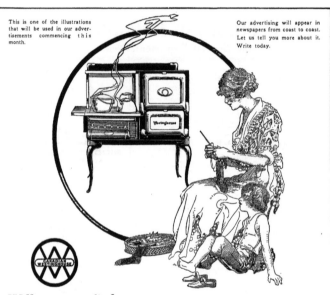

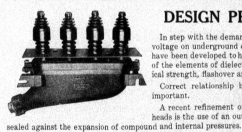

No. 8                                    Toronto, April 15, 1922

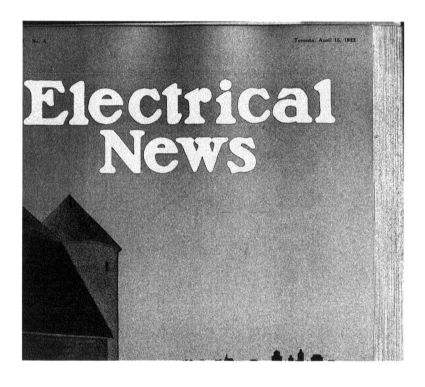

# Electrical
# News

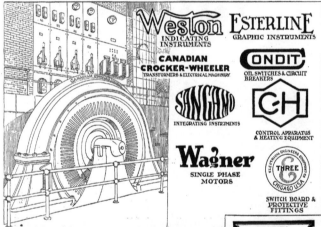

# *Northern Electric*
# Box-Line Fixtures are—

**Individually Packed**
**Photo - Labelled**
**Easily Identified**
**Quality Products**

## Seventeen Styles to Select From

This new plan of packing each of these Quality Fixtures in an individual carton and labelling the carton with a photograph of the fixture itself and all descriptive detail, saves unnecessary handling of the fixture and helps you serve customers quickly and intelligently.

You don't have to unwrap the package to see what it contains.

The fixtures remain protected until you are ready to wire them. After wiring return them to their cartons for delivery to the job of installation.

The finish of every fixture is protected and your shelves are always neat and clean.

*Write our nearest house for catalog and details of this "Picture-on-Package" Plan.*

## *Northern Electric Company*
### LIMITED

| | | | |
|---|---|---|---|
| MONTREAL | TORONTO | WINDSOR | CALGARY |
| HALIFAX | HAMILTON | WINNIPEG | EDMONTON |
| QUEBEC | LONDON | REGINA | VANCOUVER |

### *"Makers of the Nation's Telephones"*

**MANUFACTURING**
Manual Telephones
Automatic Telephones
Wires & Cables
Fire Alarm Systems
Power Switchboards

**DISTRIBUTING**
Construction Material
Illuminating Material
Power Apparatus
Household Appliances
Electrical Supplies
Power & Light Plants
Marine Fittings

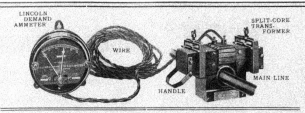

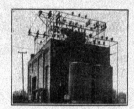

# A B C
## "ALCO"

# $110.<sup>00</sup>

### Retail Price

## The Best Dolly-Type Washers are ABC

You will always find a market for low-priced washers of fine quality, and you can get this business with A B C Alco's.
Four models: No. 51, single tub, engine drive, $72; No. 52, double tub, engine drive, $100; No. 51-E, single tub, ¼ h.p. motor (as above illustrated), $110; No. 52-E,

double tub, ¼ h.p. motor, $138. Liberal discounts. Thirteenth year! 1-¾ inch cypress tubs, quiet underneath drive, swinging power-driven wringers, heavy angle-iron frames, peg or disc-dollies. Good, sturdy, simple washers that stand up splendidly and satisfy your customers.

### The A B C Line is Complete!

To prospects interested in a rocking-tub type of washer you can sell the A B C Oscillator at $135. Has a heavy gauge copper tub, swinging electric wringer, handsome cabinet, quiet springless mechanism. Model 80-E. It is the first high quality washer of this type and size, bearing a famous maker's name, that has been sold at anywhere near so low a price as $135.
To the rest of your prospects, sell A B C Super Electrics. Five models: No. 60, semi-cabinet, galv. tub, engine drive $127; No. 60-E, semi-cabinet galv. tub, ¼ h.p. motor $155; No. 61-E semi-cabinet, copper tub, ¼ h.p. motor, $180; No. 65-E full cabinet galv. tub, metal wringer, ¼ h.p. motor $170; No. 66-E, full cabinet, copper tub, metal wringer, ¼ h.p. motor, $200. An opportunity to concentrate all your selling efforts in washers upon one famous make—to increase your profits—to reduce your overhead sales

costs—to simplify your correspondence, your ordering, your service and your bookkeeping—to combine your shipments and get maximum discounts—that is what you are offered in the complete A B C line.

For there is now an A B C—each one the best in its class—with which to sell practically every type of prospect. And every A B C bears the guarantee of the big pioneer firm of Altorfer Bros. Company—a company known nationally for its reliability and integrity.

Today! get full particulars concerning the A B C proposition, whether interested in the whole line or in any of its components. Liberal discounts! Real co-operation! Exclusive dealerships! Address any of the undersigned.

## Altorfer Bros. Company
### *Pioneer and leading makers of Washers and Ironers*
## PEORIA    :    :    :    : ILLINOIS

## C. D. Henderson, Canadian Representative    Box No. 123    Brantford, Ont.

MARITIME PROVINCES
Blackadar & Stevens,
Roy Building,
Halifax    -    -    N.S.
ONTARIO
Masco Co., Ltd.,
78 Richmond St.,    -    Toronto

WHOLESALE DISTRIBUTORS:
SASKATCHEWAN
Sun Electrical Supply, Ltd.,
Regina

ALBERTA
Cunningham Electric Co., Ltd.,
Calgary
QUEBEC
Dawson & Co., Ltd.,
148 McGill St.    -    -    Montreal

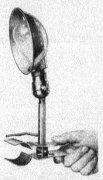

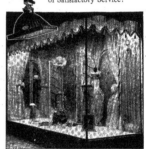

C.G.E. Twin Receptacle, Cat. No. C.G.E.
694 with Plate No. C.G.E. 698

## Doubling the Usefulness of One Outlet

IN moderate priced homes many people hesitate at the cost of an adequate number of electrical outlets, simply because the "idea" has not been sold them. Plenty of convenience receptacles are necessary for the appliances, so essential to modern housekeeping.

The market for convenience outlets is practically unlimited. The percentage of houses, even moderately well wired, is so small, that every home in your district offers a potential market for one or more C.G.E. Twin Receptacles.

The public are demanding, more and more, electric homes. Show your customers how their present houses can be converted into real electric homes by the aid of C.G.E. Twin Receptacles.

*"Made in Canada."*

# Canadian General Electric Co., Limited
## HEAD OFFICE TORONTO

Branch Offices: Halifax, Sydney, St. John, Montreal, Quebec, Sherbrooke, Ottawa, Hamilton, London, Windsor, Cobalt, South Porcupine, Winnipeg, Calgary, Edmonton, Vancouver, Nelson and Victoria.

# Wedding Presents

IT would, indeed, be hard to imagine a more suitable gift than a Hotpoint electrical device. At this time of the year, thousands of people are racking their brains, and wondering what to give for wedding presents.

Your display windows offer you a point of contact with these potential customers. They are in the market to buy—you have the merchandise for sale. The percentage of sales made from the sidewalk is surprisingly large.

Remember, one reliable electrical device sells another. Hotpoint appliances have stood the test of time. Create prestige for your store by specializing on this famous line. Order from your jobber, or direct from us.

*"Made in Canada"*

## Canadian Edison Appliance Co., Limited
### STRATFORD,   ONTARIO.

## O-B Insulators fit and are fit for every size of station

Whether it is a station that feeds a whole district or a sub for an individual consumer there are O-B Insulators that will fit in the layout.

Rigorous mechanical and electrical requirements will be well cared for. It is usually necessary to work within strict space limits—and O-B Insulators will meet that condition, too.

In addition to a wide variety of standard designs, you will find at the O-B Insulator Factory elaborate facilities for making special shapes.

*Look in Catalog No. 18
Pages 98-170*

## The Ohio Ⓑ Brass Co.
### Mansfield, Ohio, U.S.A.

Products: High Tension Porcelain Insulators; Trolley Material; Rail Bonds; Electric Railway Car Equipment; Third Rail Insulators.

# "A Washer You Can Sell and Forget"

**With All These Features :**

Tilting tub.

12-inch swinging wringer.

Galvanized iron folding tub stand.

Interchangeable dolly and disc.

Safety release on wringer.

An established reputation with more than 2200 in use in Winnipeg homes.

*Furnished in cypress and copper tub.*

We feel that we speak with authority when we say the WOODROW is

## "A Washer You Can Sell and Forget"

Thirteen years ago we started to sell electric washers, and during that time we have sold fourteen different makes.

In the spring of 1919 we sold the first WOODROW in Winnipeg, and now we have more than 2200 doing Winnipeg's weekly wash.

We have never sold a WOODROW that gave us any real trouble. Our service expense is nothing.

The WOODROW is as near perfect in construction as a machine can be.

The WOODROW is built to operate, either by gasoline engine, farm lighting plant or both direct and alternating current motors.

The new price of the WOODROW will amaze you.

*Satisfy your curiosity about this unusual machine by writing for catalog today*

# THE HOOSIER STORE

**Canadian Distributors**

**368 Portage Avenue** — — **WINNIPEG, MANITOBA**

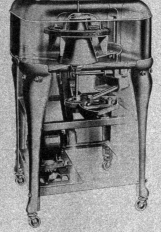

8,000 h.p. plant at Great Falls, Manitoba.

The WINNIPEG ELECTRIC RAILWAY CO. supplies

### Electric Power

industrial and other purposes of the largest manufacturers
he cities of Winnipeg and St. Boniface.

The Company is in a position to deliver dependable power to
et all demands, and invites enquiries as to

### Service and Rates

bundance of raw materials,
portation, many desirable

enings for
INDUSTRIES

will co-operate closely with you, giving
d opportunities in Greater Winnipeg.

# MPANY LIMITED
# LWAY    COMPANY

BA

A. W. McLimont
Vice-President

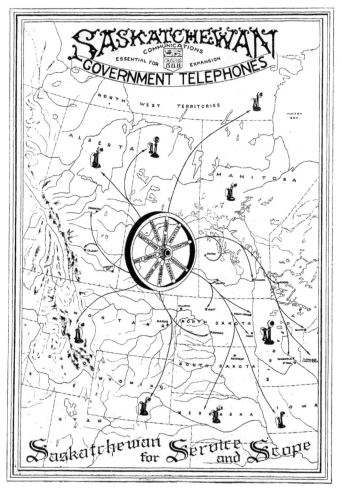

**SASKATCHEWAN TELEPHONE SYSTEMS**

SASKATCHEWAN HAS APPROXIMATELY 86% OF THE POPULATION OF CANADA AND 11½% OF THE TELEPHONES IN CANADA

SASKATCHEWAN HAS A TELEPHONE TO EVERY 8 PEOPLE AND THE FARMERS OF SASKATCHEWAN HAVE A TELEPHONE TO EVERY 9

CANADA RANKS SECOND AMONG THE NATIONS OF THE WORLD IN TELEPHONE DEVELOPEMENT WITH A TELEPHONE TO EVERY 10 PEOPLE

# Edmonton Telephone System

A public utility owned and successfully operated by the Corporation of the City of Edmonton, comprising four exchanges of Strowger type automatic equipment serving 12,900 telephones at cheaper rates than any other city of equal population.

Edmonton has 62,000 boosters for municipal ownership. The telephone system alone producing an annual surplus of $75,000.00 after taking care of all operating, maintenance and fixed charges.

Our Motto—**"Service"**

## City of Edmonton, Telephone Department
10,009—102nd Ave.          Edmonton, Alberta

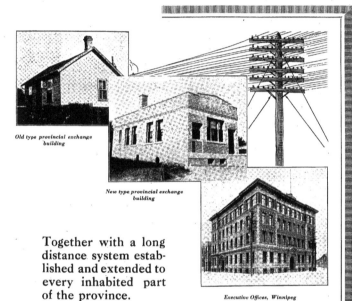

*Old type provincial exchange building*

*New type provincial exchange building*

*Executive Offices, Winnipeg*

Together with a long distance system established and extended to every inhabited part of the province.

Consisting of 11,168 miles of toll pole line and 52,367 miles of wire.

And ever continue to faithfully serve the citizens of the Province of Manitoba

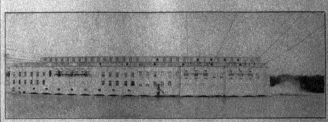

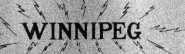

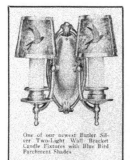

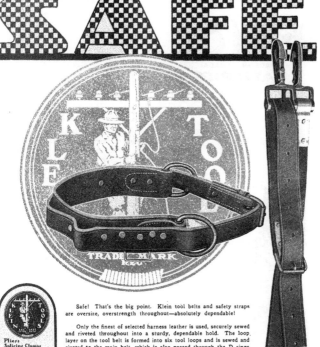

# Switch, Series 80,000

### *With Exclusive Features of Complete Protection, Complete Accessibility, and Exceptional Electrical and Structural Strength!*

Here is the safest, strongest, surest safety switch! A new Square D with radical improvements that command the attention and approval of electricians, contractors, safety experts, architects and purchasing agents! A new Square D that meets every practical requirement ever demanded of an enclosed safety switch!

**This is the Time to Act**
Write today for a Square D representative. He will show you the new radically improved Square D Switch, let you see for yourself why it is the one logical switch to stock and install, and give you the surprisingly attractive prices on the complete line from 30 amps. to 400 amps.

**SQUARE D COMPANY,** Walkerville, Ontario, Canada
*Sales Offices at Toronto and Montreal*          (s)

# Square **D** Safety Switch

# An Improved
## Square D Safety

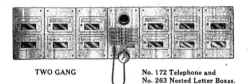

You simply push

Wire into wiremold

For its base and

Capping are rolled

Together to stay

And to make a flat pipe

That goes up in one piece

Rather than in two pieces.

Slip joints.

No threading.

*Write to-day for your copy of the Wiremold Catalogue*

# CONDUITS COMPANY LIMITED

33 Labatt Avenue,
TORONTO

602 Avenue Block,
WINNIPEG

# CONDUITS
### FOR
## INTERIOR CONSTRUCTION
## "GALVADUCT"

## "LORICATED"

*Look for this Label. It is your Guarantee of Quality.*

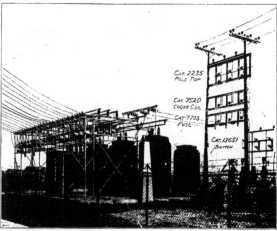

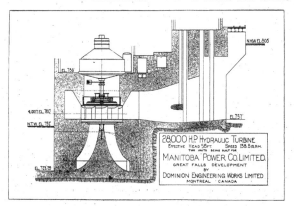

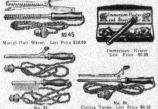

50 H.P. Link-Belt Silent Chain Drive—Canadian Cotton Co., Ltd., Hamilton, Ont. This drive is enclosed in a dust-proof, oil-retaining case. (Removed in photographing.)

A LTHOUGH great care is usually exercised in the purchase of machines, small consideration is generally given to the selection and design of the transmission.

In spite of the extremely high relative importance of the transmission and its great influence upon both cost of production and quality of product, it is often selected at random from one of the old forms of transmission, such as gears, belts and pulleys, etc., without consideration of the possibility of vastly superior results that could be obtained from the most suitable drive for the particular service.

Unfortunately, losses from such a source seldom show themselves directly to the management of a concern. Wherever conditions are such that the results produced by the transmission are recorded or noticed, Link-Belt Silent Chain will replace the older and more familiar drive in a large majority of cases.

If you desire to improve the character of work turned out by your machines or if you want to increase their output, give consideration to Link-Belt Silent Chain as the means of transmitting the power.

## CANADIAN LINK-BELT CO., LTD.
TORONTO: WELLINGTON & PETER STS.    MONTREAL: 10 ST. MICHAEL'S LANE

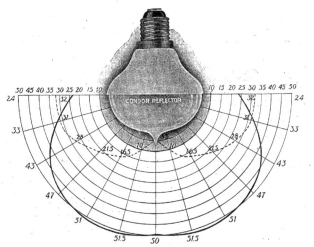

# A Better Light to Sell and Use

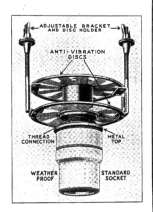

The drawing above is from a photograph showing the type of R & M Motors used throughout the mines of the Anaconda Copper Mining Co., Butte, Montana

THERE are few services which test the stamina of a motor as it is tested in the mine, for as a rule mine motors are installed in damp, out-of-the-way places where they receive the minimum attention as well as the maximum abuse. The dependable, year after year service thousands of R & M Motors are giving in leading mines throughout the world, the favor they have won among mine engineers and electricians wherever they are used, reflect the quality which is built into every R & M Motor.

While the service conditions you have to meet may not be comparable with those met in the mine, it is a comfort to know, when you have R & M Motors installed, that your power equipment will meet any test, however severe, when the occasion arises.

## The Robbins & Myers Company
## of Canada, Limited
### Brantford                              Ontario

# Robbins & Myers Motors

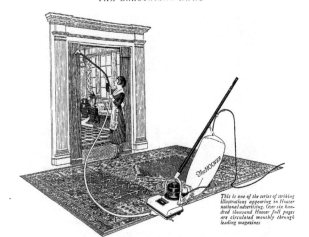

*This is one of the series of striking illustrations appearing in Hoover national advertising. Over six hundred thousand Hoover full pages are circulated monthly through leading magazines*

# Spring Housecleaning Time Is a Harvest Time for Hoover Sales

**The HOOVER**
*Converter for Attachments Is Patented*

IN order to make use of the strong suction of The Hoover for those purposes for which the machine itself cannot be used, viz.: for cleaning upholstery, curtains, hangings, books, moldings, mattresses, etc., the front end of The Hoover need only be tilted up a few inches. A converter is then easily slid under the suction opening, the machine is lowered and the converter is clamped in place. The conveniently long suction hose is easily slipped into the outlet of the converter, while any desired air cleaning tool may be attached to the open end of the hose. The dust can then be quickly collected from any location, high or low, without stooping or stretching. This converter is covered by a Hoover patent, granted March 26, 1912. Eighteen other patents protect the salient features of electric cleaner construction. Still other applications for patents are pending.

Practically everyone spends money in the spring in order to have rugs cleaned of the winter's accumulation of dirt—unless a Hoover has been in use.

Ordinarily the cost of such cleaning amounts to as much or more than the first payment on a Hoover. Yet what do people get for that money?

Nothing but a *superficial* cleaning of their rugs! And in a few weeks their rugs will be as dirty and unsanitary as before.

Do you see the favorable opportunity that this situation offers to interest an unusual number of people in The Hoover?

Take advantage of it! We will gladly assist you. For many years the Hoover organization has co-operated with dealers in conducting successful Spring Housecleaning Campaigns.

The plans we will help you to put into effect are therefore tried and proved—plans that many other dealers have tested and found highly resultful.

### The Time Is Here—Act Today!

Get into immediate touch with us! While Hoovers are year-around sellers, the spring housecleaning season is an excellent time to make your start.

Send today for a Hoover representative. Learn what a fine proposition we have for you as an Authorized Hoover Dealer.

THE HOOVER SUCTION SWEEPER COMPANY OF CANADA, LIMITED
Factory and General Offices: Hamilton, Ontario
MADE IN CANADA      BY CANADIANS      FOR CANADIANS

# The HOOVER
*It BEATS.... as it Sweeps   as it Cleans*

Generation, Transmission and Application of Electricity

For nearly thirty years the recognized journal for the
Electrical Interests of Canada.

Published Semi-Monthly By

# HUGH C. MacLEAN PUBLICATIONS
### LIMITED
THOMAS S. YOUNG, Toronto, Managing Director.
W. R. CARR, Ph.D., Toronto, Managing Editor.
HEAD OFFICE    347 Adelaide Street West, TORONTO
Telephone A. 2700

MONTREAL   -   -   119 Board of Trade Bldg.
WINNIPEG   -   -   -   302 Travellers' Bldg.
VANCOUVER   -   -   -   -   Winch Building
NEW YORK   -   -   -   -   298 Broadway
CHICAGO   -   -   Room 803, 63 E. Adams St.
LONDON, ENG.   -   -   16 Regent Street S.W.

ADVERTISEMENTS .
Orders for advertising should reach the office of publication not later
than the 5th and 20th of the month. Changes in advertisements will be
made whenever desired, without cost to the advertiser.

SUBSCRIPTIONS
The "Electrical News" will be mailed to subscribers in Canada and
Great Britain, post free, for $2.00 per annum. United States and foreign,
$2.50. Remit by currency, registered letter, or postal order payable to
Hugh C. MacLean Publications Limited.
Subscribers are requested to promptly notify the publishers of failure
or delay in delivery of paper.

Authorized by the Postmaster General for Canada, for transmission
as second class matter.

Vol. 31          Toronto, April 15, 1922          No. 8

## The Prime Need of the Electrical Industry in the Prairie Provinces Is—Organization

"The West" has always had a fascination for Canadians. It has always been characterized by so much vigor, so much optimism and, as a logical sequence, so much progressive development, that the whole continent has looked upon the "Last Great West" with envy and admiration.

It must not be supposed for a moment, however, that western progress has, in any sense, been a matter of chance. Of course, they had wonderful natural resources but, along with these, they had tremendous obstacles to overcome. The real cause of success has been their ability to "come back" in spite of all the hard knocks—the sportsman-like tenacity with which they have held on, worn out the opposing forces and finally won the game. The spirit of the real sportsman is, and has always been, the spirit of the West.

The West had, and has, wonderful natural resources and is turning them to good account. We in the electrical industry are more especially interested in the development of power and the extent of its application. For that reason we devote this issue entirely to a discussion of the various developments that have taken place in the three prairie provinces along the different electrical lines—telephones, power

plants, railways, merchandising, and organization. The information we are able to place before our readers will be found not only interesting but highly educative.

Speaking generally, the prairie provinces are not blessed with a very ample supply of water power. In this statement we must, of course, except that part of Manitoba which lies within transmission distance of the falls on the Winnipeg River—the power development in the Winnipeg district bids fair to become second only to the Niagara and the Montreal areas. But in the western section of Manitoba, in Saskatchewan and in Alberta—especially the southern portions of these provinces—the supply is far below the demand. As a compensation, however, there is coal of a fair quality—in abundance in Alberta and to a less extent in Saskatchewan. There is also crude oil and gas. All these are, therefore, playing a part in the western power development scheme.

Another outstanding feature of the western power situation is the wide distribution of small isolated farm plants. The number of these has been variously estimated from five thousand to ten thousand. They are the natural result of a shortage of water power; with the accompanying network of distribution lines and just another example of western resourcefulness.

In making a close study of western conditions, as we have been doing during the past two months in the preparation of this issue, there are certain phases that have obtruded themselves and certain suggestions we should like to make that, we think, would be for the betterment of the industry. One thing that has been emphasized time and again is the lack of general organization. The reason, we believe, is clear—the great distance between towns and cities—but the need is very urgent. There is no central station organization, no electrical engineering society, no contractors' or dealers' association such as we find in the older provinces and which have done much to hold the industry together and direct the development along right lines. The central station need could best be met, we think, by the formation of provincial associations that would tie in with similar associations in Ontario and Quebec. The same scheme would also seem to be workable for the contractors and dealers. Ontario is organized by districts, each district having its local association and executive and electing one representative to a provincial executive. If the western provinces would organize along similar lines—which seems to be the most logical scheme—there could then be formed a Do-

## Card of Thanks

*The work falling directly upon the publishers in the preparation of this special issue has been greatly lessened by the splendid assistance and co-operation from the various electrical interests. To the numerous contributors, to the officials of power companies and municipal plants, to the manufacturers, jobbers and dealers, to our Western Editor, Mr. E. H. Chapman, and to all others who assisted, we extend our warmest thanks.*

*Owing to lack of space, it is necessary to hold over several special articles for another issue.*

*The Publishers*

minion executive, made up from the various provincial executives. Organization of some kind, however, seems necessary for a continued economic development.

The other need that stands out prominently is for standardization of equipment and methods of installing it. Ontario follows regulations laid down by the Hydro Commission; Quebec follows the Underwriters' rules. The West in
this respect is very much disorganized; the matter is left in
the hands of individual inspectors who, with the best intentions and ability in the world, but because they are working
independently of one another, do not arrive at the same conclusion. This lack of standards is a handicap to everybody
and is undoubtedly retarding general progress.

We venture to suggest that if these two problems could
meet an early solution the effect would immediately be noticeable in more harmonious and economical development
all along the line.

## Western Electrical Men Strongly Favor Standard Specifications for All Canada.

The question of standard rules and regulations regarding requirements in the way of electrical material and
equipment, as noted in our last issue, has been receiving
a very considerable attention recently from a number of
Canadian electrical bodies. The prairie provinces are perhaps more interested in this question than any other section of the Dominion, inasmuch as there is an entire lack
of standardization. The viewpoint of the prairies is thus
of special interest at this time and we reproduce below
extracts from letters from representative men in the manufacturing, jobbing, and central station sections of the industry. The present moment is opportune for a discussion
of the subject by electrical men from coast to coast.

### Consumer has no Protection

"I understand that the Electrical News favors a movement for the establishment in Canada of an official central Inspection Bureau similar to the Underwriters' Laboratories of Chicago, but with extended powers to dictate
the class and design of electrical material and equipment
to be used in this country. An effort of this kind on the
part of your publication is a most praiseworthy one, and
I believe will receive the hearty support of every Canadian
interested in the electrical business.

"There is no question that if a device is unsuitable for
use in Ontario, it is equally unsuitable for use under the
same conditions in any other Province. This condition
does not now prevail. Equipment that can be used in
British Columbia may not be suitable for Ontario. Equipment suitable for Ontario may not be suitable for use in
Winnipeg. Under these conditions, the manufacturer must
consider the regulations prevailing in different localities
naturally giving precedence to that from which the largest
volume of business is obtainable. The jobber is faced
with the same problem, and the situation as a whole is
basically anything but economical.

"At present the consumer has really no protection
against buying electrical devices that are dangerous and
practically worthless. While the local rulings may forbid
the sale of any particular design of device, it is almost
impossible for the local inspector to prevent these devices
getting into the hands of users. If an official central Inspection Bureau with the proper Federal Government control were instituted, all approved appliances and supplies

would bear the official label, and in this way a uniform and
proper standard would be established."

### Industry Getting Nowhere

"I feel that the time has arrived for action to be taken
by the entire electrical industry in the Dominion, particularly the manufacturers, toward the establishment in the
Dominion of a centralized Inspection and Approval Bureau.

"At the present time local wiring inspectors have placed
on their shoulders the entire responsibility for the acceptance or rejection of practically all electrical products, and
from time to time the manufacturer is called upon to
change design of product, (and sometimes this is very
costly) to meet the requirements of a local inspector, and
whilst the product when fully changed to conform to the
inspector's requirements may do very nicely for the territory in which the inspector is operating, the same product
may not conform to the ideas of an inspector located in
some other centre.

"With conditions as they are at present you can plainly
see that methods in use are not getting the industry anywhere; consequently we are more or less working an "endless chain" proposition.

"Some of the wiring inspectors in Western Canada
recognize the inspection and approval service of the Underwriters' Laboratories of Chicago, and will accept in their
given territories electrical material approved by this body.
On the other hand, there are inspectors who absolutely
refuse to be bound by the rulings of the Underwriters'
Laboratories, and consequently reserve the right to set
up their own particular specifications, which are placed
before the manufacturer for him to follow, regardless of
the cost.

"I do not wish to give the impression that local wiring
inspectors are to be criticized for the work they are doing.
In fact, I think the industry at large is more to be criticized for not having taken steps some long time ago toward
the establishment of a centralized Inspection Bureau, whose
rulings and inspection service would be accepted by the
Fire Underwriters, wiring inspectors and others interested,
and consequently products that have passed through the
hands of such an Inspection Bureau would be saleable
throughout the entire Dominion, from Halifax to Vancouver.

"In conclusion, I am of the opinion the majority of
the wiring inspectors located in the Dominion of Canada
would welcome a move toward the establishment of an
organization such as I have referred to."

### Great Inconvenience

"With regard to the establishing of a Board centralizing the inspection of electrical material as per your editorial.

"We feel that there is a great deal of room for improvement over the present system of inspection and approval of various manufactured lines.

"In many cases in the past few years articles have
been bought by us in good faith, which were approved by
the Fire Underwriters in both the United States and Canada, and also passed the Hydro inspection in Ontario, and
yet were not approved in certain sections of the Prairies.

"Another thought that enters in here is the fact that
west of Winnipeg there is practically no inspection. A
great many times the articles inspected and passed by the
local Winnipeg inspector, for example, is a higher grade

(Continued on page 119)

# Conditions in the West Are Improving

### Emerging from Seven Critical Years of Hardship — Looking to the Future with Confidence — Winnipeg Municipal Affairs — Electric Railway Situation

#### By J. G. GLASSCO

The last seven years have been critical ones for the publicly owned utilities of Western Canada. In such places as Edmonton, Saskatoon and Regina, where coal is used as fuel, the hardships have been particularly severe. In addition to being seriously affected by the price of fuel, such places were still further handicapped on account of the fact that the major portion of the expense of running a steam utility is in the operating and maintenance costs. Hence, the soaring of all commodities and labor tended to accentuate a situation which has been aggravated by the large rise in fuel costs.

With those plants, such as Winnipeg and Calgary, that have water power, the conditions were not so bad, since the large percentage of their annual expenses are represented in fixed charges, and these did not vary.

In all places, however, where electric railway utilities were operated the high labor and material costs made satisfactory financial results impossible. Although in some cases tramway fares were largely increased to offset these conditions it is only now that the tram utilities are beginning to make two ends meet.

1920 saw the peak of labor and material costs, 1921 showing gradual decreases all around. In fact in Edmonton in 1921 the different utilities, all operated by the city,

### Winnipeg's Hydro Manager

Mr. John G. Glassco, manager Winnipeg Hydro-electric System, was born at Hamilton, Ont., in 1879 and was educated at the Public and High Schools in Hamilton, and at McGill University, Montreal, graduating with the Master of Science degree in electrical engineering. He was at first in the employ of the Montreal Light, Heat & Power Company, in 1900, but subsequently returned to the university as demonstrator and lecturer. In 1903 he went to Los Angeles, California, as superintendent of Meter and Repair Departments of the Southern California Edison Company. In 1906 he was appointed chief engineer of the Dominion Power & Transmission Company, Hamilton, Ont., and held this position until he moved to Winnipeg in 1909, as chief electrical engineer of the power construction work of the city of Winnipeg. He was appointed power engineer in 1911, in 1912 became manager of the Hydro-electric System and has occupied this position since that date. In the interval he has witnessed the growth and development of the system into a concern of which the assets exceed fifteen million dollars and where the amount of energy generated last year exceeded one hundred and twenty-five million kilowatt hours. Mr. Glassco is a member of the Engineering Institute of Canada, the American Institute of Electrical Engineers, Manitoba Club, St. Charles Country Club, Pine Ridge Golf Club and Old Colony Club. He is a consistent and enthusiastic supporter of every movement that is working for the development of the electrical industry.

showed a surplus of $351,980.00, whereas in 1920 there was a deficit of $94,867.00. The loss on the Edmonton Street Railway in 1920 slightly exceeded $200,000.00, and this was reduced to $47,795.00 in 1921. Saskatoon reports the same improvement as in Edmonton, in each case the Railway Utility being the laggard. This, of course, is to be expected as a Railway Utility's operating expense accounts for a much larger percentage of gross earnings than those of a light and power utility.

### Winnipeg Hydro Extending

For the first time in nine years the Winnipeg Municipal Plant shows a deficit. This can be accounted for by the heavy financial charges which the plant has had to carry as an additional burden on the hydro extension which has been completed and which has doubled the available electrical supply from our water power plant. The fixed charges on this extension approximated no less than 20 per cent. of our total annual expenses last year. It will readily be understood that the relatively high interest rates on the money raised during 1920 and 1921 contributed to this heavy overhead expense. The deficit was a small one of some $27,000. and in no sense materially affects the financial standing of the plant, since we carry forward a credit balance to Profit and Loss of $233,000.00, on operating account.

The Depreciation Reserve now has reached two million dollars in addition to the investment with the Sinking Fund Trustees which exceeds one million dollars, and the question arises as to how far the city is justified in increasing this reserve fund. The physical condition of the plant has been maintained at a very high standard, and some of our friendly critics contend that now with adequate reserves, which should be maintained at their present figure but not increased, further levies should be discontinued and should be applied to the reduction of rates. It is questionable, however, whether the time has arrived for a reduction in rates that are already undeniably low and which are the only thing in Western Canada that has not been increased during the last six years.

### Merchandising Policy

Another problem which has caused us a good deal of concern is the opposition that has developed on the part of some of the electrical dealers in Winnipeg to our merchandising department. We do not think this opposition is justified. Ninety per cent. of all large central stations in America are in this merchandising game and it is generally recognized now that the central station has an important and a prime function to perform in the matter of stimulating the sale of electrical appliances. Of course the difficulty arises through the difference in objectives between the central station and the dealers. In the case of the former, the prime motive is to increase the sale of electricity, whereas with the dealers their living and whole existence depends on a reasonable profit from the sale of these appliances. If, however, the central station plays the game squarely and maintains prices which will show all dealers a reasonable profit, then there should be no opposition to the central station, which, as a rule, bears a large share of advertising and which also performs the very important function of servicing and looking after customers' complaints. We have reason to believe that the opposition that developed last year in Winnipeg is gradually being overcome and that harmony and co-operative measures will obtain during the coming year.

The plant now has a surplus of 30,000 h.p., to which can be added in twelve months another 20,000 h.p. by the installation of five additional units. It should be noted that the transmission line and all switching in the generating station have been completed for the full complement of 100,000 h.p. development.

Survey work has also been started on the city's additional site at Slave Falls, and if, as expected, prosperous times come quickly to Western Canada, construction work on this site will be started within the next couple of years.

In spite of the business depression and unsatisfactory reports from commercial houses, our earnings have not only been maintained but show a healthy increase each year, the increased revenue coming from the domestic customers, with whom we have popularized electric ranges and water heaters. The power business is barely holding its own, while the commercial lighting revenue shows a small increase.

Looking to the future as we do with confidence the question arises in our minds as to whether Winnipeg, or, rather, Manitoba, will be able readily to absorb the large surplus energy which is now on hand in our case and which will be very much augmented when the new plant of the Manitoba Power Company is put into operation. This plant is expected to be completed in 1923, with an ultimate capacity of 170,000 h.p., which will bring the total power available from both plants in Winnipeg to the large figure of 300,000 h.p. According to statistics supplied by the Federal Government, Water Power Branch, in 1919, Winnipeg will not require this amount of power until 1945, but this estimate is considered to be a very conservative one, and if it is possible to induce large manufacturing establishments to locate in Manitoba, or if any progress is made in the development and use of electric power for our newly discovered mining area, even such a large amount of power should be readily absorbed. Winnipeg, itself, cannot be expected to take more than a fraction of this large surplus.

---

### Taxes and Capital

Opponents of capital must learn that enactment of a high progressive surtax in the income tax law was like the victory of Pyrrhus. Beginning at $5000, the rate increases until it becomes confiscatory for large incomes. If a man invests in a productive enterprise and receives an income of $500,000, the government takes $303,190 of it. If he earns $1,000,000, he can retain only $336,810. Should he earn $3,000,000, the government would take $1,393,000, and leave him $606,100. Courts hold that capital is entitled to a fair return. For two seasons the Soviet took the peasants' grain from them, and at the third season they refused to plant except for their own needs. Now, Russia is starving. Take away the incentive of a return and capital will not be risked in industrial enterprise. That is why new promotions are falling off so rapidly. Capital is industry's life blood, but the surtax is a halter around its neck to strangle it.---Barron's Weekly.

# Winnipeg, a Power and Industrial Centre

### A Manufacturing and Distributing City of International Importance—Great Falls Development Will Supply ample Facilities for Use of Country's Splendid Natural Resources

#### By A. W. McLIMONT.

You have asked me to let you know what I think the future possibilities of the Winnipeg district are and how they will be affected by the introduction of the new power from the Great Falls Plant of this company, which is now under construction on the Winnipeg River, and power from which is expected to be delivered in Winnipeg the latter part of next year.

While primarily a grain centre, Winnipeg is fast becoming a manufacturing and distributing centre of international importance. In the past ten years manufactured products in Winnipeg have increased 400 per cent, and as yet only the fringe of her manufacturing resources has been touched. There are numerous openings for new industries, and there is an abundance of raw material. The other essential to industrial development is Power, and this

Mr. A. W. McLimont, Vice-president and Managing Director Winnipeg Electric Railway Co., Vice-president Manitoba Power Co.

Nature has supplied in abundant quantities. In fact Winnipeg's power situation cannot be duplicated anywhere on the North American continent.

The source of the power is the Winnipeg River, which drains 55,000 square miles of lake and forest, and which guarantees to Winnipeg and vicinity a superabundance of cheap 24-hour continuous power which will be the greatest of all inducements for the location of industries. With the completion of our Great Falls development, there will be available 168,000 horse power, which will ensure the future industrial progress of Greater Winnipeg.

Western Canada produces immense quantities of grain, and to the present but little has gone through any process of manufacture in Winnipeg. Now that Winnipeg is assured of an adequate supply of cheap twenty-four hour

power it is only reasonable to believe that instead of paying high export rates on raw material and import rates on the product, we must manufacture our natural products here in the West, and no place has facilities that can compare with those of Winnipeg.

To produce the great qualities of the products of the soil that Western Canada has produced in past years machinery for tilling the soil has been very necessary and to the present nearly all of this has been imported assembled and high prices have been the result owing to long haul and high freight rates. Winnipeg, within a few years, will be a manufacturing centre for agricultural implements of all kinds.

North and east of Winnipeg and quite close to the Great Falls development are vast mineral fields now in the infancy of their development. Within a very few years large blocks of power will be required in this area both for operating the mines and the smelting and refining of the ores.

The manufacture of electric steel in this vicinity has passed the experimental stage and comparatively large blocks of power are now being utilized for this purpose. Extension of this industry on a large scale is now under contemplation and will be put into effect as soon as cheap 24-hour power is available.

I could continue indefinitely discussing what I believe to be the great future of the Greater Winnipeg district, and how Winnipeg's prosperity will be stimulated when our plant is finished, but I believe it is only necessary for me to say that the experience of every community where large quantities of hydro-electric power have been introduced is that with electric power comes capital—with capital comes new industries—with new industries come more people, and with more people comes prosperity for everyone in the community.

A deputation, representing municipalities in the Niagara district, recently visited Ottawa and waited upon the Minister of Railways, requesting that the government make improvements to the Niagara, St. Catharines and Toronto Electric Railway. The deputation stated that the roadbed of this line was in pressing need of repairs and was holding up street improvements in many of the municipalities through which it passed.

# The Power Resources of the Prairies

### Water Power, Coal, Oil and Gas Combine to Provide a
### Vigorous, Aggressive People With Ample Means for
### Developing Their Tremendous Natural Resources.

#### By Water Powers Branch, Department of the Interior, Ottawa.

Pioneers in a new country are usually at first dependent on their own muscular efforts for the power necessary for their various labors; as settlement increases and wealth accumulates it becomes possible to employ machinery driven by steam or oil engines, small towns appear supported by and supplying the wants of the inhabitants of the surrounding territory. With the passage of time and the successful development of the country, pioneer towns become cities and new towns grow up, the inhabitants of which, conscious that the pioneer stage is past, demand and obtain for themselves the amenities of civilization in the form of public utilities—water supply, drainage, electric light and power and finally street railways. These public utilities need the support of both population and industry, and industry needs power, so that the development of a new country is closely linked with its power resources.

The Prairie Provinces provide examples of all stages of civilization ranging from pioneer in the north, to the most modern in the south, and the following brief discussion of the power situation shows both how well equipped the more densely populated sections are and how great are the power resources of the whole area for future requirements.

## ALBERTA

The present thickly settled portion of the province is the south-east—this is almost entirely underlain by coal, has considerable natural gas resources now in use and good prospects of oil developments—the public utilities and manufacturing industries are largely operated by fuel and it appears must continue to be. In the north-west and north the large water-powers, the natural gas and oil fields, and vast bituminous sand deposits, will provide ample power for future development.

The province possesses available water power to the extent of 475,000 h.p. for maximum commercial development. The principal developments are those on the Bow river near Calgary with 32,380 h.p. installed but the majority of the water-power is considerably north of Edmonton and is not accessible to the, at present, most densely settled portions of the province.

### Coal

Alberta is by far the greatest of all the provinces of Canada in coal resources, possessing as she does 88 per cent of all the coal in Canada and 1/16th of the coal resources of the world. As respects Canada the comparison may be shown thus in millions of tons:—

| | |
|---|---|
| Alberta | 1,182,572 |
| British Columbia | 83,828 |
| Saskatchewan | 65,943 |
| Nova Scotia | 10,715 |

The two classes of coal in Alberta are thus:—

| | |
|---|---|
| Anthracite | 1,183 |
| Bituminous | 217,593 |
| Lignite | 963,796 |
| | 1,182,572 |

Much of this "lignite" is of high grade, suitable for steam raising and domestic use. The anthracite is found at Bankhead and along the eastern slope of the Rocky Mountains and is also reported 200 miles northwest of Edmonton. The bituminous and lignite coals are distributed over the greater part of the province including the Peace River District.

In 1920 the output of coal in Alberta was the highest on record and constituted 41 per cent of the total Canadian output.

## Oil

So far the recorded production has all been from the Turner Valley field 33 miles south of Calgary, this yielding a light oil as follows:—

| | Bbls. | Value | Per Bbl. |
|---|---|---|---|
| 1915 | small | | |
| 1916 | small | | |
| 1917 | 8,500 | $63,302 | $7.45 |
| 1918 | 13,040 | 100,004 | 7.70 |
| 1919 | 16,437 | 97,841 | 5.95 |
| 1920 | 11,718 | 75,295 | 6.42 |

indications are, however, both numerous and promising over a wide area and active investigation is proceeding.

In the south the formations are similar to those of Montana where large producing wells have recently been located near the international boundary. Mr. Ommaney, Investigation Engineer, C.P.R., states that:—

"The oil in these two great producing Montana wells undoubtedly originated from the Devonian formations which extend throughout the whole of the vast Canadian territory to the north up to the newly proven field in the far northwest."

The latter refers to the oil strike by the Imperial Oil Co. at Fort Norman on the Mackenzie river, reported at 1,000 barrels per day. This however is far north on the Alberta boundary.

Along the Athabasca river, near Fort McMurray, 200 miles north of Edmonton, are vast deposits of bituminous sands, outcropping for some 175 miles, 150 to 225 feet thick and estimated to cover an area of 750 to 1,000 square miles with probable wide extensions under heavy cover. (Dr. S. C. Ells.) Some 72 samples have been analyzed by the Department of Mines. Seepage of gas, heavy oils and tar are found throughout this district and in addition to probable wells the bituminous sands themselves are the potential source by treatment of "an enormous source of crude oil, fuel oil and their derivable by-products." (G. G. Ommaney). Six companies are reported to be drilling in this field. The known deposits are estimated to hold some 30,000 million barrels of oil.

Near Peace River Landing oil has been found in two wells in apparently commercial quantities.

## Oil Shale

Shale occurs at many points and the bituminous tar sands are largely overlaid by shale but no reports as to any extensive oil bearing shales appear to have been made.

## Natural Gas

There are a number of large gas fields in active operation. The Medicine Hat field has 33 producing wells, the Bow Island field 21, the Viking field 9, and there are several other smaller fields. The total production has been:—

|      | M. Cu. Ft. | Value       |
| ---- | ---------- | ----------- |
| 1916 | 6,094,231  | $1,113,296  |
| 1917 | 6,744,130  | 1,299,976   |
| 1918 | 6,318,389  | 1,358,638   |
| 1919 | 8,230,838  | 1,365,127   |
| 1920 | 5,633,442  | 1,181,345   |

### Market for Power

The natural resources of the Province of Alberta are large and varied and the manufacturing industries show rapid growth as illustrated by the following Dominion Census extracts:—

|          | Invested Capital | Value of Product |
| -------- | ---------------- | ---------------- |
| 1916     | $42,239,693.     | $30,592,833.     |
| 1919     | 66,673,667       | 94,855,759.      |
| Increase | 58 per cent      | 210 per cent     |

It may therefore be expected that there will be a rapidly growing demand for power.

## SASKATCHEWAN

Saskatchewan possesses available water power to the extent of 513,000 h.p. at ordinary minimum flow and 1,088,000 h.p. for maximum commercial development, but none has yet been developed. That within the settled districts in the southern portion of the province is largely on wide rivers possessing little fall, so that a large demand for power must first exist to justify the cost of the necessary dams. In the northern half of the province there are fine water powers available for future needs and good lignite south of Churchill river.

### Coal

The chief power resources for present needs is in the vast beds of lignite in the southern half of the province and it appears assured that the work of the Lignite Utilization Board of Canada will shortly: "result in the establishment of an industry of national importance" and that this fuel will be made available at favorable prices for all purposes. The process developed by the Board results in producing one ton of briquettes, equal in heating value to anthracite, from two tons of inferior fuel. In his evidence before the Special Committee of the House of Commons on the Future Fuel Supply of Canada last May the president of the Lignite Board said that the price of anthracite in Winnipeg was then $25. per ton and that: "It is difficult to think that we should rise above $12. (for briquettes), although I do not want to be quoted."

The deposits are estimated at 65,943 million tons. These occur mainly along the whole width of the extreme south and over a large area from the southwest border towards the neighborhood of Saskatoon, these areas being well served by railways; the chief production is in the Estevan district. Deposits are also reported at Lac La Ronge, south of Churchill River, and are said to be of better quality than that in the south. The quality generally is low but some of the seams being mined are from 8 to 15 feet thick.

The province is not yet a factor in the coal situation of Canada, the annual production being under 350,000 tons.

but it appears that the work of the Lignite Board will soon cause it to become so.

### Oil

While traces have been found they have not been in commercial quantities and the opinion seems to prevail that the geological structure does not warrant the expectation of this.

### Market for Power

Saskatchewan is the leading grain producing province. The southern part is of the Plain Region, but the northern part is of the Laurentian plateau, underlain by rock and having some mineral possibilities. The manufacturing development has so far been comparatively small, but as the population is now over 760,000 it is likely to considerably increase. Many large milling plants exist and development may be expected in the manufacture of leather, wood-products, bricks and pottery, flax, wool, etc. The forest resources are considerable and in addition to supplying local needs, quantities of sawn lumber are shipped out of the province.

## MANITOBA

Manitoba possesses exceptionally large and well distributed water power resources. The available power at minimum flow is 3,270,000 h.p. and the maximum commercial development would be about 5,770,000 h.p., of which some 250,000 h.p. is now developed or under construction for the supply of Winnipeg and district.

### Coal

In coal resources the province possesses about 176,000,000 tons of lignite at Turtle Mountain in the extreme south but this supply has not yet been appreciably developed and practically all coal is imported. It is probable that Manitoba will draw largely on the Saskatchewan supply when the briquetting industry is established (see remarks under "Saskatchewan") and later develop a similar industry of her own.

Of oil or gas in commercial quantities there appears little prospect as the geological formation is considered unfavorable.

Manitoba is, therefore, particularly lacking in discovered fuel resources within her own borders but has such abundant and well distributed water power that future power requirements are well provided for.

### Market for Power

The thickly settled portion of Manitoba is mainly within a very small portion of the province south of Lake Winnipeg. In population and value of manufactured products it ranks as the fourth province of the Dominion and the manufacturing progress of the principal cities is quite marked; Winnipeg now ranks (1919 Census of Industry) in all respects as the fourth manufacturing city in Canada.

This "prairie province", of which only 5% is prairie and 75% is covered by forest, produces in the north copper, gold, silver and other metals, gypsum clay, shale, limestone, timber, etc., extensive mining developments having taken place in the Le Pas and Rice Lake districts. Manitoba "must henceforth take an important place amongst the mineral producing areas in Canada, the output of copper and gold." (Commissioner of Northern Manitoba). The mineral production in 1920 was $3,900,000.

It will therefore be seen that the power demand is already large and that it is likely to increase rapidly over a wide area.

## CENTRAL ELECTRIC STATIONS IN THE PRAIRIE PROVINCES

In considering the central electric station situation in the Prairie Provinces a survey of the topographical features of the country in their relation to the centres of population will show that at present it is only on the extreme easterly and westerly boundaries of our great central plains that any extensive hydro-electric developments may be looked for. Owing to the absence of rivers possessing any considerable fall along their courses the easterly portion of Alberta, the westerly portion of Manitoba and all of the well settled part of Saskatchewan are forced to depend for the primary power for generating electricity on the generous and widely distributed supplies of native fuel, wood, coal, gas, natural and artificial and the combustible oils.

For the purpose of this review central electric stations are divided into the two main classes of generating and non-generating stations. As indicated by the name, generating stations, are those which generate power for sale, while non-generating stations are those which purchase the power they distribute from organizations operating generating stations. In some cases generating stations also purchase power from other generating stations either to supplement their own supply or to provide for the requirements of heavy peak loads. These two main classes are further divided according to character of ownership into commercial or privately owned and municipal or publicly owned stations. The generating stations are still further classified according to the type of primary power used,

into hydraulic and fuel power stations.

From the foregoing it follows that all central electric stations may be divided into the six fundamental classes of municipal hydraulic, commercial hydraulic, municipal fuel, commercial fuel, municipal non-generating and commercial non-generating. A comprehensive analysis of the central electric station industry in the Prairie Provinces along these general lines is given in the summary table. The data listed therein was collected in connection with the Census of Industry by the Dominion Bureau of Statistics in co-operation with the Dominion Water Power Branch, Department of Interior, and shows the status of the central electric station industry of the Prairie Provinces as at 1st January, 1920.

### Alberta

Alberta with a population of 581,995 has 52 central stations representing a total investment of $13,276,980. and having a total installed capacity of 78,905 horse power. These 52 stations comprise 47 generating stations representing an investment of $13,242,067, and 5 non-generating stations with an investment of $34,913.

The generating stations include 3 commercial hydraulic, 23 municipal fuel and 21 commercial fuel stations. One of the hydraulic stations maintains a fuel station which is operated continuously in conjunction with the hydro station. It is noteworthy that while only 3 or 6.4% of the generating stations depend on water as a source of power they represent an investment of $5,335,655 or 40.3% of the total capital of the generating stations and have an installed

CENTRAL ELECTRIC STATIONS IN THE PRAIRIE PROVINCES - GENERAL SUMMARY

capacity of 32,380 h.p. or 41% of the total primary power installation of the province.

The extent of the development of public ownership in the province is indicated by the fact that the investment and primary power installation in the 23 municipal fuel stations amounts to $7,395,462. and 42,818 h.p. respectively as compared with an investment and primary power installation of only $610,954. and 3,707 h.p. in the 21 commercial fuel stations.

There are three municipal non-generating stations with a total investment of $14,413. and two commercial non-generating stations with a capital of $20,500.

### Saskatchewan

Saskatchewan with a larger and more widely distributed population than either Alberta or Manitoba, possesses the greatest number of central stations, 61, but has the smallest investment, $6,758,769, and the smallest primary power installation, 42,806 h.p. As it is usually more advantageous to install small local fuel plants than to transmit electricity generated in fuel plants for any considerable distance this province has also the smallest number of non-generating stations, viz., 2.

### Manitoba

There are 29 central electric stations in the province of Manitoba, 24 of which have generating machinery installed and generate part or all of the power they distribute, the remaining five purchasing all of the power they sell from other organizations.

The total investment amounts to $16,914,922. of which $16,617,394. or 98.2% represents the value of the 24 generating stations, and $297,528. or 1.8% the value of the 5 non-generating stations.

The total primary power installation of the province amounts to 75,892 h.p. of which 72,655 h.p. or 95.8% of the whole is installed in the 4 stations depending on water as the source of their power, while the 20 fuel power stations have an installed capacity of only 3,237 h.p.

### Summary

The three Prairie Provinces have a total of 142 central electric stations representing an investment of $36,960,671. and having a total primary power installation consisting of 29 hydraulic turbines totalling 105,085 h.p., 83 steam reciprocating engines totalling 23,395 h.p., 22 steam turbines totalling 62,767 h.p. and 88 gas or oil engines totalling 7,406 h.p.

The total capacity of the electric generators installed amounts to 150,405 kv.a. of which 169 alternating current generators account for 145,976 kv.a. and 50 direct current generators for 4,429 kw.

Three of the hydraulic stations maintain fuel power auxiliary equipment consisting of 9 prime power units totalling 20,450 h.p. and 9 electric generators aggregating 13,100 kv.a.

As the population of the southerly portions of these three provinces advances in numbers and prosperity it may be expected that there will be a corresponding increase in the number of fuel plants and a considerable extension in the transmission systems of the hydraulic power organizations. As the population spreads northerly water powers at present beyond the range of economic transmission of electrical energy will be developed to meet the needs of the advancing population.

## Winnipeg Electric Railway Company's Uphill Fight

### By our Winnipeg Correspondent

Public utility companies throughout Western Canada, as indeed throughout all other parts of the country, are still feeling the stress of war-time operation and the aftermath of those unusual conditions. It is very encouraging to note, however, that many of these companies are making a good recovery and are in a stronger position today than they have been since 1914.

In this connection the record of the Winnipeg Electric Railway Company is worthy of particular mention. Harassed by jitney competition, by mounting costs of operation and withal a critical state of public opinion against it, the early part of 1917 found this company in a very perilous condition. Revenues of the company had dropped off during the previous three years at an alarming rate; and owing to the difficult conditions the company was forced to defer considerable maintenance.

In the fall of 1917 A. W. McLimont, one of the leading traction experts on the continent and a man of wide experience, particularly in the work of rehabilitating public utility properties, was appointed general manager of the company and almost from that point on the fortunes of the company took a decidedly improved turn. Within six months of taking charge of the property Mr. McLimont succeeded in effecting an agreement with the city council whereby jitneys were eliminated. A few months later the car fares were increased, giving the company more needed revenue to meet the higher operating costs, while an extensive programme of track and rolling stock rehabilitation was put into effect, with a resulting improved car service. Incidentally Mr. McLimont installed the three wire system for the mitigation of electrolysis thus relieving a condition which had been a cause for controversy between the company and the city authorities for some time.

But perhaps the most remarkable reform accomplished under Mr. McLimont's comparatively short regime has been the change in public opinion towards the company. Everywhere in Winnipeg there is to be found the desire on the part of the public to co-operate with the street railway company and boost its service. Undoubtedly, this change can be traced in a great measure to the open policy which Mr. McLimont has adopted in his dealings with the civic authorities and the public. In the newspaper columns and through the medium of the company's bulletin distributed on the street cars, Mr. McLimont has consistently discussed the various problems concerning the utility, frankly telling the public of the difficulties in the way and pointing out how the removal of the difficulties would be in mutual interests. The public have responded in every instance and have undoubtedly appreciated the manner in which Mr. McLimont has made good his promises and kept faith with them.

Another factor, and not by any means the least important in the remarkably successful administration which has characterized this company during the past few years, has been the manner in which the company has made itself a part of the public life of the city, as represented in its employees taking an important and prominent part in all movements for the welfare of the community. It is the policy of the company, inaugurated by Mr. McLimont, to get behind all such movements and link up with every organization which aims to make for a bigger, better and brighter Winnipeg.

Down-stream side of Power House, Winnipeg Municipal Plant at Point du Bois

# Winnipeg City's Hydro-electric System

### Wonderful growth in last decade—Kilowatt hours generated in 1921 five times greater than in 1912—A detail summary of generating and distributing equipment and plant

#### By E. V. CATON, Chief Engineer

The power house of the Winnipeg Hydro-electric System is situated at Point du Bois, on the banks of the Winnipeg River, 78 miles north-east of Winnipeg. Access to the site is obtained by 22 miles of standard gauge railway, owned and operated by the Department, connecting up with the Canadian Pacific Railway at Lac du Bonnet.

The Winnipeg River, with its tributary, the English River, has a drainage area of 52,000 sq. miles. The principal storage basins are the Lake of the Woods, 1500 sq. miles; Rainy Lake, 330 sq. miles; Lac Seul, 340 sq. miles. The total lake area in the watershed is 5,650 sq. miles.

Due to the nature of the country throughout the watershed the variations of flow, in the natural state, were comparatively low, minimum flow of 8,000 sec. feet, and maximum of 100,000 feet having been recorded. The annual variation is, however, much less than this and with the system of storage and control worked out by the "Water Power Branch" of the Dominion Government, a minimum flow of 20,000 sec. feet is assured.

At the power house site the river falls over a series of rapids which in their natural state gave a fall of 33 to 28 feet in a distance of approximately a quarter of a mile. By means of a dam across the head of these rapids a total head of 43 to 48 feet has been obtained and a pondage area of seven square miles. The dam across the head of the rapids consists of a rock fill dam extending 700 feet from the east bank, connecting up with a concrete spillway section 550 feet long; a second spillway 480 feet long spans a slight depression a short distance west of the main dam, while a third section 245 feet long and curving down stream forms part of the canal wall immediately above the intake. The forebay canal, on the west side of the river, extends 1200 feet down stream and conveys the water to the power house. A spillway, 225 feet, is placed in this wall below the intake, as shown. Across the south end of the canal and connected to the west shore by a wing wall is the power house.

The power house, a reinforced concrete structure 523 feet long, acts as a gravity dam at the end of the canal.

The water entering the power house, passes first through the trash racks and then through head gates into open wheel pits. The rack structure and head gate are covered in by a galvanized iron roof carried on steel trusses.

The power house is divided into three bays. viz: the wheel pits, generator room, switch and transformer rooms, the generator foundations being carried over the draught tubes. Sixteen wheel pits are provided and each pit will accommodate a 7,000 h.p. double runner horizontal water turbine. A 20 ton electrically operated crane runs the whole length of the turbine bay and extends outside the building on a gantry, below which is a spur line from the railroad which allows the loading and unloading of cars with a minimum of handling.

The eight original units were each provided with individual head gates operated off counter shafts by two 15 h.p. motors. The other units are operated with a travelling gate operating mechanism, which can place a gate into any required pit. This mechanism is driven by a 15 h.p. motor, mounted on the carriage. The original gates are timber sheathed on a steel structural frame and made in one piece, operated by screw hoists. The new gates are made in four parts, entirely of steel. In addition to the head gates, provision is made to install stop logs on the upstream side of all gates, if necessary.

The generator room is provided with a 40 ton electrically operated crane, running the length of the building. At the west end is an unloading bay and the standard gauge track runs from the outside into the bay a sufficient distance to allow of a car being placed entirely within the building. The basement is used for cable runs, pumps, etc.

The switch bay is laid out into a number of floors and compartments to accommodate the transformers, oil switches, bus bars, lightning arresters, control gallery and the necessary electrical apparatus. The west end, opposite the unloading bay, accommodates a fully equipped machine shop, offices, etc. The basement of this bay is used for storage rooms, etc., there being at the extreme west end a fireproof oil room having a capacity of 10,000 gallons of oil and a complete oil treating plant.

Between wheel pits 7 and 8 the generator room is set back to provide room for two water turbines in steel casings;

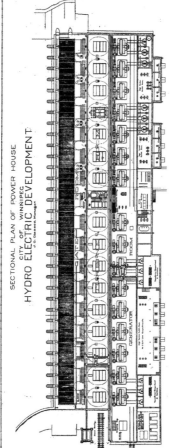

SECTIONAL PLAN OF POWER HOUSE
CITY OF WINNIPEG
HYDRO ELECTRIC DEVELOPMENT
J. G. Glassco, Manager

each turbine drives a 310 kv.a., 3 phase, 60 cycle, 230 volt alternator for supplying the auxiliary light and power.

### Water Turbines

There are at present installed eleven main units, consisting of:

Five—5,200 h.p., 164 r.p.m. Boving two runner horizontal type.

Three—6,800 h.p., 138½ r.p.m. Escher Wyss manufacture.

Three—7,000 h.p., 150 r.p.m. Boving manufacture.

Two—400 h.p., 400 r.p.m. Boving manufacture, for the auxiliary service machines.

All machines are provided with oil operated servomotors and actuating governors, the three 7,000 h.p. units being provided with the Woodward actuator type governors.

Oil for the governing system is supplied under pressure by three rotary gear pumps, driven by 40 h.p. induction motors. The system is arranged on the open tank principle and each machine has its own storage tank.

Duplicate high pressure lines are run the whole length of the building and arrangements made to connect each machine with either line, the discharge being carried to sump tanks in the basement, from which the oil is drawn.

A small air compressor with storage tank allows the oil level in the individual storage tanks to be adjusted as required. The pressure is maintained constant by the automatic unloaders on the pumps. Oil strainers are provided in the suction of each pump and provision is made to allow of the oil being treated as required. This is arranged for by having two sump tanks and a connection for a De Laval centrifugal separator, so that by cutting out one tank its contents can be thoroughly treated.

### Generators

The eleven main generators now installed consist of:

Five—3000 kv.a., 6600 volt, 3 phase, 60 cycle, 164 r.p.m., by Vickers, Limited.

Three—5000 kv.a., 6600 volt, 3 phase, 60 cycle, 138½ r.p.m., by Canadian Westinghouse Co., Ltd.

Three—6500 kv.a., 6600 volt, 3 phase, 60 cycle, 150 r.p.m. by Canadian General Electric Co., Ltd.

All generators are provided with their own direct connected exciters. The 5400 kv.a. machines are provided with temperature indicating coils, placed in the winding and connected to instruments on the control gallery. All bearings are self-lubricated but provision is made for water cooling, if necessary.

In addition to the above, two 310 kv.a., 230-volt, 3 phase, 60 cycle alternators, by the Swedish General Electric Company, are installed to supply power for local service in the building.

### Transformers and Switches

There are at present installed four banks of transformers, and provision is made for an additional bank when required. At the present time the following transformers are installed:

Six—3,000 kv.a., single phase, oil insulated, water cooled, 6600/66000 volt.

Three—5,000 kv.a., do.

One—9000 kv.a., three phase, do.

All are manufactured by the Canadian Westinghouse Company.

In addition, two 300 kv.a., 3 phase, 6600/230/115 volt oil cooled transformers are installed for local service work, one of which is supplied by the Canadian General Electric Company and one by the Moloney Electric Company. All single phase transformers are connected in banks of three and in delta on both high and low tension windings.

Cooling water is taken from a pipe header, connected to all of the wheel pits, and the discharge is run directly into the tail race.

For treating the oil, both a blotting paper and centrifugal filter are provided.

The control board is located on a gallery, in the centre of the power house. The operating board is of the bench type, all instruments and relay board being vertical. All switches are electrically operated from a storage battery.

The 6600 volt switching is located all in one bay on the generator floor level. The switches being arranged in a series of bays, each bay containing the switches to control four machines and one bank. Duplicate switches and buses are provided throughout. All switches are installed in concrete compartments with fireproof doors.

The switches for the 6600 volt control are of the G.E. H-6 type and provided with the sub-compartment disconnecting switch on the machine side and wall mounting type on the bus side.

The high tension switches are installed on a mezzanine floor, each line and bank in its own compartment. High tension switches are provided on each bank and tie switches by means of which any bank or line can be interconnected through a transfer bus running the whole length of the power house.

Lightning arresters are installed on all lines, on two of the lines these are of the electrolytic type with the impulse gap, by the Canadian Westinghouse Company, and on the other two the oxide film type with sphere gaps, by the Canadian General Electric Company.

The control and instrument circuits are wired up with multi-conductor lead covered cables, carried on angle iron racks. All leads passing through a test board before being connected to the various meters, controllers, relays, etc. The individual conductors are wrapped with colored braid and a

definite color scheme is maintained throughout the system.

All machine cables are cambric insulated, flame proof, braided cables, supported on a special structural steel and asbestos covered bracket.

The auxiliary apparatus in the power house includes a 150 kw. motor-generator set, an air compressor, storage battery, sump pump and other necessary equipment.

### Transmission Line

Four lines on two sets of steel towers and built along a

Showing general plan of development

private right-of-way owned by the Department, run between the power house and the city.

On one line the towers are spaced 600 feet and consist of alternate braced and flexible towers. The conductors are arranged on a six foot equilateral triangle, four conductors being on one plane and the other two conductors vertically above.

The other line of towers, which is spaced an average 400

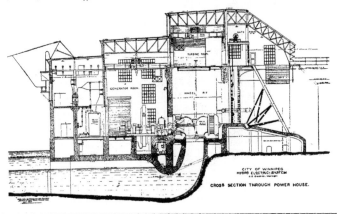

CITY OF WINNIPEG
HYDRO ELECTRIC SYSTEM

CROSS SECTION THROUGH POWER HOUSE.

feet, has braced, flexible and lattice towers, depending upon the location. The wires are spaced vertically and a 3/8 steel ground wire is strung along the top.

Cables are 278,600 cm. plain aluminum; pin type insulators are used except on dead ends and angles, where suspension type insulators in strain position are used.

All footings are concrete and over large stretches are on piles.

At all angles, dead ends and special crossings, special heavy towers are used. The longest span on the line is 1,100 feet.

At each patrol station the lines are carried through disconnecting switches to allow of the lines being tested in sections if necessary. At Tyndall, 38 miles east of Winnipeg, a special switch tower has been erected, which, in addition to allowing the lines to be sectionalized, enables any one line to be connected to one of the others. It also serves as a tap off point for a small substation located there, which supplies current to the villages of Beausejour and Tyndall.

Communication between the city and the power house is provided by a telephone circuit, strung on separate poles erected along the right-of-way. This line is supported on 8,000 volt insulators and is connected every five miles to a telephone booth, provided with instrument and test switches. Drainage coils are connected at each end of the line and at the centre.

Cottages for patrol men are provided at four points along the line, adjacent to the sectionalizing switches.

### Terminal Station

The terminal station in Winnipeg is a two storey and basement brick and structural steel building on the south bank of the Red River. The 66,000 volt power is stepped down to 12,000 volts and distributed underground to the four city sub-stations and overhead to Transcona and other adjoining municipalities.

On the top floor are placed all of the 66,000 volt switching, lightning arresters and high tension transfer bus. The general arrangement and connections of the switching are similar to that at the power house.

On the ground floor are placed the 12,000 volt switches and transformers. The switches are arranged in two rows, back to back, running the whole length of the building. On the south side are the transformer pockets and south of this is an unloading bay into which any transformer may be run and along which runs a 20 ton electrically operated crane. A spur line from the railroad runs the full length of this bay.

In the basement are located all of the 12,000 volt cables and the control cables; also pumps for the cooling water of the transformers, oil tanks and oil treating equipment.

At the present time the following transformers are installed:

Six—2700 kv.a., single phase, oil insulated, water cooled, 66,000/12,000 volt transformers, by the Canadian Westinghouse Company.

Three—2700 kv.a., do., by the Canadian General Electric Company.

Three—5000 kv.a., do., by the Canadian General Electric Company.

Connected in delta on both high and low tension sides.

There is also installed for the local service: One—500 kv.a., 12000/220 volt, three phase, oil insulated, self cooled, transformer, by the Packard Electric Co.

Cooling water taken from the city mains is circulated over a cooling tower of the natural draught type. All switches are of the electrically operated, remote controlled type, operated from a small storage battery.

The control room is an extension bay at the centre of the north side of the building. On the top storey are located the control boards. All lines, banks and condensers are controlled from a bench board, the instrument and relay being on vertical panels built immediately in front of and plainly visible from any point of the control desk. Directly behind this board is placed the recording instrument board.

Across the room from the bench board is a vertical panel board controlling the 12,000 volt outgoing feeders and behind this another vertical board on which are mounted the feeder watt hour meters.

A two panel board having mounted on it two Westinghouse automatic regulators for the synchronous condensers is placed convenient to the main bench board.

A gallery leads off the control room into the high tension switch room and from it all parts of the high tension room are clearly visible.

On the floor below the control room are the cable rooms, to which all of the control and instrument leads are

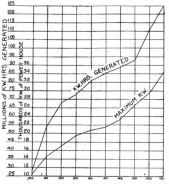

Maximum load and annual output of City of Winnipeg Hydro-electric System Power House, Point du Bois, 1912-1921.

brought and from there distributed to the various panels.

All instrument leads go direct to test links on the lower panel of the instrument and relay boards.

In these rooms are a series of panels, each panel corresponding to a panel in the control room. All control cables are brought to these panels and pass through fuses and then on up to the corresponding panel in the control room. By this means any faulty control lead can be easily and quickly located and trouble is limited to the one circuit.

As in the power house all control leads are multi-conductor, rubber insulated, lead covered cables, all conductors are distinctly marked in colors and a consistent color scheme is maintained throughout.

In the terminal station are installed two 6,000 kv.a., synchronous condensers, which are used to control the line voltage by varying the power-factor. These motors are wound for 6600 volts and are each supplied through a 6,300 kv.a., three phase, oil insulated, water cooled transformer,

# Winnipeg Municipal Hydro-Electric System from Numerous and Different Angles

Various views of the Winnipeg Hydro-electric System's operations: A—City terminal station. B—May St. sub-station. C—Ten of eleven generators installed at Point du Bois. D—McPhillips St. sub station. E—King St. sub-station. F—Part of H.T. switching, terminal station. G—12,000 volt switching, terminal station. H—6600 volt starting switches for synchronous condensers. I—Trash racks and head gates, power house.

connected to the 12,000 volt main bus. The sets are self starting, the starting voltage being supplied by low voltage taps on the 6600 volt secondary. Electrically operated, interlocked switches are installed, which make the starting up practically foolproof. An oil pump is supplied to flush the bearing at starting. These sets are started up and controlled from the main control board, no other attention than a man to start up the oil pump and see that the oil rings are picking up being necessary.

### Sub-stations

The 12,000 volt power is carried by lead covered, paper insulated cable, drawn in underground conduits to the various sub-stations. All cables are 250,000 cm., three core type, the number to each station depending upon the requirements.

King Street sub-station, situated in the centre of the business district, supplies the downtown and business areas. The 12,000 volt current is stepped down to 2400 volts and distributed partly underground and partly overhead. In addition to the a.c. supply a three wire d.c. supply at 250/500 volts is given for elevator and similar requirements.

The station equipment consists of nine 1,000 kw. single phase, oil insulated, 12000/2400 volt, self-cooled transformers, connected in banks of three and in delta on both high and low tension sides. Three 500 kw. synchronous motor-generator sets, each set consisting of a 2400 volt synchronous motor driving a 500/250 volt, 500 kw., d.c., three wire generator. All a.c. switching is of the electrically operated remote controlled type. A 500 kw., three-phase, 12,000/220 volt transformer supplies power for light and heat to the station and the main office and show room, located in the same building.

McPhillips Street sub-station is situated in the northwest end of the city, and supplies a mixed industrial and domestic load. The twelve thousand volts current is stepped down to 4,000 volts for distribution by six 500 kw. and three 1,000 kw., single phase, oil insulated, self cooled, 12,000/24,00 volt transformers, connected in banks of three; in delta on the high tension side and star on the low.

In addition to the above a bank of three 500 kw. transformers is connected delta-delta to supply 2,300 volt power to the city pumping station adjoining the sub-station. Both electrically operated and hand operated remote controlled switches are used. Single phase induction regulators, three per feeder, are connected in all outgoing feeders.

A large part of the street lighting system is supplied from this station by means of 7.5 amp. series regulators. There are also four 12,000 volt overhead feeders leaving this station, two of which go to Scotland Avenue sub-station, one to a large industrial load, and one to the village of Stony Mountain, twenty miles north.

Scotland Avenue sub-station, situated in the south-west end of the city, supplies almost entirely a domestic load. The twelve thousand volt power is stepped down to four thousand volts by six 500 kw. and three 1,000 kw. oil insulated, self-cooled, single phase, 12,000/2400 volt transformers, connected in delta on the high tension side and star on the low tension side.

All feeders are equipped with single phase induction regulators, three per feeder. All switching is electrically operated, remote controlled, the 4,000 volt feeder circuits being equipped with self-restoring relays.

From this station power is supplied to the Manitoba Power Commission. The 12,000 volt power is stepped up to 60,000 volts by three 1,000 kv.a. and three 500 kv.a., single phase, oil insulated, self cooled transformers, connected

delta-delta. Power is conveyed on steel tower lines to Portage la Prairie and other points in the Province.

Several constant current transformers are installed in this station for street lighting purposes.

McFarlane Street sub-station is situated in an annex to the main terminal station and supplies a miscellaneous power and domestic load over the north and north-east end of the city. The 12,000 volt power is stepped down to 2,400 volts by six 1,500 kv.a., single phase, oil insulated, water cooled transformers, connected in delta on high and low tension sides. All switching is of the electrically operated remote controlled type, the control board being placed on the main terminal station operating gallery. It is the intention to

Mr. E. V. Caton

change this station over to 4,000 volts in the near future.

May Street sub-station is entirely given up to street lighting equipment. Power is taken at 2,400 volts and supplied to thirty-six 50-light mercury rectifier sets, 6.6 amp., which are used to light the main streets and down-town parts of the city.

Sub-stations are also located in the town of Transcona, supplying the town at 2,200 volts, and the Canadian National shops and elevator at 12,000 volts. Another sub-station is at Stony Mountain. At Birds' Hill an outdoor sub-station supplies the village and a large number of houses and golf clubs in the vicinity.

The curve showing the growth of the load on the system is of interest, as it will be seen that in spite of the war and business depression, the load is still increasing appreciably. The low cost of power and the vigorous publicity campaign put on by the Department has resulted in a large cooking load, over 5,000 electric ranges and 1,000 water heaters being connected to the system.

With its present plant the city has over 20,000 h.p. still available and at a small additional cost can supply an additional 30,000 h.p. in a very few months. Looking to the future, the rights to Slave Falls, a power site less than six miles below the city's present development, have been secured. From 60,000 to 70,000 h.p. can here be developed at a low cost and a preliminary survey is now being made.

The Quebec Gazette announces the incorporation of the Lake St. John Light & Power Company to generate and distribute electric energy for light, heat and power, within the counties of Lake St. John, Charlevoix, Saguenay and Chicoutimi, in the province of Quebec. The firm is capitalized at $250,000. Head office: Montreal.

# The Winnipeg Electric Railway Co.

### A Short Sketch of This Pioneer Industry Which Has Played Such an Important Part in the Development of the City and of the Province

By JOHN WHITSELL, Manager W. E. R. Co.

Undoubtedly one of the principal factors in the meteoric development of the city of Winnipeg has been the Winnipeg Electric Railway Company. The pioneer in supplying gas, electricity and transportation to the community, the company has progressed in advance even of the city. The history of the company can be briefly sketched. In the year 1881, A. W. Austin incorporated the Winnipeg Street Railway Company, and the first horse-drawn car was operated on Main Street on October 24th of the next year. The first electric street car to be run in the city was run from Main Street Bridge to River Park, in July, 1891. It was operated by Mr. Austin's company, which owned River Park. Work had been commenced on this line in 1890. Two years later the Winnipeg Electric Street Railway Company obtained its charter, and made an agreement with the city of Winnipeg.

The first work done by this latter company was the construction of a line on Main St. north of C.P.R. track to Selkirk Avenue to the Exhibition grounds, and that line was formally opened on September 5, 1892. In this same year, the company which operated the horse-car line, brought suit against the Winnipeg Electric and also the city of Winnipeg, claiming that it had exclusive rights, but the Privy Council finally decided the matter in March, 1894, holding that the horse car franchise was not an exclusive franchise.

In the meantime, before the decision was handed down the electric company had built electric street car lines on

Main Street, paralleling the horse car lines. These latter were in the centre of the street and the electric lines on either side. Thus is explained why the present car tracks on Main Street are so far apart. The attached cut shows four tracks on Main Street. It was taken from a photograph which was procured at the time that the quartette of tracks was a reality.

In 1894 both the systems were being operated; the horse cars and the electric cars were of course in active competition. A rate war was inevitable, and on February 4, 1894, the electric company were selling 50 tickets for a dollar and 12 for a quarter. After the Privy Council had given its decision, however, the Winnipeg Street Railway Company was bought out by the Winnipeg Electric Street Railway Company, and the horse cars ceased their operation.

In those days Main Street and Portage Avenue were block paved. The street cars were about one-third the size of the present modern car. There were seven miles of street car tracks in the city and the population was 28,000. The number of employees on the company's payroll did not exceed 50; the annual operating costs of the company was under $50,000.00 and the invested capital $225,000.00. By the year 1900 the trackage had been increased to 16 miles on which were operated 36 single truck cars.

#### Period of Rapid Development

Then followed a period of rapid development both for

Winnipeg Electric Railway Co., Pinawa Channel. Interior generating Station

The Hydro-electric development of the W.E.R. Co. at Pinawa Channel on the Winnipeg river and, on the left, the company's manager, Mr. John Whitsell

the company, and the city. Millions of dollars of new capital were put into the property until to-day the company's total investment exceeds twenty-five million dollars. From 7 miles of trackage in 1893 the company now has 112 miles of track within the city limits and 60 miles of track in the suburbs. Approximately 350 cars, of the most modern type of rolling stock are in service and the number of employees is close to 2,000.

With regard to the development of its electric utility, steam was the motive power which generated the electricity first used by the company to operate its street cars. From the small steam plant built on the banks of the Assiniboine River, the company has built and today owns and operates a large power station at Mill Street with a capacity of 12,000 h.p. and a number of sub-stations in various parts of the city. The capacity of the Mill Street plant was soon taxed to the utmost and with a rapidly

increasing business the company realized that power generated from such a plant would be too costly a proposition, especially when within a few miles of the city, water power was running waste in almost unheard of quantities of horse power. As a result of investigation the company built its present large hydro-electric plant at Pinawa on the Winnipeg River, which was completed in 1906 and which has a capacity of 35,000 h.p. Although at the time that this was built, Winnipeg people wondered at the large development and speculated as to how it could possibly be absorbed, the demands on the company for power in the following ten years was so great as to necessitate securing further sites on the Winnipeg River, and in this connection it is interesting to note that the company's future supply is assured by the development of the Great Falls plant, now under construction by the Manitoba Power Company, Limited.

A relic of the days when Winnipeg had two Tramway Systems

Engine Room, Provincial Power House, Regina

# New Power House for Gov't. of Saskatchewan

A new power house has just been built by the Government of the Province of Saskatchewan to supply power, light and heat to the Legislative Assembly and to the other Government buildings, located nearby. It was designed by the Provincial Architect, M. W. Sharon, who can be complimented on erecting a power house harmonizing with the beautiful building adjacent to it.

The whole of the electrical work was done by the Sun Electrical Company, Ltd., of Regina, under the direction of its engineer, Garnett E. Perry. Four 250 h.p. Goldie & McCulloch boilers, which are situated in the east end of the building, supply steam for operating the six horizontal, single cylinder, 125 I.H.P. Leonard & Sons engines; direct connected to 75 kw., 250/125 volt, compound wound, 3-wire, d.c. generators, running at a speed of 277 r.p.m., conveniently arranged in two rows of three, in the well ventilated and illuminated engine room.

The huge, unsightly stack of the original power house has been replaced by two large induced draft fans, operated by 15 h.p. Westinghouse variable speed motors, which preserves the outward beauty of the building lines and retains efficiency.

Being a low voltage 3-wire d.c. plant, with six generating units, the control board is necessarily massive and expensive. It was designed and manufactured by the Canadian General Electric Company, at their Peterborough plant. The board is made up of nine two-piece panels of 2¼ inch black marine finished slate, ninety inches high.

The generator control panels (six in all), 36 inches wide, mounting: two 400 amp. ammeters and shunts; one field rheostat; one 4-point voltmeter receptacle; two 400 amp. D.P.S.T. knife switches; one 200 amp. S.P.S.T. knife switch; two recording watthour meters, and two D.P. 2-coil, 400 amp. circuit breakers.

The four poles of these breakers are tied together electrically, so that if one side should go out the other would immediately follow, otherwise the balancing compensator would be endangered. These breakers are equipped with no-voltage, overload and reverse current relays, the tripping coils being placed only in the main poles. The circuit breakers are placed in the lines beyond the series fields.

There are three feeder panels—two for lighting and one for power—24 inches wide, on which are mounted the knife switches, controlling the various feeders. All the fuses, for feeder switches, are mounted 18 inches in the rear of the main board, on separate panels.

The two voltmeters (one connected to the bus-bars and the other to 4-point receptacles to assist in equalizing the incoming machines) are mounted on brackets over the top of the centre panel.

While the front of this control board is very imposing, the manufacturers are to be commended on the ingenious manner in which their engineers planned the mounting and connecting of the bus-bars to the various pieces of apparatus and switches, mounted on the panels, which completes in a very workmanlike and substantial manner the fine finish of the main board.

In order to handle the load in the Assembly buildings, a great number of heavy lead covered, paper insulated, cables were necessary, and, owing to the distance of transmission being great (over 1200 ft.) it was found advisable by the engineer, Mr. S. S. Kennedy, of Winnipeg, who planned the electrical installation for the Government, to dispense with conduit and adopt 3/8 inch steel channels, mounted on angle iron supports, on the wall of the tunnel; also, all cables from generators to the switchboard were handled in a similar manner, making it very convenient at all times to replace a defective cable without disturbing others.

This electrical contract, amounting to $50,000, was taken by the Sun Electrical Company, Ltd., of Regina, who were at the same time handling the large electrical installation in the new Mental Hospital at Weyburn, Sask. This company deserves much credit for the very successful manner in which it has done its work. Its success as a construction company is directly due to the very capable manner in which the business of the company is handled by J. R. Young, managing director; to its construction engineer, Garnett E. Perry, and to its staff, whose efficiency can be seen by the quality of the work exhibited in this and other jobs of a similar nature undertaken by the company.

Showing General Scheme of Manitoba Power Company's Development

# The Next Big Western Power Scheme

### Manitoba Power Company Actively Developing Great Falls, Winnipeg River—Contracts Let for Two Complete 28,000 H. P. Units—Complete Capital Cost Will be less than $80 per H. P.

Active construction work is now under way on the Great Falls Development on the Winnipeg River by Manitoba Power Company Limited, after considerable preliminary study and investigation. Work is going forward rapidly and excellent camp quarters for the large force of men necessary for the work have been constructed, comprising up-to-date mess-halls and kitchens, bunkhouses and hospital, together with administration buildings and offices. A complete construction plant has been erected, including rock crushing plant, mixing plant, concrete chuting equipment, storehouses, machine, blacksmiths' and carpenters shops and a complete yard and track layout for the handling of the large quantities of materials involved in the work.

A standard gauge railway, 13½ miles in length, has been constructed, connecting the Canadian Pacific Railway tracks at Lac du Bonnet with the power site at Great Falls. This railway is now being extended a distance of three miles, from the power site to White Mud Falls, lower down the river. The road is well ballasted, is laid with 60 lb. steel and compares favorably with other branch line roads in this section of the country.

Over 1200 lineal feet of cofferdam, varying in depth from ten to thirty-five feet have been constructed, and in order to expedite the work a temporary boiler plant was installed and the power house site, enclosed by the cofferdam, was unwatered by steam pumps. Steam drills were also put in service for rock excavation.

A 60,000 volt double circuit transmission line, eighteen miles in length, supported on steel towers forty-five feet high, has now been constructed from a point on the Pinawa-Winnipeg line of the Winnipeg Electric Railway Company to the power site, and connection has been made to the City of Winnipeg Hydro-electric System's line for power supply during construction. This line will ultimately form one link of the permanent transmission system, connecting the new development with the existing water power plant on the Pinawa Channel, and with the sub-stations in the City of Winnipeg. Electric power was made available over this line early in February, and all the construc-

tion plant is now being operated electrically. A temporary transformer house has been constructed at Great Falls, which contains one bank of 600 kw. transformers and one bank of 1,500 kw. These latter transformers, while they are now being used for construction purposes, are primarily designed for station service use in the power house. Power is being distributed at 2200 volts and a double circuit wood pole line has been run from this transformer house to White Mud Falls, where a sub-station containing about 700 h.p. in 2200 volt motors operating compressors, derricks, etc. has been installed. With electrical operation at Great Falls, the work of rock excavation has been greatly facilitated, and to date approximately 30,000 cubic yards of solid rock have been excavated for the power house draft tubes and tail race. This constitutes about one-half of the total excavation required.

#### General Layout

The power house proper is located at Island No. 2, where the dam to be constructed will hold the head waters above the power house at Elevation 808, thus forming a pond or peak load reservoir of approximately 2,000 acres in extent, providing a deep, wide and extensive body of water reaching upstream a distance of five miles and flooding out the existing rapids, thus obviating the trouble so often experienced in these northern latitudes from frazil ice, which forms in swift, turbulent water.

The dam at Island No. 2 will raise the present water level forty-six feet, and investigation has disclosed the reach of the river below the power house to be from thirty to ninety feet in depth, this level being maintained by a rock reef across the river at White Mud Falls. By excavating a channel 200 ft. wide, 20 ft. deep and about 1800 feet in length through this reef, the entire reach of the river between White Mud Falls and the power site will be lowered ten feet or more, thus making a total operating head of fifty-six feet, for which head the turbines are designed. The rock excavated from this channel will be hauled by railway to the power site and used in the construction of the rock-fill dam across the East Channel, and will also supply crushed stone for the concrete.

At the site selected for this development the river is divided into two channels by an island, upon which the central section of the dam is located. It is possible, therefore, to close the West Channel by cofferdam, as has been done, passing all of the water through the East Channel. The gates and wheel pit openings in the power house in the West Channel, which will be completed first, can then be used to pass the discharge of the river while construction is under way in the East Channel.

## The Dam

As shown on Fig. 1 the dam, from the west shore to the power house, will be of the usual solid concrete non-overflow type, approximately 750 feet in length with a maximum depth of seventy feet. From the end of this dam to the island the power house bulkhead constitutes the dam. A curtain wall along the upstream face of the intake walls of the power house, extending six feet below the forebay level, will prevent ice and other foreign matter from entering the wheel pits. Adjoining the easterly end of the power house a wing wall, one hundred feet in length, will extend up the river parallel to the current. In this section a skimming weir for discharging ice and other floating materials will be included.

The first 260 feet of dam on the island will be similar in design to the non-overflow dam referred to above. The

The section of the dam between these sluices and the easterly end of the island will be 200 ft. in length and of non-overflow section, similar in design to the west wing wall.

The dam across the East Channel will be of rock-fill section, 1000 feet long, and with a maximum height of seventy feet and crown ten feet wide. The slope of the rock will be one to one. The upstream face of this dam will be faced with graded material consisting of broken rock, crushed stone, gravel and clay. The clay blanket will be built to a slope of 2½ to 1 and will be ten feet wide at the crown, and to prevent sliding a rock toe-fill will be placed at the lower edge of the fill.

The section on the east bank, 1100 feet in length, will be an earth embankment, with twenty foot crown, built to a slope of 1½ to 1 on the downstream side and 2½ to 1 on the upstream side, and will contain a concrete cut-off core wall.

## The Power House

The power house will be 110 ft. wide, 380 ft. long and have a maximum height of 145 feet. The foundations will be of reinforced concrete and will carry a structural steel superstructure with brick curtain walls provided with steel sash. The roof will be of gypsum block tile covered with standard tar and gravel roofing.

The power house proper will contain six vertical turbo-

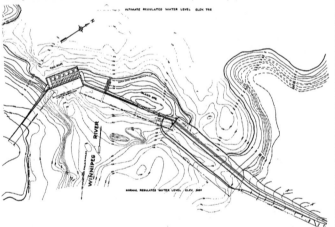

General Layout of Dam and Power House, Manitoba Power Co.

next 400 feet will be of the overflow or free spillway type, with crest at elevation 808 and with provision for flash boards to raise the head water. Four steel sluice gates, 50 ft. wide and 30 ft. deep, will be located on the crest of the island. These sluice gates are designed to pass the full flow of the river under maximum flood discharge conditions. The

generators with the usual intake racks, stop-logs, head-gates, etc. At the entrance to the scroll case for each unit, three pairs of steel head gates will be provided. Gate openings will be sufficient to pass the flow required for each turbine at a velocity of approximately 4.5 feet per second. Stop-logs will also be provided to facilitate inspection and maintenance of the gates. The screens, consisting of heavy steel bars, six inches apart and supported on a steel framework attached to the sub-structure, will be provided for preventing debris from entering the wheels.

Fig. 2 shows a cross-section of the power house, from which the relative arrangement of the main equipment can be noted. The switchboard and control room will be located on a gallery at the westerly end of the power house, beneath which will be located the governor and oil pumps, air compressors, water pumps and other auxiliary equipment.

### Initial Development

The initial installation is to include a power house building complete to accommodate three units, with screens, stop-logs, gates, operating mechanism, cranes, machine shop, heating and other necessary equipment for the successful operation of the plant. The substructure of the balance of the power house for the accommodation of the remaining three units will be completed only to such a point as is necessary in order that future extension can be made without unwatering expense.

The electrical installation includes two generators, the third to be installed in the space prepared when required, two banks of transformers and switching equipment for delivering the power to the outgoing lines. The dams and other permanent works will be constructed for the complete installation. The hydraulic installation includes the necessary turbines, governors, etc. for two units.

### Hydraulic Equipment

**Turbines:**—The turbines will be of the single runner, vertical shaft, diagonal or propeller type, and will develop 28,000 h.p. when operating under ahead of fifty-six feet and running at a speed of 138.5 r.p.m. Under these conditions the guaranteed efficiency is eighty-seven per cent. The run-away speed of these turbines when operating under full gate opening and under an effective head of fifty-six feet will be about 315 r.p.m. The full weight of the revolving part of the turbine, including hydraulic thrust, but not including the generator field is about 600,000 lbs. The shipping weight of each turbine with its accessories is 725,-000 lbs. The water enters the turbine through concrete scroll cases and will be guided into the movable guide vanes by individual stay vanes, imbedded in the concrete at their

upper and lower ends. The stay vanes will be provided with foundation bolts and adjusting screws at the upper end for adjusting the cast iron pit liners before pouring the concrete. The runner will have six blades, cast separately, and securely bolted to the runner hub. The operating gear for the guide vanes will be of the "off-set" type, which differs from the usual construction in that the guide vane levers are offset from the radial position, and as the vanes close the levers turn in a direction still further away from the radial position, resulting in a toggle action. This action greatly improves the speed regulation of the unit, due to the fact that at large gate openings the movement of the guide vanes is greatest and the amount of this movement

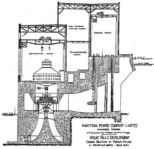

MANITOBA POWER COMPANY LIMITED
WINNIPEG, CANADA
GREAT FALLS DEVELOPMENT
Cross Section of Power House

Cross Section through Generator

decreases as the vane closes. Furthermore, as the guide vanes close the torque increases, reaching a maximum value in the closed position where the hydraulic load is also a maximum. The links connecting the guide vane levers with the operating ring are of the double-shearing type.

**Governors:**—The governors will be of the double floating level type, with relay belt drive from the main shaft. They

Construction Views of Manitoba Power Company's Development

are provided with hand control mechanism and the usual accumulator tanks, interconnecting piping, tachometer, etc., and electric motor for remote control. The governor will adjust the gates without "hunting" when the speed varies one-half to one per cent from normal, and they will be so adjusted as to open or close the turbine gates in not more than three seconds time. With full load change the time of gate movement is three seconds, and the percentage speed change, load off to zero is twenty-five per cent and load on from zero is thirty-two per cent.

**Draft Tubes:**—The draft tubes will be formed in the concrete and will be of the 'Centercone' spreading type. The centercone will extend upward to an elevation just below the runner and the upper end of this cone will be provided with a cast iron cap which will extend through the lower end of the runner cap, forming a suitable pedestal for supporting the rotating element of the turbine when the thrust bearing is dismantled. Another advantage of carrying the centercone up to the runner discharge is that it avoids all flow close to the axis, where the tendency to form vortices is greatest.

### Electrical Equipment

**Generators:**—The generators will be of 21,000 kw. capacity, of the vertical type, running at 138.5 r.p.m. This rating is based on full load temperature rise of 60 deg. C. by thermometer, or 70 deg. C. by thermo-couple, with surrounding air at 25 deg. C. The machines will have a guaranteed efficiency of 97.1 per cent and reactance of 25 per cent. The weight of the machine is 480,000 lbs. with flywheel effect of 10,000,000 WR².

These machines will generate three phase, sixty cycle, alternating current at 11,000 volts, and will be provided with 150 kw. direct connected exciters.

**Thrust Bearings:**—The entire revolving element of the unit will be supported by a Kingsbury thrust bearing mounted on top of the generator stator. This bearing has an approximate capacity of 950,000 lbs. at a speed of 138.5 r.p.m. and is provided with water cooling coils and the usual valves, thermometers and indicating devices which are customarily furnished with these bearings.

**Transformers:**—The transformers will be 7,000 kv.a. each, single phase, water cooled, 11,000/60,000 volts, with taps in the high tension winding to give 10,000 volts reduced capacity in the low tension windings. The guaranteed full load efficiency is 98.9 per cent, regulation at full load power factor 1.5 per cent and impedance approximately 9 per cent. The temperature rise at full load until constant is 55 deg. C.

The insulation test, high voltage to low voltage windings and core is 121,000 volts and from low tension windings to core 25,000 volts. The voltage across full winding is twice the normal.

Each transformer requires approximately 1,020 gallons of oil, weighing 8,800 lbs.; the weight of the transformer, less oil, being 25,000 lbs. The height from the leads is 172 ins. and from the cover 130 ins.; floor space covered being 93 ins. by 55 ins.

**Switching Equipment:**—Each generator will be connected directly to the low tension side of its transformer bank, without any intermediary oil circuit breakers or bus bars, which virtually makes each generator and its transformer bank one unit. The 11,000 volt cables from the generators to the transformers are carried in fibre conduits.

The high tension oil circuit breakers connecting the transformers to the high tension bus bars and line are of standard G.A.3 type, 73,000 volts, with interrupting capacity

of 6600 amps., and each switch is provided with automatic overload trip.

### General

The design of the plant is modern in every respect and in physical dimensions and quantity of water used the turbines are the largest for which contracts have ever been placed. The type of unit finally adopted is that recommended by a Water Wheel Committee consisting of five prominent hydro-electric engineers and the design of turbines was submitted by one of the largest Canadian turbine manufacturers, who has done very creditable work in developing a machine of such high speed and power output, resulting in a machinery cost of less than $14.00 per h.p. including turbines, generators, exciters, thrust bearings, transformers and switches. The capital cost for the initial installation, including all permanent works for the complete development is less than $80.00 per h.p. and the cost for the complete plant will be less than $60.00 per h.p. which is undoubtedly the lowest unit cost per horsepower yet recorded on this continent for a plant of this magnitude.

The plan of development is in accordance with the general scheme of the Water Power Branch of the Dominion Government, and forms the principal link in a chain of potential developments extending from Lake of the Woods to Hudson's Bay, aggregating several million horsepower.

The contractors for this work are Fraser-Brace Limited, who have successfully completed other large construc-

Mr. F. H. Martin

tion jobs of a similar nature, including the large dam for the International Nickel Company at Turbine, Ont., the La Loutre storage dam for the Quebec Government, the paper mill, dam and power house for Price Bros. at Chicoutimi, Que., and the Cedars Rapids Manufacturing and Power Company's development on the St. Lawrence River. In the execution of the above works the contractors have gathered together a very efficient organization which has been transferred to Great Falls, together with their complete construction plant, thus ensuring the successful construction of the Great Falls development.

Sir Augustus Nanton is president of the Manitoba Power Company Limited and Mr. A. W. McLimont is vice-president and general manager, under whose direction all of the work is being carried on by a loyal and efficient engineering and construction force. Mr. Julian C. Smith is supervising engineer. Messrs. L. J. Hirt, R. S. Lea and C. O. Lenz are consulting engineers and Mr. F. H. Martin is chief engineer and designer.

Power House No. 1 Calgary Power Company

Mr. F. J. Robertson

# Water Power in the Calgary District

### Two Developments by Calgary Power Company—Regulation a Big Problem—Supply to City of Calgary and Private Industries

The hydro-electric system of the Calgary Power Company, Limited, has been in continuous operation since the year 1911. The two plants are situated in the foothills of the Rockies some fifty miles west of the City of Calgary. No. 2 plant, Kananaskis Falls, is at the junction of the Bow and Kananaskis rivers, while No. 1 plant, Horseshoe Falls, is two miles down stream.

The Horseshoe Falls plant was completed in 1911 and the following year work was commenced on Kananaskis Falls plant, the completion of which gave a total installed capacity of 31,000 H.P.

The dam at No. 1 plant is of solid concrete and is provided with inspection tunnel. Flood water is handled by means of Stoney type sluice gates, operated under oil pressure; by four stop log openings and 140 feet of free spillway. The flood water equipment has been called upon to discharge 24,715 c.f.s., this somewhat abnormal flow occurring in the spring of 1916. It has been clearly shown that during flood periods the principal source of trouble to be anticipated is not so much the great and rapid increase in the quantity of water to be handled as the vast amount of drift wood which accompanies it. Entire fir trees, measuring in some cases up to fifty feet in length, which have been uprooted and brought down with the flood, require to be disposed of before any serious jam has been allowed to form.

### No. 1 Power House

The equipment at No. 1 power house consists of two 6,000 h.p. Wellman Seaver Morgan turbines, coupled to two 4,000 kv.a., 12,000 volt, 3-phase generators; two 3,750 h.p. Boving turbines, coupled to two 2,500 kv.a., 12,000 volt 3-phase generators, and two water-driven and one motor-generator exciter, any of these exciters being capable of supplying all four generators. The operating head of the turbines is 70 feet.

The transformers consist of four 3,000 kv.a., 12,000 to 55,000 volt, O.I.W.C. transformers, the transmitting voltage to the company's terminal station in Calgary being 55,000. All the switching equipment at No. 1 plant, including the

Power House No. 2, Calgary Power Company

remote control desk, is of Canadian Westinghouse Company manufacture.

#### No. 2 Power House

The dam at No. 2 plant is of solid concrete, and is provided with inspection tunnels, but it differs from No. 1 dam in that flood water is controlled wholly by means of stop logs, which are handled by an electrically driven winch. The discharge capacity with all stop logs removed, and water at normal operating level, is computed at 46,100 c.f.s.

The water is led to the gate house by means of a

Bow River at Horse Shoe Falls, June 1915

canal, excavated from solid rock, while the control to the turbines is through Tainter type gates.

The power house equipment consists of two vertical units of 5,800 h.p. each, manufactured by the Canadian Allis-Chalmers Company; two 12,000 volt generators of Swedish General Electric Company make, and one motor-driven and one water-driven exciter. The switchboard and switches are of Canadian Westinghouse Company manufacture.

#### Transmission

Duplicate transmission lines of No. 0 stranded aluminum connect the two generating stations, similar lines feeding the large plant of the Canada Cement Company at Exshaw, and also connecting the company's terminal station at Calgary. The equipment at this point consists of step-down transformers to 12,000, 2,400 and 600 volts, the city of Calgary taking part of their requirements at 12,000 volts and distributing from the company's station at 2,400 volts for local requirements.

All transmission lines are fully protected by means of electrolytic lightning arresters, which are placed in both power houses, at the Calgary terminal station and at the Exshaw cement mill sub-station.

#### Regulation

In addition to constructing the above hydro-electric plants to meet the requirements of the district, the Calgary Power Company has in operation a dam at Lake Minnewanka, 30 miles distant, the immediate object being to raise the elevation of this lake twelve feet, thereby impounding 44,000 acre feet of storage water, which is drawn upon during the low water period.

When the dam was constructed it was realized by the company that it would ultimately be necessary to build a power plant at this site, and the dam was therefore designed with this end in view.

#### A Heavy Investment

The company's total plant investment is in the region of $5,000,000 and there can be no doubt that the completion of the plants described has contributed in a great measure to the rapid development of industrial activities in the Province of Alberta.

The company's engineers have been steadily engaged during the last few years in collecting all data in connection with future power projects, and in this connection, invaluable aid has been rendered by the Dominion Water Power Branch. It is fully recognized that the time is not far distant when the present system will require to be extended to meet the rapidly growing demand for power in Alberta. All the schemes considered have, of necessity, been examined from the viewpoint of their effect on the present system, which may be considered the nucleus of hydro-electric development in the Province.

Mr. F. J. Robertson, who had a wide experience in both steam and hydro-electric plants, is the company's general superintendent.

The Quebec Gazette announces the incorporation of Mertel Motors, Ltd., with a capitalization of $20,000. The new firm is authorized to manufacture and deal in electrical machinery, appliances, etc. The head office of the company is at Montreal.

Calgary Power Co.,
Horse Shoe Falls
Development

# City of Calgary Power Department

### Steam Plant Supplements Purchased Hydro-electric—
### Special Furnace Arrangements Allow use of Lignite Coal

By J. F. McCALL, Superintendent

The city of Calgary receives its supply of electricity from two sources; hydro-electric and steam. The hydro-electric power is supplied by the Calgary Power Co. from their plant at Kananaskis Falls, about 50 miles west of Calgary. The steam plant is owned and operated by the city. It is situated in Victoria Park on the bank of the Elbow River. The power house building, which is constructed of pressed brick and concrete is 250 feet by 133 feet. A fire wall, the full length of the building, separates it into two parts, the engine room 50 ft. wide, and the boiler room 83 ft. wide.

In the boiler room there are 20 Babcock and Wilcox water tube boilers:—twelve with 3580 sq. ft. of heating surface, four with 2833 sq. ft. of heating surface, and four with 6080 sq. ft. of heating surface. The boilers are set in batteries of two, on each side of a central firing aisle. All are equipped with Babcock and Wilcox superheaters and chain grate stokers.

The induced draft plant consists of two 106 in., two 120 in. engine driven fans, and two 120 in. fans connected by silent chain drives to two 3 speed a.c. motors. These fans are connected to two 8 ft. by 75 ft., and one 10 ft. by 90 ft. steel stacks. The flues are all underground.

Coal is unloaded into a track hopper, outside the building. From this hopper it is carried by a tray conveyor to the main bucket conveyor. The bucket conveyor elevates the coal to the overhead steel bunkers, which have a capacity of 600 tons. The same conveyor takes away the ashes, which are loaded into the bucket conveyor in the conveyor tunnel, and dumped into a tray conveyor, which in turn dumps an ash hopper of 15 tons capacity. The coal handling machinery has a capacity of 50 tons per hour. From the overhead bunkers coal is delivered to the stoker hoppers by 1000 lb. weigh larries.

### Engine-Room Equipment

The engine room equipment consists of:—

One 900 h.p., 3-crank Robb-Armstrong engine, 300 r.p.m., direct connected to a Dick Kerr Co., 600 kw., 550/600 volt d.c. generator.

One 750 h.p., 3-crank Robb-Armstrong engine, 300 r.p.m., direct connected to an Allis Chalmers Bullock, 500 kw., 550/600 volt, d.c. generator.

One 1500 h.p. synchronous motor, direct coupled to a 1000 kw., 550/600 volt, d.c. generator, 400 r.p.m., supplied by Dick Kerr Co., England.

One 1800 kw. and one 5000 kw. Parsons type steam turbines, connected to 60 cycle, 2300 volts, 1800 r.p.m. generators manufactured by Allis Chalmers Co., Milwaukee.

One 2500 kw. Curtis Rateau type steam turbine, manufactured by Belliss Morcom, Birmingham, Eng., connected to 60 cycle, 2300 vo't, 1800 r.p.m. Vickers generator.

The exciters are one 100 kw., 120 volt, 290 r.p.m., steam-driven unit, Robb-Armstrong engine, Can. Westinghouse generator; one 50 kw., 120 volt, 290 r.p.m., steam-driven unit, Robb-Armstrong engine, A. C. B. generator; one 100 kw., 120 volt, 520 r.p.m., motor-generator unit.

### Condensers

The main units are equipped with surface condensers, the condensers with their respective air pumps being placed in the basement directly under the units which they serve. Condensing water is supplied from the Elbow River by two 5000 gallon per minute, Mather & Platt centrifugal pumps at 35 ft. head, driven by 2300 volt C.G.E. motors, and one 3500 gallon per minute centrifugal pump built by John Inglis Co., driven by a 3300 volt Swedish General Electric motor.

The condensate from the turbines is pumped through a recorder, and from there flows by gravity to either of two open feed water heaters.

Feed water is supplied to the boilers through a Venturi meter by two Jeansville, 3-stage centrifugal pumps driven by Terry steam turbines at 2750 r.p.m., by one Worthington vertical duplex pump, size 12 in. by 8½ in. by 10 in., and by one Fairbanks Morse vertical duplex pump, size 10 in. by 6 in. by 12 in.

### Distribution Lines

Power from the hydro-electric plant is transmitted to Calgary at 55,000 volts, over two separate transmission

Mr. J. F. McCall

lines to the company's terminal station in East Calgary. At this station the current is stepped down to 12,000 volts for distribution to the city's high tension lines. Power is supplied to the city over 12,000 volt lines as follows:—one line leading to No. 2 sub-station, Ogden, two lines leading to central power station, and a ring main circling the city to supply the different sub-stations. The central station is also connected to the sub-stations by 12,000 volt feeders.

The electrical controlling equipment at the central station consists of a 12,000 volt and a 2300 volt bus situated in the north end of the engine room basement. Each phase of these buses is placed in a separate concrete compartment, and the outgoing and incoming leads are separated by concrete barriers. Normally the buses work as a complete unit, but it is possible by the use of sectionalizing switches to cut out sections of the bus without interrupting the service. The two buses are connected together through three 3000 kv.a., 12,000/2300 volt, 3 phase transformers. The switching equipment for the main buses is placed directly over the bus on the engine room floor. There are 17 C.G.E., type F.H., 20,000 volt circuit breakers.

### Main Control Board

The main control board is mounted on a concrete and steel gallery, 70 ft. by 12 ft., supported by iron columns. This board consists of 25 black slate panels. It is equipped with the standard instruments, including power factor recording, and integrating watt meters, frequency meter, automatic synchronizing equipment, and overload and reverse current relays for the steam turbines. At the back of this board is an auxiliary board containing the feeder relays, and the control equipment for the 125 ampere hour storage battery, used for switch control. Below the main board is a skeleton bus structure for the local feeder bus. This bus is fed through a set of reaction coils from the main 2300 volt bus. All feeder switches are controlled from the main board by hand operated remote control switches. All outgoing power and light feeders are equipped with voltage regulators, which are placed in the basement, as is also a nine panel control board for house service and street lighting. This board was manufactured and set up by the city electrical department for the design and under the supervision of Mr. R. A. Brown, electrical superintendent.

### The Sub-stations

Sub-station No. 1, situated at 9th Avenue and 7th Street contains the control feeders supplying the C.P.R. locomotive and car shops at Ogden.

Sub-station No. 3, has one 1500 h.p. synchronous motor direct connected to a 1000 kw., 550/600 volt, d.c. generator, and one 1,000 h.p. synchronous motor direct connected to a 750 kw., 550/600 volt, d.c. generator. Both these machines were supplied by the Canadian Westinghouse Co.

### Purchased Power

Power is purchased form the Calgary Power Co. on a horsepower basis, the normal load purchased being approximately that required from midnight to 6 a.m. Under this system we are able to shut down our steam turbines from midnight to 6 a.m.; it also gives us a very high load-factor for our purchased power. From 6 a.m. to midnight the steam plant takes all the peaks above the hydro-electric normal load with a certain reserve for emergency purposes. At times, owing to ice trouble or snowslides in the mountains, and to lightning storms during the spring and early summer, the steam plant may be called upon at short notice to take a considerable portion of the load. The plant capacity is sufficient to handle the total city load for a reasonable length of time.

### Burning Lignite Coal

The opening up of the Drumheller coal fields made available a considerable quantity of lignite slack, if it could

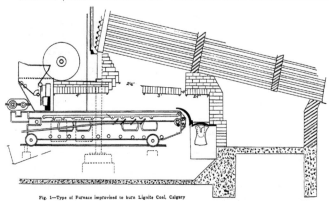

Fig. 1.—Type of Furnace improvised to burn Lignite Coal, Calgary

West, has two 1500 h.p. auto-synchronous motors direct connected to two 1000 kw. 550/600 volt, d.c., generators for street railway service. These machines were manufactured by the Swedish General Electric Co. This station also contains the necessary 12,000/2300 volt transformers, and equipment for the control of light and power for this section of the city.

Sub-station No. 2, contains one 750 h.p. synchronous motor, driving a 500 kw. 550/600 volt, d.c. generator supplied by the Dick Kerr Co., England. This station also

be burned with satisfactory results. This fuel averages about 11,500 B.t.u. per pound of dry fuel, and contains from 12 per cent to 20 per cent moisture. Efforts were made to burn this fuel on chain grate stokers for some time without success. Owing to the excessive moisture, it was impossible to ignite the coal until it had passed about three feet into the furnace. The solution of the problem depended on the radiation of sufficient heat back towards the coal gate of the stoker to dry the moisture out of the coal. After several trials the type of furnace illustrated in Fig. 1 was ob-

, taited, and has since given very satisfactory results.

In the alterations to this furnace the bridge wall was moved back 18 inches, and an 18 inch sector was removed from the sides of the stoker, and the stoker moved this

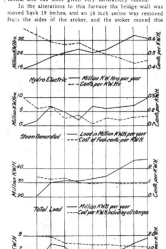

: Fig. 2—The Story of Calgary's Power Load

distance under the boiler; with this change a better arrangement of arches was obtained.

This furnace has given very satisfactory results, and it is possible to burn coal containing 23 per cent moisture, working the boilers at 110 per cent of rating. Under ordinary operating conditions 125 per cent of builders' rating is obtained with coal containing 16 per cent moisture and in emergency as high as 175 per cent rating has been obtained.

The question is often asked, "Do not the flames passing through the 2 ft. 8 in. opening between the arches, have a destructive effect on the tubes?" The answer is "No." The records of tube renewals in the plant show that in five years with 20 boilers only two tubes have had to be replaced.

### Historical

The city started their first electric light plant in 1905 with one 250 kw. generator direct connected to a Robb-Armstrong Corliss cross compound engine, and two B & W water tube boilers. This plant was situated on 9th Avenue West. The plant immediately began to grow, and in 1909 all the available space of that location was occupied, and early in 1910 it was seen that further extension would be necessary at an early date. The present location was chosen as the site of the new plant, and immediate steps were taken to get the plant under way. The new plant started in the spring of 1911, and shortly after power was obtained from the hydro-electric company. The last six months of 1911 was a period of transition. During this time the old station was being dismantled, and the machinery that was considered suitable was being erected in the new station. This was practically completed when the new station took over control of the operation in January 1912.

The growth of the electrical industry in Calgary from that date is indicated by the curves. The primary object, which the power department has in view at all times is continuity of service, with cost as a secondary consideration. Interruptions to the service are few and of short duration, while at the same time the operating and maintenance costs have shown a gradual decline. The present plant was designed and erected under the supervision of the writer, the superintendent of the Power Department.

# How Calgary's Power is Distributed

### Proportioning of Generating and Operating Expenses —Some Splendid Examples of Underground work.

By R. McKAY, Superintendent Electric Light

Calgary, situated at the gateway to the Kicking Horse Pass, on the main line of the C. P. R. about 80 miles east of that famous pleasure resort,—Banff, and surrounded by a fertile country recognized the world over for its grain growing and cattle raising possibilities, offers to manufacturers and industrial enterprises, in addition to low power rates and suitable industrial sites, all the advantages that could be obtained anywhere in the Dominion, together with local advantages peculiar to this section.

Calgary's public utilities are municipal enterprises, and are representative of what can be accomplished under public control. Previous to 1905 the rate for lighting purposes was 18 cents per kw.h. The city began operating their electric light plant in 1905 and the rates were reduced to 14 cents

per kw.hr. for lighting and 10 cents per kw.hr. for power. To-day lighting rates go as low as 2.8 cents per kw.hr. and power as low as ¾ cents per kw.hr. The total capitalization of the electric light and power department is $3,949,480.22 In addition to this $369,393.57 has been spent in improving the plant and equipment, and has been financed from surplus revenue.

Hydro power is secured from the Calgary Power Company, whose generating site is located about 52 miles west on the Bow River at the Horse Shoe and Kananaskis Falls, and is transmitted to Calgary by two transmission lines at a difference of potential of 55,000 volts to the company's sub-station located within the city limits, where it is distributed to the city's various sub-stations at 12,000 volts.

The standard primary distribution voltage is 2800, and the standard voltage for power service is 220 volts, 3 phase, 60 cycle. All primary feeders are underground to the center of distribution, then overhead—with the exception of the business portion of the city, where all distribution is underground.

The underground system consists of 116.8 duct miles of

**$831,829.67 — How it Goes — 1920**

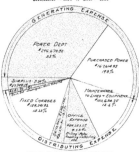

Calgary Municipal Electric Light Dept.

vitrified clay; 375 standard manholes 6 x 7 x 6; and 25 transformer manholes 9 x 14 x 7. These manholes are drained to gravel. It has been found unnecessary to drain to sewer owing to the porous nature of the soil. 25137 feet of 3 conductor, 2/0 cm. paper insulated, lead covered cable is used in carrying 13,000 volt current to sub-stations located in the center of the city, 49625 feet of 300,000 cm., 3 conductor, paper insulated, lead covered cable is used for primary dis-

**$940,380 — How it Goes — 1921**

Calgary Municipal Railway System

tribution, and 25,034 feet of secondary paper insulated, lead covered cable. All arc circuits are underground from the sub-stations to the center of distribution and 29,204 feet of

2-conductor, 6,000 volt, paper insulated, lead covered cable is used for this purpose. In addition to the foregoing there are street railway feeders and returns, police patrol, and fire alarm services, all of which to a certain extent are underground.

Underground primary switching and sectionalizing is done through the use of G. & W. primary sectionalizing boxes, and Westinghouse, C. G. E. and Conduit manhole type oil circuit breakers. On secondary distribution G. & W. six way 600 amp. secondary boxes are used, in addition to a 6 way secondary box which is made up locally.

176 C. G. E. 6.6 amp. ornamental magnetite lamps are used in lighting the main business portion of the city and 2,000, 6.6 amp., 250 c.p. to 1,000 c.p. lamps are used in lighting up the avenues and streets of the city. Ordinary mast arms and goose necks are used, and in the majority of cases converted pendant magnetite and a.c. arc lamp cases are used as fixtures. Five light ornamental standards are installed on several of the streets and avenues, but it is anticipated that they will be changed in the near future to single light standards of higher unit capacity.

The ring system is used in feeding outlying districts. This ring is 16 miles long, 12,000 volt, 3 phase, 60 cycle, and five sub-stations are fed off it. No. 1. sub-station in the center portion of the city is fed underground by two 12,000 volt, 3 conductor, paper insulated, lead covered cables.

Connected to the city lines are 16,184 lighting services. 20,129 connected h.p. motor load, (in addition to this load all the water works pumping station equipment is electrically driven and amounts to about 1500 h.p.), 682,801 lbs.

Typical Transformer Manhole, Calgary

of copper is used for overhead primary distribution. 21,550 kw. in service transformers, and 290 miles of pole line.

During the year 1921 8,528,649 kw. hrs. were sold for lighting purposes, 15.594,966 kw. hrs. for power purposes. 8,240.010 kw. hrs. to the street railway. and 2,825,920 kw. hrs. were used for arc lighting. The department will show a surplus of about $30,000.00 for the year ending December 31st, 1921.

Power rates vary from ¾ cent to 2 cents per kw. hr.; the lighting rates from 3 cents to 6 cents per kw. hr.; heating 1-¾ cents to 2 cents per kw. hr. A minimum charge is made. but no standby or other charge is imposed. A discount of 10 per cent. is allowed on all accounts for prompt payment.

# Electrical Supply in City of Saskatoon

### Has Established an Enviable Record for Continuity of Service. Total Cost of Production 2.77 cents per Kw.h.; Average Selling Price 3.28 cents.

By J.R. COWLEY, City Electrical Engineer

Saskatoon, in the province of Saskatchewan, is a city of about 25,800 inhabitants, its population having increased from 113 in the year 1903. The demand for electricity has kept pace with the increase in population, the original power plant constructed in 1907 having to be replaced by one of considerably increased capacity in 1912. The present plant is housed in a building of fireproof construction, located on the north bank of the South Saskatchewan river and situated within half a mile of the centre of the city.

#### Engine and Boiler Rooms

The engine room contains three turbo-generators of 5,000, 3,200 and 2,300 kw. rating as well as two 75 kw. turbo-exciter units, all power being generated at 2,300 volts, 60 cycle, two phase. The condensing equipment is located in a basement extending the full length of the engine room, each turbine having its own surface condenser along with the necessary air and condensate pumps; the latter being in most cases motor driven.

Eight 500 h.p. Babcock and Wilcox boilers, arranged in batteries of two, supply steam for the turbines, each boiler being fitted with B. & W. chain grate stokers arranged in a special furnace setting for the combustion of lignite slack coal. This coal carries about 16 per cent of moisture with a B.t.u. value of 9,500 and the results obtained in combustion have been very satisfactory. The steam pressure is 150 lbs. superheated 100 deg. F., and each boiler is fitted with a $CO_2$ recorder, differential draft gauge and steam flow meter.

Coal is handled by means of a bucket conveyor with ovehead storage and gravity feed; ash removal is by means either of the conveyor or by steam suction.

Each battery of boilers has also its own economizer, the boiler feed being handled by turbine and reciprocating pumps from two open type water heaters.

Induced draft is used and is provided for by two motor driven and two steam driven fans, any one of which is capable of producing draft for four boilers.

#### Switchboard

The original switching equipment installed in 1912 had to be entirely replaced in 1916. The present switchboard is of the remote controlled type operated from a benchboard situated on a gallery overlooking the engine room. All high tension switchgear is contained in concrete cubicles, the generator switches being located in the basement and the selector and feeder switches on the engine room floor. Provision has been made for 12 feeders of which 10 are in use at present. The feeders are arranged in groups of four, each group being separately protected by selector switches fed from the main bus bars, which are in duplicate.

For traction purposes there are installed two 300 kw. motor-generators and one 600 kw. Bruce Peebles convertor all of which are controlled from a 7 panel traction switchboard situated on the switching gallery. The two 300 volt switching in this case is done by means of manual remote control switchgear.

#### Percolating Water Supply

Cooling water for the condensers is pumped from the river by means of centrifugal pumps installed in a pump house located on the river bank, the capacity of same being 5,600, 5,000, and 3,600 gallons per minute respectively, all

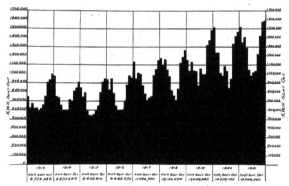

Chart Showing Output in Kw.h. per month from 1913 to 1921—City of Saskatoon

Interior of Power Plant.
City of Saskatoon

being direct driven by 2,200 volt motors.

The design of the intake well will show some features which may be of interest, and in this connection the general design is shown on sketch plan. Originally the circulating water was pumped from the well shown in the centre, which drew its supply from a 24 in. intake pipe extending into the river, but with increased demands for water this installation became inadequate and the difficulties experienced from shortage of water and the necessity of frequent cleaning out of well and condensers due to large deposits of sand and foreign matter made it desirable to increase the capacity as well as make arrangements to cope with the condition of the river water which at certain periods of the year carried large quantities of sand and other foreign matter in suspension.

In order to obviate these troubles an outer well was constructed and divided into two halves, each half separately fed from a new duplicate intake pipe. Gate valves were installed at the entrance to the outer well and openings for valves and screens were made in the inner well wall at a

Boiler Room, City of Saskatoon Power Plant

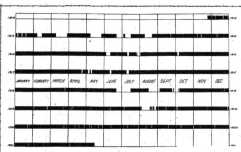

Performance Chart of 3,200 Kw. Generator, Saskatoon. Non-stop runs—October 6th, 1916, to April 22nd, 1917—June 12th, 1917, to March 24th, 1918.

point furthest away from the intake; suction drop legs being also installed so as to pull water directly from the outer wells when desired.

By this method the water flows through each or both halves of the outer well at a low velocity thus depositing its sand; while foreign matter is caught on the screens before entering the inner well. The result has been very successful in operation as water may be drawn from either of the three wells, and accumulated deposits of sand can be removed at leisure without affecting continuous operation.

The institution of a closed syphon system has also reduced the working pressure of the pumps from 70 to 50 feet of total head.

### Maintaining Continuous Supply

The city plant has been particularly fortunate in the maintaining of a practically continuous supply of current since 1913, and as a matter of interest a performance chart of the Westinghouse 3,200 kw. turbo-generator indicates the remarkable reliability of the modern steam turbine. The longest non-stop run of this unit was from June 12th, 1917, to March 24th, 1918, and the current generated during the period amounted to 9,943,650 kw.h.

In order to relieve this unit and provide a greater factor of safety, an additional 5,000 kw. Westinghouse turbo-generator was installed in the early part of 1921, this unit superseding a 750 kw. reciprocating set installed in 1912.

The difficulties in the installation of this unit were increased owing to the removal of the solid concrete foundation of the old set, and the fact that the lowering of the basement level necessitated the underpinning of the engine room walls by reinforced concrete.

All work in connection with the installation of this unit was carried out by the power house staff, the erection being supervised by the Westinghouse company's erecting engineer.

The steam consumption guarantees of this unit are as follows:—

|  |  |
|---|---|
| 5000 kw. | 14.95 lbs. |
| 4000 kw. | 14.5 lbs. |
| 3750 kw. | 14.55 lbs. |
| 2500 kw. | 15.55 lbs. |
| 1250 kw. | 19.7 lbs. |

Tests carried out since the installation have shown the consumption to be rather better than that guaranteed.

During the past year the total kw.h. sent out amounted to 15,394,380, of which 14,072,845 were sold at an average price of 3.28 cents per kw.h. sent out. The total cost of production, inclusive of distribution, and fixed charges, amounted to 2.77 cents or an overall profit of $77,735.07.

The number of meters in service at the end of 1921 were as under:—

|  |  |
|---|---|
| Domestic Light | 5591 |
| Domestic Power | 325 |
| Commercial Light | 1273 |
| Commercial Power | 367 |
| Total | 7,556 meters |

The following table shows the distribution of current sold during 1921 as well as a comparison to the revenue obtained:—

|  | Kw.h. Sold | Revenue | Percentage of Total kw.h. Sold | Percentage of Total Revenue | Revenue per kw.h Sold |
|---|---|---|---|---|---|
| Domestic Light | 1,558,910 | 127,555.42 | 11.1 | 25.5 | 8.18 |
| Commercial do. | 2,105,865 | 145,090.34 | 15.1 | 29.0 | 6.89 |
| Domestic Power | 607,190 | 16,733.89 | 4.3 | 3.4 | 2.76 |
| Commercial do. | 1,425,956 | 70,771.74 | 10.1 | 14.1 | 4.96 |
| General Light | 75,717 | 4,836.44 | .54 | .9 | 6.38 |
| City Accounts Power | 88,850 | 2,547.68 | .63 | .5 | 2.87 |
| Street Lighting (Metered) | 301,215 | 9,064.01 | 2.14 | 1.8 | 3.02 |
| Street Lighting (Arcs) | 655,560 | 23,544.15 | 4.65 | 4.7 |  |
| Pumping Plant | 1,317,981 | 18,356.79 | 8.56 | 3.6 | 1.5 |
| Street Railway | 2,034,580 | 30,518.70 | 14.45 | 6.1 | 1.5 |
| Manufacturing (Flour Mills etc.) | 4,001,021 | 52,475.07 | 28.43 | 10.4 | 1.31 |
|  | 14,072,845 | 500,994.23 | 100 per cent | 100% |  |

South end of turbo-generator room, Moose Jaw

# Moose Jaw Plant is New and Modern

### Following fire in 1912 fine new equipment and buildings make this one of the most economical plants in the Middle West

By J. D. PETERS, Superintendent

The Corporation of Moose Jaw first entered into the light and power business in 1904, and no better evidence of the rapid development of the West can be found than is represented by the growth of this business.

The plant with which the city first started in the central station industry consisted of a 100 kw. engine-driven (belted) 2 phase, 2300 volts, 60 cycle generator, and two boilers of an antiquated type. In the years following, up to 1910, this equipment was replaced by a 225 kw. direct connected, Westinghouse engine-driven set, and two B. & W. water-tube boilers, and one 225 kw., Allis-Chalmers Bullock generator, driven by a Robb-Corliss engine; also a 500 kw. General Electric vertical type steam turbine was added, along with two additional boilers.

In May, 1912, this plant, with the exception of boilers, was destroyed by fire, and a new plant built upon the same site, operating one unit in less than ten weeks after the fire, but the real construction work was carried out after service was re-established. The new plant was very much more modern than the old, and comprised one 500 kw., and one 1,000 kw., General Electric, 3 phase, 2300 volt, 60 cycles, 3,600 r.p.m. steam turbo-generators, which are equipped with surface condensers, steam and motor driven exciters, etc., and four 250 h.p. Babcock and Wilcox boilers, equipped with superheaters and chain grate stokers. Sturtevant economizers and induced draft fans were also installed.

### New Building and Equipment

The new building was of brick and steel construction and fireproof throughout, and laid out for three turbo-generator sets, of 3,000 kw., total capacity. In 1913 and 1914, the third unit, consisting of a Willans & Robinson turbine and Siemens generator of 1500 kw. capacity and 3,600 r.p.m., was installed. At about the same time four Babcock and Wil-

cox boilers of 375 h.p. capacity, each, were added, and also a coal and ash handling plant, consisting of tipple, crusher, and gravity bucket type conveyor, which conveyed the coal from the spur track about 90 ft., from the building and elevated it to the overhead coal bunkers. This same conveyor was used to deliver ashes from the ash pits to an ele-

Mr. J. D. Peters

vated ash bin outside the building. No further extensions were made to the plant until the year 1919, when the 500 kw. turbo-generator set was taken out and one of 3300 kw. capacity, also of General Electric manufacture, 3600 r.p.m., was installed in the same space, bringing the total generator capacity up to 6000 kw., but no additional boilers were added. During the same year, steam-jet ash conveyors, supplied by the American Steam Conveyor Corporation, were

| Year | Output of Plant | Yearly Load Factor | Total Revenue | Number Accts. |
|---|---|---|---|---|
| 1911 | 1,427,500 kw.h. | 13.9 per cent | $105,578.39 | 1979 |
| 1912 | 2,141,250 kw.h. | 21.9 | 128,111.94 | 2960 |
| 1913 | 3,762,900 kw.h. | 29.0 | 195,236.39 | 3450 |
| 1914 | 3,739,900 kw.h. | 30.0 | 179,428.36 | 3126 |
| 1915 | 4,266,800 kw.h. | 30.4 | 162,494.00 | 3489 |
| 1916 | 5,468,720 kw.h. | 34.7 | 186,812.41 | 3760 |
| 1917 | 6,367,790 kw.h. | 36.2 | 207,604.51 | 4037 |
| 1918 | 7,155,610 kw.h. | 36.1 | 229,965.62 | 4184 |
| 1919 | 7,348,830 kw.h. | 36.4 | 270,560.82 | 4602 |
| 1920 | 8,624,500 kw.h. | 36.3 | 310,072.52 | 4759 |
| 1921 | 10,875,800 kw.h. | 40.0 | 367,868.63 | 4910 |

also installed to relieve the gravity bucket conveyor of this work and reduce the labor required in handling ashes.

### The Switchboard

The switchboard was supplied by the Canadian General Electric Company in 1913, and also all additions made since that time. It is a very simple arrangement of remote mechanical control with no duplication of bus bars or switches, and no disconnecting switches, and has proven very satisfactory. The fact that in eleven years of continuous operation there has been only one interruption of service due to switchboard trouble, bears this out very well. As the plant capacity increased it was of course necessary to install circuit breakers of higher rupturing capacity.

### The Street Railway Supply

The street railway is not owned by the city and up to February, 1921, the railway company produced their own

View of Boiler Room, Moose Jaw Light and Power Department

power. At that time, the ci.y entered into a contract to supply this power, and a 500 kw., 2300 volt synchronous motor-generator set was installed by the company in the city plant.

The auxiliary equipment also includes Weir boiler feed pumps, a Hoppes metering feed-water heater, Bailey coal meters and a Refinite water softening plant.

The Table above shows not only the rapid growth of this business but also a marked improvement in load conditions since the year 1910.

While this remarkable increase in plant output is partially accounted for by increase in population, as indicated by the increase in the number of accounts, the most of it is due to increased sales for power purposes. During the year 1921 the sale of electric energy was divided as follows:—

Commercial and Domestic Lighting ....31.2 per cent
Electric Railway Operation .......... 11.1 per cent
Manufacturing Purposes ............ 40.0 per cent
All other Purposes ................... 17.7 per cent

Electric cooking and heating is rapidly becoming an important part of the business, over eleven per cent of the gross revenue for the year 1921 being derived from this source. Very favorable rates are given for these purposes.

The city's schedule of rates is designed as nearly as possible, to give rates which are equitable as between different classes of consumers, and to give the consumer the benefit of high load factor and diversity in the use of energy.

Western coal is used entirely, principally from the Lethbridge district and very satisfactory results are obtained, the average fuel $cost$ in the year 1921 (the highest for several years) being 1.15 cents per kw. hour sent out.

The street lighting system consists of 315 one thousand candle power, 20 ampere series gas filled lamps, and 48 five-light standards, 75 and 100 watt multiple lamps being used on these standards.

### Light and Power Department

The Light and Power Department has been very successful, financially, during its existence and in addition to giving very low rates, considering the population served, and the geographical location, it has been able to provide interest, sinking fund and depreciation reserves, and has turned over to the city, surpluses which at the end of the year 1921 total about $90,000.00

The total investment to December 31st, 1921, was approximately $1,050,000.00.

During the last thirteen years, the construction and operation of the plant and distribution system has been carried out under the direction and supervision of the writer.

## Electric Show, May 1-6

Our readers are again reminded of the Twin City Electrical Exhibition to be held in the Auditorium, Kitchener, Ont., during the week of May 1 to 6. Mr. Geo. A. Phillips is manager of this Exhibition and will gladly give information to inquirers. The object of the electric show is chiefly that the housewife, the farmer and the factory manager may become thoroughly acquainted with the almost innumerable electrical appliances that are now on the market, and which have been created to bring comfort, economy and prosperity to their users.

# Edmonton Utilities Municipally Owned

### Steam generated energy for power, light and street railway service
### —A total of 13,000 Kw. a.c. and 1150 Kw. d.c.—New unit illustrated

By W. J. CUNNINGHAM, Superintendent Power Plant

The power house and pumping station are municipally owned and are under the direct supervision of the power house superintendent. The buildings are situated on the north bank of the North Saskatchewan River, between 101st and 104th streets and are 270 feet apart. The plants have been altered and extended from time to time since first built as the needs of the city required. The original plant was placed in operation some time in 1891, the first part of the present pumping station was built in 1909 and the present power house about 1908. The old original buildings have been since abandoned, no building extensions being made to the power house or plant between 1913 and 1920.

Electrical energy is generated in the power house for street railway service and general lighting and power, including the power supplied to operate two high lift pumps in the pumping station. Steam is supplied from the power plant boilers to operate the steam pumps in the pumping station, and the circulating water for the condenser equipment in the power house is obtained from the low lift pumps in the pumping station so that the two stations are interdependent.

The building is about 140 x 256 feet and one storey in height, with a basement under the engine room. Walls are brick with brick wall between the engine room and boiler room. The engine room is concrete on steel beams.

### Main Generating Units

The general supply from the generators to the station bus bars is 3 phase alternating current at 2,300 volts, 60 cycles, and direct current 550 volts. The main units are— one 5000 kw., a.c., 3,600 r.p.m., separately excited, Canadian General Electric generator, direct connected to Canadian General Electric-Curtis turbine, with surface condenser and steam driven auxiliaries; one 4000 kw., a.c., 1,800-r.p.m., separately excited Siemens generator, direct connected to Wil-

lans-Curtis-Parsons turbine with surface condenser and motor driven auxiliaries; one 2000 kw., a.c., 3,600 r.p.m., separately excited Siemens generator, direct connected to Willans-Curtis-Parsons turbine with surface condenser and motor driven auxiliaries; one 2000 kw., a.c., 3,600 r.p.m., separately excited Westinghouse generator connected direct to Westinghouse turbine with surface condenser, and steam driven auxiliaries and motor driven condensate pump; one 750 kw., d.c., 550 volt, 285 r.p.m., compound wound, interpole Siemens generator, direct connected to Belliss & Morcom vertical triple expansion engine with surface condenser and motor driven auxiliaries, and two 400 kw., d.c., 550 volt, 350 r.p.m. Crocker-Wheeler generators, direct connected to Belliss & Morcom vertical compounnd engines with common condenser of the surface type and steam driven auxiliaries.—Total 13,000 kw. a.c., and 1,150 kw. d.c.

### Exciters

The exciters consist of one 100 kw., d.c., 125 volt, 500 r.p.m., Electric Construction generator connected direct to Howden compound, non-condensing engine; one 75 kw., d.c., 125 volt, 275 r.p.m., General Electric generator direct coupled to Goldie Ideal tandem compound non-condensing engine; one 1000 kw., d.c., 125 volt, Westinghouse generator direct connected to 2,300 volt Westinghouse induction motor— Total 275 kw. rated capacity.

### Motor Generator Sets

These include one 300 kw., d.c., 550 volt, Westinghouse generator, coupled direct to a Siemens 2,300 volt induction motor; one 250 kw., d.c., 550 volt General Electric generator, connected direct to 2,300 volt General Electric induction motor;—Total 550 kw. rated capacity.

### General Arrangement

The main generating units and boiler room are arranged

New Turbine Installed in Power Plant of the City of Edmonton

back to back separated with a wall. The condensers and auxiliaries are placed in a basement below the engine room floor.

### Boilers

There are 16 Babcock & Wilcox water tube boilers, aggregating 7,200 h.p., working under a pressure of 165 lbs., half of which are rated at 400 h.p. each and the other half 500 h.p. each. They are equipped with superheaters designed to give 125 to 150 degrees superheat when the boilers are operating at normal rating.

Firing is done with eight Babcock & Wilcox chain grate stokers in the case of the 500 h.p. boilers and the drive is arranged for two banks of four boilers. The equipment for

to 5 feet 7 inches at the bottom; they are used in connection with the Pratt system. The third, for the Sheldon plant, is an ordinary steel stack, 80 feet high above the firing floor and 6 feet diameter. The main flues are constructed of concrete and run north and south under the boilers.

### Handling of Fuel and Ashes

Two coal bunkers, one for the 500 h.p. boilers and one for the 400 h.p. boilers, are placed above the boilers central with the firing aisle. They have a total capacity of 1000 tons and are provided with chutes for supplying the stokers and hand fired boilers. The coal is delivered to the power house by rail and is unloaded to either of the two hoppers by a clam shell bucket locomotive crane. From the hoppers the

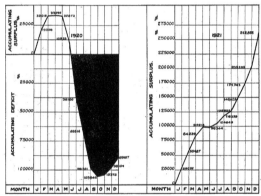

Edmonton Public Utilities—Showing accumulating Surplus or Deficit 1920-21

each bank consists of a 10 h.p. motor and standby engine belted to a countershaft from which motion is given to the stokers. Two of the 400 h.p. boilers are provided with Westinghouse chain grates driven by an engine to a countershaft common to both boilers. The remaining six 400 h.p. boilers are hand fired. The boilers are double banked and face towards a central firing aisle.

### Draft

The draft is produced mechanically and the equipment for the 500 h.p. boilers consists of a cold air Pratt ejector draft plant per bank of four boilers equipped in each case with a 45 inch Sirocco fan direct connected to a 90 h.p. motor, 550 volt, d.c., variable speed. The eight 400 h.p. boilers are served by a double fan Sheldon induced draft plant and each fan is driven by a 10 x 12 inch engine. Both the Pratt plants are located in a fan room at the north end of the boiler room and the Sheldon plant in a room at the south end.

The stacks are three in number, built of steel plates; two are of the Venturi type, 68 feet 6 inches high above the firing floor and 9 feet 10 inches diameter at the top, tapering

coal is delivered to the bunkers by way of a tray and bucket conveyor. The equipment includes a coal crusher driven by motor. The removal of the ashes is by way of the basement under the boiler room floor.

Water is delivered for condensing purposes from the pumping station, delivered into a main header. The discharge from the condensers is into a main header with connections to the river and the sedimentation basin. The boilers have a source of supply from the city mains and direct from the pumping station.

A Paterson's water treating plant of the soda and lime type, treats the make up water and the condensate from the reciprocating engines. An open heater using exhaust steam is also provided for heating the water.

Four Weir feed pumps each capable of delivering to the boilers 7,540 gallons of water per hour when supplied with steam at 150 lbs. per square inch are installed. The feed lines connecting with the boilers form two complete loops either of which supplies water from the pumps.

### Pumping Station

The pumping station is situated east of the power house.

is one storey in height and has an area of 12,400 sq. ft. The low lift pumps are situated in a pit about 40 feet deep at the south end of the building to bring them down to about the normal level of the river.

Low lift pumps include one W. H. Allen single stage centrifugal pump with a capacity of six million gallons per day against a head of 55 feet, direct connected to 105 h.p. Allen compound engine, condensing; one John Inglis two stage centrifugal pump with a capacity of six million gallons per day against a head of 80 feet, direct connected to 150 h.p. Belliss & Morcom compound engine, condensing; one W. H. Allen single stage feeder centrifugal pump with a capacity of 18 million gallons per 24 hours against a head of 55 feet, direct connected to 307 h.p. Allen compound condensing engine.

# Don't Operate Appliance Department

### Edmonton municipality believes in leaving merchandising in the hands of the dealers—Gives both dealers and customers every co-operation

#### By W. J. MURPHY, Superintendent Light and Power Dept.

The field of operations of the Electric Light and Power Department is the distribution system, the street lighting and all commercial details relating to the consumer.

The class of power is 60 cycle, three phase, 2300 volts, purchased at power plant switchboard and distributed as 2200 volts and 220 volts, three phase power; 220/110 volt single phase lighting. The feeder from the power plant to the sub-station is 6600 volts.

The consumers connected on January 1st, 1922, were as follows: Three phase power, 399; lighting and others, 15,-738. Total consumers 16,137.

Connected customers represent approximately twenty-five per cent of the entire population, i.e., one in four; approximately one hundred per cent of the houses adjacent to the electric light pole lines are connected for electric light service.

Three phase motors connected total 7,733 h.p.; meters for all consumers.

The rates charged are as follows:

#### Lighting

1-300 kw.h. at 8 cents per kw.h.
301-1000 kw.h. at 7 cents per kw.h.
1001-up kw.h. at 6 cents per kw.h.
Minimum charge one dollar per month.

#### Three-phase Power

(General three-phase power other than primary, 2200 volts).

1-300 kw.h. at 3 cents per kw.h.
301-1000 kw.h. at 2.5 cents per kw.h.
1001 and up at 1.5 cents per kw.h.
No bill issued for less than 3 h.p.; minimum charge, $2.25 per month.

#### Domestic Heating and Cooking

Same as 3-phase power, above.
Minimum charge:
Up to 2 kw. total connected load, $1.50 per month.
Over 2 kw. total connected load, $2.00 per month.

These rates are all subject to a discount of 5 per cent. Special rates are given where energy is consumed in wholesale quantities.

The street lighting lamps in service are: Tungstens, 1,319; pendant arc, 38; Whiteway, inverted arcs, 249; Total, 1,606.

The total energy purchased at the power plant switchboard for the twelve months ended December 31, 1921, for light and power consumers, and street lighting, was 16,792,500 kw.h.

#### New Feeder System

Construction will be commenced this year on the first section of a 13,200 volt feeder system and two additional sub-stations. Plans at present are for out-door sub-stations with probable change to regular indoor sub-stations, etc., when conditions warrant the expenditure at some future date.

Estimates have been approved for electric light pole lines extensions this summer, in various parts of the city.

Mr. W. J. Murphy, Edmonton

which will provide electric light service to about 150 houses not served at present.

#### Financial

The department has for many years consistently shown a surplus after paying all operation and maintenance items and providing for all capital charges, such as sinking fund, interest, and depreciation. For the twelve months of 1921, this net surplus amounted to $114,200.

A regular programme of advertising is part of our general policy, such advertising covering goodwill, information about commercial details, electrical appliances and similar subjects. For this advertising we use the local press, a civic bulletin in the street cars, and also the reverse side of our electric light bills.

#### Merchandising Electrical Appliances

The department does not operate a merchandise department or store or sell appliances. We endeavor to give all information possible, and encourage the customer to apply to us for same. We will also assist the customer in selecting suitable appliances for their needs, but for actual purchasing we refer them to the regular stores.

General view of Turbine Room, Regina

Mr. E. W. Bull

# Electricity in Saskatchewan's Capital

### Fine record of continued growth and low generating costs in face of mounting expenditures—Some interesting views of the plant

By E. W. BULL, Superintendent

The City of Regina arranged for operating its own electrical utilities and purchased a small privately owned plant and distributing system in 1904. This plant was old and consisted of a 100 h.p. Brown engine, belted to two 375 kv.a., 1100 volt, 133 cycle, single phase alternators and a more modern plant was designed by Mr. Jno. Galt, C. E., who equipped the whole city with sewerage, waterworks and electric light.

The plant located on the corner of Broad St. and Dewdney Ave. commenced operation in March, 1905 and consisted of a 300 kw., 2200 volt, three phase, 60 cycle, Westinghouse generator, direct connected to a 150 r.p.m. Inglis Corliss engine. This plant, although considered modern at the time, was arranged for non-condensing operation and operated from dark until daylight and was too large for the load. During 1907 a regular twenty-four hour service was inaugurated and motors commenced operating on printing presses and various machines in the city. To economically carry the light load a 100 kw. Westinghouse generator direct connected to a 277 r.p.m. Goldie Ideal engine was installed and a condenser was put in using city water and a cooling basin for circulating water.

There followed a growth of business and in 1909 additional equipment consisting of a 500 kv.a. Westinghouse low pressure turbine (the first to operate in Canada) was installed, using the exhaust of the 300 and 100 kw. engines.

In 1911 a. street railway power supply was met by installing a 400 kw., 600 volt, Belliss Morcom, Siemens unit, running at 350 r.p.m., and additional B & W boilers to supply steam. A 1500 kw. Willans and Robinson, Siemens steam turbine unit was also installed to meet the regular a.c. load and a spray system for cooling circulating water was installed.

The electric load continued to grow very rapidly and in 1912 and 1913 a new plant with site near a body of water, was considered necessary and the foundation for building and machinery was completed in the fall of 1913 and a 1300 kw. motor-converter (LaCour patent) was installed in No.

1 plant converting 2200 volt a.c. power to 600 volt d.c. power for railway operation. This new plant is located at the corner of Winnipeg St. and Douglas Ave. on the shore of Wascana Lake and was to carry a load of 16,000 kw. and to have all essential parts of plant in duplicate and is fitted with economizers and underfeed stokers.

The engine room was designed for six machines and

Main Switchboard, Regina Municipal Plant

boiler room for a total of sixteen boilers, fitted with mechanical stokers.

At present four machines are installed, consisting of 2-1500 kw. turbines, (one moved from No. 1 Power House), 1-3000 kw., all Willans & Robinson, Siemens units and a 5000 kw. General Electric Curtis turbine, leaving space for two more machines.

As Regina is located very far from any reliable source of water power a special effort to secure economical generation of power from coal has been made and the expensively produced energy has been distributed over a system

Chart of Output of Electric Light and
Power Department, Regina

View of Boiler Room, Regina

made as efficient as consistent with the cost of producing power. The results obtained from No. 2 plant are such that a kilowatt hour is generated with a heat consumption of 27,000 B.t.u. in coal as fired.

The distributing system uses the old power house (No. 1) as a sub-station for street lighting and street railway power as it is located close to the centre of load as well as supplying all street railway power and street lighting from this point and the motor converter is used for power

factor control. This distributing system consists of 125 miles of pole lines and supplies over 8000 consumers, 650 street lamps which light 81 miles of streets.

The amount of energy sent out from the plant as well as that registered on various classes of consumers' meters is shown in the following table, which covers eight years' records.

The McDonald & Willson Lighting Co., Ltd., 313 Fort St., Winnipeg, Man., has been awarded the contract for electrical work on a school being erected at Windsor Park, St. Vital, Man., at an estimated cost of about $42,000.

**Table I—Energy Generated, Etc.**

| Year | Kw.h. sent out from plant | Fuel Cost per ton | Fuel | Cost in cents per kw.h. | | | | |
| | | | | Total generating | Total distribution | Overhead charges | Total cost | Total revenue |
|---|---|---|---|---|---|---|---|---|
| 1914 | 9,315,355 | $6.99 | 1.41 | 2.03 | .51 | .87 | 3.41 | 3.67 |
| 1915 | 8,255,275 | 5.25 | .77 | 1.18 | .41 | 1.42 | 2.93 | 3.58 |
| 1916 | 8,902,242 | 5.00 | .72 | 1.04 | .31 | 1.41 | 2.76 | 3.40 |
| 1917 | 11,396,674 | 6.10 | .93 | 1.32 | .50 | 1.04 | 2.66 | 3.10 |
| 1918 | 12,004,705 | 7.00 | 1.18 | 1.83 | .54 | 1.03 | 3.20 | 3.08 |
| 1919 | 13,148,860 | 7.98 | 1.11 | 1.62 | .54 | 1.07 | 3.02 | 3.22 |
| 1920 | 14,626,275 | 9.09 | 1.53 | 2.33 | .39 | .99 | 3.71 | 3.58 |
| 1921 (10 mos.) | 13,264,770 | 9.09 | 1.20 | 1.85 | .15 | .93 | 3.12 | 3.42 |

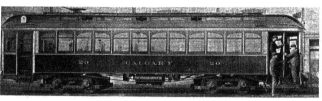

Fig. 1—Standard Type of One-man Car used in Calgary, Alta.

# Electric Railway Operation in West

### Western systems, in common with rest of world, hard hit in past five years—The one-man car as the economic solution

By R. A. BROWN,
General Manager Calgary Street Railway, Light and Power

In 1917 the Calgary Municipal Railway started operating one-man cars. The existing single and double truck cars were rebuilt at a small cost for one-man operation according to patents owned by Supt. McCauley. Since that time we have had no experience which would make us regret the change, in fact, we believe it has been the economic solution of our difficulties.

It seems to be the opinion of railway officials in many large cities that one-man car operation is suitable only for systems, or parts of systems, on which traffic is light. The safety car is responsible, I believe, in a large degree for this opinion. This car which has been developed primarily for one-man operation is a single truck car and accomodates a small number of passengers. By decreasing the headway between cars railway operators plan to handle rush hour traffic. But the wages of motor-conductors is a big item in railway operation costs, and of late years it has become a very serious item, so that putting on more cars does not solve the problem.

We believe there is a limit in density of traffic which one-man cars can handle expeditiously, but would perhaps place that limit somewhat higher than other officials. We feel that we are a step in advance of systems using the safety car. The photographs show the long double truck comfortable riding cars used in Calgary. They handle the traffic at all hours of the day at schedule speeds from 9 to 11 miles per hour. During the evening rush hour a very high percentage of the traffic originates in the down town section, and cars are loaded to full capacity just as rapidly as two man cars were loaded.

The rapid loading of cars is necessary for the success of one-man operation, and it would perhaps be of interest for me to tell how it is done in Calgary. In the first place all the cars have an entrance door and an exit door. Passengers can leave at the same time that other passengers are boarding the car. Long vestibule cars have both doors in the side of the vestibule, while short vestibule cars have one door in the side and the other in the nose of the vestibule as the photograph shows. Making the cash fare 10 cents and selling tickets 4 for 25 cents and 18 for $1.00 has resulted in reducing the cash fares to an insig-

nificant amount; this relieves the motor-conductors of a large amount of change-making. Tickets can be purchased at other places than on the cars. The abuse of transfer privileges has been eliminated as much as possible. This all helps to relieve the motor-conductors of a large amount of work. Doors are operated by the mechanical arrangements shown in the figure.

While the safety features of the safety car are not to be criticized we have had no accidents which can be traced directly to the lack of safety features on our own cars. The rear vestibule is provided with an emergency exit which can be controlled from the front vestibule.

Table I gives the schedule speeds of some of the routes in Calgary.

### Table I

|  | Miles per hour |
|---|---|
| Belt Line | 9.3 |
| Ogden Line (Rural) | 14.0 |
| South Calgary | 10.2 |
| Killarney | 10.5 |
| Sunalta | 8.12 |
| Manchester & North Hill | 10.6 |
| Red & White | 10.8 |
| White Line | 9.8 |
| Red Line | 7.8 |
| Tuxedo Park | 9.65 |
| Riverside | 8.65 |
| Sunnyside | 10.05 |
| Bowness (Rural) | 14.00 |

It will be seen that they are quite high in a good many cases. The Belt Line for instance—a very heavy traffic line—operates at 9.3 miles per hour.

We have answered many questionaires for other cities on the subject of one-man car operation. Some officials seem to be worried about the operation of the cars over railroad crossings. Calgary is fortunate in having subways or overhead bridges at points where the street railway crosses steam railroads. There are only two or three grade crossings, and these are over unimportant steam road switching tracks. I see no reason why the one-man car should not be as safe at grade crossings as the two man car, if efficient trolley guards are used to prevent the trolley from coming off.

As to the matter of accidents, our statistics show that one-man car operation reduces the number of accidents.

The circular graph shows the distribution of operating expenses and fixed charges. We believe a study of the amounts will satisfy the critics of municipal ownership that taxes, depreciation, and other fixed charges are being properly cared for.

### Miscellaneous Statistics 1921

| | |
|---|---|
| Miles Operated | 2,899,036 |
| Hours Operated | 295,820 |
| Passengers Carried | 15,629,275 |
| Revenue per Car Mile | 32.437 cents |
| Oper. Exps. per Car Mile, inc. Fixed Charges | 32.231 cents |
| Oper. Exps. per Car Hour | $ 3.158 |
| Cost of Power per Car Mile | 3.498 cents |
| Average Fare per Passenger | 5.827 cents |
| Average Daily Receipts | $ 2,576.38 |
| Average Daily Oper. Exps. | 2,559.94 |
| Percentage of Oper. Exps. and Fixed Charges to Revenue | 99.4% |
| Total Revenue | $ 940,380.01 |
| Total Expenses | 934,380.01 |
| Approximate Surplus | 6,000.00 |

While this article was intended to cover railway operation in the prairie provinces time has not permitted us to include any other than the Calgary system. This we feel is fairly representative of the systems in vogue in most western cities.

The railway buys power from the municipal power department at the following rates:

Service Charge, $2,300.00 per month.
1 cent per kw. hr. up to 400,000 kw. hrs. per month.
¾ cent per kw. hr. for all power over this amount.
In 1921 the average rate per kw. hr. was 1.26 cents.

The power department operates five sub-stations from which direct current is supplied to the railway. The sub-stations were so located as to help mitigate electrolysis, cut down the amount of feeder copper required, improve voltage regulation, and reduce distribution losses to a minimum. The sub-stations also distribute a.c. energy for light and power, including street lighting. The rate for railway power is so low that the power economies of the safety car do not assume such importance in Calgary as they would in cities with high power costs.

Calgary has a rather novel system of electrical distribution designed on the sectionalizing scheme. The switches used are C. G. E. type S. W. 4 automatic sectionalizing switch. Normally the trolley and feeder system are tied together, but in case of trouble the switches automatically isolate the section affected. This gives us the advantage of a sectionalized system with all the copper economy advantages of a tied in system. The switches work very well, and have given no trouble to date.

The line equipment failure have been reduced noticeably by a more frequent inspection and replacement of

trolley wire. Most failures occur at intersections and under subways. By using special trolley wire at these points—wire that has high tensile strength and good wearing qualities—we plan to reduce failures to a minimum.

Calgary is in the same position as a great many western cities in the matter of track. We have 83 miles of single track serving a city of 65,000 people. The real estate booms of early days spread the city over a large territory—41½ square miles in our case. This is a tremendous handicap to the running of a financially successful railway.

Utilities Commissioner, Smith Calgary

Most of the track was laid in a short period of years so that rebuilding of track is badly needed. We have already had to replace some of the special track work with solid cast manganese steel construction.

In the matter of track bonding we believe that a well bonded track is better, and certainly less costly, than many tons of return copper cables. An 80 lb. rail is equivalent to a 750,000 cm. and a double track of 4 rails is equal approximately to 3,000,000 cm. of return cable. If the joint resistance is made to four feet of rail, this means that a mile of double track is equal to a return cable of 3,000,000 cm. capacity a mile plus 10 per cent long. We use insulated copper return cables, of course, but we should need a great many more than we have were it not for a well bonded track. Good bonding is the best safeguard against electrolysis.

The type of bond used varies with conditions. Where there is paving we use the short flexible bond welded to the

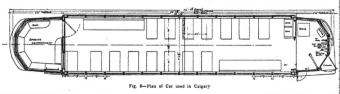

Fig. 2—Plan of Car used in Calgary

rail head, but wherever possible we use the long 4/0 Ohio"
Brass Co's. bond with the steel terminal and weld it to the
base of the rail on each side of the fish plate. The latter
type of bond does not interfere with work on the fish plate
and is down out of the way where heavy truck wheels
cannot rip it off.

At special track work the rails and manganese steel
castings are all bonded together and reinforced by long
leads which tie the ends of intersections together electrically.

Fig. 2 is a plan of the car used in Calgary, showing
the seating arrangement, the location of the fare box, and
door operating mechanism in the front vestibule. The rear
vestibule is fitted up as a smoking compartment. The
cars are heated with Peter Smith hot air heaters using coal.
Last winter Cutler-Hammer electric space heaters were installed in the smoking compartments. With a low rate
for electric power the question of heating the cars entirely
by electricity with the thermostat method of control has
been considered and it is the intention of the department
to equip a few cars and try them out next year.

The following historical data will serve to show the
cost, growth, and operating success of the department
since its inception.

## Construction and Equipment Costs to December 31, 1921

| | |
|---|---:|
| Organization | $ 4,525.51 |
| Engineering Superintendence | 6,213.25 |
| Right of Way | 4,876.00 |
| Track R. Way Construction | 1,580,475.15 |
| Electric Line Construction | 166,169.11 |
| Real Estate used in Operation of Road | 32,167.14 |
| Building Fixtures used in Operation of Road | 62,781.40 |
| Investment Real Estate | 782.00 |
| Shop Tools & Machinery | 5,806.61 |
| Cars | 418,617.09 |
| Electric Equipment of Cars | 160,063.95 |
| Miscellaneous Equipment | 13,574.79 |
| Miscellaneous St. Openings | 5,000.00 |
| Losses on Sales of Debentures | 84,922.12 |
| | $ 2,565,967.12 |

## Operation of Road Showing Passengers Carried Gross Revenue, Gross Expenses, Etc.

| Year | Passengers Carried | Gross Revenue | Gross Expenses | | Balance |
|---|---|---|---|---|---|
| 1909 | 1,594,928 | 64,317.04 | 72,229.73 | Dr. | 7,912.69 |
| 1910 | 6,720,086 | 220,472.28 | 191,147.66 | | 29,324.62 |
| 1911 | 10,578,130 | 380,239.35 | 342,451.53 | | 37,787.82 |
| 1912 | 15,986,658 | 624,998.84 | 625,629.03 | Dr. | 630.19 |
| 1913 | 17,287,860 | 767,891.14 | 754,832.89 | | 13,058.25 |
| 1914 | 16,308,279 | 702,530.26 | 698,698.66 | | 3,831.60 |
| 1915 | 13,909,298 | 561,683.18 | 561,291.96 | | 391.22 |
| 1916 | 14,519,256 | 605,634.18 | 576,912.09 | | 28,722.09 |
| 1917 | 14,006,152 | 582,553.97 | 561,061.45 | | 21,492.52 |
| 1918 | 14,871,290 | 637,579.66 | 644,697.90 | Dr. | 7,118.24 |
| 1919 | 16,918,251 | 834,413.40 | 792,180.45 | | 42,232.95 |
| 1920 | 17,061,356 | 929,700.15 | 941,335.78 | Dr. | 11,635.63 |
| 1921 | 15,829,275 | 940,380.01 | 934,380.01 | | 6,000.00 |
| | 175,420,789 | 7,862,393.46 | 7,606,849.14 | Cr. | 182,841.07 |
| | | | | Dr. | 27,296.75 |

Accumulated Surplus $ 155,544.32

| | |
|---|---:|
| Passenger Cars with Motors | 83 |
| Scenic Cars | 1 |
| Baggage Cars | 1 |
| Trail Cars | 6 |
| Freight Cars | 3 |
| Auxiliary Cars | 1 |
| Sweepers | 2 |
| Derrick Motor Cars | 1 |

### Miles of Track

| | Miles |
|---|---:|
| 1st Main Track | 68 |
| 2nd Main Track | 15 |
| Total | 83 |

# Western Men Favor Standard Specifications

(Continued from page 80)

and therefore a higher priced article than the one which
we are forced to ship further West in order to compete in
price. This in several lines causes a duplication of stock.
From the manufacturers' standpoint it means work and
hardship to them as well as to ourselves.

"We do feel that some scheme whereby inspection and
approval at one point could be established, would be a
step forward, not only from the standpoint of the quality
of merchandise marketed and the safety of the ultimate
consumer, but for the convenience of the jobbers and
manufacturers."

### All Interests will be Best Served

"I understand that some organized bodies in Eastern
Canada have considered the advisability of taking up in
a serious way the question of creating an approval board
for the electrical industry which would have Dominion
wide powers.

"In my opinion this would be a step in the right direction inasmuch as such a board or body could eliminate
some of the practices that are in effect today, due, I
believe, largely to the lack of such a body, and which are
certainly not economic.

"It seems to me that approval of appliances and apparatus can only be properly done (on the basis of economic service to the buying public) at the source of supply
and not at the many hundreds of points of use.

"Generally speaking, our source of supply of electrical
materials and equipment is geographically limited to a small
area, while the points of ultimate use of such are only limited geographically by the boundaries of our country.

"Any plan then that is built on the idea of every civic
center having its own electrical inspection bureau is, in
my mind, wrong economically when those local or civic
bureaus are allowed to each exercise its own judgment as
to giving or withholding approval of material and equipment.

"Rather should such local or civic bureaus be limited
to the work of inspection and checking to see that approved
materials and equipment have been used under conditions
that are safe and economic and to that extent only should
such local bodies be legislative.

"To illustrate, we have certain items of equipment
that carry the approval of the Board of Underwriters and
of The Ontario Hydro Commission and yet which are not
approved for use in Winnipeg proper. These same articles
can, because they carry the approval of the two mentioned
wide power bodies, be used in St. Boniface and other
civic centers which adjoin Winnipeg. This results in duplicate stocks in both dealer and jobber stocks in Winnipeg.

"While volumes might be written on this subject, it
seems to me that the case for an approval board having
national powers can be summed up by saying "All interests
will be best served when some properly constituted central
body employing the most capable engineering talent having access to the best available testing facilities can exert
a nation-wide influence by co-operating with all manufacturers and designers to the point of reducing the life and
property hazard and simplifying and standardizing equipment and material for use by both the trained and lay
public."

"Under such conditions the electrical industry can
hope to some day take a place in the forefront of those
businesses which have an economic reason for being."

# Telephone Development in Saskatchewan

### Eight times the number of rural customers and five times the number of long distance calls, represents progress of ten years

By W. WARREN
Manager  Saskatchewan  Government  Telephones

Telephonically speaking, it may be said that the Province of Saskatchewan was undeveloped previous to 1908. In the early part of that year the Telephone Acts were passed, and in the month of June, the actual work of organizing the Provincial Telephone Department was commenced.

At this time the only systems in operation were owned by the Bell Telephone Company and the Saskatchewan company. The Bell plant consisted of a long distance system following the C. P. R. main line from the Manitoba boundary to Regina and Lumsden connecting up the towns along this route, and one or two other short sections of long distance leads, together with a few exchange systems. The Saskatchewan company operated a system from Regina and Moose Jaw, via Rouleau, Weyburn and Estevan to North Portal on the International Boundary, giving service to intermediate points. With the exception of a very few municipally and privately owned systems, the above may be said to be the total telephone development in the province by the middle of the year 1908, when the Provincial Telephone Department commenced functioning as an operating company, with a staff of twelve employees.

On May 1st, 1909, the Bell company's plant in the province was purchased, and on July 1st, the same year, the Saskatchewan company's system was taken over; a number of other smaller systems were gradually absorbed, the largest outstanding privately owned system, namely Saskatoon, being taken over early in 1912.

#### Doubled in Eighteen Months

By the end of 1909 or within approximately eighteen months of starting operating, the Department had more than doubled the long distance pole and wire mileage which previously existed in the province. At that date the system consisted of 20 exchanges, 100 toll offices, 1132 miles of poles and 1640 miles of metallic circuit.

Long distance, exchange and rural systems, were from this point gradually developed, but it may be said that the maximum development has occurred during the last decade, that is from 1911 on, At the beginning of the fiscal year, 1911, the total stations in operation were 9152, including 3500 rural subscribers. There were in operation at this time, 33 exchanges, 143 toll offices, and 1647 miles of l. d. pole lead, 2203 miles of metallic l. d. circuit, 142 rural companies were operating, the total distance of pole lead built by these companies being 3491 miles.

The population of the province at this time was 492,432, giving a rate of one phone to approximately 53.7 of the population. From this period on very rapid developments were made.

#### Exchanges

The rapid development of the larger centres necessitated the use of the most efficient and very latest type of equipment, with the result that when Saskatoon was purchased, steps were immediately taken to convert it from the three to the two wire system of automatic operation, and this exchange was opened for service under the new system in October, 1912. From this time on an average of one ex-

change has been converted annually to a more modern type of equipment.

In June, 1912, the common battery exchange system in Regina was destroyed by cyclone, and a temporary building had to be erected and emergency equipment installed. The opportunity was taken, in the circumstances, to erect a new exchange for housing automatic, and automatic service was cut in at this point on January 17th, 1914. During the early part of this year, the exchanges of Weyburn and Yorkton were cut over from magneto to common battery service. In the fall of 1914, a new exchange was erected in Prince Albert for housing of automatic equipment. On February 4th, 1915, this exchange was cut over from magneto to automatic service. During 1916 Estevan exchange was built and cut over from magneto to common battery service. In 1918 Swift Current exchange was converted from magneto to automatic, and in 1919 Moose Jaw was cut over from the common battery to automatic.

#### Community Automatic Exchange

During 1920 a community automatic exchange was put into commission at Qu'Appelle. This was the first of its type installed in Canada. A.P.A.X. equipment was installed in Regina and Saskatoon, and rural automatic service was instituted in Regina. During the past year North Battleford was cut over from magneto to automatic, and about one dozen new exchange buildings erected, capable of housing automatic equipment at any time that may be decided to install same.

The small exchange development has also kept pace, as is evidenced by the fact that at the present time there are 289 government owned exchange systems in the province. 579 cities, towns, villages and hamlets within Saskatchewan are on direct l. d. communication over the government l. d. systems, 289 being exchanges, 159 toll offices, and 103 having service over rural lines, and 28 are exchange systems either owned municipally or by independent companies.

Larger exchanges, where conditions warrant, are operated by the department, but the most of the smaller points are operated on a commission basis. In larger rural centres rural companies have in a great number of cases erected buildings for the housing of the central office equipment, and operate both the exchange and rural systems as commissioned agents for the department. This system has proved very satisfactory.

The urban population of the province in 1911 was 131,365 according to Dominion census figures, and estimating it as 196,390 at the present time, the phones per head in urban communities has increased from 1 per 23.6 in 1911 to 1 per 5.8 at the present day. Taking the five cities in this province having a population of over 5,000, viz. Regina, Saskatoon, Moose Jaw, Prince Albert and Yorkton, with a total population of 92,064, the phones are at the rate of 1 per 3.5, of population, which is conclusive proof of the fact that urban development is not confined to the larger centres only, but is most uniformly distributed over all urban communities. The above condition, as far as available statistics are concerned shows a most unusual condition of development for

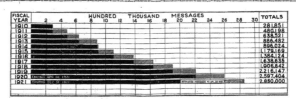

| FISCAL YEAR | HUNDRED THOUSAND MESSAGES | | | | | | | | | | | | | | | TOTALS |
|---|---|---|---|---|---|---|---|---|---|---|---|---|---|---|---|---|
| | 2 | 4 | 6 | 8 | 10 | 12 | 14 | 16 | 18 | 20 | 22 | 24 | 26 | 28 | 30 | |
| 1910 | | | | | | | | | | | | | | | | 281.851 |
| 1911 | | | | | | | | | | | | | | | | 480.198 |
| 1912 | | | | | | | | | | | | | | | | 638.521 |
| 1913 | | | | | | | | | | | | | | | | 886.482 |
| 1914 | | | | | | | | | | | | | | | | 896.024 |
| 1915 | | | | | | | | | | | | | | | | 1.179.169 |
| 1916 | | | | | | | | | | | | | | | | 1.384.124 |
| 1917 | | | | | | | | | | | | | | | | 1.638.635 |
| 1918 | | | | | | | | | | | | | | | | 1.905.842 |
| 1919 | | | | | | | | | | | | | | | | 2.216.147 |
| 1920 | | | | | | | | | | | | | | | | 2.597.404 |
| 1921 | | | | | | | | | | | | | | | | 2.850.000 |

Chart showing yearly increase and totals of long distance messages, Saskatchewan

a large operating company, as the normal development in larger centres is usually out of all proportion to the smaller community centres taken on a total population basis over a large area.

### Long Distance

The long distance development has also been very rapid, and at the present time there are 5,842 miles of pole lead and 15,552 miles of metallic circuit. This is mainly composed of No. 10 N. B. S. copper, No. 12 N. B. S. and No. 6 B. W. G. There is a very small proportion of 9 and 12 B.W.G. iron circuit still in use, but this represents considerably less than one per cent of the total circuit in the province.

In the early part of 1921, a new feature in automatic dialling was adopted, namely a two way automatic dialling service between Moose Jaw, Regina and Saskatoon. This was not the first dialling circuit installed in the province, as one way dialling had been in operation early in 1914, but no attempt to put in a two way dialling system previous to date mentioned above had been made.

### Rural Systems

Previous to 1913 the development of the rural system throughout the province was not very rapid. In this year a new rural telephone act was passed incorporating rural telephone companies, the special features of this Act are, that all companies incorporated under this Act are governed by the provisions of the Rural Telephone Act and the Company Act, and are subject to the Regulations of the Department of Telephones. They have the option of selecting their own telephone area and point of exchange connection. The majority of the resident land-owners within the area selected must agree to have telephones, and a list must be submitted of all those not requiring them. The Rural Branch of the Department then draws up plans, to be built according to the Department's standard specifications, and if all the provis-

ions of the Act are complied with, sanction the organization of the company. The company is then in a position to call for tenders for construction, and all tenders received must be submitted to the Department, where, if the amounts are reasonable, same are approved, with a recommendation to the local government board who have the power to veto or sanction the borrowing of the monies required. Rural telephone bonds are considered very good security, in that they are redeemable in fifteen years or less, and are guaranteed as a tax against all quarter sections passed by pole lead. The necessary legislation has been passed to enable the municipalities to collect these taxes in the same way as general taxes levied against the land.

In the case of leads paralleling for short sections the first lead built has the right to taxes, unless the second company is installing the phone on any quarter section passed, when the taxes on that quarter section only are transferred to the new company. Provisions are also made for the rendering of annual reports and for auditing.

### Rapid Rural Development

Under the above Act the rural development within this province has been most rapid. It is safe to say that in no other territory under any other system have such marked developments been made in this class of service. At the present time there are 1172 rural companies operating, with 53,702 miles of poles and 87,265 miles of metallic circuit and 60,020 telephones. The total capitalization of the rural companies within the province is roughly fourteen and three quarter million dollars, the approximate acreage served being twenty-two and one half million. The approximate number of quarter sections taxed under this Act are one hundred and forty thousand, the average tax levied being fifteen dollars and seventy cents per quarter section where a phone is installed, and, ten dollars and thirty-five cents where

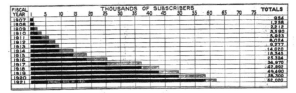

| FISCAL YEAR | THOUSANDS OF SUBSCRIBERS | | | | | | | | | | | | | | | TOTALS |
|---|---|---|---|---|---|---|---|---|---|---|---|---|---|---|---|---|
| | 1 | 5 | 10 | 15 | 20 | 25 | 30 | 35 | 40 | 45 | 50 | 55 | 60 | 65 | 70 | 75 | |
| 1907 | | | | | | | | | | | | | | | | | 954 |
| 1908 | | | | | | | | | | | | | | | | | 1.258 |
| 1909 | | | | | | | | | | | | | | | | | 2.212 |
| 1910 | | | | | | | | | | | | | | | | | 3.590 |
| 1911 | | | | | | | | | | | | | | | | | 5.913 |
| 1912 | | | | | | | | | | | | | | | | | 8.024 |
| 1913 | | | | | | | | | | | | | | | | | 9.277 |
| 1914 | | | | | | | | | | | | | | | | | 14.020 |
| 1915 | | | | | | | | | | | | | | | | | 19.345 |
| 1916 | | | | | | | | | | | | | | | | | 25.324 |
| 1917 | | | | | | | | | | | | | | | | | 36.970 |
| 1918 | | | | | | | | | | | | | | | | | 42.892 |
| 1919 | | | | | | | | | | | | | | | | | 49.490 |
| 1920 | | | | | | | | | | | | | | | | | 58.300 |
| 1921 | | | | | | | | | | | | | | | | | 62.020 |

Chart showing yearly increase and totals of rural telephones in Saskatchewan

Typical
Exterior
and
Interior
Views of
Saskatchewan's
Telephone
System

Above—Saskatoon Exchange, showing power plant.

Left Centre—Prince Albert Exchange Building.

Right Centre—Moose Jaw Exchange Building.

Left Below—Regina Exchange, showing Selector Bays.

phones are not installed. The average rental charged by rural companies to their subscribers for maintenance and operation is fifteen dollars.

In the winter of 1916-17, a school was started giving short three week courses to enable candidates selected by the rural companies to obtain as much information as possible on elementary methods of maintenance, and to date, 485 men have passed through this school.

Taking the present developed area, as from the International boundary to Township 54, an area of roughly 118,600 square miles, the present rural development not counting urban telephones shows a development of one phone for every 1.9 square miles of territory. If, however, forest and Crown lands, Indian Reserves, area covered by lakes and other unproductive lands within this area is not counted, the development shown would be approximately one phone per square mile of territory.

The rural population of Saskatchewan is given in the Dominion Census figures for 1911 as 361,067 and estimating the figures for this year as approximately 565,000, we find that in 1911 the rural development was one phone per 100.5, whereas, to-day it is one phone per 9.1. Considering the fact that they each have an outlet, via long distance to practically every town and village situated over a populated area of 1,000,000 square miles, we think it is safe to say that the development, as far as rural communications are concerned, in the province, are absolutely unique.

### Operation

On account of the province being almost entirely dependent on agricultural pursuits, small communities have sprung up at intervals of from eight to ten miles, creating rather unusual conditions as far as long distance messages are concerned, in that, the proportion of very short haul calls form a much greater proportion of the long distance business carried, than in most operating companies, there being a proportionate decrease in the long haul business handled. The conditions governing the service being somewhat different from other companies of approximately the same size, independent methods of handling the business, have been, to a very large extent, developed. The natural developments along these new lines have resulted in a reduction in operating expense from year to year, together with an increase in the efficiency of the service given.

Special cordless circuits have been installed to speed up the operating, and over 10 percent of the l. d. lines are operated as one or two way dial lines, circuits have been arranged so that two or more stations can dial in on the same line, all the features of manual operating being retained, and the transmission equivalent of the lines not being increased beyond the cord circuit equivalent which has been eliminated.

An additional circuit has been developed for automatic switching of lines at existing switching stations to reduce work to a minimum. With these methods, it has been found that an increase in efficiency generally has been obtained compared with the old operating methods, as distant exchanges immediately become a multiple office organization with respect to the automatic centre.

Due to the very rapid developments made in wireless communication in the last few months, a special study is being made of this important branch of communication work with a view to augmenting existing means of communication by applying the most recent engineering developments where same can be used advantageously for the betterment of the existing service,

The number of employees has increased from the original twelve to approximately six hundred and fifty, and the installation of telephones to ninety-five thousand six hundred and sixty-four, composed at December 31st, 1921, of 17,187 automatic, 1,667 common battery, and 12,940 magneto phones owned and operated by the government. 62,020 rural phones connected to government exchanges or telephone systems. 930 phones installed in 103 villages by rural companies, and connected to the government long distance system. 970 additional phones connected to points at present

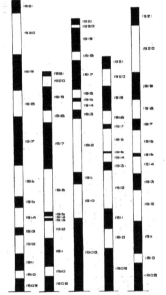

Chart showing yearly development of government system

not reached by the Department's l. d. service. The census figures just available give the population of the province as 761,390, which on the above figures shows one phone per eight of the population. During the last decade the phones per rural population have increased from one phone per 100.5 of the population to one per 9.1 and for the urban centres from one per 23.6 to one per 6.8 of the urban population. Taking this rural and urban development into consideration, Saskatchewan as a unit of comparison with the development of the telephone industry in other countries will be found to compare very favourably with any other country.

# Alberta's Policy of Rapid Extension

## Has Spent Over Twenty Millions on Her Telephone System
## — Fifty Thousand Subscribers and 37,000 Miles of Lines

### By R. B. BAXTER, General Superintendent.

If the statement seems paradoxical nevertheless it is fact that Alberta's great telephone development was foreshadowed many years before the geographical division of the North West Territories of Canada, forming the present province of Alberta, was made.

Only eight years after Graham Bell's historic triumph in transmitting human speech over a wire, citizens of Western Canada were experimenting with the new electrical wonder.

Alexander Taylor, agent for the Dominion Government Telegraphs, brought two telephones from England in 1884. One he established in the Factor's office of the Hudson's Bay Company trading post and the other in his own office in the village of Edmonton, half a mile away. The experiment, it is reported, met with indifferent success but it served to awaken intense interest in the new means of communication and early in the following year two more telephones were ordered by Dr. H. G. Wilson and Judge Rouleau.

Railways had not yet reached the Far West and delivery by ox cart was slow. Interest in the telephone was keen and speculation rife among the pioneers as to the possibilities of communication by telephone over any considerable distance. Mr. Taylor was a strong advocate of long distance telephony and to prove the possibility arranged an experiment.

Richardson, the Dominion Government telegraph agent at Battleford, in what is now the province of Saskatchewan was instructed to stop the ox-cart train, get one of the two telephones being freighted across the prairie and connect it up at his end of the telegraph line.

The operators at the various offices between Edmonton and Battleford were then instructed to cut out their batteries at midnight on a given date. All complied except one who went hunting and forgot.

On delivery of the other telephone in Edmonton, Taylor set up the connection and at midnight of the appointed day the test was made.

Both Taylor and Richardson tried to signal each other but the generators were too weak and so their repeated efforts ended in failure. Failing to signal by use of the generator they next proceeded to shout "Hello" into the transmitters and at last each heard the other. Fragmentary conversation was carried on for ten or fifteen minutes, long enough to vindicate the faith of Alexander Taylor and demonstrate the possibility of telephoning over long distances.

Two years later, in 1886, the Dominion Government telegraph service established the first long distance telephone line by converting the telegraph line between MacLeod and Lethbridge, giving fair service for a long time.

### First Exchange in 1886

The first exchange was installed during the same year in the home of Alexander Taylor at Edmonton and had four subscribers. Growth was very slow for in 1894 the four had grown only to fourteen.

1887 saw the advent of the Bell Telephone Company of Canada. They had been pressed for some months by the citizens of Calgary, then the distribution center of what is now Alberta, and finally established an exchange of forty subscribers in that city in the fall of that year. At first the exchange was a failure due to lack of competent help but gradually improved and with the growth of the frontier town and repeated changes to more modern equipment the city of Calgary has the largest and most up-to-date exchange in the province.

### Long Distance in 1891

The first telephone line built expressly for long distance purposes was erected in 1891 by the Mormon community between Lethbridge and Cardston. The scheme was engineered by the resident head of the Mormon Church, Bishop Card whose name is perpetuated in Cardston, the headquarters of the Mormon people in Alberta. In the same year Lethbridge received exchange service by the installation of an office with some seventy-five subscribers.

By this time the telephone had passed into almost com-

Mr. R. B. Baxter

mon use and exchanges and long distance lines were being installed and erected all over the country.

It is of passing interest to note that a vanished concern, known as the Alberta Telephone Company, gave long distance service between Blairmore, Coleman and Frank in the Crow's Nest Pass, without the erection of poles. The line was strung on growing trees and while breaks were frequent the service seems to have been maintained to the satisfaction of the users.

In 1905 the Province of Alberta came into being with all the rights and privileges of provincial autonomy; Immigrants were flocking into the new land of promise and spreading settlement farther and farther afield.

Pressure, growing more and more insistent, had been brought to bear on the new provincial legislature and effected the desired result in 1906 when it was decided to enter the telephone field by the erection of lines and exchanges with public funds.

## Government Ownership

The first line was built in the fall of 1906 and cut into service in April 1907 between Calgary and Banff. It is claimed that this is the first government owned long distance line in America.

Under the stimulus of this new competition the Bell

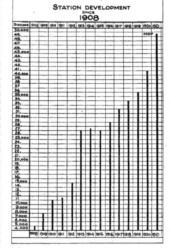

Company became active and the network of lines began to spread rapidly.

In the summer of 1907 negotiations commenced between the government and the Bell Company which resulted in the purchase by the government in April 1908 of the entire Bell interests in the province.

The telephone system was operated by the Department of Public Works until 1912 and while growth was steady the phenomenal expansion of the system may be said to have really commenced in that year with the creation of the Department of Railways and Telephones.

The policy of the government has been from the first to extend the utility to the rural community as rapidly as possible, at the same time making due provision for the fast growing towns and cities and the necessary channels of communication between them.

The policy of rapid extension of the rural system has done much to aid the development and settlement of the country. Every farm home with a telephone is connected by long distance lines over the length and breadth of the pro-

vince. Almost twenty thousand miles of standard line are in active use for rural purposes giving first-class service, almost impossible under private ownership, but an essential factor in the public welfare and development of a young and rapidly growing country.

Monopoly has been aimed at with the result that with the exception of the local system owned and operated by the city of Edmonton, the Dominion Park System, owned by the Federal government at Banff and two or three isolated farmer lines, the entire telephone system of the province is publicly owned.

Nearly twenty thousand farms and over thirty thousand homes and business places in the towns and cities are connected by telephone and the long distance lines are second to none in the Dominion. They cover a territory six hundred and fifty miles long from the International Boundary to Calling Lake, twenty miles north of Athabasca and three hundred miles wide from the Saskatchewan boundary to the mountains. In addition there is a small section at Grande Prairie and Peace River not yet connected with the main system.

Alberta is the possessor of the first carrier current sys-

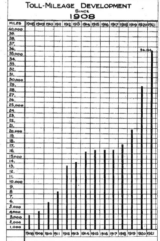

tem in Canada and the fifth in the world. It was installed in the summer of 1921 with two channels between Calgary and Edmonton and is giving excellent service.

The automatic telephone is used exclusively in the three larger cities, Calgary, Lethbridge and Medicine Hat, while common battery is installed in the bigger towns and magneto offices in the smaller.

# Manitoba Has Fine Telephone Service

### One Telephone to Every Nine of Population —
### Automatic Service Finds Favor in Larger Centres

#### By J. E. LOWRY, Commissioner of Telephones.

Telephone service in Manitoba dates back to the year 1882 at which time exchanges were established at Winnipeg, Brandon and Portage la Prairie. The system at that time and until the year 1908 was owned and operated by the Bell Telephone Company of Canada. In 1907 the Manitoba Government entered into negotiations with the Bell Telephone Company and in January 1908 took over the entire Manitoba system. At this time, there were 8,792 telephones in use in Winnipeg and 5,219 telephones in use throughout the balance of the province.

Under government ownership, telephone service in Manitoba has progressed, expanded and developed to its present high state of economical efficiency, adopting in the course of its growth practically every means of increasing the transmission of commercial and social intercourse. Its facilities have been extended to the smallest and remotest towns in the province, not with regard to cost or with an idea of profit, but for the sole purpose of serving the people irrespective of their wealth or position.

To attempt to estimate the value of the telephone service to the citizens of Manitoba in dollars and cents, would be a stupendous undertaking for with the countless benefits which the people of the town and country are enjoying because of the widespread development of local, rural and long distance service are added, an estimate of its intrinsic worth to the country at large, is a sheer impossibility. If one class of people has benefitted more than another, it is the farmer. Instead of the dreary isolation which in former times they sometimes had to endure, they are now linked together in a vast neighborhood by the wires of the Manitoba Telephone System, and when they desire to communicate with friends at a distance it is unnecessary to make a long uncomfortable trip; they are also able to keep in constant touch with their work and during harvest and threshing when repairs are needed in a hurry they find the telephone a most useful servant—and best of all, the service is very reasonable. Considered from every point of view the telephone is the farmer's best investment and fortunate for him was the day when the Manitoba government assumed control of the telephone business in the province and developed the system, which in a few short years has provided service in nearly all parts of the province.

### Developed on a Large Scale

The development of the telephone system in Manitoba during the past few years has been on a large scale and at the present time the service extends to practically all the settled towns in the province, and in the rapid development which has taken place, every precaution has been taken to use only the best material and up-to-date construction methods, with the result that to-day the people of Manitoba have a telephone system second to none. The lines and equipment of the system are built standard in every particular and it is the aim of the management to keep it at a high state of efficiency by the employment of the best available skilled labor.

Manitoba, with a population of 613,008 people has 68,000 telephones in use and of these, 64,752 are direct subscribers

to the government system, the balance being connected with municipal systems and private companies—these private systems, however, have direct connection with the government long distance lines and therefore enjoy, with the government subscribers, all the benefits which accompany connection with the provincial system.

The growth of the system has been continuous, considered as a whole. The following will show the number of telephones in use at the various periods during the past 14 years:

|          |      | Winnipeg | Provincial |
|----------|------|----------|------------|
| January  | 1908 | 8,823    | 5,219      |
| December | 1909 | 12,758   | 8,654      |
| December | 1913 | 26,089   | 19,650     |
| December | 1918 | 29,263   | 23,044     |
| December | 1921 | 38,811   | 25,942     |

To accomodate the large number of subscribers, there is a total of 116 exchanges throughout the province.

In addition to the Manitoba government having provided the people with a first-class telephone service at rea-

Mr. J. E. Lowry

sonable rates, large sums of money have been spent to provide a long distance telephone service as the following will show:

Net work of long distance lines of the system comprises 2,556 miles of pole line and 20,366 miles of wire.

### Automatic Telephone Service

In the building up of a modern telephone system, advantage must be taken of improved methods of operation and the use of advanced ideas. The Manitoba Telephone System has not lost sight of this fact, and the time having arrived when additional switchboard facilities were required to take care of the continuous growth of the city of Winnipeg, a central automatic exchange was installed to serve the subscribers of the business section of the city. Automatic exchanges were also installed to serve the subscribers of St. Boniface and Norwood. In the Winnipeg portion the present equipment capacity is 11,900 lines. The Norwood exchange present equipment has capacity of 1,700 lines.

Additional equipment is being added from time to time to take care of the increase in applications for telephone service. Plans have been made to change the service of two more manual exchanges at Winnipeg for automatic service and it is expected that the change will be completed early in the year 1923.

The city of Brandon is a full automatic service, including rural lines numbering some 480 rural subscribers.

The Manitoba Government has provided a modern telephone service to the people of the province. It is the people's own telephone system and a credit to the province as a whole.

# Edmonton Also Has Automatic System

### This Progressive City Has Invested Two and a Quarter Million Dollars and Shows a Substantial Profit Every Year

#### By R. CHRISTIE, Superintendent.

The Edmonton Telephone System is a municipal enterprise owned and operated by the corporation of the city of Edmonton.

The original plant was of the local battery magneto type switchboard operated by a private company until purchased by the city corporation in the year 1905 when there were about 600 subscribers' lines in service.

The citizens were rightly of the opinion that this magneto type of equipment was obsolete and would very soon become inadequate owing to increasing population so they began investigating the various types of modern telephone apparatus with the view to installing an up-to-date telephone exchange.

After visiting several cities in eastern Canada and the United States the representatives of the city looked into the

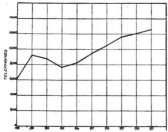

Edmonton Municipal Telephone System—Telephones in service 1912-1921

possibilities of machine switching telephone equipment and were convinced that automatic telephony was to be the system of the future.

Having investigated all types of automatic equipment, it was decided to install the Strowger system manufactured by the Automatic Electric Co., of Chicago, who completed the installation of a 1,000 line switchboard in August 1908, when subscribers' lines were cut over from the old magneto switchboard to the automatic.

Owing to the popularity of the automatic it very soon became necessary to increase the capacity of the switch-board so that in 1910 there were 2,000 lines in operation.

During the next four years the system was rapidly being enlarged because of the greatly increasing population and the subsequent development of the city. In 1913 two new exchanges were put into service and the Strathcona exchange became a part of the Edmonton system, when amalgamation of the two cities took place.

During the war period the telephone situation remained stationary and very little construction work was done until the year 1919, when the demand for service became so great that it was found necessary to erect a new Main Exchange building and install another 1,000 line addition of the latest 2-wire type. This addition was cut into service in August 1921 and owing to the numerous connections of new subscribers plans are being prepared for further additions to three exchanges during the present year.

Beginning in 1905 with less than 500 telephones, the Edmonton municipal system to-day has 11,700 lines with 12,900 telephones in service, which with an estimated population of 62,000 people, averages one phone for 4.8 of the inhabitants or 100 phones for every 480 persons.

The traffic handled by the equipment amounts to 182,000 calls each working day or an average of 15.7 calls per line per day.

The outside plant consists mainly of aerial and underground cable construction varying in size from 50 pairs to 800 pairs, making a total cable mileage of 161 miles with a total single wire mileage of 62,024 miles.

The total capital investment to-day is two and a quarter million dollars. The service rates; domestic $30.00 per year and commercial $55.00 per year, are equal to, or even better than any city of equal population on the American continent. The system being a municipally owned utility has shown a substantial profit for several years. In 1920 this surplus amounted to $60,000.00, 1921, $78,000.00 and every indication goes to show that 1922 will be an even better year than any previous one.

---

### Ask the Editor

Do you, Mr. Subscriber, write the Editor of the "Electrical News" when you want information? If not, you are missing service to which you are entitled.

Always delighted to assist our readers.

# Merchandising Problems and How They are Being Solved

### A Discussion of Many Topics. Exchange of Viewpoints Generally Leads to Solutions. The West is a Vast Field for Appliance Sales. Central Stations, for the most part, Co-operate, but do not Compete with the Electrical Merchants

## The Central Station a Factor in Selling Appliances

#### By JIM. SWAN

If the central station did not exist there could be no place for the electrical manufacturer, jobber or contractor-dealer. These three classes are entirely dependent upon the central station for their existence. All development in the electrical industry has been built around the activity begun by the central station.

The central station is the pivot upon which every merchandising wheel of the electrical industry turns. All development carried on by the central station is bound to assist, very materially, manufacturer, jobber and dealer alike.

In any community with a large population and where the cost of current for domestic purposes is but a fraction of the cost of other fuels, then the demand for appliances for the use of that cheap power will continue to grow as the people become educated to its use.

Up till the present time the merchandising of electrical appliances has been left almost entirely to the central station, more especially with regard to the merchandising of the electric range. The reason for this is, that most central stations look upon the electric range as the supreme load builder. It was therefore necessary for the central station to assume the responsibility of getting it in large numbers into the homes of the people and then keeping them 100 per cent. satisfied with real service. There is nothing that will cause dissatisfaction more quickly than poor service. Fundamentally, the idea is to keep a customer satisfied — with service.

#### Financing Difficulties

Most contractor-dealers on account of their inability to invest money in a stock of ranges or to finance time payments have been placed at a considerable disadvantage and therefore the central station was left with no other alternative than to get into the merchandising field for the safety of their own business and the unquestioned benefit to the public.

Central station development has been aptly termed the thermometer that indicates the degree of prosperity of the entire electrical industry. Can we in our mind's eye picture what Canada would be like to-day if we were to take from it the electrical industry? To a very great extent it is true that the progress of civilization is measured by the progress of electrical development. Transportation, home comforts, efficiency of production and the great possibilities of the future all have their springs deep-rooted on the solid foundation so well and truly laid by the central station.

There is a large undeveloped field in this western country. There remaineth yet much land to be possessed, and if agriculture is to be extended, factories built and homes electrified, it can only be done by real effort, coupled with enthusiasm and determination, together with a spirit of co-operation between central station and contractor-dealer.

#### Low Rates and Abundant Energy

The very low rates which have been established in many communities and the abundant surplus of energy that is now available in many of these communities should be the means of getting every electrical man into line and going forward to success.

The selling lines of electrical appliances have broadened remarkably within recent years so that it is now possible to place within any store an all electrical display that can easily equal and outclass many of the finest dis-

Mr. J. Swan, manager Winnipeg Hydro Appliance Department

plays shown by the leading shops in other lines. The electrical trade, however, has not as yet realized that fact and many of them are content to carry on in dismal surroundings of their own making, while they deprecate the fact that business is not coming their way. It is in this respect that the central station is once more an important factor, making an honest endeavour to place the electrical industry in the fore-front and raising a standard that will be the means of wholesomely stimulating the electrical trade.

One of the duties of a central station is to help the contractor-dealers make money. Some central stations are doing this in a real way, speeding up sales by consistently keeping the public thinking electrically by a sustained advertising effort which is an important factor in successful merchandising.

### Helping Contractor-Dealer

Another way the central station is helping the contractor-dealer to make money is by giving him the installation work on the range and water-heater sales effected by the central station. If the contractor-dealer is doing a real job then it is without question the duty of the central station to turn all that work over to him. During 1920 the Winnipeg Hydro-electric System gave approximately $30,000 worth of wiring and installation work to the contractor-dealers.

The same methods are practiced in other cities and are helping to keep the spirit of co-operation alive between central station and dealer.

It is not good policy for a central station to sell electrical appliances at less than manufacturers' list, for if this is done it is sure to undermine any good fellowship or co-operation that may exist between the two. Every electrical appliance sold by the dealer counts on the credit side of the central station ledger and means continued revenue for the central station, whereas the dealer's profit is obtained by selling above cost and he is entitled to a legitimate margin from the sale of all appliances.

### Immense Possibilities

The possibilities looming up before us are immense. Canada is electrically unsaturated. If the City of Winnipeg is any criterion the scope is unlimited as there are only 10 per cent. of the people enjoying the great pleasure of eating food that has been cooked electrically, while only 2 per cent. are taking advantage of the cheapest and most efficient method of heating water. At the very low rate now obtainable in most cities for domestic purposes, the aim of every central station and dealer should be the installation of an electric range and water heater in every home.

Another field that might be exploited is the washing machine field. If it is true that only one washing machine is placed on the market for every eight automobiles it is high time we were up and doing as there is no greater necessity in a home than a good clothes washer. Facts like the above should compel us to do a little human stock-taking so that we might discover just where the fault is, and what is the reason for the belated condition of the electrical industry. Surely the range, water-heater and washing machine are as necessary in the modern home, as are the typewriter and flat top desk in the modern office.

### Moloney's New District Manager

The Moloney Electric Company of Canada, Limited, beg to announce that Mr. Frank H. Girdlestone, manufacturers' agent of Winnipeg has discontinued his agency business and has accepted a position as Moloney district manager, making his headquarters in their new offices, 27 May St., Winnipeg. Mr. Girdlestone will look after the Provinces of Manitoba and Saskatchewan. Transformer shipments for these Provinces will be made from the company's Winnipeg warehouse.

## Why the Central Station Should not be a Factor in Selling Appliances in the West

By L. B. DICKSON, Vice-President McDonald & Willson Lighting Co. Ltd., Winnipeg

Municipally owned central stations should not enter into competition with the citizens of the municipality in which the utility is situated in any other business than that for which the utility was specifically created. It is to be assumed that the business for which a civic hydro plant is created is the purveying of power at as near cost as possible to the citizens whose money is invested in the concern.

To go further than the purveying of power to the citizens is to invade a field which was not included within the powers granted to the plant when the citizens brought it into being. To enter into the merchandising field in electric appliances is an invasion of private business fields no more legitimate than if the municipality were to enter upon the business of supplying boots and shoes, gramophones, or ready made clothing to its citizens.

Were a municipality to propose to provide its citizens with clothing, or any other object of merchandise usually sold in stores, there would be an instant upheaval of sentiment that would quickly put the offending administration out of business. The cry would instantly be raised that the municipality was using the citizens' own credit and their own taxes to diminish or destroy the citizens' business which is the source of taxation.

Citizens of Winnipeg voted on the money bylaw to build a power plant for the purpose only of supplying its citizens with electrical energy with the understanding that this energy would be supplied at one third the rate being charged at that time. It was not the understanding of the tax payers that the City Hydro would enter the merchandising field, nor does their present charter give them that privilege. At the time the charter was granted, there were very few appliances for domestic use on the market, such as the electric range, which is about the only appliance used in the home that is any considerable factor in creating a load for the central station.

At the present time there are in western cities electrical contractor-dealers with attractively equipped show rooms for the display and demonstration of electric ranges and washing machines, vacuum cleaners, and all other electrical devices used in the home, installed at very considerable expense to themselves; and the contractor-dealer feels that as a citizen and tax-payer, paying rent and business tax and all the other levies from which the municipality directly or indirectly derives its revenues, and which the municipally owned plant with its appliance department is not called upon to pay, he is working under a very great disadvantage. Apart entirely from the unethical proposition of using his own money to take business away from him, he feels that owing to the features of the situation above mentioned he has to meet a very unfair competition.

The management of the Winnipeg central station has on various occasions expressed its willingness to co-operate with the contractor-dealer. If it wishes to carry out this declared intention, it should adopt the principle now existing in Edmonton, for instance, where the city plant maintains a showroom for displaying electrical apparatus supplied by the various dealers in the city, and referring their prospects to the dealers, so that the sales may go through the proper channel.

The statement is reported to have been made by the manager of a central station that the civic enterprise did not have to pay any dividends. That is the case; and it is another example of the unfair competition to which the contractor-dealer is exposed. How long would it be possible for a contractor-dealer to continue in business if he were to do business for nothing?

In the Hydro News Bulletin of February, 1922, the Hydro makes the following statement: "We objected to and will continue to oppose all profiteering schemes."

In that statement the contractor-dealers feel that there is a distinct slur on themselves, whether intentional or otherwise, and that such statements are anything but clean competition; even supposing the civic owned plant had the right to compete at all. To-day there are only two dealers in Winnipeg who were in the business fourteen years ago. Numbers of dealers have started up in Winnipeg in recent years, but very few have been able to remain in the business. Such a condition does not look like profiteering.

In conclusion, it might be noted that the Winnipeg Hydro is the only municipal central station in the three prairie provinces at the present time selling electrical appliances in competition with its own ratepayers.

## Selling the Electrical Industry to the Public in the Prairie Provinces

By L. M. COCHRANE President of Cochrane, Stephenson & Co. Ltd., Winnipeg

A few years ago the public of Western Canada had not much more than a passing interest in the electrical industry, and those working in the industry itself devoted their best efforts toward the sale of strictly supply and appliance and maintenance material. Electrical merchandising as it is today was thought a possibility of the future, but little attention or time was given to selling the industry to the public. Progress in the development of things electrical has been rapid. So much so, that we have been forced to realize we are faced with a problem which must be solved by the electrical industry and no other. Our problem can be termed "Selling the Electrical Industry to the Public." It cannot be stated that progress toward this goal has not been made, and results to date have amply proven the public to be receptive, and only waiting for the industry to take advantage of its opportunity.

To get our sales message to the public is our big problem, and it is really a community service that has got to be performed by the co-operative effort of the manufacturer, jobber, contractor-dealer and central station. Our great need today is to get the interest of the consumer as greatly centred in things electrical as in home furnishings, talking machines and automobiles. There are thousands of people who go shopping every day in the week, or to put the matter another way, who go sight-seeing in the stores. As compared with other lines of business the retail electrical store gets but a very small percentage of these shoppers. They do not deliberately pass up the electrical shop—instead, they have not been sold to the idea of shopping or sight-seeing in electrical stores.

### Public Doesn't Know

Something new in the electrical industry is developed and put on the market, yet oftentimes it is months, and in some cases much longer, before the consuming public are generally familiar with it. If you do not believe this, start quizzing some of your friends outside of the industry on this point, and see how well posted they are on things electrical.

For instance, how many people do you know who have a fractional horsepower motor (better name, a kitchen motor) for polishing, buffing, grinding and a dozen and one other odd jobs it will do? Ask them if they own one, and will find the majority never heard of it before, and will no doubt tell you they thought motors were used only in machine shops, or manufacturing plants.

The public will not seek us and ask to be told the wonders of the electrical industry. Instead, we must seek them. In every city of Western Canada there are clubs and organizations to whom we have a story to tell, but we are not making speed records in getting to them with our propaganda.

As an example: There are several well-organized community clubs in the city of Winnipeg that meet at frequent periods, always in the evening. It is commonly known that one of their problems is to get timely and interesting subjects. Is there an opportunity here for the electrical industry? Thousands of new buildings are going to be erected in Western Canada in the next few years, and it is to the owners of these new buildings the electrical merchants have a message to tell. These prospective owners have got to be sold to the electrical industry before they build if things electrical are to play the part they should in these new structures. The industry has important sales work to do

Mr. L. M. Cochrane

with architects, associations, home owners, clubs, realty men, building contractors' associations and many other similar organizations.

### Special Propaganda Required

It is by working publicly with these bodies that real progress can be made, and the masses reached. To accomplish this, the right kind of publicity and propaganda has got to be planned. The manufacturer, jobber, dealer and central station must never lose sight of the fact that they are setting out to sell the consumer, and thus plan their advertising so that it will appeal to the layman. A large percentage of the so-called consumer advertising as gotten out to-day goes into the inventory bonfire of the jobber, and the dealer.

The dealers, central stations and jobbers can get along with a limited supply of bulletins, catalogues, pamphlets, etc., written in the usual terms of the industry and for use among the trade, but a new standard has got to be set in consumer advertising, and more of it supplied.

It is our business to create the desire of the consumer for things electrical, and to make this task as easy as possible. We will never do it by sending out consumer advertising loaded to the gunwales with "volts, amps, watts" and other terms that are Greek to the average buyer. In short, where is there anything creative in this kind of advertising? Let us feature the convenience, comfort and service of the article, and so create the desire to own it. Many electrical appliances are being featured in this manner—thousands are not.

Propaganda of an educational nature to interest the layman, and for use with clubs and associations, has got to be available. In the United States great progress has been made in this direction, and no doubt a large part of their propaganda is available for our purpose. The industry in Canada, however, should be capable of developing our requirements, and it is primarily up to the manufacturer and the jobber to take the lead in developing the kind of publicity that will appeal to the consumer.

In Western Canada, Winnipeg and Vancouver are the only two cities in which an effort of any kind has been made to sell the electrical industry to the public, and even in these two cities the surface has not yet been scratched.

A Western Canada Electrical Association, composed of every branch of the industry, namely jobber, dealer, and central station, and supported by the co-operation of the manufacturer, is a vital necessity. Such an organization should have as its goal nothing short of seeing that every worth-while central station district in the West has its electrical club or publicity committee (large or small) busy selling the electrical industry to the public.

When the industry starts co-operating as a unit, and sees fit to take advantage of the opportunities that surround us, then will the retail electrical shop start to get its share of the shoppers who are now passing by.

## Selling Equipment and Appliances in the West—A Different Proposition

By FRED. E. GARRETT
Sales Manager Great West Electric Co.

A jobber's salesman covering Western territory has a much different proposition from a salesman in the East for the reason that a salesman in Ontario, for instance, is never very far from his home office, whereas the salesman in the West has sometimes hundreds of miles between customers and is at all times at least a night's run from his home. Then again the salesman travelling in Saskatchewan wanting to get in touch with his firm in Winnipeg by telephone has to be careful how long he talks and how often or his telephone bill soon runs away with the profits of the order he may be conversing about.

Hotel and railway accommodation is another matter that gives the salesman in the West great concern as it is an important part of his life. He spends the most of his time, when not selling goods, on the train or in an hotel. Travelling accommodation on the main lines of the transcontinental railways is the best to be had but in the branch lines is not so good. I remember one time travelling on a mixed train and it averaged five miles per hour, including stops, for a distance of forty miles—fortunately for me this branch line was no longer, because on the prairies you do not find restaurants planted very thick. Salesmen selling electrical lines are perhaps more fortunate than those carrying such lines as groceries and dry goods, because our electrical man does not have to touch the small towns his brother salesman has to and he, therefore, lives at the better hotels. The poor devil calling on small towns of about a dozen buildings quite often finds hotel conditions much like I experienced some years ago. I stayed over night in a boarding house, which the natives called an hotel. The manager reported a full house, meaning that his rooms were all occupied, and believe me they were; about 2 a.m. I moved out into the livery barn and finished the night with the horses. Taking it, however, on the whole the salesmen are not badly used and usually if he finds business pretty good he will put up with some inconvenience.

### Western Optimism

The average dealer in the West is a natural optimist and if a line of goods looks saleable he will usually order a sample anyway, but of course the salesman must at all times watch his customer's stock to see that none of his line is sticking on the shelves and if he is the right kind of a salesman he will go to some trouble after hours in trying to help the customer sell the slow moving goods.

Different houses use different methods of merchandising their lines and jobber salesmen should watch that they do not get into the rut that seems evident in other lines. I refer to houses selling general lines of hardware, furniture, etc. Their salesman carries a catalog of very large proportions which is crowded with everything that a small town wants and it takes so much time to go through this catalog that before he gets well started the customer loses interest and it would seem that the salesman has not got a real chance to show his ability unless he concentrates on one or two lines and forgets the rest, which he can hardly do.

### Dependence on Crop Conditions

A feature of business in the West that we are continually reminded of is that business has to rely a great deal on crop prospects. For instance, in Southern Saskatchewan and Alberta the farmers have had no crop to speak of for three years and the consequence is that the business men in towns in that district have a hard time financing. The farmers give notes, payable after the crop is taken off, and if there is a poor crop, there is usually a fair crop of notes to be renewed. But, even at that, the people in the districts mentioned are today looking forward to the present year with renewed hope and it must be admitted that grit of this kind will win out in the end.

Some years ago a crop failure covering a large piece of territory affected the whole West but the acreage has increased to such an extent that a partial failure is not felt so much.

Conditions such as mentioned call for aggressive salesmen who have the personality to keep the dealer from being down-hearted.

### The Electrical Business

The electrical business as a business has improved in the past few years to this extent, that the average electrical dealer is today really selling appliances and supplies, whereas in the past the consumer stated what he or she wanted and all the dealer did was to take the money.

More improvement has taken place in the West in the past year which I believe has been due to a more friendly competition between contractor and dealer. Some dealers will say that business is bad and that when a job comes up that prices are cut so close that there is no use putting in a tender. This is partly true but you will find the successful contractor today does not take business at a loss and at the same time he is securing contracts. This is because he has taken as his slogan "Service;" I mean by that real service and the public will recognize it if the contractor will maintain a standard of service at all times.

Business conditions in the West are as yet of course in a serious state, but an improvement is in sight and all that is required is that all jobbers and dealers must keep to the front their most optimistic manner because one pessimist can do more harm than a hundred optimists can undo. Jobber and manufacturers' salesmen can do a great deal of good right now carrying to the dealer the gospel of more and better business.

### Look for a Busy Season

Prices of electrical supplies are now reduced to such an extent that this Spring construction should be carried on, under almost a normal basis, and as there is a large amount of money that will no doubt be loaned out for building purposes, there is no reason why we cannot look forward to a busy season.

It is very noticeable during the past year that towns of any size are raising money to install their own electric plant, or, as is the case in Manitoba, are appealing to Provincial Governments for financial aid to secure Hydro power. This shows that the public recognize the need of cheap power and light and are willing to pay for it. If the general public recognize this, the electrical dealer should be on his toes all the time because towns in the West are composed of retired farmers to some extent and it must be demonstrated to them the ultimate saving by the use of electric power for household uses. He already knows the economy and comfort of using electric light but to increase the electric business in the West the consumer in the small town must be educated. Take for instance a four cent cooking rate and compare it with using a coal or wood stove; he has been using this kind of stove for years and years and if we want him to use electricity, we must put up strong arguments to offset the years of service the old stove has given him.

### Farm Plants

The Farm Plant Lighting business is a branch all in itself; at the same time a large number of dealers in the West are using this as a side line. When an organization has been built up to sell farm plants alone, their business suffered considerably this last year and as there were not many plants sold to farmers the accessory end of business was equally as hard hit, but there is no doubt that when this farmer sees his spring wheat growing above the ground he will begin to look forward to a bumper harvest and again become optimistic.

One noticeable feature of the last year's operations was that the percentage of failures of electrical dealers in the West as compared to other lines of endeavor was very small which shows that the average electrical contractor has developed to such an extent that he is in the business men's class instead of being just a concern hanging out his shingle.

At a recent demonstration of appliances in a dealer's store the large crowd of attentive listeners which thronged the show-rooms demonstrated that the public want to know more about electric appliances and I would like to say in conclusion that we must all pay more attention to the selling end of our retail business and keep up the attention of the public. Financial conditions will improve shortly and if we have made a good job of educating the user of appliances the sales chart will look real interesting. The same applies to general supplies for building purposes, and education towards using more convenient outlets will boost the supply business if our information is correct that hundreds of houses will be erected this year in Western Canada.

## What Electrical Manufacturers Can Do to Help the Retailers

### By D. F. STREB

This is a question on which much can be said both pro and con.

All attachment plugs and the plugs entering the appliance should be standardized so that less would be necessary to carry in stock and so these could be manufactured in larger quantities at less initial cost.

All appliance repair parts should be lower in price, the present price for these parts if added together amount to almost twice the cost of the appliance. The discount on these parts is also in most cases less than the discounts on the appliance and in all cases too low to do business on, as anyone can readily see. 25 per cent is the usual discount, which gives a gross profit of 33-1/3 per cent out of

Mr. D. F. Streb, Proprietor of The Electric Shop, Saskatoon

which freight and express must be paid leaving a net gross profit of possibly 28 per cent which is less than the cost of doing business. In the matter of discounts on the appliances, these are also too low and barely enough to cover overhead expense.

They tell us to raise the list price but this is not practical, as all retailers would not be on the same basis. We would also point out that in this western country the freight and express add greatly to the laid down cost, so that the list price should be higher or the discount larger than what is needed in the East or near the seat of manufacture.

In the matter of advertising more of this should be done in the home magazines and daily papers, there being only a few manufacturers who help the retailer in this way, the balance expecting the retailer to put their appliances across, which without their help is expensive and almost impossible, we have learned from experience to our sorrow.

There are a number of other matters that could be taken up, I would rather discuss these at a joint meeting of manufacturers, jobbers and contractor-dealers, which I would like to be called, to take place either in Regina or Saskatoon and to cover Manitoba, Saskatchewan and Alberta. I trust that the few thoughts which I have here stated will be taken up and an attempt made to remedy.

## Electrical Industry Needs More Than Slogans

By M. E. DEERING
Winnipeg Manager, Northern Electric Co.

Everyone seems quite willing to accept as a matter of course the idea of an athlete who intends entering a contest going into physical training. It occurs to me that if we in the electrical business could see our business in the light of a contest, the preparation for which needs careful mental training, we would be able to develop quite a few winners, as well as a large number of healthy runners up.

Business for the next few years is going to be a contest, and not a mere physical or financial contest, but one that calls for mental effort to a greater degree than at any time since the slump of 1907. We have heard a great deal in recent months about dull business, and more recently still considerable has been said along the lines of "Boost Business", "Quit Knocking", "Business is as Good as We Think it is", etc., and while as slogans these are all good, we need more than mere slogans. That Electrical business, which is or-

Mr. M. E. Deering

ganized on sound business principles and the human element in which is intelligently striving to find more and better ways of serving the buying public, has nothing to fear from the future. In spite of all the so-called dull business, restricted buying, severe credits, etc., the man who is devoting his entire time to doing business in an intelligent way is finding business to do, while he who is devoting himself to talking of poor business, the cut prices of his competitor, etc., is finding nothing else to do but talk. Except in such extreme years as 1918, this will always be true to a greater or lesser degree, and because I know that some people are going to buy electrical material this year, I believe that it is better to devote myself to the job of selling them than it is to waste time talking about those who won't buy.

1922 in Western Canada will see just that amount of electrical business that we in the industry create, and if we all take stock of ourselves and our wares, put our houses in order and go after business in a clean business-like manner, we will find that there is lots of good business to be had.

The field of illumination, for instance, has hardly been touched, and no matter what you or your competitor may have sold a purchaser in the past, he is still a prospect for more and better illumination.

In the appliance line we have the dishwasher which has yet to be really exploited, while the water heater at least on Winnipeg rates offers a possible sale to every household.

In spite of all that you may think as a result of your competitor's past cut prices, the fact remains that as a general rule the owner of the new home can be made to see the advantage of wiring for a range and convenience outlets if these be intelligently demonstrated.

If you have prepared for the overwhelming demand that is now existent for radio equipment you are enjoying the fruits of your own intelligent effort, but if not, then you are probably devoting a portion of each business day to complaining about your inability to get delivery, satisfactory prices, etc.

If you would do a large and successful electrical business, study that business and its every relation to the buying public, and you will find that your business grows as you grow mentally.

## General Indications of a Betterment in Conditions

By G. R. WRIGHT
Winnipeg District Manager, Canadian General Electric Co.

While business throughout the Prairie Provinces is still offering only in small volume, there are general indications of a betterment in conditions; dealers' stocks are low; outstanding accounts are gradually being liquidated; materials being used are on a present-day cost basis; necessary economies have been effected, and the general operating efficiency of business concerns as a rule improved.

There is an increasing tendency from appliance sales and the sale of electrical specialties to enlarge the volume for the contractor-dealer, and he is rapidly learning to take advantage of the possibilities open to him.

A moderate amount of Government building will be undertaken in Saskatchewan this year. The indications are that in Manitoba the building program will be as large as last year, and that probably the number of small houses built will be greater.

Municipal expenditures are being restricted, but as a

general rule the municipalities have no accumulated stocks of material for extensions of which the majority are in need.

Those concerns whose capital expenditures must be planned two or three years ahead, are showing their faith in the country by improving their facilities where necessary to meet the natural growth of business which has been experienced in the past. Active construction is proceeding on the plant of the Manitoba Power Company. As a result of this large development of cheap power, the prospects for electrical business in the city of Winnipeg and its vicinity are enormously increased.

While conditions in Manitoba and Saskatchewan have been under a period of severe readjustment, and while it is probable that this year there will not be a large volume of business offering, it is gratifying to feel as we do that the lowest ebb has been reached, and we confidently expect that trade will again be on a normal basis in the comparatively near future.

# Do Electrical Dealers Lack Pep? Are They Neglectful of Opportunity?

### By a Farm-Plant Enthusiast

One of the really astonishing things in connection with the "Electrical Trade" to one who has been more or less interested in it for a number of years is the indifference, or shall we call it apathy, with which the legitimate electrical dealers throughout the Dominion of Canada as a class are overlooking or passing up the opportunities available in connection with the sale and installation of the Individual Electric Plant for the farm, country church, store, summer home, etc., which exist all around them in many cases, and in others are simply waiting for some energetic salesman to hustle around and secure.

There is no doubt that the "farm field" during the next few years will be one of the big fields for electrical merchandising, as the farmer is to-day a far more modern man than he was a few years ago, and not only appreciates, but wants the comforts, conveniences, and labor saving devices the city dweller enjoys through the use of electricity; and what is more he is going to have them whether he purchases them through electrical dealers, or through some free lance agent, wide-awake enough to go after him. The fact remains that no matter who gets him, he is well worth going after.

### The Logical Man

By experience and training the electrical dealer and contractor is the logical man to handle the power and light problems in his community; in fact he has a very distinct advantage in every respect over the implement man or other agent usually found selling the farmers their electrical equipment except, shall we say, the inclination to get out and hustle for the business in which, by the way, there is an attractive profit and a large field.

Hydro activities in the various provinces, especially in Ontario, are reaching out to the farm to a limited extent, and thereby creating a considerable amount of business for local dealers, but the fact remains that it will be very many years before the vast majority of the farmers of Canada can possibly be supplied from central distributing stations, if they will ever be. In the meantime thousands

of them scattered over the country are first class prospects for the sale and installation of a private system.

Doubtless the more or less crude equipments of some years ago, by courtesy called electric plants, with the constant servicing required, and at the best, uncertain service rendered, to a large extent caused the electrical dealer to fight shy; but the modern plant is a far different proposition, designed and built to give efficient service and not subject to the temperamental vagaries of its predecessors, and the selling of which does not let the dealer in for a lot of worry and trouble.

### Easy to "Service"

The really modern plants are giving highly satisfactory service on thousands of farms and other private installations in the United States and Canada; so, providing a good plant is selected, there is no cause for worry as far as service is concerned, as that has been reduced to a minimum.

Many dealers and contractors will say they are too busy with town and city work, and have not the time to take on such a proposition, but the big business man of to-day is the one who in the years gone by, when he found himself in the enviable position of having all the business he could handle, did not stop and say "I will leave well enough alone" but began to expand in order to take care of all the additional business he could get, and kept up the expansion to keep pace with the possibilities offered. On the other hand the man who either does not look for, or else passes up, opportunities for increasing his business will always be one of the "little ones," and is not liable to make much business progress as the years roll by.

### Farmer More of a "Shopper"

Generally speaking, selling the farmer is a little different from selling the town or city man, as the farmer is more of a "shopper" than the latter, does not make up his mind quite as fast and always buys more on confidence; but that feature is one to be desired, as when a particular dealer is established in the confidence of the farmer, he may rest assured that he is going to get about all of his business, and very likely that of his friends and relatives, as such things are discussed much more freely in the country than in the city and a lot of mighty good advertising is usually the result of satisfactory dealings with the farmer.

This big business opportunity lies right at your door, as we venture to say that there is not a village, town, or city in Canada surrounded by a prosperous farming district in which a little hustling on the part of a live dealer will not uncover a considerable number of first class prospects for a high grade reliable electric power and light plant. A very important feature is the fact that the sale of a plant invariably leads to the purchase of the various equipment and appliances to be used in connection with it, not to mention the wiring and fixtures, which in themselves are considerable items.

All of this business belongs logically to the electrical dealer, and can be secured by him if he is willing to make an effort to get it instead of letting it go by default to the implement man, lightning rod agent or the various other classes that seem to have been getting the cream of the business so far, simply because they were alive to the possibility and enterprising enough to get out and "cash in" on it.

### Like Automobile Business

In 1910 the automobile business was beginning to clamor for recognition, and many who thought they saw possibilities in the sale of motor cars, and were able to make a start in a small way, have now a large and prosperous business; such examples are to be found in nearly every town and city. The main thing is to have vision and foresight to realize that a new era has dawned, and the determination to "get in on it" while the "getting is good," instead of wishing afterwards that you had.

The farm plant business is in about the same position to-day that the automobile business was in just a few years ago, and the expansion gives promise of being every bit as great in so far as the rural population is concerned; the invention and adaptation of the many electrical appliances for the lightening of labor and the increasing of comfort and pleasure makes electricity a valuable and economic servant of the farmer. It takes care of the many small chores that in the past have constituted the drudgery on the farm, and particularly wearing on the life of the female part of the average farm house by the use of the old fashioned broom, wash tub, churn, water pumping and carrying, flat iron, lamp filling and cleaning, and other numerous small jobs, that are gradually sapping their health and vitality, and all of which can be made easy and pleasant by the aid of that wonderful servant electricity.

There is no field that offers such a wonderful opportunity for practical demonstration and sales promotion, attended with commensurate profits, and there is no person in such an advantageous position to take full advantage as the electrical dealer. Let "Do it Electrically" be your slogan, let the farmers in your district know in no uncertain manner that you as a practical electrical man, are the logical person to supply and advise on all electrical matters and moreover are prepared to supply both equipment and advice, also service, and the reward will be yours.

The thousands of automobiles in use by farmers to-day are there because some hustlers got out and sold them after showing the farmers that they could use them to advantage in their business and social life—just think of the sales possibilities of the farm plant that will give the farmer and his family pleasure, comfort, service and safety all the year round, no matter what weather conditions may be, as against limited service of the motor car—and you will not be lacking in convincing sales talk.

It is only a matter of putting the proposition fairly before him, and when you have sold him, and he is getting satisfaction from his purchase, the whole community knows it, is to a large extent influenced, and each succeeding sale comes easier.

There are several reliable farm plants now on the market, all nationally advertised and more or less familiar to the farmer in a general way. He needs the service such a plant will give him, and is waiting for some one to sell him. Will it be you, Mr. Contractor-dealer? Or are you going to hand this little gold mine over to the machine agent or to men of this type who have no knowledge of and no sympathy with the electrical industry, except that it brings them profits—profits that might and should go into the pockets of the electrical contractor and dealer.

The Art Electric Company, Guelph, Ont., have outgrown their old premises and have moved to larger quarters, at 104 Wyndham St. Messrs. C. C. Rasher and R. Armstrong are the proprietors.

# The Electric Merchants' Opportunity—The Public Demanding Wireless

More surprises have been supplied by the electrical industry, probably, during the last 25 or 30 years than in all other industrial operations combined. Each time we think we have reached the limit of our surprises and then along comes another one more startling than any that have gone before. Take the present day development in wireless telephony that has jumped into prominence almost overnight. Five years ago it was a brave man who prophesied that "sometime" we might be able to hear across considerable distances without the use of conducting wires—and in the short interval equipment has been so perfected and simplified that it is almost within the reach of every home to "listen in" to the programs of music, news, and speeches being broadcasted from any one of a number of large sending stations at various points on the continent. We said "almost," and, generally speaking, there seems to be only one reason today why many thousands more of these wonderful sets are not in use, and that reason is —the factories can't supply them fast enough.

Viewed from the outside, it may look as if our people had gone crazy over wireless, but to the electrical man it is merely an over-due recognition of another commercial application of this wonderful form of energy. The man on the street probably looks upon this situation as a passing fad; the electrical man believes it to be as much a permanency—as much an essential in our daily life from now on—as the electric lamp, which revolutionized the science of illumination, just as wireless will revolutionize the science of the distribution of thought and expression.

The electrical dealer is, doubtless, due for a little temporary disappointment—because he cannot fill the orders that come to him. However, the meantime should be spent in perfecting his plans for distribution. This desire on the part of the public has come to stay. From the electrical merchant's point of view, it represents a source of revenue that will yield maximum results from minimum effort.

In connection with the inability of factories to supply the demand, Mr. T. E. Menzies, president of the A. C. Gilbert-Menzies Co., states that this condition is universal. Mr. Menzies has visited, personally, a number of the largest factories and jobbers in the United States and considers that a dealer would be well advised to figure on at least three months' delay, in starting his wireless supply business. Owing to the inability to keep pace with the demand for detector and amplifier tubes, the Mineral Sets will be on the market first, although in this connection the bigger manufacturers are sold out for the time being on account of being unable to get head phones. In regard to the Gilbert factory, Mr. Menzies states that though their sales staff has not been selling radio sets for four months previous to March 1, yet they are sold up two months in advance. Canada, however, will get her fair quota, based on weekly production, and Mr. Menzies is arranging to fill present orders starting June 1, shipping in instalments until the customer's order is completed.

The whole situation certainly indicates that the electrical contractor-dealer will be well advised in placing his orders immediately. Present indications are that the business will be at its height in the autumn of the present year.

## The Manitoba Electrical Ass'n.

By MORRIS E. DEERING

I have often been asked "Do we in the Electrical Industry really need to be organized; do we need to have an Association, in order to be successful?"

Before attempting to answer such a question, I should like to make it plain that I believe the question is generally put without, what seems to me, a thorough understanding as to the proper and best functions of an organization of men in the electrical industry. Whether or not this seeming lack of understanding of proper functions is due to the history created by such past organizations as we may have had I cannot say.

Let us get a few fundamentals fixed in our minds concerning this industry of ours, and from there attempt to reason out the necessity, or lack of it, for the proper organizing of our industry. I believe that in the minds of all those who have given any serious thought to the matter that it is an accepted fact that our electrical industry renders a greater variety of services to mankind than any other single industry to-day. Just as man succeeds in finding more and better ways of causing electricity to serve him so will the industry grow. Growth of the industry is dependent on the ability of those connected with it to carry on intelligent research, development and educational work. This implies co-operation to a degree that is probably unequalled in few, if any, other similar industries.

How far could you, Mr. Central Station Manager, travel along your road if the engineer, the manufacturer, the jobber and the contractor-dealer were not constantly co-operating with you to provide more and better ways of enabling humanity to consume that "juice" you have to dispense?

How far could you, Mr. Manufacturer and Mr. Jobber, go if the central station man and the contractor-dealer refused to be interested in anything more than electric light? Suppose the contractor-dealer absolutely refused to bother with your time-payment plans and paid no attention to the new devices and appliances you know to be good.

How long would you last, Mr. Contractor-dealer, if the engineer, the manufacturer, the jobber and the central station all refused to listen to your demands for more satisfactory devices, more popular rates, etc.? Suppose that on all of your jobs there was absolutely nothing to wire for except lights.

Those responsible for carrying on the work of The Manitoba Electrical Association in Winnipeg have some such ideas as the above in mind. They look forward to the time when every member will see the advantage of taking his full share of the burden in carrying out the various plans that are from time to time mapped out, all for the betterment of the industry. When each man in our industry finds within himself a desire to help the industry or help himself, then will organization in Western Canada be found to be worth while.

The Manitoba Electrical Association, through their weekly electrical page, have, as can be proven by actual records of sales, improved the Electrical business in Winnipeg, and yet that page has largely been the result of the efforts of one or two men. How much more successful could it have been to date if every electrical merchant both wholesale and retail had taken full advantage of the opportunity to tell his story to the public, first through paid advertising and then through write-ups in the news space of the page. The use of the latter space is free and is not confined to advertisers alone.

Again, the Manitoba Electrical Association has given two demonstrations of the Electric Home before public bodies, and on each occasion the interest shown in convenience outlets and electric appliances proved that the public are waiting for the electrical man to show them the way to purchase wisely.

Those who criticize associations the most are generally found to be the least active in association matters. If every one in the industry could be aroused as to the amount of good that can be accomplished by a united effort to boost business, good business would surely follow. Electrical business, through electrical stores, can only become more than a mere saying when through co-operative effort those in the electrical industry prove to the buying public that it is to their interest to buy electrical goods from an electrical merchant. The old story of "If a man build a better mouse trap the world will beat a path to his door" holds good for us. The Manitoba Electrical Association, which to-day has a membership of 188, hopes by slow and sometimes painful progress to so weld its membership into one unified whole whose every energy is bent toward more and better public service as to make of our industry in Western Canada a business of which we can all be proud.

### Turnover Can Be Increased 25 to 50%

Taken at random from letters received during the past month, from the most prominent central station men in the prairie provinces, we print the following extracts regarding the possibilities in the West for increased turn-over of electrical appliances. They all speak with conviction:

By making all attachment plugs and receptacles standard and interchangeable, business could be increased considerably.

Turn-over of electrical goods could be increased 25 per cent.

Turn-over of electrical goods could be increased 50 per cent.

At the present time I believe that the opening is good for household appliances, especially cooking.

Washing machines and vacuum cleaners offer opportunity for live salesmen.

Turn-over of electrical goods could be increased 15 per cent.

Turn-over of electrical goods could be increased considerably by more display of goods.

Turn-over of electrical goods could be increased by greater co-operation, which is absolutely necessary.

Many electrical washers might be placed.

There should be a large increase in the sales of stoves, ranges, etc.

Think turn-over of electrical goods could be increased 25 per cent by greater efforts and co-operation by manufacturers and dealers.

Believe that turn-over of electrical goods could be increased considerably, especially if electrical washing machines were pushed.

The turn-over of electrical goods could be increased possibly 50 per cent by greater efforts and co-operation by manufacturers and dealers.

# Interesting Views and Personalities

### Further Information Regarding the Activities of the Prairie Provinces and the Personnel Behind Them.

### The Moose Jaw Electric Store

The city of Moose Jaw, Sask., has an aggressive electrical firm, known as the Moose Jaw Electric Store, Ltd. They came into existence about a year ago and took over the city retail and contracting business of the Moose Jaw Gas & Electric, Ltd., which had been in business for some time and dealt largely in farm lighting plants. These two firms are at present located on the same premises, so that between them they cater to all classes of trade, keeping a specialty salesman on the road all the time selling washing machines and vacuum cleaners. Most of these sales are made on a monthly instalment basis.

The manager of the Moose Jaw Electric Store, Ltd., known as "The Electric Shop," is Mr. H. C. Hall. Mr. Hall is an Ontario boy, born in Bleipheim and educated in Chatham. He was first with Wm. Gray & Sons Co., Ltd.,

father of J. R. Young, is also associated with the business. Mr. Young, Sr., is a thoroughly experienced business man of many years' standing, and has had much to do with the success of the company. The superintendent of the company is Mr. Garnett E. Perry, an able business man who is well known to the trade throughout Saskatchewan.

This company is thoroughly well organized. They issue price lists regularly, and catalogues; buy in large quantities---for example, washing machines are bought only in carload lots. The accounting end of the business is very systematically arranged. A card system shows the original purchasing cost, the wholesale selling price and the retail selling price; they carry forward purchase totals from preceding years and thus have on hand a record of all purchases for the past five years. The inventory is taken from the price cards on the first of each year, the quantity of stock being also indicated on the card itself.

Mr. H. C. Hall

Mr. J. R. Young

Mr. Geo. A. Young

of Chatham, and was later manager of the Manitoba branch of Gray-Campbell, Ltd., for some time; in 1914 he was secretary-treasurer in their head office at Moose Jaw. In 1920 he started in the farm lighting game with the Moose Jaw Light & Electric Company, but almost immediately took charge of his own store, as noted above. The farm plant handled is that of the Dominion Steel Products Company, the "Dominion Light."

### The Sun Electric Co., Regina

The Sun Electrical Company, Ltd., Regina, Sask., was operated by local men since 1913. It is situated in the centre of the retail section of the city. During the intervening period of nine years the size of the premises has been doubled and the staff increased from six to twenty-five members. This remarkable progress has taken place under the management of Mr. J. R. Young, who joined the staff as a clerk in 1914, but was promoted to the position of manager the following year. Since that time the assets of the firm have multiplied five times and Mr. Young is now majority owner of the stock. Mr. Geo. A. Young,

The card thus shows the record of turnover, by years, and is in itself a copy of the last inventory—a very valuable record, indeed, this company finds.

The Sun Electrical Company's store is one of the very finest in the prairie provinces. On entering the store the customer is at once attracted by the originality of the layout. Four pillars in white support the fixture room roof. Hanging from the sub-ceiling are all kinds of fixtures. On tables, also decorated in white, are reading lamps of every design. Between the tables, at selected intervals, mahogany standard lamps are placed; their dainty shades of old rose, blue and yellow complete a very effective color scheme.

The fixture room extends back about 18 ft. Running the whole length of the inside and end walls are shelves carrying shades in glass and wicker. The lower shelf contains large etched semi-indirect bowls, each bowl covering an electric light, controlled by a switch. When a customer wishes to select a bowl, the salesman switches on the lights, one by one, and each bowl is illuminated in turn. Every article in the fixture room is marked plainly with the price.

On the right of the main entrance, and opposite to the

fixture room, three glass show cases containing percolators, vibrators, violet ray machines and all kinds of electric appliances are placed. There is also a mahogany counter at the back of the show cases and running parallel are mahogany fixtures, open below, containing lamps. Slide glass cases stand above the open fixtures. These are draped in black, as this color causes the electric nickel-ware in the interior of to stand out in relief; then further down are little square drawers which hold everything from a fuse to a push button. At the end of the counter there is a small workshop, built of beaver-board and easily moved at will; pushed against the wall with just enough room to enter, it is only large enough for one man, all minor repairs of domestic appliances are executed here. This little workshop has been a great convenience, and has helped to get through the little jobs with the dispatch that is pleasing to the customer.

At the back of the store is the office. This is open and the manager, from his desk, has a full view of the store. He thus not only can keep in touch with his staff but can see his customers. The store is decorated in blue and white throughout.

The space between the first pillar of the fixture room and the entrance on the left is reserved for washing machines and vacuum cleaners. A way of advertising the washing machine which has yielded good results is to place a machine filled with suds outside the entrance, almost on the sidewalk, in full view of the public; it is no unusual sight to see five or six or more people around the machine listening to a demonstration of its qualities. This method of advertising has been most effective as shown by the sales of this machine—twenty last month. Then there is a carpet to demonstrate the vacuum cleaner, and show how easily this special machine will remove every speck of dust. "Convince your customer that you have a good thing to sell him, and he will buy" is the motto, and it has met with good results.

Mr. Young is very proud of his store, and justly so. He says that he has frequently been told, by men who have seen the best electric stores in Canada, that there is no finer store than his in the Dominion. He adds "There are bigger ones, but none more attractive and more service like. We are pleased with it, but not satisfied. It has made us many friends, and every day the number increases."

---

## History of the Great West Electric Co.

The extraodinary change that has come over the electrical supply situation in the past eight or ten years is well illustrated in the experience of the Great West Electric Company of Winnipeg, a reproduction of whose premises at 87 King Street is shown in this issue.

The expansion of business which has come to the electrical trade generally in the period mentioned has been shared by the Great West Electric Company fully, to such an extent that the company was forced to purchase a new building in October 1920, making the move into its new quarters in December, 1921.

To illustrate the growth of the trade generally by the experience of the Great West Electric Company, it is necessary to go back a few years. In 1912 the Mainer Electric Company was organized, and in 1916 its business was taken over by the Great West Electric Company, which was organized for the purpose. At that time J. Gordon Smith was appointed manager, the entire business was

reorganized, the policy changed, and from that time to the present the record of the company has been one of continued achievement and prosperity.

At the time the new company was started the market in electrical supplies depended entirely upon the building operations. Coincident with the formation of the new company, building operations were at a standstill. New markets therefore had to be sought.

The way opened up in the realization of the cheapness of power in Winnipeg. The appliance business proved the

Home of the Great West Electric Co.

medium for the needed expansion of business. Electric ranges, washers, sewing machines, and other domestic labor saving appliances were pushed, with the result that the appliance business grew rapidly and showed splendid results, which were increasingly added to as the supply end of the business gradually came back.

In September, 1921, it was found necessary, due to increasing western business, to open a branch in Calgary to take care of the Alberta field.

The lesson of the progress made by the Great West Electric Company lies in the fact that this company took advantage of the situation, and whereas the bulk of the business prior to 1916 lay in the building being done, they recognized the possibilities of the appliance field and went after the business and got it. to such an extent that the appliance business which was at the inception of the company purely a side line has taken on itself the proportions of an industry in itself, fully as important and as profitable as the original field of supplies.

Other companies have had similar experience, with the result that the business of the electric dealer in Winnipeg has grown. his credit standing has improved. he has become a better merchant and the electrical industry has without question benefitted materially through the enterprise of this and similar progressive companies.

### The Hillas Electric Co., Edmonton

Mr. C. W. Hillas, manager of the Hillas Electric Company, Edmonton, Alta., is another Ontario boy who went West and has made a success of business. He was born at Belfountain, Peel County, but when his parents moved to Calgary in 1899 he completed his education in the Calgary High School and started in the electrical business with the North West Electric Co. This was in 1906 and he remained with this company until 1911 when the North West Electric was taken over by the Northern Electric. In that year he formed a partnership with Mr. H. H. Depew, under the

Board. His chief hobby is baseball.

Mr. Carl T. Kummen was born in Norway, (the most intensely electrified country in the world). After coming to America he studied law for two years at the Lincoln-Jefferson University, Hammond, Ind., then studied industrial history and higher accounting for two years at the University of Minnesota. In the spring of 1912, he moved to Winnipeg, and joined the firm of the Mitchell-Gray Electric Co., shortly afterwards becoming secretary of the Schumacher-Gray Co., Ltd.—"The Electric Shop". Mr. Kummen served as secretary-treasurer of the old Winnipeg

Mr. C. W. Hillas            Mr. Carl T. Kummen            Mr. J. H. Schumacher

name Depew-Hillas Electric Company, in Edmonton. In May, 1917, he bought out Mr. Depew's interest and continued the business under the present name. In September, 1920, Mr. Hillas opened up a branch store, but in June, 1921, again combined his two stores at 10041 Jasper Avenue, in the very heart of the city. The Hillas Electric Co. have handled a number of the largest contracts during the last three years. They operate a fully equipped electrical store, contracting business and a motor winding department.

### The Schumacher Gray Co., Winnipeg

Mr. John M. Schumacher, president of Schumacher Gray Co. Ltd., Winnipeg, was born in Dubuque, Iowa, and received his education at Minneapolis Public Schools, graduating from the University of Minnesota. He is a member A.I.E.E., C.E.S., and E.I.C. For a time he was superintendent of construction for the W. I. Gray Co., Minneapolis; later construction manager for the Minneapolis Electric Equipment Co.; associated three years with Chas. L. Pillsbury Co., consulting engineers; then electrical inspector for the Minnesota State Board of Control. In 1911, he became associated with the Mitchell Gray Electric Co., Winnipeg. The name of this firm was eventually changed to Schumacher Gray Co. Ltd., and the business is still run under that name. Mr. Schumacher is a member of the Rotary Club, Manitoba Electrical Assn., and for the past five years has been a member of the Board of Directors, Winnipeg Builders' Exchange.

Messrs. Schumacher Gray Co. Ltd., are one of the largest firms of electrical contractors and dealers in Winnipeg, having done some of the largest contracting work in the city of Winnipeg, Being one of the most able electrical engineers in the city, Mr. Schumacher holds the appointment of Vocational Lecturer to the Winnipeg School

Jovian League for two years, and has always taken an active part in association work.

### The Espley Electrical Co., Regina

From a very meagre start some three years ago the Espley Electrical Company of Regina, Sask., now enjoy a large share of the retail and contracting business of Regina and vicinity. Their store is located just around the corner from the main thoroughfare, in the heart of the business district. Here you will find a full stock of electrical appliances, fixtures, ranges, washing machines, and electrical supplies. They are agents for Turnbull freight and passenger elevators, Moffat ranges and Eden washing machines. Their business in electrical ranges during the past two years has been one of the largest retail turnovers in Western Canada.

Mr. Espley has long been favorably known in the Regina contracting field, and has successfully completed a very large number of the big electrical installations in this district.

### The Electric Shop, Saskatoon

Saskatoon may have suffered from a boom in land values, but it boosts one of the most wide-awake "Electric Shops" in the prairie provinces. Mr. Streb's, you say; yes, of course. Everyone in the West has heard of Mr. Streb's success.

Mr. Streb was born in Canton, Ohio, but, after a wide experience in the practical side of electrical contracting and installation, he came to Canada in 1907. For a time he was town electrician of Collingwood, where he was very largely instrumental in placing the electric light department on a paying basis, but in 1912 he went West and on

teaching Saskatoon was appointed manager of the City
Electric Company. In 1917 he bought this business out.
In 1919 he moved to his present location, and in 1921,
owing to the growth of the business, organized a limited
company, including Mr. A. A. Murphy, B. Sc., A.M.E.I.C.,
formerly of Murphy & Underwood, as his partner. Mr.
Murphy was born in Portland, Ont., and educated in
Athens High School, Queen's University and McGill University, and Pittsburgh. He is also an Associate Member

Mr. A. A. Murphy

of the Engineering Institute of Canada. In 1911 he opened
the first consulting engineer's office in the Middle West
and supervised the installation of many plants, from his
Saskatoon office, which included the towns of Wilkie, As-
siniboia, Melfort, The Pas and others. He was an en-
thusiast in the use of internal combustion engines as a
prime mover under conditions prevailing in that district.
In 1921, as already noted, he took an interest with Mr.
Streb in The Electric Shop, Ltd.

## Mr. Streb Answers Mr. Brettell re Discount on Lamps

Saskatoon, Sask.

Editor, Electrical News,—

I notice in your issue of March 15th a letter from
Mr. K. Brettell of Vancouver with reference to the
discount on Mazda lamps on a consignment basis and would
like to point out that the lamp business on this basis is
the best paying proposition that the electrical dealer has
in his store.

In the first place there is no investment in lamps,
the manufacturer carrying the stock. The discount also
on this basis is a great deal larger than on the old basis and
in addition you get an extra 5 per cent for sending in a
report at the end of each month of the amount of lamps
that you have on hand and the amount of lamps that you
have sold during the month, together with a cheque for
the amount that you have sold and for which you have
already received the money.

With reference to discounts that are obtainable under
class D contracts, i.e. contracts to consumers according to
their requirements, I would point out that this is a very
great advantage in allowing you to increase the volume
of your lamp sales. While the discount left over after
giving the consumer his required discount is not large,
on the other hand do not lose sight of the fact that your
extra discount that you obtain by increasing the volume
is on your total requirements which, as you can readily
see gives you an additional discount on the lamps which you
sell over the counter at prevailing prices, without any investment whatever.

To sum up, the advantage over the old way is that
you obtain an additional discount of about 9 per cent on
the total volume you sell, besides having the advantage
of making a small percentage on business which you would
otherwise possibly not obtain.

Very truly yours,

D. F. Streb.

## Three Well-Known Western Merchandisers

Mr. Edward W. Beard
Manager Budden-Beard & Co., Calgary

Mr. G. M. Wilkinson
Manager Cunningham Electric Co., Calgary

Mr. C. E. Dashwood
Manager Burnham-Frith Electric Co.,
Edmonton

## Successful Ideas Put in Practice

### By J. R. YOUNG

A tiny workshop, 6 ft. high, 6 ft. long, with two sides 3 ft. wide; inside is a work bench with tools suitable for repairs to irons, toasters, and other small appliances. Each side of the work bench is fitted up with shelves. Separate spaces are provided on these shelves for articles to be repaired; those that have been repaired; articles waiting for parts from the factory and, in addition, a stock of fittings—plugs, terminals, cords, etc.

The advantages of such an arrangement are that all appliance repairs are kept in order, the repair work is kept out of sight of the customers and the repairman is not annoyed by customers while working. The repairman can assist at the counter when necessary, as he is close at hand, the work-shop being placed at the end of the counter. Often a boy can be employed to do these small repairs and look after deliveries. The maintaining of such a repair booth promotes business and the Sun Electrical Co. find they are able to keep a boy busy all day on repairs in Regina, a city with a population of 34,000. Minor repairs should be kept away from the back shop; men working on large jobs do not like small work and, besides, their time is too valuable.

## The Party is Over

### By E. H. SMITH

The party is over, and it was some party while it lasted! Deflation is with us, and also with the disease of the world. This will lead to sanity and health. We all knew this adjustment had to come some time, so why feel blue and discouraged? We have been through the same thing before, and we will live through this adjustment period also.

We have the money, the credit, the factories, the labor, the material, the brains, the initiative, and we are naturally optimistic. Pessimism is a disease and follows poor circulation. Exercise is the best cure for it. Go out after the orders. Put jazz, pep, ginger into your efforts. Burn the crepe and jump on the crepe hangers with both feet. Wear out the shoe leather. You will probably get some business, and anyway it will help the shoe business.

### Let Us Find the Silver Lining

Just as surely as there is a silver lining to every rain-cloud just so surely is there a silver lining to the cloud of business depression now hovering over us.

The rain-cloud's silver lining is the good old sunshine—powerful, unchanging, shining brightly.

The depression cloud's silver lining is the fundamentally sound condition of our great country—its limitless resources, its progressiveness, its men of brains. As manufacturers and dealers of the necessities and comforts of life we have a definite, important, useful place in the development of our country, and just beyond the present there's a real, lasting demand for these necessities—electrical appliances and equipment.

Let us make our product and service better than ever before; let us sell them on a broader scale; let us get down to intensive, intelligent hard work, with full faith that better business is within our grasp if we make the effort.

Messrs. T. Bellefleur & Chas. Skinner, Willow Bunch, Sask., contemplate the installation of an electric lighting plant at that place.

## City of Edmonton—Public Utilities—Finance

### By C. J. YORATH

During 1921 the financing of the public utilities, owned and operated by the city, namely, Electric Light and Power, Street Railway, Waterworks and Telephone, has been entirely readjusted and economies have been effected, with the result that the net deficit in 1920 of $94,867.00 has been converted to a net profit in 1921 of $263,855.74.

The net profit and loss of each of the utilities for the above years, after taking care of interest, sinking fund and depreciation, is as follows:

|  | 1920 | | 1921 | |
|---|---|---|---|---|
|  | Profit | Loss | Profit | Loss |
| Power Plant, Pumping Stn. | 563 | 6,242 | 74,172.15 |  |
| Waterworks | 24,000 |  | 42,500.84 |  |
| Telephone | 60,280 |  | 77,956.56 |  |
| Electric Light | 39,728 |  | 114,200.99 |  |
| Street Railway |  | 200,191 | 308,830.54 | 44,974.80 |
|  | $111,566 | 206,433 |  | 44,974.80 |
| Net Loss |  | $94,867.00 |  |  |
| Net Profit |  | $263,855.74 |  |  |

The city council has adopted the policy of taxing the utilities as though they were privately owned, and it is estimated that this year the utilities will make sufficient profit over and above all operating costs, including fixed charges, to pay by way of taxes the sum of $320,000.00 which will assist in reducing the tax rate of the city by over 5 mills. It is hoped, by the end of the present year, that in addition to the utilities making this large contribution towards relief of taxation on property, the rates for electric light and power and water will be reduced.

## Suggestions by Western Dealer to Manufacturers

### By J. R. YOUNG

1. Every manufacturer of larger electrical appliances should have some definite policy, and should have exclusive representation in the medium sized towns in Western Canada, so that where the local dealer advertises a line he will get all the results obtainable through the advertising.

2. There should be co-operative advertising.

3. Manufacturers should supply cuts and suggestions, with texts that can be used.

4. Should supply selling helps, of a good variety, window display cards, etc.

5. Should supply information on products where manufacturer's representative does not call on the dealer.

6. The dealers will not push lines where they have not got the exclusive selling rights.

## Manitoba Power Co. Lets Contracts

The contracts for the electrical equipment for the Manitoba Power Company's plant on the Winnipeg River have been let as follows: Two ATB-52, 21,000 kv.a., maximum rated, 90 per cent power-factor, 138.5 r.p.m., 11,000 volts, vertical shaft generators with 150 kw. direct connected exciters; six transformers WC., 60 cycle, 7,000 kv.a. maximum rated, 90,000 volts; four H.T. oil switches, 90,000 volts. The generators and transformers are being supplied by the Canadian General Electric Co., and the high tension oil switches by the Canadian Westinghouse Company. The two turbines are being supplied by the Dominion Engineering Company, Montreal, and manufactured by I. P. Morris Company.

# The Prairies Boast Some of the Best
# Electric Stores on the Continent

General View of Store of the Sun Electric, Regina

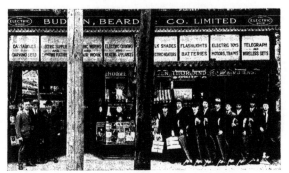

A line up that ensures certain surrender of the Calgary Housewives

# Beautiful Interior Arrangements That Attract the Customer's Attention

The Electric Shop, Moose Jaw. H. C. Hall, Proprietor

The Electric Shop, Saskatoon. D. F. Streb, Proprietor

## *Cities Vie with One Another in the Excellence of Their Merchandising Methods*

Burnham Frith Electric Company, Edmonton, Alta.

General Interior view, Cunningham Electric Co., Calgary

# Latest Developments in Electrical Equipment

The Packard Electric Co., St. Catharines, Ont., have recently developed a new piece of apparatus, in the form of a transformer temperature indicator. This indicator operates on any transformer but is more particularly designed for the service type. At the present time it is very difficult to know the actual condition under which a service transformer is operating, the first indication of trouble generally being a burnout. The new Packard temperature signal indicates the maximum temperature, the indicator remaining at that point indefinitely, until reset by hand. This temperature reading can be plainly seen from the ground, so that an inspector can determine by passing along the street just what the operating condition in any and every service transformer is at the moment, or has been in the recent past.

The indicator is in the form of a revolving frame, which

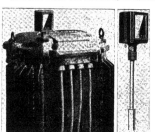

Transformer Temperature Indicator

is painted black up to a slanting line starting from the 90 deg. C. graduation; beyond this it is white. At 70 deg. C. a small portion of white appears in the upper left hand corner of the window of the metal case which encloses the frame. At 90 deg. C. all the window shows white except a small portion at the lower right hand corner.

The new device is rust-proof, weather-proof, oil-proof, bug-proof and fool-proof. It can be easily applied to any service transformer already in operation by simply drilling and tapping the cover with a ¼ in. pipe tap.

## Service Connector

Where a large number of services leave the same pole considerable difficulty is encountered in making and disconnecting services, and after a service has been disconnected four or five times it usually means that the wire gets broken, and in the majority of cases the service eventually consists of a hook of wire round the secondary, and the first wind storm that comes up usually causes interruptions to the customer's service.

To overcome this difficulty the City of Calgary Electric Light Department has designed a service connector which is giving good satisfaction and has overcome this difficulty of loose connections, and also facilitates the connection and disconnection of customer's services. This

device is made of brass and is soldered directly to the secondary, similar to the attachment of a trolley ear, but is in addition soldered. The house service is made by

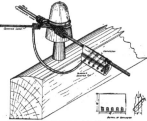

Service Connector used in Calgary

baring the service wire for about two inches, entering into a square slot and fastened by means of a tapered pin. A slight tap of a pair of pliers either makes or breaks the service.

## Banfield's Adjustable-Lite

Messrs. W. H. Banfield & Sons are now manufacturing the "Adjustable-lite" illustrated herewith. The advantage of this light is that it clamps or stands anywhere. It will clamp to a picture frame, to the head of the bed or side of

a table, top of the piano or to a dozen and one other places where concentrated illumination is desired. It is made in a variety of finishes and is suitable for use in any room in the house or in any office. The spring controlling the clamp is sufficiently powerful to ensure the lamp remaining in place. The jaws of the clamp are padded so that the furniture will not be scratched or injured in any way.

### New Drop Socket

The Tri Novelty Company, of Wilson, Pa., has now ready for the market the drop socket shown. This new socket may be connected to any lighting fixture without removing the shade. It is thus a double duty socket, giving the use of the lamp within the shade and, at the same time,

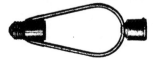

any appliance that may be attached to the lower socket. It is quickly attached or detached, being simply screwed in like a bulb. The lamp is first removed, the drop socket is inserted and the lamp replaced, leaving an extra socket below for any appliance.

### New Models of Toaster Stoves

The Equator Manufacturing Company are placing on the market three new models of their toaster stove which has become so well known to the trade. Model A-11 is an aluminum frame with 6 in. square top, consuming 550 watts, and made to sell at a very low price; with or without cord. Model A-5 measures 5¾ x 9 in. and accomodates 2 pieces

No. A-5 Full Nickel

of toast. This is built of cold rolled steel, has a highly polished nickel plate finish; equipped with ebonite handle, standard terminals, 6 ft. of cord, with attachment and connector plugs; consumes 660 watts. Model A-3 has the same characteristics as Model A-5, but measures 6 in. by 6 in. and consumes 550 watts.

### Magicoal Electric Fires

Some of the beautiful models shown by the Magicoal Electric Fires (Canada) Ltd., in their Montreal showrooms give not only combustion effect, but flame effect which is very realistic, added to which many of the models can be fitted with what is known as the "back of grate" type heater, which throws the whole of the heat out into the room although the source of heat is out of sight. This combination of fire and heater enables one to obtain on a dull chilly day in Spring and Fall an excellent imitation of a hot coal fire, not only as regards appearance but as regards heat. The switching arrangements are such that the heat may be disconnected and the fire effect left running, so that even on a dull cheerless day in summer one can get the comforting appearance of the fire without the heat if desired.

The Magicoal company are in a position to supply grates suitable for rooms containing any period of furniture and would be glad to supply to anyone interested

full particulars of their designs, heater capacities, etc. A new catalogue is in course of preparation and will be mailed on request; dealer-contractors are specially invited to communicate with the company to secure the fullest information.

### A New Eureka

The Eureka Vacuum Cleaner Company, of Kitchener, Ont., have just announced a "New Eureka" known as Model 9, and offer it to their distributing organization as their conception of a bona fide vacuum cleaner, developed to the highest possible point of efficiency, simplicity, durability and appearance. The new Eureka has many improvements. What is spoken of as the most impressive feature is the 13½ in. nozzle, tapered at the ends to increase the velocity of air at those points. Other changes include the placing of the wheels in a prominent position behind the nozzle; an improved detachable "sweep action" brush; the one-piece oil tempered steel fan; the larger dust bag, of finer material and better design; added facilities for operating the cleaner under low furniture and beds; a new handle lock bolt; shorter handle and the addition of an efficient upholstery brush to the set of attachments. The manufacturers offer the machine, with all its improvements, at the same retail price; stand behind their products with the same service, and are confident that their engineering staff, in this new machine, have met the public demand for a better and more beautiful cleaner.

### A Mov-a-long Electric Sign

The Mitchell Advertising Company, Third & Main Sts., Cincinnati, O., recently placed on the market this "Mov-a-long" electric sign, which is used chiefly in retailers' windows. It is electrically lighted and operated. Made in the form of an enamelled metal box, 9 in. high by 43 in. long with a 4 in. by 29 in. window, the device displays a lettered transparent film eleven feet in length, doubled within the box, which moves slowly to the right.

TEN WORDS THAT ARE READ ARE A T

Any advertising matter may be used, as the lettering is applied with water colors and may be removed with water and a clean cloth when a change of copy is desired, and the film used over again. The film is easily removable. A small universal motor, designed for either a.c. or d.c. current of 110 volts and equipped with adjustable speed indicator, is connected to a driving assembly for operating the film.

---

The Marchand Electrical Co., Banque Nationale Bldg., Ottawa, has secured the electrical contract on a Collegiate being erected at First Avenue, Ottawa, at an estimated cost of $700,000.

Mr. H. C. Budd has resigned his position with the Canadian General Electric Company of Winnipeg and had joined the Sales Staff of the Moloney Electric Company of Canada, Limited. Mr. Budd's headquarters will be in the Moloney Electric Company's Head Sales Office, No. 501, York Building, Toronto.

# Electricity Beyond the City Limits

### The Glamour of Light, the Luxury of Power, the Service that only Electrical Appliances Can Give, Are Within the Reach of the Country Man To-day

### By E. A. LOWDEN.

Nobody questions to-day the advantages and the benefits which electric lights afford—to those who can have them. The question has been, largely, how best to get those lights. City folks have known all about the blessings of electric lights for a long time but to folks in the country, for a long time the only answer to the question of how to have electricity in the home was—move to town.

For a long time we have been comparing life in the country with that of our brothers and sisters in the cities. Modern comforts in the cities include a wide range of conveniences which folks there have come to accept as a matter of course. In the past, those on the farms could only envy the city folks and plan for the day when they, too, could give up the farm and fly to the city with its modern conveniences.

With the development of the farm electric plant, however, many of the city comforts have come to make their abiding place in the country and country life has profited accordingly. For the farm electric plant has come to be so dependable, so efficient and so economical to own and operate that the conveniences of electricity can be enjoyed on any farm.

## A Mighty Appeal

The use of electricity on the farm has several different phases which appeal mightily to all the family. First of these will probably be the electric lights. An electric plant for farm service will not need to carry more than thirty or forty lights. These, properly placed, will give more light and immeasurably better light than could ever be enjoyed with the ordinary array of lamps and lanterns. They will be brighter, affording better living and working conditions, and they are much safer than the open flames of the old lighting systems.

On many farms and ranches, the electric power helps as much as the lights. Electric current is translated into power through the agency of the electric motor. Often the farm motor is mounted on a three-legged stand and provided with a handle so that it can be moved around from one machine to another, such as the washing machine, grindstone, churn and the like. They can be provided with an adjustable brace rod, with a hook that fits into a screw eye on the frame of the machine. This acts as a belt tightener, as well as a means of holding motor and machine firmly together.

## Carrying Water

One of the most important and the most popular uses of power from the farm electric plant is for pumping water. A good many husbands, I believe, have never realized what a burden they have left upon the shoulders of the womenfolks, where it is up to them to carry the water for house use. Think for a moment of the country homes you know, where folks live the year round, taking the conveniences of those homes—and the inconveniences—as they find them. How many of these homes do you know of, where the well is away out in the dooryard, maybe down a hill, where some

water-witch located the vein of water, years ago, with a forked stick. I have seen some of these wells a hundred feet or more away from the kitchen door, and wives and mothers in those homes had to carry the water for house use. By a simple process of arithmetic I could show you that, in instances like these, a woman will walk in the neighborhood of a hundred miles in the course of a year and carry tons and tons of water.

But that is not the point, so much as that farm homes are entertaining these out-of-date methods when they might be enjoying modern equipment and saving time—and improving health, besides. Where there is electricity at the farm home, it is possible to have a pressure water system, perhaps automatic in operation, at any rate one that will pump the water with electric power and deliver it under pressure to faucets wherever needed or store it in a pressure tank ready for use. Such a system makes possible a modern bathroom out in a country community, however isolated. It provides running water for the kitchen sink, for laundry tubs, for sprinkling yard and garden, for washing automobiles or carriages, for fire protection. It does away with pumping water for the stock, filling water troughs without any human labor or forcing it to drinking cups right in the mangers.

These drinking cups are more of an item, perhaps, in latitudes where constant stabling, or at least extended winter stabling is practiced. Dairymen who have adopted this plan, keeping pure water constantly before the dairy cattle, declare that it greatly increases the milk flow. Some say that the increase in milk thus secured will pay for a good automatic water system in less than two years with a herd of twenty cows or more.

In any event, having the water right at the turn of a faucet saves all the ordinary pumping by hand. The troughs are filled quickly and easily, and the labor thus saved from pumping can be put to good use in some other direction.

But the thing to be emphasized, however, is not that it saves time and labor for the farmer himself, to have the water pumped by electric power. It is rather that such arrangement saves those about the house from the work and trouble of procuring water for household use in the old way. And it adds so much to the life of the family—the conveniences it provides and the comforts it makes possible. Every member of the family will get benefit and satisfaction from this modern home equipment. The improved health and the increased pride which all will feel in the possession of the conveniences of running water for the home are features which in themselves are worth a considerable investment.

## Farm Plants are Low Voltage

Generally, farm electric plants are of thirty-two volts. They have been designed in this capacity because, usually, an electric storage battery is a part of the electric plant and, since the commonly accepted lead-plate battery cell has a capacity of two volts, it takes sixteen cells to make up the farm plant storage battery. A high voltage battery would take fifty-five or fifty-six cells, thus adding materially to the cost of the plant.

Where the current is to be carried any considerable distance from the plant, however, 110 volts should be used. To carry low voltage current very far requires rather large copper wires and a pretty expensive installation, though this voltage is all right for farm buildings which are grouped within a few hundred feet of each other, as they usually are.

It is hardly necessary to dwell, in the columns of this publication, upon the benefits of electric lights for the home. What we overlook, sometimes, with regard to lighting the country home is, that with electric lights we will light up the kitchen, diningroom or living room on the gray, dark days when the house is gloomy because it is cloudy or stormy outdoors. That means light and cheer inside which drives away the gloom outside. During the course of several years of farm life I do not remember that we ever lit the coal oil lamps during these "gray days" but simply crowded near the windows or strained our eyes in the inadequate light. It just wasn't done—lighting lamps in the daytime. But I know that to-day, farmers who have these farm electric plants snap on electric lights during the dark days, saving many an eyestrain and a headache on the part of the shut-in members of the family who are bound to use their eyes whether the light is good or bad.

### Light vs. Darkness

Eyestrain and headaches? Yes! Who knows but what these lights on dark days save some heartaches too. Imagine a tired mother and a house full of noisy children on a dark, stormy day on the farm. Add to that a husband and father who maybe doesn't take any too kindly to being stormbound with so much confusion. If the tension doesn't get too tight, so that something snaps and somebody's feelings get hurt along towards mid-afternoon, it is just because nerves are still too strong and hearts too brave—yet—to bend to the strain.

But if the home is brightly lighted, if cheer and comfort as manifested with electric lights can be had in the home, then mother's burden is lightened, the children forget to be "fussy" and the father can stay in the house safely all day without danger of doing violence to his own or anybody else's feelings or temper.

We don't need to dwell, much, on the advantages of electricity about the barn, except to point out the tremendous value the lights are at chore time; from the standpoint of time and labor-saving. Of course they save time and labor, for you just turn on the lights that cover the field of feeding operations and then go ahead just as you would in daylight. With the lantern, of course you have to be assured, first, that the thing is filled and the wick trimmed. Then you have to light it and see that the match stub doesn't set something afire. Then you have to get the lantern to the barn without a storm wind blowing the light out and holding up the feeding until you get lighted up and can start all over again. Next you must carry the lantern around with one hand and work with the other or get it disposed of so that it will not fall over, get kicked over or in some way set something afire—and that brings us to the question of safety. Most barn fires are caused through accidents with lanterns and matches and most barn losses include loss of livestock, feed-stuffs and farm machinery. A fire loss is a grievous wastage, something that can be avoided when the farmer's barn is eqipped with electric light, as so many barns are coming to be equipped, nowadays.

### "Which Shall It Be?"

An even balanced day of work well planned;
An evening time of work well done,
Of rest and deep content?
A home where power and light combine for
           happiness:
Where power eliminates the ceaseless round of
           drudgery,
The tedious, irksome tasks of husbandry;
Where light takes up the burden of the setting 'sun
And makes the evening hours the best of all;
Where cheerfulness and helpfulness join hands to
           lighten labor?
Or shall it be a home of gloom and discontent,
Where toil takes toll of heart and mind and soul?
Where toilers eat—and sleep—and eat—to toil again;
Where children bow beneath the weight of never-
           ending chores,
Where men and women barter love and hope for
           paltry gain?
Then choose the home where power and light do
           their full part,
Transforming, at your touch, the tedious task, the
           darkened way,
Bringing the balanced day.
The evening time of rest and deep content,
The home where happy children play;
Where laughter lives, and joy and cheerfulness;
The home where leisure lengthens out the years—
Broadens the mind and lifts life to a grander plane,
As God intends.

Outdoor Waterworks Sub-station, Calgary

# The Farm Plant and the Trade's Opportunity

### By S. F. RICKETTS

There are considerations in the marketing of electric light and power plants that make it specially interesting to the electrical trade.

The first is that the electrical trade has, as a general rule, neglected to interest itself seriously in this opportunity for business, with the result that the distributing of lighting plant and accessories has in many cases passed into the hands of garages, farm machinery firms and specially organized concerns. Considering distributing only, the above phase is not so serious to the trade, in view of the fact that the distributing of electrical accessories and apparatus still goes through the regular channels, whilst the distributing of high priced technical machinery possibly calls for a specialized organization.

Electrical dealers, however, should be concerned with the way in which the retail plant selling has been almost entirely accomplished by men who had no knowledge of electricity. Force of circumstances has been partly responsible for this as electrical dealers have not been available in small centres where there was no other electrical power available, but even where there were electrical men in business, they have, to a great extent, failed to catch the vision of the possibilities in this line, and have neglected the opportunity which is knocking at their door.

There are several reasons why it is necessary for this work to develop in the hands of the electrical trade. First and foremost is the opportunity which it offers the small town electrical dealer and contractor for wiring work and apparatus sales, and the prestige and introductions the sale of electrical units gives him with the prospective consumer. Then too, he is better equipped to advise on technical matters that crop up continually. If more of our small wiring contractors could combine some selling ability with their knowledge and experience and convey this with confidence to those who should and can install electricity, there would be a rebirth in the out of town business.

A great deal has been done during recent years to educate the prospective consuming public on the advantages of electricity for both light and power, and it is generally regarded as advisable by all modern farmers to have their houses wired when they are being built. This education has been largely carried on through the medium of farm journals and the individual plant salesman, but it would have far more weight if assistance was given by the technical man.

### Large Number of Prospects

The importance of the issue is numerically sized up in the latest survey taken by a well known journal through a questionaire. Taking Canada as a whole, they have received a reply from 20,629 readers who propose to install electricity in their homes soon. Most of these readers are living out of town, hence the deduction that the trade should be prepared to take care of this potential business. Further, when these figures are compared with additional returns, and it is found that more people intend to install electricity soon than those who replied saying they already had electricity, there is certainly food for thought. What would happen if the trade seriously believed that they could duplicate the opportunity of the last twenty years in the next five years?

In the Western Provinces there are firms who have from 2000 to 3000 actual live prospects already canvassed in each province, each prospect having been personally quoted and had a layout given him for his work. These prospects are mostly eager to go ahead and only hard times and falling prices have kept them off the market.

When one considers that the price of the plant represents about half the contract price of the job, one gets a new view-point of the opportunity. The majority of houses in the country are not wired and this alone means from $150.00 to $300.00 for each store or farm. Many are indifferently wired by contractors or local general store tinsmith and unskilled workers. Instead of No. 12 wire or heavier being used, as is necessary on 32 volt work, where the current is heavy due to low voltage, there is often the cheapest grade and size of material. Fire hazard is not usually considered, and few provinces have any rules and regulations for 32 volt work. What is the trade doing to let this sort of thing continue?

### Add the Appliance Sales

To wiring should be added the potential sale of every piece of labor-saving device on the market, for the farm is the place where the use of apparatus is appreciated.

Washing in the country cannot be sent out to the laundry; hence the washing machine is absolutely necessary. Irons are also needed, whilst the 1/4 h.p. tripod motor is the handiest piece of apparatus around the home. It drives the churn, separator and fanning mill, besides a multitude of small tools. A good practical man, however, can carry conviction further regarding the use of the individual small motor which obviates the necessity of carrying the tripod motor around and lining it up for each particular job.

The vacuum cleaner has a great future for the country, although it is not every farm that is sufficiently furnished in carpets to make the expenditure worth while.

The water pump is a piece of apparatus that works 365 days in the year, and the possibility for the sale of deep well and shallow well, automatic and semi-automatic pumps is worthy of a special article.

The sale of other apparatus, such as 32 volt fans, toasters, percolators and the multitude of special fixtures and appliances, is simply a matter of educating the user, while the "bread and butter" business of the electrical retail merchant in lamps, fixtures, fuses, etc. is assured, but is to-day largely in the hands of the hardware merchant and general storekeeper. Present electric plant users are not half equipped with these lines; therefore, there is an immense field already waiting for development and eager to absorb all the information the dealer can give on such labor-saving devices.

When the sales organization of any particular branch of the electrical industry is planned, there is always consideration given as to how far the middleman can be eliminated between the manufacturer and the consumer. It is usually found, however, that the natural course must be from manufacturers through distributor and retailer to consumer and while this course takes its toll from the final purchaser, adding much to the difficulty of making sales, the route is, nevertheless, perhaps the most economical under present conditions.

From distributor to consumer is perhaps the most interesting subject for the reader. It is here that the electrical contractor has apparently failed to complete the circuit. When the distributor starts out to line up his selling organization over a large territory, he chooses one of the following three methods:

(A) A large block agent located in a central position commanding many villages and square miles of territory; equipped with a truck and a demonstration plant and with enough money to finance his business if he is compelled to give extended credit.

(B) An agent in every town, village or trade centre, with a limited control of his territory.

(C) Retail block salesmen employed by the distributor on salary and commision with power to organize and work his individual territory in his own way.

Organizations depends entirely upon the individuals after all, and for while "A man is known by the company he keeps,"it is nevertheless a fact that "A company is known by the man it keeps." The nature of the territory covered, together with climatic conditions, also have a lot to do with the organization.

If a distributor places himself in the hands of a large block agent, he must be convinced that the man is practical, technically sound and will work his territory in season and out. It is the most remunerative to the live salesman, but it is almost impossible for a man to cover a large block of territory and effectively call on all prospects. He must have some secondary agents in the smaller centres, keeping him posted on prospective business, and this means the splitting up of commissions. There is always the possibility that another concern will come along and pick off the secondary agent by means of a larger commission, but if he is capable and his advertising is effective, he can always offset this. Climatic conditions, too, affect the power of a large block agent, for if he is weather bound for the most effective six months of the year, he certainly cannot do justice to his territory. More sales are made between the end of the harvest and the beginning of plowing than during any other period of the year. Effective work, however, can be done throughout the summer months, but the results are with the man who arrives on the spot at the phychological moment.

Organization "B," which places an agent in every small town or trade centre has much to recommend it, although requiring more attention from the distributor. While the large block agent "A" should be a good electrical man, it is not easy to find these men in the small centres. The chief reason why the local agent is desirable' is that he is in close touch with the individual prospect, knows his financial standing and can call him by his first name. The choice of agent in the small centre is limited to the usual machinery salesmen, hardware or general merchants, but their technical knowledge is limited, also their ability as salesmen is varied. More attention is required from the distributor, who is called upon to help close sales and must be constantly behind his agent keeping him at work. With this man, commission can be graduated, depending upon whether he has a demonstration plant and as to his ability to close individual sales. In other words, the distributor becomes his own block salesman and attends to wiring and service work with his own staff.

Organization "C" is only possible when the retail man has a closely settled district around a city, in which case he draws most of the commission that usually goes to an agency. If there are other practical men in the organization he need only be a salesman, but the opportunity is great for the combination of salesman-contractor working entirely on commision. He works out of head office, has no

overhead apart from his car, and can become a super-carpet-bag-Ford-plant-contractor.

Where then is the opportunity for the electrical man? As a practical salesman building up his territory, he can have an electrical store in a good centre with a plant agency for a fair sized territory. Trying to play electrical merchant to more than one concern never got a man anywhere, for no one believes that he will not have leanings in the other fellow's direction and he falls between stools. As a service man he can charge for his labour, while his travelling expenses and lost time is not so much as service from a distant distributor. He has a wonderful opportunity of picking up all the lamp and accessory business in his territory, and can indulge in demonstration campaigns that always bring results. The business is waiting for the right men to "step in" and work.

### "Service"

This perhaps is the most misunderstood term employed by the public. Gradually, education is convincing people that service costs money, and if they do not get it charged when the service is given, they must get it in another form. The automobile trade has managed to eliminate the "free service" idea, or at least reduce it to a minimum, and it is usually easy to convince the consumer that "service at cost" is more reliable than "free service" which may or may not have been covered by original commission.

One of the disadvantages of the small local agent is that he knows little or nothing about service work, either on gas or electrical machinery, consequently service devolves upon the larger block agent or back to the distributor. The latter should theoretically have nothing to do with it, and it should be the work of the agent just as garage work is carried on in every decent centre and paid for at remunerative rates.

Less troubles arise with stationary lighting plants than with automobiles, hence there is a limited call for service. Most of the calls are due to ignorance on the part of the consumer, for every service man can recall trips made to find coal oil in the gas tank, or no fuel at all; fuses blown or wires off; slight ignition trouble or loose bolts, all of which could easily be remedied by the owner if a little more interest was taken. Hence as the most effective educator is the bill, there should be a bill for "service" of any kind. There is, of course, the point where service on defective parts should be covered by the factory, but the local electrical service man need never fear charging for services rendered as it is seldom disputed, and most factories are responsible enough to take care of defects. It is a strange fact that consumers do not like service charges from men who are not practical, however effective the service, as they look upon salesmen as people who float around living on air. Possibly they confuse the "hot air" expelled with their mode of living.

---

The B. C. Telephone Company, Vancouver, B. C., are having plans prepared for an addition to their Bayview Exchange at 10th & Yew Sts., Vancouver. An addition costing in the neighborhood of $40,000 is contemplatad.

---

Messrs. Schumacher-Gray Co., Ltd., 187 Portage Ave., Winnipeg, have been awarded the electrical contract on an addition being built to the building of the Great West Life Assurance Co., at Lombard & Rorie Sts., Winnipeg, at an estimated cost of $304,000.

# The Washer You Have Waited For

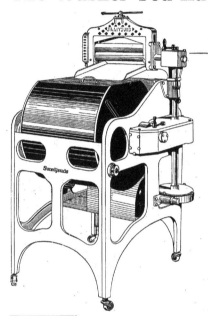

**Retail Price**

## $150

**Winnipeg and the West**

## $165

Sunnysuds was not built to create a demand but to supply one.

Here you have a dependable (note that word) standard-size, sure-action, all metal electric washer retailing at a figure that makes buying and selling reasonable.

It couldn't be better built at any price, it couldn't be built as well for a cent less.

To prove it, compare Sunnysuds with any washer at any price. If you haven't the facts we'll gladly supply them.

Dealer appointments are still being received. Have you sent in your application yet?

Wash your
Duds
in
Sunny Suds

**ONWARD MANUFACTURING COMPANY**
— KITCHENER  -  ONTARIO —

# SUNNYSUDS
## Electric Washer & Wringer

# SUPERB

*—embodying mechanical refinements*

# NEW EUREKA

## *and improvements that will carry it to even greater heights of popularity!*

Eureka Model 9 is now in the hands of Eureka dealers. The nation-wide enthusiasm which has greeted this latest product of the Eureka engineering staff is a warm tribute to the genius which has developed the refinements that make the Eureka even finer and more efficient than the winner of five international grand prizes. Each of eleven distinct improvements contribute to the rapidity and thoroughness with which the Eureka cleans, to its ease of operation, and to its ability to withstand gruelling service without failure and repair.

The nozzle has been lengthened to 13½" and its ends have been narrowed to increase the speed of the air at these points. This change now enables the Eureka to remove 95% of the clinging surface litter even without the aid of the improved "sweep action" brush. The new brush now permits a seal between floor covering and nozzle lips when clamped into position—a radical improvement representing the best combination of air and brush action. The wheels have been placed behind the nozzle—a change which not only permits easier cleaning against base

boards and in corners, but which makes it unnecessary to adjust the nozzle for the different napped floor coverings.

For convenient use under low beds and furniture, the fan case exhaust vent has been lowered 3". The enlarged dark blue bag has added materially to the cleaner's appearance and efficiency. Additional refinements as the "pocket book" clasp which locks the bag to the fan case, the new handle lock bolt, the one piece, oil tempered steel fan, and the shortened handle have all added to the beauty, simplicity, and efficiency of this superb new Eureka.

Our own conviction that this new Eureka represents an epochal achievement in the vacuum cleaner industry was substantiated and strengthened by the unbounded enthusiasm of our leading distributors when they witnessed the first demonstration of the new model. The most significant part of their approval was their prediction that they would sell fifty per cent more Eurekas than in 1921—and in 1921 Eureka was the largest selling vacuum cleaner in the world.

### *A Profitable Cleaner Agency*

To produce the maximum net profit, a vacuum cleaner must have mechanical excellence established beyond a shadow of doubt, it must be priced for mass appeal, it must be popularized by effective national advertising, and it must carry dealer discounts which leave a satisfactory profit after handling. The Eureka satisfies these four requirements. We invite correspondence from dealers regarding the dealer agency. Don't delay write us at once.

Eureka Vacuum Cleaner Company, Kitchener, Ontario

# EUREKA
## VACUUM CLEANER

# Some Legal Aspects of the Lake of the Woods and Rainy River Situation

### By EDWARD ANDERSON, K.C.

"When you realize that it is only a little over fifteen years since hydro electric energy from the Winnipeg River was first brought into Winnipeg to turn the wheels of industry in this city and in that short space of time there has been built up industries in which capital amounting to over two hundred million dollars has been invested, which industries are directly dependent upon the power produced from the Winnipeg River, you will realize what an important question this is, and how directly everyone of us and every citizen of Winnipeg is interested in it. The City of Winnipeg and the Province of Manitoba are particularly fortunate in having those tremendous possibilities of cheap power right at their doors.

The waters of the Winnipeg River come out of the Lake of the Woods—a great natural storage basin. These waters are obtained from the water-shed and from the drainage area surrounding the Lake of the Woods and Rainy Lake. In a broad way you may say the waters of the Winnipeg River come from Rainy Lake, Rainy River, and the Lake of the Woods. The water-sheds from which these waters come aggregate some twenty-eight thousand square miles. You will see what a tremendous area that covers. All these waters drain into this natural basin and flow down into the Winnipeg River, and by the reason of the conformation of territory these are created these water powers which we have already commenced to harness. It is important, therefore, to know what is the regulating or legal power over these waters. When you realize that Rainy Lake, Rainy River, and the Lake of the Woods are international waters, or boundary waters, lying between the United States on one side and Canada on the other, you can see how necessary it is that regulation should be properly enforced. The question of regulation is one which engrossed the attention of authorities for a great many years. Away back in 1887 the Dominion Government gave a grant towards the building of what was called the "Roll Away Dam" at the outlet of the Lake of the Woods, which dam had the effect of raising the water in the Lake. At that time the principal object was for the purpose of navigation, but now navigation is only a secondary matter and power the all important one. The next thing that was done that affected the levels of the Lake of the Woods was the erection of the Norman dam at the outlet of the Lake of the Woods where the Winnipeg River flows out of the Lake. It was built by a private company and shortly after an arrangement was entered into by that company and the Ontario Government whereby the Government paid $4,000.00 to the company for the privilege of regulating the stop-logs in that dam. This was to regulate the waters for navigation because even then the question of power had not become paramount. The effect of placing those dams was to raise the water in the Lake and the inhabitants of the southern shore on the Lake of the Woods in Minnesota complained through their government that a great deal of damage was done to their holdings. The result was that the question of the damage was referred to an International Joint Commission. The Commission was brought into being by joint legislation of the Dominion of Canada and the United States in 1909 whereby the question of international waters and in fact all other questions referring thereto, and disputes which might arise between the

Before Manitoba Electrical Association.

United States and Canada were to be disposed of by this International Commission.

It was a very fortunate thing that there was such a commission for such a purpose as this because if the questions had to go through the ordinary diplomatic channels they would not have been settled for a very long time and most likely settled in an unsatisfactory manner. The two governments realized that it was more than the question of the flooding of these lands which was at stake, with the result that a set of questions were submitted to this Commission, the effect of which was that they were to investigate and find out what was the most advantageous use to which the waters of the Lake of the Woods could be put. That Commission sat and took evidence and heard arguments for a period of three or four years and finally made its report. In my capacity as Counsel for the Dominion Government I had the privilege of attending before that Commission and know a great deal of what went on before it, but the most important work done before that Commission was done by the Water Powers Branch of the Dominion Government, at the head of which was Mr. Challies, and whose representative here is Mr. Atwood. That department had in mind the tremendous importance of power development on the Winnipeg River and they kept that in the forefront all the time. The question of navigation and the fisheries, and other questions came up but by reason of the energy and the foresight of the Water Power Branch at Ottawa this question of power was kept in the forefront all the time, with the result that the Commission in making its report had prominently before it the utilization of these waters for power purposes and recommended to the two governments that certain levels for the Lake of the Woods should be adopted. The Commission also disposed of the question of damages on the south shore of the lake.

They recommended that certain levels should be adopted and certain methods used to provide a continuous and dependable flow from the Lake of the Woods down the Winnipeg River. The way in which that was to be accomplished was to keep the levels of the Lake of the Woods at certain points and the report recommended certain improvements to the outlet of the Lake. It was necessary in order to get a proper flow of water that the outlet should be enlarged and certain regulating works should be put there. Those regulating works have not yet been put in but there is no question they will be put in so as to get the best use of the water in the Winnipeg River.

### Norman Dam Key to Situation

During the course of the hearing before the Commission it became evident that the strategic point or key to the whole situation was the ownership of the Norman dam. Unfortunately, as you know, governments move slowly and an ambitious and enterprising gentleman named Backus got in ahead. That created a very difficult and unsatisfactory situation so far as the water powers of the Winnipeg River were concerned, because Mr. Backus is more interested than any other single individual or combination of individuals in the waters of the Winnipeg River, Lake of the Woods, and Rainy Lake. As you can see, that created a very unfortunate condition of affairs. The Dominion Government woke up a little too late and realized that it was up to the Dominion Government to take some steps to straighten that out. The thing that brought the matter to a head was the granting by the Ontario Government to Mr. Backus of the right to develop the power site known as White Dog Falls on the Winnipeg River, and at this point let me tell you about the jurisdiction over the waters of the Winnipeg River. These waters are situated partly in Ontario and partly in Manitoba, the most important water powers of the river being in Manitoba, while White Dog Falls is the most important in Ontario. Because at confederation the natural resources of Ontario were conveyed to it, Ontario owns its own water powers and, therefore, owns and controls the waters of the Winnipeg River within the boundaries of the Province of Ontario. As the Province of Manitoba has not yet received its natural resources, the waters of the Winnipeg River lying in Manitoba were and are under the Dominion jurisdiction and have been administered by the Water Powers Branch of the Dominion Government at Ottawa. In my humble opinion, it was a fortunate thing for the people of Winnipeg and Manitoba that the waters of the Winnipeg River in Manitoba are and have been administered by the Water Powers Branch at Ottawa, and I will point out why I think that is so.

### Concession to Backus

The concession to Mr. Backus raised the question pointedly and directly, and you will see in what respect it was important there should be some jurisdiction in the Dominion Government. I understand this question is now being discussed by the Engineering Institute and that a committee is appointed to investigate and report on the question. So far as my experience and knowledge goes, I think it is of importance that the administration of those water powers should continue vested in the Dominion Water Powers Branch irrespective of whether or not the natural resources are transferred to the Province. I am quite sure that whatever Provincial Government is in power at the time that they are turned over, that government will agree with the statement I am making now. The reason why it is important that the control should remain at Ottawa is this. As soon as Mr. Backus got this concession at White Dog, and bearing in mind he had the Norman dam, it was realized at once by the people here that he had a control that was liable to be inimical and very destructive to the power users on the Winnipeg River down below.

The City of Winnipeg, the Province of Manitoba, and the Winnipeg Electric Railway Company, all got busy when this state of affairs arose, with the result that after a great deal of negotiations and discussions and representation, the Dominion Government became fully seized of the situation and the Ontario Government was made aware of the situation through Mr. Drury, the Premier. The Dominion Government and the Ontario Government agreed to introduce joint legislation to control the outlet at the Lake of the Woods, and thus control the flow of the water down the Winnipeg River. Up to that time there was no regulation, and no statute in force whereby any government could come in and take control of the situation. Those interested were absolutely at the mercy of the private interests who controlled that outlet and owned other concessions on the River. The two Governments promised to pass concurrent legislation. This was necessary because Ontario owned her water powers, while the Dominion Government had control of the water powers in Manitoba. The bill was presented to the Dominion House and passed into law. In the Ontario House for some reason it was not passed and the result, therefore, was very unsatisfactory.

### Ottawa Legislation

The legislation was passed at Ottawa, the legislation was held up at Toronto and the whole thing was in the air. The remedy adopted was this—the Dominion Government passed an independent act and in connection with that act, I can demonstrate to an absolute conclusion the necessity of having control of these waters in the hands of the Dominion Government. They passed this act which had for its express purpose the controlling of all works on the River and the outlet of the Lake. There is a very simple device by which the Dominion Government can obtain jurisdiction over Provincial undertakings and that is to pass an act declaring the particular works to be "For the general advantage of Canada." In that act they declared that all works at the outlet of the Lake of the Woods and at the English River from Lac Seul, (which has an important bearing on this question) were for the general advantage of Canada, and thereby, by that act, they became under the jurisdiction of the Dominion Government. Otherwise, as I say, there was no method of exercising that jurisdiction. Suppose, for instance, that the jurisdiction rested not with the Dominion Government but with Manitoba, there would have been no adequate means by which the Manitoba Government could have controlled the situation. All that it could possibly have control of would be that portion of the Winnipeg River situate within its boundaries and as the controlling situation was in the Province of Ontario it would have nothing to say about it at all. But by reason of the passing of this act the Dominion Government got complete jurisdiction over the matter. Among other things that the act provided for was that one of the recommendations of the International Joint Commission should be carried out—a Board appointed to have jurisdiction over the flow of water in the Winnipeg River and outlets. That Board was appointed and I understand is now functioning. By that act these waters came under the jurisdiction of the Dominion Government and the government took immediate and active steps to maintain that control.

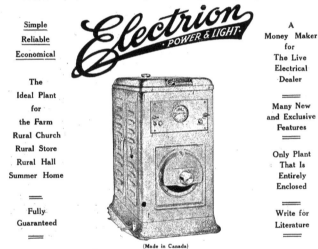

## FOR SALE

## We Are in the Market For

### S. Australian Lock Formally Opened

The Minister of Public Works for South Australia, Hon. W. Hague, recently formally opened the Blanchtown lock, the first of a series of nine locks to be built by South Australia, harnessing the Murray River and insuring a continuous navigable waterway and irrigation supply from Wentworth to the mouth of the Murray River. The work of constructing the locks was begun nearly seven years ago, but it has been delayed by floods and industrial troubles. When the whole project is completed there will be a permanent navigable waterway of 1,066 miles, and a vast increase in irrigable lands.

### Vote Against Skyscrapers

London will not emulate New York in the matter of erecting skyscrapers, if the ideas of the Royal Institute of British Architects are carried out. The architects have voted against a report of their building acts committee, advocating alteration of the London regulations to allow the erection of buildings 100 feet in height, instead of 80 ft. as at present.

### U.S. BUILDING BOOM

The value of buildings for which permits were granted last month in 141 United States cities far surpassed that of any previous February in the country's history, the total amount being $122,684,719, it was reported by Bradstreet's. This figure was a decrease of 5.3 per cent. from the lower month of January this year, during which the total reached $129,555,404, but it was a gain of 66 per cent. over February, 1921. Only one group of cities, those in New England, reported a decrease compared with February of last year. Those in the northwest showed the heaviest gain

Matches and smoking hazards are held responsible for $90,000,000 of losses, according to the report. Next comes electricity, which caused fires costing 142 per cent. higher than the corresponding month of 1921.

# GENERATORS

### In Stock for Immediate Delivery

### H. W. PETRIE, Limited

131-147 Front St. West, Toronto

## ALPHABETICAL LIST OF ADVERTISERS

# CLASSIFIED INDEX TO ADVERTISEMENTS

The following regulations apply to all advertisers:—Eighth page, every issue, three headings; quarter page, six headings; half page, twelve headings; full page, twenty-four headings.

**AIR BRAKES**
Canadian Westinghouse Company

**ALTERNATORS**
Canadian General Electric Co., Ltd.
Canadian Westinghouse Company
Electric Motor & Machinery Co.
Ferranti Meter & Transformer Mfg. Co.
Northern Electric Company
Wagner Electric Mfg. Company of Canada

**ALUMINUM**
British Aluminum Company.
Spielman Agencies, Registered

**AMMETERS & VOLTMETERS**
Canadian National Carbon Company
Canadian Westinghouse Company
Ferranti Meter & Transformer Mfg. Co.
Monarch Electric Company
Northern Electric Company
Wagner Electric Mfg. Company of Canada

**ANNUNCIATORS**
Macgillivray, G. L. & Co.

**ARMOURED CABLE**
Canadian Triangle Conduit Company

**ATTACHMENT PLUGS**
Canadian General Electric Company
Canadian Westinghouse Company

**BATTERIES**
Canada Dry Cells Ltd.
Canadian General Electric Co., Ltd
Canadian National Carbon Company
Electric Storage Battery Co.
Exide Batteries of Canada, Ltd.
McDonald & Willson, Ltd., Toronto
Northern Electric Company

**BEARINGS**
Kingsbury Machine Works

**BELLS**
Macgillivray, G. L. & Co.

**BOLTS**
McGill Manufacturing Company

**BONDS (Rail)**
Canadian General Electric Co., Ltd
Ohio Brass Company

**BOXES**
Banfield, W. H. & Sons
Canadian General Electric Co., Ltd
Canadian Westinghouse Company
G & W Electric Specialty Company

**BOXES (Manhole)**
Standard Underground Cable Company of Canada, Limited

**BRACKETS**
Slater, N.

**BRUSHES CARBON**
Dominion Carbon Brush Company
Canadian National Carbon Company

**BUS-BAR SUPPORTS**
Ferranti Meter & Transformer Mfg. Co.
Moloney Electric Company of Canada
Monarch Electric Company
Electrical Development & Machine Co.

**BUSHINGS**
Diamond State Fibre Co. of Canada, Ltd.

**CABLES**
Boston Insulated Wire & Cable Company, Ltd
British Aluminium Co., Ltd
Canadian General Electric Co., Ltd
Phillips Electrical Works, Eugene F.
Standard Underground Cable Company of Canada, Limited

**CABLE ACCESSORIES**
Northern Electric Company
Standard Underground Cable Company of Canada, Limited

**CARBON BRUSHES**
Calrbaugh Self-Lubricating Carbon Co.
Dominion Carbon Brush Company
Canadian National Carbon Company

**CARBONS**
Canadian National Carbon Company
Canadian Westinghouse Company

**CAR EQUIPMENT**
Canadian Westinghouse Company
Ohio Brass Company

**CENTRIFUGAL PUMPS**
Boving Hydraulic & Engineering Company

**CHAIN (Driving)**
Jones & Glassco

**CHARGING OUTFITS**
Canadian Crocker-Wheeler Company
Canadian General Electric Co., Ltd
Canadian Allis-Chalmers Company

**CHRISTMAS TREE OUTFITS**
Canadian General Electric Co., Ltd
Masco Co., Ltd.
Northern Electric Company

**CIRCUIT BREAKERS**
Canadian General Electric Co.
Canadian Westinghouse Company
Cutter Electric & Manufacturing Company
Ferranti Meter & Transformer Mfg. Co.
Monarch Electric Company
Northern Electric Company

**CONDENSERS**
Boving Hydraulic & Engineering Company
Canadian Westinghouse Company

**CONDUCTORS**
British Aluminum Company

**CONDUIT (Underground Fibre)**
American Fibre Conduit Co.
Canadian Johns-Manville Co.

**CONDUITS**
Conduits Company
National Conduit Company
Northern Electric Company

**CONDUIT BOX FITTINGS**
Banfield, W. H. & Sons
Northern Electric Company

**CONTRACTORS**
Seguin, J. J.

**CONSTRUCTION MATERIAL**
Oshkosh Mfg., Co.

**CONTROLLERS**
Canadian Crocker-Wheeler Company
Canadian General Electric Company
Canadian Westinghouse Company
Electrical Maintenance & Repairs Company
Northern Electric Company

**CONVERTING MACHINERY**
Ferranti Meter & Transformer Mfg. Co.

**COOKING DEVICES**
Canadian General Electric Company
Canadian Westinghouse Company
National Electric Heating Company
Northern Electric Company
Spielman Agencies, Registered

**CORDS**
Northern Electric Company
Phillips Electric Works, Eugene F.

**CROSS ARMS**
Northern Electric Company

**CRUDE OIL ENGINES**
Boving Hydraulic & Engineering Company

**CURLING IRONS**
Northern Electric Company (Chicago)

**CUTOUTS**
Canadian General Electric Company
G. & W. Electric Specialty Company

**DETECTORS (Voltage)**
G. & W. Electric Specialty Company

**DISCONNECTORS**
G. & W. Electric Specialty Company

**DISCONNECTING SWITCHES**
Ferranti Meter & Transformer Mfg. Co.
Winton, Joynes, A. H.

**DREDGING PUMPS**
Boving Hydraulic & Engineering Company

**ELECTRICAL ENGINEERS**
See Directory of Engineers

**ELECTRIC HAND DRILLS**
Rough Sales Corporation George C.

**ELECTRIC HEATERS**
Canadian General Electric Company
Canadian Westinghouse Company
Equator Mfg. Co.
McDonald & Willson, Ltd., Toronto
National Electric Heating Company

**ELECTRIC SHADES**
Jefferson Glass Company

**ELECTRIC RANGES**
Canadian General Electric Company
Canadian Westinghouse Company
National Electric Heating Company
Northern Electric Company
Rough Sales Corporation George C.

**ELECTRIC RAILWAY EQUIPMENT**
Canadian General Electric Company
Canadian Johns-Manville Co.
Electric Motor & Machinery Co., Ltd.
Northern Electric Company
Ohio Brass Company
T. C. White Electric Supply Co.

**ELECTRIC SWITCH BOXES**
Banfield, W. H. & Sons
Canadian General Electric Company
Canadian Drill & Electric Box Company
Dominion Electric Box Co.

**ELECTRICAL TESTING**
Electrical Testing Laboratories, Inc.

**ELEVATOR CONTRACTS**
Dominion Carbon Brush Co.

**ENGINES**
Boving Hydraulic & Engineering Co., of Canada
Canadian Allis-Chalmers Company

**FANS**
Canadian General Electric Co. Ltd.
Canadian Westinghouse Company
Century Electric Company
Great West Electric Company
McDonald & Willson
Northern Electric Company
Robbins & Myers

**FIBRE (Hard)**
Canadian Johns-Manville Co.
Diamond State Fibre Co. of Canada, Ltd.

**FIBRE (Vulcanized)**
Diamond State Fibre Co. of Canada, Ltd.

**FIRE ALARM EQUIPMENT**
Macgillivray, G. L. & Co.
Northern Electric Company
Slater, N.

**FIXTURES**
Banfield, W. H. & Sons
Benson-Wilcox Electric Co.
Crown Electrical Mfg. Co., Ltd.
Canadian General Electric Co. Ltd.
Jefferson Glass Company
McDonald & Willson
National X-Ray Reflector Co.
Northern Electric Company
Tallman Brass & Metal Company

**FLASHLIGHTS**
Canadian National Carbon Company
McDonald & Willson Ltd.
Northern Electric Company
Spielman Agencies, Registered

**FLEXIBLE CONDUIT**
Canadian Triangle Conduit Company
Slater, N.

**FUSES**
Canadian General Electric Co. Ltd.
Canadian Johns-Manville Co.
Canadian Westinghouse Company
G. & W. Electric Specialty Company
Moloney Electric Company of Canada
Northern Electric Company
Rough Sales Corporation George C.

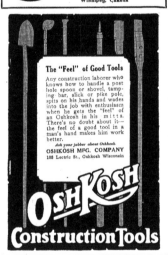

# CLASSIFIED INDEX TO ADVERTISEMENTS—CONTINUED

## CLASSIFIED INDEX TO ADVERTISEMENTS—CONTINUED

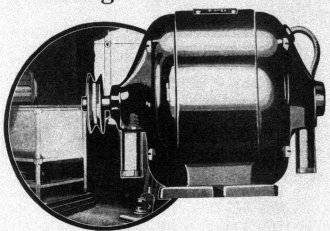

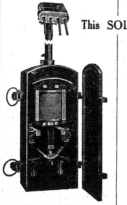

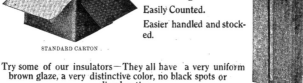

Vol. XXXI ·No. 9                                                                    Toronto, May 1, 1922

# Electrical News

### Engineering Contracting    Merchandising Transportation

*The Finest Electrical Time Equipment in the World*

## Write for Our New Catalogue

We have just printed a new catalogue of International Electric Time Systems, giving particulars of their construction, which will be invaluable to electrical contractors when tendering on this class of work. We are the only firm in Canada manufacturing these goods—all our master clocks, secondary clocks, relay equipment cabinets, etc., being made in Canada. You save duty, you avoid delay and misunderstanding and you buy in Canada. So far as we know, every tender we have made for electrical time equipment has been the lowest in price and the most satisfactory in use. Send us your name today for a copy of this catalogue.

## INTERNATIONAL BUSINESS MACHINES CO., LIMITED

**FRANK E. MUTTON,** Vice-President and General Mgr.    Head Office and Factory:    300 Campbell Ave. Toronto

For your convenience we have Service and Sales Offices in:

Vancouver,   Winnipeg,   Walkerville,   London,.  Hamilton,   Toronto,   Ottawa, Montreal,  Quebec,  Halifax,  St. John, Nfld.

*Also manufacturers of International Dayton Scales and International Electric Tabulators and Sorters·*

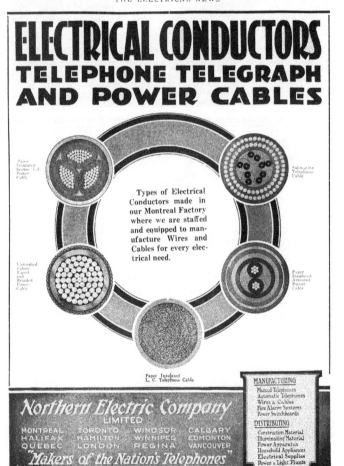

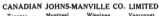

# ALPHABETICAL LIST OF ADVERTISERS

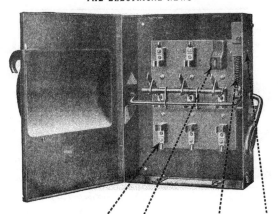

# An Improved
## Square D Safety

# Switch, Series 80,000

### *With Exclusive Features of Complete Protection, Complete Accessibility, and Exceptional Electrical and Structural Strength!*

Here is the safest, strongest, surest safety switch! A new Square D with radical improvements that command the attention and approval of electricians, contractors, safety experts, architects and purchasing agents! A new Square D that meets every practical requirement ever demanded of an enclosed safety switch!

This is the Time to Act

Write today for a Square D representative. He will show you the new radically improved Square D Switch, let you see for yourself why it is the one logical switch to stock and install, and give you the surprisingly attractive prices on the complete line from 30 amps. to 400 amps.

SQUARE D COMPANY, Walkerville, Ontario, Canada
*Sales Offices at Toronto and Montreal*

# Square D Safety Switch

# Wedding Presents

IT would, indeed, be hard to imagine a more suitable gift than a Hotpoint electrical device. At this time of the year, thousands of people are racking their brains, and wondering what to give for wedding presents.

Your display windows offer you a point of contact with these potential customers. They are in the market to buy—you have the merchandise for sale. The percentage of sales made from the sidewalk is surprisingly large.

Remember, one reliable electrical device sells another. Hotpoint appliances have stood the test of time. Create prestige for your store by specializing on this famous line. Order from your jobber, or direct from us.

*"Made in Canada"*

## Canadian Edison Appliance Co., Limited
### STRATFORD, ONTARIO.

# Not even hands are required to operate the NEW C·G·E TUMBLER SWITCH

H ERE'S a switch that responds to the sweep of a hand across the wall—no fumbling around to find the right button. The C.G.E. Tumbler Switch works like a lever, up and down, at the lightest touch. When hands are full, an elbow answers the purpose. And yet this simple device is so sturdy and reliable, that it never gets out of order.

C.G.E. Tumbler Switches are made in three styles. The flush type—applicable to every kind of residence, is finished with a black moulded handle, and has a very handsome appearance. The surface type is small, compact and unobtrusive in appearance. The all porcelain model, which is also of the surface type, is for use in damp places.

The fact, that the button is in one position for "off" and in another for "on" makes the C.G.E. Tumbler Switch self-indicating.

*"Made in Canada"*

# Canadian General Electric Co., Limited
## HEAD OFFICE ⊕ TORONTO

Branch Offices: Halifax, Sydney, St. John, Montreal, Quebec, Sherbrooke, Ottawa, Hamilton, London, Windsor, Cobalt, South Porcupine, Winnipeg, Calgary, Edmonton, Vancouver, Nelson and Victoria.

# C. G. E.
# Flush Push Switches

5 ampere three-way switch. Catalogue No. C.G.E. 3010. Hydro approval No. 108.

10 ampere double pole switch. Catalogue No. C.G.E. 2010. Hydro approval No. 108.

5 ampere single pole switch. Catalogue No. C.G.E. 1010. Hydro approval No. 108.

THE oustanding feature of this device is its simple, rugged construction. There are no delicate parts to get out of order. Most of the mechanism is manufactured from flat, steel pieces, nickel plated to prevent tarnishing or rust.

The operating spring, held in place by the central shaft, works freely under compression, thus having a much longer life than a tension spring, which has to be secured in some manner, and this usually proves to be the weakest point of the spring, consequently the switch itself.

Another important feature of this switch is the extra long stroke of the push button, which permits greater adjustment of the switch in the box, than other types of switches.

*"Made in Canada"*

# Canadian General Electric Co., Limited
## HEAD OFFICE TORONTO

Branch Offices: Halifax, Sydney, St. John, Montreal, Quebec, Sherbrooke, Ottawa, Hamilton, London, Windsor, Cobalt, South Porcupine, Winnipeg, Calgary, Edmonton, Vancouver, Nelson and Victoria.

# How will You Pay for Your Switches-- Once and for All--Or Day by Day?

Elpeco Indoor
Disconnecting Switch

D OWN at the power house the black coal or the "white coal" is working for you. It costs dollars.

Up at the machine the current is working for you. It makes dollars.

In between stand the switches. Some of them are taxing you every minute of the working day. Others are paid for once and for all.

Because Elpeco switches are designed from practical experience in the handling of high tension current and are built on an "individual" system of manufacture they save the electrical dollars that ordinary switches tax you.

By a system of grinding-in all their own the hinge and contact clips of Elpeco switches give perfect contact surface. Their double blade construction increases that surface.

The test of a switch is temperature. Think of the money wasted through some switchboards which give out enough heat to warm a house. Note how cool the Elpeco switch maintains itself in comparison with ordinary manufactured-in-quantity switches.

Remember the Elpeco trademark—the Crusader with the two shields—one against *danger* in the handling of high tension current, the other against *waste*. Elpeco's specialized knowledge on high tension apparatus and installations is available to Engineers, Contractors and Operating Executives. We solicit your correspondence.

## Protection in the future from DANGER and LOSSES

ELPECO-THE CRUSADER IN ADVANCED ELECTRICAL EQUIPMENT

Disconnecting
Switches　　Bus Supports　　Pole Top Switches　　Switch Boards

ELECTRIC POWER EQUIPMENT CORPORATION, Philadelphia, Pa.
Canadian Representives: FERRANTI METER AND TRANSFORMER MFG. CO., Limited, Toronto & Montreal

**For nearly thirty years the recognized journal for the
Electrical Interests of Canada.**

Published Semi-Monthly By

# HUGH C. MACLEAN PUBLICATIONS
### LIMITED
THOMAS S. YOUNG, Toronto, Managing Director
W. R. CARR, Ph.D., Toronto, Managing Editor
HEAD OFFICE - - 347 Adelaide Street West, TORONTO
Telephone A. 2700

| | |
|---|---|
| MONTREAL - - - | 119 Board of Trade' Bldg. |
| WINNIPEG - - - - | 802 Travellers' Bldg. |
| VANCOUVER - - - - - | Winch Building |
| NEW YORK - - - - - | 296 Broadway |
| CHICAGO - - - | 14 West Washington St. |
| LONDON, ENG. - - | 16 Regent Street S.W. |

ADVERTISEMENTS
Orders for advertising should reach the office of publication not later
than the 5th and 20th of the month. Changes in advertisements will be
made whenever desired, without cost to the advertiser.

SUBSCRIPTIONS
The "Electrical News" will be mailed to subscribers in Canada and
Great Britain, post free, for $2.00 per annum. United States and foreign,
$2.50. Remit by currency, registered letter, or postal order payable to
Hugh C. MacLean Publications Limited.
Subscribers are requested to promptly notify the publishers of failure
or delay in delivery of paper.

Authorized by the Postmaster General for Canada, for transmission
as second class matter.

Vol. 31                 Toronto, May 1, 1922                 No. 9

## Filling up the Valleys

In spite of the tremendous developments and refinements
that have been brought about in the electrical industry during
the last ten years, there still remains that old problem of peak
load, on the one hand, and the deep valley in the curves, on
the other hand. The load-factor is still low, though the overhead, the maintenance and the capital carrying charges still
work 24 hours a day. Any suggestions, therefore, towards
improving these conditions are always hailed with interest
and pleasure by the industry.

Such a suggestion comes from F. T. Kaelin, the well known
hydro-electric engineer of the Shawinigan Water & Power
Company who, in a recent address before the Engineering
Institute of Canada, spoke of the economic uses of electricity
for producing steam for heating and other purposes in off-
peak hours, thus helping to straighten out the kinky curve.
Mr. Kaelin has invented a steam generator, which is exceedingly economical in construction and, of course, practically
100 per cent. efficient as regards the transformation of electrical energy into heat. This generator has been used to good
advantage at different points in Quebec province where conditions are favorable; for example, in a pulp and paper mill
where an ample supply is available near at hand. Mr. Kaelin
cites, for example, the case of the use of electrically generated
steam over the week-end, and states that it is economically
quite feasible to turn power that was used during the six days,
—mainly for pulp grinding—into steam on the 7th day. He
stated that a recent investigation proved that the saving of
coal during the 52 Sundays and three holidays more than paid
for the installation of the steam generator. The electric

energy used during these off-peak periods would otherwise
go to waste.

The low cost of Mr. Kaelin's electric steam generator,
we understand, is due to the fact that he uses the water itself
as the resistance, varying the distance between the electrodes
as the demand for steam varies.

This idea of utilizing electricity to provide steam heat
during short periods when the plant is shut down is one that
may easily appeal to a number of industries. It is always
just about as costly an operation to keep a factory or office
building heated by coal during holidays and Sundays as it is
during the working days, and if Mr. Kaelin has a commercial
proposition to offer to different industries whereby they can
utilize electrical energy for heating purposes on the seventh
day that is used the other six days in factory production, he
has made a very important contribution to the electrical industry's economics.

## Government Extends Rural Aid

It will be remembered that at last year's session of the
Ontario Legislature an Act was passed whereby the Government undertook to bear 50 per cent of the cost of certain
rural primary distribution lines. Considered from the angle
that it places the blessing of electricity within reach of a
larger percentage of the rural community, this Act had
universal approval, but judged on a basis of legal equity,
nothing more unfair was ever placed on the Statute books.
There is no more reason why farmers should be presented
with a transmission line at half price than that the residents
along a city street should be treated in the same way. It
was an iniquitous form of legislation that got by without
serious objection, simply because every city man realizes that
life on a farm, under the best of conditions, has little enough
of luxury.

The matter has again been brought to the attention of
the country through a Bill being introduced at the present
session of the Ontario Government to extend this privilege
to municipalities that may be in a position to purchase power
from private companies (the original Act applied only to
the Hydro-electric Power Commission of Ontario). It is
not the intention of this session's Bill, apparently, to allow
any rebate to a private company that may be constructing
primary lines for rural distribution, but it provides machinery whereby a township or a group of farmers may build
primary lines and own them, the Government paying 50
per cent of the cost. This will benefit the private company
to the extent that it will not be required to put up fresh
capital for these extensions.

## Coal Sometimes Cheaper Than Water Power

With our abundance of water power in Canada, and
with many fine examples of development work carried out
at low cost, there is a natural tendency to look upon all
water power development as economical and certain to result in low power rates to the consumer. We are apt to forget that conditions vary greatly with localities; that the
size of a station has a marked effect not only on its capital
cost but also on the cost of maintenance and operation;
that the proximity of coal may reduce the cost of steam
operation to a point where water power cannot compete.

A case in point is the evidence of R. A. Ross, recently
given before the New Brunswick Power Commission. The
province of New Brunswick is just putting the finishing
touches to a 10,000 h.p. hydro-electric plant, but Mr. Ross
told the Commission that he could build a steam plant—
presumably in St. John, where this hydro-electric power will
be delivered—that would knock the spots off the hydro

plant, so far as power costs were concerned. Mr. Ross has had more than ordinary experience in the utilization of coal and his statement is thus of more than usual interest.

If this 10,000 h.p. plant had been constructed in a district far removed from an ample coal supply, no doubt the story would have been different, but it goes to show that we must not take it as a foregone conclusion that every water falls can be economically developed, whatever its size and wherever its location.

## Mayor Tisdall, Vancouver, Extols B. C. E. R. Service

Mayor C. E. Tisdall, of Vancouver told the Electric Club of that city at luncheon on March 24 that the board of aldermen had authorized him to look for an engineer of wide experience in the hydro-electric field for the purpose of advising the civic authorities on the subject of developing electric energy for Vancouver. While this had been done, the Mayor was careful to say that no definite policy had been decided. It was, however, expected that during the coming summer information on the subject would be assembled and when the council had such reports in hand, the question of a hydro-electric installation controlled by the city as against the present established service would be put up to the ratepayers. Speaking as a layman, Mayor Tisdall, in a very careful and conservative statement, pointed out that the mere existence of a waterpower did not argue for its development. The economic development of a power was a main factor in its success. Then it was still a moot question, the mayor said, as to the relative advantages of public and private ownership, as to whether a publicly owned utility can get the efficiency possible through a private corporation. He spoke also of such fundamentals as the financial statement, in which proper allowance must be made for depreciation, etc., in order to have a publicly owned project on sound basis. Financial statements, showing revenue and expenditure, might look very well until the groundwork was dug into. It was imperative to maintain an equal balance between revenue and expenditure, and incidentally the mayor pointed out the difficulties the 1922 council had just been contending with on that score. A great many people had attended at the city hall to complain of the high taxation, and when the estimates were being considered an equal number of people attended urging various expenditures. It was axiomatic that if the people asked for service someone must pay for it. That was true of all service, whether electric or otherwise.

In his introductory remarks, Mayor Tisdall commented on the development going on in the use of electricity. Time was not so far back but that many of his own age could remember when the candle was the only light. To-day the electric light was universal, as also was electric power. He predicted that electric cooking in the homes of all communities would soon be general to the same degree as light. He knew of no branch of industry or engineering which offered such an attractive future for the young man as the electrical field. Before concluding his remarks, Mayor Tisdall paid a compliment to the present electric service of the B. C. Electric Railway Company. Vancouver and its environs were well served by the present company, he said. The fact that his council was inquiring into the question of a civic hydro-electric plant was no reflection on the B.C.E.R. Co. Vancouver could not fairly compare its present rates for electric energy with that of Winnipeg for the latter city had had particularly favorable conditions in establishing its hydro-electric power plant. But Vancouver, he said, does compare favorably with most cities. Present rates were not unreasonable and "we have an abundant service."

## Utilizing a Refrigerating Plant for Heating Purposes

When Mme. Galli-Curci sang in Vancouver on March 20th the only auditorium of sufficiently large proportions to accommodate the expected audience was the Arena skating rink. Vancouver is the place where they have to make artificial ice for their skating rink, even in winter. But the cold storage atmosphere was not conducive to the best efforts of even a prima donna, to say nothing of the comfort of the audience. It was, therefore, decided to heat the Arena. The idea was to reverse the usual process by which the ice on the skating floor was formed. So the problem was put up to the engineering department of the B. C. Electric Railway Company. Here is how they solved it, the installation and test being completed in forty-eight hours—incidentally, it is the first time on record when such a method of heating electrically was ever tried. The results were entirely satisfactory, the large audience which packed the great Arena enjoying the concert with all the comfort of a theatre.

The circulating pipes for the artificial ice system are installed in parallel lines between parallel headers laid in sand under a hardwood floor, on which the ice film is ordinarily formed. There are 53,500 feet of this pipe, which is 1¼ inches in diameter. To heat these coils the method adopted was as follows :

Three sections of ¾ in. pipe, 250 feet long, were coupled in delta and immersed in the 25,000 gallon brine tank. After being heated to pre-determined value, the water was circulated by a 600 gal. centrifugal pump. Elements were supplied by a 650 kw. transformer, 2200/110 volts, coupled in parallel delta, the current on the 2,200 volt side being 90 amperes.

After heating the tank water for twenty-four hours, the temperature of the water in the tank was raised in that time to 160 deg. F. and then circulated by means of the pump through the pipe system for three hours. The following figures show the result :

Before circulating : Temp. of tank water, 160 deg.; temp. of floor, 40 deg.; temp. outside bldg., 43 deg.;

After circulating 3 hours : Temp. of tank water, 106 deg.; temp. of floor, 59 deg.; temp. outside bldg., 41 deg.

During the 30 hours' run in which the temporary heating plant was used, the power required was 9,000 kw. hours. To give a more definite idea of the heating problem solved in this original manner, the following dimensions are furnished :

| | |
|---|---|
| Height of building | 60 feet |
| Width of building | 135 feet |
| Length of building | 260 feet |
| Area of exposed walls | 47,400 sq. ft. |
| Brine tank capacity | 25,000 gals. |
| Piping system capacity | 5,000 gals. |
| Lin. ft. pipe under main floor | 53,500ft. |
| Size of pipe | 1¼ in. |

## American Institute Electrical Engineers Holding First Convention in Canada

For the first time in its history, the American Institute of Electrical Engineers will hold its mid-summer convention in Canada. Niagara Falls, at the Clifton Hotel, has been chosen as the rendezvous and Canadian members are asked to co-operate in every possible way to make the event a success. The date has been fixed as June 26-30—an ideal date, as well as location, for a little holiday.

## Toronto Section A. I. E. E. Closes the Year with Three Rousing Meetings

The American Institute of Electrical Engineers, Toronto Section, held a joint meeting with the Hamilton Branch of the Engineering Institute of Canada, in the Westinghouse auditorium, Hamilton, Ont., on March 24. The meeting was opened by Mr. H. U. Hart, chairman, Hamilton Branch E. I. C., who following a few words of welcome to the visitors, called upon Mr. Dobson, chairman of the Toronto Section of the A. I. E. E., to preside. Mr. Dobson introduced the speaker, Mr. J. S. Peters, of the Westinghouse Electric & Manufacturing Co., Pittsburg, Pa., to address the meeting on the subject "220,000 volt transmission."

Mr. Peters briefly traced the increase of voltage with the progress of electrical development since 1890 showing, by a graphic chart, that the logical point for transmission to have reached by 1923 would be 220,000 volts. He pointed out that at such a voltage the use of a solidly grounded neutral seemed imperative. He described the type of transformer which appeared most practicable for this class of service as having three windings. The primaries and secondaries would usually form a star-star connecton, while the third winding, which might be of any convenient voltage, served the double purpose of enabling a delta to be formed for stabilizing the star point, and of furnishing a circuit to which a condenser might be attached at the receiving end of the line. Single phase transformers were more practicable than three phase.

Detailed information was given as to corona, line losses, voltage regulation, the effect of ground wires, graded insulators and guard rings. The desirability of lightning arresters on 220 kilo-volt circuits was questionable. On account of the great clearances required, it was usually desirable to instal as much as possible of the equipment out of doors.

The lecture was illustrated by a generous showing of charts, diagrams and lantern slides, and was discussed from many angles by Messrs. Hart, Stevenson, H. W. Price, Dwight, Wells, C. W. Baker, Publow, Powell, Chubbuck, Cook, E. M. Wood, S. E. M. Henderson, E. W. Henderson and Borden.

The session was closed by a vote of thanks to Mr. Peters and to the Canadian Westinghouse Company, after which refreshments were served. About 50 members from Toronto were present, many of these having taken advantage of the motor-bus service which had been arranged by the Section for the occasion.

## Power Factor

Mr. H. W. Price, associate professor of Electrical Engineering, Toronto University, addressed the meeting on the subject of the "Power Factor" at the meeting of April 7th.

Professor Price had set up, on the lecture table, a very complete outfit for demonstrating the significance of power-factor on alternating current systems. This consisted in a system of loads including inductive circuits, condensers and pure resistance banks, which could be fed from an alternating source either individually or collectively. The load current passed through two sets of wattmeters, one connected to read power and the other to read re-active volt-amperes, and also through ammeters, individual and totalizing. In addition to these there were included in the circuit a direct current voltmeter and ammeter, fed from the alternating current system through an asynchronous commutator, whereby the frequency of the circuit was apparently reduced to about three cycles per minute, enabling the relative fluctuations of voltage and current to be visually observed.

With this apparatus, the speaker demonstrated with great clearness the relative phase position of current and voltage in the several types of circuits, and their effect upon the total current and power in the line. It was possible to make very plain the wasteful effect of low power factors, and to show how the lagging currents taken by unloaded transformers and other inductive apparatus could be neutralized by suitable condensers, with a consequent reduction of line current to a value representative of the energy component only of the load.

In conclusion, Professor Price demonstrated the effect of distorted current waves upon the system, showing that the harmonics which produced distortion were without effect on instruments of the wattmeter type, so that while a wattmeter connected to such a circuit continued to indicate the true power in that circuit, an instrument supplied with a quadrature voltage to give re-active volt-amperes, failed to give the true value, and entirely neglected those harmonics in the current wave which did not appear in the voltage wave. The speaker showed by a simple calculation, that the error so introduced in measurement on circuits of such wave forms as are found in practice might usually be neglected without serious results.

Mr. C. E. Schwenger of the Toronto Hydro Electric System opened the discussion and showed some interesting figures bearing on the cost of transformer magnetizing current in power purchased on a demand basis. It was brought out that it was frequently a better economic proposition to purchase transformers with comparatively large losses, and low magnetizing current, than to make efficiency the sole criterion by which they should be selected. He showed also that in districts where power was comparatively expensive, it would be wise to pay more money for the sake of obtaining transformers and other inductive apparatus having a low value of magnetizing current. Mr. Schwenger's statements were supplemented by Mr. C. W. Baker of the Packard Electric Company, who gave illuminating figures of the ratio between the magnetizing current and the cost of transformers. Messrs. Don Carlos and Borden also took part in the discussion.

## Officers for 1922-23

The annual meeting and election of officers of the Toronto section of the American Institute of Electrical Engineers took place at the Chemistry and Mining Building, University of Toronto, April 21. The reports presented, showing the result of activities during the past year, were regarded as very satisfactory and were adopted without comment. The slate of officers that had been recommended by the nominating committee was accepted as follows: S. E. M. Henderson, chairman; D. B. Fleming, secretary. Executive Committee: W. L. Amos, O. V. Anderson, L. B. Chubbock, C. H. Hopper, S. L. B. Lines and C. E. Schwenger.

Prof. A. P. Coleman, of the University of Toronto, gave an interesting address, illustrated by some very fine slides of "Rocky Mountain Trails to Mount Robson."

## Radio Forest Patrol

If present plans mature, two radio stations will be installed this season for the use of the forest aeroplane patrols at Norway House and Victoria Beach, was announced by Maj. B. D. Hobbs of Winnipeg Station of the Canadian Air Board. The outfits will be able to transmit by key only; and to receive key or phone calls. "The advantages of this method," said Maj. Hobbs, "will be that speedy communication may be made between bases in the matter of reporting fires to the district inspector of forestry, at Winnipeg, so that help may be sent to the scene of fires with the least possible delay, and for the broadcasting of reliable weather reports from the northern country."

## Re Standard Electrical Rules and Regulations Governing all Canada

Winnipeg, April 12, 1922.

Editor Electrical News,—

I am very pleased to learn by your editorial of the 1st inst. that this long deferred matter has at last received in some degree the attention that its importance deserves.

A special sub-committee of the Canadian Engineering Standards Association held a meeting on this matter in Toronto on September 8, 1920, at which I had the honor to present the views of a considerable number of inspecting interests in Canada from coast to coast outside the Province of Ontario, including representatives of the two transcontinental railway systems and the Western Canada Fire Underwriters' Association. These interests were uniformly of the opinion that the present National Electrical Code, being the universally accepted Code for the whole of Canada, with the exception of the Province of Ontario, should be accepted as the standard Code for Canada.

There has been, previous to this, a feeling prevalent in certain quarters of Canada that the National Electrical Code was a document of foreign origin and that it was derogatory to Canadians to work under it. Of course this contention is not correct as the N.F.P.A. (which may rightly be considered an international organization) has upon its committees a considerable number of Canadians, but in order to allay this feeling, the chief officials of the Association have expressed themselves as being perfectly ready to arrange for further representation of Canadians on the committees and also to loan to a responsible body in Canada the plates from which the Code is printed and to allow Canada to choose its own title—all this in order to avoid setting up what might be considered a Code varying from the present National Electrical Code.

Of course, it may be argued that the National Code is only a fire code and does not take into consideration the question of the life risk. While this was partly true at one time, this Code is gradually taking in life risk matters, and there is no doubt that in a reasonable time it will deal with this issue to a still larger extent. In the meantime, however, there is nothing to prevent a Canadian organization formulating a concise safety code dealing with the life risk and publishing this, leaving it optional as to whether it should be adopted by the various organizations dealing with such matters or not. The two codes could be adopted separately or together. Of course, we in Canada are governed to a greater degree by municipal inspection than is the case in the United States where a very considerable portion of the inspection work is performed by fire insurance interests, therefore a municipal inspector is bound to recognize the life risk and deal with it. This very fact has brought about many of the difficulties that your editorial touched upon, and until these safety to life views are standardized, this difficulty will continue to present itself and variations in rulings will naturally increase.

Yours truly,
F. A. Cambridge
City Electrician

## Revision of the Custom's Classification of Electrical Apparatus

Mr. R. H. Coats, chief statistician for the Dominion Government, has announced a revision of the Customs statistical classification of electrical apparatus and equipment, which became effective on April 1, this year. Mr. Coats' attitude has been entirely sympathetic toward the requests that have been made to him from time to time by the electrical industry to further segregate the various items of electrical apparatus and materials imported into Canada,

with the result that the present classification represents a considerable extension over previous years. Mr. Coats is also good enough to say that during the present year proper care will be taken by the authorities at Ottawa to see that other electrical apparatus which is still classified under the "not otherwise provided" heading, will be closely checked and that the principal items under this classification will be picked out, and those found to be of sufficient importance and volume will next year be lifted from the "N. O. P." list and enumerated separately.

Now that Mr. Coats has shown his interest in the electrical industry by meeting us at least half way, it is surely the least we can do to co-operate with him in every way, to the end that returns may be made as promptly, definitely and intelligently as possible. For the benefit of our readers who have not yet seen an extract of this official classification, but who may wish to make a preliminary study of it, we print the items below:

### Customs Statistical Classification—Imports— Effective April 1st, 1922.

Poles, telegraph and telephone ............ No.
Wrought or seamless tubing, iron or steel, plain or galvanized, threaded and coupled or not, 4 inches or less in diameter, n.o.p. .. $
Wire rope, stranded or twisted wire, clothes lines, picture or other twisted wire, and wire cables, n.o.p. ............................. $
Wire of iron and steel, all kinds, n.o.p. ..... Lb.
Electric vacuum cleaners .................. No.
Washing machines, domestic .............. No.
Cyclometers, pedometers and speedometers . $
Dental supplies, electric .................. No.
Steam shovels & electric shovels .......... No.
Switches, frogs, crossings and intersections for railways ............................. $
Wire, plain, brass ...................... Lb.
Copper in bars or rods, when imported by manufacturers of trolley, telegraph and telephone wires, electric wires and electric cables, for use only in the manufacture of such articles in their own factories ...........Cwt.
Copper wire, single or several, covered with cotton, linen, silk, rubber or other material, including cable so covered ................ $
Copper wire, plain, tinned or plated ........ Lb.
Time recorders and parts ................. $
Electric batteries, primary ................ $
Electric batteries, storage ................ No.
Electric heating and cooking apparatus .... $
Electric dynamos and generators .......... $
Electric fans ............................ No.
Electric fuses, fuse plugs and cutouts ...... $
Lamps, electric, arc ..................... $
Lamps, electric, incandescent bulbs ........ No.
Electric light fixtures or parts thereof, of metal .................................. $
Lightning arresters, choke coils, reactors and other protective devices .............. $
Electric meters .......................... $
Electric motors .......................... $
Rheostats, controllers and other starting & controlling devices ...................... $
Self-contained lighting outfits ............. $
Sockets, outlets & receptacles ............. $
Spark plugs, magnetos & other ignition apparatus ............................... $
Switches, switchboards, circuit breakers and parts ............................... $
Telegraph instruments, including wireless apparatus ............................... $
Telephone instruments ................... $
Transformers ........................... $
Electric apparatus, n.o.p. ................ $
Bells & gongs, n.o.p. .................... $
Wire of all kinds, n.o.p. ................. $
Drain pipes, sewer pipes & earthenware fittings therefor, chimney linings or vents, chimney tops and inverted blocks, glazed or unglazed ............................. $
Electric light carbons & carbon points, of all kinds, n.o.p. ........................ $
Insulators, electric ...................... $
Shades, lamp, n.o.p. .................... $

# The Problems of the Industrial Engineer

### A Series of Short Articles on the Everyday Questions That Arise—
### The First Article Entitled "A Morning With the General Manager"

By GEO. L. MACLEAN, E.E.
Electrical Engineer Canadian Connecticut Cotton Mills Ltd., Sherbrooke, Que.

[The industrial engineer has problems of his own, often entirely different from those of the central station engineer. In a word, one is dealing with the economical manufacture of electricity, the other with its economical application. One is selling, the other is buying; and so their problems are related to one another in much the same way as are the problems of any other buyer and seller who must bargain together.

All too rarely is the electrical-industry inclined to view matters from the industrial engineer's point of view. This, doubtless, is because we have under-estimated the importance of the position he occupies. Today he is not only responsible for the utilization of a very large percentage of the electrical energy generated, but he is also a heavy purchaser of equipment.

A problem very keenly affecting any industry that is operated electrically is the power-factor of the load. Many central stations penalize the consumer for low power-factor; it, therefore, becomes the industrial engineer's constant pursuit to maintain the power-factor of his load above the penalty point. This is not easy to do. Nor is it always easy to convince the management of the necessity of doing it, and of making the necessary outlay of capital to ensure a better power-factor.

We have been fortunate in obtaining the consent of the electrical engineer of the Canadian-Connecticut Cotton Mills, Ltd., Mr. Geo. L. Maclean, to write us a series of short articles on some of his problems and how they are handled in his organization. The first article is printed below and deals in a general way, with this question. This first article is entitled "A Morning with the General Manager." It will be followed by others dealing more specifically with various operating problems in Mr. Maclean's work—Editor.]

### A Morning With the General Manager

The G. M. came into his office one morning feeling a bit ill-tempered and desirous of an excuse to jack up someone. While looking over some bills that were O.K.'d for payment, he came across the power company's bill for the previous month with the electrical engineer's O.K. at the bottom. After looking over the bill and doing a little figuring he came to, the conclusion that all was not well on the Potomac and that his appetite for trouble was to be satisfied.

He called the engineer into his office, but, fortunately for the engineer, he had been there before and was therefore prepared for almost any kind of an argument. After a very curt "Good-morning, have a seat", the G. M. handed him the last month's power bill. "Read it over" said he, "and tell us what is the meaning of this." The engineer read as follows:—

Present reading 1,000
Previous reading 500
Kw.h. consumed 500 x 10 x 200 = 1,000,000 kw.h.
Corrected to 85% P.F. = 1,000,000 x (85 + 75) = 1,133,333 kw.h.
1,133,333 x 6 mills = $6,799.99

"Why, what's the matter?" enquired the engineer "The figures are O.K. and we used the power." "I understand that all right, but why add that 13 1/3%, why multiply by 85 and divide by 75? Don't we pay enough for the electrical

end of this outfit without increasing our power bill to suit the power company? Our electrical department costs more in proportion than any department here, besides being non-productive." This riled the engineer, who answered, "Not by any means. Your electrical department carries as much responsibility as any department in the plant. It is that department which keeps up your repairs and keeps the wheels turning. Summing it up, you are keeping your electrical department to maintain the electrical equipment and keep your plant running. In other words, it acts as an active insurance policy.

"Now I will try to explain why you pay 85/75 of your power consumption. 85 is a figure agreed upon between the power company and the consumer as unity, 75% is the power-factor of your plant." "That means nothing to me", said the G. M. "You are going a long way around to tell me something or nothing." "That may be as you see it", replied the engineer, "but, on the other hand, if I should start using technical terms of electricity with which you are not familiar, you will be farther from understading the why of this question when I am through explaining than you are at the present time, so please let me explain in a way which I am sure you will easily understand. 75% is the power-factor found to exist at our plant here." "Average power-factor?" "No, there is no such thing as average power-factor. Power-factor is a ratio; an average is a quantity; but the power company takes a test once a month and we mutually agree to abide by that test to give us our period power-factor for the month. It would take up too much of your time for me to explain why we cannot arrive at an average power-factor, and why the method we use is the best.

"Let us look at the power-factor of our plant in the light of the following illustration. Two men are pushing a truck A, the men being represented by B and C.

Fig. 1

B exerts a force of 100 pounds, C the same, the resultant of these two forces acting in the direction of D is 200 pounds. This is 100% power-factor. But suppose some obstacle causes C to move away from B as shown in figure 2. If C moves away 30 degrees away from B then the resultant of the combined forces of B and C is 200 x .9659 (which is the cosine of half the angle of 30 degrees) or 193 pounds. We arrive at this figure by using the rule that the resultant of two forces at angles to each other is the sum of the two forces multiplied by the cosine of half the angle.

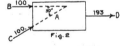

Fig. 2

"In Fig. 1 we have our voltage and current both working together, but in Fig. 2 we have put in the way of

C, which we will use to represent voltage, an obstacle called reactance, this being due to many small motors in our plant and underloaded large motors, so that C cannot push as much as B. The reason why we pay the difference between 75 and 85 is that in order for the power to reach our motors and supply the right amount of energy, the power company has to impress enough power on the line to overcome this obstacle we have imposed upon them.

"This point may also be illustrated in another way. Suppose we buy water at 110 pounds pressure and we send back pressure of 10 pounds, the water will then have to be raised to 120 in order for us to receive 110 pounds, so we have to pay for this extra 10 pounds back pressure we are imposing."

The G. M. had been following this explanation very closely and said, "Well, I guess I see it more clearly now. We cannot expect the power company to give us power when it is their business to make it for sale. But, just a minute, why do they multiply by 200?" The engineer replied, "We are using a 500 volt 5 ampere meter and in connection have to use 1,000 ampere current transformers. In this way only 1/200 part of the actual current flows through the meter." "Is there no way to correct this power-factor condition?" enquired the G. M. "Yes, there is, but as many conditions must be considered, there is hardly time this morning to take up that question."

As he left the office, the G. M. thanked the engineer for his explanation and handed him a cigar.

## Price of Copper at About Lowest Point in History

A most interesting review of the production and the maximum and minimum price of copper from the year 1860 to the present time is made by the Associated General Contractors of America, in a recent publication. The information is also shown graphically. It is noticeable that the price of copper, at the moment, is at almost the lowest ebb in its history. Note, also, that the reduction in price from the high point of 1917 is more drastic than almost any other product entering into our industrial life. At the present time, copper is shown to be slightly lower than before the war, and to those who are in the market this curve would seem to indicate that the present time is opportune for purchasing. The price curve indicates, in a general way,

that from 1860 to about 1896 the tendency was slightly downwards, but that from 1896 to the present time the tendency is slightly upwards.

The columns also indicate in a very interesting way the production of copper over the same period. Up to about 1880 copper had not attained any great prominence in the commercial world. The increased activity at that time plainly coincides with the birth of the electrical industry. During the following 35 years, as the electrical industry developed the demand for copper increased enormously. The falling off since 1918 is, doubtless, accounted for very largely by the fact that the electrical industry had over-built in the tremendous effort it had made during the war. In the last three years the production of copper has been marking time, so to speak, until the demand should catch up with the over capacity for supply.

We understand that already a noticeable increase in production has taken place, which is noteworthy as being one of the indications that the post-war readjustment, at least as regards this article, is pretty well completed.

## The Possibility and Practicability of Milton-Volt Transmission
### By C. C. CHESNEY

As is well known to electrical engineers, the principal factor limiting the maximum value of voltages for transmission purposes was that of line iinsulation. The transformer designer and builder was ever in advance of the line constructor. This is possibly due to some extent to a less difficult problem, but was principally due to the better facilities of the transformer designer to obtain at less expense by experimentation, practical working data on which to base his designs. Two hundred and fifty thousand volt transformers of substantial size have been available for experimental purposes for the past 15 years. About 10 years ago, Mr. F. W. Peek, based on data obtained from 250,000 volt tests, established certain laws governing visual corono and corona losses on transmission lines. These laws were believed to hold good for much higher voltages than they were experimentally determined. During 1921, transformers and other apparatus had been designed and built for transmission plants requiring 220,000 volts. It was, therefore, now important to know if the laws

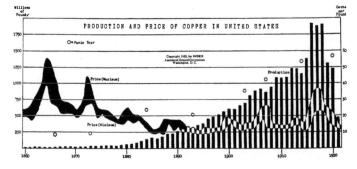

PRODUCTION AND PRICE OF COPPER IN UNITED STATES

experimentally determined at 250,000 volts would hold at much higher voltages. The purpose then, of 1,000,000 volt experiments carried on in Pittsfield during the past summer was to determine:

1. If the spark-over curves which were determined experimentally up to and including 500,000 volts, held for higher voltages.

2. If the laws developed from data obtained from 250,000 volt tests governing visual corona and corona losses on transmission lines held for voltages up to and including 1,000,000 volts.

3. To check the line insulator spark-over curve for the standard insulators and to extend it to cover the higher voltages.

The tests were made in the high tension engineering laboratory of the Pittsfield works, and were under the direct supervision of F. W. Peek, Jr.

The million volt tests were made at 60 cycles; electromotive force wave generated by the generator was approximately a sine wave. Transformers were 1,000 kv.a. in capacity and of the standard type. Our engineers estimate that the maximum error at the very high voltages in those results is not over five per cent. Tests were made on the typical electrodes that occur on transmission lines and otherwise in practice. The needle gap requires a minimum voltage for spark-over for any given sparking distance. This was well known and is due to the non-uniform field caused by the point electrodes. The sphere gap on the other hand represents as nearly uniform field as generally occurs in practice, and requires a maximum spark-over voltage for a given spacing or gap.

Visual corona tests were made on one inch, one and three-quarter inch, and three and one-half inch diameter parallel brass tube at various spacing. These tests show that the corona starting point for these very high voltages can be calculated from the laws established at the lower voltages.

The line insulator spark-over curve was found to be continuous at the high voltage.

The first 220,000 volt transformer built for transmission purposes, was built for the Big Creek Power Company, California, and has been in operation on the lines of this company for two months at 150,000 volts.

Assuming that the industrial and economical conditions would warrant the building of a million volt transmission line, these conditions are of course, extremely problematical. First, the low voltage power house would present no particular complication except that it must have the ability to handle, regulate and control the large amount of energy necessary to make such a line economical. The transformer does not offer any great difficulty in design and construction, and the design would probably be along the line of present-day practice. The bushing is probably the most difficult part of the transformer.

Third, the conductor as determined by Messrs. Faccioli and Peek, for a critical voltage of a million volts would be five inches in diameter for a minimum corona loss. Allowing for a ten per cent fluctuation of line voltage the conductor should be six and one-half inches in diameter; conductor would be hollow tube or some equivalent construction.

Four, the line insulator as has always been the case, presents the great difficulty, and from our tests should be 15 feet to 20 feet in length. From this data, therefore, you have a rough picture of the line. A six and one-half inch hollow conductor suspended by insulators 20 feet in length, and conductors spaced at least 20 feet apart. This is not an impossible engineering proposition and I am inclined to believe that it is no more formidable than

the 100,000 volt line appeared to be to the electrical engineers of twenty-five years ago. I believe that the limitations are more mechanical than electrical.

## Everybody "Listening In"

British Columbia has gone in for radiophony with true western thoroughness. In the language of a once popular, now almost forgotten rag-time melody, everybody is radiophoning now. The daily newspapers are giving publicity to the idea, which was needful to get the people into the habit of "listening in." Each of the three Vancouver dailies installed radiophone service and are now giving it a great deal of front page prominence. Attractive radiophone service programmes are billed each day and evening and sent out by each of the dailies from powerful central transmission stations. The programmes include a news bulletin of world and local daily events, short addresses from public men on important questions, and a varied musical programme. The daily papers have not only furnished lavish instructions as to how to share in the enjoyment of the radiophone service, but have as well urged their readers to install the necessary apparatus. In addition to that, reports from those having sets installed are asked for and published daily, showing the wide diffusion of the new service and the thorough practicability of this means of "broadcasting" news.

So popular have the "radiophone parties" become in the coast city, that there is a rush for placing orders to have sets installed and the necessary wiring put in houses. Hundreds of radiophone sets of all sizes and capacity from those with a 15-mile radius up to the powerful set which can catch the waves generated five hundred or a thousand miles away, are now on order and many of them are being rushed forward by express from the manufacturers in eastern centres. People are coming into the contractor-dealers, urging them to accept deposits on sets, in the hope of securing priority in having them installed. Naturally the electrical men are up on their toes to meet the demand and to take care of it. As one prominent contractor said: "This is the biggest thing that has happened in electrical circles for many a day. The industry has had something coming to it for a long time, which promised them a volume of business showing a reasonable profit. This wireless telephone development has furnished us our opportunity and it is up to us to take hold of it. My only hope is that the industry, through the manufacturers, can prevent it being "departmental-store-ized" or from falling into the hands of, say, the music trade, like the gramophone and the phonograph.

"The experience of cities in the eastern U. S. has already shown us the way, and as there is not the least question of the practicability of the system, all that remains for the electrical industry is to become thoroughly posted on all the latest appliances and methods, secure stocks promptly, and then go out after the trade that is now clamoring for wireless phones. There is not a home but will be a possible customer." said this contractor, "just as the gramophone and the phonograph went into the homes of the people."

Inquiry showed that for a set with fifteen mile radius—about the simplest form of wireless in use—the price is about $25. while for a set with 500 mile radius the price is about $110. The contractors are quoting for installation somewhat as follows:

Aerial, on top of ordinary dwelling .......... $10.00
For each outlet installed, including socket .... 5.00

These prices cover necessary wiring, and as the installation is all in multiple, but one pair of wires is necessary, even though outlets are installed in several rooms in a house.

# More Generators for Queenston-Chippawa Hydro-Electric Development

### Generator No. 1 was Described in our issue of April 1.
### No. 2, operating, and No. 3, being erected, are Identical
### with No. 1.   In the Article below we describe Nos. 4 & 5

Generators No. 4 and 5 which are being supplied by the Canadian General Electric Company, have been built at their Peterboro factory. They are 3 phase machines, each rated 16 pole, 45,000 kv.a., 187½ r.p.m., 12,000 volts, at a power-factor of 80%, current lagging. They are of the vertical shaft, revolving field type, with spring thrust bearing, upper and lower guide bearings, and with direct connected exciter.

Each generator is capable of carrying any momentary overloads to which it may be subjected, and is also mechanically capable of delivering the rated capacity of its turbine, namely 52,500 H.P. The normal speed is 187½ r.p.m., with a frequency of 25 cycles per second, and is designed for a run away speed of 347 r.p.m. It is designed to deliver its rated capacity at a normal potential of 12,000 volts, but provision is also made for carrying its rated full load amperes at any voltage up to 13,200, and it will operate under these conditions for several hours without exceeding a safe temperature limit.

The machines will operate continuously at rated load and power-factor, with an observable temperature rise not exceeding 65 degs. C. by embedded detectors. The units will meet the usual requirements with regard to wave form and parallel operation.

Each generator has been designed to withstand dead short circuits at normal voltage at the armature terminals. The leakage reactance drop with full current is 18% of the normal terminal voltage.

The efficiency of each generator at 45,000 kv.a., 80% p.f., 75 Deg. C., will not be less than 96.87%, and at 100% p. f., not less than 97.72%. These figures include all losses, including exciter losses.

For cooling purposes, free air at not more than 40 deg. C. is supplied. The temperature rise of the air in passing through the generator, when operating at rated load, is approximately 20 degs. C. The cooling air is taken into the generator at the bottom, and expelled through openings in the side of the stator frame. Openings are provided in the rotor spider rim between the sections, to allow air to pass through the space between the poles. A suitable blower is provided for exhausting the air from the chamber surrounding the stator, at a pressure equivalent to 2 in. of water.

### The Exciter

The direct connected exciter is an 8-pole, 150 kw., 187½ r.p.m., 250 volt. commutating pole machine, and has sufficient capacity for exciting the generator. when carrying a load of 45,000 kv.a., at 80% p.f., 13,200 volts.

The weights of the various parts of the generator are approximately :—

Rotor Total ......................600,000 lbs.
Stator complete ...................350,000
Base, Lower Bearing Bracket, upper
Bearing Bracket, Platforms, Bearings.
Bearing Housings, Thrust Collar and
Exciter ..........................250,000

Approximately       1,200,000 lbs.

The height of the machine from the bottom of the base to the top of the exciter bearing bracket is 28 feet 3 inches, and from the coupling face to the top of the exciter bearing bracket 33 feet 10 inches. The whole of this height does not project above the generator room floor. The height above this floor line is 13 feet 7 inches. This means that the whole of the generator stator is below the main power house floor line. The outside diameter of the stator is 24 feet 8 inches.

### The Stator

The stator, which is made in three (3) sections, has a series of radial ribs, and dovetails on the inner surface of these ribs are provided for receiving dovetail keys, on which the punchings are assembled. The punchings are built up in this vertical frame. with ventilating ducts at frequent intervals. A projecting ledge at the bottom of the armature frame is provided. which supports the armature end plates on which the punchings are built. On the top of the punchings another end plate is clamped down to hold the punchings tight. The stator frame is bolted t, the base, and provision is made for jacking the frame horizontally in all directions when setting up. The top end of the frame supports the bearing spider. The top bearing bracket is concentrically held in the frame and adjusts horizontally with it.

### The Rotor

The rotor is built up of seven (7) high grade cast steel spiders mounted on the shaft, one above the other, with their hubs contiguous. The spiders are mounted on the shaft with a press fit. A shoulder on the shaft below the spiders aligns them at the proper height. In addition to the friction of the shaft fit, heavy bolts extend through the hubs of all the spiders, holding them together. Ventilation holes between the sections are provided. On the outer periphery of the spider rims, dovetails for the poles are provided, and these slots are continuous from top to bottom.

The poles are built up of laminated sheets of steel with cast steel end plates, and the punchings are clamped by rivets runing from one end plate to the other.

The field coils are all of strip copper wound on edge. and are placed over the pole pieces, after the latter have been insulated, and before the poles are mounted in the dovetails in the rim. Heavy insulating collars are mounted at the ends of the coils.

The lower guide bearing bracket, which is of cast steel. is supported from the inside of the base. Both this bracket adjustment. and the spider adjustment. are preserved by dowel pins.

The armature coils are of the diamond type of coil. with two legs per slot. The leads are brought from the upper end of the coil where the connections between the coils are made. The coils are insulated with several thickness of mica between the turns and also on the outside. This mica is applied under the most approved processes which have been developed by this company. The coils are held in the slots by wooden wedges. and horn fibre casing is provided for mechanical protection. On the projecting end of the coil beyond the punchings, provision for tying the coil ends against strains and short circuits has been made.

### Bearings

The generators are equipped with spring thrust bearings. The latter are so designed as to carry the weight of the revolving portions of the generator, exciter and thrust bearing and the unbalanced downward thrust of the turbine, under all starting, running and stopping conditions met with in operation. The bearing is capable of sustaining and operating under a total combined weight placed on the bearing of approximately 1,000,000 lbs.

the losses and the cooling conditions. The pressure under operating conditions is within the standard allowance for these bearings of 300--400 lbs. per square inch. Copper water cooling coils are installed in the bearing housing, in order to cool the lubricating oil, thus reducing the amount of circulating oil to the minimum required to provide clean oil for the bearing. These cooling coils were given a hydrostatic pressure test. The thrust collar of the bearing is keyed to the shaft, and transmits the weight of the revolving parts to the rotating ring of the bearing. This rotating ring has a smooth rubbing surface and is so designed that a rapid circulation of oil is maintained. The stationary ring

As Generators No's 4 and 5 will appear when installed

The spring supported thrust bearing as used in these generators, and in other generators built by this company and other companies, has been developed in an effort to overcome certain difficulties experienced with other types of bearings. Of the many types in the market, the spring supported thrust is the only one which automatically adjusts itself, while in actual operation, if there is any loss of alignment due to settling of the foundations or other causes. The rubbing surfaces are in a bath of oil, and the quantity of oil circulating from the outside source, depends on

rests on springs and to insure flexibility it is made thin and has a radial saw-cut through one side. The upper surface of this ring is the stationary rubbing surface of the bearing. Sufficient springs are provided, so that the stress is only about one-half of that generally used, and the deflection under load is only about .1 of an inch. They are held in position by means of centre pins. They have been so designed that they are not stressed beyond 40% of the elastic limit of the steel. An oil well tube is provided around the shaft, which forms the inside annular wall of the oil chamber, and is welded to its supporting ring. Dowel pins are provided to keep the stationary ring from revolving. For cooling purposes, water at about 25 degs. C. is circulated through the cooling coils. When water cooling is used, 5 gallons of oil per minute are supplied to the thrust bearing, and as an additional precaution, arrangements are made to filter the oil, and cool it before recirculation. A recording thermometer is supplied, with a bulb located near the bearing in the oil.

The lower guide bearing shell is split to facilitate dismantling, and is supported in a removable spider. The opening in the lower guide bearing spider is large enough to allow the coupling on the shaft to pass through. The upper guide bearings are lined with babbitt. The usual oil deflectors are attached to the shaft to prevent oil leaking down the shaft. The drip oil from each bearing is caught in a cast iron drip pan attached to the bearing supports, from which the oil is drained by gravity. Each bearing has an independent oil feed pipe.

### The Shaft

The shaft is a hollow steel forging, with forged half-coupling, 30 feet 2¾ inches overall in length with a maxi-

mum diameter of 32 inches. A cylindrical hole, 8 in. diameter, extends the entire length of the shaft. A suitable groove
is turned in the part of the shaft above the rotor spider, for
attaching the lifting device, to enable the complete rotor to
be handled by the cranes. Shroudings are provided on the
coupling to cover the coupling bolt heads and to prevent
accidental contact with the projecting parts.

In order to investigate the temperature in the windings of the generator, copper constantan thermocouples are
provided, which are placed either between the upper and
lower coils or in the bottom of their respective slots. The
connections are made by copper constantan duplex wire, as
far as the cold junction box; from there to the recording instrument and the indicating instrument on the panel, the
connections are duplex copper wire. The indicating equip-

Armature Coil, generators 4 and 5, Queenston-Chippawa
Power Development

ment is a special design for this particular plant and is very
complete with indicating lamps and alarm contacts, besides
the instruments mentioned above.

With each generator is furnished a Tirrill voltage regulator, complete with line drop compensator.

In assembling the unit the waterwheel casing is put in
place and then grouted in. The foundations for the generator are then built up to the required height, for the assembling of the generator base. The generator proper is in a
pit. Four air inlets lead from an air supply duct through
the concrete foundations to an inlet chamber underneath
the generator. The air then passes through the generator
into the concrete pit in which the machine is placed; from
there it is collected by means of a fan, and passed into the
exhaust chamber and thence into the exhaust duct.

# Truss Arrangement of Long Span

By C. R. REID
Power House Superintendent, Shawinigan Water & Power Co.

The writer has recently had occasion to make some observations on a transmission line span which has some unusual
features. This span is 435 feet between supports and is
composed of six 500,000 circular mil aluminum conductors,
two per phase. The conductors are spaced horizontally one
foot between the two conductors of each phase and five feet
between the conductors of different phases.

The circuit of which this span is a part, is transmitting
19,000 kw. at 6,600 volts, with a current per phase of approximately 1,600 amps. The magnetic flux set up by this
current draws the two conductors of each phase together in
the central part of the span as can be observed in the
photographs. This results in a sort of truss arrangement
of the two conductors, which has a marked effect in cutting
down oscillation caused by wind. When the power is off
the line and the conductors fall apart, the oscillation is erratic
and of considerable magnitude, if there is a stiff breeze blow-

The two conductors of each phase are drawn together

ing. However, with power on the line, the oscillation is
not only greatly reduced but the period of oscillation is considerably extended. The accompanying view from a time
exposure made on a fairly windy day will give an idea of
the steadying effect of the truss arrangement of the conductors.

It is possible that in certain cases it would be desirable
to construct long spans in this truss form in order to secure
stability. A convenient arrangement would be to clamp the
two conductors together at the middle of the span and also
connect them by a jumper near each end.

# BETTER MERCHANDISING

## Electric Home Exhibition in Vancouver

### Demonstrating Every Form of Labor Saving Household Appliance

Demonstrating the electric idea in every form of labor-saving household appliance and convenience, the members of the electrical fraternity in Vancouver at the same time demonstrated to themselves the value of team-work, by their attractive electric home exhibit maintained during the two weeks of the Building and House Furnishing Exhibition from March 11th to 25th, inclusive. It was truly team-work, for every factor in the industry was represented. Central station, manufacturers, jobbers and the Vancouver Association of Electrical Contractors and Dealers united in the effort, which was successful in every way. The cash outlay to the individuals joining in the exhibit was small through the pooling of their resources. The exhibit, artistically arranged and decorated, showed interiors of three rooms, kitchen, dining-rom and bed room. In each, suitable furniture usually found in such rooms in the home; was installed through the courtesy of local manufacturers. Particular care was taken to secure proper illuminating effects and wiring for convenience outlets.

Inasmuch as the B. C. Electric Company, Canadian General Electric Co. Ltd., Canadian Westinghouse Ltd. and Northern Electric Co. Ltd., pooled their interests in this exhibit the appliances used were chosen by lot and all marks of identification on the appliances used were covered with signs which said for example 'Electric Range' rather than some particular make of electric range.

In the kitchen a range, water heater, ventilator fan, dishwasher. electric iron. cake mixer, bell transformer and electric fan were demonstrated. In the dining room a heater, vacuum cleaner, chafing dish, coffee percolator, tea- ball, tea pot and electric toaster were demonstrated. In the bed-

room a warming pad, milk warmer, vibrator, violet ray, curling iron heater and electric curling iron were demonstrated. In addition to this in the bedroom up-to-date wired furniture was a feature which attracted much attention. Bracket lamps were installed on the mirrors of the dressing table and bureau which matched the finish of the bedroom set.

Each convenience outlet was placarded. Those convenience outlets installed waist high, or higher in the walls had a card directly above the outlet. Signs on the walls with ribbon running to the baseboard outlets explained the use and purpose of each particular outlet installed in the baseboard.

Another feature which attracted considerable attention was the illuminated street number on the outside pillars of the booth.

The cost of operation of each appliance was set forth on small cards placed on or near the appliance in question.

The demonstrators were also handled in a co-operative way. The Show ran from 11 o'clock in the morning until 11 o'clock at night and it was necessary to employ one man who served from 11 o'clock in the morning until 2 o'clock in the afternoon. From 2 o'clock until 11 o'clock at night two shifts of three men each handled the exhibit. The demonstrators were furnished equally by the manufacturers, power company, and contractor dealers represented. so that a competent demonstrator was always available in each room of the exhibit to explain the proper wiring and illumination as well as the use of the appliances.

As a matter of fact the proper wiring for convenience outlets and proper illumination was stressed in the exhibit and the demonstrating of the appliances was merely incidental. As a direct result of the exhibit a local manufacturer of furniture proposes to wire a high grade line of furniture for the use of appliances. This will benefit the industry at large very greatly as this particular firm is the largest furniture manufacturer in the Province. Each piece of wired furniture sold will necessitate the installation of a convenience outlet; and convenience outlets once used in a home. sell themselves.

In addition to this, a large number of 'prospects' for

Electric Home Demonstration in Vancouver put on by Electrical Service League of B.C.

ranges and sundry small appliances, prospects for wiring for convenience outlets and for proper illumination have been obtained.

During the exhibition 10,000 pamphlets were distributed advertising the Service League as an information bureau for the prospective builder or prospective purchaser of any appliances. These pamphlets, in addition to offering the service of the Service League to the public, showed wiring plans for a two-story house based on the plans of the Toronto electrical home, a list of suggested appliances for each room in the house and a sound talk on proper wiring.

The central station is sending an illumination expert to all prospects who asked for information regarding proper illumination This man advises the type of fixture to obtain the desired illumination, but makes no effort to sell fixtures or the wiring necessary to accommodate the proposed change. When a prospect has expressed the desire to purchase fixtures as suggested the name and information is turned over to the contractor dealer who will finish up the job. If the prospect has a contractor dealer with whom he ordinarily does his electrical business, the prospect is turned over to that man, otherwise prospects are turned in in rotation to the members of the Contractor Dealers Association, regard being taken to their equipment, etc., enabling them to handle the prospective job.

Range 'prospects' are pooled and given to the manufacturers all at the same time. Prospects for particular appliances are turned over to the manufacturer furnishing that particular appliance after adequate information has been forwarded to the prospect from the office of the Service League.

The electrical home exhibit attracted more attention than any other single exhibit in the building and in the opinion of the Advisory Council of the Service League was a success in every way. It was particularly good propaganda preceding the erection of the electrical home which is contemplated in the near future.

# The Price-Cutting Jobber
## By C. D. HENDERSON*

The contractor-dealer to-day is confronted with a serious problem in the question of whom to buy from. The depression which it is to be hoped we have passed through brings the question of business ethics very much to the fore. Naturally some manufacturers and jobbers at a time like this, in their ambition to procure business, will cut below the regular market prices and the dealer is tempted very strongly to favour these concerns with his orders.

The point is—shall the dealer who has been served in a satisfactory manner for years by one or two concerns transfer his business to another jobber simply because he is offered a cut price?

Looking at the matter from a cold blooded standpoint the dealer might say that owing to the fierce competition he is forced to meet it is necessary for him to take advantage of these cut prices because his competitor across the street will likely do so and thereby have an advantage over him.

On the other hand there is no doubt of the obligation the dealer owes to his regular jobber. When times were good and materials hard to procure this jobber no doubt stood by the dealer and gave him preference over the fly-by-night dealer who sent in his orders only when he could not procure the goods elsewhere. Almost any electrical dealer has recollections of the boom years of 1917-18-19 and 20 when at times it was almost impossible to keep an adequate stock and most of us had our favorite jobber who managed in many instances to ship us materials when the market was apparently cleaned of stock. These same jobbers extended other favours to their loyal dealers in a financial way, carrying many of them for months to enable them to complete large contracts when very often financial assistance could not be procured from the banks.

There is another angle to this question which it might be well for us to consider. The dealer who buys goods on a price basis only must realize that he will eventually establish that kind of a reputation among the salesmen calling on him and will thereby lose the helpful advice which they are frequently in a position to give him. They certainly will not go out of their way to offer him any special propositions that may come their way from time to time but will favour their friendly dealers who stick by them through good and bad periods. Furthermore during good times the legitimate jobber will not be anxious for the "close buyer's" trade, realizing that the minute some price cutter comes along he will lose him again.

The problem is a very difficult one and the purpose of this article is not to offer solution but rather to start some thinking as to where to draw the line between business ethics and economic necessity.

Some dealers believing there is such a thing as sentiment in business have put the matter up to their regular jobbers very frankly by stating that while they can buy cheaper from a competing jobber they have no desire to transfer their patronage elsewhere simply on a price basis. They suggest that he sell them a little lower than the market price but still at a figure above the special cut prices. In this way the extra cost above the special cut is borne in part by the seller and in part by the buyer but the jobber retains his regular customer although his profit is cut down and the dealer maintains his reputation for fair dealing and thereby the good will thus created is mutually beneficial. The writer does not advocate this practice becoming general as it in reality sanctions price cutting. I am merely stating that I know of cases where it has bridged a difficulty.

Most of us, whether we are jobbers, manufacturers or dealers realize that price cutting in any shape or form is a curse and apart from the opinion some jobbers may have as to the attitude of the average dealer on the matter I honestly believe that 95 per cent. of the contractor-dealers of this country would gladly welcome a standard price for each and every article they buy, no matter from what jobber or manufacturer it be purchased. Then salesmanship, quality and service would predominate every transaction. The illegitimate jobber would be eliminated and the dealer would save time and money in the long run.

*President Henderson Business Service, Ltd., Canadian Manager A. B. C. Co., Brantford, Ont.

---

**Kilowatt, the Servant**

Don't work until you're tired and hot;
Leave it to Mr. Kilowatt.
He'll wash your clothes and clean your floors,
Cook your meals and do your chores.
He never complains and never sleeps.
He doesn't even stop to eat.
For a cent or two he does the lot.
Perfect servant! Kilowatt!

# The Jobber Situation

### By M. K. PIKE
Sales Manager Northern Electric Co.; President Electric Supply Jobber's Ass'n

What service does the industry require of an electrical jobber and how are the other branches of the industry endeavoring to assist the jobber to provide this service?

This question merits the close attention of everyone interested in the manufacturing, merchandising, operating or contracting field, and justifies a close analysis of our industry as it is organized to-day, to see if the jobber is being allowed to develop as the industry develops, and to learn what is being done to strengthen or to weaken his position.

There can be no argument but that the natural economic movement of electrical supplies is from the manufacturer through the jobber to the contractor-dealer or industrial consumer, and, generally speaking, manufacturers are as ready to admit this as are the jobbers themselves. As a matter of fact the very seriousness of the present day situation is due to the over anxiety on the part of the manufacturers to extend jobbers' privileges to many job-lot jobbers, who do not, never have, and in most cases cannot hope to render a real jobbing service to either the supplier or the customers.

The services expected, in fact demanded, of the jobber to-day are varied, but all are natural functions of a merchandising organization. They include carrying a large and well assorted stock of regular moving lines, some items of which turn over rapidly, others more slowly; selling by salesmen solicitation over a territory adjacent to his warehouse; handling the credits of his customers; distributing suppliers' advertising matter and tieing up with national campaigns; and last but not least, offering a consulting service on the many technical problems that arise in the course of the customer's own work.

In other words, the jobber does for the manufacturer that which the manufacturer would have to do himself were no jobbers in existence. The fact that he does this work for many manufacturers, making their products easy of access to the customers in his territory, plus the technical consulting service he maintains, makes the jobber a most important connecting link in the chain of distribution.

Surely these services can more economically be performed by a well organized jobber, with an established merchandising policy, established credit policy and trained staff of salesmen and specialists, than by the job-lot man who only caters to the buyer with a few ill-assorted lines, and then only to the buyer who is easily reached and who requires little service. Such jobbers are not creating a demand. They render no real economic service to either the supplier or the customer, and are only attempting to skim the cream off such transactions as can be handled by job-lot organizations. They are depriving the industry to-day of that which is so essential to its progress, well organized jobbing facilities—by making it impossible for organized jobbers to operate at a profit and continue to give a real economic service to both the manufacturer and the customer.

The crux of the matter is that if that portion of our business which can be taken care of with only fast moving items in stock, with no real jobbing service and no technical problem to advise on, is to be split up between real jobbers and job-lot jobbers, the jobber must curtail on his expenses. Thereby, the class of business requiring real service, special materials, credit risks, and specialists' services must suffer, and the real jobber must develop into a job-lot jobber, otherwise he cannot hope to meet the type of competition being forced upon him by the manufacturer, who is, to-day, and has been during the past two or three years indiscriminately creating new jobbers from among jobbers' salesmen or from the contractor-dealer class.

It is undoubtedly true that a new jobber added to a manufacturer's list temporarily increases that manufacturer's sales (at least to the extent of his initial stock) but does he, as a rule, create any new demand, and serve to strengthen the chain of economic distribution? He certainly does not unless he is an experienced merchandiser, knowing the service he is expected to render and is prepared to economically perform that service.

But what other effect has this indiscriminate appointing of jobbers had on the old established jobbers other than eliminating a portion of their market? It has had the very serious effect of reducing their gross profit, for the very good reason that not being expected by either the manufacturer or the customer to render the same kind of service as has been expected of the old jobber, the new jobber is satisfied with less spread and the older jobber has been left with his customers who still require a real jobbing service but with not enough spread in his schedules to enable him to render that service.

There can be no question but that the jobber of to-day must keep his house in order and improve his merchandising methods if he is to survive. If the industry is to progress, as we all hope to see it, he must be encouraged to render a real service to the industry. He should not have the considerable percentage of his market that does not require the same type of service required by the other portion, taken away from him only to be catered to by one who specializes in a very few fast moving lines, knows nothing about the real service required and cares less, has no established policy and is in business to scalp a profit without rendering an economic service to the industry.

As jobbers, we ask the other branches of the industry to pause a moment and consider that it is your problem as well as our own. We only ask a chance to perform a real service to the industry and believe we have a clearly defined field of operation, and that our existence is dependent upon getting a square deal from these looking upon us to perform a service for them.

---

Arrangements are being made for a bumper meeting of electrical contractors and dealers in London, Ont., Monday evening, May 22. Mr. V. K. Stalford, who has been acting as organizer for the Ontario Association, has the matter in charge. Contractors and dealers in the cities of London and St. Thomas, and the surrounding district, are urged to make their plans ahead so that they may meet in London on the 22nd and discuss matters of mutual interest, including organization into a District, so as to tie in with the provincial association.

---

### COME TO DEALERS' MEETING IN KITCHENER

A special Dealers' meeting will be held in Kitchener on the evening of May 5. This is the Friday of the Electrical Show week, and dealers throughout Ontario are cordially invited to arrange their plans to be present on that occasion. Mr. K. A. McIntyre will address the meeting.

A Corner of the Dominion Power and Transmission Co's Show Rooms.

# Popularizing the Electrical Idea in the City of Hamilton

### Electrical Development League, Comprising a Small Group of Enthusiastic Merchandisers, Rolling up a Campaign that Reaches It's Objective on May 13 in the opening of an Electrical Home, Built for the Purpose and Equipped as Every Modern Home Should be

The opening of an Electric Home on May 13, in Hamilton, Ont., by the Hamilton Electrical Development League will represent the climax of a campaign which has been gradually increasing in interest for several months.

It was on November 1, 1921, that the Hamilton Electrical Development League was formed, consisting of two representatives from each branch of the different electrical organizations then existing. It includes manufacturers, central station men, contractors, dealers, wholesalers and engineers.

The League was formed for the following purposes and objects :—

"To carry on educative work and by various means to create in the public mind an increased interest in electrical appliances and in the many uses of electricity; to create a realization of the need for proper and adequate wiring so that Electrical Appliances may be used conveniently and an appreciation of the requirements of good illumination.

"To carry on similiar work among architects and builders, and to arouse all members of the electrical industry to greater progressiveness and increased individual effort in order that electrical comforts and conveniences may be brought to a greater number of people and that there may be created an increased demand for electrical energy, wiring, appliances and apparatus."

On October 1st the league started the publication of an Electrical Page each Saturday in each of the daily papers. The Electrical Page consisted of 60% paid advertising and 40% news copy, with a large heading "Do it Electrically."

Beginning November 1st they started a series of prize essay contests. The winners of these contests were published on the "Do it Electrically" page. The prizes consisted of $40.00 in cash, divided as follows: First Prize, $15.00; Second Prize, $10.00; Third Prize, $5.00; Five Prizes of $2.00 each.

There were no restrictions on the winning of prizes and with few exceptions the winners were never the same.

These essay contests were used to awaken the public interest as a preliminary to the Electrical Home exhibition. The Electrical pages were continued until March 1st; the essay contests were discontinued on the same date.

The League has received a great deal of assistance in their publicity work from the manufacturers, central stations, jobbers, and contractor-dealers. Some of the essays obtained in the prize essay contests were very interesting and brought out several new suggestions for sales help.

Owing to the great amount of space being used by the industry on the Electrical Pages the League was able to decrease the amount of the budget from $10,000.00 to $5,000.00

During the Christmas holidays they distributed a very attractive poster warning the public against the danger of using candles for decorative purposes on Christmas trees, and recommending the use of electrical equipment for this purpose.

The League is now building the house for the Electrical Home. It is ideally located for this purpose on the main automobile thoroughfare through the centre of the city. There is a double street car line running past the house. They have a large illuminated sign on the property at present, informing the public of the purposes of the house. The builders advise that the Home will be completed, ready for the exhibition not later than May 10th.

# Hamilton Electric Merchants Have Some of the Finest
# Stores to be Found Anywhere

The "Hydro" is a big booster for doing it Electrically

In addition to his Association work "Joe" Cully is a real merchant

# Attractive Hamilton Store Interiors That Create the Desire to Purchase

Naturally, Harold W. Chadwick has a handsome fixture display

Fred Thornton is a booster for the Hamilton Electrical League

Hamilton League Enthusiasts gathering inspiration from the big Queenston generator

The opening date of the exhibition has been set for May 13th at 2 p.m. and at present the plans call for the exhibition to be for two weeks. The house will be fully furnished, decorated and equipped throughout, including a garage with a car, and a water softening equipment so that all water used on the premises will be chemically pure. The grounds and exterior of the house will be flood lighted.

The Hamilton Electrical Development League have carried on this campaign at the minimum expense, which has been due entirely to the splendid co-operative spirit in the industry locally, each one connected with the industry having contributed very liberally with time and effort to make the campaign a success. It is very encouraging to note the progress of this spirit of co-operation which will ultimately prove very beneficial to the industry in Hamilton. The most striking improvement is the better spirit of feeling amongst the different branches of the industry and the discontinuance of bargain sales. They have had some wonderful meetings, chock full of good fellowship, where formerly there was nothing but jealousy and knocking.

The Electrical News is authorized to say that a most cordial invitation is herewith extended by the Hamilton League to all members of the industry who may find themselves within travelling distance of Hamilton to visit the Electrical Home exhibit any time during the two weeks following the opening date, May 13. Mr. V. K. Stalford, the director of the campaign, or any of his lieutenants, will give you a hearty welcome.

At a recent meeting of the Ontario Contractors and Dealers the members were guests of the Canadian General Electric Co., at their Wallace Ave. factory and were right royally entertained

# Selling Suggestions for the Electrical Contractor and Dealer

### Sales Hints Applicable to a Business of any Size, in any Locality. Talk "Main Wires Big Enough", "Convenience Outlets" and "Electric Servants". Study the Lists Given Below Very Carefully.

Profiting by the experience gained in their first Electric Home, The Electric Home League of Toronto made certain changes in their policy of conducting their second Home, which opened on April 1 and remained open until the evening of April 22—three full weeks and four Saturdays. Continued interest on the part of the public was the determining factor in extending the period one week beyond the original period set.

One of the refinements in the organization in the second Home which has been very fruitful of results is a set of typewritten memorandum cards given to each demonstrator, covering the points it is most desirable he should bring out in explaining to the visitors the electrical features of the particular room he is in charge of at the moment. These memoranda have been prepared following the experience of the previous exhibition, and during their use have been found of great value in directing the demonstration along the proper lines and thus ensuring that it should not get out of hand, so to speak, at any point. In other words, the demonstration and talk are always given in the most concentrated form when the demonstrator follows the instructions given in his memorandum.

Electrical contractors and dealers everywhere will find the suggestions contained in these memoranda very interesting and valuable. They will not only be useful in any locality where an electric home exhibit is being put on, but each individual contractor will find the series of very material assistance in selling a better wiring job to a customer. The Toronto "League" does not offer them as perfection, but merely as something that they have found of great assistance in their own work. Suggestions as to additions or changes would be very gratefully received. The address of the secretary is 24 Adelaide St. W., Toronto. We print the data contained on these cards below.

Let us repeat that every contractor-dealer will almost certainly find something of interest and information in these memoranda. If they could be taken as a basis upon which wiring contracts are sold to the architects, or the house-owners, it would mean a tremendous improvement in the standard of our homes from the electrical point of view. Five per cent of the total cost is the ideal we must work to, and no contractor should feel satisfied with any contract he is installing unless it is meeting this standard.

### LIVING ROOM

1. Electric Home is shown by Electric Home league—a representative organization.

2. Entire electrical equipment—wiring, fixtures and appliances by **Electric Home League.**

3. Purpose — To show convenience of using electrical appliances in home properly wired.

4. Note the adequate number of **convenience outlets.**

5. So-called because of their convenience for connecting electrical appliances.

6. "Convenience outlets" are of the plug-in type—not old-fashioned screw type.

7. In building a new home, plan for sufficient number of "convenience outlets".

8. If your home is already built, they can be installed with little trouble and at nominal cost.

9. Electrical appliances selected by lot from various makers—all nameplates are removed—the appliances are displayed solely for the purpose of showing how easily they can be operated without unscrewing lamps.

10. Same provision can be applied to any house—large or small.

11. Phone—Stationary phone in kitchen. Portable desk phone can be plugged into "jacks" located in Living Room, Hall & Main Bedroom.

12. Proper lighting in the Home—have sufficient light—avoid glare. It is possible to obtain this result and still have attractive fixtures.

13. Wired tea wagon—with duplex convenience outlet and cord. Plug in anywhere. Use appliances right on the tea wagon.

In this room the well-located convenience outlets may be used for floor lamp, piano lamp, electric piano, electric cleaner, fan, tea samovar, kettle, toaster—**Make Main Wires Big Enough**—then you can have electric grate any time without re-wiring. Various styles and sizes.

### DINING ROOM

1. In the Dining Room you will want appropriate lighting. The candle effect with silver finish is quite suitable. Control switch at door.

2. Dining Room is the field of usefulness for electric appliances for the table.

3. In showing electric appliances in the Home we have endeavored to avoid extremes. We are showing only those of practical use in the home.

4. A word or two about the appliances themselves—
The Electric Waffle Iron makes deliciously crisp Southern Waffles;
In the Electric Egg Boiler, eggs are placed in the basket and thus readily handled;
For Sunday Evening the Chafing Dish prepares the Creamed Chicken;
The Electric Grill for the quick breakfast—especially if the man has to prepare it himself.

5. It is not expected that you will have all these appliances right away—you will have most of them some day. **Provide the wiring so you can use them conveniently.**

6. Three or four "convenience outlets" are located about the room so that no matter where you place the Buffet or Serving Table, appliances placed on them can be connected.

7. You can also plug in an electric heater in Spring or Fall and avoid starting the furnace too early.

8. It has been a problem to provide connections for the appliances used in the Dining Room. Cords from the fixture are unsightly and will soon break the fixture—cords from the wall you will trip over—every time.

9. The Dining Table itself is wired. One cord connects to a "convenience outlet" in the floor. The "convenience outlets" just around the edge of the table make it easy to connect the appliances. Wiring is flexible so table can be extended.

10. In similar manner it is possible to wire Buffet.

11. Floor push will call the maid from the kitchen.

## MAIN BASEMENT

1. Good lighting even in basement—enamelled reflectors direct light down where it is useful instead of letting it be wasted on ceiling.

2. Stairs well lighted— avoid falls—switch at top of stairs.

3. Electric water heater means safety and no fumes—hot water in 5 to 10 minutes—bath in ½ hour—switch in kitchen saves steps—comparatively little wiring required.

4. **Make main wires big enough.** So you can use electricity for heating water.

5. Distribution centre (fuse boxes). If your job is properly done, this part will cause you no worry.

6. Fuses are the safety valves of the electric system. If fuses blow, don't blame them for doing their duty. Have the trouble remedied.

7. Keep some spare fuses. Directory for circuits tells which one to replace, as well as the correct size.

8. Bell transformer operates bells from lighting circuit. No more batteries. Lasts a lifetime. Consumes little current and is very inexpensive.

9. We have low-priced power in Toronto. In a U. S. city of same size, power would cost 5 to 10 times as much as we pay. Yet this power is useless to you unless your home is wired properly so you can use it.

10. We have been advising you to **Make the Main Wires Big Enough.** How big? A pamphlet will be given out as you leave. It gives recommended wire sizes.

11. Wiring is usually left till very last for consideration. Then skimped in attempt to save. Often more is spent on "finish hardware"—door knobs and hinges—than on wiring.

12. **5% of your building cost will give you a good job.** Considerably less than cost of plumbing. A small price to pay for years of comfort and convenience.

13. Don't just look at the price—taking the lowest bidder—investigate—compare—see what you're getting for your money—**insist on having it done right** so that no part of it will ever have to be done over again.

## LAUNDRY

1. Electricity makes housework play. Especially in the laundry.

2. The washing and wringing are done by the electric washer leaving time free for other duties.

3. Usually best to provide outlet on ceiling for washer so cord is kept off the damp floor.

4. After washing comes ironing. The electric ironer nicely irons most of the work—the roll is turned by an electric motor—the shoe is heated electrically. The pieces are passed through thus (demonstrate). It is not even necessary to stand. A bench is usually provided.

5. For one or two pieces in a hurry or for very fine work, the electric hand iron is used for the kitchen.

6. Very little wiring inside the house is required for the electric ironer—but be sure to **Make the main wires big enough.**

7. Observe that the light in this room is just where it is needed and that the switch at the door turns it on ahead of you.

8. The cost of operating an Electric Home like this is VERY reasonable. The monthly bill for this Home would run between $5.00 and $10.00 depending upon the care with which the equipment might be used.

9. Electricity as a helper in your home WILL—

(a) Enable servants to do better work more quickly.

(b) Make it easier to secure good help because your home is electrically equipped.

10. In many homes, electric labor-saving appliances make

it possible to do without hired servants. Yet the housewife in such a home can do her work in a short time and with none of the old-time weariness.

## KITCHEN

1. Kitchen is last room you will see in the Electric Home—last but not least in importance.

2. Electricity promises brighter and happier hours in the kitchen.

3. Good lighting—centre light with switch at door—bracket over sink—no shadows—light outside with switch inside—you can see who is there before you open the door.

4. Wall Phone. Phone calls need not disturb your work—the wall phone is at hand.

5. Annunciator—indicates whether the call comes from the Dining Room or one of the outside doors.

6. Ample "convenience outlets" are provided.

7. Over the sink for exhaust fan (with switch).

8. In the baseboard for the electric dishwasher which also serves as kitchen table—no more backaches washing dishes—no more reddened hands.

9. For electric iron—in ironing board recess.

10. Duplex "convenience outlet" over shelf—for toaster, percolator or possibly grinding and polishing motor.

11. Heater switch—the heater in basement is operated from here—saving steps.

12. With an electric range—the cleanliness—comfort—safety and economy of electric cooking can be yours.

13. **Cost to operate** — A study recently made of 42 homes equipped with electric ranges, other appliances and lighting —for one year—average bill per month—only **$2.96.**

14. **Make the main wires big enough** so you too some day can save money.

## MAIN BED ROOM

1. Cheerfulness and comfort of this room is greatly increased by ample provision for electric service.

2. Duplex "convenience outlet" alongside bed—for boudoir lamp. electric cleaner, bed-warmer.

3. "Jack" for portable phone which at night can be brought up from downstairs.

4. In case of illness, lamp can be located under bed. Sufficient light for moving about room—no interference with sleep—same effect as aisles in movie theatres and Pullman cars.

5. If desired for burglar protection, a switch can be located at head of bed to turn on four lights outside doors: one at each corner of the house. Another switch can be placed in downstairs hall. Plenty of light scares away night-prowlers.

6. (South Wall) "Convenience outlet" for lamp on dresser or for portable electric heater.

7. (West Wall) "Convenience outlet" for **Wired Dressing Table,** hair dryer, massage vibrator, curling iron, portable lamp with counter-weight.

8. Good lighting—no glare, attractive fixtures—They are not disfigured by unsightly cords—"**Convenience outlets**" do **the work.**

9. Light in clothes-press. Should be a light in every one in the house.

10. Ceiling fixture controlled by switch at door. Place your switches close to doors and not behind them.

## GUEST'S ROOM

1. In the Electric Home we are not selling appliances. No quotations are given.

2. As you leave the Home you will be given a pamphlet on the last page of which is a list of contractor-dealers. Electrical information and advice will be cheerfully furnished by any of them.

3. "Convenience outlets" are also a feature of this—the guest's room.

4. (East Side) "Convenience outlet" for Violet Ray outfit or other appliances for the dresser.

5. The traveller's iron is almost indispensable for ladies away from home.

6. (South Side) "Convenience outlet" for portable heater or electric cleaner.

7. Light in clothes-press controlled by door-switch. Turned "on" and "off" with opening and closing door.

8. (South Bracket) Latest development for convenience of the public. No more bare wires sticking out. Able to buy your fixtures with a plug as you now buy an electric iron. Then bring it home—hang it and connect it at one operation. At last, you can hang a fixture like a picture.

9. Comfortable lighting in the room—appropriate fixtures—switch at the door.

10. (North Side) "Convenience outlet" alongside bed for bed-lamp or bed-warmer.

11. Periodically, the housewife wants to change the arrangement of furniture. The "convenience outlets" should be arranged with this in mind.

12. Don't trail a cord from one side of the room to the other. You may trip over it.

13. **Use "convenience outlets"—properly placed.**

### HALL AND BATH ROOM

1. "Convenience outlet" in hall for electric cleaner and electric floor polisher.

2. 3-way switches for hall lights up and downstairs.

3. Bathroom centre light controlled by switch in hall.

4. Good light for shaving by having bracket on each side of mirror. Only one light would throw the other side of face into shadow.

5. (Bath Room) "Convenience outlet" at convenient height can be used for immersion water heater, vibrator or hair dryer.

6. This is a very attractive bath room. People are learning that it is just as important to make sure of good electric wiring as it is to have good plumbing.

7. A good wiring installation can be provided for much less than the cost of the plumbing.

8. The day is not far distant when it will be very difficult to sell a house which is not properly wired.

### NURSERY

1. The nursery—a room full of delight for the kiddies.

2. Lighting—pleasant—dainty fixtures—switch at door.

3. (South Wall) "Convenience outlet" for bed-warmer or portable heater.

4. (West Wall) "Convenience outlet" for milk warmer.

5. Every "convenience outlet" in this Home is of generous capacity. This provides maximum flexibility in that you can plug any approved portable appliance into any outlet.

### SUN ROOM

1. Bright Sun Room—plenty of light—switch at door.

2. (South Wall) "Convenience outlet" for wicker table lamp and fan.

3. Electric sewing machine—pressure of the knee regulates the speed with exactness—fast or slow—the day of the tiring foot treadle is past.

4. (West Wall) "Convenience outlet" for sewing machine or portable heater.

5. You will observe that all "convenience outlets" in the Home are alike—they are standard.

6. Standard plugs have parallel blades—they are used throughout.

7. Use Duplex "Convenience outlets"—can connect two appliances at one time.

(NOTE: Show Duplex "Convenience outlet" with two standard plugs having parallel blades.)

#### 1st Floor Hall

(Given to upstairs party as they return to 1st floor Hall on their way to Dining Room. In rush hours—by traffic man. At other times—by man in Dining Room.

1. Verandah lights controlled by switch in hall.

2. "Convenience outlets" may be located on verandah for summer use—tea wagon—lamp—electric fan.

3. Hall lights upstairs and down controlled by 3-way switches.

4. "Jack" for portable phone.

5. "Convenience outlet" for torchere—electric cleaner and electric floor polisher.

### Annual Meeting Electrical Supply Manufacturers Association

The annual meeting of the Electrical Supply Manufacturers' Association, held on April 6, resulted in the following election of officers for the year 1922-23: President, F. Jno. Bell; vice-president, E. G. Mack, managing director Crouse-Hinds Company of Canada and Harvey Hubbell Company of Canada; treasurer, Geo. D. Leacock, sales manager Moloney Electric Company of Canada; general secretary, J. A. McKay. In addition, the Board of Directors includes. Messrs. H. S. Balhatchet, Benjamin Electric Company; G. W. Arnold, Boston Insulated Wire & Cable Company; A. A. McKenzie, National Carbon Company; J. J. Sorber, Economy Fuse Company; T. H. Porte, Renfrew Electric Products Company; N. S. Braden, Canadian Westinghouse Company; C. H. Keeling, Square D. Company; A. S. Edgar, Canadian General Electric Company and C. F. R. Jones, Northern Electric Company.

### Annual Meeting Electrical Supply Jobbers Association

The following officers were elected for the coming year at the annual meeting of the Electrical Supply Jobbers' Association, held in the King Edward Hotel, Toronto, on April 7: President, M. K. Pike, sales manager Northern Electric Company; vice-president, A. S. Edgar, manager Appliance Department, Canadian General Electric Co.; manager-secretary, J. A. McKay; chairman Central Division, C. A. McLean, Masco Company; chairman Eastern Division, R. J. Hiller, International Machinery & Supply Company.

### Demanded Electric Outlets Like Electric Home

Mr. R. E. Hughes, electrical contractor, of Burgessville, Ont., is in the midst of a very active season. He has just completed the wiring of a local church, a community hall, and a store. As an indication of the way the farmers are clamoring for electricity it is interesting also to note that he has just wired the buildings of three farmers in East Oxford and has 25 farmers on his waiting list. In addition to this, he is equipping six of these farms with an isolated lighting plant. Members of the industry will be interested to know, also, that two of Mr. Hughes' customers visited the Electric Home recently shown in Toronto and, on their return, demanded that their own homes be provided with many of the conveniences shown there. Mr. Hughes wires many farm homes for ranges and hot plates, installing sufficiently heavy service to take care of the iron, electric washing machine, motor for pumping water, and other modern conveniences.

# Successful Business Principles
# For Electrical Contractors and Dealers

### Showing How to Figure Overhead and Illustrating the Fundamental Principles of Better Merchandising

By LAWRENCE W. DAVIS
Special Representative of the National Association of Electrical Contractors and Dealers

It is supposed that every man who engages in the business of electrical contracting or merchandising has a definite purpose; that he is qualified to perform his part; that he is just and honest; and that he will render service in accordance with his ability. The four following elements constitute the Fundamentals of a Successful Electrical Contractor-Dealer.

1. Definite Purpose:—To make money; To serve well.

2. Character:—Aggressiveness; Intensity; Cheerfulness; Open mind; Initiative.

3. Co-operation:—Seeking the best from others; Giving the best to others; Support your national and local organization; Read your trade journal.

4. Service:—An obligation owed to yourself, to the industry, and to the public.

To gain the greatest results from such fundamentals requires organization. No one man can do it all; but he can be the leader—the active head.

#### Organization

Executive (Management):—Accounting; Financial relations; Stock; Selling; Service; Advertising.

#### Executive Management

Regardless of the size of a business the prime requisite for success lies in the complete control and knowledge of details of the business by the executive head. As the business grows to fair proportions it is essential that the executive have available at all times such information as will give him this complete knowledge without carrying the burden of the details in his head. This requires what may be properly considered the most vital organ in organization, accounting.

#### Accounting

He should provide knowledge of :—Purchases; Amount and Class of goods on hand; Cost; Overhead; Turnover.

Sales, Amount and Source:—Contracting Department; Store Sales; Fixtures; Repairs; Motors, etc.

#### Financial

Confidence of your creditors— jobbers and banks—is essential for the establishment and maintenance of credit. No other one thing can safeguard the credit of a business so thoroughly as to be able to present at the end of each month a statement showing that the executive manager has a complete knowledge and understanding of his business, even though the statement is not as strong financially as it should be. Credit is based upon confidence far more than upon financial rating.

#### Overhead

No other item affecting your business is so important as the absolute knowledge of the overhead cost of operating that business. No profit can be made upon any sale either over the counter or in construction work until the cost of making that sale, including its proportion of all the business expense, has been included in the selling price.

The following figures were taken from the result of a questionnaire submitted by the National Association of Electrical Contractors and Dealers to its 2,300 members throughout the country and represents an average, based on sales billed, which can be safely figured by contracting firms. With a strictly retail merchandising store this overhead amounts to considerably more, averaging about 30 per cent.

#### Average Overhead Percentages

| | Per cent. |
|---|---|
| Non-Productive Labor | 4.23 |
| Salaries | 8.48 |
| Rent | 1.51 |
| Light, Heat and Power | .34 |
| Postage, Tel. and Tel. | .43 |
| Advertising | 1.01 |
| Depreciation | .65 |
| Stationery and Printing | .41 |
| Incoming Freight and Express | .58 |
| Delivery Expense | .88 |
| Insurance | .62 |
| Taxes | .43 |
| Bad Debts and Allowances | .72 |
| Association Dues | .13 |
| Maintenance | .68 |
| Interest | .56 |
| Miscellaneous | 1.97 |
| | |
| Total | 23.63% |

Knowledge of your overhead in itself, however, is not sufficient. Probably no other one thing is more responsible for failures among electrical contractors than lack of knowledge of how to use overhead percentage and to apply it upon the selling price of contract work.

It must be always remembered that this percentage of overhead is the relation of the cost of operating business to the gross business done and therefore must be applied upon the selling price, and not upon the cost of a given piece of work. This may be compared to the discount off from the list price of merchandise.

For example, if the list price of a washing machine was $142 and the discount was 30 per cent, such a discount would amount to $42.60, making the cost to the dealer practically an even $100. In order, therefore, to return the selling price of the article to its listed figure we cannot add 30 per cent but rather 42 per cent to the cost price of $100.

The great mistake of the average contractor has been that after ascertaining his overhead percentage he has added it to the actual cost of a piece of work and then added to that whatever profit he has felt like obtaining upon the job.

As a matter of fact, in most cases this process, instead of showing him a net profit upon his work, has left him with nothing, or at best very little net results to show for his efforts, and we find all over the country contractors who have struggled for years and made little or no headway in developing themselves.

The accompanying table shows the proper percentages which must be added to the cost price of any job to show respective profits of from 2½ to 20 per cent with an overhead from 10 to 30 per cent.

Attention is called to the importance of studying carefully the bulletins and tables provided by the National Association of Electrical Contractors and Dealers covering this subject, as well as the many other valuable helps which members of that organization may see by studying and using the National Association's Data and Sales Book. It is urged that those who are not now members of the National Association should at once join and avail themselves of this information.

## Location of Business

The location of the contractor-dealer's business is an important factor and should be measured entirely from the standpoint of rent. Traffic, the opportunity to draw trade from the passing crowd, the nature of surrounding buildings and business, and the appearance of the store front are of greater importance than the amount of rent.

## Rent

Economy in the amount of rent paid may prove an asset or a liability, based upon the results obtained in the location selected. To secure a store in a cheap location which does not obtain local trade may make that store a liability; on the other hand, rental of many times that lower amount may prove a great asset if it is balanced by drawing features which give the store a merchandising turnover of sufficiently greater amount. Do not measure your rent in dollars and cents, but rather as a percentage of the gross business done by your store. The average rent of the contractor-dealers for 1920 was less than two per cent of the overhead.

Intensive use of:—Windows; Store arrangement; Salesmanship.

These factors determine the success of your location rather than the amount of rent.

## Stock

Choice of Merchandise: Careful consideration should be given to the quality of material or merchandise selected. These should be of the highest quality and when possible nationally advertised products backed by the reputation of thoroughly reliable concerns. See that it meets the needs of the class of trade catered to in that particular neighborhood.

## TABLE FOR FIGURING THE SELLING PRICE

| | | Overhead Percentages (Cost of Operating Business divided by Gross Sales) | | | | | | | | |
|---|---|---|---|---|---|---|---|---|---|---|
| | | 10% | 12½% | 15% | 17½% | 20% | 22½% | 25% | 27½% | 30% |
| | 2½% | 15 | 18 | 21 | 25 | 29 | 33 | 38 | 43 | 48 |
| | 5 % | 18 | 21 | 25 | 29 | 33 | 38 | 43 | 48 | 54 |
| Percentage | 7½% | 21 | 25 | 29 | 33 | 38 | 43 | 48 | 54 | 60 |
| of Net | 10 % | 25 | 29 | 33 | 38 | 43 | 48 | 54 | 60 | 67 |
| Profit | 12½% | 29 | 33 | 38 | 43 | 48 | 54 | 60 | 67 | 74 |
| Desired | 15 % | 33 | 38 | 43 | 48 | 54 | 60 | 67 | 74 | 82 |
| | 17½% | 38 | 43 | 48 | 54 | 60 | 67 | 74 | 82 | 90 |
| | 20 % | 43 | 48 | 54 | 60 | 67 | 74 | 82 | 90 | 100 |
| | | Percentage of mark-up to add to cost of labor and materials. | | | | | | | | |

Explanation:—If your cost of operating business (overhead) is 22½%, and you desire a net profit of 15%, add 60% to the cost of labor and materials.

| EXAMPLE: | | PROOF: | |
|---|---|---|---|
| Cost of labor and materials ............ | $100.00 | Labor and materials ........ $100.00=Cost | |
| Add 60% to cost ...................... | 60.00 | 22½% of $160.00 ........... | 36.00=Overhead |
| | ——— | 15% of $160.00 ............. | 24.00=Profit |
| Selling Price ...................... | $160.00 | Selling Price ........... $160.00 | |

### Buying

Two elements should be carefully considered in purchasing stock:

1. Quantities:—

Great care should be taken that the quantity purchased should be based upon the ability to dispose of it within a limited period, without being tempted to overpurchase for the sake of a seeming advantage in additional discounts.

2. Turnover:—

The advantage of a thirty-day turnover will offset even the additional discounts for quantities, when such quantity purchases require several months to make a complete turnover.

Buy of local jobbers. By so doing it is possible to keep your stock constantly filled and your turnover close to thirty days, and through the fact that less capital is required you have a greater surety of meeting all bills promptly and securing cash discounts.

Watch your stock. Prevent depreciation, waste, leakage and dead stock.

### Selling

Competent Sales People: Successful merchandising cannot be accomplished except through proper contact from the salesman or saleswoman who meets the customer. When the retail store is operated by a contractor engrossed in his contracting problems it is essential that the merchandising end of his business be handled by a competent assistant. Three adjectives may best describe the efficient sales person:—Aggressive; Attentive; Attractive.

Budget your sales plan:—Seasonable Sales; Special Sales.

Do not attempt to conduct your merchandising without a definite plan of action. Plan in advance each month, each week and each day what particular line of merchandise you are determined to push at that time. Stock allowed to remain dormant on the shelves will never sell itself profitably.

### House Canvassing

Electrical appliances have been so developed today as to have established their value beyond the experimental or introductory stage. So-called "free trials" in the homes are expensive and entail a wasteful burden of cost on the public and the dealer. Demonstration of such light appliances as a house canvasser may carry should be pushed, as well as canvassing for customers for the heavier appliances, but first demonstrations on the heavier appliances should be made in the store and such order forms adopted to be signed by prospective customers as will encourage only bona-fide purchasers, to the end that the practice of "free trials" may be discontinued.

Time payments should be limited to twelve months or less; first payments on appliance sales should be as large or larger than subsequent payments; and all time payments should have a carrying charge added to the cash price sufficient to cover interest charges and cost of collection.

### Store Selling

Store Selling:—Window displays; Store displays; Suggestions.

By suggestion is meant one of the most valuable points in merchandising, the calling attention of customers who are in your store to other lines of goods than that which they came in to buy.

All goods should be marked with uniform price tags, with all selling prices in plain figures. Price cutting and discounts under the list or recognized prices are undesirable and unprofitable, and should be used only as a last resort to move dead stock.

### Advertising and Display

Co-operative Advertising:—Creative; Educational.

Co-operative advertising has a distinctive mission in itself, which cannot be accomplished through individual advertising. It is designed to create new business and new fields for selling through the education of the public to higher standards in the use of electricity, and better understanding of the advantages to be derived from using electrical appliances:

Individual Advertising:—Newspapers; Letters and Inserts; Store Arrangement; Window Displays; Attention, Interest, Desire, Action.

These four elements must be present in any window display to make it have a drawing power that will produce sales. The window must attract the attention of passersby through movement, light or color; must interest them either through its attractiveness or its novelty; must make them desire to possess some article which they see in the windows; and, finally, if the window display be worth anything in dollars and cents it must induce them to enter the store and buy.

### Advertise Systematically

Budget your advertising:

Recommended for retail business of all kinds:—

3% to 5% of gross business.

Minimum recommended for contractor-dealers: 2% of gross business annually, divided as follows:

Annually—

Co-operative Advertising ........... ½%
Individual Advertising:
  Newspapers,
  Mailing list,
  Window displays,
  Posters and trim
     ......................... 1½%

The annual advertising plan should be divided into seasonable campaigns and the total fund budgeted to fit this plan. A $30,000 business should spend at least $1,000 in advertising annually, which may be divided into $250 for co-operative advertising and $750 for individual advertising. Window displays and store trim are included in individual advertising, and should have money and thought devoted to them.

# Annual Meetings and Other Activities of Contractors and Dealers' Associations

On May 1 the office of the Electrical Co-operative Association, Province of Quebec, will be moved to 61 McGill College Avenue, between Sherbrooke St. & Burnside Place, just above St. Catherine St., Montreal.

### French Section Elects Officers

The French Section of the Contractor-Dealers Association, at its annual meeting, elected the following officers and directors: N. Simoneau, president; W. Rochon, vice-president; J. N. Mochon, secretary; D. Vanasse, treasurer. Directors: J. O. Beaulieu, E. Hodge, M. Pelletier, L. E. Simoneau, O. Tardif, J. N. Tremblay and H. Truchon.

### Licensing of Quebec Contractors

A delegation representing both the French and English Sections of the Contractor-Dealers Association, Montreal, recently interviewed the Deputy Minister of Labour, Mr. Louis Guyon, and urged government action in the matter of putting into operation the Act respecting the licensing of electricians, and inspection of electrical work, in the province of Quebec. The Deputy Minister expressed sympathy with the desires of the deputation, and the Association is looking forward to action by the Government at a very early date.

### Mr. J. A. Anderson Again President

The English Section of the Contractor-Dealers Association, Montreal, held its annual meeting recently and elected the following executives: J. A. Anderson, president; E. J. Gunn, vice-president; Louis Kon, secretary-treasurer; J. M. Walkley, R. S. Muir, M. R. Henry, H. Vincent and C. E. Barrett. Messrs. St. Amour, Rochon and Pelletier are the delegates of the French Section on the Contractors' and Jobbers' Arrangement Committee; Messrs. Anderson, Walkley and Gunn represent the English Section. This committee maintains a very helpful point of contact between these different sections of the industry.

### Ontario Executive formally takes Control

A joint meeting of the provisional directors of the Ontario Association of Electrical Contractors & Dealers, and of the newly elected representatives to the Ontario Executive Committee was held in the Association secretary's office, 24 Adelaide St. W., Toronto, on April 5. At this meeting the affairs of the Ontario Association were formally turned over to the permanent executive committee which had been elected, and is constituted as follows : V. B. Dickeson, Windsor District; George Bremner, Kitchener District; W. W. Stuart, Guelph District; J. H. Sandham, Niagara Pena. District; W. G. Jack, Hamilton District; E. A. Drury and Harry G. Hicks, Toronto District.

Mr. Harry G. Hicks was elected chairman, Mr. J. A. McKay was secretary and treasurer, and Mr. F. W. Wegenast continued as counsel.

The next meeting of the Ontario Executive Committee was arranged to take place at 10 a.m. at the Royal Connaught Hotel in Hamilton, May 17th. It was decided to postpone the convention until Fall in order to give the new executive an opportunity to get into harness.

The Licensing Bill now before the Legislature was discussed in detail with the solicitor present and a list prepared of changes desired by the Association. Then the entire matter was placed in the hands of the solicitor for immediate action.

The New Executive Committee is composed of very able and representative men, and the members of the Ontario Association may expect an aggressive administration. Much of the delay and inconvenience recently has been due to the fact that the new machinery had to be started.

In resigning the Ontario chairmanship and passing on this important work to his successor, Mr. McIntyre spoke as follows : "I desire to thank my many friends for their hearty support during the past five and one-half years— to wish for the Ontario Association a future of great usefulness, and to wish for each of you many years of prosperity.

"Some districts are still unrepresented on the Ontario Executive Committee for the reason that there is not yet in those districts a sufficient membership to warrant organizing an official district. Will members in unorganized districts please bend their efforts toward securing more members so that you may, at an early date, secure your district organization with representation on the Executive Committee."

---

Kitchener Electrical Exhibition

May 1 to 6

---

### Hamilton Association Keeps Busy

A meeting of the Hamilton District Electrical Association was held Wednesday, March 29th, in the Board Room of the Y.M.C.A. Dinner was served at 6 P.M. This meeting was well attended by the electrical contractors, jobbers and manufacturers of Hamilton. A report was received from the executive committee on the Licensing of Electrical Contractors and Journeymen Electricians. The speaker of the evening was Mr. E. S. Jefferies, electrical engineer for the Steel Company of Canada, and president of the Hamilton Radio Society. Mr. Jefferies gave a very interesting and instructive talk in detail on radio apparatus and wireless telephony. He was assisted by Mr. George Crawford of the Hamilton Technical School. By means of a radio set a wireless talk was received from Dr. Culver of the Canadian Independent Telephone Co. at Toronto. From 9 P.M. until 9.45 P.M. wireless concerts were heard from Detroit, Chicago and Pittsburgh. At the close of the evening members were given an opportunity of listening to Trans-Atlantic messages and Washington Time Signals.

The Sun Electrical Company, Limited, Regina, Sask., have secured the electrical contract for the Prince Albert Jail. The net amount of the contract is $11,000.00, This building is being built by the Saskatchewan Provincial Government and the general contract is close to half a million dollars. The basement is now completed and the entire building is expected to be finished before winter.

Mr. D. A. Hills, 1039 College St., Toronto, has secured the electrical contract on a parish hall being erected at Avenue Road & Dupont St., Toronto, by the Church of the Messiah, at an estimated cost of $60,000.

A company to be known as the M. S. C. Radiophone Corporation of St. Thomas has been organized at St. Thomas, Ont., for the manufacture of radio telephones. It is understood that two types of equipment will be manufactured, a moderate priced one for the amateur and a higher priced one for commercial use.

## Miscellaneous Trade Notes

A. H. Winter-Joyner, Ltd., have moved their Montreal office from 701 New Birks Building to Room 606 in the same Building.

Messrs. Dawson & Co., Ltd., are moving from 148 McGill St., Montreal, to the ground floor of the Southam Building, 128 Bleury St.

The Perkins Electric Co., formerly at Phillips Place, Montreal, are moving to new offices and showrooms at 347 Bleury St., where they will carry a complete stock of fixtures and supplies.

The Associated Sangamo Electric Companies are distributing a little folder entitled "Generating electricity in 1870." The point is made that in short space of 50 years the electrical industry has developed from what was little more than an electro-plating operation into the second largest basic industry.

The annual report of the Walkerville Hydro-electric Commission, just presented by Mr. M. J. McHenry, manager of the Commission, to the town council, showed that the Walkerville System held third place in the province in the matter of total value of sales. Another interesting feature of this report was the fact that some 200 electric ranges had been installed in the homes of the citizens during the year 1921.

> ### Kitchener Electrical Exhibition
> ### May 1 to 6

## New Books

"Principles of Electrical Engineering," by William H. Timbie and Vannevar Bush, is the title of a recent publication by John Wiley & Sons, New York. This book is the outcome of experience in teaching engineering students and is designed to present the basic principles of electrical engineering in a rigorous but understandable way, problems and examples being freely used. The chapter headings will indicate the scope of the work:

1. The Electrical Engineer; 2. Electric Units and Electric Circuits; 3. Electric Power and Energy; 4. The Computation of Resistance; 5. Electrolytic Induction; 6 The Magnetic Circuit; 7. The Magnetic Field; 8 Induced Voltages; 9. Magnetic Properties of Iron and Steel; 10. Generated Voltage; 11. Force on a Conductor; 12. Conduction through Gases; 13. Dielectrics; and an appendix of useful tables. Well illustrated, 498 pages, 5¼ in. x 7⅝ in, brown cloth binding. Price, $1.00.

## Papers on Acoustics

"Collected Papers on Acoustics," Harvard University Press, Cambridge, Mass. This is a collection of papers written by the late W. C. Sabine, Professor of Mathematics and Natural Philosophy in Harvard University. The subjects covered are as follows: (1) Reverberation; (2) The Accuracy of Musical Taste in Regard to Architectural Acoustics. The Variation in Reverberation with Variation in Pitch; (3) Melody and the origin of the musical scale; (4) Effects of air currents and variation of temperature; (5) Sense of loudness; (6) The correction of acoustical difficulties; (7) Theatre acoustics; (8) Building material and musical pitch; (9) Architectural acoustics; (10) Insulation sound; (11) Whispering galleries: also an appendix dealing with the measurement of the intensity of sound and the correction of the room upon the sound. The book is printed on heavy stock, mat surface paper, and well illustrated. 280 pages 7½ in. by 10½ in. Stiff red cloth binding. Price $4.00.

## New Trade Publications

The Hydro-electric Power Commission of Ontario has issued Specification No. 16, which covers the construction of and tests on electric ranges. Copies of this Specification may be obtained by addressing the Commission at 8 Strachan Avenue, Toronto.

The Ohio Brass Company are distributing an illustrated folder, entitled "Sixty per cent. increase in trolley wire life."

The Taylor-Campbell Electric Co., of London, Ont., manufacturers of electrical supplies, have a catalogue, which they are distributing on request, covering their line of equipment, which includes outlet and service boxes, switch plates, cutouts, rosettes, etc.

Spielman Agencies, Montreal, are distributing a booklet entitled "Insulating Varnishes and Compounds and their Practical Application." This booklet is full of useful information and helpful hints, not only with regard to Griffiths Brothers' products, but also with regard to insulating materials in general and the protection and finishing of coils, electrical machinery, etc.

The Wagner Electric Manufacturing Co., St. Louis. Mo., are distributing an attractive bulletin, No. 130, entitled "Light for motion picture projection."

The Kingsbury Machine Works, Frankford, Philadelphia. Pa., are distributing Catalogue "C" on Kingsbury thrust bearings. This catalogue has been completely re-written and includes much new matter of considerable value to engineers in general. Particular attention is called to the following features: Standardization of dimensions; power loss guarantee; combined vertical thrust and journal bearings; the Queenston development; propeller thrust bearings; actual power loss determinations; standardization of power losses.

## News From the West

Mr. Clive Planta, formerly secretary of the Vancouver Electric Club, was the speaker at the weekly club luncheon on Friday, April 7th, on the occasion of his return to the Coast from a two or three month's sojourn in Toronto. He bore with him a fraternal greeting from the Toronto Electrical Club, which Secretary Lightbody was requested to acknowledge suitably. There were two leading topics which Mr. Planta dealt with in his address. The first was that of the Electric Homes' campaign. Mr. Planta had been given special attention by the Toronto committee, in view of the fact that he was anxious to tell the industry in Vancouver just how the excellent results had been obtained. He was shown over the second of the Electric Homes, and given full data as to details of preliminary arrangements, management of the exhibit, and methods of instructing the public who visited it. This information, noted carefully at the time of his visit, the speaker summarized and explained to the club members. It was a timely topic for Vancouver electrical people are very much interested in demonstrating electricity for the home.

Illuminating problems were dealt with in a highly instructive manner by Mr. Planta. Disclaiming any pretentions to be an "illuminating engineer," he gave some of the practical results worked out in many store premises in Toronto, where scientific lighting methods applied had increased the illuminating effect, at a reduced cost for light bills, and at the same time removed many of the defects from naked lighting. This, he said, was one of the livest subjects he had found in his experiences in the east, and he looked on it as one of the greatest fields for the practical electrical engineer to-day.

# The Latest Developments in Electrical Equipment

### A Pump That Handles Semi-solids

The choking of pumps operating on semi-solids is an unpleasant experience with which most engineers are familiar and the consequent cost of removing the obstruction, the burning out of motors in electrically driven pumps and the slipping and burning of belts in belt operated installations, are serious items which pump manufacturers have long been endeavoring to eliminate. The Unchokeable Pump, Limited, of London, England, whose Canadian agents are Jones & Glassco Reg'd., of Montreal and Toronto (Bank of Hamilton Building,) have made these problems their special study and as a result of many experi-

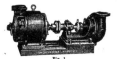

Fig. 1

ments have evolved a type of pump which they claim ensures that whatever is capable of passing through the suction must be swallowed and discharged, and handled in a better and more economical manner than could be the case when hand labor is employed. The "Unchokeable Pump" is constructed to withstand the work inseparable from the constant pumping of sludge, fibres, semi-solid and solid matter. In principle the pump operates upon ordinary centrifugal lines but in internal detail it differs from any other. No screens are necessary, for large objects that pass through the suction are ejected at the delivery. It is particularly designed to pass unscreened sewage, rags, cotton waste, fibre and all kinds of pulp and trade effluent

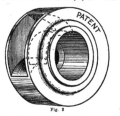

Fig. 2

from textile work in general. It has been found impossible to make the pump choke with solids or semi-solids, no matter how quickly or in what quantity they may be drawn into it.

Fig. 1 shows a pump complete with motor drive, and Fig. 2 shows the patent forms of impelling. The pump shown is motor driven with flexible couplings but belt drive and vertical pumps are also supplied. Parts are interchangeable and the impeller can be taken out and replaced. The bearings are ring oiled. Pumps of this type have been used for the handling of coal, gravel, rock ore, sand, etc., for leaves, sewage of chemical composition, and for fluids, whether containing acids or alkalis.

The Unchokeable Pump, Limited, is supplying 6 in.

vertical pumps, with special impellers, for sand and small stones, for the Mersey Docks and Harbor Board, England. These are to be used for throwing up excavated matter containing small stones to a height of 120 ft. These pumps if used to throw to a height of 60 ft., can handle stones up to 5 in. diameter.

### A Convenient Lamp Rack

The lamp-rack illustrated below is the property of the Sun Electrical Co. of Regina, Sask., and is one of the neatest display schemes seen in that province. This rack is suspended immediately above the lamp selling counter. The portion of the rack which holds the lamps may be easily revolved. The lowest lamp of the group is always lighted which is accomplished by a series of contacts within the bottom of the rack. Each lamp has a small price card stating the wattage, type and price. The front sign which reads "Edison Mazda Lamps" remains stationary and does

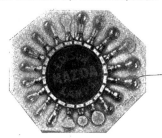

not revolve with the rack. The rack, aside from the sign and metal parts, is made of hardwood finished in mahogany to match the other fixtures. The company states, "We have found this rack extremely useful in selling lamps. We can serve a customer very rapidly as we can give instant comparison of all the popular sizes and types of lamps. We find with this rack that the sale of special kinds such as Daylite, bowl frosted or mill type lamps is greatly increased. We find that breakage is reduced to a minimum as there is no necessity to handle lamps except in testing and packing."

### Ibbotson Electric Co., Shoal Lake

Mr. A. R. Ibbotson, proprietor of the Ibbotson Electric Company, Shoal Lake, Man., sends us an interesting letter. Mr. Ibbotson was formerly superintendent of the town of Souris (Man.) electric light & power department, and for a number of years also maintained an electric shop in that town. Later he was superintendent of electric light in Shoal Lake, but since March, 1920, has been in business for himself installing electrical work, distributing supplies and fixtures, installing isolated electric plants, etc. He has two sons in the electrical business, the eldest of whom was overseas, in the Wireless Section of the Royal Canadian Flying Corps. Mr. Ibbotson writes that the municipality does not carry electrical supplies, which is another evidence that his company is supplying the demand for electrical appliances and equipment in a satisfactory manner.

### Flash Light Magic Lantern

The Fordell Projector Company, 317 East 34th St., New York City, has placed a flashlight Magic Lantern on the market. This device projects on a screen "views" from standard movie films; these being obtained, in a large assortment of views, by the manufacturers through an arrangement with a film distributing company. The apparatus con-

sists simply of a flashlight battery mounted to an ebonized baseboard. A reflector and lamp unit containing a small metal slide carrier and lens all contained in a metal projecting apparatus, combine to produce practically fifteen times the light usually obtained from an ordinary flashlight battery, according to the maker. This patented feature makes it possible through the pushing of a button to enlarge a 1-in. film view up to a 9-ft. picture.

### The "Sumbling" Washer

A new washing machine just placed on the market is the "Sumbling," manufactured at 7 St. Mary's St., Toronto. Machines are made in different sizes so as to include everything from the largest institution to the smallest home. A feature of these machines is a patented single belt reverse movement, which is said to ensure uniformity of rotation in both directions, reduces friction and wear to a minimum.

economizes in belting, etc. The claim is made that their smallest washing machine for semi-commercial work is the only machine in small compass that will consistently stand the strain of commercial washing. Special features named by the company are: fluid-tight joints, reversible wringer, special belt tightener and an oiling system that is easily accessible even while the machine is in motion.

### Folder on Shurvent Fuses

A thoroughly vented renewable fuse, recently developed, is discussed and illustrated in Folder 4472, entitled "Shurvent Protection," which has just been published by the Westinghouse Electric & Manufacturing Company. The folder explains the application and design of fuses for the protection of low-voltage circuits up to 600 volts for both alternating-current and direct-current.

### A Heater for Bathrooms

The Electric Bathroom Heater, here illustrated, is manufactured by the William H. Jackson Company, 2 West 47th St., New York. The register grill is of cast brass, but may be had in either nickel, silver or gold. The heater may

be had in various capacities, though generally supplied with a 1 kw. rating. It is placed in a recess in the tile of the bathroom wall, the illustration showing the neat effect produced. The snapswitch shown above controls this heater. All hazards due to accidental groundings appear to be removed by installing heaters in this way.

### They Appreciated Mr. Skelton

Mr. J. Skelton, the veteran salesman of the Hoover Suction Sweeper Company, withdrew from that company on April 1st, in order to devote his entire energy to the manufacture and sale of the floor waxer and polisher which he recently placed on the market. The occasion was made the "excuse" for one of those enjoyable "Get-together" suppers of Hoover retail men, to which they all look forward with real anticipation. Mr. Thos. Kelly, Canadian sales manager, presided. Next to the menu, the feature that pleased everybody most was the presentation of a very handsome table lamp to Mr. Skelton, as an appreciation of the very kindly feeling that had always existed between Mr. Skelton and the Hoover Company, and as a permanent reminder of their regret at losing him. Mr. Skelton, with his seventy-odd years—but looking only half of it— and who is generally referred to as "the grand young man of the electric sweeper industry"—made a very happy reply, in which he further emphasized the kindly relationship that had existed for a dozen years—and would always exist— between the officials of the Hoover Company and himself.

The dinner was held in the King Edward Hotel and was attended by some dozen or more of the company's agents, including Mr. R. N. Connor who had also been with the Hoover Company for some years but is now withdrawing to associate himself with Mr. Skelton in the manufacturing business.

## Motor driven air compressor for garage

For service in garages, tire repair shops, filling stations and similar places, a line of electrically operated air compressor outfits have been developed by The Auto Compressor Company, Wilmington, O. The outfits are furnished for operation on single or polyphase alternating currents, 60 cycles, 110 and 220 volts, and for 220, 110 and 32 volt d.c. circuits. The two-stage outfit, Model A-25, illustrated, has a low pressure cylinder of 2½" bore and a high pressure cylinder of 1-5/16" bore. The stroke is 2½". The compressor is air cooled, operates at 350 r.p.m., and has a displacement of 2.48 cubic feet per minute. The single stage outfit, Model E-68, has a compressor of 2" bore by 2½"

stroke. It operates at 550 r.p.m. and has a displacement of 2.5 cubic feet per minute. The tank for both outfits is 16" diameter by 48" and has a capacity of 38 gallons. An automatic diaphragm pressure switch cuts the motor in at a pressure of 120 pounds and cuts it out when the pressure reaches 150 pounds. The air is filtered to free it of oil and other impurities before it enters the tank. The two-stage outfits for single phase a.c. circuits are equipped with Robbins & Myers ¾ h.p. Type "R" repulsion-induction motors. Outfits are also furnished with polyphase induction motors and with direct current motors for polyphase and direct current circuits. The single-stage outfits are equipped with ½ h.p. motors. Each outfit is furnished complete with motor, tank, compressor, automatic switch, pressure gauge, filtering apparatus, belt, all connecting piping, and 25 feet of air hose with chuck. Either outfit can be installed in a floor area of 49½" x 20". The net weights are 450 and 380 pounds respectively for the two-stage and single-stage outfits.

## Dominion Engineering Works, Ltd., get large contracts

As noted briefly in our last issue, the Dominion Engineering Works, Limited of Montreal, Canada, have been awarded the contract by the Manitoba Power Co., for two 28,000 horse power vertical shaft, single diagonal-type runner, hydraulic turbines which will be installed in the Great Falls Plant during the latter part of the present year.

Although a young company—having been established in 1920—they have already attained an enviable position in the industrial field, more especially in the manufacture of large hydraulic turbines, Johnson valves, paper machines, etc., which, by reason of their size, are beyond the capacity of the average plant. Among the important contracts completed or under construction by this company may be mentioned : Two 11,2000 h.p. turbines for the Cedars Plant of the Montreal Light, Heat and Power Co., being the largest dimension turbines in the world; two 22,000 horse power turbines for the Laurentide Power Co.; one 41,000 h.p. turbine for the Shawinigan Water & Power Co., the cast steel scroll casing of which is the largest yet built in this or any other country; two 72 in. x 166 in., 1000 feet news machines for the Laurentide Paper Co.; one 20 ft. Johnson valve

for the Shawinigan Water & Power Co., etc. In addition to the larger units above mentioned, the company also specializes in units of smaller powers; among those already constructed may be mentioned : One 1500 horse power turbine for the Montreal Light, Heat & Power Co.; four 3900 horse power turbines for the Spruce Falls Co., Limited; one 3550 horse power, horizontal, double runner turbine for the International Nickel Co., of Canada, Limited.

Speed regulation is one of the most important factors in connection with hydraulic turbine operation for the generation of electrical energy, the exacting requirements of electrical engineers making it necessary to operate at practically constant speed. In order to accomplish this successfully, it has been found advisable to design a special type of governor, the manufacture of which requires mechanical skill of the highest order. Several governors of this manufacture are now in successful operation. One thus sees the wide range of work this company is capable of handling, varying as it does from the larger and more massive parts of turbines to the intricate and minute parts of a turbine governor and it is not difficult to realize that their capacity for the manufacture of all classes of work be it large or small, must be practically unlimited.

It is a tribute to the sagacity and foresight of the founders of this company that, within a period of three years, they have organized and equipped an industrial establishment capable of manufacturing the largest and most massive machinery yet developed. That they are an asset of nation-wide importance can readily be understood when it is mentioned that, previous to their inception, all machinery of the sizes above mentioned, had to be imported. They are licensees in the British Empire of the well known firm of W. Cramp & Sons Ship & Engine Building Co., of Philadelphia, Pa., whose improvements in the signs and manufacture of hydraulic turbines have been a contributing factor in the rapid development of this type of machinery. Standing as we are at the commencement of an era of intense development of our natural and other resources, it is fair to assume that the Dominion Engineering Works, Limited, by reason of their capacity to manufacture work of this magnitude, are destined to assist materially in this development.

The Canadian Line Materials, Ltd., Toronto, are distributing an attractive catalogue, well illustrated, describing their products, which include pine brackets, specialties, cross arms, street lighting fixtures, ground rods, anchor rods, anchors and racks.

---

## Midsummer Meeting in Niagara Falls

At a meeting of the Executive Committee of the A. M. E. U., held in Toronto on April 20, it was decided to hold the mid-summer convention, as usual, at the Clifton Hotel, Niagara Falls, Ont. The date was fixed as June 22, 23 and 24. A tentative program was arranged, which will include papers on street lighting, construction accounting, accident prevention and health promotion. A banquet will be held on the evening of the 22nd, to be followed by a radio demonstration. The program will, of course, include a trip over the Chippawa-Queenston development, on Saturday morning.

# Electric Railways

## Modern Street Car Lighting

### Proper Illumination Almost as Essential in Street Cars as in Merchandising Establishments, to Attract Business

#### BY J. A. SUMMERS*

Business men have found that light draws trade. This has been so definitely proven that few would attempt to contradict it. We see the results on every side; the crowd gathers where the bright lights are; they stop and examine the brightly lighted show window; they drift unconsciously to the brightly lighted theatre entrance or restaurant. If the street cars are made bright and attractive it is perfectly logical to assume that human nature will react in the same manner with regard to street cars as it does in other lines.

That the public does react in this manner was definitely shown recently in the subway here in New York. On some of the local trains alternate cars had about twenty-five per cent more light than the others and it was observed that the brighter cars were occupied before the others. Experiences of this kind indicate that it is entirely logical to apply merchandising methods to attract passengers.

I have heard some say that it is not necessary to attract passengers. This may be true in some cases, but in the large majority of cases the railway company is exerting every effort to attract the public to its cars.

There is severe competition in many places between the street cars and jitney buses and it behooves the traction company to make their cars as attractive as possible to hold their prestige. The statement that people have to ride anyhow regardless of conditions is not tenable. It is true that a great many are absolutely dependent on the trolleys to get to and from work, but the profits of the company depend largely on the number of people who can be induced to ride who are not forced to ride. Brightly lighted cars are without question a powerful aid in attracting these casual riders, and hence, the proper illumination of cars should receive very serious consideration.

It is unfortunate but nevertheless true, that even at the present time, illumination is often an afterthought or an annoying necessity to be disposed of as expeditiously as possible. Proper illumination cannot be provided by simply ordering several circuits of small lamps to be installed, and thus dispose of the matter without further thought or attention. I am glad to say that most of the large companies realize the importance of illumination and make very thorough investigations of lighting conditions, some of which have been reported in papers before the Illuminating Engineering Society.

There are a great many variables to be considered in planning the lighting of a car. The interior finish of the car, the voltage regulation, arrangements of the seats, and the standard of lighting in the community have a very decided bearing on the selection of the proper lighting equipment.

The interior finish makes a great difference in the re-

*A paper presented before the Illuminating Engineering Society.

sulting illumination. This is true regardless of the reflector equipment. A car with a dark finish is likely to appear dark even though the photometer shows that the illumination is adequate. The psychological effect of such a car is depressing and the passengers feel that there is insufficient illumination,—hence the unjust criticism. A gloomy car also suggests uncleanliness and frequent resentment is expressed by passengers because of being compelled to ride on "dirty, gloomy cars." This feeling does considerable harm to the railway company not only by causing the antagonism of the public but by repelling the short-haul passenger.

A flat white ceiling with light color woodwork is of course most desirable from an illumination standpoint, but a light tinted ceiling, which many companies prefer, is quite satisfactory. Such a finish will add 20 to 40 per cent to the illumination secured with the dark green ceiling so often seen, and naturally, the car will appear much brighter and more cheerful. If a gray tint is used, the paint should be mixed with vermillion and emerald green and reduced with white until the proper tint is secured. A gray that is mixed with lamp black has a much higher absorption than a warm gray mixed as above.

The shape of a ceiling also has a decided bearing on the illumination. The old ceiling with the deep half decks is rapidly being displaced with a curved ceiling extending in one sweep over the width of the car.

This is a much better arrangement from a lighting standpoint than the old one because the light is not bottled up in the narrow area between the half decks. This arrangement also provides better light on the advertising cards.

The amount of light necessary and the arrangement of the outlets are matters that have received considerable attention and aroused a great deal of discussion. It is a relatively simple matter to state the minimum requirements for safety and for reading without eyestrain, but this does not really satisfy the conditions from an economic standpoint. The standard of lighting in offices, stores and industrial plants has risen materially in the last few years and this naturally influences the public's demand for higher intensities in street cars.

An average of 3.5 to 4.0 ft-candles in a car was considered very good lighting several years ago, while now there is a strong tendency to go at least 10.0 ft-candles and even higher so as to conform to the new industrial standards. This is not an unreasonable intensity, and can easily be secured with proper reflectors with an expenditure of power of about two watts per square foot of floor area.

One must not forget the fact that the 10 ft-candles measured in an empty car at rest and with lamps operating at normal voltage will be reduced to not over 6.0 foot-candles when the car is crowded and the voltage lowered as much as it is on the majority of trolley lines. The cross reflection which adds up in the empty car is also entirely absorbed in a crowded car, and due allowance should be made for this fact when estimating the desired intensity.

A much higher intensity is necessary to read comfortably in a street car than in a home or office, because the vibration and jolts of a rapidly moving car cause a rapid movement of the reading matter which is hard to follow without great eyestrain under low intensities. Many tests on the speed of vision have shown that an intensity of 10 ft-candles is necessary to read rapidly changing print.

One row of lamps on the center line of the car is the most economical way of placing the units, and an illumination test in an empty car would show a satisfactory distribution of light. The results are also fairly satisfactory to the passengers if the seats are arranged across the car. In this case the standing passengers do not cast objectionable shadows on the seated passengers. However, if the seats are arranged longitudinally the results are often unsatisfactory whenever there are standing passengers. The arms and hands holding on to the overhead straps effectually bar the light from reaching the seated passengers. In the latter case, therefore, use should be made of two lines of units, placed just inside the edge of the seats. In all cases medium density opal reflectors should be used and in no case should they be spaced more than 5.0 ft. apart. If the units are placed under the half-deck it is of prime importance to use reflectors as a standing passenger is so close to the lamps that to him a bare lamp is exceedingly annoying. Bare lamps in this position also detract from the advertising value of the cards under the half-deck because of the glare.

Many accidents have occured to passengers alighting from street cars because of inadequate light at the steps. Persons are temporarily blinded when going from a brightly lighted platform to the dark steps and many serious falls have resulted. Several methods have been satisfactorily tried to solve this problem. One is to utilize one circuit just inside the door to operate when the door opens. Another method is to set a lamp in a recess at each end of the step, each lamp operating in series with the circuits inside the car. The latter method is the better one when the construction of the car permits its use.

The principal points to consider in designing the illumination of a car, either new or when remodeling, is the value of a light finish on the ceiling, the use of a proper reflector, spacing of the lamps not to exceed five feet, and an intensity high enough not only to make comfortable reading but to attract passengers to a well lighted cheerful car.

## An Ontario Act Respecting Construction of Municipal Radials

A Bill was recently introduced by Premier Drury in the Ontario House, entitled "An Act respecting the construction and operation of municipal electric railways."

The Bill provides that on the request, expressed by resolution, of the corporations of two or more municipalities, situate in any locality in which electrical power or energy may be supplied by the Hydro Commission, the Commission, as the agent of such corporations and at the expense of such corporations, may investigate the cost of constructing, and the advisability of constructing, the desired railway.

A section in the Bill enables such corporations to enter into an agreement with each other for the construction, equipment and operation of an electric railway.

The agreement must be submitted to a vote of the electors qualified to vote on money bylaws, the bylaw to recite the estimated cost of the work, the annual maintenance and sinking fund charges, the portion of costs to be borne by each corporation, and the estimated probable revenue.

The Bill also outlines the conditions under which money may be raised for the prosecution of this work, and for the appointment of an Association to care for the construction, equipment and operation of the railway.

In effect, the Act means that any group of municipalities may build radials if they are willing to undertake to bear the capital expenditure and the proper maintenance. The Hydro-electric Commission must approve of the work, and

power to operate the system must be purchased from the Commission. In short, it appears to be the wish of the Government to assist municipalities in every possible way to build hydro-radials where there is a reasonable guarantee that they will be a financial success. The Government, however, disclaims any responsibility whatever as regards either capital cost or maintenance.

## Electric Railway Publicity

Canadian electric railway men will be interested in the illustration herewith. It is one of the most attractive displays in the city of St. Louis, and is a continuous advertise-

Advertising Electric Railway Service

ment for the Illinois Traction Company. The semaphore raises, the headlight lights up, the wheels begin to turn, the track moves back and the car apparently makes headway.

The British Columbia Electric Railway Company recently placed all their street cars in Victoria, B. C., under one-man control.

A new trackless trolley bus has been delivered at Walkerville, Ont., and is to be used on Lincoln Road.

## Quebec R. L. H. & P. Annual

At the annual meeting of the Quebec Railway, Light Heat & Power Company, held recently, a number of changes were made in the Board of Directors, which is now composed as follows: E. A. Robert, Hon. Lorne C. Webster, J. N. Greenshields, H. G. Valiquette, Col. J. E. Hutcheson, K. B. Thornton, C. G. Greenshields, Hon. Geo. E. Amyot, Hon. Adelard Turgeon, Hon. D. O. Lesperance and A. C. Barker. At a directors' meeting, the following officers were elected: E. A. Robert, president; Hon. Lorne C. Webster, vice-president; W. J. Lynch, general manager; A. Lemoine, secretary; and R. A. Wilson, treasurer.

# Current News and Notes

**Brockville, Ont.**

To take care of the increased load brought about by the establishment of the Eugene F. Phillips 'Electrical Works at Brockville, Ont., the Hydro Commission are rebuilding their line between Brockville and Morrisburg, increasing the voltage from 26,500 to 44,000.

**Coleman, Alta.**

The town council of Coleman, Alta., contemplates the installation, in the near future, of an electric light and power service.

**Emerson, Man.**

The town of Emerson, Man., contemplates the installation of a new electric plant this spring.

**Fort Frances, Ont.**

A press report states that the International Radio Development Company has been formed and will instal a large radio station at Fort Frances, Ont.

**Hull, Que.**

Stanley Lewis, 63 Metcalfe St., Ottawa, Ont., has been awarded the electrical contract on a $50,000 apartment house being erected at Aylmer Road, Hull, by Mr. Geo. E. Hanson, 11 Front St., Hull.

Mr. E. Martel, Wright St., Hull, Que., has been awarded the contract for electrical work on an addition being built to the building at 131 Main St., Hull, owned by Mr. Ls. Bertrand.

**Joliette, Que.**

Mr. Fortunat Giroux, formerly superintendent of the municipal electric plant at St. Jerome, Que., left on May 1 to assume the position of electrical engineer & general manager of the municipal system at Joliette, Quebec.

**London, Ont.**

Messrs. R. H. Cunningham & Co., Walkerville, Ont., have secured temporary premises at 351 Glebe St., London, Ont., where the manufacture of electric furnaces will be carried on.

**Mimico, Ont.**

Messrs. Warren & Fordyce, Mimico, Ont., have been awarded the contract for electrical work on a public school building being erected at Mimico, Ont., at a cost of $70,000.

**Montreal, Que.**

The Canadian Westinghouse Co., 285 Beaver Hall Hill, Montreal, has been awarded the contract for electrical equipment in connection with the electrical pumping station at Pointe St. Charles.

**New Glasgow, N. S.**

The Nova Scotia Power Commission has been asked to submit a proposition to provide power to Pictou, Stellarton and New Glassgow, by developing Malay Falls, on the East river, Sheet Harbor. A cost of 1½ cents per kw.h. to the towns, has been suggested.

**Oakville, Ont.**

The town council of Oakville, Ont., recently decided to proceed with plans for the installation of an auxiliary generating plant, which would make the town independent of Niagara power in the event of a power tie-up. The proposed plant will probably cost in the neighborhood of $32,000, and will probably be operated by gasoline engines of 300 horse-power.

Mr. B. E. Sprowl, Oakville, Ont., has secured the contract for electric wiring on a $60,000 residence being erected at Oakville by Mr. J. Allan Ross, and a $50,000 residence being erected on the Lake Shore Road, Oakville, by Mr. H. G. Kelly.

ed on Glidden Ave., Oshawa, by the Oshawa Housing Commission.

**Ottawa, Ont.**

The Dalhousie Electric Co., Dalhousie St., Ottawa, Ont., has been awarded the contract for electrical work on a new cooling plant being erected on Besserer St., Ottawa, by the Swift Canadian Packing Company.

Mr. E. Headley, 645 Echo Drive, Ottawa, Ont., has been awarded the contract for electrical work on a Mission building recently erected in Billings Bridge District, Ottawa, by the St. Matthews Anglican Church.

**Point Grey, B. C.**

Mr. Ernest J. Pierce, 3530—28th Avenue W., Vancouver, has secured the electrical contract on an $86,000 addition being built to the Prince of Wales School, Point Grey, B. C.

**Riviere du Moulin, Que.**

La Cie Electrique, Chicoutimi, Que., has been awarded the electrical contract on a cold storage plant being erected at Riviere du Moulin, Que., by Mr. J. H. Potvin, of that place.

**Saskatoon, Sask.**

The Wheaton Electric Co., Saskaoon, Sask., has secured the electrical contract on a Chemistry Building, for the University of Saskatchewan, being erected at an estimated cost of $602,000.

**Swan River, Man.**

The town council of Swan River, Man., are having an electric lighting plant installed there this spring.

**Vancouver, B. C.**

The Johnson Electric Co., Ltd., 1633 Vennes St., Vancouver, B. C., has been awarded the electrical contract on a store being erected at 2756 Granville Ave., Vancouver, by Mr. A. V. Lewis, 1202 Seventh Ave.

**Windsor, Ont.**

Provision is made for an electric range in each flat of a 16-flat apartment house for which plans have been drawn by Mr. G. P. Jacques, architect, Windsor. This building will be erected on Victoria Ave., Windsor.

The Essex Electric & Construction Company recently opened a store at Chatham & Ferry Sts., Windsor, Ont., where they will carry on an electrical contractor-dealer business. They have an up-to-date stock of electrical equipment and supplies.

**Oshawa, Ont.**

Messrs Furdy Co., 82 Simcoe St. S., Oshawa, Ont., has secured the electrical contract on ten residences to be erect-

The Erie Electrical Equipment Co., of Johnstown, Pa., announce that S. R. Burd has been appointed acting sales manager for the company, to succeed H. A. Selah. The company was referred to in error in our issue of April 1 as The Electrical Equipment Company.

## FOR SALE

One of the best electrical construction business in Western Ontario City. Established fifteen years. Splendid opportunity for real practical man. Apply Box 806, Electrical News, Toronto.

## FOR SALE

75 H.P. Westinghouse Motor, 2200 V, 60 cycle 3 phase, induction type. Complete with starting panel and oil switch. Soo, Ont. Inquire F. J. Hathway, Soo, Mich.    8-11

Electrical Draftsman requires situation with power or industrial company. English public school education and technical training. Experienced in design of D.C. and high and low tension A.C. switch and controlling gear, also in the layout of static transformer and D.C. Traction substations. Excellent references. Box 843, Electrical News, Toronto.    9

POSITION WANTED—Electrical Engineer. technical and commercial training, over twenty years' practical experience with large power and industrial companies in operation, construction and maintenance of hydro-electric plants, sub-stations, transmission lines, distributing systems and motor installations, desires position with power or industrial company. Present position electrical engineer-manager of Street Railway, and Light and Power System. Excellent reasons for desiring change. Box 652, Electrical News, Toronto.    17-tf

A by-law authorizing the expenditure of $12,000 for extensions and improvements to the electric light system of Summerland, B. C., will, in the near future, be submitted to the ratepayers for their approval.

Mr. Jos. Aubut, until recently with the Dominion Rubber Company, Ltd., St. Jerome, Que., has resigned to accept the position of electrical engineer & superintendent of the municipal electric department, St. Jerome. Mr. Aubut commenced his new duties on May 1.

Messrs. MacKie & Driscoll, King Square, St. John, N. B., have been awarded the contract for electrical work on a $12,000 addition to be built to the Main St. Baptist church, St. John.

---

## MacLean Building Reports Limited

MacLean Building Reports will give you accurate, advance information on every building and engineering contract of consequence in the Dominion.

These Reports are issued daily and reach subscribers in ample time to bid on the work or submit prices for the machinery, equipment, materials or supplies required.

Hundreds of firms are deriving much financial benefit from the use of MacLean Building Reports. Tell us what territory you cover and put it up to us to show how we can help you get more business. Be sure to write today for rates and free sample reports.

**MacLean Building Reports Ltd.**

346 Adelaide St. W.   -   TORONTO
119 Board of Trade Bldg. MONTREAL
348 Main St.   -   WINNIPEG
212 Winch Bldg.   -   VANCOUVER

---

### NEW AND USED

# MOTORS

In Stock for Immediate Delivery

| H.P. | Phase | Cycle | Volts | R.P.M. |
|---|---|---|---|---|
| 100 h.p., 3 ph., 25 cyl., 550 volts, | | | | 710 r.p.m. |
| 75 h.p., 3 ph., 25 cyl., 550 volts, | | | | 480 r.p.m. |
| 60 h.p., 3 ph., 25 cyl., 550 volts, | | | | 750 r.p.m. |
| 52 h.p., 3 ph., 25 cyl., 550 volts, | | | | 720 r.p.m. |
| 50 h.p., 3 ph., 60 cyl., 550 volts, | | | | 970 r.p.m. |
| 40 h.p., 3 ph., 25 cyl., 550 volts, | | | | 720 r.p.m. |
| 30 h.p., 3 ph., 25 cyl., 550 volts, | | | | 750 r.p.m. |
| 30 h.p., 3 ph., 25 cyl., 550 volts, | | | | 1500 r.p.m. |
| 25 h.p., 3 ph., 25 cyl., 550 volts, | | | | 750 r.p.m. |
| 15 h.p., 3 ph., 60 cyl., 550 volts, | | | | 1750 r.p.m. |
| 15 h.p., 3 ph., 25 cyl., 220 volts, | | | | 720 r.p.m. |
| 15 h.p., 3 ph., 25 cyl., 550 volts, | | | | 750 r.p.m. |
| 15 h.p., 3 ph., 25 cyl., 550 volts, | | | | 700 r.p.m. |
| 7½ h.p., 3 ph., 25 cyl., 550 volts, | | | | 1450 r.p.m. |
| 7½ h.p., 3 ph., 25 cyl., 550 volts, | | | | 725 r.p.m. |
| 7½ h.p., 3 ph., 25 cyl., 550 volts, | | | | 700 r.p.m. |
| 5 h.p., 3 ph., 60 cyl., 550 volts, | | | | 1750 r.p.m. |
| 3 h.p., 3 ph., 60 cyl., 550 volts, | | | | 1750 r.p.m. |
| 3 h.p., 3 ph., 60 cyl., 550 volts, | | | | 1120 r.p.m. |
| 3 h.p., 3 ph., 25 cyl., 550 volts, | | | | 1500 r.p.m. |
| 3 h.p., 3 ph., 25 cyl., 550 volts, | | | | 1400 r.p.m. |
| 3 h.p., 3 ph., 60 cyl., 550 volts, | | | | 1750 r.p.m. |
| 2 h.p., 3 ph., 25 cyl., 550 volts, | | | | 1460 r.p.m. |
| 2 h.p., 3 ph., 25 cyl., 550 volts, | | | | 1400 r.p.m. |
| 1 h.p., 3 ph., 25 cyl., 550 volts, | | | | 1500 r.p.m. |
| 1 h.p., 3 ph., 25 cyl., 110 volts, | | | | 1420 r.p.m. |
| 1 h.p., 3 ph., 25 cyl., 220 volts, | | | | 710 r.p.m. |
| ½ h.p., 1 ph., 60 cyl., 110 volts, | | | | 1725 r.p.m. |
| 1/6 h.p., 1 ph., 60 cyl., 110 volts, | | | | 1725 r.p.m. |
| 1/6 h.p., 1 ph., 60 cyl., 110 volts, | | | | 1150 r.p.m. |

Machinery      Supplies

Ask for our latest stock list

## H. W. PETRIE, Limited

131-147 Front St. West, Toronto

---

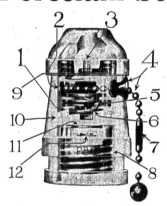

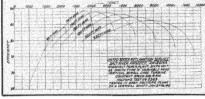

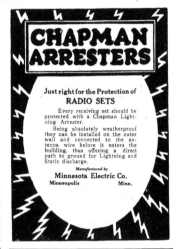

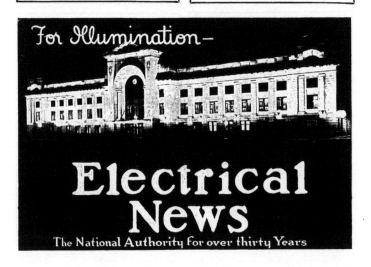

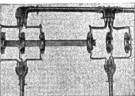

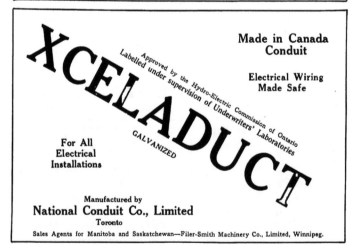

# CLASSIFIED INDEX TO ADVERTISEMENTS

The following regulations apply to all advertisers:—Eighth page, every issue, three headings; quarter page, six headings; half page, twelve headings; full page, twenty-four headings.

# CLASSIFIED INDEX TO ADVERTISEMENTS—CONTINUED

## CLASSIFIED INDEX TO ADVERTISEMENTS—CONTINUED

# Directory of Engineers
## and Special Representatives

Electrical
Mechanical
Chemical

Testing
Research
Patents

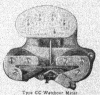

Vol. XXXI No. 10

Toronto, May 15, 1922

# Electrical News

### Engineering Contracting — Merchandising Transportation

## "Lincoln Demand Meter" Hits a
## Homer—Scoring "Watt Hour Meter"

The correct method of billing customers for the current they use, is illustrated by the above play in baseball.

The watt hour meter (with which you measure your customers' actual consumption) gets on base. The Lincoln Meter (which enables you to accurately measure each customer's **demand**) hits for a home run scoring himself and for the watt meter.

When you measure a customer's watt-hours of consumption, you only get "on base". You must measure demand also to "score" him—to bill properly.

If you measure watt hours and estimate or neglect demand, you may be measuring 10% of a customer's bill. You then guess at, or neglect, 90% of the bill—in the case of a hydro-electric plant, your fixed charges may amount to that much of your entire cost of furnishing service.

Put a Lincoln Demand Meter on your customers' lines, **with** the watt hour meter. Bill on a **100% measured basis.** Know that you are distributing equitably your fixed charges.

Write for the section (or sections) of the Lincoln Meter catalog which interests you. SECTION I is a non-technical treatise on why demand should be measured; SECTION II, the technical treatise on demand measurement; SECTION III, Installation; SECTION IV, the Lincoln Graphic Demand Meter; SECTION V, the Lincoln method of Volt-Ampere Demand Measurement (Power Factor recognition); and Section VI, the Lincoln Split-Core Transformer and Demand Ammeter.

### The Lincoln Meter Co., Limited
### 243 College Street, Toronto, Canada
Cable "Meters" Toronto    Phone College 1374

Hand now on left varies with load and shows this load at any time of day.

Hand now on right moves only to right and indicates maximum load since last resetting.

Lincoln MAXIMUM DEMAND meter

# ALPHABETICAL LIST OF ADVERTISERS

# The Safest Switch You

Individual bases permit removal of any blade or jaw in 3 minutes

## Square D

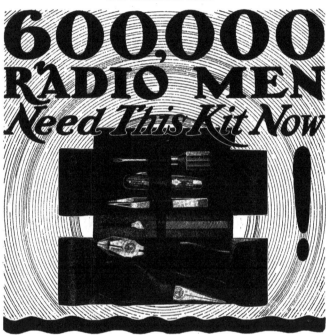

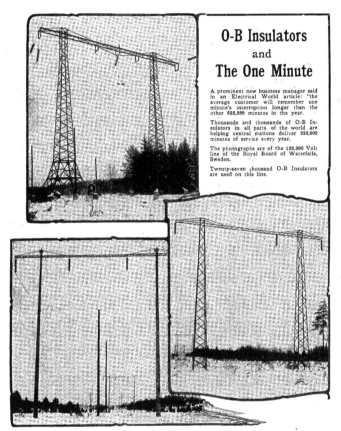

## O-B Insulators
### and
## The One Minute

A prominent new business manager said in an Electrical World article: "the average customer will remember one minute's interruption longer than the other 525,599 minutes in the year.

Thousands and thousands of O-B Insulators in all parts of the world are helping central stations deliver 525,600 minutes of service every year.

The photographs are of the 132,000 Volt line of the Royal Board of Waterfalls, Sweden.

Twenty-seven thousand O-B Insulators are used on this line.

# The New Hotpoint Water Heater

SHAVING water in one minute from turning on the current.

Sufficient hot water for bath in 15 to 20 minutes.

Cost of operating—assuming cost of current to be 1¾ cents per Kw. hr. and the heater operating on full—3¾ cents per hour. But, only on wash days is it necessary to operate on "high" for any length of time.

This is the new Hotpoint Water Heater with "The Metal Clad Element." Just as the filament in an electric lamp is positively protected from the air and so from oxidation, the element in the Hotpoint Water Heater is similarly protected. The chances of a burn-out are practically eliminated.

*" Made in Canada "*

## Canadian Edison Appliance Co., Limited
### STRATFORD, ONTARIO.

# A New Market for An Old and Tried Product

*Five Amp. Tungar Battery Charger*

Electrical dealers can broadcast some good news to the thousands of owners of wireless telephone and telegraph sets. The Tungar battery charger puts an end to the annoyance and expense of taking storage batteries to a service station for charging.

Connected to any alternating current lighting circuit the Tungar charges the battery quietly, quickly and at a small cost for current. An overnight charge once or twice a month will keep the battery in perfect condition, lengthening its life and increasing its efficiency. An added selling point for the Tungar is that it can also be used for charging automobile starting and lighting batteries and to furnish direct current for a variety of electrical experiments.

The Tungar will be promoted this year by extensive advertising in radio papers. It is dependable, easy to sell and offers a satisfactory profit.

Charges one 6-volt lead cell battery at 5 amperes, one 12-volt battery at 3 amperes or one 18 volt battery at 1.5 amperes.

There is also a 2 amp. size which will charge a 6-volt battery at 2 amperes. or a 12 volt battery at 1 ampere.

*"Made in Canada"*

# Canadian General Electric Co., Limited
## HEAD OFFICE  TORONTO

Branch Offices:  Halifax, Sydney, St. John, Montreal, Quebec, Sherbrooke, Ottawa, Hamilton, London, Windsor, Cobalt, South Porcupine, Winnipeg, Calgary, Edmonton, Vancouver, Nelson and Victoria.

## More Than Half a Million City and Town Homes In Canada Without An Electric Fan.

### YOUR OPPORTUNITY

The above figures do not take into consideration the number of stores, theatres, restaurants, barber s shops, billiard rooms, office buildings, factories, etc., that offer an immense potential market for Electric Fans.

## Sell Them C.G.E. Electric Fans

Thirty years' experience is behind the manufacture of C.G.E. Electric Fans. People not only want a fan that will keep them cool, but one that will harmonize with their furnishings. The soft green finish, and the polished blades of the C.G.E. Electric Fan will harmonize with every type of furnishing and, in fact, add a fine decorative touch.

Being prepared is the secret of success in Merchandising Fans. Directly the first hot spell strikes your district dress your windows with C.G.E. Electric Fans, and advertise in your local papers. We will gladly supply you with ready-prepared newspaper advertisements and any other co-operative material you may desire.

# Canadian General Electric Co., Limited

## HEAD OFFICE TORONTO

Branch Offices: Halifax, Sydney, St. John, Montreal, Quebec, Sherbrooke, Ottawa, Hamilton, London, Windsor, Cobalt, South Porcupine, Winnipeg, Calgary, Edmonton, Vancouver, Nelson and Victoria.

MADE · IN · CANADA

# The "Jeffersonlite"

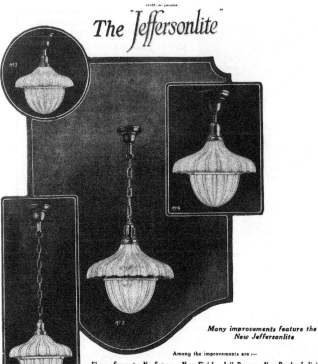

### Many improvements feature the New Jeffersonlite

#### Among the improvements are :—

**Flange Support— No Screws—New Finish—Jeff Bronze— New Bowl—Individual Cartons--New Adjustable Socket Holder--New Hangers--New Package--New Prices.**

We extend an invitation to all the Electrical Trade to visit our Bay Street Show-rooms at any time when in Toronto.

FACTORY & HEAD OFFICE
**388 CARLAW AVE**
TORONTO

DOWN-TOWN SHOW ROOMS
**154 BAY ST. TORONTO**

*Jefferson*
MADE IN CANADA
*Glass*
**COMPANY LIMITED**

285 BEAVER HALL HILL
MONTREAL
272 MAIN STREET
WINNIPEG
510 HASTINGS ST. W.
VANCOUVER

THE electrician who has a number of office appliances to keep running, finds this task easy with devices which are equipped with R & M motors. R & M fractional horsepower motors give the same high grade dependable service the larger sizes give out in the factory.

It is a noteworthy fact, too, that devices equipped with R & M motors are nearly always built on a quality basis throughout. Manufacturers of the better quality appliances naturally turn to the motor which is in keeping with their manufacturing standard when selecting motors to power their machines.

Electrical men who prefer R & M motors on the machines they have to service, find it worth while to pass this thought along to the office manager, purchasing agent and any others who have a say in the selection of these appliances.

## The Robins and Myers Company of Canada, Limited

Brantford                    Ontario

# Robbins & Myers Motors

*Many hundred thousands of Hoover advertisements are circulated each month through leading periodicals that cover Canada*

# Let the "Housecleaning Backaches" of Spring
## Help You to Greater Profits

**The Brush May Be Easily Adjusted**
*(Another Patented Hoover Feature)*

THE soft bristles of the Hoover brush, carefully selected to produce the best carpet cleaning results, will gradually wear, just as all brushes do. Eventually any brush must be lowered or become useless until replaced. The adjustment of the Hoover Beating-Sweeping Brush is simple. No tools are required; a woman can easily disconnect the belt, pull put the brush guard, and lower or raise the brush. Thus The Hoover may be kept at highest cleaning efficiency; the user is also saved the cost of new brushes. Covered by Hoover patents granted December 9, 1919, and January 20, 1920. Over fifty other patents project the salient features of electric cleaner construction. Other applications for patents pending.

Now, while men keenly recall the bare floors at home occasioned by sending out the rugs for cleaning—

Now, while women's backs still ache from moving heavy furniture and from performing old-fashioned dusting and cleaning—

Make your start as an Authorized Hoover Dealer!

Let us put one of our trained Hoover salesmen at work in your locality at this psychological time.

He will go into homes where rugs have just been cleaned by the usual methods— and prove by a Hoover demonstration that those supposedly clean rugs are in reality *not* clean!

Women will welcome him—they will be glad to hear how The Hoover can relieve them forever of such housecleaning drudgery as they have so recently been forced to endure.

Your co-operation will bring you the same splendid profits that Authorized Hoover Dealers everywhere are enjoying.

Join forces with us now—our years of experience are at your command, ready to help you start off on the right foot and continue to sell Hoovers in profitable number the year around.

A small space in your store, a small investment in Hoovers—*on a guaranteed sale basis,* twelve profits a year through monthly turnovers of your stock—that is our proposition, if you will give us your hearty co-operation.

Investigation does not obligate you in any manner. Send today for a Hoover representative.

THE HOOVER SUCTION SWEEPER COMPANY OF CANADA, LIMITED
Factory and General Offices: Hamilton, Ontario

*The* HOOVER [MADE IN CANADA]

*It BEATS—as it Sweeps—as it Cleans*

Generation, Transmission and Application of Electricity

For nearly thirty years the recognized journal for the
Electrical Interests of Canada.

Published Semi-Monthly By

## HUGH C. MacLEAN PUBLICATIONS
LIMITED.

THOMAS S. YOUNG, Toronto, Managing Director
W. R. CARR, Ph.D., Toronto, Managing Editor
HEAD OFFICE  -  347 Adelaide Street West, TORONTO
Telephone A. 2700

MONTREAL  -  -  119 Board of Trade Bldg.
WINNIPEG  -  -  -  -  302 Travellers' Bldg.
VANCOUVER  -  -  -  -  -  Winch Building
NEW YORK  -  -  -  -  -  296 Broadway
CHICAGO  -  -  14 West Washington St.
LONDON, ENG.  -  -  16 Regent Street S.W.

### ADVERTISEMENTS
Orders for advertising should reach the office of publication not later
than the 5th and 20th of the month. Changes in advertisements will be
made whenever desired, without cost to the advertiser.

### SUBSCRIPTIONS
The "Electrical News" will be mailed to subscribers in Canada and
Great Britain, post free, for $2.00 per annum. United States and foreign,
$2.50. Remit by currency, registered letter, or postal order payable to
Hugh C. MacLean Publications Limited.
Subscribers are requested to promptly notify the publishers of failure
or delay in delivery of paper.

Authorized by the Postmaster General for Canada, for transmission
as second class matter.

Vol. 31            Toronto, May 15, 1922.            No. 10

## The Hardware Store as a
## Sales Medium for Electrical Appliances

A hardware contemporary, writing on the subject of
hardware merchants handling electrical goods, has some interesting things to say regarding the merchandising ability
of electrical dealers, and incidentally points out some characteristics that may not be palatable but which, at the same
time, it may well be advisable that electrical merchants should
take to heart.

The article in question points out that "this question has
been discussed and haggled over by electrical dealers for
years—while hardware merchants are doing the selling."
This is a good hint and it should spur us on to work harder.
Another statement is that it is a "question of the survival
of the fittest between electrical contractor-dealers and hardware men." Still another, that owing to the "numerous
jibes at the hardware trade" communication with the electrical manufacturing interests, through the medium of the
Ontario Retail Hardware Association, is being made.

All this is interesting and helpful. The last quotation
indicates that electrical manufacturers are not finding the
hardware store as satisfactory a medium as formerly, or as
they expected, and are withholding patronage. This is a
foregone conclusion of course, but it is happening sooner
than most of us expected. The fact is that electrical manufacturers never seriously considered using the hardware
store as a permanent sales medium. It has been mutually
satisfactory to do so temporarily—while the electrical merchant was getting his second wind, so to speak—but for an
ultimate outlet for his merchandise it is safe to say, we be-

lieve, that such a thing was never contemplated by the
electrical manufacturer. Our contemporary hits the nail on
the head when he says "survival of the fittest." Everybody
admits that the out-and-out electric merchant, when he becomes efficient, is the "fittest" medium through which to
sell electrical goods. The industry realizes his problem, and
wonderful progress is being made. Already two or more
good stores are to be found where a year or two ago there
was one poor one—and we've only begun.

No one disputes the possibility of hardware stores selling
electrical goods, or clothing, candy, haberdashery, millinery
or groceries. However, it doesn't work out in the latter
cases, apparently, and neither can they make a success of
the electrical business. Why? They don't understand it,
that is all. The average hardware man never will understand the electrical business, or the terms, or the principles
underlying the installation and operation of electrical equipment and appliances. There will doubtless be cases where
hardware stores will maintain an electrical department and
keep a skilled man to look after it—that, in the smaller
towns especially, may well be feasible and desirable—but
where real electrical stores are available it is as logical that
the electrical business should go to them as that watch repairs should go to a watchmaker—rather than a blacksmith.

* * *

Another statement in this article is that "hardware merchants are top-notchers in merchandising." If the word
"hardware" were added, after "merchandising," we could
heartily agree. We believe hardware men are excellent merchandisers; they have had generations of experience. They
are shrewd too. Realizing the attractive appearance of electric goods and the glamour that surrounds anything connected with electricity—in the public mind—they were quick to
see the value of these goods as a shop front for their own
rather unattractive stock. That is the end of the interest
most hardware merchants take in electrical equipment, however. They are not good electrical merchants. They have
no sympathy with the electrical industry. They don't understand the uses of the goods they sell. They are unable to
arouse enthusiasm in their customers over their electric
purchases. They are poorly equipped to give service. In a
word, they are a make-shift as electrical merchandisers and
are merely used because they seem to promise better results,
temporarily, than the drug, millinery or grocery store, so
filling up the interval while we are developing real electrical
merchants. The "fitness" of the trained electrical man, who
specializes in electrical matters, cannot be questioned by
comparison.

* * *

The article concludes with the sentence: "The policy of
'the' survival of the fittest' is quite acceptable to the hardware trade of Canada."

That is the proper spirit. They are making hay while
the sun shines, realizing that it can't shine all the time. Perhaps by the time the electrical merchandising trade gets on
its feet some other attractive form of merchandise will have
been developed to help the hardware man dress up his windows. We sincerely hope so.

But the lesson running through all of this is plain for
the electrical industry. The essence of the whole matter is
speed. Hardware stores, with the best of intentions, cannot
merchandise electrical goods. The only solution is the establishment of good electric stores, manned by efficient merchandisers, in every city and town in Canada. We're getting
on, but we're going too slow. The larger cities have some
fine electric stores but we're lacking in the smaller places.
A case in point is a town of 4,000 inhabitants where the
only "electric store" in the place has one of its windows filled
with bicycles and the other with motor accessories, tools,

etc. The only indication, either inside or outside, of it's being an electric store is a cardboard sign about a foot square with the name on it "Blank Electric Company." Who ever heard of a hardware store in a town of 4,000 occupying a dirty back corner in a motor repair shop? Not in the last 1,000 years, at least. It is up to the electrical industry as a whole to remedy these conditions without delay, if for no other reason than that the people are being deprived of conveniences and service they have a perfect right to in this modern age.

## Cater to the Large Contracts But Don't Forget the Small Ones

An interesting letter has just been received from the West, commenting on our Prairie Number. The comments are complimentary and refer particularly to the value of the opinions expressed by the various branches of the trade, as the outstanding feature. The letter also points out, however, that these opinions are confined almost entirely to dealers who handle large contracts, operate with a large staff and do things, in general, in a big way.

Our correspondent goes on to say that there is another section of the retail branch of the industry that, in the bulk, contributes largely in buying power, and where the conditions are not very satisfactory. He refers to the contractor-dealers in the small towns and the small dealer in the larger towns. He speaks more particularly of the matter of quantity discounts, which the manufacturers and jobbers are reluctant to change even though the account may be one where discounts for cash are accepted.

Two large sales campaigns are cited as illustrations. The larger retailers with a known turnover find these sales advantageous, whereas the small dealer finds himself, after the sales campaign is over, with a considerable percentage of his stock still unsold. The hardware man immediately attaches a reduced price tag and, until his stock is all gone, makes it impossible for the electrical merchant to compete in price, thus giving the electric shop a black eye.

Another matter brought out is the difficulty encountered in handling devices where the resale price is advertised. Carrying charges to isolated points eat into the profits to such an extent that the margin is so small as to prohibit any profit by the merchant who studies his turnover and his costs.

The letter also again attacks the problem of manufacturers and jobbers selling direct to consumers and quotes particulars of a fan sale, showing that the customer had been supplied at a lower price, actually, than that quoted to the dealer. The letter concludes:

"In view of the greater sales of lamps and appliances by the smaller stores,—business which the jobbers cannot handle in any other way, would it not be playing the game if they would put these sales through legitimate channels, without undue checks on the quotation I refer particularly to the matter of fans, with their approximate 5 per cent. discount.

" As organization of the trade in general cannot, in my opinion, be consummated for a number of years, chiefly owing to what has been termed "Back-door members;" i.e., men who don't realize the term "costs", there seems to be no other way for the dealer than that of pushing the most popular priced line, of admittedly less quality, but with higher margin. This is unfortunate, as the cheap device, with its comparatively short life, will always be a deterrent to the industry instead of raising the standard, which we all so much desire."

There is no doubt that the problem of the small merchant is a difficult one. He does not appeal to the manufacturer and wholesaler on account of his limited purchasing power. In the end, however, will the industry not be farther ahead by co-operating in every reasonable way with the small dealer, helping him to compete with the hardware man, protecting him so that he may make a living margin and educating him so that he may grow into a bigger and better merchandiser?

## Distribution Practice in a Sister Colony

It is always of interest to know how our sister colonies are carrying on electrically, even though we may not be willing to admit that their methods are better than ours. Mr. F. A. Cambridge, city electrician of Winnipeg, Man., forwards us a letter just received from the manager of the Electricity Department of Christchurch, N. Z. Mr. Cambridge had explained to Mr. Hitchcock, in a previous letter, the scheme of electrical distribution which has been more or less standardized in Canada, and the reply, which outlines the New Zealand scheme, is very interesting, indeed. It runs in part as follows:

"I note the voltage and method of your distribution system. Also your remarks covering the increased danger in choosing a higher voltage. Possibly I did not make the position quite clear in my letter. The New Zealand Government, some six or seven years ago, adopted, as standard low tension distribution, the three phase 4-wire method, having 400 volts between phases, and 230 volts between phase and earthed neutral. This system is now extensively installed both in Christchurch and in other cities, and has given excellent results, while the dangers you speak of as possible have certainly not eventuated in practice. The ability of manufacturers to produce electrical appliances wound for 230 volts, and the general education of the public to take care in using electrical apparatus, have undoubtedly made the 230 volt apparatus safe here. What is not realized by engineers using the 110/220 volt distribution, is the very great saving in distribution at 400 volts. In this city, we have no pole transformers—owing to the fact that, for ordinary domestic supply, 400 volt distribution enables us to have our transformer groups up to a mile or two miles apart. The transformer installations are erected in substations of substantial capacity, and are all served by a ring main system of 11,000 volt underground cable. In many houses, two or even three phases, with the neutral, are led in through conduit to the distribution board, and no trouble in this connection has ever been experienced. Whether the service to a residence be single phase, two or three phase, each phase is fused on the outside of the house with a porcelain plug fuse, thus fully protecting the leads between the entrance and the house service panel.

"Where more than one phase is to be led to a cooking outfit, no flex is used, conduit wiring being carried right up to the terminals of the apparatus. The risk of the 400 volt potential difference in a flex does not, therefore, arise.

"For a similar reason, the question you raise regarding switches does not occur, no 400 volt switching being done. All the elements are wound for 230 volts; all the switching and control is done at 230 volts. The only point raised is that the two phases are led into the same apparatus, which gives rise to a potential difference between different parts of the equipment of 400 volts. All the smaller classes of apparatus, such as toasters, irons, grill-stoves, hot plates, immersion heaters. kettles, coffee percolators and the innumerable types of small equipment now on the market, are already in use extensively here at 230 volts and are perfectly satisfactory."

Messrs. Peter & Sylvester, Stratford, Ont., have secured the contract for electrical work on a $43,000 addition being built to the Birmingham St. school, Stratford.

## Federal Approval of Electrical Appliances

Brantford, Ont.

Editor Electrical News,

I have noted with a great deal of interest the comments in recent issues of your paper, suggesting the need and advisability of Federal inspection and approval of electrical appliances, and it appears to me that the organized electrical interests in Canada, could not devote energy to a better cause than the support of Federal inspection very much along the lines suggested.

While we are favored by living in a country where the purchasing public are enthusiastic regarding the uses of electrical appliances, and while we are attaining splendid results by organization, let us not lose sight of an issue of such benefit as properly constituted Federal inspection. Under present competitive conditions there is liable to be a tendency toward price cutting at the expense of the appearance, durability, and efficiency of the article. Such a condition if allowed to continue, unchecked, will result eventually in a growing scrapheap of electrical appliances, and beneath them will be buried much of the consumer's good-will that the industry cannot afford to lose. Therefore it appears to me that not only should a device be subject to criticism from the standpoint of life and fire hazard, but that the material used, the standard of workmanship, durability, and the general efficiency of design, are outstanding features which should receive more earnest consideration, if we are going to safeguard the friendly attitude of the public, on which the success of our industry is based.

Regarding local inspection and its bearing on electrical appliances, it is quite true that manufacturers are often asked to make expensive alterations to meet inspection at one point, while at some other point the same alterations are not acceptable. The impression conveyed to the mind of the buyer, moving between these points, does not tend to his support of the particular appliance, and instead of the confidence resulting from uniform inspection, he entertains an element of doubt and uncertainty, which is not conductive to the welfare of the industry. While it is admitted that many useful suggestions are received from local inspectors, there are instances where meeting some of their requirements would result in seriously impairing the life, and efficiency of the product, with the result that the manufacturer either passes up that territory, or refuses to meet the requirements, thereby often incurring the non-support of the inspector, which is not the condition of important harmony suggested by uniform Federal inspection.

Yours truly

J. A. McDonald

## Manitoba Power Co.'s Development

On April 27th. Mr. T. W. MacKay, assistant chief engineer of the Manitoba Power Company Limited, gave a very interesting talk before the Manitoba Electrical Association at the bi-monthly luncheon in Manitoba Hall, on the new hydro-electric development on the Winnipeg River at present under construction by his company.

In the course of his talk Mr. MacKay drew the attention of the members to the wonderful asset which Manitoba enjoys in the Winnipeg River, as a source of cheap and abundant water power and described the river and its drainage basin in detail, with special reference to the control of the discharge of the river by the Water Power Branch of the Dominion Government.

Reference was then made to the salient features of the new development, during which the method of construction being followed was very concisely explained. Attention was also drawn to the immense amount of material involved in the work and the procedure for handling the same.

Mr. MacKay also touched on the period of financing the project, which extended from 1914 to 1921, when arrangements were finally completed, and stated that great credit was due to the vice-president and general manager of the company, Mr. A. W. McLimont, for the successful completion of this difficult task.

In closing, reference was made to the benefits accruing to the electrical industry from the new development, not only from the standpoint of materials required for construction, but also on account of the immense quantities of appliances and equipment which will be required by consumers when the plant is in operation.

## Industrial Lighting Campaign

Owing to various causes, principal of which was the late date at which the new display headquarters of the B. C. Manufacturers Association was ready for occupation, the proposed Industrial Lighting Campaign, which has been prepared by the electrical industry in Vancouver, will be postponed until later in the year. It is possible that August or September will see the exhibit staged and the campaign begun. It is intended to be educative for the factory, shop and mill owner, instructing him in the real saving effected in cost of operation and production by correct lighting. The actual saving possible in the monthly light bill by the use of improved methods and appliances will also be shown in the campaign of demonstration.

## A Correction

In an article entitled "The Power Resources of the Prairies," which appeared on page 84 of our April 15 issue of the Electrical News the first two lines of the fourth paragraph read "The province possesses available water power to the extent of 475,000 h.p. for maximum commercial development."

This should have read "The province possesses available water power to the extent of 475,000 h.p. at ordinary minimum flow and 1,138,000 h.p. for maximum commercial development."

---

R. N. Dicer, Electrical Contractor, 1256 Pender St., W., Vancouver, has been awarded the contract for the electrical work on the new electric bakery being erected by Shelly Bros., Vancouver.

## Coming Conventions

National Electric Light Association, at Atlantic City, N. J., May 15-16-17-18-19.

Canadian Electric Railway Association, at the Drill Hall, Quebec City, Que., June 1-2-3.

Canadian Electrical Association, at the Chateau Laurier, Ottawa, June 15-16-17.

Association of Municipal Electrical Utilities, at the Clifton Hotel, Niagara Falls, Ont., June 22-23-24.

American Institute of Electrical Engineers, at the Clifton Hotel, Niagara Falls, Ont., June 26 to 30.

# Problems of the Industrial Engineer—II

## Explaining the Effects of Low Power-Factor to the General Manager

### By GEO. L. MACLEAN, E.E.
#### Electrical Engineer Canadian Connecticut Cotton Mills Ltd., Sherbrooke, Que.

"How's the world hitting you?" enquired the General Manager of the Electrical Engineer, as he strolled into the office one afternoon just as the engineer was in the middle of a transmission line problem. "Not too bad," replied the Engineer. "Lay off a while," said the G. M., "and explain to me as simply as possible how we can correct this low power-factor condition? What does this low p.f. do anyway besides making us pay out money for nothing? Does it affect our plant materially, I mean, and injure our electrical equipment?" "Yes," replied the engineer, "its disadvantages are many."

"Although the wattless current"—"What's that, wattless current?" "The watt," replied the Engineer, "is the product of volts multiplied by amperes. 100 volts of direct current times 10 amperes equals 1,000 watts, or one kilowatt, but in an alternating current of three phase (such as we use) 100 volts times ten amperes times V3=1,730 kilovolt amperes. Then 1.730 kilovolt amperes times our power-factor (.75) equals 1,397 watts or 1.297 kilowatts. The effective difference between 1.73 and 1.29 is our wattless or idle current, a current that has no component part, being out of phase. Although it puts very little load on our system the heating effect is the same though it were in phase and a component part. Therefore, it causes heating of our electrical apparatus which in turn causes deterioration of the insulation and consequently motor burn-outs, which keep our expenses up. This low power-factor also causes poor regulation of line voltage, this in turn decreases the torque or turning power of our induction motors and also increases the magnetizing current—or we will call it 'no load current.'

"As an illustration we will say that the low power-factor causes a certain percentage of increase in magnetizing current and we will have to pay for this increase in current, for it is a component part and effective. From this also you can readily see it increases our cost from the power bill end and from the practical end by increasing our maintenance cost.

"I will explain how this low power-factor would affect the first cost of an installation. We are going to install a connected load of 1,000 h.p. of induction motors ranging in size from 10 to 50 h.p. Our connected load factor, or the ratio of employed h.p. to our connected capacity, is 70 per cent, or we will use only 700 h.p. of our 1,000 available h.p. Our p.f. is 75 per cent, we will require transformer capacity of 695 kw., as 700 h.p.=522 kw. and this figure divided by .75 (our power-factor) equals 695 kw. These figures," explained the Engineer, "are taken offhand, and I am not considering stand and other apparatus.

"But from this explanation you will understand that if we had a power-factor of unity, 100 per cent, or even one of 85 per cent, it would make a great difference. Let us say that we have apparatus to allow us to bring our power-factor to 100 per cent and that it will pay us to install this equipment,—then we would install a transformer bank of 522 kw. capacity. Our first cost of installation would be reduced by the difference between the cost of a 695 kw. capacity bank of transformers and a bank of 522 kw.

"Also, suppose we have a bank of 600 kv.a. transformers and our power-factor is 75 per cent,—then we will have a capacity of 450 kw. or 603 h.p. We are now running these transformers full load and would have to enlarge our plant to increase production. We desire to make such changes as would require 400 h.p. more. Our 600 kv.a. bank of transformers is, not large enough, so we, instead of buying new transformers, install apparatus to bring our power-factor up, getting two results in one installation, viz; cutting out the power factor penalty and saving the cost of another bank of 400 kv.a. transformers.

"Could we not overload our transformers a little?" asked the G. M. "Yes, but it is poor policy to take chances, although there are instances where transformers are run at high temperatures over long periods. Manufacturers, however, do not recommend that practice. Without your transformers or spare ones, you would lose more in production than would be gained. It is a case of 'Penny wise and pound foolish.'

"Another point in connection with this," said the Engineer, "is that at unity power-factor your transmission line copper would be cut down considerably as low power-factor plays an important part in line calculation."

"You have taken up so much time explaining the evils of low power-factor," said the G. M. as he looked at his watch, "that I have not time now to learn how we can eliminate it, but I am glad for the explanation of its evils and at the first opportunity we will discuss how to get an ideal condition, or what was it? 'Unity Power-Factor,' said the G. M. as he bid the Engineer "Good afternoon."

## Radiophone Sales Campaign in Vancouver

The electrical industry in Vancouver is making every effort to handle the radiophone equipment business properly. In an effort to stabilize the sale of materials, sets and parts for radio installations, a radio campaign has been set on foot, encouraged by leading manufacturers, jobbers and the potential distributors, the contractor-dealers. As the result of a preliminary meeting held at the call of Mr. Rey E. Chatfield, secretary manager of the Electric Service League, when the situation was canvassed from all angles, a decision was reached to put on a short course of educative addresses to the salesmen employed by electrical dealers.

While it was agreed that it was certainly most desirable to sell all radio equipment through the established trade, the fact that it was new to most of the salesmen made it necessary to provide them with ample knowledge to sell intelligently. The only way then to control this business was through people who knew how to sell the equipment. For the first course, three lectures were arranged. The dates set were Tuesday, April 25th. Thursday, April 27th and Tuesday, May 2nd. For the first address. Mr. L. Mayne. radio engineer for the Marconi Wireless Telegraph of Canada. The second was to be supplied by a member of the staff of Canadian Radio Services, Ltd. and the third by the representative of C. B. Harrison & Co., agents for radio manufacturers. Merchandising phrases, description of suitable apparatus, and methods of explaining to customers apparatus suited to their requirements or desires, were to be covered in the course. It is possible that later an extended course will be given.

So interesting is the radio equipment trade becoming that it is now the intention of the electrical industry to put on a co-operative advertising campaign for the merchandising of radio sets and supplies. In this connection it is pointed out that cheap sets are being stocked and sold by department stores in the city, and that other dealers outside the electrical trade are making plans to cut in also.

# Flat Rate Operation of Water Heaters

## Lower Wattages Give Satisfaction—An Analysis of Heater Performance Under Test

By J. A. McDONALD
Manager Thermo Electric Ltd., Brantford.

The evidence of the customer who enjoys and the central station that supports a flat rate, as applied to domestic electric water heaters, seems to point to the fact that this method meets requirements to the satisfaction of the parties concerned. Heaters of this class depend for their success on continuous operation, and a reasonable period in which there is no demand for hot water, which is usually furnished by the night run. Therefore the time element does not enter into discussion, but the required wattage to successfully meet the requirements of the majority of customers is important. Considering the input in watts at which these heaters should operate, the diversity in hot water demand, and the variation in sizes of domestic boilers, there has not been a standard wattage that would meet requirements within a safe margin and eliminate unnecessary waste of energy. Because of these conditions it has been found advisable to adopt a certain degree of flexibility in the wattages of heaters for this purpose. While it is true that domestic boilers vary in capacity from 15 to 40 gallons, and even higher, the greater majority are between 30 and 40 gallons, the 30 gallon size being enough in evidence to regard it as standard for any ordinary domestic installation.

### Satisfactory Wattages

For boilers varying from 30 to 40 gallon capacity satisfactory wattages seem to vary from 700 to 1,000. It has not been found advisable to recommend heaters of lower wattage than 700 for boilers of 30 gallons, except in cases where the demand for hot water is exceptionally low, when 600 watts may meet requirements. Experiments conducted under field conditions have shown that it is advisable to keep the boiler at a temperature above 140 degrees, and preferably at 160 degrees, the latter requiring the addition of almost 50 per cent cold water before it can be used with comfort. Heaters of lower capacity than 700 watts do not seem able to take care of radiation and hot water demand, and maintain a satisfactory 30 gallon boiler temperature.

The 1,000 watt heater can be recommended for boilers of 30 gallon capacity where the demand appears excessive, and boilers under ordinary demand, upward of, and including, 40 gallon. One of the leading Canadian central stations that has several hundred 1,000 watt heaters operating continuously advises that this size gives excellent service on 40 gallon boilers, for with this size of boiler the reserve in volume of hot water is able to take care of a greater demand than the 30 gallon boiler will allow, thereby giving to the customer a greater volume of delivered hot water without additional expenditure of energy.

Because heaters operating continuously give instantaneous hot water, once they have the boiler heated, the time taken in heating needs no consideration. Next to the proper wattage, the important issue for the customer to ascertain as nearly as possible is the number of gallons of hot water that heaters of accepted capacity will deliver. While it is admitted that few customers have any idea of the actual gallons of hot water used in their households, the fact of having before them the number of gallons delivered will afford them an opportunity to judge their requirements fairly closely by comparison.

### Service Gallons

Only in exceptional cases can the hands be placed in water at a temperature of 115 degrees Fahrenheit. Owing to the boiler temperature being usually higher than this it is necessary to add cold water to the water delivered by the boiler before it can be used with comfort. This addition of water naturally increases the volume of water used, but does not take this increase from the boiler. This increased volume of water we will call service gallons, so that the number of gallons delivered by the boiler, and the number of gallons available for use may be distinguished.

### Purpose of the Chart

It must not be assumed that the chart is intended to represent actual domestic demand. Its exact purpose it to give fairly reliable information regarding the maximum number of gallons that heaters of the capacity designated will supply, the variation in boiler temperature in relation to hot water demand, and a comparison of the performances of the two heaters. These heaters were both of the immersion type and were installed in the 1" opening in the side of the boiler, commonly used for connection to the furnace coil. Owing to the fact that this opening is placed about one-third the distance from the bottom to the top of the boiler, it can be roughly stated that the heaters were heating directly only 20 gallons of water, and indirectly, 10 gallons. The test was made with exposed boilers, operating under ordinary field conditions, and covered a run of two days and two nights, the demand period being confined to the hours from 7 a.m. to 6 p.m.

The table below the chart shows the hot water gallons delivered and the consequent service gallons, the service gallon temperature being not less than 110 degrees. The full lines represent both the room and boiler temperature during the 1,000 watt run, and the dotted lines similar temperatures for the 700 watt heater.

### Heater Performance Analyzed

The 700 watt heater starts with a temperature of 46 degrees the evening preceding the first day. At 7 a.m. the following morning a boiler temperature of 155 degrees is reached. Referring to the table below we find the temperature drop starting at 8 a.m., due to the withdrawal of 20 gallons of hot water from the boiler. In order that this quantity may be comfortably used, 9 gallons of cold water are added, resulting in 29 service gallons. The chilling effect of the 20 gallons of cold water results in a temperature of 85 deg. at 10 a.m., when the temperature starts to rise and at 1 p.m. reaches a temperature of 115 deg., when 18 gallons are drawn off, which on account of their low temperature require the addition of only two gallons of cold water. The boiler is again chilled to 77 degrees but rises to 108 degrees at 5 p.m., when 20 gallons are delivered.

Starting at 60 degrees, or room temperature, at the end of the first day's demand, a temperature of 166 degrees is reached at 7 a.m. the following morning. The higher temperature reached on the morning of the second day is due to the higher starting temperature the night before. It should be mentioned here that the rate of rise as shown on the no demand period on the chart does not represent the true rate of rise of the heaters. The actual rate of rise becomes less as the temperature increases. The lines of temperature rise in the no demand period are not intended to show the actual performance of the heaters but merely the existence of the no demand period and the different temperatures arrived at.

Starting at 7 a.m. with a boiler temperature of 165 degrees, 6 gallons of hot water are drawn, which at this temperature are equivalent to 11 service gallons. At 8 a.m. an equal amount at 155 degrees is delivered, and the effect of the replacing 12 gallons of water is seen in the falling temperature until 10 a.m., when it again rises, reaching 130 degrees at 11 a.m. Here six gallons are again drawn, resulting in a falling temperature. A temperature of 125 degrees is reached at 1 p.m., when another 6 gallons is delivered. The withdrawal of similar amounts at 2 and 4 p.m. brings down the temperature to 115 degrees at 5 p.m. At this point the heater has delivered an amount of hot water only two gallons less that that delivered on the first day, and the greater part at a higher temperature. Also, there are 18 gallons at service temperature still in the boiler.

Owing to the fact that the foregoing explanation of the action of the 700 watt heater will assist in understanding the performance of the 1,000 watt heater, it will not be necessary to follow its table and the resultant temperature variations. It is natural that the temperature rise should be both higher and faster than in the case of the 700 watt heater,

TEMPERATURE VARIATION     HOT WATER SUPPLY

ELECTRIC IMMERSION HEATERS
700 AND 1000 WATT CAPACITY

DURATION OF RUN    1000 WATT  45 HOURS   700 WATT  45 HOURS

AVERAGE TEMPERATURE OF WATER IN MAINS 44 DEGREES FAH

1000 WATT TEMPERATURES ———— 700 WATT TEMPERATURES.......

No of Gallons of Hot Water Supplied
700 WATT HEATER

| U. S. GALLONS | | | | | |
|---|---|---|---|---|---|
| TIME | HOT | SERVICE | TIME | HOT | SERVICE |
| 8 A M | 6.0 | 11 | 7 A.M. | 6 | 11 |
| 1.30 A M | 18 | 8.0 | 8 A.M. | 6 | 9 |
| 3.30 A M | | 8.0 | 11. A.M | 6 | 9 |
| | | | 1. P.M. | 6 | 9 |
| | | | 2 P.M. | 6 | 9 |
| | | | 3. P.M. | 6 | 9 |
| TOTAL | 18 | 69 | TOTAL | 36 | 69 |

1000 WATT HEATER

| | | | | | |
|---|---|---|---|---|---|
| 7 A.M | 6 | 13 | 7 P.M | 6 | 13 |
| 8 A M | 6 | 7 | 10 P.M | 6 | 13 |
| 11 A M | 6 | 14 | 1 P M | 6 | 14 |
| 1 P M | 6 | 14 | 3. A.M | 18 | 44 |
| 3.30 A M | 18 | 44 | | | |
| TOTAL | 42 | 141 | TOTAL | 36 | 141 |

700 WATT HEATER SUPPLIED, 74 GALLONS HOT AND 138 SERVICE
1000  "    "    "    "    —    83  "    "    "  1262  "

Graphic results of tests on 1000 and 700 watt heaters

even considering the fact that it operates two hours less and at a lower room temperature for the entire run. The advantage of high temperature is evidenced by the fact that at 5 p.m. of the second day's run of the 1,000 watt heater, 46 gallons are available at service temperature.

### Radiation

The action of the factor of radiation is clearly evident in the rate of temperature rise of the heaters at different boiler temperatures. Noting the rate of temperature rise between 2 and 5 p.m. on the first day of the 1,000 watt run,

when an advance is made from 80 to 120 degrees, it will be seen that it is noticeably fast. Comparing this with the rate of the same heater at the same time on the second day it is apparent that it is much slower while the operating temperature is between 145 and 160 degrees. This difference is due to the increase in the rate of radiation at higher temperatures. Comparing the rate of rise of the two heaters at the same boiler temperature, the rise of the 700 watt heater from 7 to 8 a.m. on the first day and that of the 1,000 watt heater from 4 to 5 p.m. on the second day, it will be clear that the 700 watt curve is soon due to flatten out, while the 1,000 watt still maintains a satisfactory rise. This is of course suggestive of the reason for the unsatisfactory service of lower wattage heaters, which may be summed up in the fact that there is not enough energy to overcome radiation at water temperatures sufficiently high to give the customer satisfactory service.

### Diversified Demand

For further purposes of comparison it may be noted that the two heaters show four different daily demands. On the first day the 700 watt heater is called upon to supply the maximum demand early in the day, and the 1,000 watt heater supplies a minimum demand in the morning and a steadily increasing demand as the day passes. During the second day, the 700 watt heater is filling a demand of smaller quantities at more frequent intervals, and the 1,000 watt heater supplies a heavy demand in the morning and late afternoon and a light demand in the middle of the day.

### Conclusion

Those who become familiar for the first time with the actual quantity of water that can be supplied by a continuous operating low wattage heater, usually express surprise that the volume secured is much greater than they anticipated. Heaters of proper wattage for their individual boilers appear able to deliver per day a quantity of hot water equal to the capacity of the boiler, and in service gallons a quantity easily twice the boiler capacity, depending of course upon the temperature at which the boiler is operated. There does not seem to be even the remotest possibility of the adoption of a standard rule that will reasonably apply to the hot water requirements of the majority of customers. Two households with apparent similar conditions often differ widely in their hot water demand, and households where the demand appears to be excessive often use less than those where the reverse is suggested. So far the problem of the correct recommendation of wattages, for water heaters, to meet any definite requirement in the domestic field, remains unsolved. While it will always be a matter difficult of exact solution, it is safe to assume that when the product is in more general use, and dealers and public become better acquainted with its performance, many phases of the problem will become adjusted. Heaters of the capacity described are giving satisfactory service in boilers of the capacity recommended, and while the chart shows maximum quantities, careful measurement of domestic installations show that such demands are rarely if ever reached, and that the average seems to be between 40 per cent and 60 per cent, occasionally reaching 75 per cent.

---

### The United Electric Supply Co.

Mr. I. Herbert, who has been carrying on an electrical contracting business with headquarters at 46 Cameron St., Toronto, has now become associated with Mr. A. Fralich and formed the United Electric Supply Company, to carry on a general wholesale distributing business in fixtures, appliances, radio apparatus, etc. The building at 300 College St., Toronto, is being remodelled for the new company's purposes and will be opened for business about May 15.

## Prominent Consulting Electrical Engineer Heads Dominion Alloy Steel Corporation

Members of the electrical industry will be interested in noting the appointment of W. B. Boyd, M.I.E.E., as president of the Dominion Alloy Steel Corporation, Ltd., of Sarnia, Ont. Mr. Boyd is well known. He was born in Markham, in the county of York, Ontario, fifty years ago and after attending Collegiate there entered the Armour Institute of Technology, Chicago. He was only 19 years old when he was placed in charge of the electrical installation of a number of the buildings at the World's Columbian Exposition, in Chicago, and later became chief electrician in charge of lighting and power, as long as the Exposition remained open. For a time he was electrical and mechanical engineer with the Westinghouse Electrical & Mfg. Company. From there he became assistant chief electrical engineer of the Illinois Steel Company and in 1901, was appointed chief electrical engineer of the Dominion Iron & Steel Company. He became associated, in 1906, with a number of companies in which Sir William Mackenzie was interested, and held a number of important positions, including that of chief engineer to the Toronto Railway Company; chief engineer to the Electrical Development Company; also to the Toronto & Niagara Power Company and the Toronto Electric Light Company. He still holds the office of chief electrical engineer of the Toronto & York Radial Railway Company and the Sudbury-Copper Cliff Suburban Electric Railway Company. In later years Mr. Boyd has confined his attentions chiefly to consulting work, more particularly in connection with such important industrial con-

W. B. Boyd, M.I.E.E.
President Dominion Alloy Steel Co.

cerns as the Howard Smith Paper Mills, the Maple Leaf Milling Company, etc. He is also consulting electrical engineer for the Ontario Railway & Municipal Board. He is a member of the Electrical Section of the Canadian Engineering Standards Association and of the Institution of Electrical Engineers of England.

Associated with Mr. Boyd are a number of well-known men on the Board, including Sir William Mackenzie; H. R. Jones, formerly president of the Alloy Steel Corporation; Geo. A. Simpson, formerly sales manager of the Steel Company of Canada; C. H. Wills formerly chief engineer of the Ford Motor Company; W. A. Black, president of the Ogilvie Flour Mills, Ltd., and vice-president of the Kaministiquia Power Company; B. H. McCreath; J. J. Mahon and Col. William McBain.

The new Company plans to commence building operations in about two months. The contract will be let in the meantime for the new plant. This plant will be entirely self-contained and will contain two 75 ton open-hearth furnaces as well as a 20 ton and a 6 ton electric furnace; also rolling mill, blooming mill, sheet mill, etc., and ultimately, though not immediately, a blast furnace. The capacity planned within the near future is 100,000 tons annually of alloy steel, such as is used in the manufacture of automobiles, tools, farm implements, etc. The initial demand for power is estimated at 15,000 h. p. This will be supplied by the Hydro-electric Power Commission of Ontario.

This is the first plant of its kind to be established in Canada. The imports of alloy steel have now reached the annual amount of 100,000 tons. Most of this is used in the Sarnia district, and the new company anticipate that within 60 miles of their plant there will be a demand equivalent to their total output. It is indeed gratifying to note the prominent part taken by electrical men and electrical energy in establishing this very important industry.

## Electric Club of Toronto Annual

At the annual meeting of the Electric Club of Toronto, held on Wednesday, April 26, the treasurer's report showed a satisfactory surplus on hand and the report of the secretary indicated that the attendance during the past season had been the largest in the history of the Club—well over a hundred, on an average. The following officers were elected for the year 1922-23: President, E. M. Ashworth; vice-president, S. L. B. Lines; secretary, T. R. C. Flint; treasurer, R. A. L. Gray. Executive: F. Jno. Bell; W. A. Bucke; H. A. Cooch; Geo. Hogarth; R. T. Jeffrey; Frank Kennedy; Geo. Paton; L. P. Stiles and E. B. Walker.

## Big Discrepancy in Bid Prices

The electrical contract for the Chemistry Building, University of Saskatchewan, at Saskatoon, Sask., was recently let to the Wheaton Electric Co. of that city for $20,-930. The bids tendered for this work ran as follows: $33,100; $33,000; $30,000; $26,975; $25,850; $22,470; and $20,930. As we have repeatedly pointed out, there is too great a discrepancy in the bids of electrical contractors—a difference existing, in this case of $11,170. between the highest and lowest tender, or a percentage of 35. This compares with a difference of $36,097.45 in the tenders of the plumbing and heating trade (with an expenditure approximately eight times that of the electrical) or a percentage of 19. The general contract tenders varied only by $34,905. between the highest and lowest bids, representing a discrepancy of 7.5 per cent.

## Who Said Electric Homes are Only for Large Cities?

In the account of the One Man Electric Home given elsewhere, a noteworthy fact is the low cost of the whole exhibit, $150.00. This proves that any dealer can put on an Electric Home—any man who is sold on the electrical idea himself and who is willing to devote to the matter a little energy and brain power. It is necessary to secure the co-operation of a builder and a furniture dealer, but this can be done by explaining to them the advertising value of such a display. There should be an Electric Home in every town in Canada this year.

Messrs. Chipman & Power, consulting and supervising engineers, Toronto, have been retained by the town of Oakville, Ont., to make a report on a suitable auxiliary for operating the water pumps in the event of a tie-up in Niagara power. A capacity of approximately 300 h.p. will be required, which will be a single unit engine, either gasoline or Diesel type.

# Merchandising Appliances and Supplies

### Give Dealers a Reasonable "Spread"—Leave Appliances on Trial—Allow Extended Terms of Payment to the Customer —"Sell" Ourselves first—Use Manufacturers' Helps.

By GEO. R. ATCHISON*
Merchandise Manager Southern Canada Power Co.

In making up this paper on the merchandising of electrical appliances and supplies, I have not attempted to deal, in any great detail with the figures involved in the potential possibilities, but rather with some of the present problems that confront us, which must be solved if growth is to be hastened.

#### The Sale of Appliances

First, dealing with the sale of appliances:—with the exception of the flat iron, the idea must be sold and the demand created before the sale can be effected. This practically brings the sale of electrical appliances into the "Specialty Class," such as pianos, cash registers, adding machines, etc. It would hardly seem that this fact is recognized by the trade in general. The selling of specialties is expensive and requires a competent organization to handle. An outside soliciting staff is required to whom a liberal compensation must be paid, if the right type of man is to be attracted to the line. An expense of 10% can be figured on for outside solicitors alone, and in some cases this may run as high as 15%. When this initial expense is added to the regular expenses incurred in connection with the upkeep of an electrical store, it has been found that an average overhead of 26% on the selling price is a fair figure. When it is considered that a discount of 25/10% on electrical appliance purchases is that generally allowed to the dealer, it can be seen that there is very little margin left for profit, contingencies, or building up of a reserve to take care of any depression. We can, therefore, hardly hope to attract many dealers to the merchandizing of electrical appliances, until a more liberal schedule of discounts is put into effect by manufacturers and jobbers, particularly as the stock turnover in this line is not very rapid, four times per year being considered good. As an illustration of this let us take a fan, a 16 inch oscillating fan selling for $44.00, costing the dealer $33.00, to which must be added the sales tax, bringing the cost close to $34.00, and where the dealer has to pay transportation charges, the cost will be approximately $35.00. Figuring his cost of doing business at 26%, the total cost of the fan, including the selling expenses stands at about $46.00, resulting in an actual loss of $2.00, on each sale. When it is further considered that fans are somewhat of a "gamble" in this district, it can readily be understood that there is not much incentive to carry a stock of them.

Let us take another article, such as a washing machine retailing at $170.00. The invoice cost to the dealer will average $136.00, to which must be added sales tax, bringing cost up to $138,00, and when he dealer has transportation charges to pay, the cost will average $143.00 laid down in the store. To this cost must be also added the following; salesman's commission at 10%, $17.00; cartage and time of helper delivering machine to prospect's house $3.50, making a direct cost of $162.50, which leaves a margin of $7.50 to take care of other expenses, and any service calls required on the machine during the guarantee period, and this, mark you, without taking into consideration the compensation to which the dealer is entitled. Can, therefore, the dealer hope to build up a profitable business on any such basis?

*Before Montreal Section, Canadian Electrical Association.

These are only two examples, but they are practically typical of the electrical appliance business.

#### Dealer Requires a Large "Spread"

Obviously, then a larger spread must be allowed by manufacturers or jobbers, if this business is to be developed, and the present seems to be an excellent time to put such a policy into effect, when prices on appliances generally are on the downward grade. Where it is not possible for the manufacturer to decrease the list price, for the benefit of the consumer, and increase the spread to the dealer, manufacturers might allow the list price to remain as it is and pass on any benefit of reduction to the dealer in the shape of larger discounts. This is one of the problems which require careful investigation and correction.

#### How to Sell the Idea

Another problem is that generally speaking Electrical Appliances cannot yet be considered as "over the counter merchandise". The flat iron is the exception to this, and the vacuum cleaner becoming more popular and more easily sold, may also be partially considered as an exception. But, before the sale of many of the other appliances can be effected, the demand must first be created by selling the idea, and the problem is how to sell the idea to the public and stimulate the demand. This can be done in several ways, namely, local newspaper advertising; outside solicitation; attractive window displays; attractive stores; distribution of circulars which are furnished free by the manufacturers, etc., it being necessary of course, in this section to have all advertising matter both in French and English; store demonstrations; country fairs; and exhibits are known to have good results. Advantage should also be taken of having electrical appliances placed in educational institutions, domestic science schools, etc., where advantages and conveniences of appliances are learned by the students. These methods of selling the idea and creating the demand are necessarily somewhat slow, and can be measured to some extent by the number of dealers and others aggressively engaged in the merchandizing of appliances. What is needed is something that will help to sell the idea quicker and increase the demand, so that a rapid but substantial growth in the business may take place. When we have thoroughly sold the idea of electrical labor saving devices to the public, and not till then, will we have a busy time taking care of the demand. When merchandizing is properly organized and handled along the lines suggested there must be a very profitable business for all factions in the industry.

#### Campaign in Local Papers

If a co-operative advertising campaign was instituted principally in the local newspapers to sell the idea of electrical appliances in an educational way, the demand could possibly be stimulated very rapidly. One feature of this campaign would be the reservation of a full page in the newspaper at least once a month for advertising, supplemented by reading matter supplied by the industry and which the newspapers would undoubtedly be glad to insert free of charge. The cost of this campaign could be borne proportionately by the manufacturers, central stations, jobbers and dealers; while in the aggregate this cost would represent a fairly large amount, it would, when divided over those interested, inflict no great hardship on the individual. This campaign should be handled in such a way that no particular manufacturer would endeavor to sell his goods only, but that all should group together to sell the idea; say as an example of cleaning rugs, draperies, furniture, etc., with a vacuum cleaner doing away with dirt and dust, making more healthful living conditions, removing the irksome task of using a broom, etc. Electrical range manufacturers could sell the idea of cooking electrically with all its conveniences and advantages; fixture manufacturers, the idea of better lighting; washing machine manufacturers the idea of washing clothes the electrical way, quicker, easier and better, Irons, toasters, grills, etc., could be advertised in the same manner.

Vacuum cleaner manufacturers have started a co-operative advertising campaign in the United States through the leading newspapers endeavoring to sell the idea of using a vacuum cleaner for cleaning rugs, etc., and abolishing the old fashioned way of cleaning. These "ads" are 12 inches by 7 inches, and I understand will appear at least once a month. There is no mention made in the "ads" of any particular manufacturer or make of cleaner, and in fact the cut of the cleaner has been arranged so that only the handle and bag of the cleaner can be seen. A campaign conducted along these lines here should bring beneficial results.

The Electrical Home is, as we know, very much along the lines of co-operative advertising and should do a great deal towards familiarizing the public with the desirability of wiring their homes properly and with sufficient convenience outlets to use the electrical labor saving appliances to best advantage.

### Salesmen Wanted

Another problem we are confronted with is the securing of salesmen to retail electrical appliances. The training of salesmen should have the careful consideration of the industry in general. A salesman handling electrical appliances must be possessed of vision, and be able to present his sales arguments in a convincing manner. He must be prepared to meet with many difficulties in closing sales, and be strong enough to overcome all of these. The compensation paid to electrical salesmen should be sufficient to attract high class type of men. The most satisfactory method of compensation seems to be on a salary and commission basis. Straight commission is frequently used, but this has the disadvantage of the salesman being inclined to feel that his time is his own, and that he may employ it as he thinks best. He is inclined to consider that he is not on the company's pay roll, as a regular member of the staff, and that the same devotion to his work should not be expected of him as is generally given by salaried employees.

### Proper Remuneration

At this point I will revert to one of the points in the beginning of my paper regarding the necessity for a greater spread to the dealer between his cost and selling price. As an example, we might take the case of piano sales: pianos in most instances must be regarded as a specialty, and somewhat of a luxury. They sell on the average for say, $400.00. To sell any article of this value must evidently not be a very easy matter. Yet we know that pianos are sold in very large numbers. How is this possible? The piano salesman does not make a sale every day, but when he is successful in placing his wares with a customer, the transaction results in a remuneration to himself sufficient to offset his previous non-success, or times when sales were not forthcoming. Consequently, he is encouraged to keep pegging along, and like the constant dripping of water wearing away the stone, his constant endeavor results in a more or less steady output of pianos.

Could, therefore, the piano dealer continue to have his employment men capable of bringing in a steady volume of orders and conduct the business along the lines he does, unless the spread between the cost price and selling price was sufficient to pay his employees the tempting remuneration required to take care of his expenses, and net him the return he is entitled to. It is, therefore, readily seen that the piano manufacturer helps the piano dealer in no small measure.

### Highest Quality Only

It should be the policy of all employed in the merchandising of electrical appliances to sell only such appliances as are of the highest quality, and which will give the purchaser every satisfaction, under proper operating conditions. Appliances which are not up to a reasonable standard of quality should not be handled by any electrical store, as they will act as a boomerang and retard the trade generally. It should be a moral obligation on the part of the dealer due to himself, to the industry and to the purchaser to sell only such appliances as will create satisfied users, who will pass on to their friends their feeling of satisfaction at the advantages and conveniences received by them from the use of electrical labor saving devices in their homes. A satisfied user is one of the best advertisements we can have, and will be the means of helping increase the demand. Many leads can be secured from them that will result in sales. Generally they will only be too pleased to give a list of names of those likely to be interested in their desire to have their friends share in the same advantages and conveniences.

### Leave Appliances on Trial

In many instances where the dealer is endeavoring to effect a sale, he must be prepared to place the appliance in the prospect's home on trial, so that the prospect may learn for himself the many advantages and conveniences in the use of the appliance. Generally speaking, if an appliance is placed on trial in the home of a bona fide prospect, the sale is practically made, so that the likelihood of the dealer having left on his hands an appliance depreciated as a result of such demonstration may be considered remote. Demonstration in the prospect's home is probably the best way to secure a sale and should be encouraged to the fullest extent. The salesman making the demonstration should be thoroughly familiar with the appliance and be prepared to answer all questions pertaining to it in an intelligent manner.

### Extended Terms of Payment

On the higher priced appliances, such as washing machines, vacuum cleaners, electric ranges, etc., the dealer must be prepared to allow extended terms of payment to the customer if he hopes to make sales in any volume. Payments are usually extended over twelve months, with an initial cash payment on delivery of 10% of the total price. 10% is usually added to the standard list price to cover this accommodation. This addition will take care of interest on money outstanding and allow for extra book-keeping expense. Appliances sold on extended term payments should be covered by a lien form, and in this Province, the landlord must be notified by registered mail that the appliance is the property of the dealer until fully paid for. Where the dealer cannot carry these extended payment sales conveniently, arrangements can be made to have some financing company carry his paper, at a figure not exceeding 10%.

### Manufacturers "Helps"

The manufacturers are generally very willing to supply dealers with advertising circulars, and attractive window displays. Every advantage should be taken of this privilege in such a way as to bring results and justify the manufacturers' expense in supplying this free of charge. Circulars can be sent out with the monthly bills, and also enclosed with parcels.

The manufacturers are also generally prepared to supply electros for use in newspaper advertising. This should also be taken advantage of and where possible the dealers' "Ad" should tie in with the manufacturers'. In any event, wherever possible, the distribution of circulars, window displays and newspaper advertising should tie in together, and be further supplemented by special effort on the part of the salesman on the particular appliance being featured.

### More Confidence & Co-operation

If we are to develop the business to the extent we hope, we shall have to have more confidence in each other and more co-operation amongst the different factions is necessary. Instead of looking on one another as competitors in the strict sense of the word, we should feel that each and every one is doing his part towards developing the business and that by the law of average, those legitimately entitled will secure their share of business being offered.

Each faction of the industry has a function to perform in the development of the business, and the growth will be helped by a sincere spirit of co-operation throughout. Through co-operation, acquaintances are made, and that confidence established which is so necessary in any business. There should be a bond of understanding between the different factions, the function of each realized and appreciated. The difficulties in the business are not confined to any one particular branch.

The central station is anxious and willing to co-operate towards the development of the business and can be of inestimable value to the other groups. It should be the policy of all engaged in the electrical industry to boost electrical affairs in general and without detriment to any particular faction, whenever possible, so that the public may have confidence in the industry as a whole.

For instance, manufacturers, jobbers and dealers can show their faith in the industry by endeavoring to create good will towards the central station by explaining to the best of their ability the desire of the central station to supply service to the community in a satisfactory manner, explaining, if possible, some of the difficulties which are encountered in so doing.

As the growth of the merchandizing of electrical appliances will largely be measured by the growth of the central station, it obviously is imperative that this spirit of goodwill be fostered to the full. On the other hand, the central station must be prepared to reciprocate with manufacturers, jobbers, dealers and contractors in the solution of their problems and co-operate with them in every manner possible.

### Central Stations and Merchandizing

The question has frequently been raised in regard to the central stations conducting a merchandizing business, but there can really be no legitimate objection to this providing the merchandizing is carried out on a strictly ethical basis. The central station should sell at list prices, and conduct this end of their business on exactly the same basis as a dealer has to do in order to make a profit. When this is done, practical experience has proved that the fact of the central station being in the merchandizing business has, through their aggressive selling and advertising efforts in that territory, stimulated the demand for appliances and

wiring which is shared by all. On the other hand, where the central station had dropped out of this end of the business, it has been proved that the dealers' sales have dropped off very materially. There is, therefore, a distinct advantage to the dealer in territories where the central station is in the merchandizing business, conducted in an ethical manner.

We must realize that the selling of generators, motors and electrical equipment to central stations and large industrial establishments, where the demand generally speaking does not have to be created, and where we are dealing with men who know what they want, is quite different than that of selling appliances and supplies where the demand must be created and where we are dealing with people who are not familiar either with the mechanical or electrical features of the appliances we are endeavoring to sell. We must, therefore, recognize these two different phases of sales conditions and adapt ourselves in either case to the selling methods necessary to bring in results through which all the industry will benefit.

### Reduced to Figures

Just let us think of the tremendous possibilities in figures, and we will have more conception of what they mean.

Our Province has a population of approximately 2,-800,000. Using a figure of five as being the average number of persons per family, gives us an approximate total of 560,000 homes, of which 200,000 are now users of electric service. This leaves 360,000 houses still unwired. Taking an average cost of wiring each house as $60.00, and $40.00, for fixtures, we have a total of $36,000,000. of business. As the central station must spend approximately $40.00 for each new customer taken on, to cover material right back to the generating station, if these 360,000 prospects could be supplied with central station service, it would mean an additional amount of business of $14,400,000. Where the central station might not be able to supply some of the 360,000 prospects, they would then be prospects for isolated lighting plants, representing where there were any great number of them, a large amount of money.

Now let us consider the possibilities in appliances. We already have 200,000 users of electric service, of whom we are safe in saying only a few have appliances in their homes. These we may, therefore, consider as immediate prospects, and together with the 360,000 to be developed, gives us a total of 560,000, to whom we can endeavor to sell our appliances. Each of these should be a prospect for appliances to the value of, say, $500.00 made up as follows: electric range, $125.00; washing machine, 135.00; vacuum cleaner, $60.00; irons, toasters, $12.00; air heaters, $10.00; water heater, $30.00; electric sewing machine, $60.00; fan, $20.00 percolator & teapot, $35.00; heating pads, $15.00; etc. etc.

In the aggregate this amounts to $100,000,000 possible sales to present users of electric service; and to those to be developed $180,000,000, or a grand total, including wiring, central station equipment and appliances of $330,000,000; and this without taking into consideration the growth which is bound to take place in our Province.

Therefore, if we can get our message across in an intelligent and convincing manner, we can divert money into electrical channels, which is now being spent on automobiles, pianos, phonographs, etc.

### Selling Ourselves

Before, we can hope, however, to get our message across, we must be first of all 100% sold ourselves on what these labor saving appliances will do, and what they mean to the home. One way to accomplish this is to encourage

ances in his own home, and if he is able to tell the story from actual experience, a very much more convincing message can be given, and one which will carry more weight.

In conclusion, the electrical labor saving appliances we are trying to sell to all users of electricity will help to make duties easier, and more pleasant; the household tasks which are now irksome will be lightened, and the work done more easily, more quickly, more efficiently, more economically, and more pleasantly. Better and more healthful living conditions will result. The duties of the store, office and factory will be done easier and better with a pronouncer saving in both time and money.

When we ourselves are convinced that these wonderful benefits are actually secured through the use of our electrical labor saving appliances, we can be very sincere in our efforts to place them in every home and know they will give value in full measure for every dollar invested.

Mr. H. G. R. Williams has opened up a business at 12 South Main St., Welland, Ont., as an electrical contractor and merchant. These premises are temporary, however, on account of the difficulty of securing a more suitable location. M. Williams hopes, in the near future, to be moving into more suitable premises.

Shall Be Done", secures, despite apparently unsurmountable difficulties, a certain blue vase which he had been commissioned to obtain? This tale of a true fighter has lifted many a salesman from the rut and set his feet on the road to success. It remained for The Hoover Suction Sweeper Company, however, to apply "Bill Peck's" principles to a large selling organization.

This company has set aside a certain quota of Hoover cleaners for each of its salesmen, and its district and divisional managers. Those whose sales equal or better their quota figures will qualify as "Go-Getters" and will receive a handsome blue vase, symbolic of "Bill Peck's" feat and their duplication of it, and in addition all "Go-Getters" will be taken to the annual Hoover convention this summer at North canton, Ohio, at the company's expense.

The entire Hoover organization is greatly enthused over the contest and every manager and salesman is putting forth every effort to beat his quota and win the blue vase. Under the co-operative plan which is in effect between the Hoover company and its dealers, whereby the company's salesmen sell directly for the dealer, it is expected also that many electrical stores handling the Hoover will benefit from the contest in the increased sales that the salesmen, working earnestly to win the title of "Go-Getter", will make.

## Special Showcase Lighting

One of the problems of store lighting, which has vexed the electrician as well as the storekeeper, is that of satisfactory lighting for show cases. There is either too little light or else the glare of a sufficiently strong lamp dazzles the customer without properly illuminating the display in the show case. A recent installation in the stores of David Spencer Limited, Vancouver, is giving excellent service with a maximum of luminosity in all parts of the showcases. There were fifty in all of these showcases in which the reflectors were installed. The cases are wired with the lights and reflectors at the top in front. The effect is to direct the lights to the side walls as well as downward, diffusing the illumination evenly and lighting the entire display of goods in the case. The reflectors illustrated are the National X-Ray type and were supplied by the Northern Electric Company.

# A "One Man" Electric Home

Sometimes a man gets an idea that looks good, but lacks the punch to put it over. And other times, we find men with lots of punch and few ideas. The happy combination puts over stunts that look mighty difficult to the pessimists—so we've got to call Mr. R. Simpkin, of Weston, Ont., a happy combination. He has accomplished the task of putting on an Electric Home all by his lonesome.

Quite some time ago Mr. Simpkin came to the conclusion that though the people of Weston were ahead of the times in many other ways, their ideas of the value of electrical appliances might be changed to his advantage. So he began to cast around, and as a result:—

1. He secured, gratis, from a builder, a house for his purposes.

2. He persuaded a local firm to furnish the said house.

3. He filled the house with electrical appliances, including a radio set.

4. He put on an Electrical Home Exhibit.

As the house had already been wired for the builder, by himself, there was no outlay for that item. Newspaper advertising, demonstrating staff salaries and his own time were the only expenses, current being supplied free by the local Hydro Commission. Mr. Simpkin figures the whole cost of the exhibit at about $150.

The display included the usual appliances, wired furniture and electric kitchen. As there is no gas supply in Weston, the range and water heater attracted more than usual attention.

Before and during the exhibit invitations were sent out by mail; newspaper space was used and this publicity was supplemented by cards tacked up in conspicuous positions.

Visitors came in a steady stream. The bungalow was open for 10 days, and the demonstrators were kept busy from 10 a.m. to 10 p.m. A happy circumstance was that at the time newspaper radio concerts were available.

Results in the concrete form of orders for wiring, appliances and radio sets, are coming in fast. Mr. Simpkin is more than pleased with his effort.

## What Seven Contractor - Dealers
## —Working together—can accomplish

# The Electric Show in Kitchener, Ontario

### The biggest Thing of its Kind in recent years—Stimulated Public interest and demand for Electrical Appliances

Was the Kitchener Show a success?

Ask the electrical Contractor-Dealers.

Kitchener is a city of about 20,000 inhabitants. There are seven electrical dealers and, "Hydro" supplies the juice.

Nothing unusual about that?

Why, no, not exactly, but when you come to look below the surface there is something unusual about Kitchener. How does the following line-up appeal to you?

1. Every one of the seven contractor-dealers is a member of the local Contractor-Dealers' Association and of the Ontario Contractor-Dealers' Association.
2. Every one of them is on speaking terms—friendly; in fact—with every other.
3. Every one of them is a Co-operator.
4. V. S. McIntyre, the Hydro manager, gave substantial backing.
5. "Pop" Phillip, who put the show on, worked about 24 hours a day for the last two months, co-operating with everybody and—it's his own secret how he manages it—getting everyone to co-operate with him.
6. Kitchener is a real wide-awake little city whenever there's business going.
7. The manufacturers and the Ontario Hydro co-operated, too.

Briefly, those are the reasons why the Show was such a huge success.

We said Kitchener had a population of about 20,000. Well, during the week of May 1-6 fifteen thousand of them visited the Show, inspected the electrical goods, asked questions and (a goodly number of them) carried home appliances. A lot more gave their names to the contractor-dealers, who will keep in touch with them till their wants are supplied. The dealers are all enthusiastic about the amount of business that has been stirred up.

We also said the contractor-dealers co-operated. There was never anything like it before in the industry. The seven different contractor-dealer firms occupied one large booth in the centre of the hall—the same booth—pooled their interests—sold their various merchandise over the counter—sold one another's merchandise—and apportioned the profits. Just think of it!

How could it fail to be a success?

The pictures on the following pages tell the rest of the story. There were 60 exhibitors in all, representing equipment on display to the value of perhaps $300,000. The attendance was highly satisfactory. Not only has it been estimated that three out of every four people in Kitchener visited the Show, but there were visitors from outside points all over Ontario. Mr. Philip, with whom the idea of the Show originated, understands the psychology of the show business. He doesn't believe in giving anything away for nothing. His idea, rather, is to sell something for a price and give value. The booths for the week cost $65. each. The entrance fee was twenty-five cents. Did this keep anybody away? Mr. Philip thinks not. On the

contrary, he believes that it was the means of bringing in a number of substantial people who would not have come otherwise, and it kept the disinterested class away—the people who were merely looking for entertainment. Mr. Philip estimated that at least 90 per cent of those who visited the Show were interested in finding out something more about electrical appliances and what they will do.

The contractor-dealers of Kitchener have still another laudable characteristic. They believe in themselves. They believe that electric stores are the proper places to purchase electrical goods and they believe in telling the people about it in plain courteous English. One of the best pieces of propaganda work at the Show was a little booklet gotten out by these seven dealers, which reads as follows:

---

### Do It Electrically

Through the combined efforts of these Kitchener and Waterloo Electrical Contractor-Dealers supporting the plan of an electric show, has been made possible the presentation of this beautiful electric exhibition to you, and at a considerable expense.

The prime object is promotion of the electric idea, exhibiting and demonstrating all the newest, latest, and most dependable electrical appliances on the market.

These Electrical dealers are men of experience and you can rely on their advice, as being dependable.

#### Buy Your Electric Wants From an Electric Store

You then deal with the electric trade direct and if you require any repairs or service, they are in a position to intelligently serve you and give you satisfaction.

The natural outlet for electric merchandise is through an electric store and not through merchants who merely handle electric goods as a side-line, for the exploitation of temporary profit. When you need service they are not in a position to intelligently adjust your troubles, and then you have to call in your electric dealer. For that reason he is fairly entitled to the business in the first place.

The dependable electric dealer does not select goods for mere profit and sale, but his long experience enables him to select the best electric merchandise, backed up by manufacturers of unquestionable standing and you are assured of getting the best, and that is really what you want.

Let us serve you and you will be pleased with the results.

Yours for better service.

Kitchener and Waterloo Electrical
Contractor-Dealers
—Signed;

Doerr Electric Co.
Electric Service Co.
Mattell & Bjerwagen
Ellis & Howard, Limited
C. F. Schmidt
Star Electric Co.
Waterloo Electric Shop

# 44

THE ELECTRICAL NEWS

# Contractor-Dealers of Kitchener Practice
# Gospel of Co-operation

# Some Attractive Corners in the Recent Kitchener Electric Show

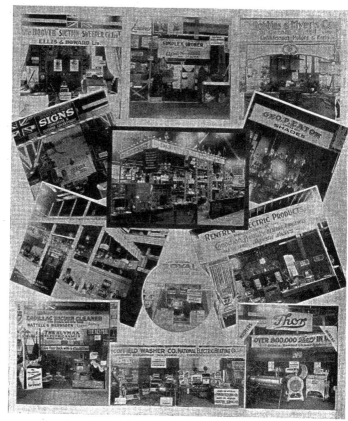

Top row—Hoover Suction Sweeper Co; Canadian Ironing Machine Co.; Robbins & Myers Co. Second row—Eureka Vacuum Cleaner Co; Central booth occupied by the Seven Contractor-Dealers of Kitchener; Geo. P. Eaton. Third row —Galt Electric Fixture Co.; Royal Electric Suction Cleaner; Renfrew Electric Products, Ltd. Bottom row—Cadillac Vacuum Cleaner & Klymax Washer Co.; Coffield Washer Co.; National Electric Heating Co.; Thor Washing Machine Co.

Everything on exhibit was for sale and it would appear that a considerable amount of business was done right on the spot. As already noted, seven local dealers pooled their interests in a large central booth and made a vigorous bid for business. They seemed thoroughly well satisfied with results.

Of course Sir Adam Beck was on hand to open the

"Pop" Philip

Show. It must have been a matter of keen satisfaction to Sir Adam, to recall that almost exactly twenty years ago the first meeting to consider and discuss the terms of a union of hydro municipalities had been held in the same city and in the same hall. Never has the short period of twenty years witnessed so great a development as that shown by the electrical industry during the last two decades. Sir Adam considered this the greatest electrical exhibition of its kind ever held in the province.

Not only were the contractor-dealers well satisfied with results but so were the manufacturers. The Show brought

in contractor-dealers from all over Ontario, and a number of manufacturers' representatives, in conversation with the Electrical News expressed their entire satisfaction with the number of sales they had made to visiting contractor-dealers. Such expressions as "Well satisfied"; "Very good results"; "Well satisfied with results"; "Made several sales each day", etc., were heard in practically every booth.

### The Baseball Match

There was a baseball match, too, on Friday afternoon. Business was a little slack immediately after luncheon, so a picked team of visiting exhibitors challenged the local electrical baseball team. The game was fast and very furious and the scorer soon lost count. The question of who won the match has not yet been raised, but it was a delightful afternoon and everybody who was there was glad of it.

### The Exhibitors

The following is a list of the exhibitors:

Ingersoll Machine & Tool Company; A. H. Winter-Joyner Company; Canadian Shade Company (two booths); Robertson Manufacturing Company; Square D Company; George P. Eaton; Robbins and Myers; A B C washer Company; (C. D. Henderson) Masco Company; Canadian Ironing Machine Company; Apex Sweeper; Elliot Machine Company; Marconi Company; Baetz Bros. Specialty Company (two booths); Onward Manufacturing Company; Renfrew Electric Products, Ltd.; Easy Washer Company; Canadian Edison Appliance Company; Hydro Electric Power Commission of Ontario; (four booths); Bluebird Corporation, Ltd.; Charles Branston, Hoover Suction Sweeper Co.; Beatty Bros.; Hughson Neulife Company; Cadillac Vacuum Cleaner Company, Empire Brass Company, Coffield Washer Company, National Electric Heating Company; Continental Electrical Company; Canadian Westinghouse Company, 1900 Washer Company; Twin City Signs; Thor Washer Company; Galt Electric Company (two booths); Moffat Stoves (six booths); and the Twin City electrical dealers: Waterloo Electric Shop, C. F. Schmidt. Star Electric Company. Mattell and Bierwagen. Ellis and Howard. Electric Service Company and Doerr Electrical Company.

---

The Sun Electric Company, Yates St., Victoria, B. C., has been awarded the electrical contract on a Club House being erected at Colwood, B. C., by the Colwood Golf & Country Club.

Some of the latest developments in Electric Fans

# The Architect and the Contractor

### The Architect is a Vital Sales factor from Start to Finish of the job
### —An indispensable link between Manufacturer, Contractor and Owner

#### By our Montreal Correspondent

"It is just a waste of time to try to do anything with the architects" is the almost universal verdict of the electrical contractor-dealers, whenever business conversation happens to drift on closer co-operation between the electrical contractor-dealers and the architects. The architects, perhaps, are the most difficult professional element in the construction industry to be drawn into intimate relations with the electrical interests, but there are other causes of the difficulty than those ordinarily attributed by the average electrical contractor-dealer—the snobbishness, because of his university education and the A.R.I.B.A., or a few other capital letters appended to his name.

There is no doubt that some of them can be accused of overestimating their own importance due to the reasons mentioned above, but, it is just as true of some men in other professions and occupations. A friend of mine, an honest, hard working chap, was recently introduced to a prominent business-man by one of their mutual acquaintances as Mr. So and So, K. C. Mr. Prominent Business-man was most courteous and hospitable towards my friend, but when asked how long he had been a King's Counsel and my friend informed him that he is just an electrical contractor and not a lawyer, and that his K.C. means a membership in the Knights of Columbus, the hospitality and most of the courtesy vanished. Mr. Prominent Business-man discovered that he had very urgent business to attend to and parted company with the electrical contractor So and So, K. C.

#### The Situation As it Exists

Let us analyze the situation as it actually exists, and discuss how we can improve the situation to bring about closer relations between the architects and contractor-dealers. The architect is the most important sales factor in the field, for in all matters he advises, and in matters outside the ken of the client, he decides.

In matters where the client has preconceived ideas about material and equipment—though these cases are not very numerous as yet in matters of electrical installations—the architect becomes the salesman either for or against these ideas. At all events, nothing of material importance can go into a job unless the architect considers that it is the best and most suitable that can be obtained for the money his client places at his disposal.

Any attempt to confine the effort of selling to the owner directly, to the exclusion of the architect is going to fall down. The architect's procedure is to take the more or less vague idea of the client, plus a more or less fixed sum of money, and evolve a tentative proposition visualized by sketch drawings showing the proposed arrangement and design. This is followed by the preparation of scale working drawings, specification, etc., leading up to the letting of the general contract. Up to this point both the architect and the owner have been planning for the things most desirable, but now they must fit these to the budget, which for many products and materials means elimination as well as substitution.

Those that remain are the ones on which the architect has been thoroughly sold.

Finally comes the supervisory stage, when the architect follows through the job, to satisfy himself that his specifi-cations are lived up to and to solve the various unforeseen problems apt to arise during the process of actual construction.

#### Have We Done Our Part?

At each of these stages the work, material and effect are discussed by the architect and the client, but have the electrical manufacturers and contractor-dealers done their part towards familiarizing either of them with what they have to offer?

The thoroughness with which the architect has been sold is often reflected during the supervisory stage of the work, and if inferior products are being used as substitutes instead of those specified, it is because the architect has not been thoroughly sold and has not been given the opportunity to acquire the knowledge of the thing that he has specified.

The architect is a vital sales factor from start to finish of the job, an indispensable link between the manufacturer, the contractor and the owner. The architect is perpetually selling this product or that, in proportion as he has previously been sold on the merits of the product, plus his own past experience or that of other architects whose names carry weight. He is singularly like other men, his own business being essentially a selling proposition, supported by the knowledge of ideals, economics and mechanics of construction products together with ability to analyze, judge, apportion, combine and correlate them. It must be remembered also that he is surrounded by a multiplicity of detail, much of which is altogether foreign to his artistic inclinations; therefore he must be given facilities for getting detailed information pertaining to such requirements when he needs it.

During the recent "Modern Electrical Home" exhibition in Montreal, several architects, visitors to the Home, admitted that such an educational campaign is one of the best means of establishing electrical installations and equipment, in the public mind, among the primary essentials of modern structures. The architects nevertheless have to be brought to the point of co-operating with the electrical industry in achieving the sought-for results, despite the opinion of most of them that as soon as the public are sufficiently educated to demand proper electrical installations and equipment they, the architects, will meet their clients' demands.

#### The Immediate Duty

The immediate work of the electrical industry, in which all the branches are vitally interested, is to simultaneously educate the public to the wider use of electricity and to offer facilities to the architects to familiarize themselves and to keep posted with the ever increasing advantages and progress in electrical illumination, heating and labor saving devices.

The Electrical Co-operative Association, Province of Quebec, has recently sent a letter to some two hundred architects in the province soliciting their suggestions how to best bring them and the electrical industry into closer contact, and it is planned to have later in the fall joint luncheon meetings between their representatives and those of the association to develop closer co-operation. Arrangements are also being made to maintain, in a convenient and suitable uptown place, an exhibit of electrical supplies, which would be accessible to the architects for inspection.

# The Manufacturer Can Often Assist the Dealer With New Sales Ideas

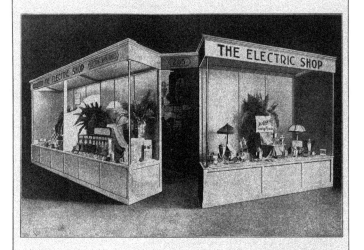

At a recent meeting of Electrical Contractors and Dealers at the Wallace Ave. Plant of the Canadian General Electric Co., the Company displayed a model store, both interior and exterior, as a help to the members. Some very excellent ideas were demonstrated.

## Enacted a Year Ago, the Act Has Remained a Dead Letter. Representation by Electrical Organizations Have Won Over the Government to Action

Electrical Contractor-Dealers in the Province of Quebec are rehearsing the ever popular song "There are Smiles that make us Happy, there are Smiles that make us Blue." Undoubtedly some feel happy and some very, very blue.

The provincial Government of Quebec published in the official Gazette on the 21st of May that on March 19th there was assented to and, by proclamation of the Lieutenant Governor in Council put into force, on May 23rd, an act amending the Revised Statutes, 1909, respecting the protection of public buildings against fire, II George V, chapter 75, popularly known as the "Electrical Licensing and Inspection Act." All this happened and was dated A. D. 1921.

Persistent rumors state that the system of selling electrical appliances was decided upon by the Government as the model for putting that law into operation. It was truly enforced on the installment plan. For over a year following the proclamation of that act, nothing was done by the Government whatever to protect the loss of property and lives from improper electrical installations and to prohibit incompetent men doing electrical wiring work.

The Electrical Co-operative Association, Province of Quebec and the lately reorganized Montreal Contractor-Dealers Association, both French and English, during the past few months made very strong representations to the Government officials directly connected with application of that act, regarding the urgent necessity of putting it into practice. This coupled with the recent disastrous fires in the Province which were attributed, and in some instances proven to be caused by, defective electrical wiring, had the desired effect.

### Being Put Into Force

The Electrical Licensing and Inspection laws are at last being put into force. The preliminary work in connection therewith is already attended to, and it is hoped that within the next few months there will be no electrical contractor or journeyman without a license and a certificate of competency wiring as much even as a hen-coop.

To begin with, the new law will be put into practice in Montreal and the adjoining municipalities and in the city of Quebec; the rest of the Province will be taken care of at a little later time.

The near-beer, so common in other parts of the Dominion, is an absolute stranger to the citizens of the Province of Quebec, but the near-electrician can be found everywhere. He has, of course, a most detrimental and discouraging effect on the legitimate contracting business from every point of view, at the same time lowering among the public the idea of value and quality in electrical work.

### Near-Beer Electricians

Those who have the qualifications to be electrical contractors or journeymen wear the smiles that make them happy; the near-bear variety electricians wear the "blue" kind of smiles. The latter will have to reconcile themselves with the fact that the electrical contractor-dealer and journeyman field is not a dumping ground for everybody and nobody, and that one must be skilled in this important trade and responsible in business to remain in it.

There is no doubt that by the time the next Legislature meets many improvements in the act will be sought; one of them will be the widening of its scope to include private dwellings as well. At last however a beginning is being made, which with a wide awake and efficient chartered Contractor-Dealers Association, with branches all over the Province, may serve to considerably improve the electrical contracting business and to assure safe electrical wiring to the public.

### "§ 2.—Installation of Electrical and Heating Systems

"3789b. Every new installation, either for light, heat or motive power, as well as every heating system, in public buildings, must be submitted for the approval of the chief inspector of industrial establishments and public buildings, and to the examiners appointed for that purpose.

"3789c. Except in the cases hereinafter provided for, no person or company shall, carry on any business, undertake or work at the installation of wires, conduits or apparatus for the transmission of electricity, for producing light, heat or motive power, in this Province, as a contractor or as a journeyman electrician, unless such person or company has obtained a license from the examiners appointed for that purpose.

"3789d. Every installation of any kind of heating system in a public building already installed or which may hereafter be installed, must be approved by one of the inspectors of public buildings, who shall give a certificate to that effect to the owner of the building. Such certificate must be constantly kept posted up in a place indicated by the inspector.

"3789e. No electrical installation in a public building in the Province, for the transmission of light, motive power or heat, can be made or altered otherwise than by a person or under the supervision of a person duly authorized and having a license to that effect.

"3789g. The chief-inspector of public buildings may, with the approval of the Minister of Public Works and Labor, declare any electric installation or any heating system already in a public building, to be defective, and order the necessary alterations to be made, and, in default of compliance with the said inspector's orders to that effect, the owner shall be liable to the penalties provided by article 3782 and 3783.

### "§ 3.—Examiners

"3789h. 1. The Minister of Public Works and Labor may appoint a board of examiners consisting of three members, who must be competent electricians, not under twenty-five years of age, and having at least five years experience as journeymen electricians. The persons so appointed must be able to speak and write French and English correctly.

2. The duties of such officers shall be as follows:

a. to examine all electric and heating installations submitted to them;

b. to examine all those desiring to become electricians, to issue certificates of competency and grant licenses;

c. to hold examinations in such places as the Minister of Public Works and Labor may be pleased to select;

d. to draw up a programme for the examinations, prepare forms and other documents for the same, collect fees, keep registers and facilitate the inspectors' work as much as possible.

"3789i. No apprentice, laborer or person not provided with a certificate of competency, shall have the right to put in electric installations, except as assistant under the immediate direction of a Journeyman electrician with a license.

"3789j. Every company, association or person whose place of business is outside the Province of Quebec, and who wishes to undertake or furnish electric installations, under the provisions of this section, must appear before the board of examiners and obtain a temporary license allowing him to continue his operations during the time required for completing his contract. Such license shall expire as soon as the work is finished.

"3789k. No certificate or license issued under this act or the regulations may be transferred or conveyed; and every such license and certificate may be suspended or cancelled by the board of examiners for sufficient reasons. Such suspension or cancellation shall, however, be subject to appeal to the Minister of Public Works and Labor, and his decision shall be final.

"3789l. Every license issued to electrical installation companies or contractors, must be posted up in the offices of such contractors or companies, and every journeyman electrician, moving-picture operator or holder of a special license, must always carry a copy of his certificate on his person. Any omission to post up the license or neglect to carry the certificate required by the regulations, shall be prima facie proof of lack of qualification.

"3789m. Proof of the fact that a contractor, company, or association employs an unlicensed person for an electrical installation, or that such installation is done contrary to the regulations adopted to that end, or that the license was obtained under false representations, shall be considered sufficient cause, under the provisions of article 3789k, for cancelling the license of such contractor, company, corporation or association.

"3789n. Certificate shall be issued for the year, and must be renewed yearly between the first and fifteenth of May of each year.

Licenses of contractors, companies, corporations or associations may be issued at any time, but ten days notice must be given to the examiners.

"3789o. Certificates shall be given for the year and be renewed annually between the 1st of May and the 1st of October of each year. The rates for fees, as regards the examination of the installations mentioned in article 3789h, shall be based on the percentage of the value of the installation, to wit: one-half of one per cent, with a minimum of ten dollars. The rate for special licenses issued to non-resident contractors doing business within this Province shall be one per cent on the value of the contract with a minimum of fifty dollars.

## "§ 4.—Certificates and Licenses

"3789p. 1. Five license forms shall be issued, designated as follows:

License A, which may be issued to any person who has satisfactorily passed the examination prescribed for journeymen electricians, and has filed an application to be registered as a contractor or master electrician in the examiners' office and paid the fee prescribed by this section;

License B, which may be granted to any company, association, corporation or firm doing or wishing to do business as contractor for electrical installation, provided one of the members of the said association, company, corporation or firm, or at least one person in its employ, holds a certificate of journeyman electrician given by the examiners, and that the fee for license has been paid;

License C, which may be given to a journeyman electrician, having at least five years experience, and who, after passing his examination successfully and complying in every respect with the prescriptions contained in the forms prepared by the examiners, has paid the fee prescribed by this section;

License D, which is that authorizing a person to take charge of a moving-picture machine. Every person applying for this license must be not less than eighteen years old, pass an examination before the examiners, obtain a certificate of competency and pay the fee hereinafter prescribed.

The operator is especially required to keep a copy of the license granted him posted up in a conspicuous place;

License E, which is the special license authorizing a person with a knowledge of electricity and employed in a public building, to do work in connection with the repair and maintenance of electrical installations in the said public building.

The person applying for such special license must pass an examination before the board of examiners.

2. Every person operating a machine driven by electricity, such as winches, derricks, travelling cranes, or any other machines more or less dangerous to the operators, workmen or the public, must obtain a license.

"3789t. 1. This section and the regulations enacted thereunder shall not affect the work in electrical stations or their branches where electric power is generated, either by a public service corporation or a municipal service, where the work is done by the employees under the control and direction of the officers of the said corporation or municipal service;

2. The following are excluded from the effects of this section and of the regulations:

a. telephone and telegraph installations where power is supplied by primary galvanic wires;

b. locomotives, cars and tramway systems operated by a public service;

c. the installations of arc lamps, used for lighting streets and public roads and operated by a public service;

d. the lights and glass bulbs, used in private houses, and the installation or preparation of carbons in arc lights in public streets.

## "§ 6.—Examinations

"3789u. Every person wishing to obtain a certificate either as a journeyman electrician, an operator of a moving-picture machine, special operator or as being in charge of hoisting apparatus as electrician, must send an application to the board of examiners on a form supplied to him for the purpose. He must give information regarding the duration of his service in his present employ and also give the board of examiners satisfactory information regarding his conduct and sobriety.

"3789v. The code known as the "National Electric Code" shall serve as a basis in drafting the examination programmes, as well as the forms and questionnaires to be used by the examiners upon the examination of candidates for certificates as electricians;

The examiners may also require from the candidates a practical as well as a theoretical demonstration in the installation of electric power in the buildings mentioned in this section.

"3789w. Every stationary engineer holding a first or second class diploma has the right to do improvement and repair work on electrical apparatus, but only in the workshops or factory where he is regularly employed.

## "§ 7.—Penalties

"3789x. The following are liable to the penalties provided by article 3782, namely: every firm, company, corporation or person contracting for electrical installation work for the production of light, heat or motive power, without a license or without being under the supervision of an electrician holding a diploma under this section or the regulations.

"3789y. The following shall be liable to the penalty provided by article 3782, namely: every journeyman electrician, operator of a moving picture machine, as well as every person in charge of repair or maintenance work in the buildings mentioned in this section comprised under the name of "Special License E," as well as every person in charge of the operating of electric machinery mentioned in the regulations, who neglects or refuses to pass his examinations before the examiners, or who, without holding the certificate required by law and by the regulations, contracts for electrical work either in connection with the installation of wires or other electrical apparatus, or for the operation of the said electrical machines.

Inspection of the electrical installations and work as yet is confined to public buildings, which according to the act mean and include: churches and chapels, or buildings used as such, seminaries, colleges, convents, monasteries, school-houses, public or private hospitals, orphan asylums, infant asylums, charity workrooms, hotels, boarding houses, capable of receiving at least fifteen boarders, theatres, halls for public meetings, lectures or amusements, buildings for the holding of exhibitions, stands on race courses or other sporting grounds, buildings in parks, skating rinks, rooms for showing moving pictures, buildings of three stories or more over the ground floor occupied as offices, stores employing at least ten clerks and court houses.

This act should prove of interest not only to the contractors in the Province of Quebec, but to contractors of Ontario as well, as occasionally large jobs in this Province are being successfully tendered for by outside concerns.

## The Development of Wireless-V

### A Series of Short, Interesting Articles Covering Wireless Progress to Date

By F. K. D'ALTON

From the time that wireless communication was first applied it has been the aim of scientists, who worked with these waves, to bring to perfection the radio telephone. Most of the improvements made in wireless telegraph systems have brought the telephone closer and it might be well to list again the steps which we have already described.

The earliest detecting devices indicated the presence of the waves but did not give their characteristics. Then came the magnetic, crystal and electrolytic detectors giving the necessary quantitative response, and being self-adjusting. These were followed by the valves giving clearer and louder signals which by amplification through valves were increased to any desired intensity as sound.

The spark transmitter gives the damped wave which has in it a definite note determined at the sending end and therefore could not be modulated by the voice to such an extent that the inherent note would be overcome. The continuous wave alternator or arc transmitter overcame this difficulty but there was not a suitable telephone transmitter available which could carry the large currents necessary for giving good radiation and working several miles and there were no further developments toward the wireless telephone until it was found that the oscillating three-element valve was capable of generating high frequency currents and of radiating waves of radio frequency at the same time being conveniently controlled by an ordinary telephone transmitter working through a small "modulation" transformer on the grid of this same valve.

The telephone transmitter carrying a small direct current is capable of varying this current in accordance with sound waves from the voice or a musical instrument. This varying current is carried through the primary winding of the modulation transformer and in consequence the voltage induced in the secondary winding has variations corresponding exactly with the changes in primary current. Now, a three electrode valve is connected as follows:—

Between filament and grid are connected the secondary winding of the modulation transformer in series with a coil of wire known as the "grid coil".

Between the filament and plate are connected a source of voltage (d.c.) and another coil known as the plate coil. Variations in plate current, through the magnetic coupling of coils cause changes in grid voltage and the reverse is true through the action of the valve.

The oscillating current resulting from this combination is made to build up a current in a third circuit consisting of another coil placed near to the plate coil, and the aerial as capacity. The consequence is that a high frequency current flows into and out of the aerial. The grid voltage is varied by the

### STATIONS WITHIN REACH OF TORONTO'S RADIO FANS

| Time signals:— | Wave length (meters) | Hour received | Type of Emission | Distance in miles |
|---|---|---|---|---|
| Arlington, Va. | 2500 | 12 a.m., 10 p.m. | Spark | 340 |
| Annapolis, Md. | 17000 | 12 a.m., 10 p.m. | c.w. | 430 |
| San Diego, Cal. | 10000 | 3 p.m. | c.w. | 2200 |
| San Francisco, Cal. | 5000 | 3 p.m. | c.w. | 2200 |
| Balboa, Panama. | 6800 | 3 p.m. | c.w. | 3100 |
| Weather:— | | | | |
| Toronto | 600 | 10.05 p.m. | spark | 10 |
| Midland | 600 | 10.05 p.m. | spark | 110 |
| Market reports, sport results, world news:— | | | | |
| Pittsburg, Pa. | 330 | 6, 7, 8.00 p.m. | radiophone | 235 |
| Music:— | | | | |
| Marconi, Tor. | 1200 | 8-10 p.m. Tues. | radiophone | 10 |
| Can. Ind. Tel., Tor. | 450 | 8-9 Mon. Thurs. | radiophone | 7 |
| Pittsburg, Pa. | 330 | 8.30-9.30 p.m. | radiophone | 235 |
| Newark, N. J. | 360 | 8.15-9.20 p.m. | radiophone | 345 |
| Springfield, Mass. | 360 | 8.15-9.20 p.m. | radiophone | 340 |
| Chicago, Ill. | 350 | 8.00-11.00 p.m. | radiophone | 450 |
| Church Services on Sundays:— | | | | |
| Pittsburg, Pa. | 330 | 10.45-12 a.m. | radiophone | 330 |
| | | 7.45-9.00 p.m. | radiophone | 330 |

The relative signal strength obtained by various transmitters and receivers, without amplifiers is shown in the following table:—

| Transmitter | Receiver | Signal strength |
|---|---|---|
| Spark | Crystal | 1 |
| Spark | 3 electrode valve (not oscillating or regenerating) | 3 |
| Spark | Valve (regenerating) | 10 |
| Spark | Valve (oscillating & regenerating) | 75 |
| Buzzer modulated C.W. | Valve, not oscillating or regenerating | 36 |
| Arc—C.W. | Valve (oscillating) | 300 |
| Oscillating Valve—C.W. | Valve (oscillating) | 600 |
| Radiophone | Valve (regenerating) | 125 |

voice through the agency of the telephone transmitter and modulation transformer and the energy in the radiated wave varies accordingly. The amplitude of the wave follows another wave which is a duplicate of the voice wave.

A wireless telephone could be detected on any receiver giving quantitative response—several of which have already been described.

The radio telephone then became a reality.

Its popularity is rapidly increasing as it does not require that the operator at the receiving end know anything of the conventional codes used in telegraphy, he only needs to have a fair knowledge of how to tune his receiving devices.

There are many methods of connecting the wireless telephone transmitter, only one of which is described here and it is usual to find a much more elaborate scheme in use, especially in the higher powered stations.

The scheme we have outlined is one in which the radiated wave is modulated by the voice controlling the grid voltage. Another good arrangement causes the voice to control the power supply; in still another scheme the voice controls the plate voltage. These have their advantages each one being found best in its own field and between certain limits of wave length and power.

Manufacturers of radio apparatus find the wireless telephone most useful in advertising their goods and entertaining their customers. They reach directly the very people with whom they do business and the people who understand the usefulness and advantages of wireless apparatus.

Foremost in this, is the Westinghouse Company who operate four wireless telephone stations—namely at East Pittsburgh; Newark, N. J.; Chicago, Ill.; and Springfield, Mass. From these stations every evening there go out concerts lasting for one, two or three hours, and in addition announcements of baseball, football and tennis scores, world news of interest, stock market and live stock reports, police reports of cars stolen, children lost, etc., and on Sundays the services of various churches, complete, including the sermon.

President Harding's speech on Armistice Day was sent out from Pittsburg, Pa., and in the evening a special Armistice Day address by Mr. W. E. MacGregor at Pittsburg. This is but one instance of special radiations. Musical concerts are sent out regularly by two stations in Toronto, both companies being manufacturers and distributors of wireless apparatus, and these are received at distances of seventy-five to one hundred miles.

The radiations from all of the Westinghouse stations come in very nicely in Toronto and are heard by about fifty amateur receiving stations in the city. It is quite the custom now for an amateur to put a "loud speaker" horn, etc., onto his receiver and entertain his friends in the evenings by concerts from these wireless telephone transmitting stations in the United States.

In addition to the music, etc., there are a few other conveniences. Time signals are sent by many commercial and navy stations at various hours of the day thus providing an accurate means of regulating your time pieces.

Weather reports are sent out from local stations on the Great Lakes enabling the listener to know in the evening the probabilities for the morrow.

The following telegraph and radio telephone stations are used regularly by the writer whose receiving station is located about ten miles northwest of the city of Toronto:

Certain types of aerial have directional effects in both transmitting and receiving. By making use of this fact, it becomes possible to give ships at sea their bearings, to direct them through fog while some distance out of port, and also to direct aeroplanes to their hangars or to particular stations

which they are anxious to reach at night or while visibility is low during the day.

The chief remaining advantage in the spark stations rests in its ability to interfere over a great range of wave length thus making "S.O.S" calls at sea more certain of reception.

## Gears of Woven Fibre

Gear cutting materials in the past have been more or less standardized, and the introduction of a new compound material has aroused interest among manufacturers and users. The base of this material consists of vulcanized woven fibre. This is combined with a condensation product made of phenol and formaldehyde. The combination is hardened and becomes inert, insoluble and infusible. It is sold under the trade name of Condensite Celoron, and the following summary of claims made in its behalf is of interest :—

(1) It is unaffected by ordinary solvents and by most acids and weak alkaline solutions.

(2) Possesses high dielectric strength.

(3) Is resilient and shock resisting.

(4) Is workable by any machining process.

### Physical Properties

The material has been subjected to engineering tests with the results indicated below :—

Tests show that Condensite Celoron has a tensile strength of 8,000 to 10,000 pounds per square inch. Its compressive strength is approximately 40,000 pounds per square inch, laminae horizontal, and 2,500 pounds per square inch laminae vertical. The transverse strength is approximately 23,000 pounds per square inch, laminae horizontal; and 25,-000 pounds per square inch, laminae vertical.

Modulus of elasticity (the load per square inch divided by the elongation per inch) is approximately 1,550,000 pounds on the average of 10 readings.

Specific gravity 1.3 to 1.4.

Brinnell hardness (by applying standard formula to the measurement of the impression made by a 10 mm. steel ball held under pressure of 500 kg. for 15 seconds) on natural face 45, on sawed face 36,40.

Coefficient of expansion .000017 in. per degree Fahr.

With absorption, test pieces 1 in. thick, .46 in 24 hours. 1.08 in one week. Absorption measured in percentage of original weight.

### Little Alteration of Standard Practice

Gear cutting from Condensite Celoron involves no radical change in the practices usually followed. It should be mentioned that the maximum thickness of a single sheet is 3 inches, and that for gears over 3-inch face, two or more pieces fastened together must be used.

Under a steady load, one way rotation, the maximum fibre stress on the key way may be safely estimated at 3,000 pounds per square inch. In extreme cases the use of a larger key may help, or end plates may be necessary. On standard gears the use of shrouds is not recommended. Gears of larger diameter in proportion to their face may because of high pressure on the key, require end plates or metal hub to relieve the fibre stress. Gears of small diameter and larger bore may require end plates to prevent distortion when the key is driven.

There are two hard and fast rules in connection with the installation of gears of this type, which are that they must always mesh with cut metal gears—never with other non-metallic gears having uncut, cast or badly worn teeth and that in every case the face of the mating gear should be equal to or greater in width than the Condensite Celoron gear.

# Province of Quebec Adopts Single Service Entrance in All New Buildings

Some time ago the Electrical Co-operative Association, Province of Quebec, on suggestion of its central station members, took steps towards the introduction of a single service entrance for any one class of service in new, remodelled and, where possible, to apply in existing buildings. The reasons for it were economy, safety and elimination of much too frequent disfiguring of buildings which instead of adding to the appearance of the community, were becoming eye-sores indeed

The Committee drafting these regulations consisted of L. A. Kenyon, engineer, of electrical distribution, Montreal Light, Heat & Power Cons.; N. L. Engell, superintendent of distribution, Montreal Public Service Corporation; E. A. Stanger, distribution engineer, Southern Canada Power Co.; H. A. Elliot, representing the Shawinigan Water & Power Company; Geo. Templeman, engineer, City of Montreal. Electrical Commission; J. N. Mochon, electrical examiner, Provincial Government of Quebec; C. M. Tait, chief electrical inspector, Canadian Underwriters' Electrical Inspection and Louis Kon, secretary-manager Electrical Co-operative Association Province of Quebec.

The Architects Association and the Builders' Exchange had, on that Committee, their representatives in advisory capacity, the former being represented by Frank R. Foster, treasurer of the Architects Association and the latter by John Quinlan, building contractor and Harry Vincent, electrical contractor.

As a result, the following rules were adopted and are being printed for distribution by the Fire Underwriters' Association as a supplement to the Code, effective from 1st of June this year.

## Wiring Rules

Overhead wires from poles to building are objectionable in many ways and their number should be kept at a minimum. As a general rule there should be only one set of lighting or power service wires running from the pole to the building. If a building is already wired and additional apparatus is installed arrangements should be made to connect the new work to the service already in, increasing the capacity of this service if necessary. If the service already in does not comply with the present specifications it is recommended that a new service of sufficient capacity for both the old and the new loads be installed.

The service wires attached to the building are without fuse protection up to the point where they reach the fuses on the service switch. An accidental ground or short circuit on this portion of the wiring generally results in the wires and the conduit burning out from the point where the trouble occurs to the end of the conduit where the wires from the pole are connected, and a fire may follow For this reason the amount of wires attached to the building and unprotected by the service switch fuse should be reduced to a minimum and the rules governing the installation of these wires should be strictly adhered to.

## Service Switch

The service switch must, where practicable, be located in the basement. If there is no basement the service switch must be located in some other public part of the building if there is one, and must be so placed as to provide sufficient space for meter installations.

The service switch and fuses must be enclosed in a place where the wires enter the building. The service box must be of approved design, arranged to be operated from the outside of the enclosure, equipped with a locking or sealing device, and shall be so marked as to indicate without opening the enclosure whether the switch is in the "on" or "off" position.

The service switch must not be located in a coal bin, clothes closet, pantry, bath-room, or in any room that is liable to be locked, or in a location where the passage to same is liable to be locked.

The service switch cabinet must be fitted with a short conduit nipple to take the wires running to the meter. This nipple shall be set in either side or top of the cabinet, but not in the bottom, and must be fitted on outer end with lock nut and bushing with which to secure it in the meter trim.

### Service Wires

The service wires whether from overhead or underground must be installed in conduit.

There must be no splices in the service wires between the service cutout and the point of connection with the wires from the pole or manhole.

Only one set of wires for each class of service may be run from the pole line to the building.

Service wires running across up or down building walls must be installed in conduit.

Conduit work on services must be waterproof. Junction boxes must not be used if they can be avoided. If they are necessary they must be of the waterproof type with the covers gasketed.

The service conduit must be placed on the outside of the building but may be embedded in concrete or other fire proof walls.

The conduit must be equipped with a service head. The wires must leave the fitting through separate insulated holes which are turned in a downward direction so that water cannot enter. The service entrance fittings must be of such design that water cannot enter.

### Grounding

All service conduits must be grounded. The ground conductor must be at least equivalent to No. 8 B & S gage copper (where the largest wire contained is not greater than No. 0 B & S gage) and need not be greater than No. 4 B & S gage (where the largest wire contained is greater than No. 0 B & S gage).

#### Grounding A. C. Services

All alternating current services must be grounded on the line side of the main switch and fuses.

The ground conductor for a.c. services must not be smaller than No. 8 B & S gage copper wire, nor smaller than one-fifth the current capacity of the wire to which it is attached, except that it need not be larger than No. 0 B & S gage.

The ground conductor for the service must be used as the ground wire for the conduit, or for any other purpose.

#### D. C. Services

Direct current systems must not be grounded at the individual services or within the building served.

In every building whatsoever now existing where electricity is at present or may hereafter be used for lighting purposes, the wires transmitting such electricity shall be so put in said building that they can be connected inside of such building with the underground conduits.

C. C. Carter and E. Bretell, prominent in the electrical industry in Vancouver, have been elected members of Vancouver Rotary Club, as representatives of the industry.

# The Latest Developments in Electrical Equipment

The Ingersoll Machine & Tool Company, of Ingersoll, Ont., have recently placed on the market the "Baby Grand" portable ironing machine. The manufacturers claim it is designed, both as to size and price, for the average household, its special features being light weight, simplicity of

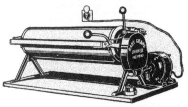

The "Baby Grand" portable ironing machine

operation and small dimensions. This machine is 3 ft. by 15 inches in size and weighs 75 lbs. It is gear driven by standard motor. Pressure upon the ironing shoe can be varied up to 500 lbs. The ironer is heated by either gas or electricity.

### The "Lamp Ette" Hangs Anywhere

The "Lamp Ette" hanging lamp device here illustrated is now manufactured in Canada by the Crown Electrical Manufacturing Co. of Brantford, Ont. This is a portable

electric bracket lamp attached to a silk brocade braid, on the other end of which is an ornamental plaque which acts as a counterweight. This lamp may be had in three colors of silk braid—gold, rose and blue— and may be used as a bed lamp, a piano lamp, a chair lamp—in fact, may be hung over any convenient object and used for any purpose.

### The "Nutype" Shadeholder

The Canadian General Electric Co. Limited, is manufacturing a new one piece, brass, three arm shadeholder, Cat. CGE 2 "Nutype" for use with 2¼ standard shadeholders and patent has been applied for. The shadeholder is of sturdy design with three arms supporting the shade rigidly at three distinct points 120. apart. The arms are reinforced by lengthwise embossings which extend upwards well into the socket head band and downwards over the lip of the shade into the cupped sections taking the shade holding screws. The clamp-

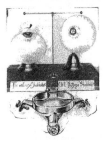

ing band encircling the socket head is reinforced by turning in the edges giving the effect of a 1/16 in. flange all around ,and is so shaped that the shadeholder is universal and can be readily used with either the standard round bead or the threaded bead socket shell. The clamping screw on the bead band is of generous length to allow for any amount of variance in diameters of beads and its length does not allow it to fall out with the clamping band opened to the fullest extent necessary to slip on any socket. This easy and positive method of attaching shades to standard sockets has made this device very popular with the trade.

The illustration shows a test made with the "Nutype" shadeholder versus the ordinary shadeholder. The test was made for a period of twenty-four hours, and as will be seen, the ordinary shadeholder was not able to support the heavy glass shade, which sagged down considerably.

---

Fred E. Garrett, who for the past six years has been associated with the Great West Electric Co., Ltd., Winnipeg, has resigned. During the past three years Mr. Garrett has held the position of sales manager of this concern. Prior to joining the staff of the Great West Electric Co., he spent six years with the C. G. E., at Winnipeg, looking after the city sales. Mr. Garrett has not definitely determined his future plans.

---

As was to be expected, books on radio are now making their appearance. The D. Van Nostrand Company have just issued one under the title "Radio Phone Receiving," which is described as "A practical book for everybody." The various topics discussed are as follows: 1. How Radio Telephoning is accomplished; 2. Tuning the simple receiving circuit; 3. Receiving the waves by crystal detectors; 4. The vacuum tube; 5. Amplifying the music or speech; 6. Regenerative and Heterodyne reception, and, 7. Radio telephone broadcasting. The book is well illustrated, 180 pages, 4½ in. by 7½ in. in size, bound in green cloth. Price $1.50 net.

### A New P & S device

The flat-back wall receptacle has gradually been losing ground because of the growing popularity of outlet-plates. These plates when used for new work, usually have a stud extending beyond the surface of the wall, and in old-house work the plate is set on the wall surface. Another cause is that outlet boxes, in general, are rarely set flush or true with the wall surface. Therefore the contractor has found it necessary to make up special brackets and canopies, where a wall receptacle would have been preferable and cheaper.

raised to conform to the natural position of the hand when ironing, and the whole iron is very carefully balanced. Another feature is the asbestos pad placed above the element, which directs and retains the heat in the bottom of the iron, requiring less current to operate. The heat is evenly graded from the hot point towards the back. There are also several mechanical features, including monel metal contact pins, which are very important in the life of an iron.

All metal BR base and socket       New Northern Electric Iron

Pass & Seymour, Inc., Solvay, N. Y. have met this demand in their new "All Metal" BR base. This is an all-metal base for every conceivable form of concealed outlet: old work, loom plates, ceiling plates, boxes for metallic and non-metallic flexible conduit. It covers them all, and when combined with the various bodies, makes a full line of wiring devices. The standard finish of BR is brush brass. The bridge inside is a full inch above the skirt and this base will therefore cover any projection such as pipe-ends or studs that extend one inch from the wall of ceiling surface. A complete series of adapters has been developed to secure this base to all forms of outlets.

### Allen Engineering Co. Activities

The Allen Engineering Company, Hamilton, has been retained to take care of the factory inspection and testing of the recent contracts let for the manufacture of the electrical equipment for the Manitoba Power Company. This same organization has just finished handling two very large contracts at Schenectady and Pittsburg for installation in Switzerland and France; the equipment consisted of 150,000 volt transformers, switching, insulation and line material. The Hydro-electric System of the City of Winnipeg also retained this company in a similar capacity for the recent extension to their power house, covering the inspection and testing of three 6,500 kv.a. generators, transformers and line material. Likewise the Laurentide Power Company has had this company look after their 2—18,500 kv.a. generators and switching recently shipped from Hamilton.

### Northern Electric "Quality" Iron

The iron is without doubt, the most widely used electric appliance and there are so many types made and in use, that it is sometimes supposed the appliance is so perfect as to make impossible further improvements in construction or design. That this is not so, is claimed for the new Northern Electric Quality Iron. Its design is distinct, in many points, from that of any other iron made. It has a long torpedo shaped nose and extended back, making it more convenient to use both ends of the iron; it has round corners at the back which prevent catching and tearing ruffles, tucks, pockets, etc; the heel is rounded—valuable when ironing cuffs and collars, lace and other fancy articles; the handle is

The Cutter Company, Philadelphia, are distributing a booklet describing, with illustrations, a variety of remote control circuit breakers of both motor and magnetically operated types, together with other forms of apparatus especially adapted to central station requirements.

The latest model "Eureka" sweeper

## The Canadian Electric Railway Convention

### Everything in Readiness for Meetings in Quebec City on June 1, 2 and 3

Very complete preparations have been made for the exhibit of electric railway equipment and accessories which will constitute one of the most important features of the Canadian Electric Ry. Ass'n. which meets in Quebec on June 1, 2 and 3. Thirty six booths have been fitted up by the Committee of Exhibits under the chairmanship of Mr. R. M. Reade, every one which has been taken, and the finest exhibit, devoted entirely to electric railway equipment, that Canada has ever seen, will be shown.

For the proper accommodation of these exhibits, the Drill Hall is being utilized. The session will be held upstairs in the same building. The officers of the Canadian Electric Railway Association are as follows: Honorary president, Mr. Thos. Ahearn, president Ottawa Electric Company; honorary vice-president, Mr. Geo. Kidd, general manager British Columbia Electric Railway Company; president, Mr. G. Gordon Gale, vice-president and general manager Hull Electric Company; vice-president, Maj. F. D. Burpee, manager Ottawa Electric Railway Company. Executive: The officers named above: Messrs. R. M. Reade;

Mr. G. Gordon Gale, president
Canadian Electric Railway Association

Quebec; H. H. Couzens, Toronto; C. C. Curtis, Cape Breton; C. L. Wilson, Toronto; E. P. Coleman, Hamilton; A. W. McLimont, Winnipeg; W. S. Hart, Three Rivers, Que., and Col. G. C. Royce, Toronto. Mr. A. Gaboury is treasurer, and Mr. H. E. Weyman, auditor.

Once more Canadian Electric railway men are reminded that this association is now no longer devoted solely to the interests of private companies. It is an association, in actual fact, of electric railway systems of Canada. The program indicates that some very helpful papers will be read and these will undoubtedly be followed by still more

valuable discussions, but the best part of the convention is always the interchange of general information, resulting from meeting men in the same line of business in a social way. This Convention actually represents the first opportunity that Canadian electric railway men have ever had of all getting together. We suggest that every man who decides to take advantage will only be consulting his own interests. The Convention Program is given herewith, as is also the complete list of exhibitors:

### PROGRAMME FOR C.E.R.A. ANNUAL MEETING

#### Wednesday, May 31st, 1922

3.00 p.m. Meeting of the Executive Committee. Preliminary meeting of Associate Members; Registration at Chateau Frontenac.

#### Thursday, June 1st, 1922

9.00 a.m. Registration at Drill Hall.

9.30 a.m. Address of welcome by His Honor the Lieut. Governor Sir Charles Fitzpatrick, and His Worship Mayor Samson of Quebec, followed by an inspection of the exhibits.
Business Session.—Minutes of last meeting; Address of president; Report of secretary; Report of treasurer; Report of special committee; Appointment of nominating committee; General business; Inspection of exhibits.

1.00 p.m. Get-together luncheon. Chateau Frontenac. $1.50. Ladies, Members and Associate Members; Short address by prominent guest.

2.30 p.m. Valuation of Street Railway Assets. Their Maintenance and Depreciation. — By Dr. Louis Herdt, consulting engineer, Montreal. Discussion; H. E. Weyman, Levis County Ry. Co.

2.45 p.m. Sight seeing around Quebec by Trolley. (Ladies.)

4.00 p.m. The Modern Street Railway Motor. — By J. K. Stotz, Canadian Westinghouse Co. Discussion; W. G. Gordon, Canadian General Electric Co., W. R. McRae. Toronto Transportation Comm.

5.00 p.m. Inspection of Exhibits.

7.30 p.m. Association banquet. Chateau Frontenac. $5.00; Chairman, Honorary President. (Ladies. Members & Associate Members.)

10.00 p.m. Dancing; Band on Dufferin Terrace.

#### Friday, June 2nd, 1922

9.00 a.m. Inspection of Exhibits.

10.00 a.m. Motor Busses and Trackless Trolleys. — By D. E. Blair, Montreal Tramways Co.

11.30 a.m. "The Romance of the Rail and Power," by courtesy of the Canadian Westinghouse Co.

12.30 p.m. Inspection of Exhibits.

2.30 p.m. Welded Track Joints. — By E. B. Entwistle, Loraine Steel Co., Johnstown. Pa. Discussion: W. F. Graves, Montreal Tramways Co.

3.30 p.m. Moving Picture Exhibit, by courtesy of the National Safety League and the Dominion Wheel & Foundries Ltd.

4.30 p.m. Trip to Quebec Bridge by Boat. Light Refreshments. (Ladies. Members & Associate Members.)

6.00 p.m. Inspection of Exhibits.

9:00 p.m. Annual Reception & Ball, $2.00, Chateau Frontenac.

**Saturday, June 3rd, 1922**

9.00 a.m. Inspection of Exhibits.

10.00 a.m. Unfinished business; election of Officers; trip to Montmorency Falls & Ste. Anne de Beaupre; trip to Saguenay.

**List of Exhibitors**

Canadian Railway & Marine World; Kenfield-Davis Publishing Co.; Canadian Westinghouse Co.; Ohio Brass Co.; Taylor Electric Truck Co.; Ottawa Car Mfg. Co.; Canadian General Electric Co.; Lyman Tube & Supply Co.; Dominion Wheel & Foundries, Ltd.; Railway & Power Engineering Corporation; National Pneumatic Co.; Nichols-Lintern Co.; Canadian Car & Foundry Co.; Witherow Steel Co.; Consolidated Steel Co.; Canadian Street Car Adver-

tising Co.; Don. M. Campbell, Mech Eng.; Transit Equipment Co.; McGuire-Cummings Co.; Cleveland Armature Works; Lindsley Bros., (Canadian) Company; The Tool Steel Gear & Pinion Co.; Allen General Supplies, Ltd.; Southam Press, Ltd.; Canadian Cleveland Fare Box Co.; Universal Lubricating Co.; The Arthur Power Saving Recorder Co.; Canadian National Carbon Co.; Standard Underground Cable Co.; Canadian Brill Co.; J. G. Brill Company; Eugene F. Phillips Electrical Works; Canadian Steel Foundries, Ltd.; Hunter Joint Block Co.; Chestham Electric Switching Device Co.; Nachod Signal Co.; United States Steel Products Co.; C. E. A. Carr Company; The St. Louis Car Co.; Dawson & Co., Ltd.; J. A. Everell Co.; Northern Electric Co.; Sarnia Bridge Co.; Ontario Safety League; Vickers Limited; The Electric Railway Journal; The Electrical News.

# Locomotives for the Chilean State Railways

## Fifteen Road Freight and Seven Switching Locomotives Under Way—Some Data on Designs and Specifications

Work on the electric locomotives being built for the Chilean State Railways is rapidly progressing. The cabs for the first eight of the fifteen road freight locomotives have been delivered by the Baldwin Locomotive Works to the Westinghouse Electric & Manufacturing Company for the installation of the equipment. There will also be seven switching locomotives.

An outline of the road locomotive is shown in an accompanying illustration. The cab is of the box type, carried on two articulated trucks, each having three driving axles with direct geared motors. The estimated weight is 226,000 pounds. The locomotives will operate at 3,000 volts direct current. This locomotive rates 1,680 h.p. at 3,000 volts and will be able to develop a maximum of 3,200 h.p. for short periods. With natural ventilation the locomotive will deliver for one hour a tractive effort 27,950 pounds at a speed of 22.6 miles per hour at 3,000 volts. The continuous capacity of the locomotive with forced ventilation is 20,880 pounds tractive effort at 24.8 miles per hour. The maximum speed is 40 miles per hour. The general dimensions and estimated weights of the locomotive are as follows:

**Table 1**
**Dimensions and Weights of Road Freight Locomotives**

| | | |
|---|---:|---|
| Track gauge | 5 ft. | 6 in. |
| Length over buffers | 49 ft. | 10 in. |
| Length over cab | 38 ft. | 0 in. |
| Total wheel base | 37 ft. | 0 in. |
| Rigid wheel base | 13 ft. | 9 in. |
| Height—top of rail to cab roof | 12 ft. | 7 in. |
| Height—top of rail to clerestory | 13 ft. | 10 in. |
| Width over cab sheets | 10 ft. | 0 in. |
| Height of coupler | | 41 in. |
| Wheel diameter | | 42 in. |
| Weight of complete locomotive | | 226,000 lb. |
| Weight of mechanical parts | | 140,000 lb. |
| Weight of electrical equipment | | 86,000 lb. |
| Weight per driving axle | | 37,670 lb. |

The road locomotives will operate over the 116-mile route between Santiago and Valparaiso and the 28-mile branch between Las Vegas and Los Andes. The heaviest grade is 2.35 per cent for 12 miles from Llai-Llai to La Cumbre. The maximum curvature is 11 degrees. There are six tunnels in the electrified zone.

The present main line freight trains average 550 short tons. They are operated with a single steam locomotive, except on the heavy 12 mile grade southbound, and on a northbound grade, of 6.8 miles. On these two sections a

steam helper is now used to maintain speeds of from 10 to 14 miles per hour.

One electric locomotive will haul a trailing load of 770 short tons in either direction between Valparaiso and Santiago without assistance except on the Tabon grade. On level tangent track the speed with such a load will be 35 miles an hour at 3,000 volts. The average running speed on the Tabon grade will be approximately 24 miles per hour. The time saved by the elimination of delay to take fuel and water and by the higher running speed will shorten the time of a trip from 4 to 5 hours in each direction.

These locomotives are equipped with Continental spring buffers and M. C. B. couplers, arranged to take attachments for chain couplers temporarily. The two six-wheel trucks are connected at the inner ends by a mallet hinge. The bar-type cast steel side frames are located outside of the wheels and connected by cast steel bumpers and cross-ties. The semi-elliptic driving springs over the journal boxes on each side are connected to one another by equal beams. The ends of each set of three driving springs connected thus are attached to the side frames through coil springs.

The 38-foot box type cab, including an engineman's compartment in each end and a central equipment compartment, is carried on center pins located approximately over the midpoint of each rigid wheel base. One center pin is restrained both longitudinally and laterally and the other in the lateral direction only, which permits free longitudinal movement of the cab relative to one truck.

The locomotives are equipped with Westinghouse air brakes, the standard of the Chilean Railways. The air brake, type 14-EL is interlocked with the regenerative brake so that the latter may be supplemented by service application of the train brakes, if desired, without applying the air brake to the locomotive drivers.

Current is collected by spring-raised, air-lowered pantagraphs, controlled by compressed air and arranged to be mechanically locked in the lowered position.

Individual switches mounted in banks establish the main circuit connections. Each switch is a complete unit in itself and may be removed without disturbing adjacent switches. Compressed air controlled by electro-magnetic valves is used to operate the switches. For certain circuits where no current is broken and for low voltage, cam switches are used. These also operate by compressed air controlled by

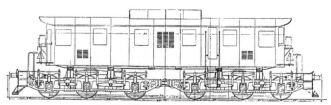

General design of Freight Locomotive

electro-magnetic valves. The cam group comprises a number of switches mounted on a single shaft, connected through a rack and pinion to a double acting air piston.

Each axle of the locomotive is driven by a Westinghouse No. 350-D motor, wound for 1500 volts and insulated to operate two in series on 3000 volts. The nominal rating of this motor on short field is 280 h.p., at 155 amperes and 1500 volts. Field control is secured by the use of two separate field windings on the main poles. The motors are geared directly to the axles with a ratio of 3.94 to 1. The gear is of the flexible type.

A motor-generator set provides low voltage power to compressors, blowers, control equipment and lights. This set has a single frame and two armatures carried by a common shaft. The 3000 volt motor is a bi-polar double-commutator machine. The continuous rating of the generator is 35 kw. at 92 volts.

A master controller is located in each engineman's compartment to provide double end operation, the same master controller being used for both motoring and regenerative braking. This controller provides 50 control notches in acceleration so that tractive effort variations are small, thereby permitting exceedingly smooth handling of trains.

The control is Westinghouse HLFR providing speed combinations by varying the grouping of the motors to give one-third, two-thirds, and full speed. Field control gives three additional speeds. Transition from one motor combination to another is made by the shunting method.

For regenerative braking, the main motor armatures are arranged for the same combinations as when motoring and the motor fields are separately excited by the motor-generator set. The range of speed in regenerative braking will be from 8 to 30 miles per hour.

### The Switching Locomotives

The switching locomotives, an outline of one of which is shown in an accompanying illustration, will be the last ones built. The cab is of the steel type and is carried on two swivel trucks. On each truck are mounted two motors driving direct through standard helical gears. The estimated weight is 136,000 pounds. The control is arranged for double end operation.

The nominal rating of this locomotive is 560 h.p. With 3000 volts, and natural ventilation, the tractive effort for one hour is 19,600 pounds at a speed of 10.6 miles per hour and the continuous capacity is 11,400 pounds at 12.7 miles per hour. With 25 per cent. nominal adhesion the starting tractive effort is 34,000 pounds. The maximum speed is 35 miles per hour. For short periods the equipment is capable of developing 1,000 h.p. In view of an expected increase in traffic these locomotives will be capable of handling trains of 1,200 short tons in yards with level tracks.

The following table gives the general dimensions and estimated weights of the locomotive:

### Table II

Dimensions and weights of switching locomotives.

| | | |
|---|---|---|
| Track gauge | 5 ft. | 6 in. |
| Length over buffers | 40 ft. | 0 in. |
| Length over central cab | 17 ft. | 0 in. |
| Length over hoods | 27 ft. | 0 in. |
| Total wheel base | 27 ft. | 4 in. |
| Rigid wheel base | 8 ft. | 6 in. |
| Height—top of rail to cab roof | 12 ft. | 3 in. |
| Width over cab sheets | 10 ft. | 0 in. |
| Height over coupler | 36 7/16 in. | |
| Wheel diameter | 42 in. | |
| Weight of complete locomotive | 136,00 lb. | |
| Weight of mechanical parts | 86,000 lb. | |
| Weight of electrical equipment | 50,000 lb. | |

The trucks are of the rigid bolster equalized type with rolled steel frames located outside of the wheels. A center pin is located at approximately mid-length of each rigid wheel base. The central cab has an engineman's stand in each end and control apparatus centrally located and suitably protected.

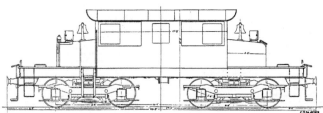

General design of Switching Locomotive

# Current News and Notes

**Ft. William, Ont.**

The Fort William Electric Company has recently been formed and will carry on an electrical contractor-dealer business at 107 South May St., Fort William. The principals of the new firm are Mr. H. G. Earley, formerly of the Ft. William Pulp & Paper Company, and Mr. W. H. Cowley, formerly of the Spanish River Pulp & Paper Company. Mr. Earley & Mr. Cowley were attached to the 8th Battalion and the 58th Battalion, respectively, of the Canadian Expeditionary Forces and saw considerable service in France.

**Hull, Que.**

Mr. M. Gravell, 36 Laval St., Hull, Que., has been awarded the contract for electrical work on a store building being erected at 90 Main St., Hull, by Mr. Donat Paquin of that place.

**Lachine, Que.**

The general contract for a power house, to cost in the neighborhood of $22,000, has been let by the Lachine city council to Messrs. Fournier & Freres, 26 St. James St., Montreal, Que.

**London, Ont.**

The Private Bills Committee, of the Ontario Legislature, recently reported favorably on the London street railway Bill, which will enable the street railway company to charge a straight five cent fare instead of the present rate of seven ordinary tickets and nine workingmen's tickets for a quarter. Children under 12, will, of course, be allowed to travel on children's tickets. The Bill was amended, however, to provide that the rate of fare may be changed by the Railway and Municipal Board, should it be deemed advisable at any time, on the application of either the city council or the railway company.

**Maxville, Ont.**

Mr. R. J. Desjardin, Maxville, Ont., has been awarded the contract for electrical work on a new store being erected on Main St., Maxville, Ont., by Mr. W. D. Campbell, of that place.

**Moncton, N. B.**

Mr. Geo. W. McCall, 72 George St., Moncton, N. B., has been awarded the contract for electrical work on residences to be erected in various parts of Moncton, by the Moncton Housing Commission, at a cost of approximately $100,000.

**Ottawa, Ont.**

Contracts for the following equipment have been let by the Ottawa Hydro-electric Commission, Ottawa, Ont.: Transformers, Canadian Crocker-Wheeler Co., Ltd., St. Catharines; Meters, Ferranti Meter & Transformer Mfg. Co., Toronto; Wire, Messrs. Garvock & Goddard, Ottawa.

Mr. H. L. Allen, 272 Bank St., Ottawa, Ont., has secured the contract for electrical work on a $150,000 addition being built to the University of Ottawa, Ottawa, Ont.

**Peterborough, Ont.**

The city council of Peterborough, Ont., recently requested that the Ontario Hydro-electric Commission reduce the street railway fares from seven cents straight, to four tickets for a quarter. This request was not granted, the Commission pointing out that the volume of traffic did not warrant the reduction.

**Point Grey, B. C.**

Messrs. Farr, Robinson & Bird, 546 Howe St., Vancouver, have secured the electrical contract on a $40,000 residence being erected at 3590 Cypress St., Point Grey, by Capt. A. Montague Tulk.

**Quebec, Que.**

Mr. Fortunat Gingras, 34 St. Augustin St., Quebec City, has been awarded the contract for electrical work on a school to be erected at Signal & St. Sauveur Sts., Quebec City, to cost in the neighborhood of $75,000.

**Riviere A Pierre, Que.**

The Shawinigan Water & Power Company, Ltd., Shawinigan Falls, Que., contemplate the installation of an electric lighting system in the town of Riviere A Pierre, Que., in the near future.

**St. Catharines, Ont.**

The Clifford Electric Co., St. Catharines, Ont. has secured the electrical contract on an addition being built to a school in Orchard Park district, St. Catharines, by the trustees of Grantham S. S. No. 7.

**St. John, N. B.**

The New Brunswick Gazette announces the incorporation of the Maritime Radio Corporation, Ltd., with a capitalization of $24,000. The new firm will manufacture and deal in radio and wireless equipment. Head office: St. John, N. B.

**Sault Ste. Marie, Ont.**

Mr. R. C. Farmer, 36 Albert St. W., Sault Ste. Marie, Ont., has been awarded the contract for electrical work on a theatre on Albert St., Sault Ste. Marie, that is being altered into a store and apartment building for Mr. L. Palumbo, 184 Schrieber St.

**Sherbrooke, Que.**

M. J. A. Choquette, 125A King St. W., Sherbrooke, Que., has been awarded the contract for electrical work on an office building being erected by the city council, on Wellington St., Sherbrooke, at an estimated cost of $75,000.

**Stratford, Ont.**

The Bennington Electric Co., Stratford, Ont., has secured the electrical contract on a new Isolation Hospital, and an addition being built to the General Hospital, Stratford, Ont., at an estimated cost of $100,000. Also the electrical contract on the new Memorial Baptist Church being built at Downie & Norfolk Sts., Stratford.

**Thessalon, Ont.**

The ratepayers of Thessalon, Ont., recently voted favorably on a by-law authorizing the expenditure of $50,000 for Hydro-electric development, of approximately 200 h.p., on the Little Rapids of the Thessalon River.

**Victoria, B. C.**

Messrs. Hawkins & Hayward, 1607 Douglas St., Victoria, B. C., have secured the electrical contract on the Royal Jubilee Hospital recently erected in that city. It is understood the contract covers electric wiring, bells, telephones and fixtures.

**Weston, B. C.**

Mr. R. Simpkin, Weston, Ontario, has secured the contract for the electrical work on an apartment house recently erected in Weston, Ontario. This includes 5 suites of five rooms each, with electric ranges, and mantels.

### Electrification Data

A interesting publication just out is Westinghouse Electrification Data, Vol. III No. 2. In this issue the foreword points out that electrification of railroads is conservation, not only of material things, but also of human energy. This is followed by an abstract of a letter on Standards for Railroad Electrification addressed by George Gibbs to the Electrification of Railways Advisory Committee in England. This letter indicates that power should be developed at 25 cycles and the use of both alternating and direct current on the trolley continued.

The "Future of Railroading is Electrification" is discussed in the third chapter of the publication. Such important existing electrifications as the Hoosac

Tunnel, the Norfolk & Western Railway, the Pennsylvania Railroad and the Chicago, Milwaukee & St. Paul Railway are cited as examples to prove this contention and a resume is given of the electrification plans of South American and other foreign countries.

The fourth chapter is descriptive of the electrification of the Paulista Railways of Brazil, outlining the reasons for the change in motive power and giving some detail of the locomotive equipment. Chapter five is a brief storey of the electrification of the Chichibu Railway of Japan, the Baldwin-Westinghouse locomotives purchased for this road forming the original introduction to Japan of American electric locomotive operation on an electrified steam railroad.

The last illustration in the book shows graphically the growth of steam railroad mileage electrified and electric locomotive tonnage in heavy traction service in the United States and Canada from 1905 to date.

The Diamond State Fibre Company of Canada Limited, have just issued a new catalogue (No. 20) on Condensite Celoron which fully describes its properties, gives technical data, etc., and is illustrated. This catalogue should prove of great interest to manufacturers who are looking for an absolutely waterproof material that will also carry out all the other functions of a vulcanized fibre material.

Pass & Seymour, Inc., Solvay, N. Y., have issued a bulletin under date of

April 24, 1922, Form 1426, describing 1". & S. standard Elexits. Complete instructions for properly installing side wall and ceiling outlets for Elexits are also given in Pass & Seymour Junior catalogue. No. 26.  ,

---

---

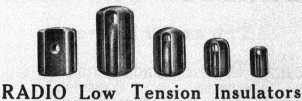

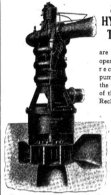

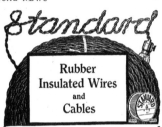

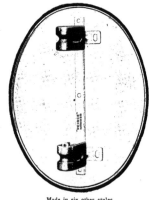

# CLASSIFIED INDEX TO ADVERTISEMENTS

The following regulations apply to all advertisers:—Eighth page, every issue, three headings; quarter page, six headings; half page, twelve headings; full page, twenty-four headings.

# CLASSIFIED INDEX TO ADVERTISEMENTS—CONTINUED

## CLASSIFIED INDEX TO ADVERTISEMENTS—CONTINUED

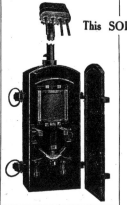
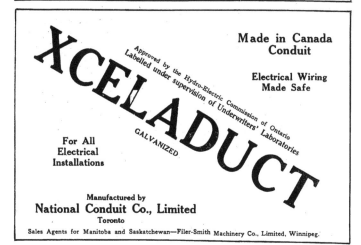

Vol. XXXI No. 11                                    Toronto, June 1, 1922

# Electrical News

**Engineering Contracting** · **Merchandising Transportation**

# COPPERWELD

## Combines the Conductivity of Copper with the Strength of Steel

Tests show that Radio Frequency Currents are carried entirely by the "skin" of the Wire. The "core" performs no current carrying service. Copperweld is therefore the Ideal Radio Antenna Wire.

### Copperweld Antenna Wire

Has a pure copper exterior permanently welded to a solid steel core without the slightest loss of current carrying properties. This "skin" is sufficiently thick to prevent nicking.

Put Copperweld on your counter and give instant satisfaction to All Radio Customers.

No measuring, packing, or explaining necessary; simply wrap up a carton of the desired length, 100, 150 or 200 feet—full directions for use are printed on the back.

Handy Packages of 100, 150 or 200 foot coils.

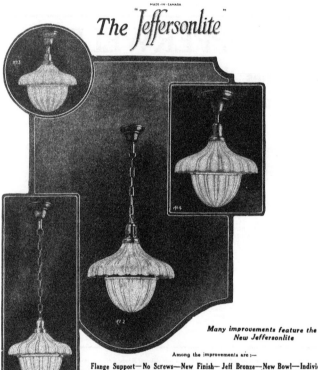

# You buy it by the pound
# He uses it by the yard

TAPES differ surprisingly in yardage per pound—an important factor in economical buying.

Inferior tape is usually short on yardage because it must be made extra thick to stand any strain whatever. The maximum yardage of Johns-Manville Tapes and splicing compounds is at once a guarante of quality and economy.

Johns-Manville Jomanco Friction Tape for general outdoor work, Johns-Manville No. 3 Friction Tape for motor leads and switchboards, Johns-Manville No. 5 Friction Tape for inside work, Johns-Manville White Friction Tape, Johns-Manville Armature Tape.

Johns-Manville Rubber Splicing Compound for ordinary use, Johns-Manville Alpha Splicing Compound for extra strength and puncture resistance, Johns-Manville Brooklyn Splicing Compound for low voltage work.

**CANADIAN JOHNS-MANVILLE CO. LIMITED**

Toronto  Montreal  Winnipeg  Vancouver
Windsor  Hamilton  Ottawa

# JOHNS-MANVILLE
## Electrical Tapes

Through
**ASBESTOS**
and its allied products

Electrical Materials
Brake Linings
Insulations
Roofings
Packings
Cements

Fire
Prevention
Products

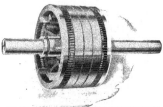

# ALPHABETICAL LIST OF ADVERTISERS

# The Standard by which all other Electric Irons are Judged

Over a quarter of a million Hotpoint Irons have been sold to Canadian housewives.

But, there are thousands of women who are still loyal to the old sad iron. To these women, the Hotpoint Iron is as much a specialty as aeroplanes are to you. To the non-user the Hotpoint Iron must be presented as a great labor saving specialty—an innovation.

Forget your inside point of view, present the Hotpoint Iron from the standpoint of the woman who has yet to experience the comfort and utility it affords.

The old time "pep" and enthusiasm, behind your selling efforts, will sell Hotpoint Irons as readily to-day, as it did several years ago. Feature Hotpoint Irons in your advertisements. Display them in your windows. You will be agreeably surprised at the results. Let us know if we can be of any assistance.

### "Made in Canada"

# Canadian Edison Appliance Co.,Limited
## STRATFORD, ONTARIO.

# "A Little Matter of Convenience"
## The New C.G.E Appliance Switch
### CAT. Nº 106

Thousands of electric irons, toasters, percolators and other appliances in daily use offer a wonderful market for this convenient little switch.

It is packed in a very attractive merchandising container—suitable for counter display. A show card attached to the carton illustrates in a striking manner the convenience of having all heating goods equipped with these switches.

After selling an iron, toaster or percolater, it is a very simple matter to draw your customer's attention to the display carton on the counter, and to suggest that you attach one of the switches to the new appliance, the result being two sales instead of one. The door is then open for still further sales, as the customer will be in a receptive mood towards having other appliances similarly equipped.

Acorn Pendent Switch
Cat. No. C.G.E. 306

Porcelain Pendent Switch
Cat. No. C.G.E. 206

*"MADE IN CANADA"*

# Canadian General Electric Co., Limited
## HEAD OFFICE  TORONTO

Branch Offices:  Halifax, Sydney, St. John, Montreal, Quebec, Cobalt, Ottawa, Hamilton, London, Windsor, South Porcupine, Winnipeg, Calgary, Edmonton, Vancouver, Nelson and Victoria.

# C. G. E.
# Service Boxes

The new line of C.G.E. Steel Service Boxes consists of 30 Amp. 125 Volt, House Service type, Cat. No. C.G.E. 100A, made in three capacities, 30, 60, 100 Amp. 125/250 Volt, triple pole, Solid Neutral Boxes—externally operated.

Cat. No.
C.G.E. 100

## To Create Prestige

Use C.G.E. Wiring Devices throughout. No better guarantee of a satisfactory job can be given, than to state that C.G.E. material has been used exclusively.

Cat. No.
C.G.E. 3030

Cat. No.
C.G.E. 3060

Cat. No.
C.G.E. 3100

Cat. No. C.G.E. 100A is a double pole box, switch fused with 125 volt Plug Fuses at the hinge-end, mounted on a Porcelain base. The externally operated handle is so arranged that by use of a small commercial padlock or sealing wire, the box may be locked closed, with the switch operative for Supply Companies Active Service; or the box may be locked closed with the switch locked open for Discontinuance of Service. With the box open the switch may be locked open for Wireman's protection when working on the circuit. The steel box is entirely die made, and consequently has a neat and pleasing appearance. It is completely equipped with grounding terminal, suitably placed ½" and ¾" knockouts, and wiring diagram showing proper and approved method of installation.

## Externally Operated

Cat. Nos. C.G.E. 3030, 3060 and 3100, 30A, 60A, 100A. T. P. 125/250V. Service Boxes are designed with separate units for each pole of the switch, and the metal parts are made removable for replacement of burned clips, etc. The centre or neutral poles are equipped with terminal lugs for a solid neutral connection with removable plate for meter testing. Knockouts are suitably placed, and ample wiring space allowed within the boxes. The Switches can be locked open with the boxes closed by means of stops and latches for padlocks. These three service switches can be readily converted into double pole 250 volt devices if desired.

*"Made in Canada."*

# Canadian General Electric Co., Limited
### HEAD OFFICE 🌀 TORONTO

Branch Offices: Halifax, Sydney, St. John, Montreal, Quebec, Cobalt, Ottawa, Hamilton, London, Windsor, South Porcupine, Winnipeg, Calgary, Edmonton, Vancouver, Nelson and Victoria.

# C-G-E
## Ornamental Lighting Installation, Sunnyside Boulevard, Toronto

The installation comprises Octagonal Bronze Lanterns complete with opalized, stippled, glass panels and refractors, giving maximum illumination on the road surface.

The system is fed from out-door type constant current transformers; 6.6 ampere secondary.

Series transformers are installed in the ground by the side of each standard, and the current stepped up from 6.6 ampere to 20 ampere, enabling the use of 20 ampere high efficiency series Mazda Lamps.

All equipment mentioned above, manufactured by

# Canadian General Electric Co., Limited
### HEAD OFFICE ⊕ TORONTO

Branch Offices: Halifax, Sydney, St. John, Montreal, Quebec, Cobalt, Ottawa, Hamilton, London, Windsor, South Porcupine, Winnipeg, Calgary, Edmonton, Vancouver, Nelson and Victoria.

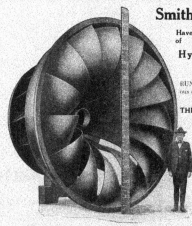

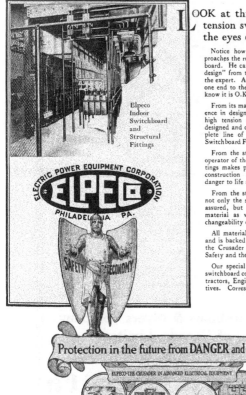

Generation, Transmission and Application of Electricity

For nearly thirty years the recognized journal for the Electrical Interests of Canada.

Published Semi-Monthly By

## HUGH C. MacLEAN PUBLICATION
### LIMITED

THOMAS S. YOUNG, Toronto, Managing Director
W. R. CARR. Ph.D., Toronto, Managing Editor
HEAD OFFICE - 347 Adelaide Street West, TORONTO
Telephone A. 2700

MONTREAL - - 119 Board of Trade Bldg.
WINNIPEG - - - - 302 Travellers' Bldg.
VANCOUVER - - - - Winch Building
NEW YORK - - - - - 296 Broadway
CHICAGO - - 14 West Washington St.
LONDON, ENG. - - 16 Regent Street S.W.

### ADVERTISEMENTS
Orders for advertising should reach the office of publication not later than the 5th and 20th of the month. Changes in advertisements will be made whenever desired, without cost to the advertiser.

### SUBSCRIPTIONS
The "Electrical News" will be mailed to subscribers in Canada and Great Britain, post free, for $2.00 per annum. United States and foreign, $2.50. Remit by currency, registered letter, or postal order payable to Hugh C. MacLean Publications Limited.
Subscribers are requested to promptly notify the publishers of failure or delay in delivery of paper.

Authorized by the Postmaster General for Canada, for transmission as second-class matter.

Vol. 31          Toronto, June 1, 1922          No. 11

## A Uniform Standard
## A Great Aid to the Electrical Industry

The Canadian electrical industry will have to be more alert than ever or the efforts that have been made to improve the standard of electrical equipment and installations will go for naught. Reports from the United States indicate that a German-made electric iron has recently been placed on the market at a price which enables the retailer to sell it to the ultimate consumer at $1.00, complete with attachment plug and four feet of flexible cord. Mr. T. H. Day, president of the Western New England Association of Electrical Inspectors states that a recent test of these irons showed that every sample was more or less defective. The voltage to which the irons should be connected is not shown, but the tests indicated a wattage of 950. Mr. Day states that none of the samples remained in circuit long enough to permit the test to be completed, the heating element breaking down in from 3 to 35 minutes. When taken apart the device was shown to consist of a wooden attachment plug, a flexible cord considerably below standard, and a heating element of sheet-mica matrix wound about with iron wire, and very poorly insulated.

It is not surprising that some of these irons should percolate through into Canada, and we understand that a number of them—or something very similar—have appeared in the province of Quebec.

There is a Dominion Act to the effect that all goods imported into Canada must be marked with the name of the country of origin, but this Act has never been enforced and advantage has been taken of this fact to bring in irons and other equipment of inferior quality. It is a very serious matter, therefore, that, instead of enforcing this Act, Finance Minister Fielding, in his recent Budget speech, announces that the Act will be repealed, as the electrical industry had hoped that its enforcement would afford some measure of protection against cheap foreign goods being brought into Canada. It looks as if this situation might not have been thoroughly understood by our finance minister.

Another announcement in the Budget speech is to the effect that duty on goods coming in from foreign countries would no longer be based on their value in depreciated currency, the point being made that under this provision the German mark, for example, was given a value entirely out of keeping with its actual value and that as a result, instead of a duty of 35 per cent, Canada has been charging as high as 1,000 per cent.

These two items will have a direct bearing on the importation of German and other cheaply constructed electrical goods into Canada. However, to the layman, the argument that these goods are cheap will not carry so much weight as the argument that they are flimsy and hazardous to operate. Our safest course, therefore, appears to be the adoption of a universal, efficient system of inspection.

This brings us back to the question that has been discussed at some length in our recent issues. The province of Ontario is the only province, apparently, where these cheap, worthless goods cannot gain a foothold. The reason is that it has a compulsory inspection and standard requirements. If all the Canadian provinces had the same arrangement, it would be infinitely better for the electrical industry as well as for the people we are endeavoring to serve.

If the new Budget is thus the means of speeding up the activities of the electrical industry to the end that we may have a universal Code of requirements, and uniform inspection, it will at least have served a useful purpose. We believe, also, it would be well if the industry would take an immediate opportunity of pointing out to Mr. Fielding the unfair position in which we shall be placed by the provisions outlined in his Budget speech, as mentioned above.

* * *

## Suggest Calling Conference
## of Provincial Representatives

At the last meeting of the main committee of the Canadian Engineering Standards Association, held in Ottawa, the secretary, Mr. R. J. Durley, reports that the work of the sub-committee on the Canadian Electrical Code was discussed and it was decided to forward to the Minister of Trade and Commerce a memorandum outlining the present situation in regard to this question, and request him to convene a meeting of representatives of the various provincial governments to discuss the possibility of interprovincial agreement, and acquaint the provincial governments with the difficulties now experienced by the electrical industry. The following resolution was unanimously adopted:

"RESOLVED that the Memorandum on the proposed Canadian Electrical Code be approved, and that the secretary be directed to forward it to the Minister of Trade & Commerce, calling his attention to the unsatisfactory situation at present existing in Canada due to the lack of uniformity in the requirements for electrical appliances and construction in the various provinces, and requesting the Minister to consider the advisability of calling a conference of representatives of all the provincial governments with this Association to discuss the matter with the view of obtaining interprovincial agreement and action."

# The Application of Electricity in the Manufacturing Industry

## A Typical Rubber Tire Plant, for Example, Uses Equipment Valued at Half a Million Dollars

By a process almost as imperceptible as it has been persistent, the industries of the world, small and great, purely local or of world-wide application, industries in their innumerable varieties, have all become dependent—absolutely tied hand and foot—to electric power. Think of the many industries that could not continue at all without electricity! Can you think of a single isolated case where, if not already used, electricity would not introduce economy, speed and refinements quite impossible without it?

It is often said that the percentage cost of power entering into the total cost of a manufactured article is so small as to be negligible. It certainly is so in comparison with its value to the industry it operates. But when considered from the point of view of dollars and cents, many of our manufacturing concerns are an exceedingly important factor to the equipment salesman. It is not at all unusual to find a half million dollars tied up, in an industrial plant, in transformers, motor-generators, motors, switching and other auxiliary equipment, with annual purchases of anywhere from $50,000 to $100,000.

With this issue we begin a series of articles more particularly designed to show the varied uses of electricity in the different industries. At the same time these articles will demonstrate the extent of the industrial field as a market for electrical equipment. The first article deals with the manufacture of rubber tires, and the Goodyear plant is taken as a type.—Editor.

The extent to which electricity is used and the effects which follow a cessation of its supply, in manufacturing, render not inapt its title "the life fluid of industry." The further carrying of the metaphor into the visualization of cables as arteries and of the power house as the heart of the whole system, is not strained: in fact, the similarity is but emphasized by the examination of a modern plant such as that of the Goodyear Tire and Rubber Company.

The principal factors considered in planning the factory layout were (1) the most efficient power distribution with the least apparatus, comprising the installation of the highest quality and safest equipment obtainable and at the lowest possible cost; (2) "safety first," that the entire system should comply with all safety requirements, and (3) flexible installation, adaptability of the system to changing conditions, motor re-location and future additions.

### Power House

In this case the power house is of brick and steel construction. It is approximately forty feet high and is divided into four rooms, namely, the boiler room, 58 by 88 feet; pump room, 40 by 66 feet; motor generator room, 36 by 66 feet, and transformer room, 22 by 66 feet. Basements are included under the pump room and the boiler room. The plan is so arranged that future extensions can be made without interfering in any way with the operation of the present plant.

The roof is constructed of reinforced concrete, plastered inside, and embodies a supplementary layer of 2 inches of wood, to prevent sweating. Over this is laid four ply of felt, the latter being covered with slag. A monitor is carried, with apparatus for opening and closing the sash. Steel

columns, concrete floors, steel sash and steel stairs are features that strike the eye of the visitor as indications of modernity and fireproofness,—in fact the building is considered entirely fire-proof.

A conception of the quantity of electricity that enters this power house is rendered easier by the statement that many towns with hundreds of inhabitants do not require as much for their entire needs. The total consumption for April 24 was 29,000 kw.h. for 24 hours. This large amount of current is supplied by the Hydro Commission from a nearby sub-station, the lines entering the power house at the east end in the transformer room. The main disconnecting switches are installed immediately under the point of entrance of the lines to the building, and are operated from a gallery 25 feet above the main floor. The lightning arresters are of electrolytic type and are installed on the gallery floor.

### Switchboard and Transformers

The 2200 volt switch structure supports a bus bar which feeds the oil circuit breakers for the 2200 volt feeders to the factory. Each breaker is furnished with a separate disconnecting switch. The bus bar is fed through a main oil breaker with no voltage release. All the oil breakers are d.c. controlled, the control switches being installed on a switchboard in the motor generator room. Lightning arresters are installed at the end of the switch structure for the protection of the switching equipment.

The greater number of the smaller motors used throughout the factory are supplied from a bank of four 150 kv.a. Westinghouse oil-cooled transformers. These are located in the same room as the power line inlet and take power from the high tension buses at 2200 volts, and feed various 550 volt

# Modern Industry Owes its very Existence
# to the Application of Electric Energy

Typical views of electrical equipment in Goodyear plant.—Upper left—Transformers and bus bars; Upper right—Direct current switch board; Middle left—75 h.p., d.c. motor, driving calender; Middle right—Mill line drive, 400 h.p., 2200 volt a.c. motor; Lower left—Motor generator set (for direct current supply;) Lower right—Individual motor drive on vertical milling machine.

circuits through their secondaries. The usual disconnecting switches are used.

Evidence of the skill with which the power requirements of the company have been foreseen and provided for,
is given added illustration in the layout and equipment of
that section of the power house which contains the switchboard, power-factor compensating and emergency equipment. A factory load naturally embraces a large number
of small motors and in general the resulting power-factor
is lower than either consumer or power company desires.
The Goodyear company have solved this problem very
efficiently by the installation of a Westinghouse synchronous
motor of 510 h.p., which improves the power-factor anywhere
from 15 to 20 per cent. and results in a considerable saving
in power bills.

The synchronous motor is used to drive a d.c. generator
of 1,400 amps. capacity, supplying various d.c. apparatus
in the factory. Some of the rubber handling or tire making
operations require the use of variable speed drive and it is
apparent that d.c. motors are more suitable for this purpose,—a fuller mention of the d.c. applications will be made
hereinafter.

A steam-driven, turbo-generator, d.c. set comprises
the emergency equipment, and is of sufficient capacity (300
kv.a.) to handle the d.c. control apparatus and emergency
lighting.

The alternating current switchboard is located on a
gallery in this room and comprises nine panels. Individual
lines are run from several heavy motors into this switchboard and each can be checked for power-factor and frequency through special meters. This arrangement enables
the exact effect of intermittent load to be determined and
though some mill operations are "born that way," so to
speak, and no amount of adjustment can alter their nature,
nevertheless it is worth while to know exactly what does
drag down the power-factor. Phase reading jacks are embodied in each panel.

Besides the heavy motor lines, floor lines, elevator lines,
factory lighting and incoming lines are represented on this
switchboard, and four graphic meters are employed for the
registration of voltage frequency, power factor and watts.

Direct current distribution is looked after by a switchboard of six panels located on the same gallery as the a.c.
switchboard. Each panel is equipped with the usual d.c.
meters and control apparatus.

An air compressor is installed in the generator room
and is driven by a 2,200 volt, 100 h.p., Westinghouse motor.

The supply of current for factory lighting is furnished
by three 40 kv.a., 2,200 to 110 and 220, transformers.

## Transmission to Factory

All the factory feed lines are enclosed in armoured conduit and pass from the power house through an underground
duct to a brick cable vault in the basement of the main
building. Provision for extension is again evident here, as
the extra capacity provided will be sufficient for some years
to come.

It may be mentioned here that even the steam end of
the power plant is not able to get along without some help
from electricity. The electric lighting is of course taken
for granted, but looking a little further we find that in the
basement of the pump room there is installed a motor
driven, vertical, centrifugal, bilge pump, which is used for
raising all drains about the power house from a sump in the
basement to the outside sewer. It is automatically operated,
the electric control for same being installed on the main
floor above.

## Applications in the Mill

In dealing with the applications of electricity in a mill
it would be other than usual to assume that the large motors,

the heavy current users, were not the most important and
the most interesting. It appears, however, that the sum
total of power used by small motors is not greatly below that
of the big fellows, and their aggregate influence upon the
power factor is sufficient to warrant the greatest care in
the choice of conditions, sizes, operation, etc.

## Heavy Motors

The heavy motors, as mentioned before, each have individual lines to the switchboard and use a.c. current, at
2,200 volts. Their main use is on the mixing mills (machines
which work the rubber up with various compounds by knead-
ing it between large steel rollers) and the arrangement is such
that the mill can be stopped in less than ⅛ of a revolution.
This is accomplished with the aid of a magnetic clutch, the
only break in a direct connection of the electric motor to
the mill. It carries a brake on either side and is operated
by a safety device. A control board with d.c. control is
supplied for each mill. The mixing mill apparatus presents
an imposing appearance and leaves an impression of power
on the observer that is missing in the contemplation of more
delicate machinery; one 400 h.p. motor is used to drive a
set comprising three 60 in. and one 84 in. mills, and the total
equipment of the three sets represents a big investment.

## Application of Direct Current Motors

The sizable amount of power, that an examination
shows to be taken by d.c. motors, as may be conjectured, is
used on machines that require variable speed drive. It is
well known that in the application of the alternating current motor to auxiliary drive, one of the chief problems
has been the development of satisfactory equipment for
variable speed service. The direct current motor has very
satisfactory inherent characteristics for such application and
its use in certain operations is almost essential; in the Goodyear factory, the major portion of the d.c. current is used
for driving the calender machines which handle sheets of
the rubber and require frequent variations of speed. Satisfactory results are obtained from the equipment, which
consists of Westinghouse motors and Westinghouse and Cut-
ler Hammer control apparatus.

Doubtless many of us have wondered how the various
treads are produced on motor tires and have hazarded a
guess that the process involves the use of some sort of mold.
This is indeed the case, and the consumption of molds is
such in this particular factory that it is found advisable to
make them on the spot. The pattern of the tread on the
mold is cut by a scriber very similar to a planer tool, carried on a swinging arm of a special machine. Naturally
very sensitive control is required and this is furnished by a
d.c. motor with suitable speed adjusting equipment. Another
application of comparatively small d.c. motors is made on
the tire building machines, which also require very sensitive control. In this operation the operator builds the tire,
practically layer upon layer, and the necessary revolution of
the whole is accomplished by the attached motor; the
machine is being continually started and stopped, resulting
in the contacts clicking all the time and convincing anyone
that a d.c. motor, in this case, is the right thing in the right
place.

A machine which gives certain fabric material a bias
cut, is also equipped with d.c. drive. The same is applied
to the "tubers," which mold bits of rubber into workable
chunks. These latter are equipped with electrically heated
dies. The slitting machines are also included in the d.c. motor
drive category.

## Grouping of Motors

An analysis of the various motor drives in the factory
shows that some use is made of the group system in certain

departments where a number of machines are placed that are easily adaptable to such an arrangement. In this regard a conservative judgment would dictate that the peculiarities of each case determine the choice of individual or group drive, the economy of first cost and the saving of space by the use of overhead motors being well known. An instance of this occurs in the tube room, where overhead motors, taking alternating current at 550 volts, are used, and in the machine shop; in the latter case, the boring mills have individual drive,

Several medium size ventilating fans are installed and are equipped with 5 or 7½ h.p., 550 volt, a.c. motors. A rather unusual ventilating feature is embodied in a horizontal fan which does most of that work for the second and third stories. This fan is enclosed in a built-in recess, and measures about 10 ft. in diameter. It is driven by a 35 h.p. motor and its installation has taken the hardship from some of the more dusty work.

### Elevators

The supply lines for the elevators are run in to the switchboard and by this means a fairly accurate check of power consumption can be obtained. Five 29 h.p., 550 volt, a.c. motors are used for these purposes and the quantity of power that is used varies considerably. It is apparent that there are a great many variables entering into the question, such as the economy of operation, number of stops, weight carried, etc., hence the range of power consumption. An elevator may make as many as 20 miles of travel per day so that even in a day's time the total energy consumption is considerable. The elevators in this case have platforms of approximately 8 by 11 feet and capacities of 6,000 pounds. They are installed in towers built outside the main walls.

### Total Motor Installation

The fact that electrical power is more efficient as a means of doing work than are hand methods is well demonstrated by the following list of installations in the Goodyear factory. The fractional horsepower motor has evidently come into its own in this case, as is evidenced by the fact that there are shown about 25 different uses for them. Their combined reactance is surprising, in fact it is somewhat of a standing joke with the power plant superintendent to put a power-factor meter on his lighting circuit for the benefit of visitors. The resulting reading is generally a longer way from unity than they expected, but he invariably clears the air with the explanation that he has about thirty fractional horsepower motors on the line.

The total connected load in the Goodyear factory amounts to 4,153½ h.p. (March 10, 1922,) including both d.c. and a.c. applications. The total is subdivided as follows:—

| Number | Voltage | Horsepower. |
|---|---|---|
| 32 | 110 a.c. | 16 |
| 34 | 125/250 d.c. | 379½ |
| 141 | 550 a.c. | 938 |
| 10 | 2,300 a.c. | 2,820 |
| Total 217 | | 4,153½ |

It will be noted that of this total of 217 motors, by far the larger number are of the 550 volt a.c. type, and this fact is an indication of standard practice for industrial motors of medium powers.

The 32 fractional horse power, 110 volt a. c. motors are used for varied purposes as ventilating, driving sewing machines and peeling potatoes in the cafeteria. The 34 direct current motors, taking 125/250 d.c., and ranging from 1 to 50 horsepower, are used mainly on the calenders and tire building machines. Advantage is taken of the ruggedness of alternating current induction motors by their application to the great majority of the operations; there are 141

of these general utility motors, 550 volt a.c., ranging from 1 to 30 horsepower.

### Lighting

Nearly all the factory lighting is carried out with 100 watt nitrogen-filled lamps. The greater part of the operations are such that a great intensity of illumination is not required and standardization is possible. Where any doubt exists, the problem is readily solved with the aid of a Westinghouse Foot Candle Meter, which gives direct readings,

### Maintenance

The west end of the factory basement is devoted to a complete electrical shop and store room. A staff of 9 or more men is regularly employed on installation and maintenance, and the principle that this department works on is to eliminate breakdowns, as far as possible by preventing their causes. The factory is provided with an Autocall system and by this means it is possible to call an expert to the scene of trouble in a very short space of time.

A very complete set of testing instruments is carried and the use of these contributes largely to the smooth running of the organization. A motor may be working underloaded or overloaded, or there may be some other abnormal condition, and unless tests are made, a burnout is the first indication of trouble. A proper set of instruments costs money but in the long run the results are worth while. The Goodyear company have portable a.c. and d.c. voltmeters and ammeters, with their necessary adjuncts of shunts and potential transformers, so that any current or voltage used in the factory can be measured. A polyphase wattmeter is also provided and a megger is available for testing for grounds and dielectric strength. A complete record of each motor is kept, which consists of a card index file, showing the rating, department, repairs, etc.

Repairs of the smaller apparatus are carried out at the factory, larger stuff has to be sent outside.

The overwhelming preponderance of electrical devices and the large consumption of power in this factory are signs of the times; electricity is a prime necessity of modern industry,—here we have an organization employing hundreds of persons and the functioning of the whole arrangement would not continue for a moment if its electrical power were cut off,.

\* \* \*

The Goodyear Rubber factory as described above is only one of a number in Canada, though it is one of the largest producers of rubber tires. Collectively, the tire industry is a field well worth cultivation by the electrical salesman. Below we give a summary of the capital expenditure and annual disbursements of the six largest tire factories in Canada. The total capital expenditure is almost two and a half million dollars and the sum total of annual purchases well over a hundred thousand. In active years, we know for a fact, this latter sum has gone very much higher—perhaps a quarter of a million. The following figures are exceedingly interesting. They cover, perhaps 90 per cent of the Canadian output:

| Factory | Capital Value | Annual Expenditure | Annual kw.h. used |
|---|---|---|---|
| 1. ...... | $ 250,000 | $ 12,000 | 4,800,000 |
| 2. ....... | 500,000 | 17,500 | 7,800,000 |
| 3. ....... | 630,000 | 73,700* | 3,500,000 |
| 4. ....... | 500,000 | 15,000 | 5,000,000 |
| 5. & 6 .... | 500,000 | 15,000 | 5,000,000 |
| | $2,380,000 | $ 133,200 | 26,100,000 |

*Including charges for current.

Musquash Development—Generating Station and surge tanks, under construction

Musquash Development—Sub-Station at Fairville

# New Hydro Plant in New Brunswick

### Unique Installation Just Completed at Musquash River, Near St. John—Actually Two Plants in One—The Work of the New Brunswick Power Commission

The New Brunswick Electric Power Commission have issued an interesting report on the Musquash hydro-electric development, which has just been completed, describing and illustrating various phases of the construction and design of this plant. The Musquash River has three branches, known as the West, the East and the North-east, the East being centrally located between the other two. The station is located on the channel of the East branch, but only the West and the North-east branches have been developed. The plant is quite unique in design, therefore, in that two pipe lines, fed from rivers more than two miles apart, converge at a common point, combining to operate generators in the same power house.

The general plan is shown in the accompanying illustration. The working head at full load on the West branch is 117 feet, and on the North-east branch, 95 feet, after allowing for pipe line losses and tidal effects. Assuming 80 per cent. efficiency of generation, the first will develop 1450 continuous horsepower and the second 1550 continuous horsepower. The pipe line from the West branch to the power house is 8 feet inside diameter, 7,400 ft. long. That from the East branch is 10 feet inside diameter, and 2935 feet long. The latter will operate two turbines and the former one turbine, so that in the one power house there will be, practically speaking, two developments, providing a two-fold reliability and insurance against interruption.

### East Branch Intake Dam

This dam, including the intake, is 1,500 feet long and has a maximum height of 78 feet in the channel. From the western end, 874 feet of concrete gravity section, including spillway, is built upon granite ledge foundation well across the stream to the high bank on the East side. From this point an earth dam with concrete core wall extends to the intake. This intake and core wall at both ends are on solid rock, and the central part of the core on good clay hardpan. East of the intake a concrete core has been carried one hundred feet into the natural bank and earth embankment graded. There are 14,675 cubic yards of concrete in this structure and 15,350 cubic yards of embankment. The dam is provided with a stop log sluiceway eight feet wide to a depth below

the level of the intake. This will permit of lowering the water to that depth, of running trash from the pond, and in the future, if any lumber should become available on the waters of the stream a log sluice can be built through the dam. There will be a pondage of 1,800 acres above this structure. The plant can be operated 24 hours continuously at average capacity, and this pond would be lowered less than three inches. It is not intended to depend on this pondage, however, for storage, except for daily regulations. Any considerable lowering of this water would reduce the head and consequently the power. Provisions have been made on the spillway of the dam so that two feet of flash boards can be placed during the decline of a flood. The water so impounded would be drawn down to further conserve storage.

To impound the water at the level of this dam eight earthen dams had to be constructed in various depressions near the lower end of the pondage. These, as well as the main structure, have been completed. 30,950 cubic yards of embankment were required for these side dams.

### West Branch Intake Dam

The main dam is located upon the site of the old driving dam, known locally as Scott Falls. A concrete overflow section 224 feet long is built across the stream bed. It is 55 feet high at the deepest point in the channel. On the East side an earth wing 1050 feet long has been constructed with a concrete core wall, and on the west side a similar construction 700 feet long, making the total length 1974 feet. There are 8432 cubic yards of concrete and 38,962 cubic yards of earth embankment in this structure. The water has already reached a level within four feet of the overflow and the structure appears to be 100% efficient thus far.

Three side dams were required in this case also. These were built of earth, rip-rapped on the water side, as are all the earth dams, and contain 21,463 cubic yards of embankment. The area of the pond will be 500 acres.

A canal 900 feet long extends from a point near the eastern end of the main dam to the concrete intake house located between two solid rock banks.

### Storage Dams on the East Branch

Dams at the foot of Loch Alva will raise the water 19

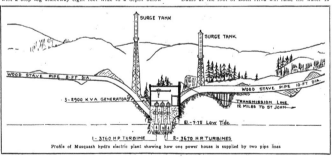

Profile of Musquash hydro electric plant showing how one power house is supplied by two pipe lines

Musquash Development—East branch intake dam

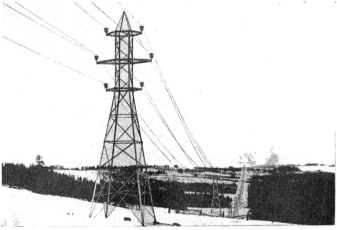

Musquash Development—Transmission line

Musquash Development—View of east branch head dam, intake and 10-ft. pipe line

Musquash Development—West branch intake and gate house, under construction

feet above summer level. The main dam is a timber structure, built on solid rock, and provided with spillway and two gates for regulating the flow. There will be 1,250,000,000 cubic feet of water impounded, enough to operate the plant at full capacity 83 days. Three side dams were necessary to close gullies near the lower end of the lake. In all four dams there are included 34,537 cubic yards of earth embankment, 3,723 cubic yards of rip-rap and 173,000 feet board measure of timber. The country adjacent having been fire swept several years ago, no timber was available at the site, and had to be hauled over four miles from the railway. The level of this lake has now been raised about ten feet and the dams are giving entire satisfaction.

Other storage sites on this stream which will be developed as required include: Eagle Lake, Deer Lake and Belvidere Lake.

### Storage Dams on the West Branch

At Log Falls, just at the upper end of the flowage from the intake dam, a yellow birch timber dam has been constructed, thirty feet high and about 400 feet long. The timber was cut from Crown Lands nearby and sawn at the site last spring. Two small earth side dams were also built. Owing to the nature of the gravel bank on the West side some trouble has been experienced in getting a satisfactory connection between the structure and the natural bank, and a crew is now working to extend the structure further into this material.

There will be 900,000,000 cubic feet of water impounded in this flowage.

In these structures 6,121 cubic yards of earth embankment and 335,000 feet of timber are used.

It is the intention to construct during the coming summer a dam at the foot of Sherwood Lake, further up the West Branch, to store 650,000,000 cubic feet of water. There is in addition 300,000,000 cubic feet of storage that can be controlled at a low cost on Seven Mile Lake.

All the work on the dams now complete has been done by the New Brunswick Contracting and Building Co., Ltd., the intake dams under their first contract and the storage dams under a contract signed in February, 1921, after calling for tenders in the case of the Loch Alva Dams.

The Log Falls dam was awarded without tender, as the price of sixty dollars per M. feet for timber erected in place including stumpage rate and fastenings, was considered as low as could be expected. Furthermore it was essential that this work be undertaken promptly so that the timber could be hauled on the ice. As it turned out the last timber was hauled only a couple of days before the ice on the flowage became unsafe.

### Intakes

At both the East and West head ponds solid rock was uncovered at the sites of the intake houses, particularly good conditions being found at the West Branch site where the intake house has been placed in a rock cut, only short wing walls at either side being necessary. Each intake is equipped, with an intake gate, operated by geared hoisting equipment at the entrance to the wood stave pipe line, trash racks of large area to prevent debris entering the pipe; and emergency stop logs at the mouth of the intake. These stops can be lowered into position or raised in their guides by a pair of travelling chain hoists suspended from runways attached to the roof I beams. The intake house may be dewatered, and repairs to the intake gate or filler gates effected. In the body of each intake gate are small filler gates, by which the pipe line may be filled before the intake gate itself is raised. The stop log hoists may also be used to withdraw sections of the trash racks for cleaning or repair, and sufficient floor space has been provided on the operating floor to allow of all stop logs being housed, when not in use, inside the building, so that minimum labor is required in handling. The windows are of steel sash, each window being provided with a ventilator.

### Pipe Lines

The pipe lines from the intakes to the surge tanks near the generating station are constructed of B. C. fir wood stave, held by steel bands spaced at short intervals of a few inches. The staves are milled radially, so that when drawn up by the steel bands, the faces of the longitudinal joints press together, forming a perfect water tight joint. To prevent leakage at the butt joints the staves are notched and fitted with steel tongues. This also allows for slight expansion and contraction of the wood. The pipes are carried on wooden saddles laid on an even grade a considerable length of the latter, in the case of the West Branch pipe line, being of close corduroy where the sub-soil is marshy.

### Surge Tanks

In order to prevent excessive variation in the pipe lines due to sudden increase and falling off of load on the generators, with corresponding change in the quantity and therefore velocity of the water at the turbines, each pipe line is connected near the generating station to a large surge tank of special design. The capacity of the tank on the East Branch pipe line is approximately 285,000 gallons, and that on the West Branch pipe line 215,000 gallons. The successful contractors for the supply and installation of these tanks and the steel pipe distributors between the tanks and the turbines were the Horton Steel Works, Limited, of Bridgeburg, Ontario.

### Hydraulic and Electrical Equipment

The generating units are of the vertical type, the turbines and auxiliary equipment being supplied by the S. Morgan Smith Company of York, Pa., the governors by Messrs. Lombard of Ashland. Mass., and the generators by the Canadian General Electric Company of Peterborough, Ontario. Each generator is of 2,900 kv.a. capacity at 13,200 volts, and carries above it and directly connected to it on the same shaft, an exciter capable of supplying direct current sufficient to excite two generators. The provision of exciters of this capacity not only ensures continuity of service in the event of failure of one exciter, but also materially simplifies the regulation of the voltage of the generators.

### Transmission Line

The transmission line consists of two separate circuits of aluminum cable, each circuit being of sufficient capacity to carry one-half of the maximum output of the generating station. These circuits are carried on fabricated towers of galvanized steel, erected on concrete footings where rock is not available.

The crossing of about one mile of muskeg presented some difficulty, and it was found necessary to drive piles to a depth of from fourteen to sixteen feet to rock or impenetrable strata, on which the concrete cap, constituting the footing for the towers, were poured in position. The spans for this portion of the line were increased to slightly over 1,000 feet to reduce the number of footings, and steel reinforced aluminum cable has been used in place of all aluminum. A special design of towers with a wide base was adopted at this point.

### Fairville Switching Station

The switching station at Fairville, is built of granite-faced concrete blocks. A steel superstructure is mounted on the roof of the switching station to accommodate emergency air break switches, for the purpose of sectionalizing the transmission line in the event of trouble occurring, should the line at any future time be extended beyond St. John towards the east.

was
wood
a few
rows
items
: pre-
l and
t ex-
arical
trable
pipe
or dry.

: lines
trura-
three-
line is
: tank
: flant
d that
scena-
tanks
of the
bridge-

be far-
i. Mac-
Messrs.
by the
orough,
: 13,000
on the
corrent
of ex-
service
fly am-
rs.

coasts of
acity to
aerating
avers of
re rock

extended
plen to
improve-
ding the
pumps for
over 1,000
reinforced
num. A
quired at

granite-
anted on
mergency
he nuns-
uould the
i towards

Musquash Development—General view of east branch intake dam and gate house

Musquash Development—West Branch intake dam; concreted overflow section

# The History of the Development of Electric Welding*

### By J. M. F. WILSON, B. Sc.

The arc fused metal is very much the same in both cases and hence we infer that very similar results would be obtained from both electrodes in practical welding.

Coated electrodes gave very much the same results, but did not include any of the covered electrodes on the market. Their specification for the coating paste was as follows:— graphite—15 grams; magnesium—7.5 g.; aluminum—4 g.; magnesium oxide—65 grams; calcium oxide—60 g.; sodium silicate—120 c.c., 40 degrees Be.; water—150 c.c. This was sufficient to coat 500.

Tests on the tensile strength of specimens made from the arc fused metal gave practically similar results (ultimate strength 50,000 lbs., from over 60,000 lbs. as electrodes and elongation fell from 14 to 8%) with properties very like low grade cast steel. Nothing was gained by coating the electrodes. Tests on specimens made from fused metal where the beads had been deposited across the piece instead of along the specimen showed a marked decrease in strength to 40,000 lbs.

### Electrode Constituents—Their Effect

In view of the authority of the above results it is interesting to compare J. S. Orton's figures for two specimens (1) R, according to the Welding Committee's specifications, and (2) W, containing a much greater percentage of C and Mn. R would be similar to B of the Bureau tests.

|     |        | C   | Si  | Mn   | Tensile Strength |                   |
|-----|--------|-----|-----|------|------------------|-------------------|
| "R" | Before | .17 | .14 | .57  |                  | % Composition     |
|     | After  | .12 | .10 | .23  | 57,300           | before and after  |
| "W" | Before | .39 | .12 | 1.01 |                  | fusion.           |
|     | After  | .23 | .02 | .84  | 76,200           |                   |

Note the retention of C and Mn in the weld, and that the greater the carbon-manganese content in the electrode the greater the percentage left in the weld. Microphotographs showed a smaller amount of oxide and slag inclusions with the high carbon manganese wire and tests gave a much higher tensile strength (see above table). The ratio of C to Mn should be 1 to 3. Further tests by H. R. Pennington on electrodes with still greater carbon manganese content throw additional light on the subject.

|                       | C   | Mn   | Analysis of a      |
|-----------------------|-----|------|--------------------|
| Before                | .85 | 11.2 | deposit on mild    |
| After (electrode neg.)| .63 | 8.11 | steel boiler       |
| After (electrode pos.)| .83 | 8.11 | plate.             |

Note the greater loss in carbon with the electrodes negative, hence to retain the carbon it is usual practice to make the electrode positive with Mn electrodes. Depositing the same wire on a carbon steel rail gave in the analysis a reduction of carbon from .99 to .71 per cent. and Mn from 10.51 to 10.19 per cent., that is a smaller loss in manganese on the carbon steel rail than on mild steel boiler plate.

The electrodes were Wanamaker Coated Electrodes No. 9, the coating preventing vaporisation of the Mn and C constituents. These particular electrodes are designed for rebuilding worn rails, frogs and crossings, the deposit being applied diagonally to the traffic and finished up against a fire brick or carbon block on the flanged side of the rail. To take advantage of the peculiar properties of 12 per cent.

*This is the conclusion of an article the first part of which appeared in the April 1st. issue and which has been unavoidably delayed in publication.

manganese steel (e.g., sudden cooling makes the metal very hard and ductile, slow cooling makes it brittle) in track work, a small stream of water is directed on the deposited metal until the heat retained in the part will just dry or vaporize the water.

Tests of Mn bars showed that if heated for one hour to 2100 degrees F. and then "toughened" by cold water, they lost strength (93,200 lbs. breaking strength), while if heated for five minutes only and then cooled they gave a test of 157,350 lbs. (ordinary 12 per cent. Mn steel gives 108,500 lbs.) From this we infer that the time for the welding cycle (a few seconds when cooled by water) would not cause any structural change in the metal.

### Automatic Welding

Automatic welding has been found to be adaptable to any form of weld from butt welding to deposition of metal on worn surfaces of shafts, wheels, etc. The speed of welding can be increased from 2 to 6 times that by hand. The weld is absolutely uniform and it has been found possible to weld any kind of steel from .1 to .55 per cent. carbon content. High speed tungsten steel has been successfully welded to cold rolled shafting, using Bessemer wire as electrode material. This kind of wire is considered as impossible for hand work, on account of sputtering, but it is just wires of this type which cut much more deeply into the work than Norway iron wire, with its quiet arc. In fact it is now possible to study the effect of the arc and the electrode without the personal equation. Some of the characteristics of the arc as the arc lengthens, an entirely automatic machine has been designed by the General Electric Co. This regulates the rate of feeding the electrode, which is now wound on a drum and gripped between a set of rollers connected through gearing to a small d.c. motor, the field of which is separately excited, and the armature connected across the arc, so that it will decrease in speed with the lower voltage due to a shortened arc, and increase in speed with the longer arc.

### Position and Handling of Electrode

It is now possible to watch the arc for long periods and amongst the observations it has been found that a variation of 5 degrees in the inclination of the electrode will make all the difference to the success of the weld. About 15 degrees from the perpendicular produces good work and with some materials the electrode should drag or point towards the welded part. H. D. Morton points out that in spite of the fact that indicating meters show practically constant readings, oscillograph records show a succession of short circuits, due probably to globules of the molten wire between the electrode and the work, the voltage falling to zero with an increase in current. The longer the arc, the fewer the short circuits. With 1/8" electrodes and 150 amperes, the globules are deposited at the rate of about two per second, as is also shown by ammeter records. As electrodes fuse at the rate of .2" per second, the electrode would have to be timed to be fed forward 1/10" after each globule to maintain a uniform arc length. The automatic machine is evidently able to respond to this.

### Alternating Current Welding

This is a more recent method of welding, its development being no doubt delayed by the fact that most of the welding was done in railway shops where d.c. was available, and that an operator has to be capable of maintaining a much shorter arc than with d.c. and takes longer to train. Again, it was contended that 75 per cent. of the heat of the d.c. arc was developed at the positive terminal and therefore the welding could never be as good in a.c. work where the amounts of heat were equal at both terminals. However, in this city it has been shown that as good a.c. work as d.c. is possible and that the point of view of the difference of heat is very much a fallacy. Quoting from the Waters Arc Welding Corporation's pamphlet, of the three factors which go to make up the heat of the arc, (1) contact resistance between the arc and negative electrode, (2) resistance of the arc, and (3) contact resistance between the arc and the positive electrode in d.c. work, the resistance of the arc accounts for 95 per cent. Of the remaining .5 per cent., the positive electrode takes 75 per cent., or about 3¾ per cent., as against 1¼ per cent. at the negative. In the a.c. of course these amounts are equal, 2½ per cent. at each electrode, so that the extra heat developed at the work by the positive terminal of a d.c. arc amounts to very little.

The simplicity of the apparatus appeals also in its favor, the weight being about half the d.c., and the current control being obtained without any external regulating apparatus, a transformer of the leakage type being all that is necessary. It is claimed that the a.c. arc gives better penetration than d.c. Before the welding committee Mr. Wagner reported that the frequency of the circuit did not make much difference, arcs being held at frequencies ranging from 12½ to 500 cycles, but of course much more easily the higher the frequency. He also found the speed of a.c. and d.c. very much the same, and that bare electrodes gave excellent results.

It is, however, so much easier to hold the arc with flux coated electrodes that they are almost invariably used and as a coating of sodium silicate (water-glass) is easily available, these need not add greatly to the cost.

### Operating Features of A.C. Welder

Against the a.c. welder it is contended that 170 amperes a.c. is required to give the same heating value as 140 d.c. and that the design of the transformer to produce a short circuit current equal to the normal welding current makes the p.f. as low as 35 per cent., hence the central station engineer is not likely to favor the type, but in these days of static condensers this ought easily to be remedied.

The recent three phase welders are attempts to equalize the load on the three phases and reduce the drop on any one phase of the system, but unless the welding transformer has some different characteristic due to its leakage factor it was demonstrated as long ago as 1892 by Steinmetz that it was impossible to take single phase power from either two or three phase systems and distribute the power equally over the several phases. On a three phase system one of the usual methods of producing single phase current is to connect two of the secondaries of the three transformers in open delta and join them in series with the third in reverse delta. Analysis by vectors will show that this simply results in putting a single phase load on two of the mains with the current in the third zero, gaining nothing over a single phase connection and losing much in simplicity and efficiency.

Another point put forward against the use of a.c. is the high voltage on open circuit causing a life hazard. On one tested in the city this was 135 volts, but there are now available on the market transformers like the Electric Arc Welding Company's which provides an auxiliary coil to limit the voltage to about 40 volts until the arc is struck.

### Conclusion

I have tried to indicate briefly the three prime factors for successful welding, (1) equipment (2) material (3) skill. As regards the equipment I have shown that it is essential to have a machine which responds quickly to the rapid changes in the resistance of the arc, and that a.c. equipment is as successful as d.c.

As regards material there must be relationship between the electrode and the work; that as the electrode arc process is essentially a casting, whatever is used must be treated to respond to the characteristics of the material; for example if it is a high percentage manganese steel that is used it must be given the same treatment a manganese casting would go through (rapid cooling, etc.), and it must be protected from the action of the atmosphere by coatings which will not allow its physical characteristics to be altered in its passage through the arc.

Finally as regards skill, not only do we mean the handling of the job by men of steady nerve, but the supervising skill to know that the weld is a good one, and the services of a welding engineer not only to select the proper material for the work, but to attend to the weld designs from a previous knowledge by test and research.

An instance of the resourcefulness of the radio fans was picked up only recently. A "radiophone party" was made up at a Vancouver home for an evening session, and while the necessary telephone equipment was on hand, no aerial or house wiring had been put in. The expert of the party "hooked on" to the wire clothes-line running to a high pole in the back yard, and in spite of the fact that it had been raining a few hours before, very good results were obtained.

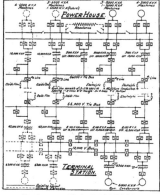

Wiring diagram of the Winnipeg Hydro Municipal System

# Problems of the Industrial Engineer——III

## Explaining to the General Manager the Different Methods for Correcting the Power-factor

### By GEO. L. MACLEAN, E.E.
#### Electrical Engineer Canadian Connecticut Cotton Mills Ltd., Sherbrooke, Que.

"There are four methods used to correct low power-factor," said the Engineer to the General Manager, who was asking how the low power-factor of his plant could be corrected.

"The phase modifying system, which has many disadvantages and consists of supplying a proper voltage to the collector rings of a phase wound motor is undesirable for many engineering reasons; this system is more generally used in European countries than here. A rotary convertor can also be used for power-factor correction but we will discuss the two most practical and common methods which are with (1) a synchronous motor, sometimes called a synchronous condenser or synchronous compensator, which is a revolving machine, and (2) a static condenser, which is stationary and although working on a different principle furnishes the same results.

"To illustrate how this correction can be brought about we will have to use a few simple diagrams. Assume our load to be 500 kw. at 75% power-factor and that we wish to raise the power-factor to 90%. In Fig. 1, the line BC represents our 500 kw. To find out what our wattless current, represented by AC, equals, we must first find out what our kv.a. load is, represented by AB. Our power-factor is 75%; then the line AB equals 500÷.75 or 666 kv.a., or we are supplied by the power company with 666 kv.a. to enable us to use 500 kilowatts. We must now find out what AC is, and this determined from the relation $AC = \sqrt{AB^2 - BC^2} = 450$ (approx.)

Fig. 1    Fig. 2

"In Fig. 2 we will have to find out the wattless energy of our 500 kw. load at 90% power-factor. Proceeding as before, AB now equals 550 kv.a., or at 90% power-factor the power-company are supplying us only 50 kv.a. over our 500 kw., but at 75% power-factor they are supplying us with 166 over our 500. I am explaining this so you can see how and why the change in power-factor affects our cost of power. In this case AC equals 230 wattless energy. A little study of these two figures will show that by cutting down our wattless energy we also shorten the line AB, which decreases the angle ABC.

"Getting back to our calculations,—we must supply to our system a component equal to the difference between our wattless current at 75% and 90%, or 450−230=220, and we will have to impress a leading kv.a. of 220 kv.a. on our system.

"Our first step must be to consider costs. Will the saving overbalance the cost, maintenance, depreciation, and interest on money invested in an apparatus to correct this power-factor? We will assume that it does.

"We will install a synchronous motor. If a synchronous motor is used instead of a static condenser, although the calculations are the same for the one as for the other, the motor will take the place of the large induction motor now driving the air compressor. It must, however, be large

enough to supply mechanical power to the compressor and not take full load current, for we will have to run this motor with an over-excited field and take full load current on the machine, but not full mechanical load, for the machine has also to supply a leading kv.a. of 220 to the system.

"A synchronous motor is practically a generator, although some changes will have to be made to meet conditions of the system. It requires an experienced attendant and, unlike an induction motor, is not flexible on the line, as any unusual conditions cause it to fall out of step and shut down, whereas an induction motor will adapt itself to voltage and frequency changes.

"Static condensers are stationary, and first cost per kv.a. runs higher, but their shortcomings are somewhat offset by the following facts: (a) low losses, (b) no rotating part, (c) less attendance, (d) operates at a low temperature, therefore, does not heat up operating room, (e) can be easily put on or taken off line. Indications from the number of static condensers under manufacture are that they may replace the synchronous motor as a power-factor corrector."

The G. M. on bidding the Engineer good day authorized him to purchase the necessary equipment for correcting the low power-factor.

## Largest High Tension Oil Circuit Breakers

The Westinghouse Electric & Manufacturing Company has just shipped to the Pacific Gas & Electric Company an oil circuit breaker for the 220,000 volt Mt. Shasta development; it is for use on a 220,000 volt system having a solidly grounded neutral. The breaker is good for a 350 kilovolt wet test. The gross shipping weight, including oil, is 90,000 pounds. The net weight of the unit, erected on foundation, is 75,000 pounds. The breaker requires 2,000 gallons of oil for each pole and the net weight of the breaker, exclusive of oil, is 30,000 pounds. The height of the breaker from the ground to the top of the bushing is 17 feet, 6¼ inches, and its height from the ground to the top of the tank is nine feet, 8½ inches. The long diameter of the tank is eight feet while its short diameter is five feet, eight inches.

## Home Made Radio Outfits

The Department of Commerce, Bureau of Standards, Washington, D. C., has prepared a circular, No. 120, entitled "Construction and operation of a simple home-made radio receiving outfit." This is for radio communication on wave-lengths between 600 and 200 metres, from high powered stations within fifty miles. The set may be constructed by anyone from materials which can be easily secured, the total cost of the equipment not exceeding $10. A single circuit with a crystal detector and an inductor, variable by steps, is used. Instructions are given for the construction of the detector, inductor, necessary switches, antennae and other parts. Certain parts, such as telephone receivers, must be purchased. Directions for operation are also given. Anyone interested may obtain a copy by sending five cents to the Superintendent of Documents, Government Printing Office, Washington, D. C.

The Nesbitt Electric Mfg. Co., formerly located at 95 King St. East, Toronto, are now established at 27 Melinda Street.

# 32nd Annual Convention of the Canadian Electrical Association to be held in Chateau Laurier, Ottawa, June 15, 16 and 17

Bigger than ever, more instructive and interesting than ever and better than ever, is the aim of the Executive and members of the C.E.A., in regard to their 32nd Annual Convention.

President Julian C. Smith and secretary Eugene Vinet, are still busy rounding off all arrangements for this convention, which is to take place in the Capital city, though the informal dinner and dance on the second night of the convention is scheduled to take place on the "Quebec side of the Ottawa River."

A large turnout of members is assured due to the splendid programme and interesting subjects to be placed and discussed at the convention; also because there is prevailing a general feeling that with the improving business conditions it is necessary for the leaders of the electrical industry to get together and mutually consult on expansion and improvements in carrying on the service from all points of view.

The Executive extends an invitation for plenary sessions to all those who are interested in the development of the electrical industry in Canada; the Electrical Co-operative Association, Province of Quebec, is sending out special announcements to all its members calling their attention to this convention and urging their presence.

Besides the following tentative programme, the Executive of the C.E.A. expects one of the best internationally known men in the electrical industry on the North American continent as a speaker:

**Thursday Morning June 15th, 1922**

Registration (Chateau Laurier)

General session, President Julian C. Smith in the chair.

Official opening by Mayor Plante.

President's address

Report of Secretary and Treasurer

Report of Membership Committee

  Chairman: M. K. Pike, Northern Electric Co.

Report of Public Relations Section

  Chairman: J. B. Woodyatt, Southern Canada Power Co.

Short Executive Session for Representatives of Classes "A" and "D" Members.

  Luncheon at Chateau Laurier

P. T. Davies, 1st Vice-Pres.

Eugene Vinet, Sec.-Treas.

Officers of the
Canadian
Electrical
Association
1921 - 22

L. W. Pratt, 3rd Vice-Pres.

Julian C. Smith, President

A. P. Doddridge, 2nd. Vice-Pres.

**Thursday Afternoon June 15, 1922**

Report of Commercial Section
Chairman: L. W. Pratt, Hamilton Cataract Power Light & Traction Co.
Industrial Lighting—J. H. O'Hara, Ottawa Electric Co.
Power Sales—P. R. Labelle, Shawinigan Water & Power Co.
Merchandising—Geo. L. Atchison, Southern Can. Power Co.
Talk by Mr. Webster & Exhibits.
Group Photograph in front of Parliament Bldgs.
Visit of new Parliament Buildings.

**Thursday Evening June 15th, 1922**

(Normal School or Russell Theatre)
Illustrated Lecture on "Illustration" by D'Arcy Ryan.
Mr. Ryan is one of the greatest authorities on illumination, approximately $30,000 have been spent for the preparation of that lecture which is a most gorgeous affair. We urgently request that every delegate should attend this part of the programme and bring his friends also.

**Friday Morning June 16th, 1922**

Report of Technical Section
Chairman. R. J. Beaumont, Shawinigan Water & Power Co.
Report of Meter Committee
Chairman: E. Holder, Shawinigan Water & Power Co.
Report of Electrical Apparatus Committee
Chairman: J. S. H. Wurtele, Southern Canada Power Co.
Underground Systems Committee
Chairman: L. A. Kenyon, Mtl. Lt. Ht. & Power Consl.
Report of Overhead Systems Committee
Chairman: A. P. Doddridge, Quebec Rly., Lt., Ht. & P. Co. Limited.
Report of Hydraulic Power Committee
Chairman: R. M. Wilson, Mtl. Lt. Ht. & Power Consl.
Report of Inductive Interference Committee
Chairman: J. H. Trimingham, Southern Can. Power Co.
Report of Lamp Committee
Chairman: Watson Kintner, Can. Westinghouse Co.,
Adjournment
Luncheon

**Friday Afternoon, June 16th, 1922**

Report of Accounting Section
Chairman: Lt. Col. D. R. Street, Ottawa Electric Co.
Report of Accident Prevention Committee
Chairman: Wills Maclachlan, Consulting Engineer Toronto.
Address by Chas. Scott, Manager Bureau of Safety, Chicago.
Report of Rural Lines Committee
Chairman: P. T. Davies, Southern Can. Power C.
Half an hour allowed to view various exhibits.
Trip by Tramways along Ottawa Valley.
Informal Dinner and Dance at Golf Club (on the Quebec side of the Ottawa River).
Speakers: Laurence W. Davis; Walter H. Johnson, 1st V. P. of N. E. L. A.

**Saturday Morning June 17th, 1922**

Executive Session for representatives of Classes "A" and "D" Members.
Laurence W. Davis, Special Representative of the National Electrical Contractor-Dealers' Association, who is scheduled to speak at Friday's dinner, coming from New York purposely for that occasion, will address, on Monday, June 19th, the electrical contractor-dealers of Montreal on the problems facing the electrical contractor-dealer and ways and means of meeting them. The Montreal meeting will be held under the auspices of the Electrical Contractor-Dealers' Association, Province of Quebec, Inc., and the Electrical Co-operative Association, Province of Quebec.

## Economic Electric Power Transmission

The British Aluminium Company, Ltd., are sending out from their Canadian office, 265 Adelaide St. W., Toronto, a very attractive bulletin. No. 175, entitled "Economic Electric Power Transmission." It contains some 50 pages of highly interesting and educative matter printed in a most attractive form and splendidly illustrated. The front page deserves particular mention, being a scene of a section of an Ontario Hydro high tension line, taken on a very windy day, the photo being captioned "Aluminium conductors withstanding severe weather conditions in Canada." The bulletin is in effect, a text book on aluminium.

## Electricity in Moving Pictures

Mr. W. Marshall, manager of a motion picture film exchange, well known in Western Canada, was the luncheon speaker at the Electric Club, Vancouver on April 21st. His topic was the great proportions to which the motion picture industry had attained and in developing it he stressed the value of electric energy and the many uses to which it was put in creating special effects and in producing the pictures. In the United States, he said, there are 18,000 motion picture houses and in Canada 900. The total number of people in the audiences of these theatres was 13,000,000.

## Quebec R. L. H. & P. Co. Installing Ten New Cars

The Quebec Railway, Light, Heat & Power Company, Quebec City, Que., are placing in operation, beginning about June 1st., ten new street cars, manufactured by the J. G. Brill Company, Philadelphia. These will be used on the city street railway division. The cars have been built according to the following specifications:

Seating capacity, 40; weight, car complete, 36,000 lbs.; length over body, 28 feet; length over vestibules, after alterations, 40 feet; width over all, 7 ft. 10 in.; body, wood; roof, Monitor type; underframe, wood; cables, Quebec Railway; air brakes, Westinghouse S.M.E.; car trimmings, Quebec Railway; control, Single End, K-35-G.2; fenders, H.B. Life Guards; motors, 4, 101-B2; gears, Nuttall; gear ratio-15-69-5" face; seating material, Rattan; seats, Longitudinal; heaters, Con. Car Heating Co.; gongs, Brill, "Dedenda"; hand brakes, Peacock; hand boxes, Brill "Dumpit"; hand straps, Ricco Sanitary; headlights, Crouse-Hinds.

---

**Coming Conventions**

Canadian Electric Railway Association, at the Drill Hall, Quebec City, June 1-2-3.

Canadian Electrical Association, at the Chateau Laurier, Ottawa, June 15-16-17.

Association of Municipal Electrical Utilities, at the Clifton Hotel, Niagara Falls, Ont., June 22-23-24.

American Institute of Electrical Engineers, at the Clifton Hotel, Niagara Falls, Ont., June 26 to 30.

---

## F. A. Rose Passes

The electrical industry has heard with regret of the death of F. A. Rose, of the firm of Rose & Ahearn, electrical engineers and contractors, Toronto. Mr. Rose was with the Canadian General Electric Company for over sixteen years, previous to going into business on his own account. He had travelled the country extensively and was one of the best known men in the trade.

# Better Merchandising and Cost Keeping

### "The Electrical Dealer is a Poor Credit Risk" Why? "He Doesn't Know His Costs"

There are many electrical merchants with very hazy notions of the cost of doing business. Most of them will admit that they ought to keep track of their costs and advance as a reason for not doing so such an excuse as "Have no time myself and can't afford a bookkeeper." This, of course, applies only to the man in a small way of business. Undoubtedly, there are many one-man businesses that cannot bear the expense of a clerk for the sole purpose of keeping accounts—but at the same time this is the very type of retailer who can least afford to be without a knowledge of his costs. The solution is for the owner to do the work himself—recognizing it as indispensable—or employ on his staff a man or woman who can handle; or learn to handle, the job as part-time work.

The inroads that are made by department stores and hardware stores on the legitimate field of the electrical merchandiser are due, in considerable degree, to the latter's poor trade rating, which in turn is the result of inefficient cost-keeping. It is needless to repeat here the oft-quoted arguments as to why the electrical man is the logical outlet for electrical goods. The fact remains that these other people are getting a good part of the business. They have been in business for a long time and are naturally the possessors of valuable merchandising experience. Their methods are efficient because their business is handled without guesswork. The electrical business is comparatively new and lacks what might be called trade traditions; i.e., established ideas and customs of doing business. In the natural course of events, these are a slow and steady growth—a man has run his business more or less efficiently, but his son takes it over, with a better experience, and so on. That leisurely development, however, cannot be tolerated in the electrical business in view of the sudden demand and wide market; in view, also, of the ready-made competition that is shooting holes in the legitimate profits of the electrical dealer.

The man who, in the past, has not been keeping accurate cost records has got to change his ways. A system of electrical merchandising must be—and is being—built up that will sweep outsiders off the map, and the man who would be successful must adopt this system.

As a practical illustration: "Service" is a good motto, but it is a mistake to spread service around too lavishly, with no idea of what it costs. How many merchandisers know what it cost them last year for all the little service calls they answered? Those jobs were little enough, but they cost money and the total would surprise the man who has not kept track of them. No right-minded merchandiser is going to shut down on service calls but he is going to know what they cost and arrange to cut down that item of overhead to a legitimate figure.

Then there's the question of rating, again. It is essential that the executive head of a business have, at all times, complete control and knowledge of the details of the busi-

ness, and this requires a proper accounting system. The man that can present a statement that shows his banker or jobber that he understands his business thoroughly is doing a great deal to maintain his credit.

### Apportioning Items of Cost

Granted that a man has an accurate knowledge of his total costs, there is another pitfall ahead, in that the various items may be improperly proportioned. Judgment and experience are the best guides in this matter. It is for this reason that during the past few weeks we have taken what may be called a census, in an endeavor to determine what the experience of Canadian merchandisers has been in this regard. A questionnaire was submitted to a large number of merchandisers, and two tables are given below which outline and clarify the result of this investigation.

Table I is a list of average costs, based upon the total number of replies received. It would doubtless be too much to say that if the various items of a business were either higher or lower than this average, there was something wrong; but, at least, this table provides a useful basis for investigation and comparison.

Table II represents the cost-of-doing-business figures of seven representative retailers. It will be noted that in some cases the items are not divided exactly alike. The general basis, however, is the same, a year's sales and a year's expenses.

These figures apply to conditions as they are today. There is no doubt that this overhead should decrease with an increase in volume of sales.

#### Table I
#### Average Overhead of Electrical Retail Merchandising in Canada

| | |
|---|---:|
| Wages of salesforce | 7.95 |
| Premiums | .86 |
| Advertising | 2.33 |
| Wrapping and other selling | .91 |
|     Total selling | 11.45 |
|     Delivery | 1.93 |
| Buying, management and office salaries | 4.59 |
| Office supplies, postage and other management | 1.37 |
|     Total buying and management | 5.96 |
| Rent | 3.62 |
| Heat, light and power | 1.06 |
| Taxes (except on buildings, income and profits) | .64 |
| Insurance (except on buildings) | .49 |
| Repairs of store equipment | .34 |
| Depreciation of store equipment | .902 |
| Total interest | 2.95 |
|     Total fixed charges and upkeep | 10.00 |
|     Miscellaneous | 1.504 |
|     Losses from bad debts | .71 |
|     Total expense | 31.55 |

#### Summary

The collection of this data brought out three striking facts, as follows:—

(1) A large number of dealers were not able to produce their cost-of-doing-business figures. They acknowledged that they should have had them and gladly admitted that they were anxious to see the costs of other men in the same line of business, but it was frankly beyond them to classify their own expenses.

(2) There is a wide discrepancy in the various items. Why should one man pay 14 per cent of his total revenue to

#### Table II
##### Overhead Expenses of 7 Representative Canadian Electrical Merchandisers

| Item | A* | B | C | D | E | F | G | H |
|---|---|---|---|---|---|---|---|---|
| | | | | | Per Cent of Net Sales | | | |
| Wages of salesforce | 3.00 | 13.9 | 9.30 | 5.3 | 8.55 | 10.5 | 2.97 | 5.92 |
| Premiums | | .6 | .10 | | | .9 | 1.68 | 1.11 |
| Advertising | 4.50 | 1.1 | 2.30 | .03 | 2.22 | 4.7 | 3.68 | 2.39 |
| Wrappings and other selling | 3.59 | | .22 | .9 | | 2.3 | .97 | 1.02 |
| Total selling | 11.09 | | 11.82 | 6.23 | | 18.4 | 8.50 | 10.44 |
| Delivery | | 7.0 | 1.10 | .92 | .4 | 2.5 | 1.06 | .52 |
| Buying, management and office salaries | 8.25 | 0.5 | 5.08 | 4.23 | 6.17 | 12.0 | 1.67 | 2.48 |
| Office supplies, postage and other management | | 0.8 | 0.72 | 1.18 | | 3.0 | .33 | 2.19 |
| Total buying and management | 8.25 | 1.3 | 5.80 | 5.41 | | 15.0 | 3.06 | 4.67 |
| Rent | 6.50 | 4.3 | 1.15 | 2.07 | 5.22 | 10.0 | 1.17 | 1.43 |
| Heat, light and power | | 0.1 | .42 | .22 | .88 | 3.3 | | 1.42 |
| Taxes (except on buildings, income and profits) | | | .89 | .11 | | 1.5 | .07 | |
| Insurance (except on buildings) | .07 | | .52 | .77 | .3 | .7 | | .15 |
| Repairs of store equipment | | | .20 | | | .6 | .23 | |
| Depreciation of store equipment | | | .47 | .08 | .53 | 1.4 | 2.90 | .033 |
| Total interest | 4.50 | | 3.00 | | 2.13 | 2.3 | 4.37 | |
| Total fixed charges and upkeep | | | 6.65 | 3.25 | | 19.8 | | 3.05 |
| Miscellaneous | | | 3.00 | .43 | 2.33 | | .79 | .97 |
| Losses from bad debts | | 1.9 | .30 | 1.0 | .35 | .6 | | .13 |
| Total expense | 41.00 | 30.2 | 28.67 | 17.24 | 24.12 | 41.3 | 16.52 | 19.79 |

*Figured on cost price.

his salesmen and another only 3 per cent? Is one dealer justified in paying 10 per cent for rent, when another "gets by" with an outlay of only 1.3 per cent? Can delivery expenses legitimately vary from .4 to 7 per cent? Buying, management and office salaries are difficult to regulate, but can they be shoved down from 12 per cent to .5 per cent without loss of efficiency? It should be possible for any retailer to answer these questions satisfactorily.

(3) The third point brought out is the extreme variation in the cost-of-doing-business totals—41.3 per cent down to 16.52 per cent. Undoubtedly there are inflexible circumstances, such as location, available capital, etc. that influence this total cost, but, is the extreme fluctuation indicated by this table consistent with good business?

The figures were obtained on the understanding that the names of the various dealers would not be published. It is, however, of interest to note that examples A and B in Table II are from the appliance departments of two large central stations, one in Western Canada and one in Eastern Canada; C, D, E and F are contractor-dealers, two in the West and two in the East; G and H are from two Hydro municipalities, among the largest in Ontario.

#### "Lest We Forget"

Do you remember the tremendous heat wave of 1921—how you were besieged by the public for fans—how you tried frantically to buy them from every jobber on your list—how you telephoned and wired to no purpose?

Don't be caught again this year. July and August almost invariably throw a heat fit, just as they did last summer. Think of the profits you lost by not being ready and lay in a fair stock while the weather is yet comfortable.

Use the same arguments with your customers to get them to purchase in advance. Remind them of the sleepless nights they spent, the days of discomfort and prostration. Remind your business customers of the loss of efficiency in their employees.

Learn experience from the straw hat dealer —you never catch him without summer headgear on the twenty-fourth of May—hot or cold.

### What Does a Woman Mean When She Says "No"?

"No," said the housewife—and it was with more or less difficulty that Bill managed to smile as he turned and walked off the porch. It was the twentieth time that day that he had been greeted—and dismissed too—with that word when he had mentioned the article he was trying to sell.

Bill was an inexperienced salesman—just out of school in fact—and he was trying hard to make a good beginning, so he made another attempt.

"No," said the twenty-first housewife.

"By golly," said Bill to himself after he got back to the sidewalk, "there's something radically wrong. Some of those women might have meant 'no' when they said it, but I don't believe all of them did. I'm not going to take 'no' for an answer after this!"

That was quite a while ago, when Bill first started out to sell. Today he is one of the best salesmen in his division, and if you will trace back you will find that his success started that day when he decided that he wouldn't take 'no' for an answer.

"I discovered," said Bill the other day, "that it is easier for a woman to say 'no' to a salesman that it is to say anything else. And I also discovered that 'no' has more shades of meaning and stands for more boiled-down excuses than there are letters in 'transubstantiation.' In other words, when a woman says 'no,' the chances are she doesn't always mean it. She may think she means it all right, but as matter of fact what she really means in a great many cases is any one of the following:

"'I don't understand.'
"'I'm afraid I can't afford it.'
"'I might buy later.'
"'I'm not feeling in the mood today.'
"'My husband would object.'
"'I'm going out and haven't time now.'

"When I start out to sell a prospect now I take into consideration the fact that she may be saying 'no' just as a matter of habit or as the easiest way of disposing of me. And straightway I say to myself, 'Madam, I don't believe you.' In other words I don't take 'no' for an answer.

"I always give my prospects who say 'no' another chance, so to speak, drawing them out to find what real shade of meaning is lurking in their minds. And very often I come away with the order!"

The "no" proposition is about the same whatever you may be selling to prospects in their homes,—especially when you happen to be out prospecting or making a cold canvass. If you take "no" for an answer your chances of making a great number of sales is mighty slim. If you analyze and draw out the picture behind the "no," your chances climb like stock in a "bull" market.

As has been pointed out somewhere or other, there's many a married man who would be a bachelor today if he had taken the first "no" his wife gave him as final!

—Hoovergrams

There is a lot of radio "junk" on the market that should only be found in Woolworth's 5, 10 and 15 cent stores. Don't get any of this on your shelves or hand it to your customers. It may mean a little immediate profit, but will react against you and your business in the very near future. The best radio equipment is none too good.

### The Ferguson Manufacturing Co.

The Ferguson Manufacturing Co., 174 King St., London, Ont., are now manufacturing conduit pipe fittings. These are made of close grained cast iron, carefully selected to ensure smooth castings and properly threaded holes, well japanned on the inside and electro-galvanized on the outside. The cover openings are ground flat and the screw holes well centred. The Hydro-electric Power Commission has approved these fittings, which are attractive both in appearance and price, and which can be shipped promptly from stock.

### Selling Chiefly to Contractor-Dealers

The Canadian Fairbanks Morse Co. Ltd., are dealing extensively in radio equipment and report business as being very good. Their chief trouble is getting shipments. They claim that the bulk of their radio equipment sales have been to the electrical dealers.

---

### It Cleans, Waxes, Polishes

In a recent issue we described the new floor waxer and polisher that had just been placed on the market by Messrs. Skelton & Connor, the principals of the Canadian Floor Waxer & Polisher Co., Ltd. In the interval this company has been working hard to reach a point where the manufactured supply would be equal to the demand, and believe they are now in a position to turn out machines as rapidly as the dealers can handle them. With the added operating experience they have also been able to eliminate any little minor troubles that developed and believe they have now a well-nigh perfect product.

The company has just made another very interesting announcement which will also give a great deal of satisfaction to the dealers, namely, that the retail price has been reduced from $65. to $52.50. As explained by the company, the original price was figured on a basis that would to some extent take care of organization and experimental charges. As production has increased, however, and the demand of the public has been demonstrated, they have decided it would be best to immediately cut the price to the consumer to the lowest point so as to assist the dealer in every possible way. The company states that dealers report splendid satisfaction on the part of their customers.

# Actual Figures on Electric Supply to Rural Customers

## Some Interesting Data Compiled out of the Experience of the Ontario Hydro Commission

The latest issue of the Hydro-electric Power Commission "Bulletin" contains some very interesting figures on the cost of the installation and operation of electrical equipment in a number of typical farm homes. The information is summed up in two tables, which are reproduced herewith.

The data included refers to farms in Brock township, Ontario County, Ontario, and are given for the purpose of illustrating, comparatively, what can be accomplished in any rural community installation costs are surprisingly reasonable.

Table Showing Comparative Data on Rural Electric Service

| Farm No. | I | II | III | IV | V | VI | VII | VIII | IX |
|---|---|---|---|---|---|---|---|---|---|
| **Rates** | | | | | | | | | |
| Annual Service | $60 | $60 | $36 | $36 | $60 | $60 | $60 | $60 | $36 |
| 1st Meter Rate | 7c. | 7c. | 7c. | 7c. | 7c. | 7c. | 7c. | 7c. | 7c. |
| **Annual K.W.H.** | | | | | ½ yr. | | | | |
| In lighting | 130 | 946 | 295 | 387 | 88 | 1138 | 434 | 604 | 278 |
| In power | 284 | 1219 | 702 | 347 | 119 | 681 | 398 | 514 | 278 |
| **Work Done** | | | | | | | | | |
| Bu. chopping | 1000 | 5000 | 3000 | 1500 | 175 | 4000 | 3000 | 2400 | 1500 |
| Root pulping | 1200 | 35 Hrs. | — | 3000 | 700 | 3000 | 500 | 3000 | 4500 |
| Hrs. Milking | 90 | 150 | — | — | — | — | — | — | — |
| Separating | yes | yes | yes | — | — | yes | — | — | — |
| Pumping | yes | yes | — | yes | — | yes | — | 175 hrs. | 400 hrs. |
| Feed Cutt'g | — | yes | yes | — | 15 hrs. | 60 hrs. | 30 hrs. | 20 hrs. | — |
| Total Cost | $85.12 | $196.10 | $95.15 | $79.45 | $26.48 | $151.07 | $110.57 | $126.49 | $73.78 |

Table Showing Comparative Data on Installation for Rural Electric Service

| Farm No. | I | II | III | IV | V | VI | VII | VIII | IX |
|---|---|---|---|---|---|---|---|---|---|
| **House Wiring** | | | | | | | | | |
| No. of Outlets | 18 | 23 | 21 | 29 | 12 | 41 | 26 | 28 | 20 |
| Cost of Wiring | $100.00 | $190.00 | $100.00 | $115.00 | $100.00 | $130.00 | $ 75.00 | $100.00 | $ 75.00 |
| Cost of Fixtures | $ 10.00 | $ 75.00 | $ 80.00 | $ 18.00 | — | — | — | — | — |
| **Barn Wiring** | | | | | | | | | |
| No. of Outlets | 17 | 22 | 16 | 17 | 13 | 20 | 16 | 15 | 11 |
| Motor | 1 | 1 | 1 | 1 | 1 | 1 | 1 | 1 | 1 |
| Cost of Wiring | $151.50 | $150.00 | $170.00 | $158.73 | $138.00 | $200.00 | $175.00 | $166.00 | $145.00 |
| **Motor Installation** | | | | | | | | | |
| Sizes in H.P. | 3 | 8 | 5 | 5 | 5 | 5½ | 5 | 5 | 5½ |
| Cost | $148.50 | $378.00 | $165.00 | $149.50 | $176.00 | $220.00 | $165.00 | $155.00 | $235.00 |
| **Line to Road** | | | | | | | | | |
| Length in Rods | 18 | 20 | 255 | 36 | 60 | 24 | 60 | 49 | 24 |
| Poles included | — | — | — | — | — | — | 2 | — | — |
| Cost of line | $ 30.00 | $ 50.00 | $119.34 | $ 63.45 | $265.00 | $ 49.56 | $100.07 | $ 53.87 | $ 54.34 |
| Total Cost | $440.00 | $843.00 | $634.34 | $504.76 | $679.00 | $509.56 | $515.07 | $474.87 | $509.34 |

# Electric Co-operative Leagues Performing Valuable Services for the Industry

The scope and possibility of accomplishment of Electrical Co-operative Leagues in the various Provinces of the Dominion is exceeding the expectations of even those who most favourably look upon and most ardently support this movement; in fact, if properly conducted and supported, these Leagues will perform an economic function in the changed conditions that will not only help to unify the scattered efforts, in the same direction, of the various groups within the industry and educate the public in the value of electricity as one of the most important factors in improving social conditions of communities, but will prove to be a saving of considerable value in marketing electrical merchandise.

Such movements as the recent efforts of business development of the N.E.L.A., the education campaigns of Trade Papers, in their editorial and special articles, to be made effective, must have local agencies; it is through the Electrical Co-operative Associations that this work can be carried on most effectively.

The Electrical Co-operative Association, Province of Quebec, which began its work some eighteen months ago, though working with a very limited budget as yet, has proven the usefulness of such an organization; this, despite the great handicap of bilingualism prevailing in that province, which makes the work so much more costly and tedious. Its secretary-manager, referring to the Association's third semi-annual report to the executive, under the sub-title of "The Boy is Growing," says:

"Born eighteen months ago under the leadership of some of the most prominent men in the industry, our Association has no apologies to offer for the progress in its growth. We've cut, in that time, all our teeth and are trying to crack some tough nuts; we are off the baby's chair and are kicking around like a spring colt on a meadow; in another six months we shall be ready for a larger allowance of oats to pull the load along.

"During the first half of the year 1921-1922 there were held ten meetings of the Executive. The Trade Relations Committee met every Tuesday since its first meeting on November 29th, making a total of twenty-one meetings at which matters mainly pertaining to the relations of contractor-dealers and jobbers, and infractions of the adopted Code of Business Ethics, were discussed. Sale of foreign made goods on the Montreal market, competition of fixture manufacturers with the dealers in selling to the users, the Modern Electrical Home and Entertainments had also the attention of this committee. Members have attended the meetings very well, except for the past two or three weeks, which is accounted for by the moving season somewhat increasing business activities and efforts to get business. There is no doubt that this committee has performed, and will perform, very valuable services toward popularizing the work of the Association.

"The Single Service Entrance Committee, consisting of representatives of power companies, Fire Underwriters and Government Inspection Bureaus, Builders' Exchange and the architects, the two latter bodies in consulting capacity, after several meetings have adopted rules governing the single service entrance in all the new, remodelled and other buildings, where practicable. The Underwriters' Bureau has decided to have printed instructions covering it in detail, and to enforce it from June 1, 1922.

## Organization

A great deal of time and effort was spent on organizing the English Section of the contractor-dealers, which included 14 general meetings, 3 joint executive meetings of the French and English Sections, and 3 joint contractor-dealers and jobbers meetings; during the time reported, 642 letters, not including usual meeting notices, were mailed. There is no doubt that improvement in the Contractor-Dealers' Association and acceleration of the activities of the new chartered Association is due to the assistance and work done on their behalf by the Co-operative Association.

### Window Dressing

A campaign for better window dressing was begun last Christmas and, by means of the bulletins, it is being continually brought to the attention of the dealers. No very positive results have been accomplished as yet, though some requests were received and a certain slight improvement is noticeable. It will require special effort.

### Government Licensing

The persistent requests of the Association by letters, telephone and personally directed appeals to the Deputy Minister of Labor to enforce the May, 1921, Act of licensing electrical contractors and journeymen and of inspection of electrical work, had considerable bearing on the Department of Labor taking final steps towards the enforcing of the Act. The preliminary work in connection therewith having already begun, the licenses are being already issued to those contractor-dealers who can pass the necessary examination.

### Co-operative Luncheons

The Association supplied speakers for the second Wednesday luncheons in January, February, March and April. The speakers supplied not only gave interesting addresses on various phases of co-operation, but also enabled the Association to be brought into closer contact with organizations whose goodwill and support will prove to be of great value in future work of the Association. Bulletins reporting activities of the Association are sent periodically to all the members and others connected with the electrical contracting business in the province.

### Electrical Homes

Electrical Home No. 1, with all its defects and drawbacks, was visited by 10,555 persons, at a cost of approximately .08¾ cents per visitor. A large number of "Improve on Yesterday" pamphlets were given away and a satisfactory number of requests for "Wiring the Home for Convenience and Comfort" were received and are still coming in.

We had to postpone the opening of Home No. 2 from April to September, due to the moving season, which did not permit of getting the necessary number of hosts to attend to the visitors.

The next "Home", in the Notre Dame de Grace district, will be a great improvement on the previous one, from every point of view.

The Lancashire Land & Construction Company is building a special house for that purpose, and all the committees are organized to make it a success. Besides special features of the building itself, the first electrified full suite dining and bedroom furniture manufactured in Canada will be displayed.

Gibbard & Company, manufacturers of solid walnut and mahogany furniture, in Napanee, Ont., are co-operating with our Association in this respect. There will be scientific illumination and artistic fixtures provided throughout the house, and a radio outfit installed.

### Distribution of Literature and Emblems

A number of pamphlets "Improve on Yesterday" were distributed directly among the public, giving suitable explanations as to how the home can be improved by application of electricity; all members of Quebec Legislature had sent to them copies of "Comfort and Conveniences of electricity in your Home"; some 350 copies of "Wiring the Home for Convenience and Comfort" were distributed among the contractor-dealers and the prospective builders in Montreal and the Province and 200 were sent to the architects in the Province of Quebec.

Some 300 emblems of the Association were sent to the members and can be seen in offices, stores and display windows; it eventually will have significance as the public is beginning to be familiar with it.

600 copies of Braley's "Origin of Co-operation" were distributed among those connected with the electrical industry in the Province.

### Socials

A smoking concert attended by over 400 men was held during the past winter and a dance in the Venetian Gardens after the Electrical Home was closed.

It served to bring together the electrical fraternity and to give considerable publicity to the Association among many who otherwise do not come in direct contact with the Association's work.

Besides, there is a desire among a considerable portion of the electrical men to get together occasionally for social purposes.

### Model Wiring Plan

A Wiring Committee, consisting of Geo. K. McDougall, electrical consulting engineer, and a specialist on illumination; Frank Penden, architect; E. J. Gunn and Wm. Rochon, electrical contractors and Fred. J. Parsons, fixture dealer are working out a model wiring plan for an average sized and priced dwelling according to modern requirements for electricity in a home. It will be available to all the contractor-dealers and builders of houses as a suggestion for "properly wired homes." The Lancashire Land and Construction Company is adopting that plan for wiring a number of houses built by them this season; we do not doubt that other builders will gradually follow. The same committee is preparing a set of standard electrical symbols for use in architects' and builders' plans.

### Man Who Does not Know His Costs is a Menace

No one believes more firmly than we, nor has anyone lived more faithfully up to that belief, that prices must come down as closely to the pre-war basis as possible and that to do so the manufacturer must use much inventive talent and the factory must be standardized as far as possible and production problems solved.

We also believe that what is commonly known as the market price should be set by those who can produce the best article at the lowest figure, not by those who keep on in the same old way with no ability or attempt to meet their problems by a change in their program.

If all concerns would insist on getting a fair profit, quite as many goods would be sold and business would the more quickly return to an even keel.

The chief trouble with a broken market on any given line is the danger of a price which benefits no one for long but injures many forever. In such a warfare purchasing agents become most perspicacious—almost uncanny—even going so far as to place almost unheard of discounts on an order as a tryout after the real bottom if there be any. They tell the salesman any sort of tale as to the price of the other firm and confusion becomes worse confounded.

People shop around—they turn down their best and oldest friends for three cents and even go so far as to think the old friend is not on to his job, but that the new low pricer is the Daniel come to judgment.

A great contributing cause to such a situation is the distrust that gets fixed in the head of the purchaser when high prices run a riot of debauch as in 1919-1920.

Just the same there is the other dangerous extreme witnessed today in some of the wiring-device lines—where selling below cost is the habit of the hour. Of course other less competitive lines have to bear the loss. If there be no such lines to stand such losses—the answer is go-broke-sir.

It also is rather pathetic to note on occasion the opposite policies followed by the selling and buying end in the same concern. The selling department may be fighting a noble fight against a price break-down, struggling to win a profit against hard odds, and gaining headway through service and selling brains, while all the time the buyer is endeavoring to squeeze the last ounce of profit out of the man who sells him.

Such a procedure lowers the vitality of business. It breeds suspicion, it injures electrical standards, it hurts morale, breeds tricksters and in the final analysis it does very little to stimulate business.

None of us are perfect in any particular and in times like those we have gone, and still are going, through, pesky people are bound to arise and foolish deeds be done, but if such a demoralizing condition extends too far and soaks too deep it acts as a very real barrier to the return of the better day.—Trumbull Tem

### W. Ross Hilton Joins S. W. Farber

Mr. Ross Hilton, of Montreal, Quebec, has recently become associated with S. W. Farber of 141 5th Street Brooklyn, New York, inventors of the famous "Adjusto-Lite," as factory representative in the Eastern Provinces of Canada. Mr. Hilton comes very highly recommended as he has a very wide knowledge of Canadian territory and Canadian business methods. He has travelled the Dominion for a good many years for such concerns as the Northern

**W. Ross Hilton**

Electric Company, Ltd., Crown Electric Company, Ltd., and McDonald & Willson, Ltd. It is the intention of F. W. Farber to carry out a very extensive advertising and sales campaign on the "Adjusto-Lite" throughout the Dominion of Canada. Mr. Hilton will also represent Mr. Farber on the various lines which they manufacture, such as brass and copper ware, silver holloware, smoking articles, electric portable lamps, and stationary desk sets. In selecting Mr. Hilton as their factory representative, Mr. Farber states he had in mind the desire of having a man represent them who would prove an asset to the Canadian merchants, from his knowledge and experience, and afford them a valuable marketing and merchandising counsellor.

# Cyclone Paralyzes Winnipeg Service

### Records Established in Building Transmission Lines Under Most Adverse Conditions—The Value of a Standby Plant

A serious interruption in electric service occurred in the City of Winnipeg on May 10th when a break in the transmission lines of both light and power utilities occurred. A cyclone of unparalleled force struck the country about 50 miles north-east of Winnipeg, levelling the trees, demolishing farm houses and wrecking 15 towers on both the old and new transmission lines of the Winnipeg Hydro and 11 towers on the transmission line of the Winnipeg Electric Railway. The first indication that a disaster had occurred was when all power on the company's lines was interrupted at about 12:30 p.m. tying up street car traffic and all services fed from the W.E.R. went dead. 20 minutes later all services supplied from the municipal plant were in turn affected, completely tying up all industries. At 2:05 p.m. service was partially restored on the street cars by the operation of the steam auxiliary plant of the railway company, but it was not until 24 hours later that all services were once again in operation in the city.

A great amount of damage was done to the Winnipeg Hydro lines in a very inaccessible part of the system, north of Milner. This was on the high ground known as Milner Ridge which is in a very exposed situation. The break occurred 6½ miles in from the Lac du Bonnet branch line in a country consisting mostly of muskeg. To make matters worse there had been a continuous down pour of rain for 3½ days previously, resulting in the country being practically impassable to traffic. A valiant attempt was made by the repair gang to get to the scene of operations quickly but owing to the impassable condition of the roads it was after 9 o'clock before men and material arrived at the site of operations where they already found the power house superintendent with two of his section crews on the job. The 6½ miles from the railway was traversed by men with material packed on their backs wading up to their knees in water.

The freakish nature of the cyclone was evident by the fact that towers on the west section of the track and also

Steel tower leveled by Manitoba cyclone

the trees in that vicinity were blown over to the south while on the east side of the track trees and towers were blown to the north; also the cable was blown for over 100 feet on either side of the right of way. A homesteader's log house within 200 yards of the line was completely demolished and the roof carried ½ mile away. Many other farm build-

ings in the surrounding country were practically wiped out, the homesteaders being left without buildings of any description. The collie dog shown in the accompanying photograph had his first experience of flying as both dog and kennel disappeared entirely, the dog returning 2 days later.

The work of organizing the camps supplying food was made exceedingly difficult by the continued down-pour of rain, but repair work was carried on throughout the night. It was decided to get one circuit into operation as soon as possible. While the linemen were engaged in disentangling cables the section gang dug holes and carried in light poles. A reinforcement of men arrived about 4:30 a.m. on the morning of May 11th on a special train from Winnipeg. Many of the linemen walked 21 miles to get to the scene of operations after being carried, as far as Tyndall in motor trucks where it was impossible for them to proceed any

Temporary transmission line structure following cyclone at Milner, Man.

further on account of the bad condition of the roads.

The work of erecting a temporary wooden structure over one mile in extent was practically completed by 5:20. The greatest difficulty was incurred in disentangling the cables and resplicing them where they had been cut or broken.

After lines No. 1 and No. 2 were in operation the work of diverting No. 3 and No. 4 circuits to the north of the right of way was commenced. This was necessary as these circuits were required to be put in operation before permanent repairs could be gone ahead with. The members of the staff and emergency gang went to work in a cheerful spirit and with unstinted efforts through the most trying conditions in order to give the citizens of Winnipeg service as soon as possible. Everyone was wet through, in some places it being necessary to work in water up to the waist. This, however, did not check their enthusiasm and they worked with a willingness that has earned them great praise.

The estimated cost of repairs will be approximately $15,-000 for permanent repairs while it will take at least $5,000 to take care of the cost of building the temporary structure. A shut down of long duration is an unusual experience for the City of Winnipeg but it has brought to the surface the question of building a standby plant as it is felt that industrial concerns particularly cannot afford to have their stock endangered by a tie-up of this description. The City Council are alive to the situation and have asked manager Glassco to report on the cost of building a standby plant. This will in all probability be of 83,000 h.p. capacity, consisting of 2-12,500 kv.a. steam turbo-generators with oil fired boilers. The cost of a plant of this kind would probably amount to $1,-500,000.

## Manitoba Power Co. to Deliver Power this Year

The visit of a number of prominent business men and engineers of Winnipeg to the plant of the Manitoba Power Company on May 17 was signalized by the announcement of the engineers on the ground that the first unit would be completed and delivering power to Winnipeg by Christmas this year.

The party numbering 150 men left Winnipeg by a special train, the guests of Sir Augustus Nanton, and inspected the work done which consists of the erection of an immense cofferdam on one side of the island which here splits the river. Where the diversion of the current has laid bare the bed of the river, tremendous excavation work has been done, and concrete is being poured in for the foundations of the power house. Work has been commenced on a dam across the island and from its further side to the opposite bank which will raise the level of the water 46 feet, while at White Mud Falls, two or three miles down the river, work is in progress by which the water level will be lowered 10 feet, thus giving a head of 56 feet for generating power.

After dinner, the members of the party visited every part of the work under the direction of capable guides. F. H. Martin, engineer in charge, explained the system by which expense of construction has been reduced to a minimum. Practically the entire engineering personnel was on the ground to explain the project in all its details.

### Confidence in Winnipeg

In the after dinner speeches a strong feeling of confidence in Winnipeg and its ability to absorb the whole of the power available as quickly as the development takes place was predominant. Sir Augustus Narton traced the history of the Winnipeg Electric Railway Company, and in referring to the pioneers mentioned, Sir William Van Horne, Sir William Whyte, Sir William Mackenzie, Sir Donald Mann, D. B. Hanna and J. H. Munson. He spoke well of the work of the late manager, Wilford Phillips, and praised in glowing terms the work of A. W. McLimont, the vice-president.

Mr. McLimont drew an optimistic picture to show how the power development would insure the industrial development of the city, and emphasized the need of co-operation. In closing, he paid a tribute to Mr. Julian C. Smith, supervising engineer, and the Fraser Brace Company.

### Julian Smith's Assurance

Julian C. Smith, vice-president of the Shawinigan Water and Power Company, said that it was most remarkable that the only adverse comment he had heard of the Manitoba Power Company was in Winnipeg. "Surely," he said, 'Winnipeggers do not intend to stand back and let people from Montreal and New York make this city assume the importance it rightly should?" He thought that Winnipeg was in need of confidence in their own future by the citizens themselves. Outsiders had that confidence in Winnipeg.

Major James H. Brace, also appealed for a demonstration by Winnipeg people of their confidence in the city.

He thought they should show their appreciation of the efforts of those responsible for starting this great power project by assisting and co-operating; if they did not they were not worthy of having the city progress at all.

D. H. Cooper, manager of the National Trust Company, thanked Sir Augustus Nanton for providing the opportunity for visiting the plant, which he regarded as a tremendous factor in the future development of Winnipeg as a manufacturing centre.

Travers Sweatman, president of the Board of Trade, said Mr. McLimont's message of co-operation should be strongly emphasized. The fact that outsiders had been persuaded to invest $7,500,000 in the project spoke for the confidence they had in the future of the city, he declared.

## Ahead of Last Year

Alex. MacKenzie, general sales manager of the Canadian National Carbon and Prest-O-Lite Co's., was a recent visitor to Winnipeg, spending a week with R. F. Kingsbury, manager of their Western division. Mr. MacKenzie stated that their business in this division is ahead of last year and anticipates good business for the balance of 1922.

## Appointed Representatives

The Russell-Fowler Co., have removed from 306 Notre Dame Ave., to 104 Capitol Theatre Building, Winnipeg. This firm have just been appointed western representatives for the Pittsburg Reflector and Illumination Co., Pittsburg. Pa., and Smith and Stone Ltd, Georgetown, Ont.

## Attended Annual Conference

W. H. Reynolds, western manager of the Eugene F. Phillips Electrical Works Ltd., has just returned to Winnipeg from a two weeks' trip to Montreal and Toronto, having attended the annual conference of the firm at Montreal. Mr. Renyolds states that present prospects point to more activity in the West. While he does not predict a boom, he is looking forward to increased sales from now onwards, and feels that the depression of trade that has existed, is a thing of the past.

## Contract for Oddfellows Home

Gamble and Willis Co. Ltd, 306 Notre Dame Ave., Winnipeg, have been awarded the electrical contract for the Oddfellows Home, at St. Charles, Manitoba; the total contract price of the building is in the neighbourhood of $100,000.

## Business Better than Anticipated

W. W. Robinson, Western representative of the Jefferson Glass Co., and the Crown Electrcal Mfg. Co., has just returned to Winnipeg after a two months' trip through Western Canada as far as the Pacific Coast, and reports business far better than he anticipated. The dealers appear to be very optimistic and a number of them are buying novelty lines, such as bronze table lamps, etc., in preference to staple lines. Mr. Robinson says. that the dealers are carrying very low stocks, and if the West has a good crop this year, there is sure to be heavy buying this Fall.

Messrs. Ferranti Limited, of Hollinwood. Lancashire, have lately been entrusted by the Publc Works' Department. of the New Zealand Government. with a contract for the supply of seven 4,000 kv.a., 110,000 volt, single-phase. oil-immersed force-cooled transformers. in connection with the Mangahao Hydro-electric Power Scheme."

## The Flow of Fluids Measured Electrically

The measurement of fluid flowing in a pipe presents a difficult problem; in the case of steam, this problem is extremely important to the electrical man. This is particularly the case in industry where, though the plant may be using hydro electric power, steam is used for many purposes and the boiler plant is under the supervision of the power plant superintendent. There has been perfected a device for the purpose of measuring the flow of steam or other fluid, that measures the flow directly by means of ordinary electrical instruments, and that seems to be very little known.

This measurement is accomplished by means of an electric current, which is regulated by the differential pressure of the flow. The current measured electrically represents directly the amount of fluid passing through the pipe to which the instrument is attached.

The principles involved are demonstrated diagrammatically in the figure. The instrument is designed on the U-tube principle, partly filled with mercury and balances the impact pressure of the flow in the pipe by the rise of mercury in the low-pressure side of the tube. The mercury column also forms a part of the electric circuit, as shown in the figure. This electric circuit contains a fixed external resistance $R_1$, a constant electromotive force E, an ammeter A and a watt-hour meter W. In the contact chamber C, which forms the low-pressure side of the U-tube, there are a number of conductors of varying lengths placed above the mercury column, and as the mercury rises it makes

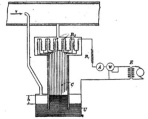

Fig. 1.

contact with one conductor after another. The variable resistance $R_1$ is subdivided by these conductors into resistance steps corresponding to the varying lengths of the conductors, so that the rise and fall of the mercury column varies the amount of resistance and thereby regulates the amount of current passing through the circuit.

The basic principle accordingly involves the laws governing the flow of fluids through pipes, along with those governing the flow of an electric current. The problem of establishing the theoretical relation between these fundamental laws offered little difficulty because of the similarity between the units of flow measurement, such as pressure and velocity, and the units of electric measurement, such as voltage and current.

As may be surmised the practical instrument is of vastly different appearance, and in some cases includes the use of an integrating feature which is of importance, since the readings from the watt-hour meter are more accurate than those taken from a recording ammeter. This feature therefore eliminates the necessity of planimetering charts and insures accurate results for any variation of flow.

When measuring the flow of steam generated by a battery of boilers the flow indicators are placed in front of each boiler, showing the momentary performance for the guidance of the fireman. At the same time, supplementary recorders connected electrically with the indicators are placed conveniently for the supervision of the chief operator.

Recently the manufacturers of water gas adopted the use of low-pressure exhaust steam for gas generation, which created an urgent demand for a measuring device to operate intermittently, varying every few minutes from zero to maximum. The electrical method of flow measurement was adopted, as this made it possible to measure successfully the steam required for the manufacture of water gas.

The main advantage of the electrical method of flow measurement is the accuracy with which the differential pressure is transmitted through a mercury column, which is free to attain the true level under all conditions of flow. Furthermore, the electrical instruments used to register the flow can be checked at any time without interfering with the operation or installation of the measuring device.

## Hydro and Dealers Working Together

The difference that has existed for some months between the Winnipeg Hydro and the Winnipeg contractor-dealers has now been overcome. Among the various objections raised by the dealers was, that the Hydro used the term, "Can be obtained from your Own Hydro" in their advertising copy. This has now been changed to "Can be obtained from Electrical Dealers or Your Hydro," and sometimes to "Visit the Electrical Dealers or Your Hydro," omitting the word "Own." Another matter that was objected to by the dealers was that of a flat rate for wiring being given by the Hydro. The latter have pointed out that the flat rate is a better proposition for everyone in the industry, and it is now being considered by the Contractor-Dealers' Association. In the month of April, the Hydro had eighty-three wiring jobs done by tenders from the various contractor-dealers; after the wiring had been completed, there was a nice surplus left over for the Hydro. The following flat rates for wiring are being charged by the Hydro:—

| | |
|---|---|
| Range only | $40.00 |
| Range and spare | $50.00 |
| Range and water heater | $60.00 |

The above prices apply only for installations in individual frame houses, and not in duplex houses or apartment blocks.

The Hydro have agreed to maintain re-sale prices; should there be an adjustment of prices necessary the executive committee of the Contractor-Dealers' Association will be notified by the Hydro, so that the matter can be discussed.

Arrangements have been made between the Hydro and the Home Appliances Manufacturing Co., Elmwood, Man., that will permit the dealers to purchase community washing machines direct from the factory should they so desire.

Mr. J. G. Glassco, general manager of the Winnipeg Hydro states that he is sincerely interested in the welfare of the Contractor-Dealers' Association and will do all in his power to establish good-will so that the electrical industry may flourish in Winnipeg.

The Ontario Association of Electrical Contractors and Dealers (Toronto District) recently announced that on and after April 1st, 1922 the wage rate for journeymen electricians will be 80 cents per hour.

# To Whom Does the "Radio" Business Belong?

### Radio Equipment Will Sell With a Minimum of Effort—It Will Turn Over Rapidly and Yield a Fair Profit—It is Electrical and has no Place in the Hardware or Department Store

### Who Does The Radio Business Belong To?

"The electrical dealers," you say, "of course."

But, who is getting this business?

Everybody—the hardware man and the department store are the chief contenders. But everybody is after it—crazy about it, in fact.

Are we, as an electrical industry, going to stand idly by and let these outsiders steal one of our most profitable lines of business? Why are they crazy about it? Because radio equipment sells at sight. Because people come clamoring for it. Because it yields quick returns.

Then they wash their hands of the whole thing and leave it to the electrical contractor and dealer to bear the brunt of any "service" that has to be rendered.

This is the way the electrical industry has been made the goat by the other trades for years past. Ever since the first electric lamp was invented we have provided the powder and shot; they have just pulled the trigger and collected the game.

They are not even decent enough to say "thank you." It has become so much a habit with them to steal the electric merchant's business that they are beginning to look upon it as their right. Read this from a recent issue of "The Dry Goods Economist." They are discussing "radio" and who should sell it, and say:

"It is a case of striking while the iron is hot—a very much overworked expression, but applicable. The stores that get the business first have the best hold on their communities. If the department store doesn't get in while the getting is best, people are going to get into the habit of buying elsewhere—in hardware stores, electrical shops, and so on. And the department store man will have to sit by and see his chance fade away or spend a good deal of extra money and effort taking away from those stores the business he could have had for nothing—or very little—had he got there first."

"Just think of it!" they say, holding up their hands in horror, "the very idea of electrical merchants having the presumption to sell electrical goods!"

Wouldn't it be too bad, though, if we should let these outsiders come in and corner one of the most profitable bits of retail business that has appeared on the horizon of the electrical merchant? Nothing has taken the popular fancy to such an extent in the history of the whole industry as has this radio. The sale of it can scarcely prove unprofitable. The turnover is rapid and the demand so great that only the minimum of "service" is expected. The department store has no more right to it—nor the hardware store—than the electrical store has a right to start selling a new fad in shoes or hosiery, or monkey wrenches or whatnot, just because the demand is great and the profit correspondingly satisfactory.

But are we going to cry over spilt milk—spilled into the coffers of the hardware and department store? Don't let's do it! Let us have backbone enough to write our past losses off as "experience" and set our teeth against spilling any more milk.

We have the solution if we will apply it. We all know what it is. In the first place, the electrical merchants must equip themselves to supply the public demand promptly and fully. In the second case, the manufacturer must stand back of the electrical merchant and give him a reasonable guarantee that irresponsible stores will not be among his competitors. Surely the electrical dealers can rise to the occasion. Surely the electrical manufacturers will stand together and support their own industry to this extent.

Quite aside from sentiment, or the question of right and wrong, there is also the question of expediency. It may not be entirely evident today, or even tomorrow, that the manufacturer will be consulting his own interest by marketing his product through the electrical merchant—the man who understands what he is selling—but a month from now, or three months from now, when the question of service comes up, extensions, enlargements and refinements, then it is certain that the electrical merchant will demonstrate that he is the essential link between the public and the manufacturer. We have no doubt in the world that it will come around to this, but the part that appeals to us as so entirely unreasonable and unfair is that every Tom, Dick and Harry in the way of a storekeeper should be allowed to handle radio outfits when they have no interest in them, beyond the profits, and when they may be handled just as quickly by the legitimate dealer who would at the same time sell the necessary amount of service. While the radio industry is young, let us start it on a proper foundation. It will develop much more rapidly and much more substantially and be the means of helping many an electrical contractor-dealer to establish himself more firmly in business so that he may carry on in other lines much more effectively.

There is another phase of the situation, too. It has worked out in the United States—and signs are not wanting that similar conditions are developing here—that a quantity of worthless material, turned out by manufacturers who don't know what they are doing, and handled by dealers who know no more than the manufacturers, is being foisted on the public. After the first flush of enthusiasm is over the public will waken up to the fact that they have not received value for their money and that they have something in the way of radio equipment that is of comparatively little use. All that will have been accomplished, therefore, will be that the radio business will have been discredited in the minds of the public and the products of reliable manufacturers, handled by reliable dealers, will be looked upon with suspicion. This, too, is a situation that seems to call for immediate action, not only that the electrical industry may be protected but that the general public as well may not become the victims of an unscrupulous group of persons who have no interest in radio, or any other electrical appliance, beyond the mere profit it will yield them today and tomorrow.

### Has Added an Electric Store

Mr. J. N. Lee, electrician, has opened an electric store at 501 College St., Toronto. He formerly had his office and storerooms at 347½ Euclid Avenue. Mr. Lee has had a wide experience in electrical matters. He was 6 years with the Toronto Hydro. in wiring and construction work, meters. etc., being detailed on exhibition work during the autumn seasons. He also spent some three years with the University of Toronto. in the Electrical Department. and for the past four years has been in business on his own account. He contemplates carrying a full line of appliances.

# Latest Developments in Electrical Equipment

### Bryant No. 651 Switch Plug

An indicating appliance switch plug has been placed on the market by the Bryant Electric Company of Bridgeport, Connecticut, and will be distributed throughout Canada by the Northern Electric Company. This plug is rated at 6 amps. 125 volts, 3 amps. 250 volts. One of its main features is the indicating switch which is made on the toggle principle. To the switch mechanism is attached a dial that clearly indicates whether the current is "on" or "off." Other features embodied include:—steel reinforcing springs which hold the self adjusting contacts firmly, thereby assuring tight connections; large terminal screws that make wiring easy; wide breaking area of switch between jaw and blade; casings are of tough composition material to stand lots of abuse;

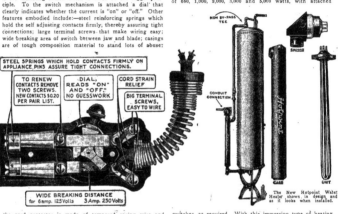

STEEL SPRINGS WHICH HOLD CONTACTS FIRMLY ON APPLIANCE PINS ASSURE TIGHT CONNECTIONS.

| TO RENEW CONTACTS REMOVE TWO SCREWS. NEW CONTACTS $0.20 PER PAIR LIST. | DIAL, READS "ON" AND "OFF." NO GUESSWORK | CORD STRAIN RELIEF |
| BIG TERMINAL SCREWS, EASY TO WIRE |

WIDE BREAKING DISTANCE
for 6 amp. 125 Volts        3 Amp. 250 Volts

the cord protector is made of tempered spring wire and stretching won't hurt it a bit; composition bushing in the end saves wear on the cord. It will take either flat or round terminals. This plug fits:—American, Edison, Northern Electric, G. E., Renfrew, National, Manning Bowman, Hotpoint. Westinghouse, Universal, Simplex, and many other makes of heating appliances.

### A New Storage Battery

H. M. Trade Commissioner. F. W. Field, describes the manufacture of a lead hydrate storage battery by Siebe, Gorman & Co. of 187. Westminster Bridge Road. London S. E. I. The following claims are made for the new battery:

(1) It has 300 to 800 per cent. greater capacity than any other battery of the same size;

(2) Its greatest superiority is attained at the highest discharge rates;

(3) It is indestructible as a battery;

(4) It is rechargeable in fifteen minutes;

(5) It cannot sulphate; and

(6) It has a very much longer life than any other battery.

While in external appearance the lead hydrate battery differs little from an ordinary lead battery, the plates are very much thicker and the density of the dilute sulphuric acid electrolyte is much greater (Sp. Gr. 1.34 to 1.35). Moreover, the plates are put into stock fully formed and are capable of yielding a discharge voltage of over 2 volts per cell directly they are put in the acid. It is stated that the hydrate paste does not vary in volume under any conditions of charge or discharge, so that the plates do not buckle, and consequently remain unaltered for years.

### The Hotpoint Water Heater

The new Hotpoint water heater that has just been placed on the market by the Canadian Edison Appliance Co., Stratford. Ont., is of the immersion type—efficient and easily replaced. The unit is so constructed that the manufacturers claim its life will be practically indefinite. It is made in sizes of 660, 1,000, 2,000, 3,000 and 5,000 watts, with attached

The New Hotpoint Water Heater shown in design and as it looks when installed.

switches, as required. With this immersion type of heating element the efficiency is practically 100 per cent, as the water, passes over the element immediately into the lagged tank, this absorbing and retaining all the heat energy. The arrangement for removing and replacing the element, should this be found necessary, is remarkably simple, requiring only two or three minutes of the electrician's time, the services of a plumber not being required at all. In a number of Canadian cities flat rates are now established for water heating purposes. When using a high efficiency heater, such as the one described above, the cost of operation often does not run over $2. or $3. a month for all the water required in an ordinary house.

### Electric Stylo pen

The Post Electric Company, 80 East Forty-second St., New York City, has placed on the market an electric pen, known as the "Stylo-electric." It is used for marking wood, leather and paper, and by the use of colored transfer papers, will burn brilliant tints into leather, celluloid, glass, hard

rubber, wood, etc. It can also be used as a miniature soldering iron, on fine work. The Stylo-electric is designed along the lines, and is about the size, of a fountain pen. It is supplied with two different points, six feet of cord, circuit tap switch, and a supply of colored transfer paper. The manufacturer states that the heat of the point can be easily regulated.

### Improved and Simplified New Attachments for Hoover Electric Cleaner

Certain changes and improvements have recently been made in the attachments designed for use with the Hoover suction sweeper. Generally speaking, these changes have been dictated by the desire of Hoover engineers to simplify the design and improve the efficiency of the attachments. The Research Laboratories at the Hoover plant claim they have developed some new ideas which will increase the durability and more than double the cleaning ability of their product. One of the most important improvements is that made in the converter, the device by which the air tools are attached, which has been redesigned to allow the hose to be inserted into the converter opening instead of the converter being inserted into the hose. The manufacturer claims that this change alone increases, to an appreciable extent, the volume of air which passes through the attachments.

It is stated that a new and improved hose will hereafter be furnished. This hose, to quote a statement from the Hoover engineering department, is made with "soft molded rubber ends and covered with a heavy quality of olive-drab colored duck." This covering is very strong and durable and is of such a color that it will not easily show soil from frequent contact with the floor.

In addition to these changes, one of the heavy fibre extension tubes has been curved, at an angle of 45 degrees, which makes it possible to dispense with the 45 degree elbow attachment, thereby reducing, by two, the number of joints in the set of attachments. The Hoover people claim that this will not only remove an obstruction to the air flow at this joint, but will also prevent the parts from becoming disconnected in use.

Changes have also been made in the general purpose brush, which has been "designed to combine the uses of the orifice nozzle, flat rubber nozzle and library brush, this one tool, this tool being separable so that it can be used as an aluminum single orifice nozzle if desired. In addition, the taper on this combination tool has been reversed, so that the tool enters the hose.

All in all, some rather important improvements appear to have been made which, together with the greater simplicity of design, are destined · to increase the popularity of Hoover attachments in the eyes of the dealers and eventual owners. We are told that every possible cleaning need in the home is thoroughly provided for.

### Radio in Unsettled Condition

Harry F. Allen, purchasing agent of the Great West Electric Co. Ltd., Winnipeg has just returned from a business trip, having visited Chicago, New York, Toronto, Hamilton, Brantford and Woodstock. He was particularly interested in radio development, but found that this line is still in a very unsettled condition. However, there is lots of room for the contractor-dealer to take care of this line, owing to the fact that it is more or less of a technical nature, and at the present time is in the hands of other people who know nothing of this line. Mr. Allen found conditions generally are getting brighter, with considerable building in progress in the Toronto district.

### Exclusive Imperial Headlight Agents

Dating from June 1, 1922, The Ohio Brass Company. Toronto, will act as exclusive sales agent of the Crouse-Hinds Company of Canada, Limited, in connection with the Imperial line of headlights. Imperial headlights are

well known for their excellent design, their durability in service and their large variety. There are Imperial headlights—including luminous arc, carbon arc and incandescent types—for every class of electric railway and mining service. The Ohio Brass has had broad experience in the marketing of headlights and is prepared to give adequate service.

### Announcement

The Fibre Conduit Company, Orangeburg, New York, has acquired the plant of the American Fibre Conduit Corporation at Fulton, New York, and the conduit manufacturing business of the Johns-Manville Company at Lockport, New York, and has appointed Johns-Manville Inc., as Sales Agent for its products, effective as of May 15th, 1922.

### Insulating and Soldering

A catalogue supplement describing their line of insulating and soldering compounds and announcing the extension of that line of products has been issued by the Westinghouse Electric & Manufacturing Company. The publication is known as 5-A, Supplement No. 2.

The Travelers Indemnity Company, Hartford, Conn., has recently included a comprehensive Machinery Policy in its list, under which insurance can be had on engines, fly wheels, steam turbines and electrical units. The company is distributing a folder describing its new policy.

The Department of Railways of the Province of Saskatchewan has issued its annual report for the financial year ended April 30, 1921, copies of which are now available. Interesting information is given regarding the financial and physical status of the Moose Jaw Electric Railway Company, the Regina Municipal Street Railway and the Saskatoon Municipal Railway.

---

We know of an electrical merchant whose cash sales have been doubled by the addition of radio equipment to his stock.

---

Mr. A. Harker has recently opened an electrical fixture and supply store at 183 Church St., Toronto. He is displaying a full line of the "Alpha" fixtures, of which he is, himself, the inventor.

---

Messrs. R. E. T. Pringle. Ltd., Tyrrell Building. 95 King St. E., Toronto, have moved to larger quarters. at 27 Melinda St., Toronto.

---

A combination stove and heating pad is being marketed by the Ingersoll Electrical Products Co., 109 W. Austin Ave., Chicago. This is a 3-heat equipment, the low heat being used as a heating pad, and the medium and high heat being suitable, chiefly, for the purpose of toasting, warming, etc. The heating element is said to be of new design and permanent and, owing to complete insulation by the use of amber mica and asbestos, all danger of short circuit or

The same company are manufacturing a metallic flexible heating pad, which is made of aluminium and bronze, nickle plated, and also has three heats. A special advantage claimed for this pad is that it may be fastened to the body in any manner; also, it may be used with wet or damp applications. Its metallic construction adds greatly to its variety of applications. without interfering with its flexibility; for example, the pad may be used by a patient or invalid in any position.

## A One-man Electric Home

The wiring of the house, the installation of the appliances and the advertising for this Home Electric were done by one man, Mr. R. Simpkin of Weston, Ont.—An example that can be followed anywhere.

### A Rheostat Regulator

The Wirt Company, Philadelphia, are manufacturing a fan and vibrator regulator. This is a rheostat, as illustrated, which can be used in every electrical home, for the following purposes: to give four speeds to single speed fans; to furnish four distinct speed controls to vibrators without a speed regulator; to give four heat controls to single heat

heating pads; to reduce the excessive speed of the spindle on fountain drink mixers; to dim table portables and floor lamps which have two or three lights, where the total wattage of the lamps does not exceed 60 watts; the regulator in this case is screwed into the baseboard receptacle, and portable plug inserted into regulator. The various controls are clicked off by simply turning the fibre ring on the regulator.

### Value—Plus Service

A trade mark which means something is that of the Sibbald Electric Co. Ltd., which believes that "Value plus

Service equals satisfaction." Any dealer who believes it and acts up to it is bound to find himself doing a good business with contented customers.

Messrs. Hawkins & Hayward, electrical engineers and contractors, have in hand the work of installing the electrical equipment in the Royal Jubilee Hospital, Victoria.

### Introducing New Motor

C. D. Ellsworth, sales engineer of the Century Electric Co., of St. Louis, Mo., recently spent two weeks in Winnipeg and is now visiting the trade in the Prairie Provinces introducing the Century automatic start polyphase motor. This motor is more or less new to Western Canada, although it has been in use in Eastern Canada and the States for the past four or five years. Mr. Ellsworth is very pleased with business conditions in Winnipeg.

---

### Sample Message Printed on the back of Monthly Bill by the Edmonton City Light and Power Department

#### Why a Monthly Minimum Charge?

So that you can have electric light, or power for a motor, just at any instant you may choose to "snap the switch," certain equipment must be held ready for service at all times. Power plant machinery and all its complicated auxiliaries, poles, wires, transformers, meters and similar engineering equipment together with a number of other technical matters, are all involved in that "snap of the switch."

This equipment and its many operating details are all items of expense, and certain more or less fixed costs exist even although during a certain period, the consumer has very small, or not any energy consumption.. For example, there are interest and depreciation charges, maintenance costs, line and transformer losses, meter reading cost, billing office expense and other items going on just the same.. Of course, adequate revenue from light and power service is necessary to meet these costs, otherwise they might soon cause an increase in general taxation. It is reasonable that the actual electric light and power users should bear these items as part of the cost of service, and this is the case in practice, the Electric Light and Power Department, being operated on an efficient commercial basis, entirely self supporting with a satisfactory margin of profit.

The details of equitably distributing these many items of costs among the various classes of service are part of the highly technical problem which rate-making involves, and it is only possible to here make a passing reference to the matter. However, it may be stated in general terms that these more or less fixed charges are averaged for the various classes of customers. The rates are carefully calculated to include these items, but if you use less than a certain amount of energy consumption each month then the minimum charge is necessary so that you pay your share of such costs for that particular class of service.

Briefly and stripped of many technicalities that is why "monthly minimum charges" exist in our own and practically all other rate schedules.

# Current News and Notes

**Birchcliffe, Ont.**

The Canada Electric Co., 175 King St. E., Toronto, have secured the electrical contract on a $60,000 school building to be erected at Birchcliffe, Ont.

**Durham, Ont.**

At a meeting held recently at Durham, Ont., the Eugenia Hydro-electric Association was formed.

**Emerson, Man.**

A by-law authorizing the expenditure of $9,000 for an electric lighting plant at Emerson, Man., was recently passed.

**Donnancona, Que.**

Mr. J. B. A. Lachance, 12½ Mont Marie Ave., Levis, Que., has secured the contract for electrical work on school building to be erected at Donnacona, Que., at an estimated cost of $43,000.

**Halifax, N. S.**

The Northern Electric Co., Ltd., Halifax, N. S., has been awarded the contract, by the Halifax city council, for the supply of street lighting equipment for the city of Halifax.

**Hamilton, Ont.**

Headquarters for the Hydro-electric System, Hamilton, Ont., have been secured by the purchase, recently, of the Bank of North America building, 12 King St. E., Hamilton. The building will be completely remodelled and will likely be ready for occupancy by September next.

Messrs. Culley & Breay, 35 King St. W., Hamilton, have been awarded the contract for electric wiring on a new school being erected on Dunsmure Ave., Hamilton, at an estimated cost of $425,000.

Mr. Jos. Hartnett, 368 King St. W., Hamilton, Ont., has been awarded the electrical contract on a $40,000 school being erected on Gage Avenue, N., Hamilton, by the Separate School Board.

Mr. Fred T. Brooks, 29 Mary St., Hamilton, Ont., who carries on a sign painting business, has added electrically illuminated signs, the letters of which will be of opalite glass.

Mr. Stanley Lewis, 63 Metcalfe St., Hull, Que., has been awarded the contract for electrical work on an apartment building being erected for Mr. G. E. Hanson, 11 Eddy St., Hull, at an estimated cost of $30,000.

Mr. Raoul Viau, Hotel de Ville Ave., Hull, Que., has been awarded the electrical contract on a music store at 55 Main St., Hull, now undergoing alterations.

Mr. E. Martel, 39 Wright St., Hull, Que., has been awarded the contract for electrical work on an addition being built to the store of Messrs. Fortin & Gravelle at 25 Main St., Hull.

**Kingston, Ont.**

Work has started on a power house to be built at King St. W., Kingston, Ont., for the Kingston General Hospital and Queen's University, the cost of which is estimated at around $50,000.

**Kitchener, Ont.**

The Star Electric Co., Kitchener, Ont., has secured the contract for electrical work on eight residences being erected on Rose Ave., Kitchener, by the Star Construction Company, of that city.

The Commercial Electric Co., London, Ont., has secured the electrical contract on an office building located on Queens Ave., London, that is undergoing alterations.

The Vincent & Say Electric Co., 344 Union Ave., Montreal, has secured the electrical contract on the Royal Arthur School to be erected on Canning St., Montreal, by the Protestant School Board, at a cost of about $146,000.

Mr. H. R. Cassiday, 255 Regent St., Montreal, Que., has been awarded the contract for electrical work on a service station being erected at St. Dennis St., between Bellechasse & Reaubien Sts., Montreal.

The Acme Electric Co., Herald Building, Montreal, has been awarded the contract for electrical work on an addition and alterations being made to the pharmacy of Mr. H. Singer, 81 Laurier Avenue W., Montreal.

Mr. J. A. St. Amour, 2171 St. Denis St., Montreal, has been awarded the electrical contract on a public garage being erected at Laurier & Esplanade Avenues, Montreal, by the Savard Motor Supply Co., 124 Laurier Ave.

Mr. G. Houle, 200 Boyer St., Montreal, has secured the contract for electrical work on sixteen residences being erected in the vicinity of Wilson & Terrebonne Aves., Montreal, by Mr. V. E. Lambert, 350 Marcil Ave., at an estimated cost of $168,000.

**London, Ont.**

The Commercial Electric Co., 489 Richmond St., London, Ont., has secured the electrical contract on a store building being erected on Dundas St., London, by the Cowan Hardware Company, London, at an estimated cost of $100,000.

**Montreal, Que.**

Mr. O. Labelle, 326 Northcliffe Ave., Montreal, has been awarded the contract for electrical work on a store and apartment building being erected at 6592 Sherbrooke St., Montreal.

J. J. Joubert Co., Limited, 975 St. Andre Street, Montreal, are constructing a 1,000 KW. transformer house, the work being done under the supervision of J. A. Burnett, M. E.I.C., consulting engineer, New Birks Building.

**Ottawa, Ont.**

Mr. Stan. Lewis, 63 Metcalfe St., Ottawa, Ont., has secured the contract for electrical work on an addition to be built to the Arlington Avenue School, Ottawa. Also the electrical contract on a $60,000 addition to be built to the Cambridge St. School, Ottawa.

The Canada Gazette announces the incorporation of the Griffin Radio Manufacturing Co. Ltd., for the purpose of manufacturing and dealing in radio equipment of all kinds and the manufacture and sale of electrical goods. The new firm is capitalized at $2,000,000. Head office: Toronto.

**Peterborough, Ont.**

Mr. Chas. Bowra, 92 Brock St., Oshawa, Ont., has secured the contract for electrical work on a factory building being erected at Hunter St. E. & Armour Road, Peterborough, Ont., at an estimated cost of $180,000 for the Western Clock Co., Peterborough.

**Quebec City, Que.**

The Quebec Gazette announces the incorporation of the Canadian Wireless & Electric Co., Ltd., with a capital-

# POWER CABLE

### 350,000 C. M. 3-Conductor 12000 volt
### Paper Insulated, Steel Tape Armoured Cable

Overall Diameter—3.62 ins.

*Built to Specification of Hydro-Electric Power Commission
of Ontario*

---

# Eugene F. Phillips Electrical Works, Limited
# Montreal

*The Oldest and Largest Manufacturers of Bare and Insulated Wires and Cables
in the British Overseas Dominions*

ization of $20,000. The new firm is authorized to manufacture and deal in electrical goods and radio equipment of every kind. Head office: Quebec City.

## Sault Ste. Marie, Ont.

A wireless station was recently opened at Whitefish Point ,Lake Superior, and it is understood that in the near future two stations, one at Detour and the other at Grand Marais, will also be opened. Ships may thus obtain their bearings from these stations in bad weather and fog, and direct their course so as to avoid dangerous rocks, etc.

## Summerland, B. C.

The ratepayers of Summerland, B. C., recently voted favorably on a by-law authorizing the expenditure of $12,000 for improvements to the municipal electric lighting plant at Summerland.

## Three Rivers, Que.

The North Shore Power Company, Three Rivers, Que., has been awarded the street lighting contract for the city of Three Rivers, by the city council.

## Toronto, Ont.

Messrs. Nason Bros., 456 Carlaw Avenue, Toronto, have secured the contract for electrical work on four stores to be erected at Donlands & Danforth Avenues, and four stores recently erected at Danforth & Greenwood Ave., Toronto, by Mr. S. Lumb, 29 Oakdale Crescent.

Mr. Jas. E. Harnan, 42 Arundel Ave., Toronto, has been awarded the contract for electrical work on a $30,000 apartment house being built on Bloor St. W., near Willowvale for Mr. R. Dale, 98 Castle Frank Road; also the contract for electrical work on four stores recently erected at Danforth & Jones Avenues, Toronto, for Messrs. Grant & Leslie, 837 Logan Avenue.

Mr. R. Mitchell, 1174 Danforth Ave., Toronto, has been awarded the contract for electrical work on four stores being erected at Danforth & Woodbine Aves., Toronto; also on eight store buildings nearing completion at Danforth & Greenwood Aves., Toronto.

Mr. R. C. Smith, 704 Shaw St., Toronto, has secured the contract for electrical work on a store and apartment building recently erected at Bloor St. & Brock Ave., Toronto, by Loblaws Groceteria, Ltd., at a cost of approximately $30,000.

Messrs. Richardson & Cross, 81 King St. E., Toronto, have secured the electrical contract on a store and apartment building being erected at St. Clair & Oakwood Aves., Toronto, at an estimated cost of $20,000.

The Western Electric Co., 347 Spadina Avenue, Toronto, has been awarded the electrical contract on a store building being erected at Danforth & Meagher Avenue, Toronto, for Mr. S. MacAluso, 1940 Queen St. E.

The Canada Electric Co., 175 King St. E., Toronto, has secured the electrical contract on the Riverdale Technical School, Toronto, to be erected by the Board of Education at an estimated cost of $600,000.

Mr. J. F. Calverly, 2603 Yonge St., Toronto, has been awarded the contract for electrical work on a store and apartment building recently erected at Yonge & Albertus Ave., Toronto, by Mr. G. E. Williamson, 2497 Yonge St.

Messrs. Cavers & Harvey, 702 Brock Ave., Toronto, have secured the electrical contract on eight stores to be erected at Yonge St. & Manor Road, Toronto, at an estimated cost of $80,000; also the electrical contract on eleven stores to be erected at Yonge St. & Cashin Road, Toronto, at an estimated cost of $110,000, and six stores recently erected on Glebemanor St., North Toronto, at a cost of approximately $60,000.

Messrs. Richardson & Cross have been awarded the electrical contract on a building at 136-140 Bay St., Toronto, that is being remodelled at an estimated cost of $25,000. Also the electrical contract on a building at 43 Adelaide St. W., Toronto, undergoing alterations at an estimated cost of $25,000.

Messrs. Beattie-McIntyre, Ltd., 72 Victoria St., Toronto have secured the electrical contract on a new church and Sunday School building to be erected at Heath St. & Lawton Blvd., Toronto, by the Christ Church Anglican Congregation, at an estimated cost of $100,000.

Mr. J. Everard Myers, Bank of Hamilton Bldg., Gould & Yonge Sts., Toronto, has been awarded the contract for alterations and additions to the electrical equipment of the Booth Memorial Building, Davisville, by the Salvation Army.

Messrs. Ramsden & Roxborough, 36 Front St. E., Toronto, have been awarded the electrical contract on a store building to be erected at 277-279 Victoria St., Toronto, at an estimated cost of $50,000.

Messrs. Harry Alexander, Ltd., 6 King St. West, Toronto, have secured the electrical contract on a $50,000 addition being built to the Toronto Conservatory of Music, 135 College St., Toronto.

Messrs. Moss & Stocks, 14 Price St., Toronto, have secured the contract for the electric wiring on a $40,000 addition to be built to the Hughes School, Toronto. Also the contract for wiring a school being erected in Moore Park, Toronto, at an estimated cost of $60,000.

Messrs. R. A. L. Gray & Co., 85 York St., Toronto, have been awarded the electrical contract on two stores being erected at Bloor St. W. & Brock St., Toronto, for Loblaw Groceteria, Ltd., 157 King St. E.

The Flexlume Sales System, Ltd., have moved from 92 Adelaide St. E. to 34 Eastern Ave., Toronto.

Mr. N. McLeod, 808 Danforth Ave., Toronto, has secured the contract for electrical work on eight residences being erected on Millbrook Crescent, near Broadview Ave., Toronto, for Dr. J. A. Gallagher, 2 May St.

Mr. Ed. Griffin, 151 Westminster Ave., Toronto, has secured the contract for electrical work on a store and apartment building being erected at Annette St. & Willard Ave., Toronto, by Mr. J. M. Bentley, 517 Runnymede Road.

Messrs. Nichol & Fagen, 111 Logan Ave., Toronto, have secured the contracts for electrical work on four stores being erected at Coxwell Ave. & Small St., Toronto, by Mr. H. W. Ormerod, 21 Chisholme Ave. at an estimated cost of $40,000.

## Weston, Ont.

The Weston Water, Power and Light Commission are installing 3/300 k.v.a. Packard transformers in their station to take care of the increased demand for both power and light. There is a heavy electric range load in Weston, and the demand for these services is steadily increasing.

## Winnipeg, Man.

The McDonald & Willson Lighting Co., 309 Fort St., Winnipeg, has been awarded the electrical contract on the building at 52 Albert St., Winnipeg, owned by Messrs. G. R. Gregg & Co., Ltd., that is to be remodelled at an estimated cost of about $25,000.

Messrs Gamble & Willis, 306 Notre Dame Ave., Winnipeg, have secured the electrical contract on a $100,000 I.O.O.F. Home to be erected at St. Charles by the Odd Fellows Grand Lodge of Manitoba.

In the list of exhibitors at the Electric Show, Kitchener, Ont., printed in our issue of May 15, the name of the McClary Manufacturing Company was inadvertently omitted. The McClary company had a very attractive display.

Pass & Seymour, Inc., Solvay, N. Y., have issued a bulletin under date of April 24, 1922, Form 1426, describing P. & S. standard Elexits. Complete instructions for properly installing side wall and ceiling outlets for Elexits are also given in Pass & Seymour Junior catalogue, No. 26.

F. S. Hunting, who has for several years been connected with the General Electric Company as general manager of The Fort Wayne division, is severing his connection with that company to become president and general manager of The Robbins & Myers Company with factories at Springfield, Ohio, Xenia, Ohio and Brantford, Ontario, Canada. C. F. McGilvray, former president of The Robbins & Myers Company, becomes chairman of the board, while W. J. Myers will continue as vice president, W. A. Myers as secretary and H. E. Freeman as treasurer. Mr. Hunting has had 34 years experience in the production and sale of small motors, and is ideally equipped for his new position with The Robbins & Myers Company.

The Gordon Electrical Co., have recently moved from 909 Main Street, Winnipeg, to their new Electrical Store, 428 Selkirk Avenue Winnipeg, Man.

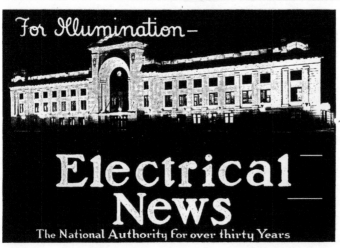

For Illumination—
**Electrical News**
The National Authority for over thirty Years

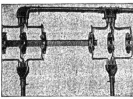

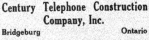

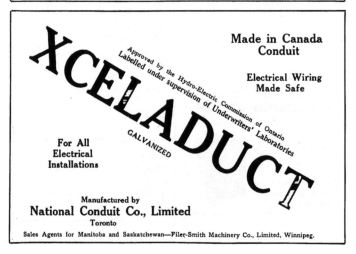

# CLASSIFIED INDEX TO ADVERTISEMENTS

The following regulations apply to all advertisers:—Eighth page, every issue, three headings; quarter page, six headings; half page, twelve headings; full page, twenty-four headings.

**AIR BRAKES**
Canadian Westinghouse Company

**ALTERNATORS**
Canadian General Electric Co. Ltd.
Canadian Westinghouse Company
Electric Motor & Machinery Co.
Ferranti Meter & Transformer Mfg. Co.
Northern Electric Company
Wagner Electric Mfg. Company of Canada

**ALUMINUM**
British Aluminum Company.
Spielman Agencies, Registered

**AMMETERS & VOLTMETERS**
Canadian National Carbon Company
Canadian Westinghouse Company
Ferranti Meter & Transformer Mfg. Co.
Monarch Electric Company
Northern Electric Company
Wagner Electric Mfg. Company of Canada
Winter Joyner Ltd., A. H.

**ANNUNCIATORS**
Macgillivray, G. L. & Co.

**ARMOURED CABLE**
Canadian Triangle Conduit Company

**ATTACHMENT PLUGS**
Canadian General Electric Company
Canadian Westinghouse Company

**BATTERIES**
Canada Dry Cells Ltd.
Canadian General Electric Co., Ltd
Canadian National Carbon Company
Electric Storage Battery Co.
Exide Batteries of Canada. Ltd.
McDonald & Willson, Ltd., Toronto
Northern Electric Company

**BEARINGS**
Kingsbury Machine Works

**BELLS**
Macgillivray, G. L. & Co.

**BOLTS**
McGill Manufacturing Company

**BONDS (Rail)**
Canadian General Electric Co., Ltd
Ohio Brass, Company

**BOXES**
Banfield, W. H. & Sons
Canadian General Electric Co., Ltd
Canadian Westinghouse Company
G & W Electric Specialty Company

**BOXES (Manhole)**
Standard Underground Cable Company of Canada, Limited

**BRACKETS**
Slater, N.

**BRUSHES CARBON**
Dominion Carbon Brush Company
Canadian National Carbon Company

**BUS BAR SUPPORTS**
Ferranti Meter & Transformer Mfg. Co.
Moloney Electric Company of Canada
Monarch Electric Company
Electrical Development & Machine Co.
Winter Joyner Ltd., A. H.

**BUSHINGS**
Diamond State Fibre Co. of Canada, Ltd.

**CABLES**
Boston Insulated Wire & Cable Company, Ltd
British Aluminium Company
Canadian General Electric Co., Ltd
Phillips Electrical Works, Eugene F.
Standard Underground Cable Company of Canada, Limited

**CABLE ACCESSORIES**
Northern Electric Company,
Standard Underground Cable Company of Canada, Limited

**CARBON BRUSHES**
Calebaugh Self-Lubricating Carbon Co.
Dominion Carbon Brush Company
Canadian National Carbon Company

**CARBONS**
Canadian National Carbon Company
Canadian Westinghouse Company

**CAR EQUIPMENT**
Canadian Westinghouse Company
Ohio Brass Company

**CENTRIFUGAL PUMPS**
Boving Hydraulic & Engineering Company

**CHAIN (Driving)**
Jones & Glassco

**CHARGING OUTFITS**
Canadian Crocker-Wheeler Company
Canadian General Electric Co., Ltd
Canadian Allis-Chalmers Company

**CHRISTMAS TREE OUTFITS**
Canadian General Electric Co., Ltd
Maxco Co., Ltd.
Northern Electric Company

**CIRCUIT BREAKERS**
Canadian General Electric Co.
Canadian Westinghouse Company
Cutter Electric & Manufacturing Company
Ferranti Meter & Transformer Mfg. Co.
Monarch Electric Company
Northern Electric Company

**CONDENSERS**
Boving Hydraulic & Engineering Company
Canadian Westinghouse Company

**CONDUCTORS**
British Aluminum Company

**CONDUIT (Underground Fibre)**
American Fibre Conduit Co.
Canadian Johns-Manville Co.

**CONDUITS**
Conduits Company
National Conduit Company
Northern Electric Company

**CONDUIT BOX FITTINGS**
Banfield, W. H. & Sons
Northern Electric Company

**CONTRACTORS**
Seguin, J. J.

**CONSTRUCTION MATERIAL**
Oshkosh Mfg., Co.

**CONTROLLERS**
Canadian Crocker-Wheeler Company
Canadian General Electric Company
Canadian Westinghouse Company
Electrical Maintenance & Repairs Company
Northern Electric Company

**CONVERTING MACHINERY**
Ferranti Meter & Transformer Mfg. Co.

**COOKING DEVICES**
Canadian General Electric Company
Canadian Westinghouse Company
National Electric Heating Company
Northern Electric Company
Spielman Agencies, Registered

**CORDS**
Northern Electric Company
Phillips Electric Works, Eugene F.

**CROSS ARMS**
Northern Electric Company

**CRUDE OIL ENGINES**
Boving Hydraulic & Engineering Company

**CURLING IRONS**
Northern Electric Company (Chicago)

**CUTOUTS**
Canadian General Electric Company
G. & W. Electric Specialty Company

**DETECTORS (Voltage)**
G. & W. Electric Specialty Company

**DISCONNECTORS**
G. & W. Electric Specialty Company

**DISCONNECTING SWITCHES**
Ferranti Meter & Transformer Mfg. Co.
Winter Joyner, A. H.

**DREDGING PUMPS**
Boving Hydraulic & Engineering Company

**ELECTRICAL ENGINEERS**
See Directory of Engineers

**ELECTRIC HAND DRILLS**
Rough Sales Corporation George C.

**ELECTRIC HEATERS**
Canadian General Electric Company
Canadian Westinghouse Company
Equator Mfg. Co.
McDonald & Willson, Ltd., Toronto
National Electric Heating Company

**ELECTRIC SHADES**
Jefferson Glass Company

**ELECTRIC RANGES**
Canadian General Electric Company
Canadian Westinghouse Company
National Electric Heating Company
Northern Electric Company
Rough Sales Corporation George C.

**ELECTRIC RAILWAY EQUIPMENT**
Canadian General Electric Company
Canadian Johns-Manville Co.
Electric Motor & Machinery Co., Ltd.
Northern Electric Company
Ohio Brass Company
T. C. White Electric Supply Co.

**ELECTRIC SWITCH BOXES**
Banfield, W. H. & Sons
Canadian General Electric Company
Canadian Drill & Electric Box Company
Dominion Electric Box Co.

**ELECTRICAL TESTING**
Electrical Testing Laboratories, Inc.

**ELEVATOR CONTRACTS**
Dominion Carbon Brush Co.

**ENGINES**
Boving Hydraulic & Engineering Co., of Canada
Canadian Allis-Chalmers Company

**FANS**
Canadian General Electric Co. Ltd.
Canadian Westinghouse Company
Century Electric Company
Great West Electric Company
McDonald & Willson
Northern Electric Company
Robbins & Myers

**FIBRE (Hard)**
Canadian Johns-Manville Co.
Diamond State Fibre Co. of Canada, Ltd.

**FIBRE (Vulcanized)**
Diamond State Fibre Co. of Canada, Ltd.

**FIRE ALARM EQUIPMENT**
Macgillivray, G. L. & Co.
Northern Electric Company
Slater, N.

**FIXTURES**
Banfield, W. H. & Sons
Benson-Wilcox Electric Co.
Crown Electrical Mfg. Co., Ltd.
Canadian General Electric Co. Ltd.
Jefferson Glass Company
McDonald & Willson
National X-Ray Reflector Co.
Northern Electric Company
Tallman Brass & Metal Company

**FIXING DEVICES**
Inventions Ltd.
Rawlplug Co., of Canada

**FLASHLIGHTS**
Canadian National Carbon Company
McDonald & Willson Ltd.
Northern Electric Company
Spielman Agencies, Registered

**FLEXIBLE CONDUIT**
Canadian Triangle Conduit Company
Slater, N.

**FUSES**
Canadian General Electric Co. Ltd.
Canadian Johns-Manville Co.
G. & W. Electric Specialty Company
Moloney Electric Company of Canada
Northern Electric Company
Rough Sales Corporation George C.

## CLASSIFIED INDEX TO ADVERTISEMENTS—CONTINUED

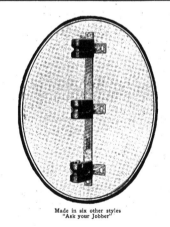

## CLASSIFIED INDEX TO ADVERTISEMENTS—CONTINUED

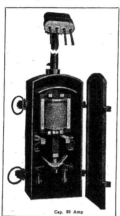

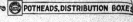

Vol. XXXI No. 12                                                    Toronto, June 15, 1922

**Electrical News**

Engineering
Contracting

Merchandising
Transportation

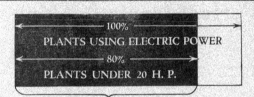

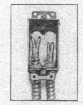

## ALPHABETICAL LIST OF ADVERTISERS

A Galvaduct Building—Y. M. C. A. at Kitchener, Ont.
Electrical Contractors—The Star Electric Co., Kitchener, Ont.

# GALVADUCT

### "GALVADUCT"

The most perfect interior construction conduit on the market. Recognized as the standard of high quality.

Always specify Galvaduct or Loricated CONDUITS

### "LORICATED"

A high-class interior construction conduit of the enameled type, proof against acid or other corrosive agents.

If your jobber cannot supply you—write us.

## CONDUITS COMPANY LIMITED
### TORONTO                     MONTREAL
**Western Sales Office: 602 Avenue Building, Winnipeg**

# LORICATED

Protective covers for line
terminals prevent shocks
and shorts. Optional.

Only authorized persons
having key can open the
box when the switch is on.

Powerful quick make and
quick break mechanism
enclosed to prevent clog-
ging by dust etc.

# Safety Switch

# The Electroplax Company, Limited
### Harry E. Corey, President

## Everything in **RADIO** Insulation !

#### Canadian Manufacturers of :—

| | |
|---|---|
| AERIAL INSULATORS | LIGHTNING ARRESTORS |
| BINDING POSTS | LAMINATED TUBING |
| CONDENSOR END PLATES | MICA CONDENSOR CONTAINERS |
| COIL MOUNTS | PANEL BOARD (REDMANOL, BAKELITE) |
| COUPLER INSULATING PARTS | RHEOSTAT BASES |
| DIALS | RADIO BOXES |
| DETECTOR BASES | ROTOR INSULATING PARTS |
| FUSE PARTS | SWITCH BOARDS |
| GROUND SWITCH BASES | SWITCH HANDLES |
| INSULATORS | SOCKETS |
| INSULATION ON JACKS | TELEPHONE EAR CAPS |
| KNOBS | VACUUM TUBE HOLDERS, ETC., ETC. |

ALSO MOULDERS OF
PHENOL RESIN PLASTICS
REDMANOL - BAKELITE - CONDENSITE
AND COLD PRESSED PLASTICS, THERMOPLAX, PYROPLAX.
ACID AND HEAT PROOF VARNISHES.

=== SYNTHETIC AMBER PRODUCTS ===

FACTORY, MOUNT DENNIS, ONT.           CITY SALES OFFICE, YORK BUILDING, TORONTO
Jct. 5985; City Office, Ade. 7971.            W. M. Davidson, Sales Manager

---

# "UNIVERSAL" VACUUM CLEANERS

Among many special features of the "Universal" Vacuum Cleaner, the most important is the combination of a motor, more powerful and with a greater suction than any other cleaner, with a patented nozzle having thread catchers on the mouth of the nozzle on both sides with air spaces between so that the air is powerfully whirled in from both sides, as well as up through the carpet and blows thread, lint, etc., off from the thread catchers up into the bag together with whatever dust and dirt is on the surface or in the carpet.

The disadvantage of brushes—unsanitary, hard and disagreeable to clean, easy to get out of order and destructive of rugs and carpets—has long been recognized.

The "Universal" does away with brushes entirely and does everything by a powerful air suction.

Manufactured by
### Landers-Frary & Clark

No. E720 Polished Aluminum Weight 12 pounds.
No. E7201 Polished Aluminum complete with Attachments 16 pounds.
Packed one in a wooden case 26½ pounds, with Attachments 30½ pounds.
Packed two in a wooden case 38½ pounds, with Attachments 43 pounds.

# Great West Electric Company Limited
### 61-65 Albert St.      -      -      Winnipeg, Man.

# KLEIN CLIMBERS

### Take no Chances—*Demand* Klein Climbers

You cannot blame the "old-timers" for demanding Klein Climbers! A good climber is a matter of life or death.

Protect your outside gangs with the climbers that have been standard for 65 years. Every single Klein Climber is carefully forged from the finest spring steel we can buy—hardened and tempered under expert supervision and subjected to rigid factory tests before sold.

The gaff itself is made of a special steel, hand tempered and tested, and securely riveted to the leg iron in such a manner that it will not come loose.

The shape of Klein Climbers, the set of the angle and the temper are all the product of years of experience, and designed for safety, ease and comfort.

Play safe. Buy Klein Climbers.

Pliers
Splicing Clamps
Sleeve Twisters
Climbers
Tool Belts
Safety Straps
Lag Screw Wrenches
Wire Grips
Tool Bags
Tree Trimmers
Charcoal Furnaces
Staysalité Torch

## Mathias KLEIN & Sons
### Established 1857        Chicago, Ill. U.S.A.

# Imperial Headlights give you---
## THE TYPE YOU NEED
## THE SERVICE YOU WANT

You'll find an Imperial Headlight which is exactly suited for your cars. There are several designs of Incandescents, of Luminous Arcs and Carbon Arcs in the complete Crouse-Hinds Imperial line. You can choose the type which pleases you most or you can outline your requirements and we will make a recommendation.

Here's the big idea to keep in mind— whatever the Imperial Headlight that goes on your cars, you're certain to get long and satisfactory service from it.

Under an arrangement consummated June 1, 1922, Imperial Headlights of The Crouse-Hinds Company of Canada Limited are now sold exclusively through

## The Ohio ⒷBrass Co.
8 Hillingdon Ave.                    Toronto, Ont.
Main Office and Works : Mansfield, Ohio, U.S.A.

Products: Trolley Material, Rail Bonds, Electric Railway Car Equipment, High Tension Porcelain Insulators, Third Rail Insulators

# National Advertising

A National Advertising Campaign, for the famous Hotpoint Iron, is now under course of preparation. Newspapers and magazines from Coast to Coast will be used, —over a million homes will be reached.

The steady demand, for the Hotpoint Iron, is already large—it's going to be much larger. We strongly urge you to secure a plentiful supply immediately.

It is our earnest desire to co-operate with our dealers to the fullest extent. We are willing to supply ready-prepared newspaper advertisements, counter display cards, literature etc., and do everything in our power to help you sell Hotpoint Irons.

*"Made in Canada"*

## Canadian Edison Appliance Co.,Limited
### STRATFORD, ONTARIO.

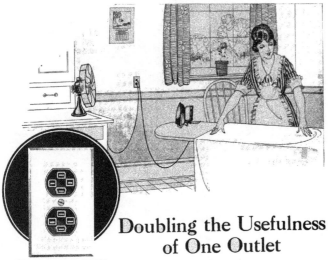

C.G.E. Twin Receptacle, Cat. No. C.G.E.
634 with Plate No. C.G.E. 695

# Doubling the Usefulness
# of One Outlet

I N moderate priced homes many people hesitate at the cost of an adequate number
of electrical outlets, simply because the "idea" has not been sold them. Plenty of
convenience receptacles are necessary for the appliances, so essential to modern housekeeping.

The market for convenience outlets is practically unlimited. The percentage of
houses, even moderately well wired, is so small. that every home in your district offers
a potential market for one or more C.G.E. Twin Receptacles.

The public are demanding, more and more, electric homes. Show your customers
how their present homes can be converted into real electric homes by the aid of
C.G.E. Twin Receptacles.

*"Made in Canada"*

# Canadian General Electric Co., Limited
## HEAD OFFICE 〈G.E.〉 TORONTO

Branch Offices:  Halifax, Sydney, St. John, Montreal, Quebec, Cobalt, Ottawa, Hamilton, London, Windsor
South Porcupine, Winnipeg, Calgary, Edmonton, Vancouver, Nelson and Victoria.

# ANOTHER BEAVERDUCT INSTALLATION

THE·NEW·MOUNT·ROYAL· HOTEL MONTREAL·

Beaverduct is a Canadian made conduit, manufactured in a modern plant under the underwriters' inspection. Its use on such an installation, as this, is a fitting testimonial of its superiority.

Good conduit can only be produced in a plant where the equipment permits only absolutely clean material being used. Electrical apparatus and plating solutions must be kept up to the mark under scientific management. The proper application of the enamel, after the galvanizing of the outside, is of vital importance. And, lastly, rigid inspection.

*"Made in Canada"*

# Canadian General Electric Co., Limited
## HEAD OFFICE TORONTO

Branch Offices: Halifax, Sydney, St. John, Montreal, Quebec, Cobalt, Ottawa, Hamilton, London, Windsor, South Porcupine, Winnipeg, Calgary, Edmonton, Vancouver, Nelson and Victoria.

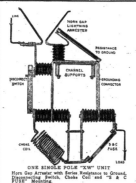

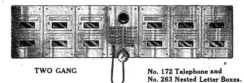

MADE-IN CANADA

# The "Jeffersonlite"

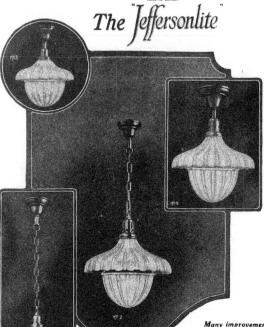

**Many improvements feature the
New Jeffersonlite**

Among the improvements are :—

**Flange Support— No Screws—New Finish— Jeff Bronze—New Bowl—Individual
Cartons--New Adjustable Socket Holder--New Hangers--New Package--New Prices.**

We extend an invitation to all the Electrical Trade to visit our Bay Street Show-
rooms at any time when in Toronto.

# WHAT ABOUT

**Thermo-Electric** Water Heating Devices are the result of painstaking experiment and research. Every factor tending to adverse or doubtful influences has been eliminated. Every condition tending to increased efficiency, durability, and satisfactory performance has been incorporated. The price is based on faultless design and material, high grade workmanship, and the built-in completeness, which secures and retains the lasting good-will of the customer.

# THE HOMES WITHOUT

**Thermo-Electric** service is an opportunity for every Dealer to add to his list of Satisfied Customers. An opportunity to enter a field of excellent possibilities, with a Product of Proven Worth. The initial installation, properly accomplished, is the final servicing. Its service to your Customer, Yourself and us, is reflected in His Satisfaction, Your support of our Product, and our increase in Sales Volume.

# HOT WATER?

**Thermo-Electric** Water Heaters are built to meet any range of requirements, Domestic or Industrial. We can place you in touch with domestic installations that have never been off the line since 1914, and are still giving the same efficient service as when installed. We have designed and placed in successful operation continuous service heaters of 25 KW. capacity, securing the same efficient performance as any of our smaller types. Heating Water Electrically is our business, our main business. We have found it a good business. So will you.

# Heated Electrically

**Thermo-Electric** Water heaters enjoy national representation, and support. A guaranteed Product with the final element of doubt removed.

**Western Sales Offices:**
**602 Avenue Block, Winnipeg, Man.**

## Place your order with your jobber

All Domestic Heaters adapted for direct conduit connection, and remote control only. Twin Element Circulation Heaters complete with three heat indicating switch. All come in individual carton.

MANUFACTURED BY

# Thermo Electric Limited
*Manufacturing Engineers*
**BRANTFORD   —   —   CANADA**

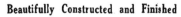

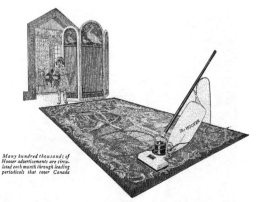

*Many hundred thousands of Hoover advertisements are circulated each month through leading periodicals that cover Canada*

# A Small Investment, Small Floor Space, Monthly Turnovers, Large Yearly Profits

**The Hoover Bag Is Always Properly Held**
*(Another Patented Hoover Feature)*

REGARDLESS of the position of the Hoover handle, and no matter whether the bag is inflated or deflated, it cannot get caught under the wheels, nor can any strain be put upon it to tear the seams or to pull the fabric out of the frame at the exhaust opening. This automatic protection is accomplished by the Hoover bag-sliding hook. When the bag is deflated, or the handle is raised, the hook slides down. As bag inflates, the hook travels upwards. The bag is also easily unhooked for emptying. Covered by a Hoover patent granted May 1, 1917. Over eighteen patents now protect The Hoover. Other applications for patents are pending.

Four important factors of the Authorized Hoover Dealer franchise are named above.

Frequent ordering is encouraged as Hoover dealers are urged to strive for monthly turnovers.

This makes it possible for a dealer to secure twelve profits a year on his investment. And, no matter whether his order is for three or three hundred Hoovers, the same discount applies.

Then, too, a dealer need devote only a small, well-located space in his store for demonstrating and displaying The Hoover.

When his machines arrive, a Hoover man—trained in the best methods of retail selling—will spend ample time with the dealer to post him on successful methods of making sales.

Display and advertising literature will be liberally provided. Furthermore, this Hoover man will continually keep in close touch with the dealer, to assist him to build up his volume.

Does not this franchise appeal to you? Largely because the Hoover organization backs up its dealers in this sincere, helpful manner, The Hoover is the largest selling electric cleaner in the world.

Without obligation, of course, will you not permit our representative to go fully into this matter with you? Please let us know when he may call.

THE HOOVER SUCTION SWEEPER COMPANY OF CANADA, LIMITED
Factory and General Offices: Hamilton, Ontario

*The* HOOVER  MADE IN CANADA

*It BEATS ··· as it Sweeps — as it Cleans*

Generation, Transmission and Application of Electricity

**For nearly thirty years the recognized journal for the Electrical Interests of Canada.**

Published Semi-Monthly By

# HUGH C. MacLEAN PUBLICATIONS
### LIMITED

THOMAS S. YOUNG, Toronto, Managing Director
W. R. CARR, Ph.D., Toronto, Managing Editor
HEAD OFFICE - 347 Adelaide Street West, TORONTO
Telephone A. 2700

MONTREAL - - - 119 Board of Trade Bldg.
WINNIPEG - - - - 302 Travellers' Bldg.
VANCOUVER - - - - - Winch Building
NEW YORK - - - - - 296 Broadway
CHICAGO - - 14 West Washington St.
LONDON, ENG. - - 16 Regent Street S.W.

ADVERTISEMENTS
Orders for advertising should reach the office of publication not later than the 5th and 20th of the month. Changes in advertisements will be made whenever desired, without cost to the advertiser.

SUBSCRIPTIONS
The "Electrical News" will be mailed to subscribers in Canada and Great Britain, post free, for $2.00 per annum. United States and foreign, $2.50. Remit by currency, registered letter, or postal order payable to Hugh C. MacLean Publications Limited.
Subscribers are requested to promptly notify the publishers of failure or delay in delivery of paper.

Authorized by the Postmaster General for Canada, for transmission as second class matter.

*Vol. 31*         *Toronto, June 15, 1922*         *No. 12*

## What Can We do to Establish An Electrical Merchant Class?

The month of June again brings us to the conventions of the two Canadian central station organizations—the Canadian Electrical Association, which includes all the privately operated companies in Canada, and the Association of Municipal Electrical Utilities, which to date has not extended its activities beyond the province of Ontario and, without doubt, as in former years, delegates will come away from the meetings deeply impressed with the value of the work each of these associations is doing. It is, we believe, unfortunate that certain discussions which had in view the holding of these two conventions at the same time and place did not bear fruit, but next year's efforts may prove more successful. The advantages appear to outweigh the objections.

Having in mind the excellent work being done by both of these associations one must be an egotist to offer criticism on the proceedings. A suggestion is never a criticism, however, and the Electrical News takes the liberty of offering one, believing that its interest in the electrical industry is so well recognized that no member of either of these associations will misconstrue anything we may say.

A chain is just as strong as its weakest link. The electrical industry is a chain made up of six main links—the manufacturer, the engineer, the wholesaler, the contractor, the retailer and the public. The strongest links are engineering and manufacturing. The weakest link—the link that is making it impossible for the whole chain to lift more than a

very small percentage of the load the larger links are capable of carrying—is the retailer.

What would we think of a building contractor operating a chain with a weak link, day after day and month after month, and being satisfied to delay progress by lifting very small loads, thus prolonging the time required to carry out his operations to two or three times what it should be? Wouldn't the stupidest contractor see the advantage of strengthing his chain so that his machinery would work to capacity?

Now, doesn't this, to a very considerable degree, portray the electrical industry? Here are two central station organizations—whose primary function, truly, is to supply electric current but whose success and existence depends on the absorption of that current, by the public, chiefly through the greater use of electrical appliances—occupying most of their time at conventions, and devoting most of their energies, to still further strengthening those phases of the industry that are already very strong. It wouldn't be a vitally serious matter if no addition were made to our knowledge of pure engineering for a couple of years, or if the present status of the art of generating and distributing electric energy should remain as it is for the time being. On the other hand, it is, indeed, a tremendously vital problem that we are so terribly weak and lacking in merchandising ability and facilities. The outlet of the industry, so to speak, is clogged and the people who would buy appliances, with very little encouragement, have nowhere to go for that encouragement. We have practically no merchant class.

Wouldn't it be in the interests of the industry if our two central station conventions were at least to give over a greater percentage of their time to the consideration of this problem of merchandising? Looking over the program of the Canadian Electrical Association one sees that of eighteen reports to be submitted in two days, one only deals strictly with merchandising. This will probably narrow down the consideration of this most important problem to about half an hour. We cannot but feel that this association would be well advised to hold a two or three day convention and consider nothing at all but merchandising—ways and means of improving the present, unsatisfactory condition of distributing electrical merchandise, through the dealer, to the public.

The same remarks hold good as regards the A. M. E. U. program, which, this year, seems to have overlooked the merchandising end almost entirely.

The crying need of the electrical industry is the development of a merchant class. Both in numbers and quality our retail stores are weak. The few good stores merely serve to emphasize the inadequacy of the others. How shall we go about it to improve this condition, which, as long as it remains, will dwarf the growth of the electrical industry? Isn't it worthy of greater consideration on our convention programs?

## Standardizing Color Signals

An important conference, called by the American Engineering Standards Committee, was held in New York on May 23 for the purpose of considering the advisability of standards regarding colors used for traffic signals. The conference included representatives from more than 50 bodies and organizations interested. A quantity of information was forthcoming regarding the extent to which signals are already being used, particularly in steam railway work. It was pointed out that signals controlling railway traffic were already completely standardized and that the proper function of the Standards Committee would be to decide on the course to be taken in connection with traffic signals, as distinguished from signals used by the railway companies. Discussion resulted in the following recommendations:

1. It is desirable that there shall be national uniformity in the use of colors for signals.

2. The unification contemplated should include the following items:

(a) The use of colored lights on all vehicles used on highways.

(b) The use of colored lights on all signals, both permanent and temporary, along highways and at curves.

(c) The use of colored lights for highway crossing signals for steam and electric railways.

(d) The co-ordinated relation of color, position, and number of lights.

(e) The co-ordinated relation to systems of flashing, moving, or other similar systems of signals.

(f) The use of color for non-luminous as distinguished from luminous signals.

(g) The use of colors for emergency exit signals, and

(h) Methods of specifying or defining colors for signal purposes.

It was decided also that the A.E.S.C. should form a Sectional Committee for the purpose of furthering the recommendations reached by the conference.

## Commencing Work on Power Plant at Indian Chute, Matachewan River

Work has just been commenced by the Matachewan Power Company on a power plant at Indian Chute, Matachewan river. In the district there are two falls in close proximity, each with an approximate head of 48 feet and a capacity, under control, of 6,000 h.p. The Matachewan Power Company will instal three units at the lower falls —Indian Chute—the first unit being already on order. It is expected this unit will be operating by next Christmas. The power will be used in the Matachewan gold camps and the Gowganda silver mines.

Construction work is in charge of H. W. Sutcliffe, of Sutcliffe & Nealands, New Liskeard, Ont., assisted by the following Board of Directors: President, Robt. Fennell, Nasmith & Fennell, Toronto; 1st. vice-president, F. J. Westaway, W. J. Westaway Co., Hamilton; 2nd. vice president,

PLAN OF
DEVELOPMENT
OF
INDIAN CHUTES

Col. Robt. Stark, president, Stark-Seabold Co., Montreal; D. W. Caldwell, Boyd-Caldwell Co., Perth; Sir H. L. Galaway, D.S.O., London, Eng.

### Well on the Way to Recovery

We are pleased to announce that J. Gordon Smith, manager of the Great West Electric Co. Ltd., Winnipeg, who has been confined to his home for the past two months. with pneumonia, is well on the way to recovery, although it is likely to be several weeks, before he is able to return to business.

---

## Business is Better

Business is better. The depression which began in the United States in 1920 and continued throughout 1921 has passed, and substantial progress has already been made toward normal activity and a new business cycle has been entered upon. Adverse conditions, as the coal strike, may temporarily retard the upward swing. Other factors, such as widespread crop failures, might even result in recession for a time, but no circumstances can alter the fact that there is now an unassailable basis for confidence in slow and steady expansion of the commercial and financial activities of this country.'

This basis for confidence is fivefold: First, there is plenty of money to be had at reasonable rates both for short-time and long-time requirements. Second, stocks of finished goods and of raw materials have been reduced to reasonable proportions. Third, commodity prices are stabilizing. Fourth, conditions in basic industries, including agriculture, are improving, and production is expanding. Fifth, gains are not confined to the United States. Conditions are improving throughout the world. Some countries constitute exceptions to this statement, but their bearing on the international situation is not great enough to alter the fact that the world outlook is better, with the United States and Canada in the forefront of improvement.—

Commerce Monthly, New York.

# The Application of Electricity in the Manufacturing Industry

### Upwards of Three Thousand Wood-working Plants in Canada
### —Only Small Percentage Fully Equipped Electrically—Millions
### of Dollars Worth of Business in the Offing

In our last issue we described the uses of electricity in the rubber tire industry. It was shown that the amount of electrical equipment required was very large, that the consumption of electric energy was of very considerable proportions, and that the very existence of the industry may be said to depend on the refinements and efficiencies that were only made possible by the flexibility of electric drive.

The present article—the second of the series—deals in a similar way with the woodworking industry. This is an industry that differs widely from the rubber tire industry. In place of a small number of large and highly developed organizations we find a large number of smaller plants distributed all over Canada, located in every city and almost every town of any proportions. In a recent report of the Dominion Bureau of Statistics, it is indicated that the number of woodworking plants in Canada is in the neighborhood of three thousand.

The demand for power in the individual woodworking plants is naturally of comparatively small proportions, but in the total it bulks up large. For example, in Ontario alone, made up of a large number of units of course, the motor installation amounts to well over 10,000 h.p. This item will be mentioned in greater detail later in the article.

In this particular industry, besides giving the benefit of

Motor-driven, Billiard Table Slate-Finisher—Brunswick-Balke-Collender Co.

increased efficiency, electricity helps retard the decay of craftsmanship. The woodworkers of Canada are artisans who still stamp their work with a measure of individuality. They do so because their processes are not altogether standardized as is done for the purpose of production on a gigantic scale. The peculiar needs of the industry have something to do with this, but a more potent factor is the economic possibility of running a large number of small plants. These small plants are scattered throughout Canada and their continued operation is largely dependent upon ready access to electric energy. Our extensive distribution of water-power,

among its other blessings, has thus helped us to get away from a deadly duplication in houses and furniture.

Over 60 per cent of the plants engaged in the manufacture of lumber into its finished products at the present day, are electric motor driven, and of the remainder a large part

Individual drive, motor belt-connected to sticker

are prevented from using electricity only by a restricted supply in the locality in which they are operating.

### Economy of Electric Drive

Woodworking machinery is especially adaptable for electric drive, because of the intermittent nature of many of the operations requiring individual drive for the highest economy. It is no uncommon thing to find small plants where such a machine as the sander, shown in Fig. 1, is in use less than 50 per cent of the time. The economy effected by the use of motor drive in such a case is considerable, as the losses due to the turning of non-effective overhead shafting account for a large part of the power consumed. Mr. R.J. Owens, in a recent article in the "Canadian Woodworker," says in this regard: "tests made in a large number of plants have shown conclusively that without motor drive the transmission losses are abnormal—in the majority of cases approximately 50 per cent of the total power consumed."

### Individual Drive

Peculiar economies can be effected by the use of individual drive in woodworking plants due to the variation in speeds required by different machines. An engineer who has had a wide experience with this type of drive claims the following advantages for it:—

1.—An increased output per machine, due to better control, constant speed and many other features.

2.—Smaller power house or transformer capacity required,

3.—Operating cost minimized.

4.—Increased safety for operators by removing shafting and belts.

5.—Better daylight due to absence of overhead shafting.

Band resaw direct-driven, using flexible coupling

6.—Permits machines to be placed to best advantage as regards flow of materials, etc.

### Motor Selection

The conditions in a woodworking plant, unavoidable on account of the material handled, demand equipment to which there is attached the least possible degree of fire hazard. It will be seen from Fig. 2, which is a normal location, that the motor is exposed to a shower of wood dust. This renders desirable, freedom from moving contacts liable to cause arcing, and requires a sturdy construction capable of withstanding the accumulation of this wood dust deposit.

A motor of the desired characteristics is the squirrel cage induction motor. It not only meets the above requirements but has grown in popularity on account of its even torque and constant speed. Where a variable speed is essential, direct current motors predominate, but latterly advances have been made in the application of alternating

Sander driven by two motors, belt-connected

current motors both of the slip ring variable speed and multi-speed squirrel cage type, that makes them applicable under nearly all conditions.

It has been found that a good deal of trouble is caused in some plants through the omission of good protective starting equipment for motors of smaller size than those provided with an auto starter. This results in single phase

operation and burnouts, which an oil starter, or oil switch would prevent.

### Lighting

In far too many cases that have been under recent observation, it has been found that practically no attention has been paid to the principles of scientific illumination. A drop cord with a bare lamp over each machine or at each workmen's bench with a few more thrown in for general illumination is often considered sufficient.

### Design Tendencies

The increased attention being given to the design of woodworking machinery has resulted in the production of some radically different designs, for which high efficiency is claimed. This is generally accomplished by greater speeds and built-in motors. The latest developments include several separate motors on each machine. In some cases the speed demanded is greater than can be obtained with

Belt Sander driven by overhead motor—Brunswick-Balke-Collender Co.

25 cycle current—a fact that opens an interesting question as to the possibilities of development of motor equipment to meet the demand.

### Many Plants Electrically Driven

The general tendency in the industry seems to be toward the complete electrification of all possible woodworking operations. That this inclination is of importance to electrical men may be judged by the results from a canvass of the woodworking plants of Ontario. An inventory was recently made of 257 representative factories and it was found that these were equipped as follows:—

Fully motor driven ................. 129
Partially motor driven ............. 46
Steam driven ...................... 82

The combined horsepower of the motors installed was found to be 10,570 made up of 547 motors. Of the firms who did not use electric power, the majority intimated that it was because of its non-availability.

### Some Plants Still use Steam

There is always a good deal of waste around a woodworking plant. Also, some of the equipment such as kilns and presses, require steam. For these reasons many plant managers cling to the old steam boiler, using their plant waste as fuel, and supplement it with coal. In many cases electric installations are being put in, large enough to take care of the excess power requirements over and above that supplied by the fuel gathered around the plant. This has been done in cases where it has been demonstrated to the satisfaction of the plant manager, that electric power can be bought more cheaply, than the same amount of power can be generated by coal in the plant.

An analysis of the information furnished by those plants which are either partially or completely motor-driven showed that 163 preferred group drive, while 21 had adopted the individual motor system.

Some of the individual installations were as follows:—

| 20 motors totalling | 200 horse-power |
| 20 motors totalling | 150 horse-power |
| 20 motors totalling | 125 horse-power |
| 9 motors totalling | 145 horse-power |
| 9 motors totalling | 70 horse-power |
| 13 motors totalling | 175 horse-power |
| 44 motors totalling | 135 horse-power |
| 14 motors totalling | 400 horse-power |

### An Outstanding Factory

The factory of the Brunswick-Balke-Collender Co., of Toronto, Ontario, is one of the most modern and complete in the matter of electrical equipment.

The machinery is entirely motor-driven requiring a total capacity of 400 horsepower. There are 95 motors, 89 of which are 550 volt, a.c.; the remaining 6 are 110 volt, a.c. This installation takes care of every possible woodworking operation in the factory, and includes the motors required for elevators and ventilation.

The switchboard arrangement is made very simple due to current being received at 550 volts from the supply company. This eliminates transformers, except for two small ones used for lighting purposes.

Group drive is used wherever possible. This is in

Group Drive—Bowling Ball Dept. Brunswick-Balke-Collender Co.

general where the machines are small, and run at the same speed, as is illustrated by the photographs. The bowling ball and cue machines, are driven in this way, and in the phonograph department, the bandsaw, cutoff saw and boring machine are grouped.

A saving is made in floor space by mounting motors on specially constructed platforms suspended from the ceiling. This method also avoids the danger of floor level belts. In cases where the construction or location of the machines prevents this method of driving them the motors are placed on the floor and the drive belt is well guarded.

The general result has been the elimination of hand

work, and the securing of added efficiency. A flat rubbing machine now does jobs formerly considered the exclusive field of the hand worker. Machines such as moulders or stickers run up to 4,000 r.p.m. and turn out quantities of work undreamed of a few short years ago.

### More Motors Needed

An examination of the facts reveals that as a whole the woodworking industry is a field of tremendous possibilities for the application of electrical apparatus. There are in Canada upwards of 3,000 factories engaged in the manu-

50 h.p. motor direct-connected to fast feed planer

facture of wood into its finished products. Based on the average shown by the 267 Ontario plants, mentioned above, they require 40 h.p. each. This makes a total of 3,000 x 40 = 120,000 h.p. required for the whole industry, in Canada.

It is not easy to strike an average as to the cost of the necessary motor equipment, but putting it roughly at $40 a horse-power, this represents a total expenditure of 120,000 times $40 = $4,800,000. This is for motors alone and does not take into consideration a large amount of auxiliary equipment, transformers, wire, conduit, condulets, lighting fixtures, lamps, switchboards, outlet boxes, and so on, that would very materially increase this sum.

This is not all new business, but much of it is waiting for the electrical salesman. There are comparatively few woodworking plants in Canada that are as fully equipped as they ought to be, and there is a large number of them that have never yet heard the pleasant hum of an electric motor. Just now, while business is none too brisk, would seem to be an opportune time to see what could be done toward placing our 3,000 woodworking plants on a basis of efficiency that can only be made possible through the more general application of electric drive.

A publication, known as circular 1579-B, containing general information and technical data for cutting micarta gears and pinions, has recently been issued by the Westinghouse Company for distribution among gear cutters. The book contains a great amount of useful information for gear manufacturers.

## Removing Weeds from Lake at Summer Resort---A Clever Invention

An interesting item comes from Mr. M. Freeman, Commissioner of Public Utilities of the city of Lethbridge, Alta., describing a weed cutting boat, which Mr. Freeman improvised to remove objectionable weeds in a small lake at a summer resort served by the municipal electric railway system of Lethbridge. Mr. Freeman furnishes the following details as likely to be of interest to any electric railway men who may be confronted with similar conditions:

The boat is a scow, of the folowing dimensions; hull, 8 ft. by 20 ft. by 18 in., (a 24 in. depth would be better for rough water.) There is a deck overhang of 2 ft. at the front and 14 in. at the stern, and the paddle boxes extend 24 in. on either side, making over-all measurement 12 ft. by 25 ft. The boat is propelled by paddles 6 in. by 18 in., 4 ft. apart and hinged to endless chains passing over 16 in. sprocket wheels, fore and aft. The motor power is a four-cylinder 60 h.p. automobile engine (though a 40 h.p. would be plenty.)

Parallel with the engine is a counter shaft, connected to the engine by chain drive, ratio 1 to 2½ of the engine. On this counter shaft is mounted an 8 in. diameter, 12 in. face pulley for driving the paddle shaft, extending across the boat near the stern and connected therewith by a 12 in. rubber belt. The paddle drive shaft carries a 20 in. pulley, 12 in. face.

Also on this same counter shaft there are mounted two 16 in. grooved rope pulleys, for 7/8 in. rope; this counter shaft forms the fulcrum for the adjustable frame which carries an endless saw with which the weeds are cut.

The saw frame is of 2½ in. pipe. The side arms, of which there are two on each side of the boat, are attached to collars on the counter shaft and fulcrum thereon as the frame is raised or lowered. These arms are sufficiently long to clear the rear end of the boat, and at the rear end carry a rectangular frame of the same size pipe, attached to which at one upper and two lower corners are band saw pulleys. The upper or drive pulley is 22 in. diameter, 1 in. face, and has a shaft to which is also attached a rope drive pulley 18 in. in diameter. A rope tension pulley is also supplied close to the counter shaft and kept in tension by a heavy coiled spring and threaded bolt, for adjustment. The lower band saw pulleys are 16 in. diameter.

The weight of the saw frame is taken care of by a length of 4 in. I-beam, fulcrumed at a point between the rear end of the boat and the engine, one end being attached to the saw frame at the rear, by wire rope, and the other end carrying a counter- balancing weight. A 3/4 in. 22 gauge, 6 point, endless band saw is used. The saw frame is so balanced that it will lower itself by gravity when released.

The boat has four rudders, one at each corner, for ready control. The engine control, steering wheel and a wheel to raise the saw are all within reach of the operator, so that only one man is necessary. The boat makes four miles per hour and cuts 8 ft. wide, with a minimum depth of 1 ft. and a maximum depth of 6 ft.

## A. I. E. E. at Niagara Falls June 26-30

Special interest is being manifested in the circuit breaker symposium which will be a feature of the meeting of the American Institute of Electrical Engineers at Niagara Falls, Ont., June 26 to 30, inclusive. The papers relate to the Baltimore tests of oil circuit breakers, which are the highest power tests ever made. The papers to be presented are. "Baltimore Oil Circuit Breaker Tests" by H. C. Lewis, Chief of Tests, Consolidated Gas, Electric Light and Power Company, and A. F. Bang, Testing Engineer, Pennsylvania Water & Power Company; "Oil Circuit Breaker Tests," interrupting capacity tests of General Electric oil circuit breakers at Baltimore, by J. D. Hilliard, General Electric Company; and "Tests on Westinghouse Oil Circuit Breakers at Baltimore," by J. B. MacNeil, of the Westinghouse Electric & Manufacturing Company.

Cutting weeds in small lake at summer resort, served by the municipality of Lethbridge, Alta.

Sunnyside Boulevard by daylight—Photo taken from boardwalk

# Beauty and Utility in Street Lighting

### Lighting System Installed on Sunnyside Boulevard, Toronto, Embodies Many Noteworthy Features — Series System Used

The city of Toronto, situated on the North shore of Lake Ontario, is naturally adapted for parks and lakeshore drives, but not until the recent past did they have a boulevard worthy of any particular note. Some years ago, the city, the Provincial Government of Ontario, and the Dominion Government, each appointed a Commissioner on the "Harbor Board," with the idea of making vast improvements along the waterfront of Toronto. In addition to improving the harbor for shipping purposes, they worked out a scheme for the construction of a boulevard along the entire waterfront of the city, and where necessary, reclaiming the land along the lake shore. Some three or four miles of this boulevard, at the west end of the city, in the Sunnyside district, is now open to traffic.

In the summer of 1920, the question of lighting this boulevard came to the forefront. The matter was one that warranted considerable thought, for although comparatively little of the boulevard had been completed at that time, a lighting system was to be decided upon which could be standardized for the whole length of the waterfront. With this idea in view, arrangements were made between the Harbor Commissioners, who were to assist in choosing a suitable lighting unit; the Property Commissioners of the Toronto city council, whose department was to pay for the installation, and the Toronto Hydro-electric System, who were to install the system and maintain it for the city. These three bodies had therefore to agree upon and determine the lighting system to be used.

After considerable time had been spent on the display of sample units by various manufacturers, it was decided, from an ornamental standpoint and from a maintenance point of view, to adopt an octagonal cast bronze lantern, the unit to be equipped with opalized stippled glass panels on the sides and top, and each unit to be complete with an 8½ inch Holophane dome refractor with a porcelain en-

Sunnyside Boulevard by night—Motorists speak enthusiastically of the splendid illumination

anielled top reflector, constituting the refractor holder, so as to obtain the maximum candle power on the road surface.

These lanterns were to be mounted on ornamental poles, the light centre to be approximately 15 feet from the ground, the units being placed 100 feet apart, opposite. The ornamental bronze lanterns and cast iron capitals were supplied by the Canadian General Electric Co. Ltd., the poles being supplied by R. E. T. Pringle; lamps by C.G.E., Canadian Laco and Canadian Westinghouse.

Another point in connection with this system which is worthy of note is the fact that this is the first series system installed in the city of Toronto. The remainder of the street lighting, which consists of some 60,000 lights, is fed from multiple circuits; but after making comparisons on lamp life, costs, and flexibility, between the series and multiple systems, the engineers of the Toronto Hydro-electric System recommended that the series system should be used on this new boulevard lighting scheme. This was then sanctioned, and the system was laid out with a C.G.E. series I.L., 1000 c.p. transformer buried in the ground at the side of each standard, stepping the current up from 6.6 to 20 amperes. On the side streets running off the boulevard, there were erected, on poles, 20 kilowatt pole type R.O. constant current transformers, to feed the system. The ultimate intention is to dig waterproof pits and install these transformers therein. At the present time there is only a 600 c.p., 20 ampere lamp used in each unit, although the capacity of the I.L. transformer is 1000 c.p. A point of interest in connection with this installation is that the entire equipment had to be supplied for a frequency of 25 cycle, as practically the whole of Toronto is supplied with 25 cycle power from Niagara. Cable was supplied by Standard Underground, Eugene Phillips and Northern Electric Co's.

The system has now been in operation several months, and has met with the entire approval of all three bodies who were instrumental in deciding on the lighting scheme. It is without question considered to be one of the most efficient incandescent street lighting systems in Canada. It is also proving a wonderful advertisement, and has created a desire among the citizens of Toronto for better lighting in the remainder of the city. The system is also being continually inspected by engineers from neighboring towns and cities, who speak very highly of it; indeed when any section of Toronto or a neighboring city is considering street lighting, the Sunnyside Boulevard offers an example of a very high standard of illumination.

From the photographs attached, it will be seen that both an ornamental and efficient system has been installed on this boulevard.

## Manager of English Electric Co. Visiting Canada

Mr. P. J. Pybus, managing director of the English Electrical Company, is a visitor in Canada. The English Electric Company of Canada, at St. Catharines, Ont., is co-operating closely with the English Electric Company and part of Mr. Pybus' time was occupied in an inspection of the St. Catharines plant. With Mr. Pybus was Mr. John Sampson, of the John Brown & Company. Both of these gentlemen are directors of Power, Traction & Finance Co., Ltd., a British concern comprising five British manufacturing and one financial company, and formed for the purpose of enabling the carrying out of complete contracts of a varied nature. This latter company will operate on the North American continent, it is understood, and will be in a position to compete for any large engineering work that may be required.

## "Americans Deserve Business; They Advertise for it"

The Toronto Advertising Club held the final meeting of its season at the King Edward Hotel, May 11. Although the attendence was small, considerable interest and discussion was aroused by the address of Russell T. Kelley, of the Hamilton Advertisers' Agency, Ltd., the principal speaker of the evening, whose subject on "Advertising from a business man's standpoint" permitted a wide expression of opinion.

Mr. Kelley touched upon such matters as national unity, the development of Canadian exports, the establishment of industrial bureau to decentralize industry and build up smaller centres; but the larger portion of his talk, and most of the discussion which followed, centred about the extension of domestic trade in Canadian made goods. The speaker referred to the large consumption here of a particular American soap, the popularity of American clothes, and to the fact that half a million dollars a day is being spent by Canadians in American resorts.

E. Sterling Dean, who was among others to comment on Mr. Kelley's remarks, stated emphatically that if American firms were getting a large share of Canadian business they were entitled to it. 'They go after it;" he said, "they advertise for it; they get results. If the Canadian manufacturer really wants the business he can get it by advertising for it. Just to satisfy myself of the extent of American advertising in Canada," said Mr. Dean, "I recently checked over the national advertising in a Canadian publication and found that sixty per cent of it was American. This may not be a fair average for all publications, but it is an indication of the aggressiveness of U. S. firms in this country. Until our own manufacturers become equally aggressive the 'Made-in-Canada' cry will remain largely inoperative." —Marketing.

## Automatic Telephone Exchanges to be Installed

The Bell Telephone Company have announced that they will begin almost immediately the erection of an automatic exchange on Main St., near Kingston Road, East Toronto. This will be the first automatic exchange installed in Toronto and will be known as the "Blackstone." A second exchange will follow, which will be located in the down-town district, to take care of additional requirements as they develop. It is not the intention to replace any of the present manual exchanges, except as equipment becomes obsolete or worn out, or as additions may be required. We understand the same policy will be adopted throughout the Dominion, the larger cities receiving first attention on account of greater pressure at those points. Fear has been expressed that with the inauguration of automatic exchanges a number of operators will be thrown out of employment. The company does not anticipate that the installation of automatics can possibly be made at a rate that would make this at all likely or possible.

---

The towns of Winkler and Plum Coulee, Manitoba, are installing electric light systems, purchasing power from the Manitoba Power Company.

A. T. Hicks, Vice-president          M. J. McHenry, President          S. R. A. Clement, Secretary

# Convention of the Association of Municipal Electrical Utilities to be Held in Clifton House, Niagara Falls, June 22, 23, 24

The regular summer Convention of the Association of Municipal Electrical Utilities will be held at the Clifton House, Niagara Falls, Ont., June 22-23-24. Municipal engineers and officials eagerly anticipate a trip to Niagara at this time of the year and it is expected the attendance will very large. A trip over the Chippawa works will as usual be one of the outstanding features of interest.

### THURSDAY, JUNE 22

**Morning.**
  Registration.
**Afternoon, 2. 30 o'clock.**
  President's Address.
  Reports.
  Address—"Public Health as a National and Industrial Asset," by C. J. O. Hastings, M. D., Medical Officer of Health for the City of Toronto.
**Evening. 6.30 o'clock.**
  Convention Dinner.
  Address—"Co-operation." by Hon. I. B. Lucas, Solicitor, H.E.P.C. of Ontario. (Tickets may be obtained at time of registration. Subscription $2.50.)

### FRIDAY, JUNE 23

**Morning. 9.30 o'clock.**
  Papers—"Modern Street Lighting From Illumination Standpoint," by R. M. Love, Street Lighting Specialist, Canadian General Electric Co., Limited.
  "Modern Street Lighting from Construction and Apparatus Standpoint," by M. B. Hastings, Secretary, A. H. Winter-Joyner, Limited.
  Discussion.
**Afternoon. 2.30 o'clock.**
  Paper—"Some Peculiar Phases of Allocation of Charges for Municipal Electric Utilities," by R. C. McCollum; Auditor Municipal Accounts, H. E. P. C. of Ontario.
  Discussion.
**4.00 o'clock.**
  Baseball Match—Commercial Members, C. H. Hopper, Capt., vs. Utility Members, O. M. Perry, Capt. Indoor equipment. "Clary" Settell, Referee.

M. J. McHenry, as president of the Association, will be Master of Ceremonies, ably assisted by A. T. Hicks of Oshawa, vice-president. S. R. A. Clement, whose indispensable services seem to have made him the permanent secretary of the association will, as usual, be at everybody's service. G. J. Mickler will sit at the receipt of customs. The program follows:

### SATURDAY, JUNE 24

**Morning.**
  Visit to the Queenston Generating Station.
  Arrangements have been made with the Hydro-Electric Power Commission to permit delegates wearing the Convention badge to go over the works at Queenston on this morning.

### LADIES

Ladies accompanying delegates to the Convention will be furnished with badges which will permit their taking part in the following programme especially prepared for them. free of charge.

### THURSDAY, JUNE 22

**Afternoon. 4 o'clock.**
  Tea will be served at the Refectory, Queen Victoria Niagara Falls Park.

### FRIDAY, JUNE 23

**Morning. 10 o'clock.**
  Motor Trip Around Niagara Falls, returning in time for lunch.
**Evening. 9 o'clock.**
  Convention Dance (informal.)

### Commercial Exhibit

Commercial Exhibit—There will be an Exhibit by the Commercial members which will be open at all times excepting during Convention sessions.

### GOLF

Playing privileges can be secured for only a limited number. Those intending to play will send their names to Mr. P. B. Yates, Manager, Public Utilities Commission, St. Catharines.

# The Electrical Contractor

### A Model Wiring Plan For Electrical Contractors—Many New Features—Electric Job Should Cost Five Per cent

Electrical contractors everywhere will be interested in a description of the wiring features of the the model Electric Home that was shown in Hamilton during the two weeks, May 13-27. This is an ideal that every contractor should work to. It not only means satisfaction for the owner but it means larger business for the contractor and bigger sales of appliances for the electric merchant.

It is high time that the wiring of a house for $50., $60. or $75. should have passed into ancient history. Few houses today cost less than $3,000 to $5,000. If the electrical contractor fails to sell himself to the extent of five per cent of this cost he has failed in his duty to the house owner, to say nothing of his duty to himself and the rest of the industry.

The Hamilton Home was well wired. It probably may be said that it is the best wired home in Canada, for its size, to date. Certain things were included that the average contractor may not be able to sell to his customers, though it is difficult to see how anyone can argue that they don't give value in convenience, comfort and saving of time and labor. The house is a model for electrical contractors to

Service Equipment Enclosed in Steel Boxes

work to, and while they may feel they are making progress towards this ideal, they should never be quite satisfied until they have reached it. Some of the most interesting details are given below:

The house was wired throughout with BX cables and rigid conduit. Service equipment was entirely enclosed in steel boxes of the safety type, mounted on steel panels; this equipment is shown in one of the figures. The meter was located in a cabinet, so placed that it could be read from the outside of the house, through a glass door; this also is shown in one of the accompanying figures; it is a great convenience in that the meter reader saves a great deal of

his own time as well as that of the housewife, to say nothing of the mud he tracks into the cellar. The service conduit was embedded in the solid brick wall.

The telephone service was also embedded in the wall in conduit; one advantage of this is that burglars are unable to cut off the service, as frequently happens with exposed wires. Five telephone outlets were conveniently located at different points throughout the Home. This admits of the telephone being carried from room to room and is a convenience out of all proportion to the cost involved.

The wiring of the Home included an illuminated porch number. Night lighting was provided in the halls by means of flush bull's eye receptacles, mounted in the baseboard in the upper and lower halls and on the stairway. Special lighting was provided over the buffet, to give show window

The meter reader does not enter the house

effect on the top of the buffet. The china cabinet was illuminated like a show case. The dining room table was wired for a number of appliances, also the tea wagon, the dressing tables in the different rooms, and one of the beds. Every closet had a light outlet.

It was something of an innovation in this Home that the convenience outlets were not placed in the baseboards, but were located about 42 in. from the floor. This arrangement has many advantages, not the least of which is that it avoids the necessity of crouching down to insert or disconnect a plug—an operation none too pleasant or comfortable to people of mature years, to whom, after all, these conveniences should make the greatest appeal.

In the Hamilton Home the kitchen was particularly attractive; a tiny breakfast room occupied an inglenook in one corner and the end of the breakfast table being in prominent view from every point of the kitchen, this was used for a couple of bull's eye switches, one of which controlled the

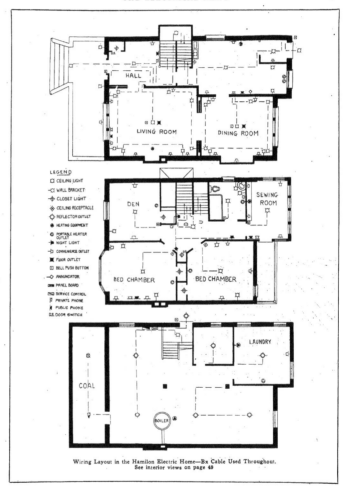

LEGEND
□ CEILING LIGHT
−◁ WALL BRACKET
−⊕ CLOSET LIGHT
✦ CEILING RECEPTACLE
◇ REFLECTOR OUTLET
◉ HEATING EQUIPMENT
◎ PORTABLE HEATER OUTLET
−◈ NIGHT LIGHT
−□ CONVENIENCE OUTLET
▨ FLOOR OUTLET
⊡ BELL PUSH BUTTON
−◇ ANNUNCIATOR
▭ PANEL BOARD
▭ SERVICE CONTROL
⌐ PRIVATE PHONE
✕ PUBLIC PHONE
D.S. DOOR SWITCH

Wiring Layout in the Hamilon Electric Home—Bx Cable Used Throughout.
See interior views on page 49

water heater in the basement and the other the equipment on the table itself.

A liberal number of ceiling brackets, of the removable and interchangeable type were installed throughout. All mantles were of the electric type, no flues being provided. All the rooms were sufficiently heavily wired to carry portable heaters of the necessary size. The porch lights were controlled by means of a key switch.

It is of interest to note that in the city of Hamilton a flat rate is given for water heaters. The electric water heater in this Home was of the two-element type— 660 watts and 2,400 watts. The 660 watt element was connected ahead of the meter—for which the power company gives a special rate of $1.50 per month. The 2,400 watt element was metered and controlled, as already noted, by means of an indicating switch in the kitchen.

The garage was wired for lighting and provided also with convenience outlets for portable lamps or heater. The ceiling outlets instead of being placed in the centre of the garage, as is usual, were placed near the sides of the ceiling —two on each side.

## Selling Better Window Lighting

The opportunity of selling better window lighting exists in most of the stores in your vicinity. Every installation or alteration in a store gives a fine chance for this business. It is not hard to get—if you can show a merchant in a tactful manner, that his competitor's windows are better lighted, you have a strong chance of landing him for an order.

A preliminary step in going after business, is to pick out the most likely prospects. Go around at night, list the poorly lighted windows and note how they compare with others.

is the man to get after, rather than a subordinate. He is the man who spends the money and he is the man to sell on the idea.

When you go to see him, carry with you a demonstration unit which you can quickly install to show color effects. Be certain to know, in advance, what your costs per unit are, so that the cost of an installation can be figured from the window dimensions.

The actual conversation with your customer can not be laid down by any hard and fast rule. Every man can not be in a courteous manner and tell them the facts. Start in with some pleasant remark, in regard to the arrangement of the store, or the briskness of trade, etc. Then get right down to business. Show him to the best of your ability, that better window illumination is a dividend payer.

Do not interrupt the proprietor if he is busy making a sale, and do not back out at the first excuse. It is a mistake, however, to hang on to the point where you antagonize him; retire before that happens and you leave the way open for another call.

Talk to the prospect in his own language, "Foot-candles," " angle of incidence," and so on are Greek to him and just about as interesting. The contractor must know the scientific end, as part of his business, but it is poor stuff to spring on the customer.

Color lighting is growing in use for show window display. New forms of lighting equipment are coming out continually. For example, the spot light tendency is now in evidence and might be worth cultivation. The "show-window-lighting-specialist" must be up-to-date, and he can only do this by keeping in touch with every new thing in his line that comes out.

A point for the contractor—showing how the position of the bulb in a reflector may affect the illumination of a window display.

This will give you a sizeable list. Pick out the cream of these, such as haberdashers, boot and shoe stores, etc., and make up your mind to concentrate on them .

A little advance publicity is the next thing. A reputation for being well posted on lighting matters is a great sales help. Put on a tip-top lighting display in your own window, or co-operate with somebody else, and have your card in the window. You can add some fancy title, as "Lighting Specialist,"—that is, if you can deliver the goods. Three or four good mailings sent out will assist. These letters should be brief and to the point—each containing some little fact on the show window lighting game. If you are concentrating on any particular type of reflector or lamp, enclose a folder showing its uses. Have your name prominently on the folder.

Now comes the actual sales call. The store proprietor

## Contractors in England Take Small Percentage, Too

In England, as in Canada, the electrical contractor does not take a sufficient percentage of the cost of house building. The following table, recently printed in the Electrical Review, gives a good idea of the disparity that exists:—

|  | Per cent. of cost labor only. |
| --- | --- |
| Bricklayers | 32.5 |
| Carpenters | 24.0 |
| Plasterers | 17.0 |
| Painters | 7.1 |
| Plumbers | 4.0 |
| Electricians | 2.4 |
| All others | 13.0 |
|  | 100.0% |

# BETTER MERCHANDISING

## Home Wiring Must be Improved

### A Complete Installation at First Saves Money in the End*

The electric wiring of the home nowadays is of much more important consideration than ever before, on account of the rapid development of electrical household appliances, and the increasing general dependence of the householder on electricity for both lighting and power.

The cost of wiring in the average home is a very small percentage of the total cost of the building, and only a few dollars stand between a poor installation and a good installation. At the present time there are more poor installations than good ones, which accounts for the fact that so many alterations are made in the wiring of the home after the owner moves in. It is quite evident that the matter of the wiring of a residence is not given enough consideration, either by the owner himself or by the general contractor, who is responsible for the wiring in about ninety-five per cent. of the installations. Certainly as much consideration should be given to the wiring of a home as to a factory or an office building, where architects and engineers are employed to give full consideration to the convenience, original cost, and cost of maintenance and operation of the equipment. The home which is in use twenty-four hours a day should certainly have more consideration and thought given to it than a commercial institution which is only in use for about one third of that time.

It has been unfortunate in many ways that the electrical wiring has been left to the general contractor, as there are two kinds of builders, the House Builder and the Home Builder. The house builder merely puts in a job with a bare amount of equipment in the building, so as to make a quick sale, no thought given any further. The home builder tries to give it full consideration and consults the owner of the house. The house builder has even gone so far as to merely provide some wiring, not even installing the switches or plug receptacles. It is just about as absurd to leave out the switches and receptacles, on an electrical wiring job, as it is to omit the painting, requiring the owner, to provide that after he occupies the house.

A concise outline follows, which the home builder can use when he consults the owner, which gives a fair consideration of the modern convenience requirements.

1. Proper ceiling and bracket outlets for each room with necessary switches.
2. Switches to control the first and second floor hall lights.
3. A liberal installation of plug receptacles for portable lamps for living room and bedrooms.
4. Provisions for an electric range in kitchen, electric grate in living room, and a number of heating plugs in living room, dining room, and bed rooms. The above plugs can be used for vacuum cleaners, heating pads, and portable electric heaters for Fall and Spring use, after the furnace has been allowed to go out.
5. Provision for an electric water heater on tank.
6. Make wiring service large enough to accommodate all the above equipment, with at least twenty-five per cent

*By J. H. Schumacher, Treas. & Mgr., Schumacher-Gray Co., Ltd., Winnipeg, Man.

additional capacity for future requirements which cannot be seen at the present time.

7. Provide for a system of call bells and annunciators, and other signal and telephone systems as may be desired.

If a home builder intends to deal honestly with his client he will attempt to consider the above items in consultation with the owner. Some builders may question the above, and say it is too expensive, but on looking into the matter a little closer they will see the amount of money it would save in future expenditures or alterations within a few years, and also have the work concealed. If outlets are not placed properly, more lamps are required to get adequate light. This in itself will cause a larger lighting bill than necessary. This fact is ordinarily overlooked—that it is light and heat you buy from the central station company, but pay for it in the form of killowatt hours, so that the more conveniently the equipment is located and the more efficient equipment used, the less the cost of operation.

Up to the present time there are city and provincial by-laws that regulate certain plumbing and heating requirement but comparatively little has been done by any authorities outside of fire considerations. It will only be a few years before some steps will be taken to make mandatory certain definite requirements so as to conserve the health, eyesight, and good tempers of those who spend most of the time in operating the household, both in city and country.

---

### Plan Now to Make Business Brisk This Summer

The passing of spring, the end of the housecleaning season and the approach of summer bring dealers to a fork in the road. There are two alternatives. You can let business drag—which it has a normal tendency to do during the summer—or you can inject pep into it and maintain a satisfactory volume of sales straight through.

Undoubtedly many dealers are planning, in one way or another, to stimulate business. Some will get an earlier start than others—which brings the thought that, although we are barely entering the summer season, right now is by no means too early to be laying plans! Indeed, there are some dealers who are already conducting campaigns as forerunners of more extensive, sustained effort later on between now and fall. Start now!

---

### The "Universal" Electric Washer

The Great West Electric Co. Ltd., Winnipeg, announce that the "Universal" Electric Washer, which they are be. ginning to introduce in Western Canada, is receiving favorable comment.

> A Hamilton contractor-dealer reports that he has received an order from a customer to install 14 convenience outlets in exactly the same manner as they were installed in the Hamilton Electric Home.

## Pacific Coast Points the Way
## in Better Merchandising

# Vancouver Electric Merchants' Stores

### Among the Finest on the Continent—Strengthening
### the Weakest Link in the Electric Chain—Central
### Station and Private Stores Working in Unison

Having sold the electrical idea to themselves, Vancouver electrical men have placed sure ground under their feet in their progress towards selling the idea properly to the public. To gather an outline impression of what the industry is doing in the ambitious coast city, there has been attempted in the following paragraphs, a symposium of some of the methods and practices adopted by a few of the contractor-dealers in Vancouver. Then to sketch in a background, in which one may discern in part at any rate, how they have sold themselves rightly on the electrical idea, it is necessary to draw attention to the enterprise which has been shown in this B. C. city in organization. First there is the Service League, which with a trained secretary-manager is Contractor-Dealers' Association. Next comes the Electric doing great educative work in the industry. Then there is the Electric Club which, once a week at least, holds a meeting at which every branch of the industry is represented, central station, the electrical engineers, manufacturer, wholesaler, dealer and electrician.

Vigorous campaign work is constantly being planned and carried on by the members of the various organizations within the industry. There is in all the effort put forth, the guiding star of co-operation among all the branches. The endeavor of the men in this industry in Vancouver is to have every man in it realize that the interests of all are identified and that only by team-work can the best results be accomplished. The industry as a whole has clear vision on that score. And the attitude of mind, joined with the energetic organization work in which all unite, is accountable for the progress the industry has made and is still making in electrical merchandising in Vancouver.

Possibly no province in the Dominion holds greater things in electrical development for the future. British Columbia is a land of great water powers. It is destined to be a community of great industrial enterprises. To-day its basic industries, such as mining, lumber manufacturing and pulp making, are great users of electric energy, and the stage of development of some of these industries is comparatively small to what it will be in the near future. For instance, the opening of just one more great mine in one district in the province will very probably mean the construction of another hydro-electric power plant of large capacity. It may be that the electrical idea—of doing things the electrical way—has a natural atmosphere in B. C. which fosters the spirit of enterprise the industry is showing.

#### Selling the Electric Range Idea

"The dominating idea underneath any selling plan for electric ranges is the housewife's pride in her home, and more especially in her kitchen," is the way Jno. Priestman of the sales department of the B. C. Electric Company puts it in describing the successful outcome of a long and uphill campaign to put the electric range over in Vancouver.

Mr. E. E. Walker, sales engineer of the B. C. Electric devised and directed the entire plan of the selling campaign which that company has waged in the past five years, and which is now demonstrating its soundness by its success in bringing in results. Every detail of the daily and monthly efforts made by his staff was closely scrutinised by Mr. Walker and careful note made of every response to those efforts, so that systematic follow-up methods might be adopted.

Discussing the situation of electric kitchen equipment to-day, as the outcome of his chief's successful plans, Mr. Priestman said; "Cleanliness, economy, better cooked food, no fuel or ashes, convenience in planning meals and housework—the final appeal of the clean kitchen—are powerful incentives to induce the ambitious housewife to invest in this modern means of increasing home comfort."

"Public sentiment is in favor of the electric range now. Little time is spent on convincing any one of the practicability of its use. The question now is mainly that of costs. On that score we are able to convince the customer of the economy of the electric range, as compared with either gas or coal. Based on the 3 cent rate per kilowatt, which is what the user of an electric range pays, we have all the data of numerous and exhaustive tests, which show that when used for cooking only, the operating cost is lower. It is an actual fact that it is easier to sell an electric range to-day than it is to sell a coal range.

"To-day in Vancouver and its vicinity there are possibly 600 electric ranges in use. That is what has been accomplished in a campaign extending over five years. In 1916 there were practically no electric ranges in use in this area. In five years from now we confidently expect that there will be 5,000 homes equipped with electric kitchens.

#### Early Efforts Bore Fruit Slowly

"It took us us a long time to make a beginning with the electric range, once the campaign had been decided on," remarked Mr. Priestman, in referring to the time more than five years ago, when Mr. Walker first planned to introduce electricity into the kitchens of Vancouver. "Our very first step was to test out all the different ranges on the market, with the desire to put before the public the best in their interests.

"Our next job was to overcome a very definite prejudice which at that time existed against the electric range. We had to convince the women folk that they could cook on them and that their use was attended with superior advantages. The plan was to demonstrate the ranges constantly in our show-rooms and at the various exhibitions wherever we had the opportunity. Actual cooking was done at all demonstrations and the ladies were invited to see all the familiar culinary processes carried out with electric heat used instead of coal or other fuel. After possibly a year of steady, if slow, progress we began to get results in a small way. Putting out two salesmen to call on likely prospects, selected from a carefully revised list, we got the first two or three ranges installed, and from that on it was comparatively easy to get testimonials and also at times to get per-

(Continued on page 50)

Vancouver Electric Shops—Upper left: Typical exhibit of electric kitchen used in numerous demonstration campaigns by B.C Electric Railway Co. Upper right: Interior view Electric Supply & Contracting Company's store, 781 Granville St.—a handsome store. Lower left: Another interior view; this is the attractive store of Percy F. Letts. Lower right: The Domestic Science class giving a demonstration at a recent Vancouver exhibition, using an electric range, supplemented by individual hot-plates.

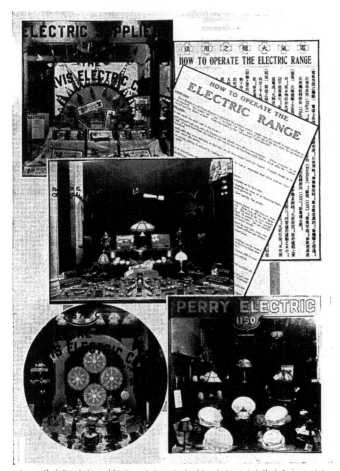

Vancouver Electric Shops—In the upper left hand corner is shown a Hotpoint window entered by the Jarvis Electric Co., in a recent all Canadian competition; it took second prize. Middle: Rankin & Cherrill's uptown shop; an electrically operated airplane was the drawing card in this window. Lower left: Jarvis Electric Company's window on the occasion of a recent Rotary convention held in Vancouver. Lower right: A special sale of ornamental lamps and shades, featured by the Perry Electric Co. Note in the upper left hand corner an explanation card in both Chinese and English on "How to operate the electric range."

Hamilton Electric Home—Above, the Kitchen; note how complete it is, with its ventilating fan, convenience outlets and well placed lights; the "table" in the corner is a dishwasher. Below, the dining room; wired table and tea wagon with many convenience outlets; note the lighting of the buffet, to the right.

mission to invite visitors to see the electric range actually installed in a home.

"In making demonstrations, and through the work of our salesmen making calls, we built up a fairly good mailing list. To-day we work on that steadily. We find we get our best results through a series of letters accompanied by selected printed matter discussing electric ranges. We have also used a monthly calendar, each one being accompanied by a short letter. This spring we have four salesmen out making calls on our list of prospects, in this way following up our mail campaign closely. These men have made a study of their subject and have been trained to talk, demonstrate and sell electric ranges. In that way we have put over the electric range in Vancouver, until, as I have said, there are now about 600 in use and the number is growing steadily.

"One consideration has undoubtedly been helpful to our sales, and that is the sentiment that there is a certain amount of prestige attaching to having an electrically equipped kitchen—every woman is proud to show such a kitchen. In that way it almost invariably follows that putting a range in a house is the means of selling two or three more in the same neighborhood, often in the same block," stated Mr. Priestman. "Another point is that in areas not served by our company's gas system, the time and labor saving and convenience of electric cooking is appreciated. In such districts, especially where there are large homes occupied by people who can afford every comfort, our sales have been good.

"In selling, we have adopted the policy of quoting a man a price for the range installed complete, including the wiring and connections. The buyer appreciates the fact that he knows exactly what the entire cost is to be, ready to turn on the current. He is saved all trouble of letting a contract for wiring after he selects a range. There are no extras, and more than that, a maintenance service is given, which we feel is somewhat generous. The manufacturer's guarantee of one year is supplemented by a ready assistance to get the household familiarized with the range and proper methods of using it,—even to the extent of an instruction card printed in Chinese characters for the Chinese cook! We have also found that quite a large number of customers appreciate our system of time payments, which we have arranged to meet the convenience of the purchaser.

"Going with the electric range has been the necessity of providing a hot water supply for the home. Many houses are now being built without a kitchen chimney at all, through the kitchen being designed exclusively for electric equipment. Therefore an electric hot water heater has to be installed. For that a flat rate of $3.00 to $4.00 per month has been given, according to the size of the heater. The rate is based on $4.00 per kw. per month. Continuous heat for the hot water tank is thus provided.

"The cost of the electric range is based on use for actual operating only, not for heating purposes. With that condition it works out theoretically the same cost as gas. The average is one dollar per month per person. That is, a family of five would find their electric range bill run $4.00 to $5.00 per month. A great assistance in selling electric ranges" commented Mr. Priestman, "has been the improvement in their appearance of late years. Then too, the installation of wiring and connections has been done so that it is practically all out of sight and the appearance of the kitchen is thus greatly improved. And after all that is the great secret—the clean, attractive kitchen—it appeals to the housewife who has to use it. On account of the capital involved in stocking and in campaigning the sale of electric ranges, in our position as the central station, the B. C. Electric Company had to pioneer the field and take the lead. But as with other

appliances, which we have almost turned over to the dealers, in time the same course will likely be followed."

### Best Thing They Ever Did, Said Mr. Brettell

"This was the very best thing we ever did," said Mr. E. Brettell of the Electric Supply & Contracting Co. Ltd., 781 Granville St., when showing the alterations made to the counter tops of the electrical supplies section of the store. Glass has replaced wood in these tops, and a shallow showcase below has displayed all different types of lamps, etc., with prices and description carefully carded. These are attractively displayed on black velvet cloth, and suitably lighted. The customer looks at the particular item desired—if it is a lamp the salesman at once flashes the switch on a similar one just above and out of the way. Then immediately behind are the cabinets carrying the stock. A decision made, the salesman has no search, the customer no delay, for the stock is carefully kept right behind where the sample is displayed.

"That system is carried out throughout our store," said Mr. Brettell. "We have found it work so well that we have made it a leading feature of our stock-keeping and display. For instance, in large cabinets below counter level, and also below the shelves in which every ornamental lamp and fixture is shown, we have ready for immediate delivery a sufficient quantity of each article to meet at once all likely demands in the course of the day's selling. Naturally, we make it a point to see that this supply is replenished promptly from our storage. The method has the additional advantage that every single fixture or lamp placed in the cabinets below the display has been looked over to see that it is in readiness for delivery."

Carrying out the idea of keeping every department in a definite place, the store is arranged so that the appliances are in one section, the ornamental fixtures in another. On entering the store, large and handsome cabinets, are met, which are devoted to all appliances. This series of cabinets is one of the latest additions to the high class fixtures which mark this modern electric shop. Built of solid mahogany, and costing close to $1,000, these cabinets demonstrate the firm's belief in adequate surroundings for proper display of electric goods. On the opposite wall are arranged shelves, in cellular form, for ornamental fixtures and lamps —one article only being displayed in each. The result is unquestionable—the full value is brought out and no confusion caused as a purchaser passes from one to the other.

### This Shop Fits Into Surroundings

Fitting your shop and business to the locality in which it is located is well worked out by Percy F. Letts, secretary of the Vancouver Contractor-Dealers' Association. His shop is in a prosperous residential district. In fact it is the very last shop at the top of Granville Street, and at the boundary of Shaughnessy Heights, into which residential area no shop premises are allowed to be pushed. To appeal to the home instinct Mr. Letts has his show-rooms so arranged that, as he says, "it simulates as closely as possible the atmosphere of the home." Being divided into three parts, all opening into each other and well provided with show windows facing the street, it has been possible to work out this idea. The corner show-room is arranged much like a large dining or sitting room. Fixtures on the wall resemble an old-fashioned suitable to display there. In the centre of the room is a large dining table. On this are arranged various lamps and fittings suitable to display there. In the centre of the room is a large dining table. On this a variety of appliances as well as different types of table lamps are arranged from time to time. They are connected up to the lighting or heating circuit by cords carried under the rug which covers the floor. Beside the table, other lamps on tall stands of metal or turned wood are set when suited to the rest of the display. In the window,

library and table lamps, household appliances and numerous
decorative features are arranged to attract the eye of those
passing in the street.

In the second show-room electric ranges, electric washing
machines and other of the larger utilities are arranged in
effective manner for demonstration. In the third show-room
the stock of electric lights and sundry supplies is concentrated, and this is the only portion of this unique shop which
is arranged in conventional "store" shape, with counter and
shelves. Mr. Letts features wood stands, turned in any
variety of cabinet woods, to match both in material, shading
and form, the furnishings of any room. Another special
feature is parchment shades, done in oils—and real paint-
ings of artistic merit are these. Possibly Mr. Letts' good
fortune and success in this attractive feature of his merchandising is due to his fortune in marriage. Mrs. Letts is an
artist in oils, of talent and training, and she does these parchment shades so well that they have been bought by customers
from Fort George to Winnipeg, and even so far away as
Virginia.

### Electric Aeroplane Was Attraction

"Our experience with that aeroplane in our window fully
demonstrated the value of getting goods out where they can
be seen and to show what they can do," said Mr. W. H.
Slater of Rankin & Cherrill, talking of the attraction provided by a toy aeroplane which formed the centre of interest
in their show window. "It demonstrates also the value of a
moving window display, especially for electrical goods, which
lend themselves to being actuated by a current, easily arranged, as was done for the aeroplane.

"The tiny plane was driven by electricity from a toy
transformer, taken off the regular 110-volt lighting circuit.
The wires which held the aeroplane suspended from the ceiling of the show window supply the current to the motor on
the plane. This motor drives the propellor and is the only
means of running the machine except for a counterbalance
at the ceiling to steady the machine and keep it flying in a
circle. If run at full speed the plane would cover about a
25-foot circle. The speed can be controlled through the
transformer.

"Perhaps the most interesting detail about the display,
from a sales point of view," said Mr. Slater, "is the fact that
directly from having the machine flying in the window we
sold out entirely a stock of these toys which had been dead
on our hands for eight years. We had three large ones,
three medium size and two smaller. If we had cared to sell
the one we still have in the window, we could have done so
a dozen times. These planes were made by the Rittenhouse
Mfg. Co., at a time when aeroplanes were immensely popular.
Later they failed to catch the popular fancy, and we carried
them until we were able to display them in the windows of
our new store. In our other store we had no suitable space
to display them."

Rankin & Cherrill is a pioneer name in the electrical
contractor-dealer circles in Vancouver. Mr. H. V. Rankin
is the original member still remaining in the firm. Two stores
are operated, both on Hastings Street. The uptown shop
features very prominently the electrical appliances and handsome electric fixtures stocked by the company. The main
store, at 55 Hastings Street, lacked sufficient window and
show-room space for the proper and attractive display of the
large stock carried. That led to the other store being taken
late in 1921.

### Uses Window Displays and Bright Lighting

"I am an advocate of window displays, attractively arranged and frequently changed," said Mr. Roy V. Perry,
manager of the Perry Electrical Company, "and on the principle of taking our own prescription, we use electric lighting

schemes freely in all possible displays. Bright lights are the
greatest of drawing cards. This is especially true of selling
electric lamps and fixtures as I have proved from experience."

Because of securing larger show window space The Perry
Electric Company has taken premises at 985 Robson Street,
and is making preparations to close its other shop at 1150
Robson Street.

## Small Stores Can Supply Large Amounts of Courtesy

Service means courtesy. It includes as well, attention
to the troubles that happen to a customer's appliances,
promptness in deliveries, the furnishing of the right materials and so forth—but back of these there has to be an unfailing courtesy that leaves the customer with the impression that he has been well treated. This can be achieved by
the small stores just as well as by the large ones,—perhaps
better. An incident that occurred recently in the store
of a man who runs the business practically by himself, indicates how the principle is carried out and partly explains,
by the way, the success of this particular business.

A lady came, somewhat diffidently, into this store the
other day. She seemed at a loss until she saw a female
clerk behind a certain counter—then she stepped out boldly,
and presently was pouring a tale of woe into the ear of the
attentive clerk. Her electric door-bell needed attention and
her explanation of the trouble was garrulous, to say the
least. The clerk was the recipient of a host of instructions,
as to what hour the man was to call, to knock loudly and
so on—and right there was where the service came in. That
whole volume of talk was listened to without the least sign
of impatience; the clerk even appeared interested and agreed
smilingly to the most minute instructions. Finally the customer went away happy, certain that her needs would be
fully looked after.

It isn't hard to tell what would have been the result if
that same woman had happened to drift into some other
stores. She would have been snapped off by a brusque
male salesman, or shooed out of the way as quickly as possible by a man too busy figuring on his wiring contracts to
devote any time to consideration of the vagaries of a door-
bell. And that would have been bad business—about the
same as telling the customer to go elsewhere. The very
fact that she was timid about coming into an electric store
and yet came in, showed that at least she had an idea of the
right place to look for electrical work. That idea should be
fostered with the greatest assiduity. The impression carried away by the customer is the thing that counts. Impressions are formed unconsciously, and a good impression
brings a customer back to a store.

The employment of one or more women clerks might
prove of benefit to some stores. Unquestionably they would
require a great deal of training, but they are generally neat
in appearance and courteous. Undoubtedly the average
woman clerk is superior to the combination wireman-sales-
man, who is often 90 per cent wireman and 10 per cent
salesman, and whose clumsy handling of customers does a
great deal to injure the trade.

# What is the Future of Radio?

## The Craze for Cheap Sets is Dying Out—Radio Equipment is Undergoing Evolution—Developments Must be Watched

We have all asked ourselves this question. We have all had it put us many times. At first it may have been a little difficult to answer. Even yet it is quite impossible to foresee what new developments and inventions may bring forth. This much, however, seems certain; if radio sets can be placed in our homes at a reasonable cost, and if they are capable of reproducing music, addresses, news, reports, etc., in such a way as to give us pleasure and profit, the public will find the means of buying them.

Radio development, we believe, will follow lines analogous to the gramophone business. In the early days the reproductions were very imperfect. They lacked "quality." Yet the wonder of the thing created a tremendous public interest and it was only those whose appreciation of quality in tone exceeded their curiosity and enthusiasm who resisted the temptation. The true situation soon asserted itself, however, and the "music box" fell into disfavor. Then came improvements and refinements over a series of years—and with these, of course, added costs—until to-day there are hundreds of thousands of homes where quite expensive phonographs are not only tolerated but enjoyed. The demand for the cheaper machines has almost vanished.

Many signs point to a similar evolution in radio equipment, though the various stages will probably follow one another much more quickly. Already, however, the novelty of receiving messages in this way is losing its hold on the people and they are saying that they want something better—something that is less a strain on the imagination and on the nerves. The craze for ten dollar sets—except with the small boy—has about died out and the average citizen is sitting back waiting for the further development of the art. He is not satisfied with present results but he is enthused with the possibilities and convinced of the value—commercial and otherwise—to such an extent that he is willing to pay anything in reason for a real radio set.

There seems to be only one conclusion to the whole discussion, viz., that just as soon as thoroughly satisfactory and reliable equipment can be placed on the market the better-to-do classes of our citizens will buy them.

Now, what is the proper attitude of the electrical dealer towards radio?

As we pointed out in our last issue, this business, logically and morally, is a part of the electrical industry. The electrical merchant is the ultimate channel through which radio equipment must reach the public. The electrical dealers themselves should assure this by perfecting their knowledge and sales methods in radio matters, and the manufacturer should support the electrical dealer against all outside retail competition. If, in the meantime, radio is demanded by the youth of our country it might not be a bad plan to let the department store and the hardware man handle it so that he will become discredited all the sooner in the public mind as a dealer in electrical equipment. However, there is a profit on the sale of this equipment due largely to its rapid turnover and it does not seem to be any sufficient reason why the electrical merchant should not have the profit simply because he realizes that the quality of what he sells is not what it should be or what it will be in the near future. It would seem to be the wisest course, however, for the dealer to watch his stock very carefully

and not be caught with a quantity of obsolete supplies on his shelves. It would be well, also, for him to keep in very intimate touch with the whole radio situation so that he may be in a position to offer sound advice to his customers where they seem inclined to appreciate it. He will thus establish himself as an authority on radio and will be the logical person to consult at a later date when this equipment has become more nearly standardized.

Without question, there is going to be a tremendous demand for radio sets when the equipment has been perfected. What man of means will hesitate to pay two or three hundred dollars for such an installation when it gives him first-hand news, entertainment and education all in one? There is another viewpoint; most well-to-do homes own gramophones, yet the phonograph itself is just the begining of the expense. Records costing hundreds of dollars are a necessary acquisition before the phonograph can give satisfactory service. It will not be so with radio equipment. The first expense will, practically speaking, be the whole expense involved.

Then think what a boon a radio service will be to the individual or family, or community, living in an isolated district. Every home under these conditions will become a prospect. The line of demarcation between city and country—which has been largely one of entertainment—will disappear.

The best word for the electrical dealer seems to be—use caution, study the problem, be prepared to take advantage of future developments.

## Some Radio Selling Points

The demand for radio supplies continues unabated. Forecasts that the business would prove to be of a very temporary nature, have been disproved, and the market goes on widening. Sufficient time has elapsed since the first rush to allow any mushroom tendencies a good opportunity to manifest themselves, and they have not done so. The public has been won over to radio, and all the electrical dealer has to worry about is keeping the trade in its proper channels, and stocking the right kind of goods.

The development of the radio demand has been much faster than was the demand for the first telephones. It, however, has been very similar. The change in status of the new convenience seems to be progressing along the same lines, at first it was distinctly a novelty, as were the pioneer telephones. At the present time it may be called a very popular luxury or fad. Eventually will come the "necessity" stage—where the telephone is now. It is not too much to say that the fore-sighted dealer will regard this stage as something that is bound to happen and will lay his plans accordingly.

### Pick Your Stock

There is such a rush of business in this field that some dealers are buying indiscriminately. The quantity and variety of the goods that can be handled by the dealer are determined exactly by his knowledge of his previous trade. Keep accurate track of your enquiries, and this will give you a line on what you can dispose of.

Nearly every electrical dealer has dabbled more or less in

radio supplies already. There was and is a good demand and the stuff had to be obtained wherever possible—the result has been that there is very little standardization and some of the stocks look pretty well mixed up. There is inevitably going to be alteration in parts, and improvements, accessories added and so on, and the man with a mixed up lot of apparatus is liable to find it left on his hands.

In regard to stocking, there is the usual tendency now, when supplies are scarce, to lay hands on as much stuff as possible. This is particularly the case when a dealer has been unable to get a previous order filled. We can picture that Mr. Bill Smith, electrical dealer, has had to turn down a couple of orders for head phones some day this week. The same afternoon along comes Mr. Highflyer of the X—— Radio Company and tells him that if he orders right away he can get three or four dozen head phones immediately. Ten to one Mr. Bill Smith orders as many as he can possibly take, not thinking of the fact that there may perhaps be improved and maybe cheaper phones on the market before he has sold this three or four dozen.

### Sales of Parts

It is a mistake to think the sale of parts is not worth catering to. There is a tendency with many dealers to neglect this end of their business. They supply what they have in stock without any painful feelings, but as for studying the demand, and laying in a well assorted stock—well, they simply don't do it. Perhaps this is a reflection of the feeling that existed when radio was confined to a few "bugs." In those days, as now, the items handled were small and inexpensive, and the buyers were exceedingly particular in their choice. The business was infrequent, anyway, and so in general the dealers passed it up as not worth the trouble.

That does not obtain to-day. Now, the people who cannot buy the more expensive assembled sets, or who wish to make up their own, for the pure joy of doing it, are numerous. The little 5-cent or 10-cent sale to-day will lead to more to-morrow. These parts are not so expensive but that a dealer can carry a pretty full line. Display them where they can be seen.

### Know More Than Your Customer

There are many demands made upon the electrical dealer in regard to the technical knowledge he is supposed to have. Besides a general knowledge of electricity, the public expects him to understand the scientific side of illumination and the latest call is for information in regard to radio. This is natural enough, and it is easy for a dealer to get to know the subject so that he can tell the customer how to put in the set and get results.

From the standpoint of efficient salesmanship, however, it is not the right thing to tell the customer what he wants to know and let him go. He should be gently led on to the consideration of better equipment. If he has already a set and is purchasing some part therefor, perhaps you can suggest another improvement. Knowledge is power, and there is a distinct change in the attitude of the customer if he finds that the dealer can give him authoritative information. In spite of the publicity that all phases of the subject have been getting, there are still some people who know absolutely nothing about it other than that you hear through phones. They are easily handled at first, but the next comeback demands a little more knowledge on the dealer's part and so on. Many men with a mechanical turn of mind, develop into real radio fans and go in for everything possible. This type demands expert knowledge on the part of the salesman; the dealer who can handle this demand has that part of the trade cornered.

### The Follow-Up Business

There are special opportunities in radio merchandising

that do not occur to the same extent in the retailing of ordinary appliances. In the great majority of cases, the first stab at the game is made by a customer with the purchase of a $25.00 or $30.00 set. The installation of this outfit satisfies him for a while, but very soon he wants better results,—besides there is the trouble of passing the telephones around to the various members of the family.

The customer generally comes back and enquires about attaching a loud speaker to the present outfit. This is the point where the dealer can sell him one of the more expensive sets. The dealer explains that the loud speaker cannot be attached to the crystal detector type, and with a little salesmanship, sends the customer away with an outfit costing up to perhaps $200.00.

The appeal of radio is to the whole family and to every family. There should be no trouble in placing radio sets in the majority of homes that use electrical appliances. The farm market has not been touched. The business is just beginning, and though we cannot look for a feverish demand to continue forever, there is no doubt that the use of radio will become almost as universal as the telephone. The securing and servicing of this business will add greatly to the prestige of the electrical dealer.

# The A. B. C. of Wireless

The transmitting equipment of a wireless station is very different from the receiving equipment. There are in use perhaps a thousand receiving sets to one transmitting set and this is some indication of which apparatus is the most complicated—and also, which is the most costly.

In general, transmitting apparatus is not of interest to the amateur. The apparatus required is expensive and needs much care and adjustment not called for in the operation of a receiving set. We do not promise, therefore, to burden our readers at the moment, with an unnecessary description of the construction or operation of sending sets. At the present time the general desire is merely to have apparatus that will receive clearly what is sent out from the commercial transmitting stations. The general attitude of the amateur is that knowing there are several excellent broadcasting stations, he wishes to understand the operation of the apparatus that will enable him to pick up the programs they send out.

### Wireless Impulses of Wave Form

Wireless communication is carried on by means of radiated waves that are produced electrically—hence the term "radio," from the manner in which the waves are given out.

When a stone is dropped into a pond, radiated waves of water are produced—the ripples spread out in concentric circles. We can see and feel these water waves, and hence know that their rate of travel is relatively slow.

Another example of wave motion is in the production of sound. These sound waves are audible and they travel faster than the water waves.

Light waves are faster still—they go at a rate of 186,-000 miles per second. We realize their existence by their effect on the retina of the eye.

Radio waves travel just about as fast as light waves. They are different from any of the other wave forms we have mentioned, however, in that they can be neither heard, felt nor seen. Their detection can be accomplished only by electrical apparatus that changes the waves received into a form recognizable by the human senses. This transformation is effected in a more or less satisfactory degree by a number of simple radio receiving sets now on the market.

### Simple Receiving Apparatus

In its most elementary form the receiving apparatus consists of five parts. These are:—

(1) An antenna, or aerial. This consists of a wire suspended in the air, between insulators, the purpose of which is to intercept the waves.

(2) An inductive coil or "tuning coil." This is a coil of wire connected through a sliding contact to the antenna. Its effect is to change the electrical character of the receiving circuit in such a way as to permit the reception of

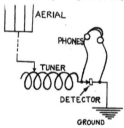

waves from any particular sending station. When the coil has been adjusted to one particular sending station, all other waves, unless very powerful, pass through the tuning coil to the ground, unnoticed. By tuning, the workable distance is much increased.

(3) A detector. This is the next link in the chain from the antenna. The most common type is the crystal detector. It consists of a metal point resting lightly on crystals of certain minerals, such as galena, through which the current passes to ground. This detector has the property of rectifying the current so that only the oscillations in one direction are effective.

(4) The phones. These render the incoming signal audible.

(5) A "ground." The system must be well grounded.

## Vancouevr to Have Electric Home Exhibit

An Electric Home Campaign has been successfully launched by the Electric Service League of Vancouver. Through a co-operative arrangement with a number of building contractors and supply houses, a handsome residence is being erected in Shaughnessy Heights, one of the most attractive residential districts. The location of the home is such that it is readily accessible to all who may desire to visit the exhibit. Architects Townley & Matheson designed the residence which is now being erected under the general supervision of Messrs. Rush & Reed, contractors.

It is expected the construction of this Electric Home will be completed early in August. While the building is under construction the electric wiring is to be installed to provide for every demand of the most recent advances in electrical service for the home. The work is to be carried out by members of the League. The home is to be handsomely decorated and furnished throughout. Every electrical appliance in modern use will be installed and demonstrated during the period the home is open for exhibition.

While the home "atmosphere" is sought in the building, its interior fittings and exterior surroundings, the Electrical Service League is taking special care that the Electric Home impression will be predominant. In the wiring, in addition to the necessary separate circuits for lighting and service, the arrangement of switches has been most carefully designed. Remote control for almost every part of the house

lighting is to be provided, so that the utmost convenience is given for turning on the current in various rooms. In the principal bedroom, designed for the owner of the house, a master switch will throw on every light in the house if desired. In addition, this switch also controls a series of concealed lights under the gables, which flood the approaches to the house with light. It would be an awkward and startling reception for any untimely visitor.

There will be electric fireplaces in one or two of the living rooms, and every provision made for using electric appliances in bedrooms, dining room and kitchen. The laundry arrangements will also include electrical equipment. The kitchen will be equipped with an electric range, water heater and other features. An electric washing machine, wringer and dryer will be placed in the laundry.

At present it is anticipated that the exhibition will be announced for late August or early September. In the meantime the members of the Vancouver League are busy on details of all the plans for the adequate presentation of electric services in the modern home.

## E. W. Playford Opens Office in Montreal

Mr. E. W. Playford has resigned his position as assistant district manager of the Canadian General Electric Co. Ltd., Montreal, to engage in business for himself as manufacturers agent, and has opened an office in the New Birks Building, Montreal. He will represent the Dominion Carbon Brush Co., of Toronto and the Dominion Engineering

E. W. Playford

Agency Limited, of Toronto, who are Eastern Canadian representatives for the well known I. T. E. circuit breakers, Sundh automatic control equipment, Garfield moulded insulation, Berry's compound, etc. etc. Mr. Playford has been with the electrical industry in the Province of Quebec for the past twenty years, eleven years of which has been spent with the Canadian General Electric Co. Ltd., having held the positions of construction engineer, district engineer and later assistant district manager. He was previously connected with the R. E. T. Pringle Co., the Montreal Light Heat & Power Co., and for many years was on the electrical engineering staff of the Dominion Textile Co.

### Mr. Geo. P. Eaton goes to Lincoln Electric Co.

Mr. Geo. P. Eaton, who has been with the Chamberlain & Hookham Meter Company since 1911, and latterly also with the Lincoln Meter Company—as Ontario sales manager for both companies—has severed his connection with these companies and is now Ontario salesman with the Lincoln Electric Company of Canada, manufacturers of electric welding machines and motors.

# Winnipeg an Active Electrical Centre

The Winnipeg Hydro is installing underground, two 12,-000 volt feeders from the terminal station on Rover Street to No. 5 sub-station in Fort Rouge. The total length of cable is 52,000 feet. It is being manufactured by the Northern Electric Co., Montreal, at a cost of $95,000. A portion of this cable is being installed in conduits that form part of the underground distribution system. For about 14,000 feet, however, the conduits are being laid by the Carter-Halls-Aldinger Co., at a cost of $25,000. The total cost of the new feeders, including conduit construction, cost of conduits, man-hole covers, sewer connections, etc., will be approximately $135,000. Arrangements have been made at two points in the feeders for sectionalizing through disconnecting switches which will facilitate testing in case of trouble. On the completion of these feeders the system of sub-stations will then be inter-connected by a loop feeder system.

### Winnipeg Hydro May Install Steam Standby

Mr. Glassco, manager of the Winnipeg Hydro, has stated that during informal discussions in connection with the installation of a standby plant, the general opinion of the committee seems to be in favor of building a plant at an early date. The main question is the size of the plant that should be installed. In all probability one of about 13,500 h.p. capacity would be put in to start with, with the building and other equipment large enough to take care of at least one additional unit of a similar kind. All evidence gathered to date confirms the original estimate of about $75.00 per kw., that is about $56.50 per h.p. It is pointed out, however, that as the building and various equipment would have to be large enough to take care of the ultimate capacity of the standby plant, the cost per horse power for the first unit would be higher, the above being the estimated cost per horse power for the complete installation.

### Attended Western Sales Conference

A. S. Edgar, manager of the Supply Department of the Canadian General Electric Co., Toronto, spent a few days in Winnipeg, on his return from Vancouver, where he attended the Western Supply Sales Conference. Mr. Edgar was accompanied on the trip by Geo. R. Wright, district manager, Winnipeg. The conference was held at the Vancouver Hotel, Vancouver, from the 8th to the 13th May, and was attended by A. P. Horner, district manager, Calgary; D. R. Logan, manager of lamp sales section, Toronto; S. E. H. Smith, Western supply sales inspector; and H. Pim, district manager, Vancouver.

### A Change in Distribution Will Improve Service

Conforming with the policy of the Hydro Department of keeping ahead in all the latest improvements in electrical service, changes are being made in the distribution system which will increase the efficiency of the service by giving much steadier voltage, free from the fluctuations which sometimes occur when heavy loads are suddenly put on the lines. The system in certain sections of the city will be changed from what is known as the 3 wire connection to 4 wire star connection common neutral. The result of this change will be that without any additional cost for material they will be able to distribute three times the amount of energy as formerly. For example, in one district that has already been changed they are now able to distribute 27,000 h.p. where only 9,000 h.p. was formerly carried. The cost of installing

sufficient equipment to accomplish this end on the old delta system would probably have run to $50,000 whereas by changing to the new system the cost was $7,600 only. By their keeping in touch with the most approved methods the citizens are assured efficient service at the lowest possible rates.

### Representing Eastern and U. S. Lines

Mr. Fred E. Garrett, who recently resigned the sales managership of the Great West Electric Company, has opened a sales office in the Confederation Life Building, Winnipeg. He intends to represent some Eastern and U. S. manufacturers, selling the jobbing trade from Fort William to the Pacific Coast. As he has had a great deal of experience in the electrical business in all branches he is qualified to be of assistance to both dealer and jobber in selling his lines. Mr. Garrett advises that as some of his lines are not as yet completed, he is not able to say what all the lines are, but he is ready to announce the following: Ingersoll Machine & Tool Co., Ingersoll, Ont., (Baby Grand Ironer); The Continental Electric Co., Toronto, (Royal

Fred. E. Garrett

Vacuum Cleaner); L & N Manufacturing Co. Ltd., Montreal, (Insuladuct & Loomduct); Trenle Porcelain Company, East Liverpool, Ohio, (Porcelain Products); Holophane Company, (Illumination).

### A Regular Western Visitor

H. S. Balhatchet, vice-president of the Benjamin Electric Mfg. Co., of Canada Ltd., Toronto, was a recent visitor to Winnipeg, spending four days with Mr. Reg. Smith, western manager of the firm. Mr. Balhatchet makes annual trips to Winnipeg, in order to keep in touch with his larger distributors.

### Russell-Fowler Changes

The Russell-Fowler Co., 104, Capitol Theatre Building, Winnipeg, have disolved partnership; the business will be continued by Ray Fowler, at the same address; the name of the firm will not be changed.

The Winnipeg Hydro have placed a contract with Messrs. Julian & Henry, Winnipeg, for the painting of the towers on the old transmission line. This work is now commenced and will be completed before Oct. 15th. The cost of this work will be approximately $9,000.

# The Latest Developments in Electrical Equipment

### New Type Lightning Arresters

The Westinghouse Electric & Manufacturing Company has developed a new type of direct current electrolytic lightning arresters, known as "Type AR", for car or station use on railway, power and lighing circuits. These arresters, for voltage applications up to 3,800 contain one to 12 cells. Each cell consists of two aluminum plates immersed in a suitable inorganic electrolyte and supported from a porcelain cover clamped by a zinc ring to a glass jar with a gasket placed between the porcelain cover and the glass jar. Hollow concentric cylinders made from sheet aluminum form the plates, the outer cylinder or plate being punched and upset at

D. C. Electrolytic Arrester

frequent intervals in order to allow free circulation of electrolyte within the cell. Balancing resistors are used with arresters of more than one cell. These resistors cause each cell to take its proper portion of the line voltage and thereby tend to keep the aluminum hydroxide films equally formed. The arresters are "floated" between line and ground so that a leakage current of only a few millamperes passes continually. This leakage current serves to keep the film upon the aluminum plate or plates in proper order.

The product is capable of passing a surge current of approximately 1,000 amperes at double normal voltage when the arrester is functioning and one arrester should be used for each 500 kilowatts of feeder bus, rotary converter or motor generator capacity to which the arrester is connected. Any voltage in excess of normal line voltage is discharged promptly through the arrester

---

### New Bell Ringing Transformer

The Hydro-electric Power Commission of Ontario requires that the circuit of a bell ringing transformer include a double pole cut-out fused with not over three ampere fuses. In order that a householder may meet this requirement with the least amount of wiring trouble in installing one of these necessary conveniences, the Canadian General Electric Co., has just brought out an a.c. 5 watt, fused, bell ringing transformer in which the transformer is combined with a double pole plug fuse, porcelain covered, cut-out in one unit. To install this device it is only necessary to remove the dry batteries from the bell circuit, mount the fused transformer with wood screws to the wall or ceiling, attach the lighting circuit wires to the cut-out terminals marked "110 volts" and the bell wires to the lower terminals marked "Bell" and the job is completed—meeting all the requirements of any inspection department. The C. G. E.

fused transformer is made both 25 cycle and 60 cycle covered by Cat. Nos. CGE113 and CGE114 respectively, and while the term bell ringing feature only is treated, there are many other uses to which it may be put, such as lighting low voltage lamps (6 to 14), night lights, or wherever low voltage alternating current is desired. The device is all porcelain, glazed white, with approximate over-all dimensions of 4-1/8 in. L. x 3-1/4 in. W. x 3-1/8 in. H. Due to its compactness, neatness in appearance and ease of wiring it should become a very popular appliance with the trade.

### Rail Bonds for Arc Weld Application

Two new rail bonds, for arc weld application, have been developed by The Ohio Brass Company, Mansfield, Ohio. They are known as Type AW-7, for base of rail application and Type AW-8 for ball of rail application They are shown in Figs. 1 and 2 respectively. The AW-8 bond consists of 2 copper strands formed into a U and welded into steel terminals which, in turn, are welded to the rail. The AW-7 bond is the same in principle; it has a single strand, of any desired length, welded into suitable steel terminals. It is a proven fact that steel to steel is the easiest kind of electric welding and the new bonds conform to this principle. The steel terminals are slightly rounded and provide a wide angle or scarf in which to build the weld. The steel is heavy and will not be burned through by the arc. A copper sleeve, which surrounds the strands at the terminal, shields the wires and absorbs vibration.

Fig. 2—AW 8 Bond

Fig. 1—AW 7 Bond

### Heavy Duty Electric Range.

A new venture on the part of the Langley Electric Mfg. Co. Ltd., Winnipeg, is the manufacture of the heavy duty Electric Range shown herewith. Orders have been received from the Provincial Government for the Deaf and Dumb Institute, Winnipeg; and the Nurses Home, Brandon. The current consumption of these ranges is 14 kw., each range having a cooking surface of 24 in. by 36 in. The container element is cast iron, with the elements so arranged that they

are accessible and easy to repair. The frame is of heavy brass tubing, nickle plated. This is said to be the first heavy duty range to be built without an oven; the oven is separate and manufactured by MacNab & Roberts Ltd., Winnipeg. This range is suitable for homes, hospitals, or any concern handling a large number of meals, as it is capable of cooking for 250 people. It is made up of four separate elements.

### The Alpha Fixture Company's Showrooms

The Alpha Fixture Company has opened showrooms at 183 Church St., Toronto, where they display a very attractive new line of fixtures, known as the "Alpha." This is described as "an entirely new idea in fixture making and a 95 per cent Canadian product." The distinctive feature of the Alpha fixtures is that they are constructed largely of wood, thus making them easily adaptable to all furniture finishes so that they blend with whatever decoration scheme may be desired.

### The Electroplax Company

A new manufacturing concern, The Electroplax Company, Ltd., with a factory in Mount Dennis, Ont., has opened a city sales office in the York Building, Toronto, with Mr. W. M. Davidson, sales manager, in charge. This company manufacture "Redmanol Bakelite" and Condensite phenol resin plastics. They will specialize on electrical insulation and moulded products, wireless insulating equipment, synthetic amber products, and acid and heat proof varnishes and lacquers.

The principals of the firm are: President, H. E. Corey, late of the Standard Oil Company of Pennsylvania; Vice-president, Dr. L. V. Redman, president of the Redmanol Chemical Products Co., of Chicago, 2nd vice-president, P. A. Thompson, of Nesbitt Thompson and Co., investment brokers of Montreal.

B. P. Corey of the Corson Oil Co., Petrolia, Ont., and V. R. Smith actuary of the Confederation Life Co., are Directors.

The British Aluminium Company announce that, beginning May 25, their address will be 592 King St. West, Toronto, instead of 263 Adelaide St. West.

### Transformer Load Indicator

The transformer load indicator of the Westinghouse Electric & Mfg. Co., illustrated below, is a device arranged for mounting in place of the drain plug or, preferably, in the side of the tank well a few inches below the oil level, in a distributing transformer, for the purpose of indicating whether any predetermined temperature has been reached or exceeded. When this predetermined temperature is reached, a yellow flag, or semaphore, drops into view to give a visible indication of the fact. The distinctive color and considerable area of this semaphore make it readily seen by the patrolman, from the street.

This device can be used for several purposes, the most important being to indicate when the transformer is over-

loaded and should be changed for a larger one. Another use is in determining whether transformers operating in parallel are properly dividing the load. The device can also be used to obtain more efficient loading of transformers, since underloaded units can be detected by setting the indicator to trip at a low temperature, in which case the transformer should be replaced by a smaller one.

The indicator consists, in general, of two parts—the indicating mechanism and the plug. The indicating mechanism consists of a base plate, tube, bi-metallic strip, adjusting rig, and cover. The tube fits the inside of the plug and has a threaded piece at the end, which engages the nut in the plug. By this means the indicating mechanism can be inserted in the plug with a turn or two, after the transformer is hung, and can then assume a vertical position without any danger of falling out. The bi-metallic strip set in the end of the tube projects beyond the base plate far enough to hold the flag in position until, through an increase in temperature, this bi-metallic strip bends sufficiently to unlatch the flag, which drops by gravity into the position where it may be observed from the ground. The flag is reset by simply pushing it back into place.

### What the Dealers Say About Radio Merchandising

"Department stores are cutting into the business."

"Plenty of enquiries, but few sales."

"Selling lots of parts, principally head phones."

"Can not get enough."

"The kids come in and ask about prices but have not struck much real business."

"Our down town store is doing well but we have cut that department out of our uptown place."

"I could sell all kinds of phones and bulbs but simply can't get them."

"The manufacturers' terms do not suit me and I am not handling that line."

"There is some price cutting on antenna wire. No. 14, bare, has been sold at 40 cents a hundred."

"Little demand for better class of equipment."

## Siluminite

The George C. Rough Sales Corporation, Montreal, are now handling, for Canada, a new British insulating material, —"Siluminite"—which possess qualities and characteristics of value to the manufacturer. It is supplied in the form of sheets, short rods, tubes, etc., and can be readily molded during its manufacture to any form or shape desired. It is of a dense black color and takes a fine polish. It is not affected by heat. It is not brittle nor has it any softening points. Immersion in alkali, hot oil or boiling water leaves it unchanged. It is not affected by atmospheric changes. This material is thus claimed to be an effective substitute for slate and marble for switchboards, panel boards, etc., as having very high electrical resistance, thinner sheets can be used, thereby reducing handling and manufacturing costs. Its weight is approximately one pound to 15 cu. in. Its mechanical strength is said to be very great and to excel in this respect porcelain, mica, fibre, ebonite, wood, slate, marble or the molded compounds. It is being manufactured in sheets of standard sizes 48 in. square thickness from 1/8 in. to 1/2 in.; and in sheets 30 in. by 40 in., in thickness from 9/16 in. up to 2 in. It may also be had in all molded forms and shapes.

## Selling Electric Ranges Scientifically

The Canadian Westinghouse Co. Ltd., Winnipeg, have been holding a very successful demonstration of their electric range, at the show rooms of the Winnipeg Hydro, for the past month or so. The demonstration has been in charge of a young lady who has had considerable experience in teaching Domestic Science, her mornings are taken up visiting the homes of purchasers and prospective purchasers and the afternoons by baking, in the Hydro show rooms. The company report very good results so far. About a week after a range has been installed in the customer's home ,the demonstrator pays a visit to enquire if the purchaser thoroughly understands the range; if not, she is only too glad to give the required information. This is followed up about a week later by a telephone call, asking if the purchaser is now familiar with the range. In practically every case this is answered in the affirmative.

## New Northern Electric Halifax Offices

The Northern Electric Company announce that their Halifax premises have been moved from 67-69 Upper Water St. to 86 Hollis St. The new location is right in the heart of the wholesale district. The building is owned by the Eastern Telephone & Telegraph Company, and the Northern Electric will occupy one half of the lower floor—approximately 40 ft. by 80 ft. and, as warehouse space, all of the second floor. The company reports that business is decidedly on the increase in the Maritime provinces. They recently secured the contract for the installation of a complete new street lighting system for the city of Halifax, which it is expected will be in operation about October 1.

## Using Ferranti Transformers

In connection with the Mangahao hydro-electric development being carried out by the New Zealand Public Works Department, Ferranti Limited announce that they have the contract for seven 4000 kv.a., single phase transformers. The transformers are for use in banks of three, forming two 3 phase groups, each 12,000 kv.a., 110/10 500 volts, 50 cycles. They are of the core type, oil immersed, with forced oil circulation. Oil-filled terminals of special Ferranti design are to be supplied for the 110,000 volt sides. The illustration shows the first of these transformers to be completed.

## Campbell Mfg. Co.

The Campbell Manufacturing Company (formerly Taylor-Campbell Electric Company) are located at 526 Adelaide St., London, Ont., where they have a well equipped factory for manufacturing their various lines, which include enclosed entrance safety switches, all types of steel boxes and cabinets, cutouts, octagon outlet boxes, box covers, flush switches boxes, tight fitting covers, ground clamps, lugs, switch plates, plug cutouts, fuse plugs, rosettes, knobs, tubes, etc. Mr. E. L. Campbell is general manager of the company. It is one of this company's characteristics that they are unusually careful in the inspection of their goods, as they desire to give the best possible service and believe that this is only maintained by manufacturing their goods the greatest care.

## Trade Note

Mr. G. Wilkinson, who operates an up-to-date electrical store at 3031 Dundas St. W, Toronto, has now installed a wireless outfit on his roof, gives concerts every night and carries a full line of wireless equipment in his store.

The Murphy Electric Co., 603 Sayward Bldg., Victoria, B. C., has been awarded the contract for electrical work on a warehouse of Wm. N. O'Neil & Co., situated at 555 Yates Avenue, Victoria, that is undergoing alterations.

The McDonald & Willson Lighting Co., 309 Fort St., Winnipeg, has been awarded the electrical contract on a $170,000 film exchange being erected at Ellice & Hargrave Sts., Winnipeg, by Mr. M. Chechik, 375 Alfred Ave.

The Levvy Electric Company, Ltd., 493 Portage Ave., Winnipeg, have been awarded the electrical contract on the Montreal Trust Building, Portage Ave. & Main St., Winnipeg, that is being altered for a bank, by the Royal Bank of Canada, at a cost of about $35,000.

Messrs. Hawkins & Hayward, 1607 Douglas St., Victoria, B. C., have been awarded the electrical contract on an addition and alterations that are being made to the F. W. Woolworth store at Douglas & View Sts., Victoria, at approximate cost of $55,000.

Mr. W. H. Hughes, 24 Duncaton Ave., Toronto, has been awarded the contract for electrical work on six stores being erected on Danforth Ave., near Woodbine Ave., Toronto, at an estimated cost of $50,000, for Mr. E. J. Furniss, 23 Glenfern Ave.

The Vancouver Harbor Commission, which at present operates a steam locomotive on its two miles of industrial trackage on Granville Island, has decided to electrify the entire trackage on the Island.

Harry Alexander, Ltd., 6 King St. W., Toronto, has been awarded the electrical contract on a $60,000 addition to be built to a Lodge at Dovercourt Road, north of Bloor St., Toronto, the property of the Sovereign Hall Company.

The King Electrical Supply Company, formerly of 300 College St., Toronto, has moved to 288 College St., a few doors east of their old stand, where business will be carried on as before in contracting and supplies.

## Eighteenth Annual Convention Canadian Electric Railway Association Most Successful in History

One must be a true realist to be able to report the business proceedings of a convention under conditions such as surround one during a trip to historic Quebec: a most gorgeous trip down the St. Lawrence from Montreal to Quebec city, May, the ever inspiring month of the year just fading into the even more delightful time; the quaint and picturesque streets of Quebec, recalling the days when even a thought of electricity might have been sufficient to cause one's burning at the stake; the green lawns and boulevards of many shades, with sweetly fragrant and multi-colored plots and beds of tulips, narcissus, geraniums, forget-me nots and lily of the valley, flanked by bushes of lilacs lining almost the whole way from Chateau Frontenac to the Drill Hall, where the convention was held and exhibits displayed; and in addition to all this the ever present and limitless hospitality of the entertainment committee and especially of Mr. and Mrs. W. J. Lynch and Mr. and Mrs. R. M. Reade.

The prevailing spirit of the convention was one of decided optimism as to the development of the country and settling down to serious, real and stabilized and improving business, signs of which are not so much contained in posters like "Prosperity around the corner" and professional optimists' verbose and loud, though meaningless and baseless speeches, as in businesslike analysis of existing conditions and consideration of factors designed to bring the country back to its normal level.

Great stress was also laid on the fact that in street and suburban transportation improved business is possible of achievement at all times if the public is properly catered to and good service given. In the opinion of many present, in the matter of equipment, companies are not considering sufficiently the desirability of basing the size and carrying capacity of the unit upon satisfactory and efficient operation during 90 per cent of the daily service under normal conditions, as against the principle of efficiency and ignoring economy in order to handle temporary overloads during 10 per cent of the time in service.

Sir Charles Fitzpatrick, the Lieutenant Governor of Quebec, welcomed the delegates at the inaugural session of the convention, inviting, on behalf of Lady Fitzpatrick, all the delegates and ladies to attend a garden party in the Government House on the closing day of the convention, the King's birthday. Mr. Gordon Gale, president of the association, presided.

In the course of his address, the Lieutenant Governor said: "I consider no gathering of greater importance than yourselves, and there are no problems more difficult of solution than those of transportation, light, heat and power.

"In Canada we have no coal to speak of save in Cape Breton and in the Rocky Mountains, and we largely depend on the United States. Under these circumstances, we are solving the problem by utilizing our great water powers, and the heating of water by the direct action of electric current would contribute still more to the success of mine, field and forest."

### Exhibits

Following Mayor Samson's speech, who in a very appropriate manner welcomed the delegates on behalf of the city of Quebec, Mr. Gale thanked His Honor and His Worship for their kind attendance and also their gracious remarks, and the meeting adjourned to the right wing of the drill hall, which was given over to exhibits; it was formally opened by the Lieutenant Governor, the military band playing delightful selections while the delegates and visitors inspected the exhibits. The Exhibition Committee, consisting of Messrs. R. M. Reade, Quebec R'y. Light, Heat and Power Co.; D. M. Campbell, Preston, Ont.; A. Gaboury, Montreal Tramways; Peter J. Quinn, Quebec R'y. Light, Heat and Power Co., were recipients of many well deserved congratulations. It was decided at the next convention to surpass even this year's exhibit and to improve and enlarge from year to year the exhibit feature of the convention.

### Reports

Following an inspection of the exhibits the delegates returned to the Assembly Hall, and the business of the day commenced. The president, Mr. Gordon Gale, vice-president and general manager Hull Electric Railway Co., gave his annual report, in which he said that as a result of the policy decided upon last year, important improvements have taken place. The active membership during the past year has increased 35 per cent, while the associate membership made a gain of 160 per cent.

Relations have been established between the Canadian association and the American Electric Railway Association, the latter of which was represented at the convention by Mr. R. I. Tod, Ex-President of the A.E.R.A., Mr. J. W. Welch, its secretary, and a number of other prominent officials.

Mr. Gale referred to the value of the Exhibition as an adjunct of the convention.

The Secretary's report was delivered by Mr. L. E. Moreland of Hull. He showed a membership of 35 companies as against 27 last year. The associate members are included in 37 companies.

The honorary treasurer, Mr. A. Gaboury, superintendent of the Montreal Tramways, showed a credit balance in the bank which was highly satisfactory.

### Social Functions

The first day of the convention a get-together lunch was held in Chateau Frontenac attended by the members, associate members, ladies and guests.

At night a banquet was held, at which over 800 people sat down, under the chairmanship of Col. J. E. Hutcheson, general-manager of the Montreal Tramways. Besides a princely meal and a splendid musical program, which in reality was a concert, the presence of such still famous names as "Veuve Cliquot," "Heidesick," "Mumm" and "Pommery" helped to create a most enjoyable atmosphere.

Following the toast to the King, Major F. D. Burpee, the newly elected president, proposed the toast "Our Guests." This was responded to by Mr. J. A. Welch, secretary of the American Electric Railway Association. Mr. F. S. Neale, Toronto, proposed the toast to the ladies; Mr. Chas. L. Wilson in responding stated that at the first meeting of the C.E.

R.A., there were only sixteen members in all, while at the present convention there were no less than sixty-two ladies registered.

Col. Hutcheson requested Mr. C. D. Henry of Indianapolis, to propose the toast to the Association. Mr. Henry, a former president of the A.E.R.A., stated that it had been his privilege to be connected with the Association for a number of years in the United States and added that the good work of the Canadians had been known, his reason for attending the convention being a desire to meet the Canadians face to face and to join with them in the development of the industry. Nothing was better for industry than that all work together. Individual efforts, no matter how strong, could not accomplish what a united effort could. Co-operation, is the secret of success.

Mr. Gordon Gale, responding to the toast, drew a comparison between the first meeting of the association when some sixteen members were present, and the present convention where there were some three hundred registered. He

interest to all the delegates and caused the most discussion. There were two particular reasons for this: the newness of the subject, which is rapidly becoming of great importance and concern to the tramways companies of this continent, and is resulting in an extensive research, mainly theoretical as far as the trackless trolley is concerned, as the application of this means of locomotion is hardly even in a pioneering stage either in the United States or in Canada; secondly, the form in which Mr. Blair prepared his paper—a form which if adopted as a standard by all conventions would develop business-like exchange of ideas covering the subject in a thorough and orderly manner without any waste of time. Mr. Blair, after giving a concise presentation of his subject by statements, pertinent to the most salient points of his subject, developed his paper by a series of questions grouped under six captions or divisions. Such an arrangement enables the participants of a convention to express themselves in a few sentences on the points they are most interested and familiar with without rambling all over the subject. The

Major F. D. Burpee, president

Mr. E. A. Robert, Hon. president

Mr. H. H. Couzens, vice-president

also dwelt briefly on the ultimate aim of the Canadian Electric Railway Association.

Dancing and general hilarity held sway till late after midnight.

The second night of the convention was rendered no less enjoyable by a ball and a late dinner. Those two chief events of the social program of the convention were supplemented by afternoon tea parties, sight-seeing trips by street cars and boat, including a visit to the famous Quebec Bridge.

### Papers

The first paper to be read and discussed was "Valuation of street railway assets, their maintenance and depreciation" by Dr. Louis Herdt, consulting electrical engineer, of Montreal. Discussion was led by Mr. H. E. Weyman of the Levis County Railway, Co. Other papers were: "The Modern Street Railway Motor" by J. K. Stotz, Canadian Westinghouse Co., the discussion being led by W. G. Gordon, transportation engineer, Canadian General Electric Co.; "Welded Track Joints" by E. B. Entwistle, Loraine Steel Co., the discussion was led by. W. F. Graves, chief engineer, Montreal Tramways Co., and "Motor Buses and Trackless Trolleys" by D. E. Blair, superintendent of rolling stock, Montreal Tramways Co., the discussion was led by W. M. Manz, of the Ohio Brass Company.

The last mentioned paper was undoubtedly of unusual

general opinion of the convention on this subject was that both electrically and gasoline driven buses should be operated by tramways companies, but considered only as adjuncts and supplementary means of service. That they are applicable and desirable for short cuts, temporary hours traffic and where laying of tracks is an objection either temporarily or permanently; and that a thorough study should be given to the problem of gasoline driven and the trackless trolley buses.

On the strength of information gathered at many points in the United States and Canada and his personal observations in regard to the efficiency and economy of maintenance of gasoline engines, including a year's study of the subject, Mr. Blair seemed inclined to believe that whenever and wherever the electric street railway companies have to resort to the use of buses, the gasoline driven vehicles will prove preferable from certain points of view. He made it plain, however, that he did not give his opinion as authoritative, due to the comparatively limited material one has. as yet, at one's disposal for a study of this subject, as well as due to the recent adoption of the idea of trackless trolleys on this side of the water.

### Discussion

The conditions of street railway industry during the last few years brought about a situation where any method that will offer an operating solution for certain conditions where traffic is not dense, is looked upon with great interest. The trackless trolley in Europe has been in service for some ten or fifteen years, operating under various conditions and with various types of overhead and collection equipment. Practically all the trackless trolley lines on the continent are operating with a variety of current collection equipment ranging from that of four wheel truck to various combinations of slider, both for single and double pole. In England alone the practice was standardized on a two-pole, two-base collection construction. The latter is the only type that has been able to operate over a wide range of roadway and at sufficient speed to maintain an economic schedule. The same system is in operation in running the buses in Shangnai, China, the equipment being installed by an English concern.

### Past Practice the Basis of Bus Construction

Mr. Manz in leading the discussion, covered very thoroughly the overhead construction, supplementing his remarks by very interesting and instructive slides of trackless trolleys in operation and many diagrams of overhead construction. Past practice was the basis of bus construction and motor management. The absence of grounded return and the necessity for providing a positive and negative feeder within a short distance of one another make the problem of current collection at first seem rather difficult, until a careful analysis of the fundamental conditions are made.

However, it is necessary, and it is quite possible and practical, to lay out the overhead with standard overhead materials and secure the positive operation necessary.

W. R. Robertson, general superintendent of the Railway Department of the Hydro-electric Power Commission of Ontario, gave a most interesting account of the study made by him, and material gathered for it, in connection with the trackless trolley put into operation in Windsor, Ont., about a month ago.

He was emphatic in his contention that electric street railways companies should encourage proper development of trackless trolleys, where the motive power of the buses is electric current as against the use of gasoline engine. Following a very careful analysis of both means of bus locomotion he was ready to believe that electrically operated buses are, for many reasons, more desirable and economical than their competitor—the gasoline bus.

### Operation of Trolley Buses in Windsor

The reason for installing trolley buses in Windsor was the realization of the fact that certain districts were not properly served with transportation facilities. The use of trolley buses was decided upon to take care of this service until such time as the density of traffic warranted the construction of rail lines. In making the traffic survey of the district now served by the trolley bus it was found that one district has a population of approximately 4,700 people residing within 500 feet of the then-proposed and now-in-operation line; the other two districts had a considerably larger number of inhabitants, varying from 6 to 11 thousand, but yet not sufficient to justify the expense of a permanent tram line.

The line now in operation is between one and a half and one and three quarter miles in length. In determining the type of bus that would be used both the gasoline and the electrically driven bus type were very seriously considered. From the best information that could be secured the finding was that the economical life of an internal combustion engine, such as used on the gasoline trucks is 125,000 miles. With the electric motor the average yearly car-mileage in street car service is 35 to 50,000 miles, and the life and economy of operation appear to be governed entirely by the the obsolesence of the type, as a motor even after operating

20 years with ordinary maintenance is often in as good condition as when first put into operation.

With respect to the cost of operation, from the best information that could be gathered, gasoline when used as fuel costs in the neighborhood of from 5 to 7 cents per bus mile. The power consumption on the type of bus adopted in Windsor is about one kw.h. per bus mile. Another consideration is that electric street railway companies have as employees men experienced to handle electrical repairs economically without having to provide any further facilities.

The overhead construction essentials are simplicity, safety, lightness, with a maximum strength for a minimum weight; the collector must be flexible enough to take care of wire variations, and compensate for the swaying of the bus on account of poor roadway surface. The bus must be able to diverge, running at normal speeds, to permit the passing of other vehicles on the street; it also must operate to the curb line in each case to pick up and discharge passengers.

The buses in Windsor have a running speed of 25 miles per hour on level roadway, with about 450 volts. In order to make a complete turn a bus requires a radius of approximately 31 feet.

Since the inauguration in Windsor of the trolley buses on Lincoln Road, on a 15 minute headway, not a trip has been lost. There have been no dewirements of the trolley poles; loops are provided at each end of the line, so as to prevent any delay in wyeing. The public, the operators and the management are very well pleased with the new system, the operation being practically noiseless, which makes them so much more desirable for operation on the streets of residential districts.

### Jitneys Cut into Revenue

Mr. Palmer of Baltimore, in contributing to the discussion, stated that his company found itself confronted with the jitneys and other "Free-Lancers" in the transportation business taking away from them of from $300 to $500 per day of revenue, which made it necessary to give a supplementary service by means of buses. He mentioned that the engines used on gasoline buses are of the truck type, which are neither economical nor speedy enough to be satisfactory; but that there is being now developed a special type of gasoline engine for passenger buses, which may make that kind of bus more desirable than a trackless trolley.

A feature of the convention was the large number of splendid and educative films shown.

The officers elected for the coming year are as follows: hon. president, E. A. Robert, president Montreal Tramways Co.; hon. vice-president, W. C. Hawkins, president Dominion Power & Transmission Co., Hamilton; hon. council, Col. J. E. Hutcheson, Montreal; T. Ahearn, Ottawa; Geo. Kidd, Vancouver; Acton Burrows, Toronto. President, Maj. F. D. Burpee, general manager Ottawa Electric Railway Co.; vice-president, H. H. Couzens, general manager Toronto Transportation Commission. Executive: A. P. Coleman, Hamilton; C. L. Wilson, Toronto; A. W. McLimont, Winnipeg; R. M. Reade, Quebec; W. L. Weston, Toronto; D. W. Houston, Regina; A. Eastman, Kingsville, Ont.; H. E. Weyman, Levis, Que.; W. R. Robertson, Toronto. The hon. treasurer is T. Hart. Three Rivers, Que.; Auditor, Col. Geo. C. Royce, Toronto.

### Appointed Selling Agent

The Ward Leonard Electric Company, of Mount Vernon, manufacturers of Vitrohm Ribohm resistors and electrical control appartus, announce the appointment of Joseph E. Perkins, 113 East Franklin Street, Baltimore, Md., as their selling agent for Maryland, Virginia and that part of Pennsylvania identified as the Susquehanna Valley, as far north as Harrisburg.

# Electric Resistance Welded Rail Joints

## A paper read before the Canadian
## Electric Railway Association Convention

### By E. B. ENTWISLE*

### Electric Resistance Welded Rail Joints

When girder rails were first introduced to the electric street railway industry in the United States in 1884 there was also introduced the problem of joints, as this question had been treated in an altogether different manner in the use of the flat, or "stringer" rail joint in the early years of street railways.

It required a dozen years of use of the girder rail to prove that its life depended not so much upon the chemical proportions of the steel or the quality of its supporting ballast, as it did upon the character of the joints. Rails were removed from the track, not because the rail as a whole had reached the limit of wear, but on account of the failure of the joints, and the measure of value of the rail was cut down in a rapidly increasing ratio the longer the defective joint remained in the track.

We all remember the lighter rail of 30 to 60 pounds per yard in the early days of the electric traction, as compared with the larger rail, heavier splice bars and heat-treated bolts of the standard construction of to-day, made necessary by the heavier cars and equipment. The evil of the joint still remains, although great improvements have been made in the various types of bolted, cast-weld and electrically welded joints.

The type of joint this paper has to deal with is the electric resistance joint of The Lorain Steel Company, the successors of The Johnson Company of Pennsylvania.

This form of welding first took place about 1894 and was designed to unite two rails with a bar 7 inches long on each side of the web, with contact against the web at the extremities of the bar. The current was passed through these contacts and welded them to the web. The first of this type of welding was done in Boston on what was known as the Providence Rail, the joint then in use being a casting which received the foot of the rail, with a wedge driven along one side to hold rail and casting together. Later and more successful designs of electrically welded splice bars replaced the first type, and with few alterations has been the standard for twenty-five years.

To go back for a moment into the history of continuous track. It was not without much time, labor and expense that the fact was demonstrated that 7 inch or 9 inch girder rail laid in paving in long stretches of 1,000 feet or over, would remain in the paving under the extremes of heat and cold. An allowance for expansion had always heretofore been provided. In order to determine this question a stretch of 6 inch rail, 1,000 feet long, laid in macadam and operated over by the Johnstown Traction Company, was united at each joint by two bars 1¼ inches thick, 2½ inches wide and 5 feet long; with 18 turned bolts of 1¼ inch diameter. Examinations for expansion or contraction in the extremes of heat and cold were daily made, covering over six months of time. None being found upon the most careful measurements, the welding of rails with the ends abutting was begun, and continues to be successfully prosecuted.

The bar-welding with head support, if done on new rail with no splice bar holes, is 18 to 22 inches long, 3¼ inches wide, 1½ inch thick, of mild steel. If on old rail the bars are long enough to extend beyond the old bar holes. To prepare the bars for welding they are heated and placed under a drop-hammer which forces a pear-shaped portion of the

metal above the surface of the bar at each end, and a strip 1¼ inch wide across the bar in the middle. These three areas are the sections welded. A head-support about 4 inches wide is welded to the bars at the time the middle weld is made. This is also welded to the side of the rail head, and to the web. The equipment consists of a sand-blast outfit to clean the ends of the rails; a car to carry the current generator and air compressor, and a car equipped with grinding apparatus. The process, after the rails to be welded are brought to line and surface and the heads are abutting, is to hold the bars against the rail web with two powerful clamps, send the electric current through the contacts and fuse the bars and the rail together. The centre weld is first made, then the end welds. After the weld is made and the current turned off, the side pressure against the weld is maintained until the critical temperature is passed. This pressure is equal to about 45 tons on an area of slightly less than four square inches. This form of electric welded joint has probably reclaimed more track and brought it to its greatest use, than any other form of joint in use. It comes nearer to eliminating the joint than other types, as the rail heads are brought tightly together by the contraction of the bars and when failures occur the rail breaks in the web beyond the weld, and this is accounted for sometimes by the contraction of the rail behind the weld, the accumulating force of the contraction of a long rail centralizing at the break.

For twenty-five years this process has been carried on until now there have been welded over 800,000 joints distributed among the principal cities of the United States, and many repeat orders testify to the estimation of its value held by those who know it best. A repeat order for the 15th consecutive year of one of the largest traction companies has just been entered.

The process has been successfully applied to open tee rail construction on the Elevated Railway of the Myrtle Avenue line, Brooklyn. Here the track is exposed to rapid changes in temperature of the air, without the aid of the earth to absorb and dissipate the accumulated heat. An experimental installation of 370 joints was made in December, 1914, January, 1915, and a recent inspection of this track reveals that four joints have been removed on account of breakage through the rail, one other showing fracture through the welded bar, and eight other joints slightly defective due apparently to not filling up the openings between rails with shims when the weld was made. This installation after 7½ years of service is still carrying traffic with a minimum cost as to joints. No other type of joint ever used on the elevated railway can show a record comparable with this, and there is on no other elevated structure a more nearly noiseless track than this stretch of welded track.

In welding old track it is especially important to close open-joints with shims prior to welding. In new track it is equally important, and it is the fact that to drive the rails together as close as you may, the shrinkage of the bars after the last weld will draw the rails together still tighter. All welded joints of whatever character are subject to the strains of expansion and contraction, and all track is subject to vibration. Under heat stresses we get expansion, and with cold comes tension, and accelerated vibration due to the increase in tension. A joint to withstand

---

*Chief Engineer Lorain Steel Co.

these conditions must be as strong in tension as the rail itself. Breakages in this form of joint have been in the rail and not in the bars.

## Failures

Joint failures can be accounted for on the hypothesis of power shortage during the moment of welding, on overheating, on imperfect alignment and surfacing of the rail prior to welding, and on contraction of the rail localizing at the joint. A few percentages of failures will serve to give an idea of the worth of this form of joint simply as a maintenance proposition. An installation of 1943 joints in 1917-18 on 7 inch new girder rail's, had 5 breaks. A later installation of 3833 in '19-'20-'21 has no breaks to date. This is a 1-10 of one per cent, record.

Six miles of 56-lb. tee rail track laid in macadam with one rail uncovered to the ties, was welded ten years ago, and an inspection this month reveals no broken joints.

Another large installation covering about 17 years, for the Public Service Corporation of New Jersey, required 130,-158 joints. There has been welded in Chicago since 1907 a total of 184,423 joints on the standard Chicago rail, which weighs 129 pounds per yard and is 9 inches high. Of these the percentage of breaks during the first six years was 2.74, and the average for nine years was reduced to below 1.5 per cent. This improvement, to quote from the report of the Board of Supervising Engineers, Chicago Traction, was brought about "by the combined efforts in improving the workmanship of the weld and by the more uniform conditions of power supply which became available as the work proceeded."

## Butt Weld

The Lorain Company is now operating under the Jacob's patents on a butt weld. This process contemplates bringing the milled ends of the rails together and welding them without the aid of bars. This was the process aimed at by the predecessors of the Company back in '93-'94, but their engineers were unsuccessful in achieving what apparently today is the 100 per cent. joint.

The butt-weld process consists essentially in a method of heating the end, or abutting surfaces of the rails to be joined, by having them in contact with a liquid flux which is heated to a high degree by the passage of a comparatively small electric current through it, and a powerful clamp against the rails, operated by a hydraulic pump, for forcing the rails together, end to end, after the ends have been brought to a welding heat. The apparatus is mounted on two trolley cars, the first of which contains a 30 kw. motor-driven alternating current generator, an air compressor, and a furnace for melting the flux, and a re-heating torch. A switch-board with all necessary meters and regulating apparatus is also provided.

In the process the rails are spiked to a few ties to hold them to gauge and to insure a fair surface and alignment, and an opening of about ¼ inch between the ends. The rail clamp and cross-head are then adjusted so as to automatically bring the two rails in line with each other. A spacer is set between the rail ends and a preliminary pressure of 5,000 pounds per square inch is applied before the molds are set in place. When the molds are set the cable terminals from the generator are clamped to the cross-heads and the current is passed through the flux, which has been previously melted and poured into the mold. The current in its passage through the flux heats the ends of the rails to a high degree above the welding heat of steel, and when this point is reached, which requires about 3½ minutes, the pressure is applied, the rails are forced together and simultaneously the current is cut off. At about a half minute before a welding heat is reached, the current reaches its maximum when

about 1,650 amperes are taken at about 40 volts. No oxidation in the weld is possible, as the rail ends are completely immersed in the flux and no air can reach them.

Immediately upon applying the full pressure the top mold is taken off and the hot metal which has extruded at the weld is hammered across the head, thus producing a forging effect to obviate any tendency that might exist to soften the rail at the exact point of weld.

Scleroscope tests show that the portion which has been heated and hammered is actually somewhat denser than the parts of the rail that have not been heated. The pressure holding the rails together is held until the weld has cooled below its critical temperature, when the re-heating torches are applied and the flame directed to the centre of the web, directly against the weld, and it is re-heated to about 1,500 deg. Fahr. This relieves any strains that may have been set up due to the end pressure or other causes.

About 5/8ths of an inch of each rail is combined in the weld, the extruded metal being ground off the head and tram. As a continuous operation about three welds an hour can be made on 9 inch rail.

The first experimental butt welds were made in Pittsburgh, Pa., in December, 1919, and in Johnstown in the summer of 1921, on the tracks of the Johnstown Traction Company. Ninety-three welds on 134-lb. 9 inch girder rail were made, making two rails each 2,700 feet long. About one-half of this track was sprung into a curve and was not paved in for several weeks. Yet there was no distortion under constant operation, and no breaks to date.

In Boston, 504 welds were made in 1921 on four different rail sections. In Philadelphia 402 welds on 9 inch, 141 pound rail were made within the last year. In this process, due to the single track operation in that city, the welding apparatus was set on a motor truck, the rails being laid on top of the paving outside of the track, which was being operated over by the railway company. Sections from 1,000 to 1,500 ft. long were welded, and were afterward set over into the track in their permanent position. The advantage of the process is due to the fact that the entire cross-section of the rail is welded, without the addition of any other metal. While special conditions require special treatment, there are no difficulties which cannot be surmounted, and they are more than compensated for by the permanency of the result. There has been but one failure in over 500 joints welded in Boston, and five joints in over 400 welded in Philadelphia, and none in Johnstown where 93 joints were welded, a total of 999 with six failures.

---

The Board of Commissioners of Public Utilities of the Province of Nova Scotia is at present engaged on the valuation of the property of the Nova Scotia Tramways and Power Co. Limited. The company will have the option, on the completion of the valuation, to make application to the board for approval of such rate schedules as will yield a net return of 8 per cent. on the value of the investment in the various departments, as found by the Board.

---

## Coming Conventions

Canadian Electrical Association, at the Chateau Laurier, Ottawa, June 15-16-17.

Association of Municipal Electrical Utilities, at the Clifton Hotel, Niagara Falls, Ont., June 22-23-24.

American Institute of Electrical Engineers, at the Clifton Hotel, Niagara Falls, Ont., June 26 to 30.

# Current News and Notes

**Annapolis Royal, N. S.**

Mr. S. Rippey, Annapolis Royal, N. S., has been awarded the electrical contract on a town hall being erected at that place.

**Boissevain, Man.**

Mr. N. Johnson, who has been superintendent of the electric light plant of the Melita Milling Co., Melita, Man., has received the appointment of superintendent of the municipal electric light plant for the town of Boissevain, Man.

**Fort William, Ont.**

During the year 1921, in the city of Fort William, it is estimated that about 190 electric ranges were installed. Electric water heaters, to the number of about 190, were also installed during that period.

**Guelph, Ont.**

Mr. W. Stuart, Quebec St., London, Ont., has been awarded the contract for electrical work on two stores to be erected on Quebec St., London, for Mr. F. Kloepfer, of London.

**Hamilton, Ont.**

The Electric Supply Co., Ltd., 65 St. James St. South, Hamilton, Ont., has been awarded the electrical contract on 14 stores, apartments and theatre to be erected near King St. E. & Main St., Hamilton, at an estimated cost of $300,000, by Deta Properties, Ltd., Hamilton. Also the electrical contract on a $25,000 bank building to be erected at King & Main Sts., Hamilton, by the Bank of Hamilton.

The Salisbury Electric Co., 112 King St. W., Toronto, has secured the electrical contract on a building at 55-59 King St. E., Hamilton, that is to be remodelled at an estimated cost of around $120,000.

Mr. J. Dynes, Prospect St., Hamilton, Ont., has secured the contract for electrical work on five stores being erected at King and Spadina Ave., Hamilton, for Messrs. McKay Bros., of that city.

**Kitchener, Ont.**

The Star Electric Co., Kitchener, Ont., has secured the electrical contract on six residences being erected at Albert & Filbert Sts. and Hohner Ave., Kitchener, by Messrs. Schmalz & Peterson.

**London, Ont.**

Mr. J. H. Pollock, 307 Clarence St., London, Ont., has been awarded the contract for electrical work on a building on Queens Avenue, London, owned by Dr. Geo. McNeill, that is being remodelled for offices.

**Montreal, Que.**

Messrs. A. A. Giddings & Co. Ltd., 660 Dorchester St. W., Montreal, have been awarded the contract for transformers for the $8,000,000 hotel being erected by the Mount Royal Hotel Company at Peel & St. Catherine Sts., Montreal.

Mr. W. Rochon, 454 Lafontaine Park, Montreal, has been awarded the contract for electrical work on a $60,000 apartment house being erected at Roy & Alfred Sts., Montreal, by F. X. Bissonette, 341 Drolet St.

**Oakville, Man.**

The Oakville Electric Co. Ltd., has been formed, with a capitalization of $20,000, to carry on an electrical contractor-dealer business, etc., in the town of Oakville, Man.

**Ottawa, Ont.**

Mr. E. G. Tressider, 5th Avenue, Ottawa, has been awarded the contract for electrical work on a store building being erected on Bank St. South, Ottawa, by Mr. Ainslie Green, Hope Chambers, Ottawa.

The Canada Gazette announces the incorporation of the Radio Sales Co. Ltd., with a capitalization of $75,000. Head office, Montreal.

The Success Electric Products Company, Ltd. has recently been formed, with a capitalization of $50,000. Head office, Ottawa, Ont.

**Port Arthur, Ont.**

About 106 electric ranges, and about the same number of electric heaters, were installed in Port Arthur homes during the year 1921.

The first of the rebuilt one-man cars for use in Port Arthur, Ont., was recently given a try-out on the streets of that city, with entirely satisfactory results.

**Port Colborne, Ont.**

Messrs. Sherk & Kennedy, Port Colborne, Ont., have secured the electrical contract on a $34,000 school to be erected at Humberstone, Ont., in the immediate future.

**Quebec City, Que.**

Messrs. Goulet & Belander, 190 Richardson St., Quebec City, Que., have secured the electrical contract on St. Sacrement school, to be erected on St. Foye Road, Quebec City, at a cost estimated to be in the neighborhood of $70,000.

**St. James, Man.**

Messrs. Gamble & Willis, 306 Notre Dame Ave., Winnipeg, have secured the electrical contract on a $96,000 school being erected on Sutherland St., St. James, Man.

The St. James Electric Co., 207 Roseberry Ave., St. James, Man., has been awarded the electrical contract on two stores to be erected on Portage Ave., near Roseberry St., Winnipeg, for Dr. V. M. Leech, 249 Canora St., Winnipeg.

**Souris, Man.**

Mr. R. Logan, electric and town engineer of the town of Souris, Man., for the last three and a half years, is leaving shortly to start in business for himself. He has purchased the storage battery business of J. Dubick, Souris, and will also do electrical contracting in the district, handling electrical appliances and lighting plants as well.

**Toronto, Ont.**

Mr. M. Nealon, 9 Glen Morris Ave., Toronto, has secured the contract for electrical work on a $125,000 church and rectory to be erected on Roncesvalles Ave., Toronto, by the Parish of St. Vincent de Paul.

Messrs. Taylor Bros., 25½ Norwood Ave., Toronto, have secured the contract for electrical work on twenty bungalows now being erected on Strathmore Blvd. near Monarch Park Ave., Toronto, by Messrs. Churchill & Morrell, 699 Woodbine Ave.

Mr. A. Mitchell, 1174 Danforth Ave., Toronto, has been awarded the contract for electrical work on 12 pair of residences recently erected at Glebeholme Blvd. & Caithness Ave., Toronto, at an approximate cost of $100,000, by Messrs. Grimshaw Bros., 117 Danforth Ave.

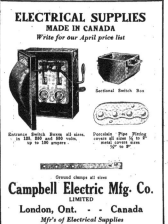

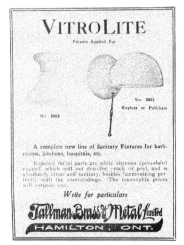

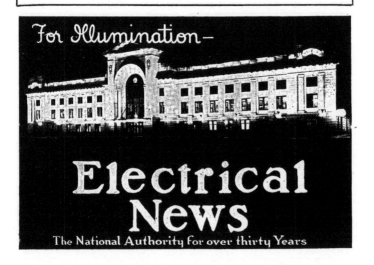

# CLASSIFIED INDEX TO ADVERTISEMENTS

The following regulations apply to all advertisers:—Eighth page, every issue, three headings; quarter page, six headings; half page, twelve headings; full page, twenty-four headings.

**AIR BRAKES**
Canadian Westinghouse Company

**ALTERNATORS**
Canadian General Electric Co., Ltd.
Canadian Westinghouse Company
Electric Motor & Machinery Co.
Ferranti Meter & Transformer Mfg. Co.
Northern Electric Company
Wagner Electric Mfg. Company of Canada

**ALUMINUM**
British Aluminum Company.
Spielman Agencies, Registered

**AMMETERS & VOLTMETERS**
Canadian National Carbon Company
Canadian Westinghouse Company
Ferranti Meter & Transformer Mfg. Co.
Monarch Electric Company
Northern Electric Company
Wagner Electric Mfg. Company of Canada
Winter Joyner Ltd., A. H.

**ANNUNCIATORS**
Macgillivray, G. L. & Co.

**ARMOURED CABLE**
Canadian Triangle Conduit Company

**ATTACHMENT PLUGS**
Canadian General Electric Company
Canadian Westinghouse Company

**BATTERIES**
Canada Dry Cells Ltd.
Canadian General Electric Co., Ltd
Canadian National Carbon Company
Electric Storage Battery Co.
Exide Batteries of Canada, Ltd.
McDonald & Willson, Ltd., Toronto
Northern Electric Company

**BEARINGS**
Kingsbury Machine Works

**BELLS**
Macgillivray, G. L. & Co.

**BOLTS**
McGill Manufacturing Company

**BONDS (Rail)**
Canadian General Electric Co., Ltd
Ohio Brass Company

**BOXES**
Banfield, W. H. & Sons
Canadian General Electric Co., Ltd
Canadian Westinghouse Company
G & W Electric Specialty Company

**BOXES (Manhole)**
Standard Underground Cable Company of Canada, Limited

**BRACKETS**
Slater, N.

**BRUSHES CARBON**
Dominion Carbon Brush Company
Canadian National Carbon Company

**BUS BAR SUPPORTS**
Ferranti Meter & Transformer Mfg. Co.
Moloney Electric Company of Canada
Monarch Electric Company
Electrical Development & Machine Co.
Winter Joyner Ltd., A. H.

**BUSHINGS**
Diamond State Fibre Co. of Canada, Ltd.

**CABLES**
Boston Insulated Wire & Cable Company, Ltd
British Aluminium Company
Canadian General Electric Co., Ltd
Phillips Electrical Works, Eugene F.
Standard Underground Cable Company of Canada, Limited

**CABLE ACCESSORIES**
Northern Electric Company
Standard Underground Cable Company of Canada, Limited

**CARBON BRUSHES**
Calebaugh Self-Lubricating Carbon Co.
Dominion Carbon Brush Company
Canadian National Carbon Company

**CARBONS**
Canadian National Carbon Company
Canadian Westinghouse Company

**CAR EQUIPMENT**
Canadian Westinghouse Company
Ohio Brass Company

**CENTRIFUGAL PUMPS**
Boving Hydraulic & Engineering Company

**CHAIN (Driving)**
Jones & Glassco ●

**CHARGING OUTFITS**
Canadian Crocker-Wheeler Company
Canadian General Electric Co., Ltd
Canadian Allis-Chalmers Company

**CHRISTMAS TREE OUTFITS**
Canadian General Electric Co., Ltd
Masco Co., Ltd.
Northern Electric Company

**CIRCUIT BREAKERS**
Canadian General Electric Co.
Canadian Westinghouse Company
Cutter Electric & Manufacturing Company
Ferranti Meter & Transformer Mfg. Co
Monarch Electric Company
Northern Electric Company

**CONDENSERS**
Boving Hydraulic & Engineering Company
Canadian Westinghouse Company

**CONDUCTORS**
British Aluminum Company

**CONDUIT (Underground Fibre)**
American Fibre Conduit Co.
Canadian Johns-Manville Co.

**CONDUITS**
Conduits Company
National Conduit Company
Northern Electric Company

**CONDUIT BOX FITTINGS**
Banfield, W. H. & Sons
Northern Electric Company

**CONDUIT PIPE FITTINGS**
Ferguson Mfg. Co.

**CONTRACTORS**
Seguin, J. J.

**CONSTRUCTION MATERIAL**
Oshkosh Mfg., Co.

**CONTROLLERS**
Canadian Crocker-Wheeler Company
Canadian General Electric Company
Canadian Westinghouse Company
Electrical Maintenance & Repairs Company
Northern Electric Company

**CONVERTING MACHINERY**
Ferranti Meter & Transformer Mfg. Co.

**COOKING DEVICES**
Canadian General Electric Company
Canadian Westinghouse Company
National Electric Heating Company
Northern Electric Company
Spielman Agencies, Registered

**CORDS**
Northern Electric Company
Phillips Electric Works, Eugene F.

**CROSS ARMS**
Northern Electric Company

**CRUDE OIL ENGINES**
Boving Hydraulic & Engineering Company

**CURLING IRONS**
Northern Electric Company (Chicago)

**CUTOUTS**
Canadian General Electric Company
G. & W. Electric Specialty Company

**DETECTORS (Voltage)**
G. & W. Electric Specialty Company

**DISCONNECTORS**
G. & W. Electric Specialty Company

**DISCONNECTING SWITCHES**
Ferranti Meter & Transformer Mfg. Co.
Winter Joyner, A. II

**DREDGING PUMPS**
Boving Hydraulic & Engineering Company

**ELECTRICAL ENGINEERS**
See Directory of Engineers

**ELECTRIC HAND DRILLS**
Rough Sales Corporation George C.

**ELECTRIC HEATERS**
Canadian General Electric Company
Canadian Westinghouse Company
Equator Mfg. Co.
McDonald & Willson, Ltd., Toronto
National Electric Heating Company

**ELECTRIC SHADES**
Jefferson Glass Company

**ELECTRIC RANGES**
Canadian General Electric Company
Canadian Westinghouse Company
National Electric Heating Company
Northern Electric Company
Rough Sales Corporation George C.

**ELECTRIC RAILWAY EQUIPMENT**
Canadian General Electric Company
Canadian Johns-Manville Co.
Electric Motor & Machinery Co., Ltd.
Northern Electric Company
Ohio Brass Company
T. C. White Electric Supply Co.

**ELECTRIC SWITCH BOXES**
Banfield, W. H. & Sons
Canadian General Electric Company
Canadian Drill & Electric Box Company
Dominion Electric Box Co.

**ELECTRICAL TESTING**
Electrical Testing Laboratories, Inc.

**ELEVATOR CONTRACTS**
Dominion Carbon Brush Co.

**ENGINES**
Boving Hydraulic & Engineering Co., of Canada
Canadian Allis-Chalmers Company

**FANS**
Canadian General Electric Co. Ltd.
Canadian Westinghouse Company
Century Electric Company
Great West Electric Company
McDonald & Willson
Northern Electric Company
Robbins & Myers

**FIBRE (Hard)**
Canadian Johns-Manville Co.
Diamond State Fibre Co. of Canada, Ltd.

**FIBRE (Vulcanized)**
Diamond State Fibre Co. of Canada, Ltd.

**FIRE ALARM EQUIPMENT**
Macgillivray, G. L. & Co.
Northern Electric Company
Slater, N.

**FIXTURES**
Banfield, W. H. & Sons
Benson-Wilcox Electric Co.
Crown Electrical Mfg. Co., Ltd.
Canadian General Electric Co. Ltd.
Jefferson Glass Company
McDonald & Willson
National X-Ray Reflector Co.
Northern Electric Company
Tallman Brass & Metal Company

**FIXTURE PARTS AND FITTINGS**
Banfield, W. H. & Sons.

**FIXING DEVICES**
Inventions Ltd.
Rawlplug Co., of Canada

**FLASHLIGHTS**
Canadian National Carbon Company
McDonald & Willson Ltd.
Northern Electric Company
Spielman Agencies, Registered

**FLEXIBLE CONDUIT**
Canadian Triangle Conduit Company
Slater, N.

**FUSES**
Canadian General Electric Co. Ltd.
Canadian Johns-Manville Co.
Canadian Westinghouse Company
G. & W. Electric Specialty Company
Moloney Electric Company of Canada
Northern Electric Company
Rough Sales Corporation George C.

# CLASSIFIED INDEX TO ADVERTISEMENTS—CONTINUED

## CLASSIFIED INDEX TO ADVERTISEMENTS—CONTINUED

## Dealers Who Handle
## The Westinghouse Line of Fans

are assured that they are selling a line that will give unqualified satisfaction to their customers.

Their attractive, symmetrical design appeals at once to the users—then you can point out their other good qualities such as their low cost of maintenance, an oiling once a season keeps them in good condition for years; their strong guards; their protecting felt bases covering the entire bottom of the fans; their perfect oiling system; their adaptability for either desk or bracket mounting.

### Canadian Westinghouse Co., Limited, Hamilton, Ont.

| | | |
|---|---|---|
| TORONTO, Bank of Hamilton Bldg. | MONTREAL, 285 Beave. Hall Hill | OTTAWA, Ahearn & Soper, Ltd. |
| HALIFAX, 105 Hollis St. | FT. WILLIAM, Cuthbertson Block | WINNIPEG, 158 Portage Ave. E. |
| CALGARY, Canada Life Bldg. | VANCOUVER, Bank of Nova Scotia Bldg. | EDMONTON, 211 McLeod Bldg. |

REPAIR SHOPS :

| | | |
|---|---|---|
| MONTREAL—113 Dagenais St. | VANCOUVER | WINNIPEG—158 Portage Ave. E. |
| TORONTO—366 Adelaide St. W. | 1090 Mainland St. | CALGARY—316 3rd. Ave. E. |

# Westinghouse